DICTIONARY
OF
BIOLOGY

ENGLISH / GERMAN / FRENCH / SPANISH

BY

GÜNTHER HAENSCH
AUGSBURG

AND

GISELA HABERKAMP DE ANTÓN
BARCELONA

ELSEVIER SCIENTIFIC PUBLISHING COMPANY
AMSTERDAM / OXFORD / NEW YORK
1976

ELSEVIER SCIENTIFIC PUBLISHING COMPANY
335 JAN VAN GALENSTRAAT
P.O. BOX 211, AMSTERDAM, THE NETHERLANDS

Distributors for the United States and Canada:

ELSEVIER/NORTH-HOLLAND INC.
52, VANDERBILT AVENUE
NEW YORK, N. Y. 10017

ISBN: 0-444-99828-4

Preface

The main design of this "Dictionary of Biology" is to provide a representative selection from the huge vocabulary of the biological sciences for use by all those who work in this field, including translators and interpreters. The immense range of the subject has meant that the authors could not aim at completeness. Nevertheless, care was taken to obtain as proportionate a selection of terms as possible from all areas of biology. The limited choice of words in certain subjects such as anatomy and systematics was quite intentional. In systematics, for example, only the major groups of animals and plants, the phyla, divisions and in some cases the classes have been included. On the other hand, modern subjects like the environment, ethology and ecology have been given due consideration.

The authors wish to express their thanks to the following contributors:

Dr. *Schauer,* BLV Publishing House in Munich, who checked the German terminology and suggested additional entries

Dr. *Alfred J. Szumski,* Professor of Physiology at the Medical College of Virginia, Richmond (USA)

Dr. *Lenoch,* Translation Service at the Commission of the European Communities, Luxembourg

Frl. *R. Weisgerber,* Translation Service at the Commission of the European Communities, Luxembourg

Mr. *Derek Rutter,* Chief Translator at the British Embassy in the Federal Republic of Germany, Bonn

The authors would be grateful for any proposals for amendment or improvement of this dictionary.

Munich, January 1976 THE AUTHORS

Vorwort

Das „Wörterbuch der Biologie“ will allen, die sich, sei es als Fachleute, sei es als Übersetzer oder Dolmetscher, mit dieser Materie befassen, eine repräsentative Auswahl aus dem ungeheuer reichen biologischen Wortschatz bieten. Wegen des beschränkten Umfanges konnten die Verfasser keine Vollständigkeit anstreben. Bei der Auswahl der Termini wurden alle Gebiete der Biologie möglichst gleichmäßig berücksichtigt. Die Auswahl des Wortschatzes gewisser Gebiete wie Anatomie und Systematik wurde bewußt eingeschränkt, z. B. wurden in der Systematik nur die großen Tier- und Pflanzengruppen, Stämme, Abteilungen und teilweise die Klassen aufgenommen. Andererseits wurden heute so wichtige Gebiete wie Umwelt, Verhaltensforschung und Ökologie berücksichtigt.

Für Mitarbeit sind die Verfasser besonders zu Dank verpflichtet:

Herrn Dr. *Schauer,* München, der die deutsche Terminologie überprüft und Ergänzungsvorschläge gemacht hat

Herrn Dr. *Alfred J. Szumski,* Professor für Physiologie am Medical College of Virginia, Richmond (USA)

Herrn Dr. *Lenoch,* Übersetzungsdienst der Kommission der Europäischen Gemeinschaften, Luxemburg

Frl. *R. Weisgerber,* Übersetzungsdienst der Kommission der Europäischen Gemeinschaften, Luxemburg

Mr. *Derek Rutter,* Chefübersetzer der Britischen Botschaft in der Bundesrepublik Deutschland, Bonn

Die Verfasser nehmen Vorschläge zur Verbesserung und Ergänzung dieses Wörterbuches gerne entgegen.

München, im Januar 1976 DIE VERFASSER

Préface

Le «Dictionnaire de la Biologie» souhaite offrir à tous ceux qui s'occupent de cette matière, soit en spécialistes, soit en traducteurs ou interprètes, un choix représentatif du vocabulaire immense de la biologie. La limitation d'espace ne permet pas aux auteurs de viser à l'intégralité. Dans la sélection des termes, ils ont cherché à représenter aussi uniformément que possible tous les domaines de la biologie. Dans certains, comme l'anatomie et la systématique, le choix du vocabulaire a été systématiquement restreint. Par exemple, en systématique, on n'a repris que les grands groupes d'animaux et de plantes, les embranchements, divisions et en partie les classes. Par ailleurs, on a pris en considération des domaines aujourd'hui aussi essentiels que l'environnement, l'éthologie et l'écologie.

Les auteurs remercient tout spécialement de leur collaboration:

Le Dr. *Schauer,* Editions BLV de Munich, qui a vérifié la terminologie allemande et proposé certaines additions

Le Dr. *Alfred Szumski,* professeur de physiologie au Medical College of Virginia, à Richmond (USA)

Le Dr. *Lenoch,* Service de traduction de la Commission des Communautés Européennes, à Luxembourg

Mlle. *R. Weisgerber,* Service de traduction de la Commission des Communautés Européennes, à Luxembourg

M. *Derek Rutter,* traducteur en chef de l'ambassade de Grande-Bretagne dans la République fédérale d'Allemagne, à Bonn

Les auteurs seront heureux de recevoir toute suggestion visant à améliorer et compléter ce dictionnaire.

Munich, janvier 1976 Les Auteurs

Prólogo

El «Diccionario de Biología» se propone ofrecer a todas aquellas personas que se ocupan de esta materia, sea como científicos, sea como traductores o intérpretes, una selección representativa del vocabulario inmensamente rico de la biología.
Debido a la necesidad de limitar la extensión de la obra, los autores no pueden pretender abarcar la totalidad del vocabulario biológico. En ciertos campos se hizo deliberadamente una selección restrictiva, por ej. en anatomía y sistemática. En cuanto a esta última, sólo se han incorporado los términos referente a los grandes grupos, fila, divisiones y en parte las clases de animales y plantas. Por otra parte se han tenido en cuenta campos hoy en día tan importantes como el medio ambiente, la etología y la ecología.

Los autores quieren expresar su especial agradecimiento:

al Dr. *Schauer,* colaborador del BLV-Verlag, de Munich, quien revisó la terminología alemana e hizo propuestas para completarla

al Dr. *Alfred J. Szumski,* catedrático de fisiología del Medical College of Virginia, Richmond (E.E.U.U.)

al Dr. *H. Lenoch,* del Servicio de Traducción de la Comisión de las Comunidades Europeas, Luxemburgo

a la Srta. *R. Weisgerber,* del Servicio de Traducción de la Comisión de las Comunidades Europeas, Luxemburgo

al Sr. *Derek Rutter,* Jefe del Servicio de Traducción de la Embajada del Reino Unido en la R.F.A., Bonn.

Los autores aceptarán gustosos cuantas sugerencias se les hagan para mejorar y completar el presente Diccionario.

Munich, enero de 1976 — Los Autores

Introduction

This "Dictionary of Biology" comprises:

- a quadrilingual glossary in Englisch, German, French and Spanish and
- three unilingual alphabetical indices in German, French and Spanish.

As the glossary is arranged alphabetically in English, there is no need to have an index for English. All the catchwords have been consecutively numbered and the German, French and Spanish equivalents are given after each English term.

In the English part of the dictionary, expressions comprising an adjective + substantive, such as *adaptive radiation* or similar compounds, have been entered under the first word. In the example quoted, it would be "adaptive."

The spelling of the German terminology has been based on Duden (e. g. *Kode, Nukleo ..., Zyto ...*), rather than on the customary orthography used in scientific literature *Code, Nucleo* ... and *Cyto* ...

The following abbreviations are used in the dictionary:

adj.	adjective	micr.	microscopy
anat.	anatomy	microb.	microbiology
bot.	botany	n	neuter
ecol.	ecology	neur.	neurophysiology
embr.	embryology	obs.	obsolete
env.	environment	phys.	physiology
eth.	ethology	pl	plural
evol.	evolution	pop.	popular (expression)
f	feminine	sing.	singular
gen.	genetics	sub.	substantive
geol.	geology	tax.	taxonomy
ins.	insects	US	term used in the United States
LA	term used in Latin-America		
m	masculine	zool.	zoology

~ means repetition of one or several parts of the previous expression, e. g. Dominanzwechsel, ~ umkehr (= Dominanzumkehr), active substance, ~ principle (= active principle).

THE AUTHORS

Einführung

Das „Wörterbuch der Biologie" umfaßt:

- ein viersprachiges Glossar (englisch-deutsch-französisch-spanisch)
- drei einsprachige Register (deutsch, französisch, spanisch).

Das Glossar ist in der alphabetischen Folge der englischen Stichwörter angeordnet, so daß sich ein englisches Register erübrigt. Alle Stichwörter sind laufend numeriert. Hinter jedem englischen Stichwort sind dessen deutsche, französische und spanische Äquivalente angegeben.

Im englischen Teil sind Ausdrücke aus Adjektiv + Substantiv, wie z. B. *adaptive radiation* oder sonstige Zusammensetzungen unter dem ersten Element des zusammengesetzten Ausdruckes aufgeführt, in unserem Falle unter dem Adjektiv „adaptive".

Bei der deutschen Terminologie wurde die Schreibweise nach Duden bevorzugt (z. B. *Kode, Nukleo ..., Zyto ...*), obwohl in der wissenschaftlichen Literatur auch die Schreibweise *Code, Nucleo ..., Cyto ...* gebräuchlich ist.

Im Wörterbuch werden folgende Abkürzungen verwendet:

adj.	Adjektiv	micr.	Mikroskopie
anat.	Anatomie	microb.	Mikrobiologie
bot.	Botanik	n	sächlich
ecol.	Ökologie	neur.	Neurophysiologie
embr.	Embryologie	obs.	obsolet, veraltet
env.	Umwelt	phys.	Physiologie
eth.	Ethologie	pl	Plural
evol.	Evolution	pop.	populär, umgangssprachlich
f	feminin	sing.	Singular
gen.	Genetik	sub.	Substantiv
geol.	Geologie	tax.	Taxonomie
ins.	Insekten	US	in den USA üblicher Ausdruck
LA	in Latein-Amerika üblicher Ausdruck	zool.	Zoologie
m	maskulin		

~ bedeutet Wiederholung eines oder mehrerer Bestandteile des vorhergehenden Ausdrucks, z. B. Dominanzwechsel, ~umkehr (= Dominanzumkehr), active substance, ~ principle (= active principle).

DIE VERFASSER

Introduction

Le «Dictionnaire de la Biologie» comprend:

- un Glossaire quadrilingue (anglais-allemand-français-espagnol)
- trois index monolingues (allemand, français, espagnol).

Le glossaire est rangé dans l'ordre alphabétique des termes anglais, ce qui rend superflu un index anglais. Tous les termes sont numérotés à la suite. Chaque terme anglais est suivi d'équivalents allemand, français et espagnol.

Dans la partie anglaise, les expressions adjectif plus substantif, telles que *adaptive radiation* ou autres termes composés sont rangées d'après le premier élément de l'expression, dans notre cas l'adjectif «adaptive».

Dans la terminologie allemande, on a préféré l'écriture Duden (par exemple *Kode, Nukleo . . ., Zyto . . .*), bien que la littérature scientifique écrive aussi couramment *Code, Nucleo . . ., Cyto . . .*

Le dictionnaire utilise les abréviations suivantes:

adj.	adjectif	micr.	microscopie
anat.	anatomie	microb.	microbiologie
bot.	botanique	n	neutre
ecol.	écologie	neur.	neurophysiologie
embr.	embryologie	obs.	obsolète, désuet
env.	environnement	phys.	physiologie
eth.	éthologie	pl	pluriel
evol.	évolution	pop.	populaire
f	féminin	sing.	singulier
gen.	génétique	sub.	substantif
geol.	géologie	tax.	taxonomie
ins.	insectes	US	terme utilisé aux Etats-Unis
LA	terme utilisé en Amérique latine	zool.	zoologie
m	masculin		

~ signifie la répétition d'un ou plusieurs éléments de l'expression précédente, par ex. Dominanzwechsel, ~umkehr (= Dominanzumkehr), ou: active substance, ~principle (= active principle).

LES AUTEURS

Introducción

El «Diccionario de Biología» comprende:

- un glosario en cuatro lenguas (inglés-alemán-francés-español)
- tres índices alfabéticos monolingües (alemán, francés, español).

El glosario sigue el orden alfabético de los términos ingleses, de manera que no hace falta un índice alfabético de éstos. En la parte principal del diccionario cada término va numerado. Al lado de la columna inglesa aparecen los equivalentes alemanes, franceses y españoles. En la parte inglesa las expresiones compuestas de adjetivo + substantivo, como por ej. *adaptive radiation,* u otros compuestos aparecen en el orden alfabético del primer elemento de la expresión compuesta, en nuestro caso de «adaptive».

En cuanto a la terminología alemana, hemos dado preferencia a la grafía según el diccionario de Duden (por ej. *Kode, Nukleo* . . ., *Zyto*. . .), si bien en la bibliografía científica se da también con frecuencia la grafía *Code, Nucleo* . . ., *Cyto* . . ., etc.

En el Diccionario se usan las siguientes abreviaturas:

adj.	adjetivo
anat.	anatomía
bot.	botánica
ecol.	ecología
embr.	embriología
env.	medio ambiente
eth.	etología
evol.	evolución
f	femenino
gen.	genética
geol.	geología
ins.	insectos
LA	término empleado en América latina
m	masculino
micr.	microscopia
microb.	microbiología
n	neutro
neur.	neurofisiología
obs.	obsoleto
phys.	fisiología
pl	plural
pop.	popular
sing.	singular
sub.	sustantivo
tax.	taxonomía
US	término empleado en los Estados Unidos
zool.	zoología

~ significa repetición de uno o varios elementos de una expresión compuesta anteriormente citada, por ej. Dominanzwechsel, ~ umkehr (= Dominanzumkehr); active substance, ~ principle (= active principle).

LOS AUTORES

Dictionary in four languages

Viersprachiges Wörterbuch

Dictionnaire en quatre langues

Diccionario en cuatro lenguas

	English	Deutsch
1	abdomen, belly	Bauch *m*, Unterleib *m*, Abdomen *n*; Hinterleib *m ins.*
2	abdominal, ventral	abdominal, Bauch...
3	abdominal cavity, peritoneal ~	Bauchhöhle *f*, Leibeshöhle *f*, Peritonealhöhle *f*
4	abdominal leg → pleopod	—
5	abdominal respiration	Bauchatmung *f*
6	abdominal wall	Bauchwand *f*
7	abducens nerve	Abduzens *m*, seitlicher Augenmuskelnerv *m*
8	abductor, abducent muscle	Abduktor *m*, Abziehmuskel *m*
9	aberration	Aberration *f*
10	aberration rate	Aberrationsrate *f*
11	abiogenesis → spontaneous generation	—
12	abiogenetic	abiogenetisch
13	abiotic	abiotisch
14	able to reproduce → reproducible	—
15	abnormal	abnorm
16	abnormality	Abnormität *f*
17	abomasum, read, true stomach	Labmagen *m*
18	abortive transduction	abortive Transduktion *f*
19	above threshold → supraliminar	—
20	to abscise	abwerfen (Pflanzenteile)
21	abscisin	Abszisin *n*
22	abscission	Abszission *f*, Abwerfen *n* (von Pflanzenteilen)
23	absciss(ion) layer	Trennungsschicht *f*
24	abscission zone	Trennungszone *f*
25	absolute threshold	absolute Schwelle *f*
26	to absorb	absorbieren
27	absorbent	Absorbens *n*, Absorptionsmittel *n*
28	absorbing, absorptive	absorbierend
29	absorption	Absorption *f*
30	absorption capacity → absorptive ~	—
31	absorption cross → grading up	—
32	absorption curve	Absorptionskurve *f*
33	absorption spectrum	Absorptionsspektrum *n*
34	absorptive → absorbing	—
35	absorptive capacity, ~ power, absorption capacity	Absorptionsfähigkeit *f*, Absorptionsvermögen *n*
36	absorptive capacity of the soil → suction force of the soil	—
37	abundance	Abundanz *f*
38	abyssal, abysmal	abyssisch, abyssal, Tiefsee...
39	abyssal fauna, deep-sea ~	Tiefseefauna *f*
40	abyssal zone	Abyssal *n*, abyssale Zone *f*
41	acantha → prickle	—
42	acanthocarpous	stachelfrüchtig
43	Acanthocephala, acanthocephalans, spiny-headed worms	Kratzer *mpl*
44	acauline, stemless	stengellos
45	acceleration	Akzeleration *f*, Entwicklungsbeschleunigung *f*
46	acceptor	Akzeptor *m*
47	accessory bud	Beiknospe *f*, akzessorische Knospe *f*
48	accessory cell → auxiliary cell	—
49	accessory gland	akzessorische Drüse *f*
50	accessory nerve	Akzessorius *m*, Beinnerv *m*
51	accessory pigment	akzessorisches Pigment *n*
52	accidental host	Gelegenheitswirt *m*
53	acclimatization, acclimatation; acclimation *US*	Akklimatisierung *f*, Akklimatisation *f*
54	to acclimatize, to acclimate	akklimatisieren
55	accommodation	Akkommodation *f*
56	acellular, noncellular	nichtzellig
57	acentric	azentrisch
58	acetabulum	Hüftgelenkpfanne *f*
59	acetaldehyde	Azetaldehyd *n*

	Français	Español
1	abdomen *m*, ventre *m*	abdomen *m*, vientre *m*
2	abdominal, ventral	abdominal, ventral
3	cavité *f* abdominale, ~ péritonéale	cavidad *f* abdominal, ~ peritoneal
4	—	—
5	respiration *f* abdominale	respiración *f* abdominal
6	paroi *f* abdominale	pared *f* abdominal
7	nerf *m* oculaire externe, ~ abducens	nervio *m* abducens, ~ ocular externe
8	muscle *m* abducteur	músculo *m* abductor
9	aberration *f*	aberración *f*
10	taux *m* d'aberration	porcentaje *m* de aberración
11	—	—
12	abiogénétique	abiogenético
13	abiotique	abiótico
14	—	—
15	anormal	anormal
16	anomalie *f*	anormalidad *f*
17	abomasum *m*, caillette *f*	abomaso *m*, cuajar *m*, cuajo *m*
18	transduction *f* abortive	tra(n)sducción *f* abortiva
19	—	—
20	perdre (feuilles etc.)	desprenderse
21	abscissine *f*	abscisina *f*
22	abscis(s)ion *f*	abscisión *f*, desprendimiento *m*
23	couche *f* d'abscis(s)ion, ~ séparatrice	capa *f* de abscisión, estrato *m* ~ ~
24	zone *f* d'abscis(s)ion, ~ de séparation	zona *f* de abscisión
25	seuil *m* absolu	umbral *m* absoluto
26	absorber	absorber
27	absorbant *m*, substance *f* absorbante	absorbente *m*
28	absorbant	absorbente
29	absorption *f*	absorción *f*
30	—	—
31	—	—
32	courbe *f* d'absorption	curva *f* de absorción
33	spectre *m* d'absorption	espectro *m* de absorción
34	—	—
35	pouvoir *m* absorbant, capacité *f* absorbante, ~ d'absorption	poder *m* absorbente, ~ de absorción
36	—	—
37	abondance *f*	abundancia *f*
38	abyssal	abisal, abismal, abísico
39	faune *f* abyssale	fauna *f* abisal
40	abysse *m*	zona *f* abisal
41	—	—
42	acanthocarpe	acantocarpo, acantocárpico
43	acanthocéphales *mpl*	acantocéfalos *mpl*, gusanos *mpl* de cabeza espinosa
44	acaule	acaule
45	accélération *f*	aceleración *f*
46	accepteur *m*	aceptor *m*
47	bourgeon *m* accessoire	yema *f* accesoria
48	—	—
49	glande *f* accessoire	glándula *f* accesoria
50	nerf *m* accessoire	nervio *m* accesorio
51	pigment *m* accessoire	pigmento *m* accesorio
52	hôte *m* occasionnel, ~ facultatif	huésped *m* accidental
53	acclimatation *f*	aclimatación *f*
54	acclimater	aclimatar
55	accommodation *f*	acomodación *f*
56	acellulaire	acelular, no celular
57	acentrique	acéntrico
58	acétabule *m*	acetábulo *m*
59	acétaldéhyde *m*, aldéhyde *m* acétique	acetaldehído *m*, aldehído *m* acético

60	acetic acid	Essigsäure *f*
61	acetic fermentation	Essigsäuregärung *f*
62	aceto-acetic acid	Azetessigsäure *f*
63	Acetobacter, vinegar bacteria	Essig(säure)bakterien *fpl*
64	acetone	Azeton *n*
65	acetylation	Azetylierung *f*
66	acetylcholine	Azetylcholin *n*, Vagusstoff *m*
67	acetylcholinesterase	Azetylcholinesterase *f*
68	achene, achenium	Achäne *f*
69	achlamydeous	achlamydeisch
70	achlorophyllous	chlorophyllfrei
71	achromasie *gen.*	Achromasie *f gen.*
72	achromatic	achromatisch
73	achromatic spindle	achrome Spindel *f*
74	achromatin, linin	Achromatin *n*, Linin *n*
75	achromic → coulourless	—
76	acicle, acicula	Nadel *f bot.*
77	acicular, aciform	nadelförmig
78	acicular leaf	Nadelblatt *n*
79	acicular-leaved tree → conifer	—
80	acid amide	Säureamid *n*
81	acid-base balance	Säure-Basen-Gleichgewicht *n*
82	acid dye	saurer Farbstoff *m*
83	acid-fast, ~-proof, ~-resistant	säurefest, säurebeständig
84	acid humus → raw humus	—
85	acid-nonfast, ~-nonresistant	nichtsäurefest
86	acid resistance	Säurefestigkeit *f*
87	acid-soluble	säurelöslich
88	acidic agglutination	Säureagglutination *f*
89	acidification	Ansäuerung *f*; Versauerung *f (Boden)*
90	to acidify	ansäuern; versauern *(Boden)*
91	acidity	Säuregehalt *m*
92	acidity indicator *(plant)*	Säurezeiger *m (Pflanze)*
93	acidophil(ic), acidophilous	azidophil, säureliebend
94	acoustic... → auditory...	—
95	acoustical signal, sound ~	Schallsignal *n*
96	acquired character, ~ characteristic	erworbenes Merkmal *n*, erworbene Eigenschaft *f*
97	acquired immunity	erworbene Immunität *f*
98	Acrania, Cephalochordata, cephalochordates	Akranier *pl*, Schädellose *pl*
99	acrocarpic, acrocarpous	akrokarp, gipfelfrüchtig
100	acrocentric	akrozentrisch
101	acrogynous	akrogyn
102	acropetal	akropetal
103	acrosome, perforatorium	Akrosom *n*, Perforatorium *n*
104	ACTH → adrenocorticotropic hormone	—
105	actin	Aktin *n*
106	actinobiology → radiobiology	—
107	actinomorphic, actinomorphous	aktinomorph
108	actinomorphy	Aktinomorphie *f*
109	actinomycete	Strahlenpilz *m*
110	actinomycin	Aktinomyzin *n*
111	Actinopterygii, actinopterygian fish	Strahlenflosser *mpl*
112	Actinozoa → Anthozoa	—
113	action current	Aktionsstrom *m*
114	action potential	Aktionspotential *n*
115	action spectrum	Aktionsspektrum *n*, Wirkungsspektrum *n*
116	action system	Aktionssystem *n*
117	to activate	aktivieren
118	activated sludge	Belebtschlamm *m*, Aktivschlamm *m*, belebter Schlamm *m*
119	activated-sludge method	Belebtschlammverfahren *n*, Schlammbelebungsverfahren *n*
120	activating enzyme	Aktivatorenzym *n*
121	activation	Aktivierung *f*
122	activation energy	Aktivierungsenergie *f*

60	acide *m* acétique	ácido *m* acético
61	fermentation *f* acétique	fermentación *f* acética
62	acide *m* acétyle-acétique, ~acétoacétique	ácido *m* acetoacético
63	acétobactéries *f pl*, bactéries *f pl* acétiques	bacterias *f pl* acéticas
64	acétone *f*	acetona *f*
65	acétylation *f*	acetilación *f*
66	acétylcholine *f*, substance *f* vagale	acetilcolina *f*, sustancia *f* vagal
67	estérase *f* d'acétylcholine	esterasa *f* de acetilcolina
68	akène *m*, achaine *m*, achène *m*	aquenio *m*
69	achlamydé	aclamídeo
70	non chlorophyllien, dépourvu de chlorophylle	sin clorofila
71	achromasie *f gen.*	acromasia *f gen.*
72	achromatique	acromático
73	fuseau *m* achromatique	huso *m* acromático
74	achromatine *f*, linine *f*	acromatina *f*, linina *f*
75	—	—
76	aiguille *f bot.*	acícula *f*
77	aciculaire, aciforme	acicular
78	feuille *f* aciculaire, ~ en aiguille	hoja *f* acicular
79	—	—
80	amide *m* acide	amida *f* de ácido
81	équilibre *m* acido-basique	equilibrio *m* ácido-base
82	colorant *m* acide	colorante *m* ácido
83	acido-résistant, résistant aux acides	ácidorresistente, ácido-resistente
84	—	—
85	acido-nonrésistant	no ácido-resistente
86	acidorésistance *f*, résistance *f* aux acides	ácidorresistencia *f*, resistencia *f* a los ácidos
87	acido-soluble	soluble en ácido
88	agglutination *f* acide, ~ acidique	aglutinación *f* ácida
89	acidification *f*	acidificación *f*
90	acidifier	acidificar
91	acidité *f*	acidez *f*
92	indicateur *m* de l'acidité (du sol)	indicador *m* de la acidez (del suelo)
93	acidophile, acidiphile	acidófilo
94	—	—
95	signal *m* acoustique, ~ sonore	señal *f* acústica
96	caractère *m* acquis, caractéristique *f* acquise	carácter *m* adquirido, característica *f* adquirida
97	immunité *f* acquise	inmunidad *f* adquirida
98	acraniens *m pl*, céphalocordés *m pl*	acranios *m pl*, cefalocordados *m pl*
99	acrocarpe	acrocárpico
100	acrocentrique	acrocéntrico
101	acrogyne	acrógino
102	acropète	acropetal
103	acrosome *m*, perforateur *m*	acrosoma *m*, perforatorium *m*
104	—	—
105	actine *f*	actina *f*
106	—	—
107	actinomorphe	actinomorfo
108	actinomorphie *f*	actinomorfia *f*
109	actinomycète *m*	actinomiceto *m*
110	actinomycine *f*	actinomicina *f*
111	actinoptérygiens *m pl*	actinopterigios *m pl*
112	—	—
113	courant *m* d'action	corriente *f* de acción
114	potentiel *m* d'action	potencial *m* de acción
115	spectre *m* d'action	espectro *m* de acción
116	schéma *m* d'action	sistema *m* de acción
117	activer	activar
118	boues *f pl* activées	cieno *m* activado, fango *m* ~
119	procédé *m* des boues activées	proceso *m* de activación del fango, ~ del fango activo
120	enzyme *f* activatrice, ~ d'activation	enzima *m* activante, ~ activador
121	activation *f*	activación *f*
122	énergie *f* d'activation	energía *f* de activación

123	activator	Aktivator *m*
124	active immunity	aktive Immunität *f*
125	active site, ~ centre	aktives Zentrum *n*
126	active substance, ~ principle	Wirkstoff *m*
127	active transport	aktiver Transport *m*
128	activity rhythm	Aktivitätsrhythmus *m*
129	actomyosin	Aktomyosin *n*
130	acyclic	azyklisch
131	acylation	Azylierung *f*
132	adaptability, adaptivity	Anpassungsfähigkeit *f*
133	adaptable	anpassungsfähig
134	adaptation	Anpassung *f*, Adap(ta)tion *f*
135	adapted, fitted	angepaßt
136	adapted to environment, fitted ~ ~	umweltangepaßt
137	adaptedness, adaptiveness, adaptive fitness	Angepaßtheit *f*
138	adapter molecule	Adaptormolekül *n*
139	adaptive	adaptiv
140	adaptive character, ~ trait	Anpassungsmerkmal *n*
141	adaptive colo(u)ration	Anpassungsfärbung *f*, Farbanpassung *f*
142	adaptive enzyme, inducible ~	adaptives Enzym *n*, induzierbares ~
143	adaptive hormone, adaptative ~	Anpassungshormon *n*, „Streßhormon" *n*
144	adaptive radiation	adaptive Radiation *f*
145	adaptive reaction	Anpassungsreaktion *f*
146	adaptive value, ~ significance	Anpassungswert *m*
147	adaptiveness → adaptedness	—
148	adaptivity → adaptability	—
149	adductor (muscle)	Adduktor *m*, Anziehmuskel *m*
150	adelphogamy	Adelphogamie *f*, Geschwisterbestäubung *f*
151	adenine	Adenin *n*
152	adenocyte → glandular cell	—
153	adenohypophysis	Adenohypophyse *f*, Drüsenhypophyse *f*
154	adenoid	adenoid, drüsenähnlich
155	adenosine diphosphate, ADP	Adenosindiphosphat *n*, Adenosindiphosphorsäure *f*, ADP
156	adenosine monophosphate, adenylic acid, AMP	Adenosinmonophosphat *n*, Adenosinmonophosphorsäure *f*, Adenylsäure *f*, AMP
157	adenosine triphosphate, ATP	Adenosintriphosphat *n*, Adenosintriphosphorsäure *f*, ATP
158	adenous → glandulous	—
159	adequate stimulus	adäquater Reiz *m*
160	adermin → pyridoxine	—
161	adhesion	Adhäsion *f*
162	adhesive disc	Haftscheibe *f*
163	adhesive organ, organ of attachment; holdfast *bot.*	Haftorgan *n*
164	adhesive power, adhesiveness	Adhäsionsfähigkeit *f*, Haftvermögen *n*
165	adipose, fatty	fetthaltig
166	adipose fin	Fettflosse *f*
167	adipose panniculus	Unterhautfettgewebe *n*
168	adipose tissue, fatty ~	Fettgewebe *n*
169	adiuretin → vasopressin	—
170	adjuvant	Adjuvans *n*
171	adolescence	Adoleszenz *f*
172	ADP → adenosine diphosphate	—
173	adrenal cortex, suprarenal ~	Nebennierenrinde *f*
174	adrenal cortical hormone	Nebennierenrindenhormon *n*
175	adrenal gland, suprarenal ~, ~ capsule	Nebenniere *f*

No.	Français	Español
123	activateur *m*	activador *m*, agente *m* ~
124	immunité *f* active	inmunidad *f* activa
125	site *m* actif, centre *m* ~	centro *m* activo, locus *m* ~
126	substance *f* active, matière *f* ~, principe *m* actif	sustancia *f* activa, materia *f* ~, principio *m* activo
127	transport *m* actif	transporte *m* activo
128	rythme *m* d'activité	ritmo *m* de actividad
129	actomyosine *f*	actomiosina *f*
130	acyclique	acíclico
131	acylation *f*	acilación *f*
132	adaptabilité *f*, adap(ta)tivité *f*, faculté *f* d'adaptation	adaptabilidad *f*, adaptividad *f*, aptitud *f* adaptiva, capacidad *f* de adaptación
133	adaptable	adaptable
134	adaptation *f*	adaptación *f*
135	adapté	adaptado
136	adapté au milieu	adaptado al medio ambiente
137	état *m* d'adaptation	estado *m* de adaptación
138	molécule *f* adaptrice	molécula *f* adaptadora
139	adaptif, adaptatif	adaptivo, adaptativo
140	caractère *m* adaptatif	carácter *m* adaptivo, ~ de adaptación
141	adaptation *f* chromatique	coloración *f* adaptativa
142	enzyme *m* adap(ta)tif, ~ inductible	enzima *f* adap(ta)tiva, ~ inducible
143	hormone *f* adaptative	hormona *f* adaptativa
144	radiation *f* adaptative, diversification *f* ~	radiación *f* adap(ta)tiva
145	réaction *f* adaptative, ~ d'adaptation	reacción *f* adaptativa, ~ de adaptación
146	valeur *f* adaptative, ~ d'adaptation	valor *m* adaptativo, ~ de adaptación
147	—	—
148	—	—
149	(muscle *m*) adducteur *m*	(músculo *m*) aductor *m*
150	adelphogamie *f*	adelfogamia *f*
151	adénine *f*	adenina *f*
152	—	—
153	adénohypophyse *f*	adenohipófisis *f*, hipófisis *f* glandular
154	adénoïde	adenoide
155	adénosine *f* diphosphate, diphosphate *m* d'adénosine, acide *m* adénosine diphosphorique, ADP	adenosina *f* difosfato, adenosindifosfato *m*, ADP
156	adénosine *f* monophosphate, monophosphate *m* d'adénosine, acide *m* adénosine monophosphorique, ~ adénylique, AMP	adenosina *f* monofosfato, adenosinmonofosfato *m*, ácido *m* adenílico, AMP
157	adénosine *f* triphosphate, triphosphate *m* d'adénosine, acide *m* adénosine triphosphorique, ATP	adenosina *f* trifosfato, adenosintrifosfato *m*, ATP
158	—	—
159	stimulus *m* adéquat	estímulo *m* adecuado
160	—	—
161	adhésion *f*, adhérence *f*	adhesión *f*, adherencia *f*
162	disque *m* adhésif	disco *m* adhesivo
163	organe *m* adhésif, ~ fixateur	órgano *m* fijador
164	pouvoir *m* adhérent, capacité *f* d'adhésion, adhésivité *f*	capacidad *f* de adhesión, adhesividad *f*
165	adipeux, graisseux	adiposo, graso
166	nageoire *f* adipeuse	aleta *f* adiposa
167	pannicule *m* adipeux	panículo *m* adiposo
168	tissu *m* adipeux, ~ graisseux	tejido *m* adiposo, ~ graso
169	—	—
170	adjuvant *m*	adjutor *m*
171	adolescence *f*	adolescencia *f*
172	—	—
173	surrénale *f* corticale, cortico-surrénale *f*, cortex *m* surrénal	corteza *f* suprarrenal, ~ adrenal, corticosuprarrenal *f*
174	hormone *f* cortico-surrénale	hormona *f* cortical adrenal, ~ adrenocortical, ~ corticosuprarrenal
175	(capsule *f*) surrénale *f*, glande *f* ~	(glándula *f*) suprarrenal *f*, ~ adrenal, cápsula *f* suprarrenal

176	adrenal hormone	Nebennierenhormon *n*
177	adrenal medulla, suprarenal ~	Nebennierenmark *n*
178	adrenal medullary hormone	Nebennierenmarkhormon *n*
179	adrenaline, adrenin(e), epinephrin(e)	Adrenalin *n*
180	adrenergic	adrenergisch
181	adrenocorticotropic hormone, adreno-(cortico) tropin, corticotrop(h)in, ACTH	Adrenokortikotropin *n*, adrenokortikotropes Hormon *n*, Kortikotropin *n*, ACTH
182	adrenosterone	Adrenosteron *n*
183	to adsorb	adsorbieren
184	adsorbent	Adsorbens *n*, Adsorptionsmittel *n*
185	adsorption	Adsorption *f*
186	adsorption capacity, adsorbency	Adsorptionsfähigkeit *f*, ~vermögen *n*
187	adsorption chromatography	Adsorptions-Chromatographie *f*
188	adult	adult, erwachsen
189	adult (animal)	Adulttier *n*, Alttier *n*
190	adult mo(u)lt, imaginal ~	Imaginalhäutung *f*
191	adult stage, adulthood	adulte Periode *f*, Erwachsenenstadium *n*
192	adventitious bud	Adventivknospe *f*
193	adventitious embryony	Adventivembryonie *f*
194	adventitious plant	Adventivpflanze *f*
195	adventitious root	Adventivwurzel *f*
196	aecidiospore, spring-spore	Äzidienspore *f*, Aecidienspore *f*
197	aecidium, aecium	Äzidium *n*, Aecidie *f*
198	aerenchyma, aerating tissue	Aerenchym *n*, Durchlüftungsgewebe *n*
199	aerial (growing) parts	oberirdische Teile *m/n pl*
200	aerial root	Luftwurzel *f*
201	aerial shoot	Luftspross *m*
202	aerobe	Aerobier *m*, Aerobiont *m*
203	aerobic, aerobian, oxybiotic	aerob
204	aerobiology	Aerobiologie *f*
205	aerobiosis	Aerobiose *f*, Oxybiose *f*
206	aerophyte	Aerophyt *m*, Luftpflanze *f*
207	aeroplankton	Aeroplankton *n*
208	aerotaxis	Aerotaxis *f*
209	aerotropism	Aerotropismus *m*
210	aesthetes	Ästheten *pl*
211	(a)estivation[1], summer dormancy	Sommerschlaf *m*
212	(a)estivation[2], prefloration	Knospendeckung *f*, Ästivation *f*
213	afference	Afferenz *f*
214	afferent	afferent, zuführend
215	affinity	Affinität *f*
216	to (af)forest	aufforsten
217	(af)forestation	Aufforstung *f*
218	aflagellar → atrichous	—
219	after-birth	Nachgeburt *f*, Fruchtkuchen *m*
220	after-brain → myelencephalon	—
221	after-discharge	Nachentladung *f*
222	after-potential	Nachpotential *n*
223	after-ripening	Nachreife *f*, Nachreifung *f*
224	after-stain	Nachfärbung *f*
225	agamete	Agamet *m*
226	agamic, agamous	agam
227	agamobium	Agamobium *n*
228	agamogenesis	Agamogenese *f*
229	agamogony	Agamogonie *f*
230	agamont	Agamont *m*
231	agamospecies	Agamospezies *f*
232	agamospermy	Agamospermie *f*
233	agar(-agar), gelose	Agar(-Agar) *n*, Gelose *f*
234	agar block	Agarblock *m*
235	agar culture	Agarkultur *f*

176	hormone *f* surrénalienne, ~ surrénale	hormona *f* suprarrenal
177	surrénale *f* médullaire, médullo-surrénale *f*	médula *f* suprarrenal, ~ adrenal
178	hormone *f* médullo-surrénale	hormona *f* medular adrenal, ~ adrenomedular, ~ medulosuprarrenal
179	adrénaline *f*, adrénine *f*, épinéphrine *f*	adrenalina *f*, adrenina *f*, epinefrina *f*
180	adrénergique	adrenérgico
181	hormone *f* adrénocorticotrope, (adréno)-corticotrophine *f*, cortico-stimuline *f*, ACTH	hormona *f* adrenocorticotropa, (adreno)-corticotropina *f*, ACTH
182	adrénostérone *f*	adrenosterona *f*
183	adsorber	adsorber
184	adsorbant *m*	adsorbente *m*
185	adsorption *f*	adsorción *f*
186	capacité *f* d'adsorption, pouvoir *m* adsorbant	poder *m* de adsorción
187	chromatographie *f* d'adsorption	cromatografía *f* por adsorción
188	adulte	adulto
189	adulte *m*	adulto *m*
190	mue *f* imaginale	muda *f* imaginal
191	stade *m* adulte, état *m* ~	estado *m* adulto, adultez *f*
192	bourgeon *m* adventif	yema *f* adventicia
193	formation *f* d'embryons adventifs	embrionía *f* adventicia
194	plante *f* adventive	planta *f* adventicia
195	racine *f* adventive	raíz *f* adventicia
196	écidiospore *f*	ecidióspora *f*
197	écidie *f*	ecidio *m*
198	aérenchyme *m*, tissu *m* d'aération	aerénquima *m*, tejido *m* aerífero
199	organes *mpl* aériens, parties *fpl* aériennes	partes *fpl* aéreas
200	racine *f* aérienne	raíz *f* aérea, ~ epigea
201	pousse *f* aérienne	brote *m* aéreo, vástago *m* epigeo
202	aérobie *m*	aerobio *m*
203	aérobie, oxybiotique	aerobio
204	aérobiologie *f*	aerobiología *f*
205	aérobiose *f*, oxybiose *f*	aerobiosis *f*, oxibiosis *f*
206	plante *f* aérophyte, ~ aéricole	aerófito *m*
207	aéroplancton *m*, plancton *m* aérien	aeroplancton *m*, plancton *m* aéreo
208	aérotactisme *m*	aerotaxis *f*
209	aérotropisme *m*	aerotropismo *m*
210	esthètes *mpl*	estetos *mpl*
211	estivation *f*, repos *m* estival, léthargie *f* estivale	estivación *f*, letargo *m* estival
212	estivation *f*, préfloraison *f*	estivación *f*, prefloración *f*
213	afférence *f*	aferencia *f*
214	afférent	aferente
215	affinité *f*	afinidad *f*
216	boiser	plantar bosques; aforestar (LA)
217	boisement *m*	plantación *f* de bosques; aforestación *f* (LA)
218	—	—
219	arrière-faix *m*, délivre *m*	secundinas *fpl*
220	—	—
221	décharge *f* réflexe polysynaptique, «after-discharge»	descarga *f* ulterior, posdescarga *f*
222	post-potentiel *m*, potentiel *m* consécutif	potencial *m* secundario, pos(t)potencial *m*
223	postmaturation *f*	posmaturación *f*
224	coloration *f* secondaire	tinción *f* complementaria
225	agamète *m*	agámeto *m*
226	agame	ágamo, agámico
227	agamobium *m*	agamobio *m*
228	agamogenèse *f*	agamogénesis *f*
229	agamogonie *f*	agamogonia *f*
230	agamonte *m*	agamonte *m*
231	agamoespèce *f*, espèce *f* agame	agamoespecie *f*
232	agamospermie *f*	agamospermia *f*
233	agar(-agar) *m*, gélose *f*	agar(-agar) *m*, gelosa *f*
234	bloc *m* d'agar, ~ de gélose	bloque *m* de agar
235	culture *f* sur agar	cultivo *m* de agar

236	agar gel	Agargel *n*
237	agar medium	Agarnährboden *m*
238	agar plate	Agarplatte *f*
239	to age	altern
240	age and area theory	Alter- und -Areal-Theorie *f*
241	age determination	Altersbestimmung *f*
242	age distribution	Altersverteilung *f*
243	age structure	Altersstruktur *f*
244	ageing → senescence	—
245	agenesis, agenesia	Agenesie *f*
246	agent	Agens *n* (*pl*: Agentien)
247	agglutinable	agglutinierbar
248	agglutinate	Agglutinat *n*
249	to agglutinate	agglutinieren
250	agglutination	Agglutination *f*
251	agglutination reaction	Agglutinationsreaktion *f*
252	agglutinin	Agglutinin *n*
253	agglutinogen	Agglutinogen *n*
254	aggregate	Aggregat *n*
255	aggregate fruit → composite ~	—
256	aggressive behavio(u)r	Aggressionsverhalten *n*
257	aggressive drive	Aggressionstrieb *m*, Angriffstrieb *m*
258	agmatoploid	agmatoploid
259	agmatoploidy	Agmatoploidie *f*
260	Agnatha	Agnathen *pl*
261	agonist, agonistic muscle	Agonist *m*
262	agrobiology	Agrobiologie *f*
263	A-horizon, eluvial ~	A-Horizont *m*, Oberboden *m*
264	air-bladder → swim-bladder	—
265	air breathing	Luftatmung *f*
266	air chamber *bot.*	Luftkammer *f bot.*
267	aircraft noise, aviation ~ *env.*	Fluglärm *m env.*
268	air passages → respiratory ~	—
269	air pollution, atmospheric ~	Luftverschmutzung *f*, ~ verseuchung *f*
270	air-pore *bot.*	Luftspalte *f bot.*
271	air-sac	Luftsack *m*
272	akaryotic → non-nucleated	—
273	akinesis, akinesia	Akinese *f*, Akinesie *f*, Bewegungslosigkeit *f*
274	akinetic	akinetisch
275	alanine	Alanin *n*
276	alarm call, warning ~	Warnruf *m*, Warnlaut *m*
277	alarm signal, warning ~	Warnsignal *n*
278	alarm substance	Schreckstoff *m*
279	alary muscle	Flügelmuskel *m*
280	alate, winged	geflügelt
281	albinism	Albinismus *m*
282	albino	Albino *m*
283	albumen[1], egg-white	Albumen *n*, Eiklar *n*
284	albumen[2]	Sameneiweiß *n*, Nährgewebe *n*
285	albumin	Albumin *n*
286	albuminoid	Albuminoid *n*
287	alburnum → sapwood	—
288	alcohol series	Alkoholreihe *f*
289	alcoholic fermentation	alkoholische Gärung *f*
290	aldehyde	Aldehyd *n*
291	aldehyde acid, aldehydic ~	Aldehydsäure *f*
292	aldosterone	Aldosteron *n*
293	alecithal, alecithic	alezithal, dotterlos
294	aleuron(e)	Aleuron *n*
295	aleuron(e) grain	Aleuronkorn *n*
296	aleuron(e) layer	Aleuronschicht *f*
297	aleuroplast → proteinoplast	—
298	alga (*pl* algae)	Alge *f*

236	gel *m* d'agar, ~ de gélose	gel *m* de agar
237	milieu *m* à agar, ~ gélosé	medio *m* de agar
238	plaque *f* d'agar, ~ de gélose	placa *f* de agar
239	vieillir	envejecer
240	théorie *f* de l'âge et de l'espace	teoría *f* de la edad y área
241	détermination *f* de l'âge	determinación *f* de la edad
242	répartition *f* par âge	distribución *f* por edades
243	structure *f* par âge	estructura *f* por edades
244	—	—
245	agénésie *f*	agénesis *f*, agenesia *f*
246	agent *m*	agente *m*
247	agglutinable	aglutinable
248	agglutinat *m*	aglutinado *m*
249	agglutiner	aglutinar
250	agglutination *f*	aglutinación *f*
251	réaction *f* d'agglutination	reacción *f* de aglutinación
252	agglutinine *f*	aglutinina *f*
253	agglutinogène *m*	aglutinógeno *m*
254	agrégat *m*	agregado *m*
255	—	—
256	comportement *m* agressif, ~ d'agression	comportamiento *m* agresivo, conducta *f* agresiva
257	impulsion *f* agressive	instinto *m* agresivo
258	agmatoploïde	agmatoploide
259	agmatoploïdie *f*	agmatoploidia *f*
260	agnathes *mpl*, agnathostomes *mpl*	agnatos *mpl*
261	agoniste *m*	músculo *m* agonista
262	agrobiologie *f*	agrobiología *f*
263	horizon *m* A, ~ éluvial	horizonte *m* A, ~ eluvial, superficie *f* del suelo
264	—	—
265	respiration *f* aérienne	respiración *f* aérea
266	chambre *f* à air, cavité *f* ~ ~ *bot.*	cámara *f* de aire, ~ aerífera *bot.*
267	bruits *mpl* émis par la navigation aérienne	ruidos *mpl* producidos por la aviación
268	—	—
269	pollution de l'air, ~ atmosphérique	polución *f* del aire, contaminación *f* atmosférica
270	lacune *f* aérifère *bot.*	estoma *m* aerífero, poro *m* ~
271	sac *m* aérien	saco *m* aéreo
272	—	—
273	akinésie *f*, acinèse *f*	acinesia *f*
274	acinétique, acinésique	acinético
275	alanine *f*	alanina *f*
276	cri *m* d'alarme	sonido *m* de alarma, grito *m* ~ ~
277	signal *m* d'alerte, ~ avertisseur	señal *f* advertidora, ~ preventiva
278	substance *f* d'alarme	sustancia *f* de alarma
279	muscle *m* alaire	músculo *m* alar
280	ailé	alado; pterigoto *ins.*
281	albinisme *m*	albinismo *m*
282	albinos *m*	albino *m*
283	albumen *m*, blanc *m* d'œuf	albumen *m*, clara *f* de huevo
284	albumen *m*	albumen *m*
285	albumine *f*	albúmina *f*
286	albuminoïde *m*	albuminoide *m*
287	—	—
288	série *f* des alcools	serie *f* de soluciones alcohólicas
289	fermentation *f* alcoolique	fermentación *f* alcohólica
290	aldéhyde *m*	aldehído *m*
291	acide *m* aldéhydique, ~ d'aldéhyde	ácido *m* aldehídico
292	aldostérone *f*	aldosterona *f*
293	alécithe, alécithique	alecito, alecítico
294	aleurone *f*	aleurona *f*
295	grain *m* d'aleurone	grano *m* de aleurona
296	couche *f* d'aleurone	capa *f* de aleurona
297	—	—
298	algue *f*	alga *f*

299	algal fungi → phycomycetes	—
300	alginic acid	Alginsäure *f*, Algensäure *f*
301	algology, phycology	Algenkunde *f*, Algologie *f*
302	Algonkian (era), Algonkin	Algonkium *n*
303	alimentary bolus	Nahrungsbissen *m*
304	alimentary canal, enteric ~	Verdauungskanal *m*
305	alimentary cycle → food cycle	—
306	alimentary deficiency → food shortage	—
307	aliphatic amino-acid	aliphatische Aminosäure *f*
308	alkalimeter	Alkalimeter *n*
309	alkalimetry	Alkalimetrie *f*
310	alkaline	alkalisch, laugenartig
311	alkalinity	Alkalinität *f*
312	alkaloid	Alkaloid *n*
313	all(a)esthetic	allaästhetisch
314	allantochorion, chorioallantoic membrane	Allantochorion *n*, Chorioallantois *f*
315	allantoin	Allantoin *n*
316	allantois	Allantois *f*, Urharnsack *m*
317	allele, allelomorph	Allel *n*, Allelomorph *n*
318	allelic, allelomorphic	allel, allelomorph
319	allelism, allelia, allelomorphism	Allelie *f*, Allelomorphismus *m*
320	allelomorph → allele	—
321	allelomorphic series	allelomorphe Serie *f*
322	allelotaxy	Allelotaxis *f*
323	allelotype	Allelotyp *m*
324	allergen	Allergen *n*
325	allergic	allergisch
326	allergy	Allergie *f*
327	alliance, protocooperation	Allianz *f*, Verband *m*
328	allobiosis	Allobiose *f*
329	allocarpy	Allokarpie *f*
330	allochroic	wechselfarbig
331	allochroism	Farbänderung *f*
332	allochronic, allochronous	allochron
333	allochthonous	allochthon, bodenfremd
334	allocycly	Allozyklie *f*
335	allodiploid	allodiploid
336	allodiploidy	Allodiploidie *f*
337	allogamous	allogam(isch)
338	allogamy, cross-pollination, cross-fertilization	Allogamie *f*, Fremdbestäubung *f*, ~befruchtung *f*, Kreuzbestäubung *f*
339	allogene, recessive allele	Allogen *n*, rezessives Allel *n*
340	allogenetic	allogenetisch
341	allogenic, allogenous	allogenisch
342	allohaploid	allohaploid
343	allohaploidy	Allohaploidie *f*
344	alloheteroploid	alloheteroploid
345	alloheteroploidy	Alloheteroploidie *f*
346	alloiogenesis	Alloiogenese *f*
347	allolysogenic	allolysogen
348	allometry	Allometrie *f*
349	allopatric	allopatrisch
350	alloplasm	Alloplasma *n*
351	allopolyploid	allopolyploid
352	allopolyploidy	Allopolyploidie *f*
353	all-or-none principle, ~ law	Alles-oder-Nichts-Gesetz *n*
354	allosomal	allosomal
355	allosome → heterochromosome	—
356	allosyndesis, allosynapsis	Allosyndese *f*, Allosynapsis *f*
357	allosyndetic	allosyndetisch
358	allotetraploid, amphidiploid	allotetraploid, amphidiploid
359	allotetraploidy, amphidiploidy	Allotetraploidie *f*, Amphidiploidie *f*
360	allotriploid	allotriploid
361	allozygote	Allozygote *f*
362	alternate *bot.*	wechselständig

299	—	—
300	acide *m* alginique	ácido *m* algínico
301	algologie *f*, phycologie *f*	algología *f*, ficología *f*
302	algonkien *m*	algonquino *m*
303	bol *m* alimentaire	bolo *m* alimenticio
304	tube *m* digestif, canal *m* ~	tubo *m* digestivo
305	—	—
306	—	—
307	amino-acide *m* aliphatique	aminoácido *m* alifático
308	alcalimètre *m*	alcalímetro *m*
309	alcalimétrie *f*	alcalimetría *f*
310	alcalin	alcalino
311	alcalinité *f*	alcalinidad *f*
312	alcaloïde *m*	alcaloide *m*
313	allaesthétique	alaestético
314	allanto-chorion *m*, membrane *f* chorio-allantoïdienne	corioalantoides *f*, membrana *f* corio-alantoidea
315	allantoïne *f*	alantoína *f*
316	allantoïde *f*	alantoides *f*, saco *m* alantoideo
317	allèle *m*, allélomorphe *m*	alelo *m*, alelomorfo *m*
318	allélique, allélomorphe	alélico, alelomórfico
319	allélisme *m*, allélomorphisme *m*	alelismo *m*, alelomorfismo *m*
320	—	—
321	série *f* allélomorphique	serie *f* alelomórfica
322	allélotaxie *f*	alelotaxia *f*, alelotaxis *f*
323	allélotype *m*	alelotipo *m*
324	allergène *m*	alergeno *m*
325	allergique	alérgico
326	allergie *f*	alergia *f*
327	alliance *f*	alianza *f*, protocooperación *f*
328	allobiose *f*	alobiosis *f*
329	allocarpie *f*	alocarpia *f*
330	allochroïque	alocroico
331	allochroïsme *m*	alocroismo *m*, alocromasia *f*
332	allochrone	alócrono
333	allochtone	alóctono
334	allocyclie *f*	alociclia *f*
335	allodiploïde	alodiploide
336	allodiploïdie *f*	alodiploidia *f*
337	allogame	alógamo
338	allogamie *f*, pollinisation *f* croisée, fertilisation *f* ~, fécondation *f* ~	alogamia *f*, polinización *f* cruzada, fertilización *f* ~, fecundación *f* ~
339	allogène *m*	alógeno *m*
340	allogénique	alogenético
341	allogène	alógeno
342	allohaploïde	alohaploide
343	allohaploïdie *f*	alohaploidia *f*
344	allohétéroploïde	aloheteroploide
345	allohétéroploïdie *f*	aloheteroploidia *f*
346	alloiogenèse *f*	aloiogénesis *f*
347	allolysogénique	alolisogénico
348	allométrie *f*	alometría *f*
349	allopatrique	alopátrico
350	alloplasme *m*	aloplasma *m*
351	allopolyploïde	alopoliploide
352	allopolyploïdie *f*	alopoliploidia *f*
353	loi *f* du tout ou rien	ley *f* del todo o nada
354	allosomique	alosómico
355	—	—
356	allosyndèse *f*	alosíndesis *f*
357	allosyndétique	alosindético
358	allotétraploïde, amphidiploïde	alotetraploide, anfidiploide
359	allotétraploïdie *f*, amphidiploïdie *f*	alotetraploidia *f*, anfidiploidia *f*
360	allotriploïde	alotriploide
361	allozygote *m*	alocigoto *m*
362	alterne *bot.*	alternado *bot.*

363	alternate host, alternative ~	Wechselwirt *m*
364	alternating dominance	Dominanzwechsel *m*, ~umkehr *f*
365	alternation of generation(s)	Generationswechsel *m*
366	alternative disjunction, disjunctional separation	Alternativverteilung *f*
367	alternative inheritance	alternative Vererbung *f*
368	altitudinal zone	Höhenstufe *f*
369	altricial bird → nidicolous bird	—
370	alveolar air	Alveolarluft *f*
371	alveolar duct	Alveolengang *m*
372	alveolar hypothesis, ~ theory	Alveolarhypothese *f*
373	alveolar structure	Wabenstruktur *f*
374	alveolus	Alveole *f*
375	ambient noise, environmental ~	Umweltlärm *m*
376	ambient temperature, environmental ~	Umgebungstemperatur *f*
377	ambisexual, ambosexual	ambisexuell
378	ambivalent	ambivalent
379	amboceptor	Ambozeptor *m*
380	ambulacral plate	Ambulakralplatte *f*
381	ambulacrum, ambulacral foot, ~ tube	Ambulakralfüßchen *n*
382	ameba → amoeba	—
383	ameiosis	Ameiose *f*
384	ameiotic	ameiotisch
385	ament → catkin	—
386	amentaceous plants, amentiferous ~	Kätzchenblütler *mpl*, Amentifloren *pl*
387	Ametabola → Apterygota	—
388	ametabolous	ametabol
389	amicron	Amikron *n*
390	amicroscopic	amikroskopisch
391	amide	Amid *n*
392	amination	Aminierung *f*
393	amine	Amin *n*
394	aminoacetic acid → glycine	—
395	amino acid	Aminosäure *f*
396	aminoacid activation	Aminosäureaktivierung *f*
397	aminoacid decarboxylase	Aminosäuredekarboxylase *f*
398	aminoacid oxidase	Aminosäureoxydase *f*
399	aminobenzoic acid	Aminobenzoesäure *f*
400	amino group	Aminogruppe *f*
401	amitosis	Amitose *f*, amitotische Kernteilung *f*, direkte ~ *f*
402	ammonia	Ammoniak *n*
403	amnion	Amnion *n*, Schafhaut *f*
404	amniotes	Amnioten *pl*
405	amniotic	amniotisch
406	amniotic cavity	Amnionhöhle *f*
407	amniotic fluid	Amnionflüssigkeit *f*, Fruchtwasser *n*
408	amniotic sac, ~ pouch	Amnionsack *m*, Fruchtwassersack *m*
409	am(o)eba	Amöbe *f*
410	am(o)ebocyte	Amöbozyt *m*
411	am(o)eboid	amöboid
412	am(o)eboid movement	amöboide Bewegung *f*
413	amorphous, amorphic	amorph, gestaltlos
414	amount of coverage → degree ~ ~	—
415	AMP → adenosine monophosphate	—
416	amphiaster	Amphiaster *m*
417	amphiastral mitosis	Amphiastralmitose *f*
418	Amphibia, amphibians, Bathracia	Amphibien *fpl*, Lurche *mpl*
419	amphibian, amphibious, amphibiotic	amphibisch
420	amphicarpous, amphicarpic	amphikarp
421	amphidiploid → allotetraploid	—
422	amphigastria *pl*	Amphigastrien *fpl*
423	amphigony, amphigenesis	Amphigonie *f*

363	hôte *m* alternant	huésped *m* alternativo, hospedante *m* ~
364	dominance *f* alternée	dominancia *f* alternativa
365	alternance *f* de(s) générations	alternación *f* de generaciones, alternancia *f* ~ ~
366	disjonction *f* alternative	disyunción *f* alternativa
367	hérédité *f* alternée	herencia *f* alternativa
368	étage *m* altitudinal	piso *m* altitudinal, ~ de altitud
369	—	—
370	air *m* alvéolaire	aire *m* alveolar
371	conduit *m* alvéolaire	conducto *m* alveolar
372	hypothèse *f* des alvéoles	hipótesis *f* de los alvéolos
373	structure *f* alvéolaire	estructura *f* alveolar
374	alvéole *f*	alvéolo *m*
375	bruits *mpl* ambiants, ambiance *f* sonore	ruido *m* ambiental
376	température *f* ambiante	temperatura *f* ambiente
377	ambisexuel, ambisexué	ambisexual
378	ambivalent	ambivalente
379	ambocepteur *m*	amboceptor *m*
380	plaque *f* ambulacraire	placa *f* ambulacral
381	ambulacre *m*, pied *m* ambulacraire	ambulacro *m*, pie *m* ambulacral
382	—	—
383	améiose *f*	ameiosis *f*
384	améiotique	ameiótico
385	—	—
386	amentacées *fpl*, plantes *fpl* amentifères	amentifloras *fpl*
387	—	—
388	amétabolique	ametábolo
389	amicron *m*	amicrón *m*
390	amicroscopique	amicroscópico
391	amide *m*	amida *f*
392	amination *f*	aminación *f*
393	amine *f*	amina *f*
394	—	—
395	acide *m* aminé, amino-acide *m*	aminoácido *m*, ácido *m* aminado
396	activation *f* des acides aminés	activación *f* de los aminoácidos
397	aminoacide-décarboxylase *f*	descarboxilasa *f* de aminoácidos
398	aminoacide-oxydase *f*, oxydase *f* d'amino-acide	oxidasa *f* de aminoácidos, aminoácido-oxidasa *f*
399	acide *m* aminobenzoïque	ácido *m* aminobenzoico
400	groupe(ment) *m* amino, ~ aminé	grupo *m* amino
401	amitose *f*	amitosis *f*, división *f* amitótica
402	ammoniac *m*	amoniaco *m*
403	amnios *m*	amnios *m*
404	amniotes *mpl*	amniotas *mpl*
405	amniotique	amniótico
406	cavité *f* amniotique	cavidad *f* amniótica
407	liquide *m* amniotique	líquido *m* amniótico
408	sac *m* amniotique	saco *m* amniótico, bolsa *f* amniótica
409	amibe *f*	ameba *f*, amiba *f*
410	amibocyte *m*	amebocito *m*
411	amiboïde	ameboide
412	mouvement *m* amiboïde	movimiento *m* ameboide
413	amorphe	amorfo
414	—	—
415	—	—
416	amphiaster *m*	anfiaster *m*
417	mitose *f* amphiastrale	mitosis *f* anfiastral
418	amphibies *mpl*, amphibiens *mpl*, batraciens *mpl*	anfibios *mpl*, batracios *mpl*
419	amphibie	anfibio
420	amphicarpe	anficárpico
421	—	—
422	amphigastres *mpl*	anfigastros *mpl*
423	amphigonie *f*	anfigonia *f*

424	amphikaryon, amphinucleus	Amphikaryon *n*
425	amphimictic	amphimiktisch
426	amphimixis, amphimixia	Amphimixis *f*, Amphimixie *f*
427	amphimutation	Amphimutation *f*
428	Amphineura	Urmollusken *fpl*, Amphineuren *pl*
429	amphiploid	amphiploid
430	amphitene	Amphitän *n*
431	amphitoky	Amphitokie *f*
432	amphogenic	amphogen
433	amplexicaul	stengelumfassend
434	amyelinate, amyelinic, non-myelinated, non-medullated	marklos *(Nervenfaser)*
435	amygdala → palatal tonsil	—
436	amylase, diastase	Amylase *f*, Diastase *f*
437	amyliferous, amylaceous	stärkehaltig, -führend, -reich
438	amylopectin	Amylopektin *n*
439	amyloplast	Amyloplast *m*
440	amylose	Amylose *f*
441	anabiosis	Anabiose *f*, Scheintod *m*; Trockenstarre *f*
442	anabolic	anabolisch
443	anabolism, constructive metabolism	Anabolismus *m*, Baustoffwechsel *m*
444	anaboly, anabolie	Anabolie *f*
445	anadromous	anadrom
446	anaerobe	Anaerobier *m*, Anaerobiont *m*
447	anaerobic, anoxybiotic	anaerob
448	anaerobiosis, anoxybiosis	Anaerobiose *f*, Anoxybiose *f*
449	anagenesis	Anagenese *f*, Höherentwicklung *f*
450	anal fin	Analflosse *f*
451	anal gland	Afterdrüse *f*, Analdrüse *f*
452	anal papilla	Analpapille *f*
453	analogous, analogic(al)	analog
454	analogue	Analoge *n*
455	analogy	Analogie *f*
456	to analyse	analysieren
457	analyser	Analysator *m*
458	analysis	Analyse *f*
459	analysis of variance	Varianzanalyse *f*
460	anamniotes	Anamnier *pl*
461	anaphase	Anaphase *f*
462	anaphase movement	Anaphasebewegung *f*
463	anaphoresis	Anaphorese *f*
464	anaphylaxis	Anaphylaxie *f*
465	to anastomose	anastomosieren
466	anastomosis	Anastomose *f*
467	anastral mitosis	Anastralmitose *f*
468	anatomist	Anatom *m*
469	anatomy	Anatomie *f*
470	anatonosis	Anatonose *f*
471	anatropous	anatrop
472	ancestor, ascendent	Vorfahr *m*, Ahn *m*, Aszendent *m*
473	androdioecism, androdioecy	Androdiözie *f*
474	androecium	Andrözeum *n*
475	androgen, male hormone	Androgen *n*, männliches Geschlechtshormon *n*
476	androgenesis	Androgenese *f*
477	androgynal, androgynous	androgyn
478	androgynism, androgyny	Androgynie *f*
479	androhermaphrodite	Androhermaphrodit *m*
480	andromerogony	Andromerogonie *f*
481	androsome	Androsom *n*
482	androsporangium → microsporangium	—
483	androspore → microspore	—
484	androsterone	Androsteron *n*
485	anelectrotonus	Anelektrotonus *m*

424	amphicaryon *m*	anficario *m*
425	amphimictique	anfimíctico
426	amphimixie *f*	anfimixia *f*, anfimixis *f*
427	amphimutation *f*, mutation *f* double	anfimutación *f*, mutación *f* doble
428	amphineures *mpl*	anfineuros *mpl*
429	amphiploïde	anfiploide
430	amphitène *m*	anfiteno *m*
431	amphitoquie *f*	anfitoquía *f*
432	amphogène	anfogénico
433	amplexicaule, embrassant la tige	amplexicaule
434	amyélinique	amielínico, no mielinizado, no medulado
435	—	—
436	amylase *f*, diastase *f*	amilasa *f*, diastasa *f*
437	amylacé, amylifère	amiláceo, amilífero
438	amylopectine *f*	amilopectina *f*
439	plaste *m* amylifère	amiloplasto *m*
440	amylose *m*	amilosa *f*
441	anabiose *f*	anabiosis *f*
442	anabolique	anabólico
443	anabolisme *m*	anabolismo *m*, metabolismo *m* constructivo, ~ asimilador
444	anabolie *f*	anabolia *f*
445	anadrome	anádromo
446	anaérobie *m*, anaérobionte *m*, anoxybionte *m*	anaerobio *m*, anoxibionte *m*
447	anaérobie, anoxybiotique	anaerobio
448	anaérobiose *f*, anoxybiose *f*	anaerobiosis *f*, anoxibiosis *f*
449	anagenèse *f*, évolution *f* progressive	anagénesis *f*, evolución *f* progresiva
450	nageoire *f* anale	aleta *f* anal
451	glande *f* anale	glándula *f* anal
452	papille *f* anale	papila *f* anal
453	analogue, analogique	análogo, analógico
454	analogue *m*	análogo *m*
455	analogie *f*	analogía *f*
456	analyser	analizar
457	analyseur *m*	analizador *m*
458	analyse *f*	análisis *m*
459	analyse *f* de variance	análisis *m* de varianza
460	anamniés *mpl*, anamniotes *mpl*	anamnios *mpl*, anamniotas *mpl*
461	anaphase *f*	anáfase *f*
462	mouvement *m* anaphasique	movimiento *m* anafásico
463	anaphorèse *f*	anaforesis *f*
464	anaphylaxie *f*	anafilaxis *f*
465	anastomoser	anastomosar
466	anastomose *f*	anastomosis *f*
467	mitose *f* anastérale	mitosis *f* anastral
468	anatomiste *m*	anatomista *m*
469	anatomie *f*	anatomía *f*
470	anatonose *f*	anatonosis *f*
471	anatrope	anátropo
472	ancêtre *m*, ascendant *m*	antecesor *m*, antepasado *m*, ascendiente *m*
473	androdioécie *f*	androdioecia *f*
474	androcée *m*	androceo *m*
475	androgène *m*, hormone *f* mâle	andrógeno *m*, hormona *f* masculina
476	androgenèse *f*, androgénie *f*	androgénesis *f*
477	androgyne	andrógino
478	androgynie *f*	androginia *f*
479	androhermaphrodite *m*	androhermafrodita *m*
480	andromérogonie *f*	andromerogonia *f*
481	androsome *m*	androsoma *m*
482	—	—
483	—	—
484	androstérone *f*	androsterona *f*
485	anélectrotonus *m*	anelectrotono *m*

486	anemochory, wind dispersal	Anemochorie *f*, Windverbreitung *f* (des Samens)
487	anemogamy → anemophily	—
488	anemophilous	windblütig, anemophil
489	anemophilous plant	Windblütler *m*
490	anemophily, anemogamy, wind-pollination	Windblütigkeit *f*, Windbestäubung *f*, Anemogamie *f*, Anemophilie *f*
491	anemotaxis	Anemotaxis *f*
492	aneuploid	aneuploid
493	aneuploidy	Aneuploidie *f*
494	aneurine → thiamine	—
495	angiology	Angiologie *f*, Gefäßlehre *f*
496	angiospermous	bedecktsamig
497	angiosperms	Angiospermen *pl*, Bedecktsamer *mpl*, bedecktsamige Pflanzen *fpl*
498	angle of divergence	Divergenzwinkel *m*
499	angustifoliate, narrow-leaved	schmalblättrig
500	anhydride	Anhydrid *n*
501	animal anatomy → zootomy	—
502	animal behavio(u)r	tierisches Verhalten *n*
503	animal breeder	Tierzüchter *m*
504	animal breeding	Tierzucht *f*, Tierzüchtung *f*
505	animal cell	Tierzelle *f*
506	animal dispersal → zoochory	—
507	animal ecology	Tierökologie *f*, Zoo-Ökologie *f*
508	animal experiment → experimentation on animals	—
509	animal genetics → zoogenetics	—
510	animal geography → zoogeography	—
511	animal heat, caloricity	tierische Wärme *f*
512	animal host	Wirtstier *n*
513	animal kingdom	Tierreich *n*
514	animal life	Tierleben *n*
515	animal migration	Tierwanderung *f*
516	animal parasite	Tierschmarotzer *m*
517	animal phylum	Tierstamm *m*
518	animal physiology	Tierphysiologie *f*
519	animal plankton → zooplankton	—
520	animal pole	animaler Pol *m*
521	animal protection	Tierschutz *m*
522	animal protein	tierisches Eiweiß *n*
523	animal protein factor, APF	tierischer Proteinfaktor *m*
524	animal society	Tiergesellschaft *f*
525	animal sociology	Tiersoziologie *f*, Zoozönologie *f*
526	animal species	Tierart *f*
527	animal starch → glycogen	—
528	animal systematics, ~ taxonomy	Tiersystematik *f*
529	animal toxin → zootoxin	—
530	animal world	Tierwelt *f*
531	anion	Anion *n*
532	anisogamete, heterogamete	Anisogamet *m*, Heterogamet *m*
533	anisogamy, heterogamy	Anisogamie *f*, Heterogamie *f*
534	anisogeny	Anisogenie *f*
535	anisophylly	Anisophyllie *f*
536	anisoploidy	Anisoploidie *f*
537	anisotropic	anisotrop
538	anisotropy	Anisotropie *f*
539	anlage → primordium	—
540	Annelida, annelids	Gliederwürmer *mpl*, Ringelwürmer *mpl*, Anneliden *pl*
541	annual *bot.*	einjährig
542	annual periodicity	Jahresperiodik *f*, ~rhythmik *f*
543	annual (plant)	Annuelle *f*, einjährige Pflanze *f*
544	annual ring, tree ~	Jahresring *m*
545	annular	ringartig, ringförmig
546	annular thickening	Ringverdickung *f*

486	anémochorie *f*, dispersion *f* par le vent	anemocoria *f*
487	—	—
488	anémophile, anémogame	anemófilo, anemógamo
489	plante *f* anémophile	planta *f* anemófila
490	anémophilie *f*, anémogamie *f*	anemofilia *f*, anemogamia *f*
491	anémotaxie *f*, anémotactisme *m*	anemotaxis *f*, anemotactismo *m*
492	aneuploïde	aneuploide
493	aneuploïdie *f*	aneuploidia *f*
494	—	—
495	angiologie *f*	angiología *f*
496	angiosperme	angiospermo
497	angiospermes *fpl*	angiospermas *fpl*, angiospermos *mpl*
498	angle *m* de divergence	ángulo *m* de divergencia
499	angustifolié	angustifolio
500	anhydride *m*	anhídrido *m*
501	—	—
502	comportement *m* animal	comportamiento *m* animal
503	zootechnicien *m*	zootécnico *m*
504	sélection *f* animale	selección *f* animal
505	cellule *f* animale	célula *f* animal
506	—	—
507	écologie *f* animale	ecología *f* animal
508	—	—
509	—	—
510	—	—
511	chaleur *f* animale	calor *m* animal, caloricidad *f*
512	hôte *m* animal	animal *m* huésped, hospedante *m* animal
513	règne *m* animal	reino *m* animal
514	vie *f* animale	vida *f* animal, ~ de los animales
515	migration *f* d'animaux	migración *f* animal
516	parasite *m* animal	parásito *m* animal
517	phylum *n* animal	filum *m* animal
518	physiologie *f* animale, zoophysiologie *f*	fisiología *f* animal
519	—	—
520	pôle *m* animal	polo *m* animal
521	protection *f* des animaux	protección *f* de animales
522	protéine *f* animale	proteína *f* animal
523	facteur *m* de protéine animale	factor *m* proteínico animal
524	société *f* animale	sociedad *f* animal
525	sociologie *f* animale, zoocénotique *f*	sociología *f* animal
526	espèce *f* animale	especie *f* animal
527	—	—
528	systématique *f* animale	sistemática *f* animal, taxonomía *f* ~
529	—	—
530	monde *m* animal	mundo *m* animal
531	anion *m*	anion *m*
532	anisogamète *m*, hétérogamète *m*	anisogameto *m*, heterogameto *m*
533	anisogamie *f*, hétérogamie *f*	anisogamia *f*, heterogamia *f*
534	anisogénie *f*	anisogenia *f*
535	anisophyllie *f*	anisofilia *f*
536	anisoploïdie *f*	anisoploidia *f*
537	anisotrope	anisotrópico
538	anisotropie *f*	anisotropia *f*
539	—	—
540	annélides *m/fpl*, vers *mpl* annelés	anélidos *mpl*, gusanos *mpl* anillados
541	annuel *bot.*	anual *bot.*
542	périodicité *f* annuelle	periodicidad *f* anual
543	annuelle *f*, plante *f* ~	anual *f*, planta *f* ~
544	anneau *m* annuel, cerne *m*	anillo *m* anual
545	annulaire	anular
546	épaississement *m* annelé	engrosamiento *m* anular

547	annular tracheid	Ringtracheide *f*
548	annular vessel, ring ~	Ringgefäß *n*
549	anode	Anode *f*
550	anodic rays	Anodenstrahlen *mpl*
551	an(o)estrus	Brunstlosigkeit *f*, Unbrunst *f*, Anöstrus *m*
552	anomalous	anomal
553	anomaly	Anomalie *f*
554	anorthogenesis	Anorthogenese *f*
555	anoxybiosis → anaerobiosis	—
556	antagonism	Antagonismus *m*
557	antagonist, antagonistic muscle	Antagonist *m*
558	antenna, feeler	Fühler *m*, Antenne *f*
559	antennal gland	Antennendrüse *f*
560	antennule	Antennula *f*
561	anterior pituitary, ~ lobe (of hypophysis), prehypophysis	Hypophysenvorderlappen *m*
562	anterior pituitary hormone	Hypophysenvorderlappenhormon *n*
563	anterior root → ventral root	—
564	anther	Staubbeutel *m*, Anthere *f*
565	antherid(ium)	Antheridium *n*
566	antherozoid	Antherozoid *n*
567	anthesis	Anthese *f*, Blütenöffnung *f*
568	anthocyan(in)	Anthozyan(in) *n*
569	anthocyanidine	Anthozyanidin *n*
570	anthophytes → spermatophytes	—
571	anthoxanthin	Anthoxanthin *n*
572	Anthozoa, Actinozoa	Korallentiere *npl*, Blumentiere *npl*, Anthozoen *pl*
573	anthropobiology	Anthropobiologie *f*
574	anthropochory, dispersal by man	Anthropochorie *f*
575	anthropogenesis	Anthropogenese *f*, Menschwerdung *f*
576	anthropogenic	anthropogen
577	anthropogeny	Anthropogenie *f*
578	anthropogeography	Anthropogeographie *f*
579	anthropography	Anthropographie *f*
580	anthropoids	Anthropoiden *pl*, Menschenaffen *mpl*
581	anthropology	Anthropologie *f*, Menschenkunde *f*
582	anthropometry	Anthropometrie *f*
583	anthropomorphism	Anthropomorphismus *m*
584	anthropomorphology	Anthropomorphologie *f*
585	antiauxin	Antiauxin *n*
586	antibiont	Antibiont *m*
587	antibiosis	Antibiose *f*, Widersachertum *n*
588	antibiotic	Antibiotikum *n*
589	antibody	Antikörper *m*
590	anticlinal	antiklin
591	anticodon	Antikodon *n*
592	antidiuretic hormone → vasopressin	—
593	antidote, counterpoison	Gegengift *n*, Antidot *n*
594	antidromic	antidrom
595	antienzyme	Antienzym *n*, Antiferment *n*
596	antifertilizin	Antifertilizin *n*
597	antigen	Antigen *n*
598	antigen-antibody reaction	Antigen-Antikörper-Reaktion *f*
599	antigenicity	Antigenität *f*
600	antigibberellin	Antigibberellin *n*
601	antihaemorrhagic vitamin → phylloquinone	—
602	antihormone	Antihormon *n*, Gegenhormon *n*
603	antimetabolite	Antimetabolit *m*
604	antimorph	antimorph
605	antimutagen	Antimutagen *n*

547	trachéide *f* annelée	traqueida *f* anular
548	vaisseau *m* annelé	vaso *m* anular, ~ anillado
549	anode *f*	ánodo *m*
550	rayons *mpl* anodiques	rayos *mpl* anódicos
551	anoestrus *m*	anestro *m*
552	anormal	anormal
553	anomalie *f*	anomalía *f*
554	anorthogenèse *f*	anortogénesis *f*
555	—	—
556	antagonisme *m*	antagonismo *m*
557	(muscle *m*) antagoniste *m*	(músculo *m*) antagonista *m*
558	antenne *f*	antena *f*
559	glande *f* antennaire	glándula *f* antenal
560	antennule *f*	anténula *f*
561	antéhypophyse *f*, préhypophyse *f*, lobe *m* antérieur (de l'hypophyse)	prehipófisis *f*, pituitaria *f* anterior, lóbulo *m* ~ (de la hipófisis)
562	hormone *f* antéhypophysaire, ~ préhypophysaire	hormona *f* prehipofisaria
563	—	—
564	anthère *f*	antera *f*
565	anthéridie *f*	anteridio *m*
566	anthérozoïde *m*	anterozoide *m*
567	anthèse *f*	antesis *f*
568	anthocyan(in)e *f*	antocianina *f*
569	anthocyanidine *f*	antocianidina *f*
570	—	—
571	anthoxanthine *f*	antoxantina *f*
572	anthozoaires *mpl*, coralliaires *mpl*	antozoos *mpl*, antozoarios *mpl*, actinozoos *mpl*, animales *mpl* coralíneos
573	anthropobiologie *f*, anthropologie *f* biologique	antropobiología *f*, antropología *f* biológica
574	dispersion *f* par l'homme	antropocoria *f*
575	anthropogenèse *f*	antropogénesis *f*
576	anthropogène	antropogénico
577	anthropogénie *f*	antropogenia *f*
578	anthropogéographie *f*	antropogeografía *f*
579	anthropographie *f*	antropografía *f*
580	anthropoïdes *mpl*, singes *mpl* anthropomorphes	antropoides *mpl*, monos *mpl* antropomorfos
581	anthropologie *f*	antropología *f*
582	anthropométrie *f*	antropometría *f*
583	anthropomorphisme *m*	antropomorfismo *m*
584	anthropomorphologie *f*, anthropologie *f* morphologique	antropomorfología *f*
585	antiauxine *f*	antiauxina *f*
586	antibiote *m*	antibionte *m*
587	antibiose *f*	antibiosis *f*
588	antibiotique *m*	antibiótico *m*
589	anticorps *m*	anticuerpo *m*
590	anticlinal	anticlinal
591	anti-codon *m*	anticodon *m*
592	—	—
593	antidote *m*, contre-poison *m*	antídoto *m*, contraveneno *m*
594	antidromique	antidrómico
595	antienzyme *m*	antienzima *f*
596	antifertilisine *f*	antifertilizina *f*
597	antigène *m*	antígeno *m*
598	réaction *f* antigène-anticorps	reacción *f* antígeno-anticuerpo
599	antigénéité *f*	antigenicidad *f*
600	antigibbérelline *f*	antigiberelina *f*
601	—	—
602	antihormone *f*	antihormona *f*, hormona *f* inhibidora
603	antimétabolite *m*	antimetabolito *m*
604	antimorphe	antimorfo
605	antimutagène *m*	antimutagen *m*

606	antineuritic vitamin → thiamine	—
607	antinoise campaign → noise control	—
608	antipellagrous vitamin → nicotinamide	—
609	antiperistalsis	Antiperistaltik *f*
610	antipermeability vitamin, vitamin P	Permeabilitätsfaktor *m*, Vitamin P *n*
611	antipodal cell	Antipode *f*, Gegenfüßlerzelle *f*
612	antipollution act, ~ law	Gesetz *n* gegen Umweltverschmutzung
613	antipollution campaign	Bekämpfung *f* der Umweltverschmutzung
614	antirhachitic vitamin → calciferol	—
615	antiscorbutic vitamin → ascorbic acid	—
616	antiseptic	antiseptisch; Antiseptikum *n*
617	antiserum	Antiserum *n*
618	antisterility vitamin → tocopherol	—
619	antithrombin	Antithrombin *n*
620	antitoxic	antitoxisch
621	antitoxin	Antitoxin *n*
622	antitrypsin	Antitrypsin *n*
623	antivitamin	Antivitamin *n*, Vitaminantagonist *m*
624	antixerophthalmic vitamin → axerophthol	—
625	antizymotic	gärungsverhindernd
626	anuclear → non-nucleated	—
627	Anura, Salientia	Froschlurche *mpl*, Anuren *pl*
628	anurous, tailless	schwanzlos
629	anus	After *m*, Anus *m*
630	aorta	Aorta *f*
631	aortic arch	Aortenbogen *m*
632	apetalous	apetal
633	apex (*pl* apices), tip	Apex *m*, Scheitel *m*, Spitze *f*
634	apex of a leaf → leaf tip	—
635	aphotic zone	aphotische Zone *f*
636	aphyllous, leafless	blattlos
637	aphylly	Aphyllie *f*, Blattlosigkeit *f*
638	apical bud, terminal ~	Apikalknospe *f*, Gipfel- *f*, Terminal- *f*, End- *f*
639	apical cell	Scheitelzelle *f*
640	apical dominance	Apikaldominanz *f*
641	apical growth	Spitzenwachstum *n*
642	apical meristem	Apikalmeristem *n*
643	aplanospore	Aplanospore *f*
644	aplasia	Aplasie *f*
645	apocarpous	apokarp, getrenntfrüchtig
646	apocarpy	Apokarpie *f*
647	apocrine gland	apokrine Drüse *f*
648	apodal, apodous	fußlos
649	apoenzyme, colloidal bearer	Apoenzym *n*, kolloidaler Träger *m*
650	apogamy	Apogamie *f*
651	apomeiosis	Apomeiosis *f*
652	apomictic	apomiktisch
653	apomixis, apomixia	Apomixis *f*
654	aponeurosis	Sehnenhaut *f*, Aponeurose *f*
655	apophysis	Apophyse *f*, Knochenfortsatz *m*
656	aporogamy	Aporogamie *f*
657	aposematic colo(u)ration → warning ~	—
658	apospory	Aposporie *f*
659	apothecium	Apothezium *n*
660	apparent free space, AFS	freier Raum *m*
661	appeasement gesture, ~ ceremony *eth.*	Befriedungsgeste *f*, Beschwichtigungszeremonie *f eth.*
662	appendage *anat.*, *zool.*	Anhang *m anat.*, *zool.*
663	appendicular skeleton, skeleton of limbs	Gliedmaßenskelett *n*
664	appetitive behaviour	Appetenzverhalten *n*
665	apposition	Apposition *f*, Substanzanlagerung *f*

606	—	—
607	—	—
608	—	—
609	antipéristaltisme *m*, mouvement *m* antipéristaltique	movimiento *m* antiperistáltico
610	vitamine *f* d'antiperméabilité, ~ P	vitamina *f* de la permeabilidad, ~ P
611	(cellule *f*) antipode *f*	(célula *f*) antípoda *f*
612	loi *f* contre la contamination de l'environnement	ley *f* contra la polución del medio ambiente, ~ anticontaminación
613	lutte *f* contra la pollution	lucha *f* antipolutiva, ~ antipolución, ~ contra la contaminación, ~ anticontaminación
614	—	—
615	—	—
616	antiseptique (*m*)	antiséptico (*m*)
617	anti-sérum *m*	antisuero *m*
618	—	—
619	antithrombine *f*	antitrombina *f*
620	antitoxique	antitóxico
621	antitoxine *f*	antitoxina *f*
622	antitrypsine *f*	antitripsina *f*
623	antivitamine *f*	antivitamina *f*
624	—	—
625	antizymotique	anticímico
626	—	—
627	anoures *m pl*	anuros *m pl*
628	anoure, acaudé, sans queue	anuro, sin cola
629	anus *m*	ano *m*
630	aorte *f*	aorta *f*
631	arc *m* aortique	arco *m* aórtico
632	apétale	apétalo
633	apex *m*, pointe *f*	ápice *m*, punta *f*
634	—	—
635	zone *f* aphotique	zona *f* afótica
636	aphylle, dépourvu de feuilles	áfilo, sin hojas
637	aphyllie *f*	afilia *f*
638	bourgeon *m* apical, ~ terminal	yema *f* terminal
639	cellule *f* apicale	célula *f* apical
640	dominance *f* apicale	dominancia *f* apical
641	croissance *f* apicale	crecimiento *m* apical, ~ terminal
642	méristème *m* apical	meristema *m* apical
643	aplanospore *f*	aplanóspora *f*
644	aplasie *f*	aplasia *f*
645	apocarpe	apocarpo, apocárpico
646	apocarpie *f*	apocarpia *f*
647	glande *f* apocrine	glándula *f* apocrina
648	apode	ápodo
649	apoenzyme *f*	apoenzima *m*, apofermento *m*, soporte *m* coloidal
650	apogamie *f*	apogamia *f*
651	apoméiose *f*	apomeyosis *f*, apomeiosis *f*
652	apomictique	apomíctico
653	apomixie *f*	apomixis *f*, apomixia *f*
654	aponévrose *f*	aponeurosis *f*
655	apophyse *f*	apófisis *f*
656	aporogamie *f*	aporogamia *f*
657	—	—
658	aposporie *f*	aposporia *f*
659	apothécie *f*	apotecio *m*
660	espace *m* libre apparent	espacio *m* libre
661	geste *m* d'apaisement *eth.*	gesto *m* de apaciguamiento, ~ pacificador *eth.*
662	appendice *m* *anat.*, *zool.*	apéndice *m* *anat.*, *zool.*
663	squelette *m* appendiculaire, ~ du membre	esqueleto *m* apendicular, ~ de los miembros
664	comportement *m* appétitif, ~ d'appétence	comportamiento *m* apetitivo, ~ de apetencia
665	apposition *f*	aposición *f*

666	apposition eye	Appositionsauge *n*
667	appositional growth	Appositionswachstum *n*
668	apterous, apterygotous, wingless	flügellos
669	Apterygota, Ametabola, ametabolous insects	Urinsekten *npl*, Flügellose *pl*, Apterygoten *pl*
670	aquatic	wasserbewohnend, -lebend
671	aquatic animal	Wassertier *n*
672	aquatic plants → hydrophytes	—
673	aquatic resources → water ~	—
674	aqueous, watery	wässerig
675	aqueous extract	wässeriger Extrakt *m*
676	aqueous humo(u)r	Kammerwasser *n*
677	aqueous solution, watery ~	wässerige Lösung *f*
678	Arachnida, arachnids	Spinnentiere *npl*, Arachniden *pl*
679	arachnoid (membrane)	Spinnwebenhaut *f*, Arachnoidea *f*
680	arachnology	Spinnenkunde *f*, Arachnologie *f*
681	arborescent	baumartig, baumförmig
682	arboricole, arboreal, tree-dwelling	baumbewohnend
683	arborization	Verästelung *f (Nerven etc.)*
684	Archaean (era)	Archaikum *n*
685	Archaeozoic (era)	Archäozoikum *n*, Eozoikum *n*
686	archallaxis	Archallaxis *f*
687	Archegoniatae	Archegoniaten *pl*
688	archegonium	Archegonium *n*
689	archencephalon	Urhirn *n*, Archenzephalon *n*
690	archenteron, primitive gut	Urdarm *m*, Archenteron *n*
691	archesporium	Archespor *n*
692	archetype, architype	Urtyp *m*, Archetyp *m*
693	archibenthic zone	Archibenthal *n*
694	archiblast	Archiblast *m*
695	archipallium	Archipallium *n*
696	archiplasm, archoplasm	Archiplasma *n*, Archoplasma *n*
697	area opaca, opaque area	dunkler Fruchthof *m*
698	area pellucida, pellucide area	heller Fruchthof *m*
699	areal type	Arealtyp *m*
700	areolar *bot.*	behöft *bot.*
701	arginase	Arginase *f*
702	arginine	Arginin *n*
703	arginine phosphate, phosphoarginine	Argininphosphat *n*
704	argyrophil fibre → reticulin ~	—
705	aril(lus)	Samenmantel *m*, Arillus *m*
706	arista → awn	—
707	aristate, awned	begrannt
708	aromatic	aromatisch
709	aromatization	Aromatisierung *f*
710	to aromatize	aromatisieren
711	aromorphosis	Aromorphose *f*
712	arousal reaction	Weckreaktion *f*, „arousal reaction“
713	arrest of development, arrested ~	Entwicklungsstillstand *m*
714	arrest of growth, cessation ~ ~	Wachstumsstillstand *m*, -stockung *f*
715	arrhenotoky	Arrhenotokie *f*
716	artefact, artifact	Artefakt *n*
717	artenkreis	Artenkreis *m*
718	arterenol → noradrenaline	—
719	arterial	arteriell
720	arteriole	Arteriole *f*
721	artery	Arterie *f*, Schlagader *f*
722	arthrology	Gelenklehre *f*, Arthrologie *f*
723	Arthropoda, arthropods	Gliederfüßer *mpl*, Arthropoden *pl*
724	arthrospore	Arthrospore *f*, Gliedspore *f*
725	articular, synovial	Gelenk...
726	articular capsule, joint ~	Gelenkkapsel *f*
727	articular cartilage	Gelenkknorpel *m*
728	articular cavity	Gelenkhöhle *f*

666	œil *m* à vision par apposition	ojo *m* de visión por aposición
667	croissance *f* appositionnelle	crecimiento *m* por aposición
668	aptère, désailé	áptero, apterigoto
669	aptérygotes *m pl*, amétaboles *m pl*, insectes *m pl* amétaboliques	apterigotos *m pl*, ametábolos *m pl*
670	aquatique	acuático, acuícola
671	animal *m* aquatique	animal *m* acuático
672	—	—
673	—	—
674	aqueux	acuoso
675	extrait *m* aqueux	extracto *m* acuoso
676	humeur *f* aqueuse	humor *m* acuoso
677	solution *f* aqueuse	solución *f* acuosa
678	arachnides *m pl*	arácnidos *m pl*
679	arachnoïde *f*	aracnoides *f*
680	arachnologie *f*	aracnología *f*
681	arborescent	arborescente, arboriforme
682	arboricole	arborícola, arbóreo
683	arborisation *f*	arborización *f*
684	archéen *m*	arcaico *m*
685	archéozoïque *m*, éozoïque *m*	arqueozoico *m*
686	archallaxie *f*	arcalaxis *f*
687	archégoniates *f pl*	arquegoniadas *f pl*
688	archégone *m*	arquegonio *m*
689	archencéphale *m*	arquencéfalo *m*
690	archentère *m*, archentéron *m*, intestin *m* primitif	arquenterio *m*, arquenterón *m*
691	archéspore *f*	arquesporio *m*
692	archétype *m*	arquetipo *m*
693	zone *f* archibenthique	zona *f* arquibentónica
694	archiblaste *m*	arquiblasto *m*
695	archipallium *m*	arquipalio *m*
696	archiplasme *m*, archoplasme *m*	arquiplasma *m*
697	aire *f* opaque	área *f* opaca
698	aire *f* pellucide	área *f* pelúcida, ~ transparente
699	type *m* d'aire	tipo *m* de área
700	aréolé *bot.*	areolado *bot.*
701	arginase *f*	arginasa *f*
702	arginine *f*	arginina *f*
703	arginine-phosphate *m*	fosfato *m* de arginina, fosfoarginina *f*
704	—	—
705	arille *m*	arilo *m*
706	—	—
707	plein de barbes ou arêtes	aristado
708	aromatique	aromático
709	aromatisation *f*	aromatización *f*
710	aromatiser	aromatizar
711	aromorphose *f*	aromorfosis *f*
712	réaction *f* (r)éveillante	respuesta *f* de alerta
713	arrêt *m* de développement	cesación *f* del desarrollo, cese *m* ~ ~
714	arrêt *m* de croissance	cesación *f* del crecimiento, cese *m* ~ ~, detención *f* ~ ~
715	arrhénotocie *f*, arrhénotoquie *f*	arrenotoquía *f*
716	artefact *m*	artefacto *m*
717	cercle *m* d'espèces	círculo *m* de especies
718	—	—
719	artériel	arterial
720	artériole *f*	arteriola *f*
721	artère *f*	arteria *f*
722	arthrologie *f*	artrología *f*
723	arthropodes *m pl*	artrópodos *m pl*
724	arthrospore *f*	artróspora *f*
725	articulaire, synovial	articular, sinovial
726	capsule *f* articulaire	cápsula *f* articular, ~ sinovial
727	cartilage *m* articulaire	cartílago *m* articular, ~ artrodial
728	cavité *f* articulaire	cavidad *f* articular

No.	English	German
729	Articulata	Gliedertiere *npl*
730	articulated pod	Gliederhülse *f*
731	articulated siliqua	Gliederschote *f*
732	articulation, joint	Gelenk *n*
733	artifact → artefact	—
734	artificial insemination	künstliche Besamung *f*
735	Artiodactyla, even-toed ungulates	Paarhufer *mpl*, Paarzeher *mpl*
736	artioploidy	Artioploidie *f*
737	arytenoid cartilage	Gießbeckenknorpel *m*
738	ascendency	Aszendenz *f*
739	ascendent → ancestor	—
740	ascending colon	aufsteigendes Kolon *n*, aufsteigender Grimmdarm *m*
741	ascent of sap	Saftanstieg *m*
742	Aschelminthes → Nemathelminthes	—
743	ascogonium	Askogon *n*
744	ascomycetes, sac fungi	Schlauchpilze *mpl*, Askomyzeten *mpl*
745	ascorbic acid, antiscorbutic vitamin, vitamin C	Askorbinsäure *f*, antiskorbutisches Vitamin *n*, Vitamin C
746	ascospore	Askospore *f*
747	ascus (*pl* asci)	Askus *m*, Sporenschlauch *m*
748	aseptate → nonseptate	—
749	asexual, non-sexual	ungeschlechtlich, geschlechtslos, asexuell
750	asexual reproduction, vegetative ~	ungeschlechtliche Fortpflanzung *f*, vegetative ~
751	asparagine	Asparagin *n*
752	aspartic acid	Asparaginsäure *f*
753	asperifoliate	rauhblättrig
754	assimilable	assimilierbar
755	to assimilate	assimilieren
756	assimilating tissue	Assimilationsgewebe *n*
757	assimilation	Assimilation *f*
758	assimilation pigment, assimilative ~	Assimilationsfarbstoff *m*
759	assimilation product	Assimilationsprodukt *n*, Assimilat *n*
760	assimilation ratio	Assimilationsrate *f*
761	assimilation starch	Assimilationsstärke *f*
762	assimilatory power	Assimilationsfähigkeit *f*
763	assimilatory quotient	Assimilationsquotient *m*
764	associated	vergesellschaftet
765	association, plant ~	Assoziation *f* (Pfl. soz.); Vergesellschaftung *f*
766	association area	Assoziationsfeld *n*
767	association centre (US: center)	Assoziationszentrum *n*
768	association learning *eth.*	Assoziationslernen *n eth.*
769	aster *gen.*	Aster *m gen.*
770	asternal rib → false ~	—
771	astral rays → polar ~	—
772	astrocyte	Astrozyt *m*
773	astrosphere, astral sphere	Astrosphäre *f*
774	asyndesis, asynapsis	Asyndese *f*, Asynapsis *f*
775	atavism, throwing-back	Atavismus *m*, Rückschlagsbildung *f*
776	atavistic, atavic	atavistisch
777	atmospheric humidity, ~ moisture	Luftfeuchtigkeit *f*
778	atmospheric nitrogen	Luftstickstoff *m*
779	atmospheric pollution → air ~	—
780	atomic age	Atomzeitalter *n*
781	atomic energy, nuclear ~	Atomenergie *f*, Kernenergie *f*
782	atomic fall-out → radioactive ~	—
783	atomic nucleus	Atomkern *m*
784	atomic waste → radioactive ~	—
785	atomic weight	Atomgewicht *n*
786	atoxic → nontoxic	—
787	ATP → adenosine triphosphate	—
788	atrichous, atrichic, aflagellar	geißellos, atrich
789	atrium	Atrium *n*, Herzvorhof *m*, Vorkammer *f*
790	atrophy	Atrophie *f*

729	articulés *m pl*	articulados *m pl*
730	gousse *f* articulée, ~ lomentacée	legumbre *f* articulada
731	silique *f* articulée	silicua *f* articulada
732	articulation *f*, jointure *f*	articulación *f*
733	—	—
734	insémination *f* artificielle	inseminación *f* artificial
735	artiodactyles *m pl*	artiodáctilos *m pl*
736	artioploïdie *f*	artioploidia *f*
737	cartilage *m* aryténoïde	cartílago *m* aritenoides
738	ascendance *f*	ascendencia *f*
739	—	—
740	côlon *m* ascendant	colon *m* ascendiente
741	ascension *f* de la sève, montée *f* ~ ~ ~	ascensión *f* de la savia, subida *f* ~ ~ ~
742	—	—
743	ascogone *m*	ascogonio *m*
744	ascomycètes *m pl*	ascomicetos *m pl*
745	acide *m* ascorbique, vitamine *f* anti-scorbutique, ~ C	ácido *m* ascórbico, ~ cevitámico, vitamina *f* antiescorbútica, ~ C
746	ascospore *f*	ascospora *f*
747	asque *m*	asco *m*, asca *f*
748	—	—
749	asexuel, asexué	asexual, asexuado
750	reproduction *f* asexuée, ~ végétative	reproducción *f* asexual, ~ vegetativa
751	asparagine *f*	asparagina *f*
752	acide *m* aspartique	ácido *m* aspártico
753	aspérifolié	asperifolio
754	assimilable	asimilable
755	assimiler	asimilar
756	tissu *m* assimilateur	tejido *m* asimilador
757	assimilation *f*	asimilación *f*
758	pigment *m* assimilateur	pigmento *m* asimilador, ~ asimilable
759	produit *m* d'assimilation, assimilat *m*	producto de asimilación, ~ asimilado
760	intensité *f* d'assimilation	proporción *f* de asimilación
761	amidon *m* d'assimilation	almidón *m* de asimilación
762	capacité *f* d'assimilation, pouvoir *m* assimilateur	capacidad *f* de asimilación
763	quotient *m* assimilateur	cociente *m* de asimilación, ~ asimilatorio
764	associé	asociado
765	association *f* (végétale)	asociación *f* (vegetal)
766	aire *f* associative	área *f* de asociación
767	centre *m* d'association, ~ associatif	centro *m* de asociación
768	apprentissage *m* associatif *eth.*	aprendizaje *m* asociativo *eth.*
769	aster *m gen*	áster *m gen.*
770	—	—
771	—	—
772	astrocyte *m*	astrocito *m*
773	astrosphère *f*	astroesfera *f*
774	asynapsis *f*	asinapsis *f*, asindesis *f*
775	atavisme *m*	atavismo *m*
776	atavique	atávico
777	humidité *f* atmosphérique	humedad *f* atmosférica
778	azote *m* atmosphérique	nitrógeno *m* atmosférico
779	—	—
780	âge *m* atomique	era *f* atómica
781	énergie *f* atomique, ~ nucléaire	energía *f* atómica, ~ nuclear
782	—	—
783	noyau *m* atomique	núcleo *m* atómico
784	—	—
785	poids *m* atomique	peso *m* atómico
786	—	—
787	—	—
788	atriche	atriquio
789	atrium *m*	atrio *m*
790	atrophie *f*	atrofia *f*

791	atropous, atropal	atrop
792	attractant, attractive substance	Lockstoff *m*
793	attraction cone → fertilization ~	—
794	attraction sphere → centrosphere	—
795	audition	Hören *n*
796	auditory	Gehör ..., Hör ...
797	auditory acuity	Hörschärfe *f*
798	auditory capacity	Hörvermögen *n*
799	auditory capsule, otic ~	Gehörkapsel *f*
800	auditory cell	Hörzelle *f*
801	auditory meatus (external, internal)	Gehörgang *m* (äußerer, innerer)
802	auditory nerve, acoustic ~	Hörnerv *m*, Akustikus *m*, Statoakustikus *m*
803	auditory organ, hearing ~	Hörorgan *n*
804	auditory ossicle → ear ~	—
805	auditory sense, sense of hearing	Gehörsinn *m*
806	auditory stimulus	Hörreiz *m*
807	auditory threshold, acoustic ~	Hörschwelle *f*
808	auditory tube → Eustachian ~	—
809	auditory vesicle, otocyst	Gehörbläschen *n*, Ohrbläschen *n*
810	auricle *anat.*	Herzohr *n*
811	autecology	Autökologie *f*
812	autoadaptation	Autoadaptation *f*
813	autoalloploidy	Autoalloploidie *f*
814	autoantigen	Autoantigen *n*
815	autobivalent	Autobivalent *n*
816	autocatalysis	Autokatalyse *f*
817	autochory, self-dispersal	Autochorie *f*
818	autochthonous	bodenständig, autochthon
819	autoclave	Autoklav *m*
820	autoecious	autözisch
821	autogamous	autogam
822	autogamy → self-fertilization; self-pollination	—
823	autogenesis, autogeny	Autogenese *f*
824	autogenic	autogenisch
825	autogenous	autogen
826	autograft → autotransplant	—
827	autoheteroploidy	Autoheteroploidie *f*
828	autoimmunization	Autoimmunisierung *f*, Selbstimmunisierung *f*
829	autolysate	Autolysat *n*
830	autolysis	Autolyse *f*
831	automatism	Automatie *f*, Automatismus *m*
832	automixis	Automixis *f*
833	automutagene	Automutagen *n*
834	autonomic (vegetative, visceral) nervous system	autonomes (vegetatives, viszerales) Nervensystem *n*
835	autonomy	Autonomie *f*
836	auto-orientation	Autoorientierung *f*
837	autoparthenogenesis	Autoparthenogenese *f*
838	autoplastic	autoplastisch
839	autoplasty → autotransplantation	—
840	autopolyploidy	Autopolyploidie *f*
841	autoradiography, radioautography	Autoradiographie *f*, Radioautographie *f*
842	autoregeneration	Selbsterneuerung *f*
843	autosome	Autosom *n*
844	autosyndesis	Autosyndese *f*
845	autosynthesis	Autosynthese *f*
846	autotomy	Autotomie *f*, Selbstverstümmelung *f*
847	autotoxin	Autotoxin *n*
848	autotransplant, autograft	Autoplantat *n*
849	autotransplantation, autoplasty	Autoplastik *f*, autoplastische Transplantation *f*
850	autotrophia	Autotrophie *f*

791	atropique	átropo
792	substance *f* attractive, attractif *m*	sustancia *f* de atracción, atrayente *m*
793	—	—
794	—	—
795	audition *f*	audición *f*
796	auditif	auditivo
797	acuité *f* auditive	acuidad *f* auditiva, agudeza *f* ~
798	capacité *f* auditive	capacidad *f* auditiva, facultad *f* ~
799	capsule *f* otique	cápsula *f* auditiva, ~ ótica
800	cellule *f* auditive	célula *f* auditiva
801	conduit *m* auditif, méat *m* ~ (externe, interne)	conducto *m* auditivo, meato *m* ~ (externo, interno)
802	nerf *m* auditif	nervio *m* auditivo, ~ acústico
803	organe *m* auditif, ~ de l'audition	órgano *m* auditivo, ~ de la audición
804	—	—
805	ouïe *f*, sens *m* de l'ouïe, ~ auditif	oído *m*, sentido *m* auditivo, ~ del oído
806	stimulus *m* auditif	estímulo *m* auditivo, ~ acústico
807	seuil *m* auditif, ~ d'audition	umbral *m* auditivo, mínimo *m* audible
808	—	—
809	vésicule *f* auditive, ~ otique, otocyste *m*	vesícula *f* auditiva, otocisto *m*
810	auricule *f*, oreillette *f*	aurícula *f*
811	autécologie *f*, écologie *f* individuelle	autecología *f*
812	autoadaptation *f*	autoadaptación *f*
813	auto-alloploïdie *f*	autoaloploidia *f*
814	auto-antigène *m*	autoantígeno *m*
815	autobivalent *m*	autobivalente *m*
816	autocatalyse *f*	autocatálisis *f*
817	dissémination *f* autochore	autocoria *f*
818	autochtone	autóctono
819	autoclave *m*	autoclave *m*
820	autoïque	autoico
821	autogame	autógamo
822	—	—
823	autogenèse	autogénesis *f*, ortogénesis *f* autógena
824	autogónique	autogénico
825	autogène	autógeno
826	—	—
827	autohétéroploïdie *f*	autoheteroploidia *f*
828	auto-immunisation *f*	autoinmunización *f*
829	autolysat *m*	autolisado *m*
830	autolyse *f*	autolisis *f*
831	automatisme *m*	automatismo *m*
832	automixie *f*	automixis *f*
833	automutagène *m*	automutagen *m*
834	système *m* nerveux autonome (végétatif, viscéral)	sistema *m* nervioso autónomo (vegetativo, visceral)
835	autonomie *f*	autonomía *f*
836	auto-orientation *f*	autoorientación *f*
837	autoparthénogenèse *f*	autopartenogénesis *f*
838	autoplastique	autoplástico
839	—	—
840	autopolyploïdie *f*	autopoliploidia *f*
841	autoradiographie *f*	autorradiografía *f*, radioautografía *f*
842	autorégénération *f*	autoregeneración *f*
843	autosome *m*	autosoma *m*
844	autosyndèse *f*	autosíndesis *f*
845	autosynthèse *f*	autosíntesis *f*
846	autotomie *f*	autotomía *f*
847	substance *f* autotoxique	autotoxina *f*
848	autotransplant *m*, autogreffe *f*, greffe *f* autoplastique	autoinjerto *m*, injerto *m* autoplástico
849	autotransplantation *f*, autoplastie *f*	autotrasplante *m*, autoplastia *f*
850	autotrophie *f*	autotrofia *f*

No.	English	German
851	autotrophic	autotroph
852	autotropism	Autotropismus *m*
853	autumn wood → late ~	—
854	auxanography	Auxanographie *f*
855	auxiliary cell, accessory ~, subsidiary ~	Hilfszelle *f*, Auxiliarzelle *f*, Nebenzelle *f*
856	auxin	Auxin *n*
857	auxocyte	Auxozyt *m*
858	auxospore	Auxospore *f*, Wachstumsspore *f*
859	auxotroph(ic)	auxotroph
860	avena test, coleoptile curvature bioassay	Avenatest *m*, Haferkoleoptilen-Krümmungstest *m*
861	avifauna, bird fauna	Vogelwelt *f*, Vogelfauna *f*
862	avirulent	nichtvirulent
863	avitaminosis	Avitaminose *f*, Vitaminmangelkrankheit *f*
864	avoidance reaction, avoiding ~	Meidereaktion *f*
865	awl-shaped → subulate	—
866	awn, arista	Granne *f*
867	awned → aristate	—
868	axenic culture → pure ~	—
869	axerophthol, antixerophthalmic vitamin, vitamin A	Axerophthol *n*, Xerophthol *n*, Vitamin A *n*
870	axial filament	Achsenfaden *m*
871	axial organ	Axialorgan *n*
872	axial skeleton	Achsenskelett *n*
873	axial structure	Achsenstruktur *f*
874	axilemma, axolemma	Axilemm(a) *n*, Axolemm(a) *n*
875	axillary	achselständig, axillär
876	axillary bud, lateral ~	Achselknospe *f*, Seitenknospe *f*
877	axis cylinder → axon	—
878	axocoel	Axozöl *n*
879	axon, axis cylinder, neurite	Axon *n*, Achsenzylinder *m*, Neurit *m*
880	axon reflex	Axonreflex *m*
881	axon terminal	Axonendigung *f*
882	axoplasm	Axoplasma *n*, Achsenplasma *n*
883	axostyle	Achsenstab *m*, Axostyl *n*
884	azoic era	Azoikum *n*
885	azotation	Azotierung *f*
886	azygote	Azygote *f*
887	bacciferous, berry-bearing	beerentragend
888	bacciform, berry-shaped	beerenartig, beerenförmig
889	Bacillariophyceae → diatoms	—
890	bacillary	bazillär, Bazillen...
891	bacilliculture	Bazillenkultur *f*
892	bacilliform	bazillenförmig
893	bacillus	Bazille *f*, Bazillus *m*
894	backbone → vertebral column	—
895	back cross, back crossing	Rückkreuzung *f*
896	back-cross breeding	Rückkreuzungszüchtung *f*
897	back-cross parent	Rückkreuzungselter *m*
898	back-cross ratio	Rückkreuzungsverhältnis *n*
899	back-fin → dorsal fin	—
900	background noise	Geräuschkulisse *f*, Lärmkulisse *f*
901	back mutation → reverse ~	—
902	back pollination	Rückbestäubung *f*
903	bacterial	bakteriell
904	bacterial capsule	Bakterienkapsel *f*
905	bacterial cell	Bakterienzelle *f*
906	bacterial colony	Bakterienkolonie *f*
907	bacterial culture	Bakterienkultur *f*
908	bacterial flagellum	Bakteriengeißel *f*
909	bacterial genetics	Bakteriengenetik *f*
910	bacterial host → host bacterium	—
911	bacterial spore	Bakterienspore *f*

851	autotrophe	autótrofo, autotrófico
852	autotropisme *m*	autotropismo *m*
853	—	—
854	auxanographie *f*	auxanografía *f*
855	cellule *f* auxiliaire	célula *f* auxiliar, ~ subsidiaria
856	auxine *f*	auxina *f*
857	auxocyte *m*	auxocito *m*
858	auxospore *f*	auxóspora *f*
859	auxotrophe, auxotrophique	auxótrofo
860	test *m* avoine, ~ de courbure	prueba *f* de la avena, ~ de la curvatura del coleóptilo
861	avifaune *f*, faune *f* des oiseaux	avifauna *f*, fauna *f* aviar, ornitofauna *f*
862	avirulent	avirulento, no virulento
863	avitaminose *f*	avitaminosis *f*
864	réaction *f* d'évitement	reacción *f* de evasión, ~ de evitación
865	—	—
866	arête *f*, barbe *f*	arista *f*, barba *f*
867	—	—
868	—	—
869	axérophtol *m*, vitamine *f* antixérophtalmique, vitamine *f* A	axeroftol *m*, vitamina *f* antixeroftálmica, vitamina *f* A
870	filament *m* axial	filamento *m* axial
871	organe *m* axial	órgano *m* axial
872	squelette *m* axial	esqueleto *m* axial
873	structure *f* axiale	estructura *f* axial
874	axilemme *m*	axilema *m*, axolema *m*
875	axillaire	axilar
876	bourgeon *m* axillaire	yema *f* axilar, ~ lateral
877	—	—
878	axocoele *m*	axocele *m*
879	axone *m*, cylindre-axe *m*, cylindraxe *m*, neurite *m*	axón *m*, cilindroeje *m*, neurita *f*
880	réflexe *m* d'axone	reflejo *m* de axón
881	terminaison *f* axonique	terminal *m* axónico
882	axoplasme *m*	axoplasma *m*
883	axostyle *m*	axostilo *m*
884	azoïque *m*	azoico *m*
885	azotation *f*	azotación *f*
886	azygote *m*	acigoto *m*
887	baccifère	baccífero
888	bacciforme	bacciforme, abayado
889	—	—
890	bacillaire	bacilar
891	culture *f* de bacilles	cultivo *m* de bacilos, bacilicultura *f*
892	bacilliforme	baciliforme
893	bacille *m*	bacilo *m*
894	—	—
895	rétrocroisement *m*, recroisement *m*, croisement *m* en retour, backcross *m*	cruzamiento *m* retrógrado, retrocruza *f*, retrocruzamiento *m*
896	reproduction *f* par rétrocroisement	retroreproducción *f*
897	parent *m* récurrent, ~ de backcross	progenitor *m* recurrente
898	proportion *f* de rétrocroisement	relación *f* de retrocruzamiento
899	—	—
900	fond *m* sonore, bruit *m* de fond	fondo *m* sonoro, ruido *m* de fondo
901	—	—
902	pollinisation *f* en retour	retropolonización *f*
903	bactérien	bacteriano, bactérico, bacterial
904	capsule *f* bactérienne	cápsula *f* bacteriana
905	cellule *f* bactérienne	célula *f* bacteriana
906	colonie *f* bactérienne	colonia *f* de bacterias
907	culture *f* bactérienne	cultivo *m* bacteriano
908	flagelle *m* bactérien	flagelo *m* bacteriano
909	génétique *f* bactérienne, bactériogénétique *f*	genética *f* bacteriana
910	—	—
911	spore *f* bactérienne	espora *f* bacteriana

912	bacterial strain	Bakterienstamm *m*
913	bacterial toxin → bacteriotoxin	—
914	bacterial virus → bacteriophage	—
915	bacteria-retaining filter, bacteria-proof ~, germ-proof ~	Bakterienfilter *m*, Entkeimungsfilter *m*
916	bactericidal	bakterientötend, bakterizid
917	bactericide	Bakterizid *n*
918	bacteriform	bakterienförmig
919	bacteriochlorophyll, bacteriochlorin	Bakterienchlorophyll *n*, Bakteriochlorin *n*
920	bacteriocin	Bakteriozin *n*
921	bacteriologic(al)	bakteriologisch
922	bacteriologist	Bakteriologe *m*
923	bacteriology	Bakteriologie *f*
924	bacteriolysin	Bakteriolysin *n*
925	bacteriolysis	Bakteriolyse *f*
926	bacteriolytic	bakteriolytisch
927	bacteriophage, phage, bacterial virus	Bakteriophage *m*, Phage *m*, Bakterienvirus *n*
928	bacteriostatic	bakteriostatisch
929	bacteriotoxin, bacterial toxin	Bakteriengift *n*, Bakteriotoxin *n*
930	bacteriotropin → opsonin	—
931	bacterium (*pl* bacteria)	Bakterie *f*, Bakterium *n*
932	bacteroid	Bakteroide *f*
933	baculiform, rod-like, rod-shaped	stäbchenförmig
934	baker's yeast	Bäckerhefe *f*
935	balance theory of sex determination	Balance-Theorie *f* der Geschlechtsbestimmung
936	balanced lethal	balancierter Letalfaktor *m*
937	balancer → haltere	—
938	balancing organ, balance ~, statical ~	Gleichgewichtsorgan *n*
939	balanus	Eichel *f* *anat.*
940	Balbiani ring	Balbiani-Ring *m*
941	Baldwin effect	Baldwin-Effekt *m*
942	ball-and-socket joint → diarthrosis	—
943	ballast	Ballaststoffe *mpl*
944	ballistic fruit	Schleuderfrucht *f*
945	balsam	Balsam *m*
946	band *gen.*	Querscheibe *f* *gen.*
947	bark, rhytidome	Borke *f*
948	Barr body → sex chromatin body	—
949	barreness	Befruchtungsunfähigkeit *f*
950	basal body, ~ corpuscle	Basalkörper *m*
951	basal cell	Basalzelle *f*
952	basal ganglion	Basalganglion *n*
953	basal granule, kinetosome	Basalkorn *n*, Kinetosom *n*
954	basal metabolic rate (BMR)	Grundumsatzwert *m*
955	basal metabolism	Grundumsatz *m*
956	basal plate	Basalplatte *f*
957	base analogue	Basenanalog *n*
958	base exchange, cation ~	Basenaustausch *m*, -umtausch *m*, Kationenumtausch *m*
959	base exchange capacity	Basenumtauschkapazität *f*
960	base of skull	Schädelbasis *f*
961	base pairing	Basenpaarung *f*
962	base sequence	Basensequenz *f*
963	basement membrane	Basalmembran *f*
964	basic	basisch
965	basic culture medium	Standardnährboden *m*
966	basic dye	basischer Farbstoff *m*
967	basic number, fundamental ~	Basiszahl *f*, Grundzahl *f*
968	basichromatin	Basichromatin *n*
969	basicity	Basengehalt *m*, Basizität *f*
970	basidiomycetes, club fungi	Ständerpilze *mpl*, Basidiomyzeten *mpl*

912	souche *f* bactérienne	cepa *f* de bacterias
913	—	—
914	—	—
915	filtre *m* imperméable aux germes, ~ stérilisant	filtro *m* bacteriano, ~ impermeable para las bacterias
916	bactéricide	bactericida
917	bactéricide *m*	bactericida *m*
918	bactériforme	bacteriforme
919	bactériochlorophylle *f*	bacterioclorofila *f*, bacterioclorina *f*
920	bactériocine *f*	bacteriocina *f*
921	bactériologique	bacteriológico
922	bactériologiste *m*, bactériologue *m*	bacteriólogo *m*
923	bactériologie *f*	bacteriología *f*
924	bactériolysine *f*	bacteriolisina *f*
925	bactériolyse *f*	bacteriolisis *f*
926	bactériolytique	bacteriolítico
927	bactériophage *m*, phage *m*, virus *m* des bactéries	bacteriófago *m*, fago *m*, virus *m* bacteriano
928	bactériostatique	bacteriostático
929	bactériotoxine *f*	bacteriotoxina *f*, toxina *f* bacteriana
930	—	—
931	bactérie *f*	bacteria *f*, bacterio *m*
932	bactéroïde *m*	bacteroide *m*
933	en forme de bâtonnet	baculiforme
934	levure *f* de boulangerie	levadura *f* de panificación
935	théorie *f* de l'équilibre de la détermination des sexes	teoría *f* del equilibrio de la determinación de los sexos
936	lét(h)al *m* balancé	letal *m* equilibrado
937	—	—
938	organe *m* de l'équilibre, ~ d'équilibration	órgano *m* del equilibrio
939	gland *m* *anat.*	bálano *m*, glande *m* *anat.*
940	anneau *m* de Balbiani	anillo *m* de Balbiani
941	effet *m* de Baldwin	efecto *m* de Baldwin
942	—	—
943	substances *f pl* encombrantes, lest *m*, «ballast» *m*	materia *f* fibrosa
944	fruit *m* explosif	fruto *m* espermobólico
945	baume *m*	bálsamo *m*
946	bande *f* *gen.*	banda *f* (transversal) *gen.*
947	rhytidome *m*	ritidoma *m*
948	—	—
949	infécondabilité *f*	infecundidad *f*
950	corpuscule *m* basal	cuerpo *m* basal
951	cellule *f* basale	célula *f* basal
952	ganglion *m* basal	ganglio *m* basal
953	grain *m* basal, cinétosome *m*	gránulo *m* basal, cinetosoma *m*
954	valeur *f* du métabolisme basal	tasa *f* metabólica basal, índice *m* de metabolismo basal
955	métabolisme *m* basal	metabolismo *m* basal
956	plaque *f* basale	placa *f* basal
957	base *f* analogue	análogo *m* de base, base *f* análoga
958	échange *m* de bases, ~ de cations	cambio *m* de bases, intercambio *m* ~ ~
959	capacité *f* d'échange de bases	capacidad *f* de cambio de bases
960	base *f* du crâne	base *f* del cráneo
961	accouplement *m* de bases	apareamiento *m* de bases, emparejamiento *m* ~ ~
962	séquence *f* de bases	secuencia *f* de bases
963	membrane *f* basale	membrana *f* basal
964	basique	básico
965	milieu *m* (de culture) basal	medio *m* (de cultivo) básico (o: basal)
966	colorant *m* basique	colorante *m* básico
967	nombre *m* de base, ~ basal, ~ fondamental	número *m* básico, ~ basal, ~ fundamental
968	basichromatine *f*	basicromatina *f*
969	basicité *f*	basicidad *f*
970	basidiomycètes *m pl*	basidiomicetos *m pl*

971	basidiospore	Ständerspore *f*, Basidiospore *f*
972	basidium (*pl* basidia)	Basidie *f*, Ständer *m*
973	basifugal	basifugal
974	basilar *bot.*	grundständig *bot.*
975	basilar membrane	Basilarmembran *f*
976	basipetal	basipetal
977	basket cell	Korbzelle *f*
978	basophil(ic), basiphil(ic)	basophil
979	basophily, basophilism	Basophilie *f*
980	bast → liber	—
981	bast cell	Bastzelle *f*
982	bast fibre (US: fiber)	Bastfaser *f*
983	bast parenchyma	Bastparenchym *n*
984	bastard → hybrid	—
985	to bastardize → to hybridize	—
986	Batesian mimicry	Batessche Mimikry *f*
987	Bathracia → Amphibia	—
988	bathyal zone	Bathyal *n*, bathyale Zone *f*
989	bathypelagic zone	Bathypelagial *n*
990	bearer of heritable characters	Erbträger *m*
991	becoming extinct	aussterbend
992	bee venom, ~ poison	Bienengift *n*
993	beer yeast	Bierhefe *f*
994	behavio(u)r	Verhalten *n*
995	behavio(u)r flexibility	Verhaltensflexibilität *f*
996	behavio(u)r genetics	Verhaltensgenetik *f*
997	behavio(u)r pattern	Verhaltensmuster *n*, ~weise *f*
998	behavio(u)ral analysis	Verhaltensanalyse *f*
999	behavio(u)ral trait, ~ characteristic	Verhaltensmerkmal *n*
1000	behavio(u)ral physiology	Verhaltensphysiologie *f*
1001	behaviorism	Behaviorismus *m*
1002	behaviorist	Behaviorist *m*
1003	belly → abdomen	—
1004	bending strength, flexural ~	Biegungsfestigkeit *f*
1005	beneficial insect → useful ~	—
1006	benthic, benthal	benthisch
1007	benthic zone	Benthal *n*
1008	benthos, benthic community	Benthos *n*
1009	berry	Beere *f*; Beerenfrucht *f*
1010	berry-bearing → bacciferous	—
1011	berry-shaped → bacciform	—
1012	beta-rays	Betastrahlen *mpl*
1013	between-brain → diencephalon	—
1014	B-horizon, illuvial horizon, enriched ~	B-Horizont *m*, Anreicherungshorizont *m*, Unterboden *m*
1015	biennial	zweijährig
1016	biennial (plant)	Bienne *f*, zweijährige Pflanze *f*, bienne ~
1017	bifid	zweigespalten
1018	biflagellate	zweigeißelig
1019	bifloral	zweiblütig
1020	bifoliate	zweiblättrig
1021	bifurcate, forked	gegabelt
1022	bilateral cleavage	Bilateralfurchung *f*
1023	bilateral symmetry	Bilateralsymmetrie *f*
1024	bilaterally symmetrical	bilateralsymmetrisch
1025	bile	Galle *f*
1026	bile acid	Gallensäure *f*
1027	bile duct, biliary ~	Gallengang *m*
1028	bile pigment	Gallenfarbstoff *m*, ~pigment *n*
1029	bile salt	Gallensalz *n*
1030	biliary	Gallen ...
1031	biliation	Gallenabsonderung *f*, ~sekretion *f*

971	basidiospore *f*	basidiospora *f*, basidióspora *f*
972	baside *f*	basidio *m*
973	basifuge	basifugal
974	basilaire	basilar
975	membrane *f* basilaire	membrana *f* basilar
976	basipète	basípeto
977	cellule *f* en panier, ~ à corbeille	célula *f* en cesta
978	basophile, basiphile	basófilo
979	basophilie *f*, basophilisme *m*	basofilia *f*
980	—	—
981	cellule *f* libérienne	célula *f* liberiana
982	fibre *f* libérienne	fibra *f* liberiana
983	parenchyme *m* libérien	parénquima *m* liberiano
984	—	—
985	—	—
986	mimétisme *m* batésien	mimetismo *m* batesiano
987	—	—
988	zone *f* bathyale	zona *f* batial
989	zone *f* bathypélagique	zona *f* batipelágica
990	porteur *m* d'hérédité	portador *m* de la herencia, transmisor *m* ~ ~ ~
991	en voie d'extinction	en vías de extinción, en trance ~ ~
992	venin *m* d'abeille	veneno *m* de abeja
993	levure *f* de bière	levadura *f* de cerveza
994	comportement *m*	comportamiento *m*, conducta *f*
995	flexibilité *f* du comportement	flexibilidad *f* del comportamiento
996	génétique *f* du comportement	genética *f* del comportamiento, ~ conductista
997	patron *m* de comportement, modèle *m* ~ ~	modelo *m* de comportamiento, pauta *f* de conducta
998	analyse *f* du comportement	análisis *m* del comportamiento
999	caractère *m* du comportement, ~ éthologique	característica *f* comportamental, ~ etológica, rasgo *m* de la conducta
1000	physiologie *f* du comportement	fisiología *f* del comportamiento
1001	béhaviorisme *m*, théorie *f* béhavioriste	behaviorismo *m*, conductismo *m*
1002	béhavioriste *m*	behaviorista *m*, conductista *m*
1003	—	—
1004	résistance *f* à la flexion	resistencia *f* a la flexión
1005	—	—
1006	benthique	béntico, bentónico
1007	zone *f* benthique	zona *f* bentónica
1008	benthos *m*	bentos *m*, comunidad *f* béntica
1009	baie *f*	baya *f*
1010	—	—
1011	—	—
1012	rayons *mpl* bêta	rayos *mpl* beta
1013	—	—
1014	horizon *m* B, ~ illuvial	horizonte *m* B, ~ iluvial, subsuelo *m*
1015	biennal	bienal
1016	(plante *f*) biennale *f*, plante *f* bisannuelle	(planta *f*) bienal *f*
1017	bifide	bífido
1018	biflagellé	biflagelado
1019	biflore	bifloro
1020	bifolié	bifolio
1021	bifurqué, fourchu	bifurcado
1022	segmentation *f* bilatérale	segmentación *f* bilateral
1023	symétrie *f* bilatérale	simetría *f* bilateral
1024	de symétrie bilatérale	simétricamente bilateral
1025	bile *f*	bilis *f*
1026	acide *m* biliaire	ácido *m* biliar
1027	canal *m* biliaire	conducto *m* biliar
1028	pigment *m* biliaire	pigmento *m* biliar
1029	sel *m* biliaire	sal *f* biliar
1030	biliaire	biliar
1031	sécrétion *f* biliaire	biliación *f*

No.	English	German
1032	bilineurine → choline	—
1033	biliprotein	Biliprotein *n*
1034	bilirubin	Bilirubin *n*
1035	biliverdin	Biliverdin *n*
1036	bilobar, bilobular	zweilappig
1037	bilocular	zweifächerig
1038	binary fission → bipartition	—
1039	binary nomenclature → binomi(n)al ~	—
1040	binocular microscope	Binokularmikroskop *n*
1041	binocular vision	binokulares Sehen *n*
1042	binomial distribution	Binominalverteilung *f*
1043	binomi(n)al nomenclature, binary ~	binäre Nomenklatur *f*
1044	binovular → dizygotic	—
1045	binovular twins → dizygotique ~	—
1046	binuclear, binucleate	zweikernig
1047	bioassay, biological assay	Biotest *m*, Lebendversuch *m*
1048	bioblast	Bioblast *m*
1049	biocatalysis, biological catalysis	Biokatalyse *f*, biologische Katalyse *f*
1050	biocatalyst, biocatalyzer	Biokatalysator *m*
1051	biochemical	biochemisch
1052	biochemical oxygen demand, BOD	biochemischer Sauerstoffbedarf *m*, BSB
1053	biochemist	Biochemiker *m*
1054	biochemistry	Biochemie *f*
1055	biochore	Biochore *f* (*pl* Biochorien)
1056	biocide	Biozid *n*
1057	bioclimatology	Bioklimatologie *f*, Bioklimatik *f*
1058	bioc(o)enology	Biozönologie *f*, Biozönotik *f*
1059	bioc(o)enosis, biotic community	Biozönose *f*, biotische Gemeinschaft *f*, Lebensgemeinschaft *f*
1060	bioc(o)enotic	biozönotisch
1061	biocybernetics → bionics	—
1062	biocycle	Biozyklus *m*
1063	biodegradable	biologisch abbaufähig
1064	biodemography → population ecology	—
1065	biodynamics	Biodynamik *f*
1066	bioenergetics	Bioenergetik *f*
1067	biogen(e)	Biogen *n*
1068	biogenesis	Biogenese *f*
1069	biogenetic	biogenetisch
1070	biogenetic isolation	biogenetische Isolation *f*
1071	biogenetic law, Haeckel's ~	biogenetisches Grundgesetz *n*
1072	biogeny	Biogenie *f*
1073	biogeochemical cycle	biogeochemischer Zyklus *m*
1074	biogeochemistry	Biogeochemie *f*
1075	biogeography	Biogeographie *f*
1076	biologic(al)	biologisch
1077	biological assay → bioassay	—
1078	biological clock, internal ~	biologische Uhr *f*, physiologische ~, innere ~
1079	biological half-life (period)	biologische Halbwertzeit *f*
1080	biological membrane	biologische Membran *f*
1081	biological organization plan	biologischer Bauplan *m*
1082	biological (pest) control	biologische Schädlingsbekämpfung *f*
1083	biological purification of waste water	biologische Abwasserreinigung *f*
1084	biological race	biologische Rasse *f*
1085	biological spectrum	biologisches Spektrum *n*
1086	biologist	Biologe *m*
1087	biology	Biologie *f*
1088	bioluminescence	Biolumineszenz *f*
1089	biolysis	Biolyse *f*
1090	biomass	Biomasse *f*
1091	biome	Biom *n*
1092	biomechanics	Biomechanik *f*

1032	—	—
1033	biliprotéine *f*	biliproteína *f*
1034	bilirubine *f*	bilirrubina *f*
1035	biliverdine *f*	biliverdina *f*
1036	bilobé, bilobulaire	bilobar, bilobular
1037	biloculaire	bilocular
1038	—	—
1039	—	—
1040	microscope *m* binoculaire	microscopio *m* binocular
1041	vision *f* binoculaire	visión *f* binocular
1042	distribution *f* binômiale, ~ binominale	distribución *f* binomial
1043	nomenclature *f* binominale	nomenclatura *f* binomial, ~ binaria
1044	—	—
1045	—	—
1046	binucléaire, binucléé	binucleado
1047	bioessai *m*, essai *m* biologique, test *m* ~	bioensayo *m*, ensayo *m* biológico
1048	bioblaste *m*	bioblasto *m*
1049	biocatalyse *f*	biocatálisis *f*, catálisis *f* biológica
1050	biocatalyseur *m*	biocatalizador *m*
1051	biochimique	bioquímico
1052	demande *f* biochimique d'oxygène, DBO	demanda *f* bioquímica de oxígeno, DBO
1053	biochimiste *m*	bioquímico *m*
1054	biochimie *f*	bioquímica *f*
1055	biochore *m*	biocora *f*
1056	biocide *m*	biocida *m*
1057	bioclimatologie *f*	bioclimatología *f*
1058	biocoenotique *f*, biocénotique *f*	biocenología *f*, biocenótica *f*
1059	biocoenose *f*, biocénose *f*, communauté *f* biotique	biocenosis *f*, comunidad *f* biótica
1060	biocénotique	biocenótico
1061	—	—
1062	cycle *m* biologique	biociclo *m*
1063	biodégradable	biodegradable
1064	—	—
1065	biodynamique *f*	biodinámica *f*
1066	bioénergétique *f*, énergétique *f* biologique	bioenergética *f*
1067	biogène *m*	biógeno *m*
1068	biogenèse *f*	biogénesis *f*
1069	biogénétique	biogenético
1070	isolement *m* biogénétique	aislamiento *m* biogenético
1071	loi *f* biogénétique (fondamentale), ~ de Haeckel	ley *f* fundamental biogenética
1072	biogénie *f*	biogenia *f*
1073	cycle *m* biogéochimique	ciclo *m* biogeoquímico
1074	biogéochimie *f*	biogeoquímica *f*
1075	biogéographie *f*	biogeografía *f*
1076	biologique	biológico
1077	—	—
1078	horloge *f* biologique, ~ physiologique, ~ interne	reloj *m* biológico, ~ interno
1079	demi-vie *f* biologique, période *f* biologique de demi-valeur	vida *f* media biológica, semiperíodo *m* de vida biológico
1080	membrane *f* biologique	membrana *f* biológica, biomembrana *f*
1081	plan *m* d'organisation biologique	plan *m* de organización biológica
1082	lutte *f* biologique (contre les parasites)	control *m* biológico (de plagas)
1083	épuration *f* biologique des eaux résiduaires	depuración *f* (o: purificación *f*) biológica de aguas residuales
1084	race *f* biologique	raza *f* biológica
1085	spectre *m* biologique	espectro *m* biológico
1086	biologiste *m*	biólogo *m*
1087	biologie *f*	biología *f*
1088	bioluminescence *f*	bioluminescencia *f*
1089	biolyse *f*	biólisis *f*
1090	biomasse *f*	biomasa *f*
1091	biome *m*	bioma *m*
1092	biomécanique *f*	biomecánica *f*

No.	English	German
1093	biometeorology	Biometeorologie *f*
1094	biometric(al)	biometrisch
1095	biometry, biometrics, biostatistics, biological statistics	Biometrie *f*, Biometrik *f*, Biostatistik *f*
1096	biomolecule	Biomolekül *n*
1097	bion	Bion *n*
1098	bionics, biocybernetics	Bionik *f*, Biokybernetik *f*
1099	bionomy, bionomics	Bionomie *f*
1100	biont	Biont *m*
1101	biophore	Biophore *f*
1102	biophysicist	Biophysiker *m*
1103	biophysics	Biophysik *f*
1104	biopsy	Biopsie *f*
1105	biorhythm	Biorhythmus *m*
1106	bios	Bios *m*, Biosstoffe *mpl*
1107	biosociology	Biosoziologie *f*
1108	biosome	Biosom *n*
1109	biosphere	Biosphäre *f*
1110	biostatistics → biometry	—
1111	biosynthesis	Biosynthese *f*
1112	biosystematics	Biosystematik *f*
1113	biotic	biotisch
1114	biotic community → biocoenosis	—
1115	biotic factor	biotischer Faktor *m*
1116	biotic potential	biotisches Potential *n*
1117	biotin, vitamin H	Biotin *n*, Vitamin *n* H
1118	biotope	Biotop *m*, Lebensstätte *f*
1119	biotype	Biotyp(us) *m*
1120	bipartition, binary fission	Zweiteilung *f*
1121	biped	Zweifüßer *m*
1122	biped(al)	zweifüßig
1123	bipedalism	Bipedie *f*, Zweifüßigkeit *f*
1124	bipinnate	doppelt gefiedert
1125	bipolar	zweipolig, bipolar
1126	bipolarity	Bipolarität *f*
1127	bird fauna → avifauna	—
1128	bird flock	Vogelschwarm *m*
1129	bird migration	Vogelzug *m*
1130	bird of prey, raptorial bird, predatory ~	Raubvogel *m*
1131	bird pollination → ornithophily	—
1132	bird preservation	Vogelschutz *m*
1133	bird reserve	Vogelschutzgebiet *n*
1134	bird ringing	Beringung *f* *(Vögel)*
1135	birds breeding gregariously	Koloniebrüter *mpl*
1136	birefringence, double refraction	Doppelbrechung *f*
1137	birefringent, doubly refracting	doppelbrechend
1138	birth, parturition	Geburt *f*
1139	birth-rate	Geburtenrate *f*
1140	bisexual	bisexuell, zweigeschlechtig
1141	bisexuality	Bisexualität *f*, Zweigeschlechtigkeit *f*
1142	biuret reaction	Biuretreaktion *f*
1143	bivalent, geminus (*pl* gemini)	Bivalent *n*, Geminus *m* (*pl* Gemini)
1144	bivalve, two-valved	zweiklappig
1145	Bivalvia, bivalves, Lamellibranchiata, lamellibranchs, Pelecypoda, pelecypods	Muscheln *fpl*, Bivalven *pl*, Lamellibranchier *pl*
1146	blastema	Blastem *n*, Keimgewebe *n*
1147	blastocholine	Blastokolin *n*
1148	blastocoel, segmentation cavity, cleavage ~	Blastozöl *n*, Furchungshöhle *f*, Keimhöhle *f*
1149	blastocyst	Blastozyste *f*, Keimblase *f*
1150	blastocyte	Blastozyt *m*
1151	blastoderm	Blastoderm *n*, Keimhaut *f*
1152	blastodisc, germinal disc	Keimscheibe *f*, Blastodiskus *m*
1153	blastogenesis	Blastogenese *f*
1154	blastogenic	blastogen, keimgebunden

1093	biométéorologie *f*	biometeorología *f*
1094	biométrique	biométrico
1095	biométrie *f*, biométrique *f*, biostatistique *f*	biometría *f*, bioestadística *f*
1096	biomolécule *f*	biomolécula *f*
1097	bion *m*	bíon *m*
1098	bionique *f*, biocybernétique *f*, bioélectronique *f*	biónica *f*, electrónica *f* biológica
1099	bionomie *f*	bionomía *f*, bionómica *f*
1100	biote *m*	bionte *m*
1101	biophore *m*	bióforo *m*
1102	biophysicien *m*	biofísico *m*
1103	biophysique *f*, physique *f* biologique	biofísica *f*
1104	biopsie *f*	biopsia *f*
1105	rythme *m* biologique	ritmo *m* biológico
1106	bios *m*, substances *f pl* ~	bíos *m*
1107	biologie *f* sociale	biosociología *f*
1108	biosome *m*	biosoma *m*
1109	biosphère *f*	biosfera *f*
1110	—	—
1111	biosynthèse *f*	biosíntesis *f*
1112	biosystématique *f*	biosistemática *f*
1113	biotique	biótico
1114	—	—
1115	facteur *m* biotique	factor *m* biótico
1116	potentiel *m* biotique	potencial *m* biótico
1117	biotine *f*, vitamine *f* H	biotina *f*, vitamina *f* H
1118	biotope *m*	biótopo *m*
1119	biotype *m*, type *m* biologique	biotipo *m*
1120	bipartition *f*, division *f* binaire	bipartición *f*, división *f* binaria
1121	bipède *m*	bípede *m*, bípedo *m*
1122	bipède	bípede, bípedo
1123	bipédie *f*	bipedismo *m*, bipedalismo *m*
1124	bipenné	bipinnado
1125	bipolaire	bipolar
1126	bipolarité *f*	bipolaridad *f*
1127	—	—
1128	bande *f* d'oiseaux, troupe *f* ~	bandada *f* de aves, bando *m* ~ ~
1129	migration *f* d'oiseaux	migración *f* de las aves
1130	oiseau *m* prédateur, ~ de proie, rapace *m*	ave *f* de presa, ~ de rapiña, (~) rapaz *f*
1131	—	—
1132	protection *f* des oiseaux	protección *f* de aves
1133	réserve *f* avienne	reserva *f* de pájaros, ~ de aves
1134	baguage *m*	anillado *m*, anillamiento *m*
1135	oiseaux *m pl* nichant en colonies	aves *f pl* que anidan en colonias
1136	biréfringence *f*	birrefringencia *f*, refracción *f* doble
1137	biréfringent	birrefringente
1138	part *m*, parturition *f*	parto *m*, parición *f*
1139	taux *m* de(s) naissance(s)	tasa *f* de natalidad, índice *m* ~ ~
1140	bisexuel, bisexué	bisexual, bisexuado
1141	bisexualité *f*	bisexualidad *f*
1142	réaction *f* du biuret	reacción *f* del biureto
1143	bivalent *m*	bivalente *m*, gémino *m*
1144	bivalve	bivalvo
1145	bivalves *m pl*, lamellibranches *m pl*, pélécypodes *m pl*, acéphales *m pl*	bivalvos *m pl*, lamelibranquios *m pl*, pelecípodos *m pl*
1146	blastème *m*	blastema *m*
1147	blastokoline *f*	blastocolina *f*
1148	blastocoele *m*, blastocèle *m*, cavité *f* de segmentation	blastocele *m*, cavidad *f* de segmentación
1149	blastocyste *m*	blastocisto *m*
1150	blastocyte *m*	blastocito *m*
1151	blastoderme *m*	blastodermo *m*
1152	blastodisque *m*, disque *m* germinatif	blastodisco *m*, disco *m* germinativo
1153	blastogenèse *f*	blastogénesis *f*
1154	blastogène	blastógeno

1155	blastomere	Blastomer *n*, Blastomere *f*
1156	blastomycete	Hefepilz *m*, Sproßpilz *m*, Blastomyzet *m*
1157	blastopore	Blastoporus *m*, Urmund *m*
1158	blastopore lip	Urmundlippe *f*
1159	blastospore	Blastospore *f*, Sproßkonidie *f*
1160	blastula	Blastula *f*, Blasenkeim *m*
1161	blastulation	Blastulabildung *f*
1162	bleeding *bot.*	Blutung *f* *bot.*
1163	blending character, quantitative ~	quantitatives Merkmal *n*
1164	blending inheritance	Vererbung *f* quantitativer Merkmale
1165	blepharoplast	Blepharoplast *m*
1166	blind gut → caecum	—
1167	blind spot	blinder Fleck *m*
1168	block of genes	Genblock *m*
1169	block pattern effect	Sperrmustereffekt *m*
1170	blood-brain barrier	Blut-Gehirn-Schranke *f*
1171	blood cell → haemocyte	—
1172	blood circulation	Blutkreislauf *m*, ~zirkulation *f*
1173	blood coagulation, ~ clotting	Blutgerinnung *f*
1174	blood coagulum, ~ clot	Blutgerinnsel *n*
1175	blood corpuscle	Blutkörperchen *n*
1176	blood flow, ~ stream	Blutstrom *m*
1177	blood-forming → haemopoietic	—
1178	blood gill	Blutkieme *f*
1179	blood group	Blutgruppe *f*
1180	blood group antigen	Blutgruppen-Antigen *n*
1181	blood group determination, ~ grouping, ~ typing	Blutgruppenbestimmung *f*
1182	blood island	Blutinsel *f*
1183	blood picture, h(a)emogram	Blutbild *n*, Hämogramm *n*
1184	blood pigment	Blutfarbstoff *m*, Blutpigment *n*
1185	blood plasma	Blutplasma *n*
1186	blood platelet → thrombocyte	—
1187	blood pressure	Blutdruck *m*
1188	blood protein	Bluteiweiß *n*
1189	blood sample	Blutprobe *f*
1190	blood sedimentation	Blutsenkung *f*
1191	blood serum	Blutserum *n*
1192	blood smear	Blutausstrich *m*
1193	blood space	Blutlakune *f*
1194	blood stream → ~ flow	—
1195	blood sugar	Blutzucker *m*
1196	blood supply	Durchblutung *f*
1197	blood test	Blutprobe *f*
1198	blood typing → blood group determination	—
1199	blood-vascular system	Blutgefäß-System *n*
1200	blood vessel	Blutgefäß *n*
1201	blood volume	Blutvolumen *n*
1202	to bloom → to flower	—
1203	to blossom, blossom → to flower, flower	—
1204	blossom-time → flowering period	—
1205	blue-green algae → Cyanophyta	—
1206	body cavity	Leibeshöhle *f*, Körperhöhle *f*
1207	body cell → somatic ~	—
1208	body fluid	Körperflüssigkeit *f*
1209	body heat	Körperwärme *f*
1210	body plan → structural ~	—
1211	body structure	Körperbau *m*
1212	body temperature	Körpertemperatur *f*
1213	body weight	Körpergewicht *n*
1214	bog moss → peat ~	—
1215	boiling point	Siedepunkt *m*
1216	bone	Knochen *m*
1217	bone earth, ~ salt	Knochenasche *f*, Knochenerde *f*

1155	blastomère *m*, sphère *f* de segmentation	blastómero *m*
1156	blastomycète *m*	blastomiceto *m*
1157	blastopore *m*	blastóporo *m*
1158	lèvre *f* blastoporique	labio *m* del blastóporo
1159	blastospore *f*	blastospora *f*
1160	blastula *f*	blástula *f*
1161	blastulation *f*	blastulación *f*
1162	exsudation *f* de la plante (après la taille)	lacrimación *f*
1163	caractère *m* quantitatif, ~ mêlé	carácter *m* cuantitativo, ~ mezclado
1164	hérédité *f* à facteurs multiples	herencia *f* mezclada
1165	blépharoplaste *m*	blefaroplasto *m*
1166	—	—
1167	tache *f* aveugle	punto *m* ciego, mancha *f* ciega
1168	bloc *m* de gènes	bloque *m* de genes
1169	effet *m* de combinaison en bloc	efecto *m* de combinación en bloque
1170	barrière *f* hémato-encéphalique	barrera *f* hemáticacerebral
1171	—	—
1172	circulation *f* sanguine	circulación *f* sanguínea, ~ hemática
1173	coagulation *f* sanguine, ~ du sang	coagulación *f* sanguínea, ~ hemática
1174	caillot *m* de sang	coágulo *m* sanguíneo
1175	globule *m* du sang	glóbulo *m* sanguíneo
1176	courant *m* sanguin, flux *m* ~	corriente *f* sanguínea, flujo *m* sanguíneo
1177	—	—
1178	branchie *f* sanguine	branquia *f* sanguínea
1179	groupe *m* sanguin	grupo *m* sanguíneo
1180	antigène *m* de groupe sanguin	antígeno *m* de grupo sanguíneo
1181	détermination *f* du groupe sanguin, typage *m* du sang	determinación *f* del grupo sanguíneo
1182	îlot *m* sanguin	isla *f* sanguínea, ~ hemática
1183	formule *f* sanguine, ~ hématologique, hémogramme *m*	cuadro *m* hemático, ~ hematológico, hemograma *m*
1184	pigment *m* sanguin	pigmento *m* sanguíneo, ~ hemático
1185	plasma *m* sanguin	plasma *m* sanguíneo
1186	—	—
1187	pression *f* sanguine, tension *f* artérielle	presión *f* sanguínea, tensión *f* arterial
1188	substance *f* protéique du sang	proteína *f* sanguínea, ~ hemática
1189	échantillon *m* de sang	muestra *f* de sangre
1190	sédimentation *f* globulaire	sedimentación *f* sanguínea, ~ globular
1191	sérum *m* sanguin	suero *m* sanguíneo, ~ hemático
1192	frottis *m* de sang	extensión *f* sanguínea, frotis *m* sanguíneo
1193	lacune *f* sanguine, espace *m* sanguin	laguna *f* sanguínea, espacio *m* sanguíneo
1194	—	—
1195	sucre *m* sanguin	azúcar *m* sanguíneo
1196	irrigation *f* sanguine	irrigación *f* sanguínea, riego *m* sanguíneo
1197	analyse *f* du sang	análisis *m* de sangre, prueba *f* sanguínea
1198	—	—
1199	système *m* sanguin	sistema *m* vasculosanguíneo
1200	vaisseau *m* sanguin	vaso *m* sanguíneo
1201	volume *m* sanguin	volumen *m* sanguíneo
1202	—	—
1203	—	—
1204	—	—
1205	—	—
1206	cavité *f* corporelle	cavidad *f* corporal
1207	—	—
1208	liquide *m* corporel, ~ organique	líquido *m* orgánico
1209	chaleur *f* corporelle	calor *m* corporal, ~ orgánico
1210	—	—
1211	conformation *f* du corps	constitución *f* corporal, estructura *f* del cuerpo
1212	température *f* corporelle, ~ interne	temperatura *f* corporal, ~ interna
1213	poids *m* corporel	peso *m* corporal
1214	—	—
1215	point *m* d'ébullition	punto *m* de ebullición
1216	os *m*	hueso *m*
1217	cendres *f pl* d'os	ceniza *f* ósea, sal *f* ~

1218	bone formation → osteogenesis	—
1219	bone marrow	Knochenmark *n*
1220	bony → osseous	—
1221	bony cell → osteocyte	—
1222	bony fish → Osteichthyes	—
1223	bony labyrinth, osseous ~	knöchernes Labyrinth *n*
1224	bony plate	Knochenplatte *f*
1225	bony skull → osteocranium	—
1226	bony tissue, bone ~	Knochengewebe *n*
1227	book-lung	Buchlunge *f*, Tracheenlunge *f*, Fächerlunge *f*, Phyllotrachee *f*, Fächertrachee *f*
1228	bordered pit	Hoftüpfel *m*
1229	bostryx, helicoid cyme	Schraubel *f*
1230	botanical	botanisch
1231	botanist	Botaniker *m*
1232	botany, phytology	Botanik *f*, Pflanzenkunde *f*, Phytologie *f*
1233	bouillon culture, broth ~	Bouillonkultur *f*
1234	boundary cell → heterocyst	—
1235	bouquet stage	Bukettstadium *n*
1236	brachial	Arm ...
1237	Brachiopoda, brachiopods, lamp shells	Armfüßer *mpl*, Brachiopoden *pl*
1238	brachyblast, brachyplast	Kurztrieb *m*, Brachyblast *m*
1239	brachymeiosis	Brachymeiosis *f*
1240	brachyuric → short-tailed	—
1241	brackish water	Brackwasser *n*
1242	bract	Deckblatt *n*, Tragblatt *n*, Stützblatt *n*, Braktee *f*
1243	bract scale → cover ~	—
1244	bracteole	Brakteole *f*
1245	bradytelic	bradytelisch
1246	brain, encephalon	Gehirn *n*, Hirn *n*
1247	brain case → cranial capsule	—
1248	brain mantle → pallium	—
1249	brain stem	Hirnstamm *m*, Stammhirn *n*
1250	brain vesicle	Hirnbläschen *n*
1251	branch	Zweig *m*, Ast *m*
1252	to branch → to ramify	—
1253	branchia → gill	—
1254	branchial (see also gill ...)	Kiemen ...
1255	branchial heart	Kiemenherz *n*
1256	branchial skeleton	Kiemenskelett *n*, Branchialskelett *n*
1257	Branchiopoda	Blattfüßer *mpl*, Blattfußkrebse *mpl*
1258	to break down, to degrade	abbauen
1259	"breakage first" hypothesis	Bruch-Hypothese *f*
1260	breakage-fusion bridge-cycle	Bruch-Fusions-Brücken-Zyklus *m*
1261	breakage-reunion bivalent	Bruch-Reunions-Bivalent *n*
1262	breakdown, degradation	Abbau *m*
1263	breakdown product, degradation ~	Abbauprodukt *n*
1264	breastbone → sternum	—
1265	to breathe, to respire	atmen
1266	breathing ... → respiratory	—
1267	to breed	züchten
1268	breeder	Züchter *m*
1269	breeding	Züchtung *f*
1270	breeding area, ~ ground, ~ site	Brutplatz *m*
1271	breeding behavio(u)r → incubation ~	—
1272	breeding colony	Brutkolonie *f*
1273	breeding experiment	Züchtungsversuch *m*
1274	breeding for resistance	Resistenzzüchtung *f*
1275	breeding in colonies, colonial nesting	Kolonienbrüten *n*, Brüten *n* in Kolonien
1276	breeding method, selection ~	Züchtungsmethode *f*, Züchtungsverfahren *n*
1277	breeding of new varieties	Neuzüchtung *f*
1278	breeding parasite	Brutparasit *m*
1279	breeding parasitism, brood ~, nest ~	Brutparasitismus *m*
1280	breeding plumage → nuptial dress	—

1218	—	—
1219	mœlle *f* osseuse	médula *f* ósea, tuétano *m*
1220	—	—
1221	—	—
1222	—	—
1223	labyrinthe *m* osseux	laberinto *m* óseo
1224	plaque *f* osseuse	placa *f* ósea
1225	—	—
1226	tissu *m* osseux	tejido *m* óseo
1227	sac *m* pulmonaire, phyllotrachée *f*	tráquea *f* laminar, pulmón *m* ~, filotráquea *f*
1228	ponctuation *f* aréolée	punteadura *f* areolada, poro *m* areolado
1229	cyme *f* hélicoïde	bóstrix *m*, cima *f* helicoidal
1230	botanique	botánico
1231	botaniste *m*, phytologue *m*	botánico *m*, fitólogo *m*
1232	botanique *f*, phytologie *f*	botánica *f*, fitología *f*
1233	culture *f* en bouillon	cultivo *m* en caldo
1234	—	—
1235	stade *m* «en bouquet»	estado *m* de ramillete
1236	brachial	braquial
1237	brachiopodes *mpl*	braquiópodos *mpl*
1238	brachyblaste *m*	braquiblasto *m*
1239	brachyméiose *f*	braquimeiosis *f*
1240	—	—
1241	eau *f* saumâtre	agua *f* salobre
1242	bractée *f*	bráctea *f*, hoja *f* tectriz
1243	—	—
1244	bractéole *f*	bractéola *f*
1245	bradytélique	bradítélico
1246	encéphale *m*	encéfalo *m*
1247	—	—
1248	—	—
1249	tronc *m* cérébral	tronco *m* cerebral, ~ encefálico
1250	vésicule *f* cérébrale	vesícula *f* cerebral
1251	branche *f*	rama *f*
1252	—	—
1253	—	—
1254	branchial	branquial
1255	cœur *m* branchial	corazón *m* branquial
1256	squelette *m* branchial	esqueleto *m* branquial
1257	branchiopodes *mpl*	branquiópodos *mpl*
1258	dégrader	desdoblar(se), degradar, demoler
1259	hypothèse *f* de la cassure d'abord	hipótesis *f* de "ruptura inicial"
1260	cycle *m* «rupture-fusion-pont»	ciclo *m* "ruptura-fusión-puente"
1261	bivalent *m* «rupture-réunion»	bivalente *m* de ruptura y reunión
1262	dégradation *f*, démolition *f*	degradación *f*, desdoblamiento *m*
1263	produit *m* de dégradation, ~ de démolition	producto *m* de degradación, ~ ~ desdoblamiento
1264	—	—
1265	respirer	respirar
1266	—	—
1267	sélectionner	seleccionar
1268	sélectionneur *m*	criador *m*
1269	sélection *f*	selección *f*, cría *f*
1270	zone *f* de nidification, ~ de reproduction	área *f* de cría
1271	—	—
1272	colonie *f* reproductrice, ~ de ponte	colonia *f* de cría, ~ ~ nidificación
1273	expérience *f* d'élevage, essai *m* ~	experiencia *f* de cría, experimento *m* ~ ~
1274	sélection *f* de résistance	selección *f* de resistencia
1275	nidification *f* en colonies	nidificación *f* en colonia, cría *f* ~ ~
1276	procédé *m* de sélection	procedimiento *m* de cría
1277	création *f* de nouvelles variétés	creación *f* de nuevas variedades
1278	parasite *m* de couvée	parásito *m* de cría
1279	parasitisme *m* de couvée, ~ de ponte	parasitismo *m* de cría
1280	—	—

1281	breeding season, brood period	Brutzeit *f*
1282	breeding territory	Brutrevier *n*
1283	breeding value	Zuchtwert *m*
1284	bridge-fragment configuration	Brücken-Fragment-Konfiguration *f*
1285	bright field	Hellfeld *n*
1286	bright field microscopy	Hellfeldmikroskopie *f*
1287	bristle, seta, chaeta	Borste *f*, Borstenhaar *n*
1288	bristly, setaceous	borstig
1289	broad-leaved tree → deciduous ~	—
1290	bronchiole	Bronchiole *f*
1291	bronchus (*pl* bronchi)	Bronchie *f*
1292	brood	Brut *f*
1293	to brood → to incubate	—
1294	brood body → propagule	—
1295	brood bud, ~ gemma	Brutknospe *f*
1296	brood care → care of young	—
1297	brood parasitisme → breeding ~	—
1298	brood patch → incubation ~	—
1299	brood period → breeding season	—
1300	brood pouch	Bruttasche *f*
1301	broth → nutrient bouillon	—
1302	broth culture → bouillon culture	—
1303	brown algae → Phaeophyta	—
1304	brush border	Bürstensaum *m*
1305	brushwood layer	Strauchschicht *f*
1306	bryology, muscology	Mooskunde *f*, Bryologie *f*
1307	bryophytes	Moospflanzen *f pl*, Bryophyten *pl*
1308	Bryozoa, Polyzoa, bryozoans	Moostierchen *n pl*, Bryozoen *pl*
1309	buccal, oral	Mund ...
1310	buccal cavity, mouth ~, oral ~	Mundhöhle *f*
1311	bud	Knospe *f*, Auge *n bot.*
1312	to bud	knospen
1313	bud-bearing → gemmiferous	—
1314	bud dormancy	Knospenruhe *f*
1315	bud mutation → sport	—
1316	bud-scale, tegmentum	Knospenschuppe *f*, Tegment *n*
1317	budding	Knospenbildung *f*; Sprossung *f (Fortpfl.)*
1318	buffer	Puffer *m*
1319	to buffer	puffern
1320	buffer action	Pufferwirkung *f*
1321	buffer(ing) capacity	Pufferungsvermögen *n*, Pufferkapazität *f*
1322	buffer(ing) gene	Puffergen *n*
1323	buffer solution	Pufferlösung *f*
1324	buffer substance	Puffersubstanz *f*
1325	buffer system	Puffersystem *n*
1326	buffering	Pufferung *f*
1327	bulb	Zwiebel *f* (Pfl. *anat.*)
1328	bulb scale	Zwiebelschuppe *f*
1329	bulbil	Brutzwiebel *f*, Bulbille *f*
1330	bulbous, tuberous	knollig
1331	bulbous plant	Zwiebelpflanze *f*, -gewächs *n*
1332	burdo	Burdo *m* (*pl* Burdonen)
1333	buttress-root → stilt root	—
1334	butyric acid	Buttersäure *f*
1335	butyric fermentation	Buttersäuregärung *f*
1336	byssus	Byssus *m*
1337	cacogenesis, kakogenesis	Kakogenese *f*
1338	c(a)ecum, blind gut	Blinddarm *m*, Caecum *n*, Coecum *n*
1339	caenogenesis, kainogenesis	Kainogenese *f*, Caenogenese *f*, Zänogenese *f*
1340	Caenozoic → Cenozoic	—
1341	caffeic acid	Kaffeesäure *f*
1342	calamus, quill	Federkiel *m*
1343	calcareous shell	Kalkschale *f*
1344	calcareous skeleton	Kalkskelett *n*

1281	période *f* de couvaison, ~ d'incubation	época *f* de cría, estación *f* ~ ~, período *m* de incubación
1282	territoire *m* de reproduction, ~ de ponte	territorio *m* de cría
1283	valeur *f* génétique	valor *m* genético
1284	configuration *f* «fragment-pont»	configuración *f* de fragmento-puente
1285	fond *m* clair	campo *m* brillante
1286	microscopie *f* à fond clair	microscopia *f* de campo brillante
1287	soie *f* *zool.*	cerda *f*, queta *f*, seda *f*
1288	sétacé	setáceo
1289	—	—
1290	bronchiole *f*	bronquíolo *m*
1291	bronche *f*	bronquio *m*
1292	couvée *f*, nichée *f*	nidada *f*, pollada *f*
1293	—	—
1294	—	—
1295	gemme *f*	gema *f*, yema *f* reproductora
1296	—	—
1297	—	—
1298	—	—
1299	—	—
1300	poche *f* incubatrice	bolsa *f* incubadora
1301	—	—
1302	—	—
1303	—	—
1304	bordure *f* en brosse	borde *m* en cepillo, ~ estriado
1305	strate *f* arbustive	estrato *m* arbustivo
1306	bryologie *f*	briología *f*
1307	bryophytes *mpl*	briofitos *mpl*, briofitas *fpl*
1308	bryozoaires *mpl*, polyzoaires *mpl*	briozoos *mpl*, polizoos *mpl*
1309	buccal, oral	bucal, oral
1310	cavité *f* buccale	cavidad *f* bucal
1311	bourgeon *m*, bouton *m*, œil *m*	yema *f*, botón *m*
1312	bourgeonner, boutonner	brotar, abotonar
1313	—	—
1314	dormance *f* des bourgeons	letargo *m* de las yemas
1315	—	—
1316	écaille *f* de bourgeons	escama *f* gemaria, tegmento *m*
1317	bourgeonnement *m*	gemación *f*, brotación *f*
1318	tampon *m*	tampón *m*, amortiguador *m*
1319	tamponner	tamponar, amortiguar
1320	action *f* tampon	acción-tampón *f*, acción *f* amortiguadora
1321	pouvoir *m* tampon	poder *m* tampón, ~ amortiguador
1322	gène *m* tampon	gen *m* amortiguador
1323	solution *f* tamponnée	solución *f* tampón, ~ amortiguadora
1324	substance *f* de tampon	sustancia *f* tampón
1325	système *m* tampon	sistema *m* tampón, ~ amortiguador
1326	tamponnage *m*	amortiguación *f*
1327	bulbe *m*	bulbo *m*
1328	écaille *f* de bulbe	escama *f* del bulbo
1329	bulbille *f*	bulbilo *m*
1330	bulbeux, tubéreux	bulboso, tuberoso
1331	plante *f* bulbeuse, ~ à bulbe	planta *f* bulbosa, ~ bulbífera
1332	bourdon *m*	burdón *m*
1333	—	—
1334	acide *m* butyrique	ácido *m* butírico
1335	fermentation *f* butyrique	fermentación *f* butírica
1336	byssus *m*	biso *m*
1337	cacogenèse *f*	cacogénesis *f*
1338	coecum *m*	(intestino *m*) ciego *m*
1339	cénogenèse *f*, caenogenèse *f*	cenogénesis *f*
1340	—	—
1341	acide *m* caféique	ácido *m* cafeico
1342	calamus *m*	cálamo *m*, cañón *m*
1343	coquille *f* calcaire	concha *f* calcárea
1344	squelette *m* calcaire	esqueleto *m* calcáreo

1345	calcicole, calcicolous plant, calciphyte	Kalkpflanze *f*, kalkliebende Pflanze *f*
1346	calciferol, antirhachitic vitamin, vitamin D_2	Kalziferol *n*, antirachitisches Vitamin *n*, Vitamin D_2
1347	calcification	Verkalkung *f*
1348	calcifuge (plant), calciphobe	Kalkflieher *m*, kalkfliehende Pflanze *f*
1349	to calcify	verkalken
1350	caliciform cell → goblet cell	—
1351	call note, calling song	Lockruf *m*
1352	callose	Kallose *f*
1353	callosity	Schwiele *f*
1354	callous corpus → corpus callosum	—
1355	callus	Kallus *m*
1356	caloric value	Kalorienwert *m*
1357	caloricity → animal heat	—
1358	calorie	Kalorie *f*
1359	calorie intake	Kalorienaufnahme *f*
1360	calorie requirements	Kalorienbedarf *m*
1361	calorific centre → heat centre	—
1362	calorification, thermogenesis, heat production	Wärmeerzeugung *f*
1363	calorimeter	Kalorimeter *n*
1364	calorimetry	Kalorimetrie *f*
1365	calycle, calicle	Außenkelch *m*
1366	calyptra → root-cap	—
1367	calyptrogen	Kalyptrogen *n*
1368	calyx (*pl* calyces)	Blütenkelch *m*, Kalyx *m*
1369	cambium	Kambium *n*
1370	cambium activity, cambial ~	Kambiumtätigkeit *f*
1371	cambium cell	Kambiumzelle *f*
1372	Cambrian (period)	Kambrium *n*
1373	camera eye	Kameraauge *n*
1374	camerate, chambered	gekammert
1375	cameration	Kammerung *f*
1376	to camouflage	tarnen
1377	camouflage, procrypsis	Tarnung *f*
1378	campylotropous	kampylotrop
1379	Canada balsam	Kanadabalsam *m*
1380	canalizing selection	kanalisierende Selektion *f*
1381	cancerogenic → carcinogenic	—
1382	cancerology → carcinology	—
1383	cancrology → carcinology	—
1384	cane sugar → saccharose	—
1385	canine (tooth)	Eckzahn *m*
1386	cannibalism	Kannibalismus *m*
1387	capable of flight	flugfähig
1388	capable of germination → germinative	—
1389	capable of regeneration	regenerationsfähig
1390	capacity to respond, responsiveness	Reaktionsfähigkeit *f* (auf einen Reiz)
1391	capillary *adj.*	Haar ...
1392	capillary *sub.*	Kapillare *f*, Kapillargefäß *n*, Haargefäß *n*
1393	capillary network	Kapillarnetz *n*
1394	capillary water	Kapillarwasser *n*
1395	capitulum, flower head	Blütenköpfchen *n*, ~ körbchen *n*
1396	capsular fruit	Kapselfrucht *f*
1397	capsule	Kapsel *f*
1398	carapace	Panzer *m* *zool.*, Carapax *m*
1399	carbohydrate, glucide	Kohle(n)hydrat *n*
1400	carbohydrate metabolism	Kohlenhydratstoffwechsel *m*
1401	carbohydrate reserve	Kohlenhydratreserve *f*
1402	carbon	Kohlenstoff *m*
1403	carbon assimilation	Kohlenstoffassimilation *f*
1404	carbon compound	Kohlenstoffverbindung *f*
1405	carbon content	Kohlenstoffgehalt *m*, Kohlegehalt *m*
1406	carbon cycle	Kohlenstoffzyklus *m*

1345	plante *f* calcicole	planta *f* calcícola
1346	calciférol *m*, vitamine *f* antirachitique, vitamine D_2	calciferol *m*, vitamina *f* antirraquítica, vitamina D_2
1347	calcification *f*	calcificación *f*
1348	plante *f* calcifuge	planta *f* calcífuga
1349	(se) calcifier	calcificar(se)
1350	—	—
1351	cri *m* d'appel, chant *m* ~	llamada *f* de atracción, sonido *m* de llamada
1352	callose *f*	calosa *f*
1353	callosité *f*	callosidad *f*
1354	—	—
1355	cal *m*, callus *m*	callo *m*
1356	valeur *f* calorique	valor *m* calórico
1357	—	—
1358	calorie *f*	caloría *f*
1359	apport *m* calorique	aportación *f* calórica
1360	besoins *m pl* en calories	necesidades *f pl* calóricas, requerimientos *mpl* calóricos (o: de calorías)
1361	—	—
1362	calorification *f*, thermogenèse *f*	calorificación *f*, termogénesis *f*
1363	calorimètre *m*	calorímetro *m*
1364	calorimétrie *f*	calorimetría *f*
1365	calicule *m*	calículo *m*
1366	—	—
1367	calyptrogène *f*	caliptrógeno *m*
1368	calice *m*	cáliz *m*
1369	cambium *m*, assise *f* génératrice	cámbium *m*, zona *f* generatriz
1370	activité *f* cambiale	actividad *f* del cámbium
1371	cellule *f* cambiale	célula *f* cambial
1372	cambrien *m*	cámbrico *m*
1373	œil *m* camérulaire	ojo *m* cámara
1374	cloisonné	dividido en cámaras
1375	cloisonnement *m*	tabicación *f*, tabicamiento *m*
1376	camoufler	camuflar
1377	camouflage *m*, cryptisme *m*	camuflaje *m*, cripsis *f*
1378	campylotrope	campilótropo
1379	baume *m* du Canada	bálsamo *m* del Canadá
1380	sélection *f* canalisante	selección *f* canalizante
1381	—	
1382	—	—
1383	—	—
1384	—	—
1385	canine *f*	diente *m* canino, colmillo *m*
1386	cannibalisme *m*	canibalismo *m*
1387	apte à voler, capable de ~, voilier	capaz de volar
1388	—	—
1389	capable de régénérer	capaz de regeneración
1390	capacité *f* de réaction	capacidad *f* de respuesta
1391	capillaire	capilar
1392	capillaire *m*, vaisseau *m* ~	capilar *m*, vaso *m* ~
1393	réseau *m* capillaire	red *f* capilar
1394	eau *f* de capillarité	agua *f* de capilaridad, ~ capilar
1395	capitule *m*	capítulo *m*, cabezuela *f*
1396	fruit *m* capsulaire	fruto *m* capsular
1397	capsule *f*	cápsula *f*
1398	carapace *f*	caparazón *m*
1399	hydrate *m* de carbone, glucide *m*	carbohidrato *m*, hidrato *m* de carbono, glúcido *m*
1400	métabolisme *m* glucidique	metabolismo *m* glucídico, ~ hidrocarbonado
1401	réserve *f* glucidique	reserva *f* de carbohidratos
1402	carbone *m*	carbono *m*
1403	assimilation *f* du carbone, ~ carbonée	asimilación *f* del carbono
1404	composé *m* carboné	compuesto *m* carburado
1405	teneur *f* en carbone	contenido *m* en carbono
1406	cycle *m* du carbone	ciclo *m* del carbono

1407	carbon dioxide, carbonic-acid gas	Kohlendioxid *n*
1408	carbon dioxide assimilation	Kohlensäureassimilation *f*
1409	carbon isotope	Kohlenstoffisotop *n*
1410	carbon skeleton	Kohlenstoffskelett *n*
1411	carbonic acid	Kohlensäure *f*
1412	carbonic anhydrase	Karboanhydrase *f*
1413	Carboniferous (period)	Karbon *n*
1414	carboxyl group	Karboxylgruppe *f*
1415	carboxylase	Karboxylase *f*
1416	carboxylation	Karboxylierung *f*
1417	carboxylic acid	Karbonsäure *f*
1418	carcinogenic, cancerogenic	karzinogen, kanzerogen, krebserregend
1419	carcinology, cancerology, cancrology	Krebsforschung *f*, Carcinologie *f*, Karzinologie *f*
1420	cardia	Kardia *f*, Magenmund *m*
1421	cardiac	Herz ...
1422	cardiac activity	Herztätigkeit *f*
1423	cardiac cycle	Herzzyklus *m*
1424	cardiac muscle → myocard	—
1425	cardiac rhythm	Herzrhythmus *m*
1426	cardinal point	Kardinalpunkt *m*
1427	care of young, brood care	Brutpflege *f*
1428	carnassial tooth	Reißzahn *m*
1429	Carnivora, carnivores	Raubtiere *n pl*
1430	carnivore	Fleischfresser *m*, Karnivore *m/f*
1431	carnivorous	fleischfressend
1432	carnivorous plant → insectivorous ~	—
1433	carotenase	Karotinase *f*
1434	carotene, carotin	Karotin *n*
1435	carotenoid, carotinoid	Karotinoid *n*
1436	carotid artery	Karotis *f*, Karotide *f*, Kopfschlagader *f*, Halsschlagader *f*
1437	carpal bone, wrist ~	Handwurzelknochen *m*
1438	carpel	Fruchtblatt *n*, Karpell *n*
1439	carpogonium	Karpogon *n*
1440	carpophore, carpopodium	Fruchthalter *m*, Karpophor *m*
1441	carpospore	Karpospore *f*
1442	carposporophyte	Karposporophyt *m*
1443	carpus	Handwurzel *f*
1444	carrier	Träger *m*, Trägersubstanz *f*
1445	cartilage	Knorpel *m*
1446	cartilage bone, replacing ~, replacement ~	Ersatzknochen *m*, Primordialknochen *m*
1447	cartilaginous, chondric (see also chondro ...)	knorpelig, Knorpel ...
1448	cartilaginous fish → Chondrichthyes	—
1449	cartilaginous precursor → precartilage	—
1450	caryo ... → karyo ..., nucleo ...	—
1451	caryopsis	Schalfrucht *f*, Kornfrucht *f*, Grasfrucht *f*, Karyopse *f*
1452	casein; paracasein (US)	Kasein *n*
1453	Casparian strip, ~ band	Casparyscher Streifen *m*
1454	caste	Kaste *f*
1455	castrate	Kastrat *m*
1456	to castrate, to emasculate	kastrieren, entmannen
1457	castration, emasculation	Kastration *f*, Entmannung *f*, Emaskulation *f*
1458	cat ... see also kat ...	—
1459	catabolism, katabolism, destructive metabolism, disassimilation	Katabolismus *m*, Abbaustoffwechsel *m*
1460	catabolite, katabolite, katabolic product	Katabolit *m*, Stoffwechselabbauprodukt *n*
1461	catadromous	katadrom
1462	catalase	Katalase *f*
1463	catalepsis, catalepsy	Katalepsie *f*
1464	catalysis	Katalyse *f*
1465	catalytic	katalytisch

1407	anhydride *m* carbonique, gaz *m* ~, dioxyde *m* de carbone	anhídrido *m* carbónico, dióxido *m* de carbono
1408	assimilation *f* du gaz carbonique	asimilación *f* del dióxido de carbono
1409	isotope *m* du carbone	isótopo *m* del carbono
1410	squelette *m* carboné	esqueleto *m* carbonado
1411	acide *m* carbonique	ácido *m* carbónico
1412	anhydrase *f* carbonique	anhidrasa *f* carbónica
1413	carbonifère *m*	carbonífero *m*
1414	groupe(ment) *m* carboxyle, ~ carboxylique	grupo *m* carboxilo, ~ carboxílico
1415	carboxylase *f*	carboxilasa *f*
1416	carboxylation *f*	carboxilación *f*
1417	acide *m* carboxylique	ácido *m* carboxílico
1418	cancérigène, carcinogène	cancerógeno, carcinógeno, cancerígeno
1419	cancérologie *f*, carcinologie *f*	cancerología *f*, carcinología *f*
1420	cardia *m*	cardias *m*
1421	cardiaque	cardíaco
1422	activité *f* cardiaque	actividad *f* cardíaca
1423	cycle *m* cardiaque	ciclo *m* cardíaco
1424	—	—
1425	rythme *m* cardiaque	ritmo *m* cardíaco
1426	point *m* cardinal	punto *m* cardinal
1427	soins *m pl* aux jeunes	cuidado *m* de la cría, ~ ~ ~ prole
1428	(dent *f*) carnassière *f*	diente *m* carnicero, presa *f*
1429	carnivores *m pl*, carnassiers *m pl*	carnívoros *m pl*, carniceros *m pl*
1430	carnivore *m*	carnívoro *m*
1431	carnivore, carnassier	carnívoro, carnicero
1432	—	—
1433	caroténase *f*	carotinasa *f*
1434	carotène *m*	carotina *f*, caroteno *m*
1435	carotinoïde *m*	carotinoide *m*, carotenoide *m*
1436	carotide *f*	(arteria *f*) carótida *f*
1437	os *m* carpien	hueso *m* carpiano
1438	carpelle *m*, feuille *f* carpellaire	carpelo *m*, hoja *f* carpelar
1439	carpogone *m*	carpogonio *m*
1440	carpophore *m*	carpóforo *m*, carpopodio *m*
1441	carpospore *f*	carpospora *f*
1442	carposporophyte *m*	carposporófito *m*
1443	carpe *m*	carpo *m*
1444	porteur *m*	portador *m*, sustancia *f* portadora
1445	cartilage *m*	cartílago *m*
1446	os *m* de cartilage, ~ ~ substitution, ~ primordial	hueso *m* de cartílago, ~ cartilaginoso, ~ de sustitución, ~ primordial
1447	cartilagineux, chondral	cartilaginoso, cóndrico
1448	—	—
1449	—	—
1450	—	—
1451	caryopse *m*	cariópside *f*, cariopsis *f*
1452	caséine *f*	caseína *f*
1453	assise *f* de Caspary, cadre *m* ~ ~	banda *f* de Caspary
1454	caste *f*	casta *f*
1455	castrat *m*	castrado *m*
1456	castrer, châtrer, émasculer	castrar, emascular
1457	castration *f*, émasculation *f*	castración *f*, emasculación *f*
1458	—	—
1459	catabolisme *m*, désassimilation *f*	catabolismo *m*, metabolismo *m* destructivo, desasimilación *f*
1460	catabolite *m*, produit *m* catabolique	catabolito *m*
1461	catadrome	catádromo
1462	catalase *f*	catalasa *f*
1463	catalepsie *f*	catalepsia *f*
1464	catalyse *f*	catálisis *f*
1465	catalytique	catalítico

1466	to catalyze	katalysieren
1467	catalyzer, catalyst	Katalysator *m*
1468	cataphoresis	Kataphorese *f*
1469	cataphyll, cataphyllary leaf	Niederblatt *n*
1470	catechin	Katechin *n*
1471	catecholamine	Katecholamin *n*
1472	catelectrotonus, katelectrotonus	Katelektrotonus *m*
1473	catenation	Kettenbildung *f* (Chrom.)
1474	cathepsin, kathepsin	Kathepsin *n*
1475	cathode	Kathode *f*
1476	cathode rays	Kathodenstrahlen *mpl*
1477	cathode-ray oscillograph	Kathodenstrahl-Oszillograph *m*
1478	cation, kation	Kation *n*
1479	cation exchange → base ~	—
1480	catkin, ament(um)	Kätzchen *n* *bot.*
1481	caudal fin, tail ~	Schwanzflosse *f*
1482	caudal vertebra, tail ~	Schwanzwirbel *m*
1483	caulescent	stengeltreibend
1484	cauliflorous	stammblütig
1485	cauliflory	Stammblütigkeit *f*, Kauliflorie *f*
1486	cauline	Stamm ..., Stengel ...
1487	cauline vascular bundle → stem bundle	—
1488	caulocarpous	stengelfrüchtig
1489	caulome, stem axis	Sproßachse *f*, Kaulom *n*
1490	cava, caval vein	Hohlvene *f*
1491	cave fauna, ~ animals, troglobionts	Höhlenfauna *f*, -tiere *npl*, Troglobionten *pl*
1492	cave-nesting animal, hole-nester	Höhlenbrüter *m*, Baumbrüter *m*
1493	cavernicolous, cave-inhabiting	höhlenbewohnend
1494	cavernous body	Schwellkörper *m*
1495	cavernous tissue, erectile ~	Schwellgewebe *n*
1496	C^{14} dating → radiocarbon datation	—
1497	cecidium (*pl* cecidia), plant gall	Pflanzengalle *f*, Zezidie *f*
1498	cecum → caecum	—
1499	cell	Zelle *f*
1500	cell aggregation	Zellaggregat *n*
1501	cell biology, cytobiology	Zellbiologie *f*
1502	cell body	Zellkörper *m*, Zelleib *m*
1503	cell boundary	Zellgrenzschicht *f*
1504	cell constancy, eutely	Zellkonstanz *f*
1505	cell constituent, ~ component	Zellbestandteil *m*
1506	cell culture	Zellkultur *f*
1507	cell death, necrocytosis	Zelltod *m*
1508	cell differentiation, cytodifferentiation	Zelldifferenzierung *f*
1509	cell division	Zellteilung *f*
1510	cell elongation, ~ enlargement	Zellstreckung *f*
1511	cell extract	Zellextrakt *m*
1512	cell fractionation	Zellfraktionierung *f*
1513	cell-free system	zellfreies System *n*
1514	cell fusion	Zellverschmelzung *f*
1515	cell inclusion	Zelleinschluß *m*
1516	cell layer, cellular ~	Zellschicht *f*
1517	cell lineage	Zell-Linie *f*
1518	cell matrix	Zellmatrix *f*
1519	cell membrane	Zellmembran *f*, Zellhaut *f*
1520	cell metabolism, cellular ~	Zellstoffwechsel *m*
1521	cell migration	Zellwanderung *f*
1522	cell mouth → cytostome	—
1523	cell multiplication → cell proliferation	—
1524	cell nucleus → nucleus	—
1525	cell organelle	Zellorganell *n*
1526	cell permeability	Zellpermeabilität *f*
1527	cell physiology, cytophysiology	Zellphysiologie *f*
1528	cell plate	Zellplatte *f*

1466	catalyser	catalizar
1467	catalyseur *m*	catalizador *m*, agente *m* ~
1468	cataphorèse *f*	cataforesis *f*
1469	cataphylle *f*	catafilo *m*
1470	catéchine *f*	catequina *f*
1471	catécholamine *f*	catecolamina *f*
1472	catélectrotonus *m*	catelectrotono *m*
1473	caténation *f*	catenación *f*
1474	cathepsine *f*	catepsina *f*
1475	cathode *f*	cátodo *m*
1476	rayons *mpl* cathodiques	rayos *mpl* catódicos
1477	oscillographe *m* à rayons cathodiques, ~ cathodique	oscilógrafo *m* de rayos catódicos
1478	cation *m*	catión *m*
1479	—	—
1480	chaton *m*	amento *m*
1481	nageoire *f* caudale	aleta *f* caudal
1482	vertèbre *f* caudale	vértebra *f* caudal
1483	caulescent	caulescente
1484	cauliflore	caulífloro
1485	cauliflorie *f*	caulifloria *f*
1486	caulinaire	caulinar, caulino
1487	—	—
1488	caulocarpe, caulocarpien, caulocarpique	caulocarpo
1489	axe *m* caulinaire	eje *m* caulinar
1490	veine *f* cave	vena *f* cava
1491	faune *f* des cavernes, cavernicoles *mpl*, troglobies *mpl*	fauna *f* cavernícola, troglobios *mpl*
1492	oiseau *m* qui niche dans des cavités	ave *f* que anida en cavidades
1493	cavernicole	cavernícola
1494	corps *m* caverneux	cuerpo *m* cavernoso
1495	tissu *m* caverneux, ~ érectile	tejido *m* cavernoso, ~ eréctil
1496	—	—
1497	cécidie *f*, galle *f*	cecidio *m*, agalla *f*
1498	—	—
1499	cellule *f*	célula *f*
1500	agrégat *m* cellulaire	asociación *f* celular
1501	biologie *f* cellulaire, cytobiologie *f*	biología *f* celular, citobiología *f*
1502	corps *m* cellulaire	cuerpo *m* celular
1503	limitante *f* cellulaire	capa *f* limitante (celular)
1504	constance *f* cellulaire, eutélie *f*	constancia *f* celular, eutelia *f*
1505	constituant *m* cellulaire, composé *m* ~	constituyente *m* celular, componente *m* ~
1506	culture *f* cellulaire	cultivo *m* celular
1507	mort *f* cellulaire, nécrocytose *f*, nécrose *f* cellulaire	muerte *f* celular, necrocitosis *f*
1508	différenciation *f* cellulaire	diferenciación *f* celular
1509	division *f* cellulaire	división *f* celular
1510	élongation *f* cellulaire, allongement *m* ~	elongación *f* celular, alargamiento *m* ~
1511	extrait *m* cellulaire	extracto *m* celular
1512	fractionnement *m* cellulaire	fraccionamiento *m* celular
1513	système *m* acellulaire	sistema *m* libre de células
1514	fusion *f* cellulaire	fusión *f* celular
1515	inclusion *f* cellulaire, enclave *f* ~	inclusión *f* celular
1516	couche *f* cellulaire, assise *f* ~	estrato *m* celular
1517	lignée *f* cellulaire	linaje *m* de células
1518	matrice *f* cellulaire	matriz *f* celular
1519	membrane *f* cellulaire	membrana *f* celular
1520	métabolisme *m* cellulaire	metabolismo *m* celular
1521	migration *f* cellulaire	migración *f* celular
1522	—	—
1523	—	—
1524	—	—
1525	organite *m* cellulaire	orgánulo *m* celular
1526	perméabilité *f* cellulaire	permeabilidad *f* celular
1527	physiologie *f* cellulaire, cytophysiologie *f*	fisiología *f* celular, citofisiología *f*
1528	plaque *f* cellulaire	placa *f* celular, lámina *f* ~

1529	cell proliferation, ~ multiplication	Zellvermehrung *f*
1530	cell sap	Zellsaft *m*
1531	cell structure, cellular ~	Zellstruktur *f*
1532	cell suspension	Zellaufschwemmung *f*
1533	cell theory	Zelltheorie *f*
1534	cell wall	Zellwand *f*
1535	cellobiose	Zellobiose *f*
1536	celloidin section	Zelloidinschnitt *m*
1537	cellular	Zell ..., zellular
1538	cellular protein	Zelleiweiß *n*
1539	cellular respiration	Zellatmung *f*
1540	cellular tissue	Zellgewebe *n*
1541	cellulose	Zellulose *f*
1542	cellulose membrane	Zellulosemembran *f*
1543	cellulose wall	Zellulosewand *f*
1544	celo ... → coelo ...	—
1545	ceno ... → coeno ...	—
1546	Cenozoic, Caenozoic, Kainozoic	Känozoikum *n*, Neozoikum *n*, Zänozoikum *n*, Erdneuzeit *f*
1547	centipedes → Chilopoda	—
1548	central body → centrosome	—
1549	central cylinder	Zentralzylinder *m*
1550	central nervous system, CNS	Zentralnervensystem *n*, ZNS
1551	central spindle	Zentralspindel *f*
1552	centric	zentrisch
1553	centric fusion	zentrische Fusion *f*
1554	centrifugal	zentrifugal
1555	centrifugation	Zentrifugation *f*
1556	to centrifuge	zentrifugieren
1557	centrifuge	Zentrifuge *f*
1558	centriole	Zentriol *n*
1559	centripetal	zentripetal
1560	centrochromatin	Zentrochromatin *n*
1561	centrodesm(us), centrodesmose	Zentrodesmose *f*
1562	centrogene	Zentrogen *n*
1563	centrolecithal	zentrolezithal
1564	centromere, kinetochore, spindle attachment	Zentromer *n*, Kinetochor *n*
1565	centromere distance	Zentromerdistanz *f*, -abstand *m*
1566	centromere interference	Zentromerinterferenz *f*
1567	centroplasm	Zentroplasma *n*
1568	centrosome, central body	Zentrosom *n*, Zentralkörperchen *n*
1569	centrosphere, attraction-sphere	Zentrosphäre *f*
1570	cephalic	Kopf ...
1571	cephalic index	Kopfindex *m*
1572	cephalin, kephalin	Kephalin *n*
1573	cephalization	Zephalisation *f*
1574	Cephalochordata → Acrania	—
1575	Cephalopoda, Siphonopoda	Kopffüßer *mpl*, Zephalopoden *mpl*
1576	cephalothorax	Kopfbrust *f*, Kopfbruststück *n*
1577	cerebellar	Kleinhirn ...
1578	cerebellum	Kleinhirn *n*
1579	cerebral, encephalic	Gehirn ..., Hirn ..., Großhirn ...
1580	cerebral convolution → gyrus	—
1581	cerebral cortex	Hirnrinde *f*, Großhirnrinde *f*, Kortex *m*
1582	cerebral ganglion	Zerebralganglion *n*
1583	cerebral hemisphere	Großhirnhemisphäre *f*
1584	cerebral ventricle	Hirnkammer *f*, Hirnventrikel *m*
1585	cerebralization	Zerebralisation *f*
1586	cerebroside	Zerebrosid *n*
1587	cerebrospinal fluid, CSF	Gehirnrückenmarksflüssigkeit *f*, Zerebrospinalflüssigkeit *f*
1588	cerebrum	Großhirn *n* (oft auch: Gehirn *n*)
1589	certation	Zertation *f*
1590	cerumen	Ohrenschmalz *n*
1591	cervical	Hals ..., zervikal

1529	prolifération *f* cellulaire, multiplication *f* ~	proliferación *f* celular, multiplicación *f* ~
1530	suc *m* cellulaire	jugo *m* celular, savia *f* ~
1531	structure *f* cellulaire	estructura *f* celular
1532	suspension *f* cellulaire	suspensión *f* celular
1533	théorie *f* cellulaire	teoría *f* celular
1534	paroi *f* cellulaire	pared *f* celular
1535	cellobiose *m*	celobiosa *f*
1536	coupe *f* à la celloïdine	corte *m* de celoidina, sección *f* en ~
1537	cellulaire	celular
1538	protéine *f* cellulaire	proteína *f* celular
1539	respiration *f* cellulaire	respiración *f* celular
1540	tissu *m* cellulaire	tejido *m* celular
1541	cellulose *f*	celulosa *f*
1542	membrane *f* cellulosique	membrana *f* celulósica
1543	paroi *f* cellulosique	pared *f* celulósica
1544	—	—
1545	—	—
1546	cénozoïque *m*, néozoïque *m*	cenozoico *m*, kainozoico *m*
1547	—	—
1548	—	—
1549	cylindre *m* central	cilindro *m* central
1550	système *m* nerveux central, SNC	sistema *m* nervioso central, SNC
1551	fuseau *m* central	huso *m* central
1552	centrique	céntrico
1553	fusion *f* centrique	fusión *f* céntrica
1554	centrifuge	centrífugo
1555	centrifugation *f*	centrifugación *f*
1556	centrifuger	centrifugar
1557	centrifugeur *m*, centrifugeuse *f*	centrifugadora *f*, centrífuga *f*
1558	centriole *m*	centríolo *m*
1559	centripète	centrípeto
1560	centrochromatine *f*	centrocromatina *f*
1561	centrodesmose *f*	centrodesma *m*
1562	centrogène *m*	centrogen *m*
1563	centrolécithe, centrolécithique	centrolecito, centrolecítico
1564	centromère *m*, cinétochore *m*, kinétochore *m*	centrómero *m*, centrómera *f*, quinetócoro *m*
1565	distance *f* au centromère	distancia *f* del centrómero
1566	interférence *f* du centromère	interferencia *f* del centrómero
1567	centroplasme *m*	centroplasma *m*
1568	centrosome *m*, corpuscule *m* central	centrosoma *m*, corpúsculo *m* central
1569	centrosphère *f*, sphère *f* d'attraction	centrosfera *f*, esfera *f* atractiva
1570	céphalique	cefálico
1571	indice *m* céphalique	índice *m* cefálico
1572	céphaline *f*	cefalina *f*
1573	céphalisation *f*	cefalización *f*
1574	—	—
1575	céphalopodes *mpl*	cefalópodos *mpl*
1576	céphalothorax *m*	cefalotórax *m*
1577	cérébelleux	cerebelar
1578	cervelet *m*	cerebelo *m*
1579	cérébral, encéphalique	cerebral, encefálico
1580	—	—
1581	écorce *f* cérébrale, cortex *m* cérébral	corteza *f* cerebral
1582	ganglion *m* cérébral	ganglio *m* cerebral
1583	hémisphère *m* cérébral	hemisferio *m* cerebral
1584	ventricule *m* cérébral	ventrículo *m* cerebral
1585	cérébralisation *f*	cerebralización *f*
1586	cérébroside *m*	cerebrósido *m*
1587	liquide *m* céphalorachidien, ~ cérébrospinal	líquido *m* cefalorraquídeo
1588	cerveau *m*	cerebro *m*
1589	certation *f*	certación *f*
1590	cerumen *m*	cerumen *m*
1591	cervical	cervical

1592	cervical vertebra	Halswirbel *m*
1593	Cestoda, tapeworms	Bandwürmer *mpl*, Zestoden *pl*
1594	chaeta → bristle	—
1595	chain molecule	Kettenmolekül *n*
1596	chain reaction	Kettenreaktion *f*
1597	chain reflex, reflex chain	Kettenreflex *m*, Reflexkette *f*
1598	chalaza	Chalaza *f* a) Knospengrund *m* *bot.* b) Hagelschnur *f* *zool.*
1599	chalazogamy	Chalazogamie *f*
1600	chamaephytes	Chamäphyten *mpl*
1601	chambered → camerate	—
1602	chance for survival, ~ of surviving	Überlebenschance *f*
1603	change of diet	Nahrungswechsel *m*
1604	change of function	Funktionswechsel *m*
1605	character, characteristic, trait	Merkmal *n*, Kennzeichen *n*, Eigenschaft *f*
1606	character constancy	Merkmalskonstanz *f*
1607	character displacement	Merkmalsverschiebung *f*
1608	character gradient → cline	—
1609	characteristic species, index ~, indicator ~	Charakterart *f*, Kennart *f*, Leitart *f*
1610	Charophyta	Armleuchteralgen *fpl*, Charophyceen *pl*
1611	chasmogamy	Chasmogamie *f*
1612	chasmophytes	Felsspaltpflanzen *fpl*, Chasmophyten *mpl*
1613	check-cross	Kontrollkreuzung *f*
1614	chela (*pl* chelae), pincer	Schere *f* *zool.*
1615	chelate	Chelat *n*, Chelatverbindung *f*
1616	chelating agent, chelant	Chelatbildner *m*, Chelator *m*
1617	chelation	Chelatbildung *f*, Chelatisierung *f*
1618	Chelicerata	Fühlerlose *pl*, Chelizeraten *pl*
1619	chelicere, chelicera	Chelizere *f*
1620	chemical mediator → transmittor	—
1621	chemo-autotrophic	chemoautotroph
1622	chemonasty	Chemonastie *f*
1623	chemoreception	Chemorezeption *f*
1624	chemoreceptor	Chemorezeptor *m*
1625	chemosynthesis	Chemosynthese *f*
1626	chemotactic	chemotaktisch
1627	chemotaxis, chemiotaxis	Chemotaxis *f*
1628	chemotrophic	chemotroph
1629	chemotropism, chemiotropism	Chemotropismus *m*
1630	chewing (insect)	kauend (Insekt)
1631	chiasma (*pl* chiasmata)	Chiasma *n* (*pl* Chiasmata, Chiasmen)
1632	chiasma frequency	Chiasmafrequenz *f*
1633	chiasma interference	Chiasmainterferenz *f*
1634	chiasmatypy	Chiasmatypie *f*
1635	Chilopoda, chilopods, centipedes	Hundertfüßer *mpl*, Chilopoden *pl*
1636	chim(a)era	Chimäre *f*
1637	Chiroptera, chiropterans	Handflügler *mpl*
1638	chitin	Chitin *n*
1639	chitinous	chitinös, chitinhaltig
1640	chitinous coat	Chitinmembran *f*, Chitinhülle *f*
1641	chitinous plate	Chitinplatte *f*
1642	Chlamydobacteriales	Fadenbakterien *fpl*
1643	chlamydospore	Chlamydospore *f*
1644	chlorenchyma	Chlorenchym *n*
1645	chlorocruorin	Chlorokruorin *n*, Chlorocruorin *n*
1646	chlorophyll	Chlorophyll *n*, Blattgrün *n*
1647	Chlorophyta, green algae	Grünalgen *fpl*, Chlorophyceen *pl*
1648	chloroplast, chlorophyll body, ~ granule	Chloroplast *m*, Chlorophyllkörnchen *n*
1649	choana (*pl* choanae), posterior nare, internal ~	Choane *f*, hintere Nasenöffnung *f*
1650	Choanichthyes	Choanenfische *mpl*

1592	vertèbre *f* cervicale	vértebra *f* cervical
1593	cestodes *mpl*	cestodos *mpl*
1594	—	—
1595	molécule *f* en chaîne	molécula *f* en cadena
1596	réaction *f* en chaîne	reacción *f* en cadena
1597	chaîne *f* réflexe, réflexes *mpl* en série	cadena *f* refleja, reflejos *mpl* en cadena
1598	chalaze *f*	chalaza *f*, cálaza *f*
1599	chalazogamie *f*	calazogamia *f*
1600	chamaephytes *mpl*, chaméphytes *mpl*	caméfitos *mpl*
1601	—	—
1602	chance *f* de survie	probabilidad *f* de supervivencia, ~ de sobrevivir
1603	changement *m* d'alimentation, ~ de nourriture	cambio *m* de alimentación
1604	changement *m* de fonction, ~ fonctionnel	cambio *m* de función
1605	caractère *m*, caractéristique *f*, trait *m*	carácter *m*, característica *f*
1606	constance *f* des caractères	constancia *f* de los caracteres
1607	déplacement *m* des caractères, divergence *f* ~ ~	desplazamiento *m* de caracteres, divergencia *f* ~ ~
1608	—	—
1609	espèce *f* caractéristique, ~ indicatrice, ~ type	especie *f* caracteristica, ~ indicadora
1610	charophycées *fpl*	carofíceas *fpl*, carofitos *mpl*
1611	chasmogamie *f*	casmogamia *f*
1612	chasmophytes *mpl*	casmófitos *mpl*
1613	croisement *m* témoin	cruza *f* de verificación
1614	chela *f* (*pl* chelae), pince *f*	quela *f*
1615	chélat *m*	quelado *m*
1616	agent *m* de chélation, chélateur *m*	agente *m* de quelación, ~ quelante
1617	chélation *f*	quelación *f*
1618	chélicérates *mpl*	quelicerados *mpl*
1619	chélicère *f*	quelícero *m*
1620	—	—
1621	chimio-autotrophe	quimioautótrofo
1622	chimionastie *f*	quimionastia *f*
1623	chimioréception *f*	quimiorrecepción *f*
1624	chimiorécepteur *m*, chémorécepteur *m*	quimiorreceptor *m*
1625	chimiosynthèse *f*, synthèse *f* chimique	quimiosíntesis *f*
1626	chimiotactique	quimiotáctico
1627	chimiotactisme *m*, chimiotaxie *f*	quimiotactismo *m*, quimiotaxis *f*
1628	chimiotrophe	quimiótrofo, quimiotrófico
1629	chimiotropisme *m*	quimiotropismo *m*
1630	masticateur *ins.*	masticador *ins.*
1631	chiasme *m*, chiasma *m* (*pl* chiasmas, chiasmata)	quiasma *m*
1632	fréquence *f* des chiasmata	frecuencia *f* de los quiasmas
1633	interférence *f* chiasmatique	interferencia *f* quiasmática
1634	chiasmatypie *f*	quiasmatipia *f*
1635	chilopodes *mpl*	quilópodos *mpl*
1636	chimère *f*	quimera *f*
1637	chiroptères *mpl*, chéiroptères *mpl*	quirópteros *mpl*
1638	chitine *f*	quitina *f*
1639	chitineux	quitinoso
1640	enveloppe *f* chitineuse, membrane *f* ~	envoltura *f* quitinosa, membrana *f* ~
1641	plaque *f* chitineuse	placa *f* quitinosa
1642	chlamydobactériales *fpl*	clamidobacterias *fpl*
1643	chlamydospore *f*	clamidospora *f*
1644	chlorenchyme *m*	clorénquima *m*
1645	chlorocruorine *f*	clorocruorina *f*
1646	chlorophylle *f*	clorofila *f*
1647	chlorophycées *fpl*, chlorophytes *fpl*, algues *fpl* vertes	clorofíceas *fpl*, clorofitos *mpl*, algas *fpl* verdes
1648	chloroplaste *m*, plaste *m* chlorophyllien	cloroplasto *m*, grano *m* de clorofila
1649	choane *m*	coana *f*, abertura *f* nasal posterior
1650	choanichthyens *mpl*	coanicties *mpl*

1651	choanocyte, collar cell	Choanozyte *f*, Kragengeißelzelle *f*
1652	choanoflagellates	Choanoflagellaten *mpl*
1653	cholecyst → gall bladder	—
1654	choledoch(us), common bile duct	gemeinsamer Gallengang *m*
1655	cholesterol, cholesterin	Cholesterin *n*
1656	choline, bilineurine	Cholin *n*, Bilineurin *n*
1657	cholinergic	cholinergisch
1658	cholinesterase	Cholinesterase *f*
1659	chondric → cartilaginous	—
1660	Chondrichthyes, chondrichthian fish, cartilaginous ~	Knorpelfische *mpl*
1661	chondrification	Verknorpelung *f*
1662	chondrin	Chondrin *n*, Knorpelleim *m*
1663	chondriocont, chondriokont	Chondriokont *m*
1664	chondriome	Chondriom *n*
1665	chondriomere, plastomere	Chondriomer *n*, Plastomer *n*
1666	chondriosome → mitochondrion	—
1667	chondriosphere	Chondriosphäre *f*
1668	chondroblast	Chondroblast *m*, Knorpelbildungszelle *f*
1669	chondroclast	Chondroklast *m*, Knorpelzerstörer *m*
1670	chondrocranium, cartilaginous skull	Knorpelschädel *m*
1671	chondrocyte, cartilage cell	Knorpelzelle *f*
1672	chondrogenesis	Knorpelbildung *f*
1673	chondroid tissue	Knorpelgewebe *n*
1674	chondrology	Knorpellehre *f*
1675	chondroskeleton, cartilaginous skeleton	Knorpelskelett *n*
1676	chorda dorsalis → notochord	—
1677	chorda-mesoderm	Chordamesoderm *n*, Urdarmdach *n*
1678	Chordata, chordates	Chordatiere *npl*, Chordaten *pl*
1679	chordotonal organ	Chordotonalorgan *n*, Saitenorgan *n*
1680	chorioallantoic membrane → allantochorion	—
1681	chorion	Chorion *n*, Zottenhaut *f*
1682	chorionic cavity	Chorionhöhle *f*
1683	chorionic gonadotropin, prolan	Choriongonadotropin *n*, Prolan *n*
1684	chorionic villus	Chorionzotte *f*
1685	choripetalous → dialypetalous	—
1686	C-horizon	C-Horizont *m*, Untergrund *m*
1687	choroid	Aderhaut *f*, Gefäßhaut *f*
1688	chorology	Chorologie *f*, Arealkunde *f*
1689	chromaffin	chromaffin
1690	chromatic	chromatisch
1691	chromatid	Chromatide *f*
1692	chromatid aberration	Chromatidenaberration *f*
1693	chromatid bridge	Chromatidenbrücke *f*
1694	chromatid interference	Chromatideninterferenz *f*
1695	chromatin	Chromatin *n*
1696	chromatin body, ~corpuscle	Chromatinkörper *m*
1697	chromatogram	Chromatogramm *n*
1698	to chromatographe	chromatographieren
1699	chromatography	Chromatographie *f*
1700	chromatometer, chromometer	Chromatometer *n*, Farbenmesser *m*
1701	chromatophore	Chromatophor *n*, Farbstoffträger *m*
1702	chromatoplasm	Chromatoplasma *n*
1703	chromocentre	Chromozentrum *n*
1704	chromocyte → pigment cell	—
1705	chromomere	Chromomer *n*
1706	chromomeric	Chromomeren...
1707	chromometer → chromatometer	—
1708	chromonema (*pl* chromonemata)	Chromonema *n* (*pl* Chromonemata)
1709	chromoplast	Chromoplast *m*
1710	chromoprotein	Chromoproteid *n*
1711	chromosomal	Chromosomen...
1712	chromosomal mutation	Chromosomenmutation *f*
1713	chromosome	Chromosom *n*

1651	choanocyte *m*, cellule *f* à collerette	coanocito *m*, célula *f* en collar
1652	choanoflagellés *mpl*	coanoflagelados *mpl*
1653	—	—
1654	(canal *m*) cholédoque *m*	colédoco *m*, conducto *m* biliar común
1655	cholestérine *f*, cholestérol *m*	colesterina *f*, colesterol *m*
1656	choline *f*, bilineurine *f*	colina *f*, bilineurina *f*
1657	cholinergique	colinérgico
1658	cholinestérase *f*	colinesterasa *f*
1659	—	—
1660	chondrichthyens *mpl*, poissons *mpl* cartilagineux	condricties *mpl*, peces *mpl* cartilaginosos
1661	chondrification *f*	condrificación *f*
1662	chondrine *f*	condrina *f*
1663	chondrioconte *m*	condrioconto *m*
1664	chondriome *m*	condrioma *m*
1665	chondriomère *m*	condriomero *m*
1666	—	—
1667	chondriosphère *f*	condriosfera *f*
1668	chondroblaste *m*	condroblasto *m*
1669	chondroclaste *m*	condroclasto *m*
1670	chondrocrâne *m*, crâne *m* cartilagineux	condrocráneo *m*, cráneo *m* cartilaginoso
1671	chondrocyte *m*, cellule *f* cartilagineuse	condrocito *m*, célula *f* cartilaginosa
1672	chondrogenèse *f*	condrogénesis *f*
1673	tissu *m* cartilagineux	tejido *m* cartilaginoso
1674	chondrologie *f*	condrología *f*
1675	squelette *m* cartilagineux	condroesqueleto *m*, esqueleto *m* cartilaginoso
1676	—	—
1677	chordo-mésoblaste *m*	cordamesodermo *m*
1678	cordés *mpl*, chordés *mpl*	cordados *mpl*
1679	organe *m* chordotonal	órgano *m* cordotonal
1680	—	—
1681	chorion *m*	corion *m*
1682	cavité *f* chorionique	cavidad *f* coriónica
1683	gonadotrophine *f* chorionique, ~ placentaire, prolan *m*	gonadotropina *f* coriónica, prolán *m*
1684	villosité *f* choriale	vellosidad *f* coriónica, ~ corial
1685	—	—
1686	horizon *m* C	horizonte *m* C
1687	choroïde *f*	coroides *f*
1688	chorologie *f*	corología *f*
1689	chromaffine	cromafín, cromafínico
1690	chromatique	cromático
1691	chromatide *f*	cromátida *f*, cromatida *f*
1692	aberration *f* chromatidique	aberración *f* cromatídica
1693	pont *m* chromatidique	puente *m* cromatídico
1694	interférence *f* chromatidique	interferencia *f* cromatídica
1695	chromatine *f*	cromatina *f*
1696	corps *m* chromatinien, corpuscule *m* de chromatine	cuerpo *m* cromatínico, corpúsculo *m* ~
1697	chromatogramme *m*	cromatograma *m*
1698	chromatographier	cromatografiar
1699	chromatographie *f*	cromatografía *f*
1700	chromatomètre *m*, chromomètre *m*	cromatómetro *m*, cromómetro *m*
1701	chromatophore *m*	cromatóforo *m*
1702	chromatoplasma *m*	cromatoplasma *m*
1703	chromocentre *m*	cromocentro *m*
1704	—	—
1705	chromomère *m*	cromómero *m*
1706	chromomérique	cromomérico
1707	—	—
1708	chromonème *m*, chromonéma *m*	cromonema *m*
1709	chromoplaste *m*	cromoplasto *m*
1710	chromoprotéine *f*	cromoproteína *f*
1711	chromosomique, chromosomal	cromosómico, cromosomal
1712	mutation *f* chromosomique	mutación *f* cromosómica
1713	chromosome *m*	cromosoma *m*

1714	chromosome aberration, chromosomal ~	Chromosomenaberration *f*
1715	chromosome arm	Chromosomenarm *m*, ~schenkel *m*
1716	chromosome break(age)	Chromosomenbruch *m*
1717	chromosome complement	Chromosomenbestand *m*
1718	chromosome duplication, ~ doubling	Chromosomenverdoppelung *f*, ~duplikation *f*
1719	chromosome map	Chromosomenkarte *f*
1720	chromosome number	Chromosomenzahl *f*
1721	chromosome pair	Chromosomenpaar *n*
1722	chromosome segment, chromosomal ~	Chromosomensegment *n*
1723	chromosome set	Chromosomensatz *m*
1724	chromosome theory of inheritance	Chromosomentheorie *f* der Vererbung
1725	chromosomin	Chromosomin *n*
1726	chronaxy, chronaxia	Chronaxie *f*, Kennzeit *f*
1727	chrysalis → pupa	—
1728	Chrysophyta, golden brown algae	Goldalgen *fpl*
1729	chyle	Chylus *m*, Darmlymphe *f*
1730	chyliferous vessel, lacteal	Chylusgefäß *n*
1731	chylification	Chylusbildung *f*
1732	chyme	Chymus *m*, Nahrungsbrei *m*
1733	chymosin → rennin	—
1734	chymotrypsin	Chymotrypsin *n*
1735	cicatricial tissue, scar ~	Narbengewebe *n*
1736	ciliary body	Ziliarkörper *m*, Strahlenkörper *m*
1737	ciliary movement	Flimmerbewegung *f*, Wimperbewegung *f*
1738	ciliary muscle	Ziliarmuskel *m*
1739	Ciliata, ciliates	Wimpertierchen *npl*, Ziliaten *pl*
1740	ciliate(d)	bewimpert
1741	ciliated cell	Wimperzelle *f*, Flimmerzelle *f*
1742	ciliated epithelium	Wimperepithel *n*, Geißel~ *n*, Flimmer~ *n*
1743	ciliation, ciliature	Bewimperung *f*
1744	cilium (*pl* cilia)	Wimper *f*, Flimmer *f*, Zilie *f*
1745	cincinnus, scorpioid cyme	Wickel *m* *bot.*
1746	cinnamic acid	Zimtsäure *f*
1747	cion → graft	—
1748	circadian rhythm, diurnal ~	zirkadianer Rhythmus *m*, Tagesrhythmus *m*
1749	circle of races → rheogameon	—
1750	circular chromosome → ring ~	—
1751	circular muscle	Ringmuskel *m*
1752	circulation of materials	Stoffkreislauf *m*
1753	circulation of sap → sap-flow	—
1754	circulatory system	Kreislaufsystem *n*
1755	circumnutation, revolving nutation	Zirkumnutation *f*
1756	cirrus	Zirrus *m*
1757	cis-configuration, cis-arrangement	Cis-Konfiguration *f*
1758	cis-trans effect	Cis-Trans-Effekt *m*
1759	cis-trans test	Cis-Trans-Test *m*
1760	cistron	Cistron *n*, Zistron *n*, Funktionsgen *n*
1761	citric acid	Zitronensäure *f*
1762	citric acid cycle → Krebs cycle	—
1763	cladode	Kladodie *f*, Kladodium *n*, Flachsproß *m*
1764	cladogenesis	Kladogenese *f*, Stammbaumverzweigung *f*
1765	cladophyll → phylloclade	—
1766	clan	Sippe *f*
1767	clasper *bot.*	Haftwurzel *f*
1768	class *tax.*	Klasse *f* *tax.*
1769	classical conditioning	klassische Konditionierung *f*
1770	classificatory system, classification ~	Klassifikationssystem *n*
1771	clavicle, collar-bone	Schlüsselbein *n*
1772	claw	Klaue *f*, Kralle *f*
1773	to clear	aufhellen
1774	clear felling; ~ cutting *US*	Kahlschlag *m*

1714	aberration *f* chromosomique, ~ des chromosomes	aberración *f* cromosómica
1715	bras *m* chromosomique	brazo *m* cromosómico
1716	cassure *f* de chromosomes, rupture *f* chromosomique	rotura *f* cromosómica, ruptura *f* de cromosomas
1717	stock *m* chromosomique, équipement *m* ~	conjunto *m* cromosómico, equipo *m* ~
1718	dédoublement *m* des chromosomes	duplicación *f* cromosómica
1719	carte *f* chromosomique	mapa *m* cromosómico
1720	nombre *m* chromosomique	número *m* cromosómico
1721	couple *m* de chromosomes, paire *f* chromosomique	par *m* de cromosomas, ~ cromosómico
1722	segment *m* chromosomique	segmento *m* cromosómico
1723	garniture *f* chromosomique, lot *m* ~, jeu *m* ~	dotación *f* cromosómica, juego *m* cromosómico
1724	théorie *f* chromosomique de l'hérédité	teoría *f* cromosómica de la herencia
1725	chromosomine *f*	cromosomina *f*
1726	chronaxie *f*	cronaxia *f*
1727	—	—
1728	chrysophycées *f pl*	crisófitos *m pl*, algas *f pl* pardo-doradas
1729	chyle *m*	quilo *m*
1730	vaisseau *m* chylifère	quilífero *m*
1731	chylification *f*	quilificación *f*
1732	chyme *m* (alimentaire)	quimo *m*
1733	—	—
1734	chymotrypsine *f*	quimotripsina *f*
1735	tissu *m* cicatriciel	tejido *m* cicatricial
1736	corps *m* ciliaire	cuerpo *m* ciliar
1737	mouvement *m* ciliaire	movimiento *m* ciliar
1738	muscle *m* ciliaire	músculo *m* ciliar
1739	ciliés *m pl*, ciliates *m pl*	ciliados *m pl*
1740	cilié	ciliado
1741	cellule *f* ciliée, ~ à cils vibratiles	célula *f* ciliada
1742	épithélium *m* cilié, ~ vibratile	epitelio *m* ciliado
1743	ciliation *f*, ciliature *f*	ciliación *f*
1744	cil *m* (vibratile)	cilio *m*
1745	cyme *f* scorpioïde	cincino *m*
1746	acide *m* cinnamique	ácido *m* cinámico
1747	—	—
1748	rythme *m* circadien, ~ diurne, ~ nycthéméral	ritmo *m* circadiano, ~ diurno, ~ nictemeral
1749	—	—
1750	—	—
1751	muscle *m* circulaire	músculo *m* circular
1752	circulation *f* des substances	circulación *f* de la materia, ciclo *m* ~ ~ ~
1753	—	—
1754	système *m* circulatoire, appareil *m* ~	sistema *m* circulatorio
1755	circumnutation *f*	circumnutación *f*
1756	cirre *m*	cirro *m*
1757	configuration *f* cis, arrangement *m* cis	disposición *f* cis, cis-configuración *f*
1758	effet *m* cis-trans	efecto *m* cis-trans
1759	test *m* cis-trans	prueba *f* cis-trans
1760	cistron *m*	cistrón *m*
1761	acide *m* citrique	ácido *m* cítrico
1762	—	—
1763	cladode *m*	cladodio *m*
1764	cladogenèse *f*	cladogénesis *f*
1765	—	—
1766	clan *m*	clan *m*, estirpe *f*
1767	racine *f* crampon	raíz *f* adhesiva, ~ adherente
1768	classe *f* *tax.*	clase *f* *tax.*
1769	conditionnement *m* classique	condicionamiento *m* clásico
1770	système *m* de classification	sistema *m* de clasificación
1771	clavicule *f*	clavícula *f*
1772	griffe *f*, ongle *m*	garra *f*, garfa *f*, pezuña *f*
1773	éclaircir	aclarar
1774	coupe *f* claire, ~ à blanc, ~ blanche, ~ rase	corte *m* a tala rasa, ~ a matarrasa, desmonte *m* completo

1775	clearing *micr.*	Aufhellung *f micr.*
1776	clearing agent, ~ medium	Aufhellungsmittel *n*
1777	cleavable → fissile	—
1778	to cleave	sich furchen
1779	cleavage, segmentation	Furchung *f*
1780	cleavage cavity → blastocoel	—
1781	cleavage cell	Furchungszelle *f*
1782	cleavage division	Furchungsteilung *f*
1783	cleavage nucleus, segmentation ~	Furchungskern *m*
1784	cleavage plan	Furchungsebene *f*
1785	cleavage spindle	Furchungsspindel *f*
1786	cleft foot	Spaltbein *n*, Spaltfuß *m*
1787	cleistocarp	Kleistokarp *n*
1788	cleistogamous	kleistogam
1789	cleistogamy	Kleistogamie *f*
1790	climate chamber, controlled environment ~	Klimakammer *f*
1791	climatic change	Klimawechsel *m*, ~änderung *f*
1792	climatic conditions	Klimabedingungen *f pl*
1793	climatic factor	Klimafaktor *m*
1794	climatic race, climate ~	Klimarasse *f*
1795	climatic region	Klimazone *f*
1796	climatic type	Klimatyp *m*
1797	climax	Klimax *f*
1798	climax community	Klimaxgesellschaft *f*
1799	climax forest	Klimaxwald *m*
1800	climbing plant, climber	Kletterpflanze *f*
1801	climbing root	Kletterwurzel *f*
1802	clinal variation	Merkmalsänderung *f*
1803	cline, character-gradient	Merkmalsgefälle *n*, ~gradient *m*, Klin *m*, Kline *f*
1804	clinostat	Klinostat *m*
1805	cloaca	Kloake *f*
1806	cloacal aperture, ~ opening	Kloakenöffnung *f*
1807	clonal, clonic	klonisch
1808	clonal selection	Klonzüchtung *f*, ~selektion *f*, Verklonung *f*
1809	clon(e), klon	Klon *m*
1810	close breeding → inbreeding	—
1811	closing membrane	Schließhaut *f*
1812	clostridium	Klostridium *n*
1813	clotting → coagulation	—
1814	club fungi → basidiomycetes	—
1815	club mosses → Lycopsida	—
1816	clutch (of eggs)	Gelege *n*
1817	clutch size	Gelegegröße *f*
1818	cnida → nematocyst	—
1819	Cnidaria	Nesseltiere *n pl*, Knidarien *pl*
1820	cnidoblast, stinging cell, thread ~	Nesselzelle *f*
1821	cnidocil	Knidozil *n*
1822	cnidocyst → nematocyst	—
1823	CNS → central nervous system	—
1824	coacervate	Koazervat *n*
1825	coacervation	Koazervation *f*
1826	coadaptation	Koadaptation *f*
1827	coagulability	Gerinnbarkeit *f*
1828	coagulable	gerinnbar
1829	to coagulate, to clot	gerinnen, koagulieren
1830	coagulation, clotting	Gerinnung *f*, Koagulation *f*
1831	coagulum, clot	Gerinnsel *n*
1832	coarctate pupa	Tönnchenpuppe *f*
1833	coarse grained	grobkörnig
1834	coat → pelage	—
1835	cobalamin(e), cyanocobalamin, extrinsic factor, vitamin B_{12}	Kobalamin *n*, Extrinsic-Faktor *m*, Vitamin *n* B_{12}
1836	cocaine	Kokain *n*
1837	coccus (*pl* cocci), spherical bacterium	Kokke *f*, Kokkus *m*, Kugelbakterie *f*

1775	éclaircissement *m*, clarification *f*	aclaramiento *m*, aclarado *m*
1776	éclaircisseur *m*	(agente *m*) aclarante *m*, medio *m* ~, aclarador *m*
1777	—	—
1778	cliver	segmentarse
1779	clivage *m*, segmentation *f*	segmentación *f*, clivaje *m*
1780	—	—
1781	cellule *f* de segmentation	célula *f* segmentada
1782	division *f* de segmentation	división *f* de segmentación
1783	noyau *m* de clivage	núcleo *m* segmentado
1784	plan *m* de clivage	plano *m* de segmentación
1785	fuseau *m* de segmentation	huso *m* de segmentación
1786	pied *m* fendu	pie *m* bífido
1787	cléistocarpe *m*	cleistocarpo *m*
1788	cléistogame	cleistógamo
1789	cléistogamie *f*	cleistogamia *f*
1790	chambre *f* climatique, ~ climatisée	cámara *f* climática, ~ climatizada
1791	changement *m* de climat	cambio *m* climático
1792	conditions *fpl* climatiques	condiciones *fpl* climáticas
1793	facteur *m* climatique	factor *m* climático
1794	race *f* formée sous l'effet du climat	raza *f* climática
1795	zone *f* climatique	zona *f* climática
1796	type *m* climatique	tipo *m* climático
1797	climax *m*	clímax *m*
1798	communauté *f* climacique, association *f* ~	comunidad *f* climácica, ~ clímax, ~ final
1799	forêt *f* climacique	bosque *m* clímax
1800	plante *f* grimpante	planta *f* trepadora
1801	racine *f* grimpante	raíz *f* trepadora
1802	variation *f* clinale, ~ des caractères	variación *f* clinal
1803	cline *m*	cline *m*, clina *f*
1804	clinostat *m*	clinóstato *m*
1805	cloaque *m*	cloaca *f*
1806	ouverture *f* cloacale, orifice *m* cloacal	abertura *f* cloacal
1807	clonique, clonal	clónico, clonal
1808	sélection *f* clonale, clonage *m*	selección *f* clonal, clonación *f*
1809	clone *m*	clono *m*, clon *m*
1810	—	—
1811	membrane *f* obturante	membrana *f* oclusiva, ~ obturante
1812	clostridium *m*	clostridio *m*
1813	—	—
1814	—	—
1815	—	—
1816	ponte *f*	puesta *f*
1817	abondance *f* de la ponte	volumen *m* de la puesta
1818	—	—
1819	cnidaires *mpl*	(c)nidarios *mpl*
1820	cnidoblaste *m*, cellule *f* urticante	(c)nidoblasto *m*, célula *f* urticante
1821	cnidocil *m*	(c)nidocilio *m*
1822	—	—
1823	—	—
1824	coacervat *m*	coacervato *m*
1825	coacervation *f*	coacervación *f*
1826	coa(da)ptation *f*	coa(da)ptación *f*
1827	coagulabilité *f*	coagulabilidad *f*
1828	coagulable	coagulable
1829	coaguler	coagular
1830	coagulation *f*	coagulación *f*
1831	coagulum *m*, caillot *m*	coágulo *m*
1832	pupe *f* coarctée	pupa *f* coartada
1833	à gros grains	de grano grueso
1834	—	—
1835	cobalamine *f*, cyanocobalamine *f*, facteur *m* extrinsèque, vitamine *f* B_{12}	cobalamina *f*, factor *m* extrínseco, vitamina *f* B_{12}
1836	cocaïne *f*	cocaína *f*
1837	coque *m*, coccus *m*	coco *m*, bacteria *f* esférica

1838	coccygeal vertebra	Steißwirbel *m*
1839	coccyx	Steißbein *n*
1840	cochlea *anat.*	Schnecke *f anat.*
1841	cocoon	Kokon *m*
1842	to code	kodieren, verschlüsseln
1843	code reading, reading of the genetic code	Ablesen *n* des genetischen Kodes
1844	coding, codification	Kodierung *f*, Verschlüsselung *f*
1845	codominance	Kodominanz *f*
1846	codominant	kodominant
1847	codon	Kodon *n*
1848	coefficient of relationship	Verwandtschaftskoeffizient *m*
1849	coefficient of variation	Variationskoeffizient *m*
1850	Coelenterata, coelenterates	Hohltiere *npl*, Zölenteraten *pl*
1851	coelenteron	Gastralraum *m* (der Hohltiere)
1852	coelom, deuterocoele, coelomic cavity	Zölom *n*, Deuterozöl *n*
1853	coelomic fluid	Zölomflüssigkeit *f*
1854	coelomic sac, ~ pouch	Zölomtasche *f*, Zölomsack *m*
1855	coelomoduct	Zölomodukt *m*
1856	coenobium, cenobium	Zönobium *n*
1857	coenocyte	Zönozyt *m*, Zönozyte *f*
1858	coenogamete	Zönogamet *m*
1859	coenospecies	Zönospezies *f*
1860	co-enzyme, co-ferment	Koenzym *n*, Koferment *n*
1861	cofactor	Kofaktor *m*
1862	cohesion-tension theory	Kohäsions-Spannungs-Theorie *f*
1863	cohesive force	Kohäsionskraft *f*
1864	coincidence	Koinzidenz *f*
1865	coitus	Koitus *m*
1866	colchicine	Kolchizin *n*, Colchizin *n*
1867	cold-blood	Kaltblüter *m*
1868	cold-blooded, h(a)ematocryal	kaltblütig
1869	cold injury	Kälteschaden *m*
1870	cold receptor	Kälterezeptor *m*
1871	cold resistance, ~ hardiness, winter ~	Kälteresistenz *f*, Winterfestigkeit *f*
1872	cold rigo(u)r → winter rigidity	—
1873	cold-sensitive	kälteempfindlich
1874	cold stimulus	Kältereiz *m*
1875	cold treatment	Kältebehandlung *f*
1876	Coleoptera, coleopterans	Koleopteren *pl*, Käfer *mpl*
1877	coleoptile	Koleoptile *f*, Keimscheide *f*
1878	coleoptile curvature bioassay → avena test	—
1879	coleorhiza	Koleor(r)hiza *f*
1880	colinearity	Colinearität *f*
1881	collagen	Kollagen *n*
1882	collagenic	kollagen, leimgebend
1883	collagenic fibre (US: fiber)	Kollagenfaser *f*
1884	collar-bone → clavicle	—
1885	collar cell → choanocyte	—
1886	collateral (fibre)	Kollaterale *f*
1887	collective fruit → multiple ~	—
1888	Collembola, collembolans, springtails	Kollembolen *pl*, Springschwänze *mpl*
1889	collenchyma	Kollenchym *n*
1890	collodion	Kollodium *n*
1891	colloid	Kolloid *n*
1892	colloidal	kolloidal, kolloid
1893	colloidal bearer → apoenzyme	—
1894	colloidal solution	kolloide Lösung *f*
1895	colon	Grimmdarm *m*, Kolon *n*
1896	colony	Kolonie *f*; Tierstock *m*
1897	colony foundation	Koloniebildung *f*
1898	colorability → stainability	—
1899	colorimeter	Kolorimeter *n*
1900	colorimetry	Kolorimetrie *f*
1901	colostrum, foremilk	Kolostrum *n*, Biestmilch *f*
1902	colo(u)r-blind, daltonian	farbenblind
1903	colo(u)r blindness, daltonism	Farbenblindheit *f*

1838	vertèbre *f* coccygienne	vértebra *f* coccígea
1839	coccyx *m*	cóccix *m*
1840	cochlée *f* (auditive)	coclea *f*, caracol *m* *anat*.
1841	cocon *m*	capullo *m*
1842	coder	codificar
1843	lecture *f* du code génétique	lectura *f* del código genético
1844	codage *m*, codification *f*	codificación *f*
1845	codominance *f*	codominancia *f*
1846	codominant	codominante
1847	codon *m*	codón *m*
1848	coefficient *m* de parenté	coeficiente *m* de parentesco
1849	coefficient *m* de variation	coeficiente *m* de variación
1850	coelentérés *mpl*, coelentérates *mpl*	celentéreos *mpl*
1851	coelentéron *m*	celenterón *m*, celenterio *m*
1852	coelome *m*, cavité *f* coelomique	celoma *m*, cavidad *f* celomática
1853	liquide *m* coelomique	fluido *m* celómico
1854	sac *m* coelomique	bolsa *f* celomática
1855	coelomoducte *m*	celomiducto *m*
1856	coenobe *m*, cénobe *m*, cénobie *f*	cenobio *m*
1857	coenocyte *m*	cenocito *m*
1858	coenogamète *m*	cenogameto *m*
1859	coeno-espèce *f*	cenospecie *f*
1860	coenzyme *m*	coenzima *m*
1861	cofacteur *m*	cofactor *m*
1862	théorie *f* de la tension de cohésion	teoría *f* de la tensión-cohesión
1863	force *f* de cohésion	fuerza *f* cohesiva, poder *m* de cohesión
1864	coïncidence *f*	coincidencia *f*
1865	coït *m*	coito *m*
1866	colchicine *f*	colquicina *f*, colchicina *f*
1867	animal *m* à sang froid	animal *m* de sangre fría
1868	à sang froid	de sangre fría
1869	dégât *m* dû au froid	daño *m* debido al frío
1870	récepteur *m* du froid, cryorécepteur *m*	frigidorreceptor *m*
1871	résistance *f* au froid	resistencia *f* al frío
1872	—	—
1873	sensible au froid, cryo-sensible	sensible al frío
1874	stimulus *m* du froid	estímulo *m* del frío
1875	traitement *m* par le froid	tratamiento *m* (por el) frío
1876	coléoptères *mpl*	coleópteros *mpl*
1877	coléoptile *m*	coleóptilo *m*
1878	—	—
1879	coléorhize *m*	coleorriza *f*
1880	colinéarité *f*	colinearidad *f*
1881	collagène *m*	colágeno *m*
1882	collagène	colágeno
1883	fibre *f* collagène	fibra *f* colágena
1884	—	—
1885	—	—
1886	collatérale *f*	(fibra *f*) colateral *f*
1887	—	—
1888	collemboles *mpl*	colémbolos *mpl*
1889	collenchyme *m*	colénquima *m*
1890	collodion *m*	colodión *m*
1891	colloïde *m*	coloide *m*
1892	colloïdal	coloidal, coloideo
1893	—	—
1894	solution *f* colloïde	solución *f* coloidal
1895	côlon *m*	colon *m*
1896	colonie *f*	colonia *f* (animal)
1897	formation *f* de colonies	formación *f* de colonias
1898	—	—
1899	colorimètre *m*	colorímetro *m*
1900	colorimétrie *f*	colorimetría *f*
1901	colostrum *m*	calostro *m*
1902	daltonien	daltoniano
1903	daltonisme *m*	daltonismo *m*

1904	colo(u)r change	Farbwechsel *m*
1905	colo(u)r reaction	Farbreaktion *f*
1906	colo(u)r sense	Farbensinn *m*
1907	colo(u)r solution	Farblösung *f*
1908	colo(u)r vision	Farbensehen *n*
1909	colo(u)ration	Färbung *f*
1910	colo(u)rless, achromic	farblos
1911	columella	Kolumella *f*
1912	column chromatography	Säulenchromatographie *f*
1913	columnar epithelium	Zylinderepithel *n*, prismatisches Epithel *n*
1914	combination	Kombination *f*
1915	combining ability	Kombinationseignung *f*, ~fähigkeit *f*
1916	combustion heat *phys.*	Verbrennungswärme *f* (physiologische)
1917	commensal	Kommensale *m*
1918	commensalism	Kommensalismus *m*
1919	commissure	Kommissur *f*
1920	common bile duct → choledochus	—
1921	communication *eth.*	Kommunikation *f*, Verständigung *f* *eth.*
1922	community *ecol.*	Gesellschaft *f* *ecol.*
1923	community development	Gesellschaftsentwicklung *f*
1924	companion cell	Geleitzelle *f*
1925	comparative study of behavio(u)r, ~ ethology	vergleichende Verhaltensforschung *f*
1926	compartmenta(liza)tion	Kompartimentisierung *f*
1927	compatibility	Kompatibilität *f*, Verträglichkeit *f*
1928	compensating chiasma	kompensierendes Chiasma *n*
1929	compensation point	Kompensationspunkt *m*
1930	compensator	Kompensator *m*
1931	compensatory hypertrophy	kompensatorische Hypertrophie *f*
1932	competence *embr.*	Kompetenz *f* *embr.*
1933	competition for food	Nahrungskonkurrenz *f*, ~ wettbewerb *m*
1934	competitive inhibition	kompetitive Hemmung *f*
1935	complement fixation	Komplementbindung *f*
1936	complement fixation reaction	Komplementbindungsreaktion *f*
1937	complemental air, inspiratory reserve	Komplementärluft *f*, inspiratorisches Reservevolumen *n*
1938	complemental male	Ergänzungsmännchen *n*
1939	complementary chiasma	komplementäres Chiasma *n*
1940	complementary gene, ~ factor	Komplementärgen *n*
1941	complementary nucleus, co-nucleus	Komplementärkern *m*
1942	complementation	Komplementation *f*, Komplementierung *f*
1943	complete cleavage, total ~, ~ segmentation	Totalfurchung *f*
1944	complete metamorphosis → holometabolism	—
1945	complete penetrance	vollständige Penetranz *f*
1946	Compositae, composites	Korbblütler *m pl*
1947	composite fruit, aggregate ~, compound ~	zusammengesetzte Frucht *f*
1948	compound crossing-over	Mehrfachaustausch *m*
1949	compound eye	Komplexauge *n*, Facettenauge *n*, Netzauge *n*
1950	compound leaf	zusammengesetztes Blatt *n*
1951	compound protein, conjugated ~	zusammengesetztes Eiweiß *n*, konjugiertes ~
1952	compound umbel, composite ~	zusammengesetzte Dolde *f*, Doppeldolde *f*
1953	compressive strength	Druckfestigkeit *f*
1954	concentrating culture → enrichment ~	—
1955	concentration gradient	Konzentrationsgradient *m*, ~gefälle *n*
1956	conceptacle	Brutbecher *m*, ~körbchen *n*, Konzeptakel *n*
1957	conch(a)	Ohrmuschel *f*
1958	concha, shell	Muschelschale *f*, Konchylie *f*
1959	conchiolin	Konchin *n*, Konchiolin *n*, Konchyolin *n*
1960	concrescence	Verwachsung *f*
1961	concrete	verwachsen
1962	conditional factor	Konditionalfaktor *m*
1963	conditioned lethal	bedingter Letalfaktor *m*

1904	changement *m* de couleur, ~ chromatique	cambio *m* de color
1905	réaction *f* colorée	reacción *f* coloreada
1906	sens *m* chromatique, ~ des couleurs	sentido *m* cromático, ~ del color
1907	solution *f* colorée	solución *f* colorante
1908	vision *f* des couleurs, ~ colorée	visión *f* cromática, ~ del color
1909	coloration *f*	coloración *f*
1910	incolore, achromique, achrome	incoloro, acrómico
1911	columelle *f*	columela *f*
1912	chromatographie *f* sur colonne	cromatografía *f* en (o: de) columna
1913	épithélium *m* cylindrique	epitelio *m* cilíndrico, ~ columnar
1914	combinaison *f*	combinación *f*
1915	aptitude *f* à la combinaison	aptitud *f* combinatoria, ~ de combinación
1916	chaleur *f* de combustion (physiologique)	calor *m* de combustión (fisiológico)
1917	commensal *m*	comensal *m*
1918	commensalisme *m*	comensalismo *m*
1919	commissure *f*	comisura *f*
1920	—	—
1921	communication *f* *eth.*	comunicación *f* *eth.*
1922	communauté *f* *ecol.*	comunidad *f* *ecol.*
1923	évolution *f* des communautés	evolución *f* de las comunidades
1924	cellule-compagne *f*, cellule *f* annexe	célula *f* acompañante, ~ anexa
1925	étude *f* comparée du comportement	estudio *m* comparado del comportamiento, etología *f* comparada
1926	compartimentation *f*	compartimentación *f*
1927	compatibilité *f*	compatibilidad *f*
1928	chiasme *m* de compensation	quiasma *m* de compensación
1929	point *m* de compensation	punto *m* de compensación
1930	compensateur *m*	compensador *m*
1931	hypertrophie *f* compensatoire (ou: ~trice)	hipertrofia *f* compensatoria
1932	compétence *f* *embr.*	competencia *f* *embr*
1933	compétition *f* alimentaire, concurrence *f* ~	competencia *f* alimentaria
1934	inhibition *f* compétitive	inhibición *f* competitiva
1935	fixation *f* du complément	fijación *f* del complemento
1936	réaction *f* de fixation du complément	reacción *f* de fijación del complemento
1937	air *m* complémentaire, volume *m* de réserve inspiratoire	aire *m* complementario, volumen *m* de reserva inspiratoria
1938	mâle *m* complémentaire	macho *m* complementario
1939	chiasme *m* complémentaire	quiasma *m* complementario
1940	gène *m* complémentaire, facteur *m* ~	gen *m* complementario, factor *m* ~
1941	noyau *m* complémentaire	núcleo *m* complementario
1942	complémentation *f*	complementación *f*
1943	segmentation *f* totale	segmentación *f* completa
1944	—	—
1945	pénétrance *f* complète	penetración *f* completa
1946	composacées *f pl*, composées *f pl*	compuestas *f pl*
1947	fruit *m* composé, ~ agrégé	fruto *m* compuesto, ~ agregado
1948	échanges *m pl* multiples, crossing-over multiple	intercambio *m* múltiple, entrecruzamiento *m* compuesto
1949	œil *m* composé, ~ à facettes	ojo *m* compuesto, ~ en facetas, ~ facetado
1950	feuille *f* composée	hoja *f* compuesta
1951	protéine *f* composée, ~ conjugée, hétéroprotéine *f*	proteína *f* conjugada, heteroproteido *m*
1952	ombelle *f* composée	umbela *f* compuesta
1953	résistance *f* à la pression	resistencia *f* a la presión
1954	—	—
1955	gradient *m* de concentration	gradiente *m* de concentración
1956	conceptacle *m*	conceptáculo *m*
1957	pavillon *m* (de l'oreille)	pabellón *m* (de la oreja)
1958	coquille *f*	concha *f*
1959	conchyoline *f*	conquiolina *f*
1960	concrescence *f*, coalescence *f*	concrescencia *f*
1961	coalescent	concrescente
1962	facteur *m* conditionnel	factor *m* condicional
1963	lét(h)al *m* conditionné	letal *m* condicionado

1964	conditioned reflex	bedingter Reflex *m*
1965	conducting system, conduction ~	Leitungssystem *n*
1966	conducting tissue, conductive ~	Leitgewebe *n*
1967	conduction of excitation	Erregungsleitung *f*
1968	conduction of impulse	Impulsleitung *f*
1969	conduction of stimulus	Reizleitung *f*
1970	conduction velocity	Leitungsgeschwindigkeit *f*
1971	conductivity, conductibility	Leitfähigkeit *f*
1972	conduplicate (vernation)	konduplikativ (Knospenlage)
1973	condyle	Gelenkkopf *m*
1974	cone *bot.*, strobile, strobilus	Zapfen *m* *bot.*, Strobilus *m*
1975	cone *anat.* → retinal cone	—
1976	cone scale	Zapfenschuppe *f*
1977	cone-shaped, conic(al)	kegelförmig, konisch
1978	conflict behavio(u)r	Konfliktverhalten *n*
1979	congenital, innate	angeboren
1980	congruent crossing	kongruente Kreuzung *f*
1981	conidiophore	Konidienträger *m*, Konidiophor *n*
1982	conidium (*pl* conidia)	Konidie *f*
1983	conifer, coniferous tree, acicular-leaved ~, softwood ~	Konifere *f*, Nadelbaum *m*
1984	conifer(ous) forest, needle-leaved ~, softwood ~	Nadelwald *m*
1985	conifers	Nadelhölzer *npl*, Koniferen *fpl*
1986	conjugant	Konjugant *m*
1987	conjugated protein → compound ~	—
1988	conjugation	Konjugation *f*
1989	conjunctiva	Bindehaut *f*
1990	connective	Konnektiv *n*
1991	connective tissue	Bindegewebe *n*
1992	connective tissue cell → fibrocyte	—
1993	conservation of nature, protection ~ ~	Naturschutz *m*
1994	conservationist	Naturschützer *m*
1995	consociation	Konsoziation *f*
1996	conspecific *adj.*	artgleich
1997	conspecific, species member	Artgenosse *m*
1998	constancy of species, fixity ~ ~	Stetigkeit *f*
1999	constant of sedimentation	Sedimentationskonstante *f*
2000	constituent	Bestandteil *m*
2001	constitution(al) type	Konstitutionstyp *m*, Körperbautyp *m*
2002	constitutive enzyme	konstitutives Enzym *n*
2003	constriction	Einschnürung *f*
2004	constructive metabolism → anabolism	—
2005	consubspecies	Konsubspezies *f*
2006	consumer *ecol.*	Konsument *m* *ecol.*
2007	consummatory act	Endhandlung *f*
2008	contact point	Kontaktpunkt *m*
2009	contact stimulus, haptic ~	Berührungsreiz *m*, Kontaktreiz *m*, haptischer Reiz *m*
2010	contaminant	Verseuchungsmittel *n*
2011	to contaminate	verseuchen, verschmutzen
2012	contamination	Verseuchung *f*, Verschmutzung *f*
2013	continental drift theory	Kontinentalverschiebungstheorie *f*, Drifttheorie *f*
2014	continental shelf	Kontinentalschelf *n*, ~sockel *m*
2015	continued stimulus, constant ~	Dauerreiz *m*
2016	continuity of the germ plasm, germinal continuity	Kontinuität *f* des Keimplasmas
2017	continuous area	geschlossenes Areal *n*, kontinuierliches ~
2018	continuous variability, quantitative ~	kontinuierliche Variabilität *f*, fluktuierende ~, quantitative ~
2019	contorted (aestivation)	kontort, gedreht (Knospendeckung)

1964	réflexe *m* conditionné	reflejo *m* condicionado
1965	système *m* conducteur, appareil *m* ~	sistema *m* conductor
1966	tissu *m* conducteur	tejido *m* conductor, ~ de conducción
1967	conduction *f* de l'excitation	conducción *f* de la excitación
1968	conduction *f* de l'influx	conducción *f* del impulso
1969	conduction *f* du stimulus	conducción *f* del estímulo
1970	rapidité *f* de conduction	velocidad *f* de conducción
1971	conductivité *f*, conductibilité *f*	conductividad *f*, conductibilidad *f*
1972	condupliquée (vernation)	conduplicada (vernación)
1973	condyle *m*	cóndilo *m*
1974	cône *m*, strobile *m*	cono *m*, estróbilo *m*
1975	—	—
1976	écaille *f* des cônes	escama *f* del estróbilo
1977	coniforme, conique	coniforme, cónico
1978	comportement *m* en situations conflictuelles	comportamiento *m* en situaciones de conflicto, ~ conflictivo
1979	congénital, inné	congénito, innato
1980	croisement *m* congru	cruzamiento *m* congruente
1981	conidiophore *m*	conidióforo *m*
1982	conidie *f*, conidiospore *f*, spore *f* conidienne	conidio *m*
1983	conifère *m*, résineux *m*, arbre *m* à feuilles aciculaires	conífera *f*, árbol *m* aciculifolio
1984	forêt *f* aciculifoliée, ~ sempervirente, ~ résineuse	bosque *m* de coníferas, ~ aciculifolio
1985	conifères *m pl*	coníferas *f pl*
1986	conjugant *m*	conjugante *m*
1987	—	—
1988	conjugaison *f*	conjugación *f*
1989	conjonctive *f*	conjuntiva *f*
1990	connectif *m*	conectivo *m*
1991	tissu *m* conjonctif	tejido *m* conjuntivo
1992	—	—
1993	conservation *f* (ou: protection *f*) de la nature	conservación *f* (o: protección *f*) de la naturaleza
1994	défenseur *m* de la nature	defensor *m* de la naturaleza
1995	conassociation *f*	consociación *f*
1996	congénère	congénere
1997	congénère *m*	congénere *m*
1998	constance *f* des espèces, fixité *f* ~ ~	constancia *f* de las especies
1999	constante *f* de sédimentation	constante *f* de sedimentación
2000	partie *f* constituante, élément *m* constitutif, constituant *m*	constitutivo *m*, parte *f* constitutiva
2001	type *m* de constitution	tipo *m* constitucional
2002	enzyme *m* constitutif	enzima *f* constitutiva
2003	constriction *f*	constricción *f*
2004	—	—
2005	consubespèce *f*	consubespecie *f*
2006	consommateur *m ecol.*	consumidor *m ecol.*
2007	acte *m* consommatoire	acto *m* consumatorio
2008	point *m* de contact	punto *m* de contacto
2009	stimulus *m* de contact	estímulo *m* de contacto, ~ háptico
2010	agent *m* de contamination, produit *m* contaminant, substance *f* contaminante	(agente *m*) contaminante *m*, sustancia *f* contaminadora
2011	contaminer	contaminar
2012	contamination *f*	contaminación *f*
2013	théorie *f* de la dérive des continents	teoría *f* de la deriva de los continentes, ~ de los desplazamientos continentales
2014	plateau *m* continental	talud *m* continental
2015	stimulation *f* continue	estimulación *f* contínua, ~ sostenida
2016	continuité *f* germinale	continuidad *f* del plasma germinal
2017	aire *f* continue	área *f* continua
2018	variabilité *f* continue, ~ quantitative	variabilidad *f* continua, ~ cuantitativa
2019	tordue (estivation)	contorta (estivación)

2020	contour feather	Konturfeder *f*, Deckfeder *f*
2021	contractile	kontraktil
2022	contractile root, pull ~	Zugwurzel *f*
2023	contractile vacuole, pulsating ~	kontraktile Vakuole *f*, pulsierende ~
2024	contractility, contractibility	Kontraktilität *f*, Kontraktionsfähigkeit *f*
2025	control animal	Kontrolltier *n*
2026	control experiment	Kontrollversuch *m*
2027	controlled environment chamber → climate ~	—
2028	co-nucleus → complementary nucleus	—
2029	convergence *evol.*	Konvergenz *f* *evol.*
2030	convergent evolution	konvergente Evolution *f*
2031	conversion *gen.*	Konversion *f* *gen.*
2032	coordination	Koordination *f*
2033	coordination centre (US: center)	Koordinationszentrum *n*
2034	coorientation	Koorientierung *f*
2035	Copepoda, copepods	Kopepoden *pl*
2036	coprophage	Koprophage *m*, Kotfresser *m*
2037	to copulate	kopulieren, (sich) begatten
2038	copulation	Kopulation *f*, Begattung *f*
2039	copulation tube → fertilization ~	—
2040	copulatory behavio(u)r	Begattungsverhalten *n*
2041	copulatory organ	Begattungsorgan *n*, Paarungsorgan *n*, Kopulationsorgan *n*
2042	copy-choice hypothesis	Kopiewahlhypothese *f*
2043	coracoid bone	Rabenschnabelbein *n*
2044	coral reef	Korallenriff *n*
2045	corepressor	Ko-Repressor *m*
2046	coriaceous, leathery	lederig
2047	corium → dermis	—
2048	cork, suber	Kork *m*
2049	cork cambium → phellogen	—
2050	cork cell, suberized ~	Korkzelle *f*
2051	cork layer	Korkschicht *f*
2052	cork tissue, phellem	Korkgewebe *n*, Phellem *n*
2053	corky, suberose	korkig
2054	corm(us)	Kormus *m*
2055	cormophytes	Kormophyten *mpl*, Sproßpflanzen *fpl*
2056	cornea	Hornhaut *f* (Auge)
2057	corneal reflex	Hornhautreflex *m*
2058	corneous, horny	hornig
2059	corneous layer, horny ~	Hornschicht *f*
2060	cornification	Verhornung *f*, Hornbildung *f*
2061	to cornify	verhornen
2062	corolla	Blumenkrone *f*, Korolle *f*
2063	coronary artery	Koronararterie *f*, Herzkranzgefäß *n*
2064	corpora allata	Corpora allata
2065	corpora cardiaca	Corpora cardiaca
2066	corpora pedunculata → mushroom bodies	—
2067	corpus callosum, callous corpus	Balken *m* *anat.*
2068	corpus luteum, yellow body	Gelbkörper *m*, Corpus luteum
2069	corpus luteum hormone	Gelbkörperhormon *n*, Corpus-luteum-Hormon *n*, Schwangerschaftshormon *n*
2070	corpus striatum, striate corpus	Corpus striatum
2071	corpuscle	Körperchen *n*
2072	correlation	Korrelation *f*
2073	correlation coefficient, ~ index	Korrelationskoeffizient *m*, ~index *m*
2074	cortex → rind	—
2075	cortex parenchyma	Rindenparenchym *n*
2076	Corti's membrane, tectorial ~	Deckmembran *f*
2077	Corti's organ, spiral ~	Cortisches Organ *n*
2078	cortical pore → lenticel	—
2079	corticoid	Kortikoid *n*
2080	corticosteroid	Kortikosteroid *n*
2081	corticosterone	Kortikosteron *n*

2020	plume *f* de contour, ~ tectrice, ~ de couverture	pluma *f* cobertera
2021	contractile	contráctil
2022	racine *f* tractrice, ~ de traction	raíz *f* contráctil
2023	vacuole *f* contractile, ~ pulsatile, ~ pulsatrice	vacuola *f* contráctil, ~ pulsativa
2024	contractilité *f*, contractibilité *f*	contractilidad *f*
2025	animal *m* témoin	animal *m* control, ~ testigo
2026	expérience *f* témoin	experiencia *f* de control
2027	—	—
2028	—	—
2029	convergence *f evol.*	convergencia *f evol.*
2030	évolution *f* convergente	evolución *f* convergente
2031	conversion *f gen.*	conversión *f gen.*
2032	coordination *f*	coordinación *f*
2033	centre *m* de coordination	centro *m* coordinador
2034	coorientation *f*	coorientación *f*
2035	copépodes *mpl*	copépodos *mpl*
2036	coprophage *m*	coprófago *m*
2037	copuler	copular(se)
2038	copulation *f*	copulación *f*, cópula *f*
2039	—	—
2040	comportement *m* copulatoire, ~ de copulation	conducta *f* copulativa, comportamiento *m* de cópula
2041	organe *m* copulateur, ~ d'accouplement	órgano *m* copulador
2042	hypothèse *f* de la choix de copie	hipótesis *f* de la elección de la copia
2043	coracoïde *f*	coracoides *f*
2044	récif *m* corallien, ~ coralliaire	arrecife *m* de coral, ~ coralino
2045	corépresseur *m*	co-represor *m*
2046	coriacé	coriáceo
2047	—	—
2048	liège *m*, suber *m*	corcho *m*, súber *m*
2049	—	—
2050	cellule *f* subéreuse	célula *f* suberosa
2051	couche *f* subéreuse	capa *f* suberosa
2052	tissu *m* subéreux	tejido *m* suberoso, felema *m*
2053	liégeux, subéreux	corchoso, suberoso
2054	cormus *m*	cormo *m*
2055	cormophytes *mpl*	cormófitos *mpl*
2056	cornée *f*	córnea *f*
2057	réflexe *m* cornéen	reflejo *m* corneal
2058	corné	córneo
2059	couche *f* cornée	estrato *m* córneo
2060	cornification *f*	cornificación *f*
2061	transformer en corne	cornificar(se)
2062	corolle *f*	corola *f*
2063	artère *f* coronaire	arteria *f* coronaria
2064	corpora allata	cuerpos *mpl* alados
2065	corpora cardiaca	corpora cardiaca
2066	—	—
2067	corps *m* calleux	cuerpo *m* calloso
2068	corps *m* jaune	cuerpo *m* lúteo, ~ amarillo
2069	hormone *f* du corps jaune	hormona *f* del cuerpo lúteo
2070	corps *m* strié	cuerpo *m* estriado
2071	corpuscule *m*	corpúsculo *m*
2072	corrélation *f*	correlación *f*
2073	coefficient *m* de corrélation, indice *m* ~ ~	coeficiente *m* de correlación
2074	—	—
2075	parenchyme *m* cortical	parénquima *m* cortical
2076	membrane *f* tectrice	membrana *f* tectoria, ~ de Corti
2077	organe *m* de Corti	órgano *m* de Corti
2078	—	—
2079	corticoïde *m*	corticoide *m*
2080	corticostéroïde *m*	corticosteroide *m*
2081	corticostérone *f*	corticosterona *f*

2082	corticotrop(h)in → adrenocorticotropic hormone	—
2083	cortin	Kortin *n*
2084	cortisol → hydrocortisone	—
2085	cortisone	Kortison *n*
2086	corymb	Doldentraube *f*
2087	cosmic rays	kosmische Strahlen *mpl*
2088	cosmopolite	Kosmopolit *m*
2089	costal cartilage	Rippenknorpel *m*
2090	costal pleura	Rippenfell *n*
2091	costal respiration	Brustatmung *f*, thorakale Atmung *f*
2092	costate, ribbed	gerippt
2093	cotyle, cotyloid cavity	Gelenkpfanne *f*
2094	cotyledon, seed leaf	Keimblatt *n bot.*, Samenblatt *n*, Kotyledone *f*
2095	counterpoison → antidote	—
2096	count(ing)	Zählung *f*, Auszählung *f*
2097	counting cell, ~ chamber	Zählkammer *f*
2098	coupling phase	Kopp(e)lungsphase *f*
2099	course of inheritance	Erbgang *m*
2100	to court	balzen
2101	courtship	Balz *f*
2102	courtship behavio(u)r	Balzverhalten *n*
2103	courtship colo(u)ration	Prachtkleid *n*
2104	courtship movements	Balzbewegungen *fpl*
2105	covariation, covariance	Kovariation *f*, Kovarianz *f*
2106	cover-glass, cover-slip	Deckglas *n*
2107	cover scale, bract ~	Deckschuppe *f*
2108	coxa → hip	—
2109	coxal → sciatic	—
2110	coxal gland	Koxaldrüse *f*
2111	cranial	Schädel...
2112	cranial bone, skull ~	Schädelknochen *m*
2113	cranial capacity	Schädelkapazität *f*
2114	cranial capsule, brain case	Schädelkapsel *f*, Hirnschale *f*
2115	cranial cavity	Schädelhöhle *f*
2116	cranial nerve, brain ~	(Ge)hirnnerv *m*, Kopfnerv *m*
2117	Craniata → Vertebrata	—
2118	craniology	Kraniologie *f*, Schädellehre *f*
2119	craniometry	Kraniometrie *f*, Schädelmeßlehre *f*
2120	cranium, skull	Schädel *m*
2121	crawling movement	Kriechbewegung *f*
2122	creatine	Kreatin *n*
2123	creatine phosphate, phosphocreatine	Kreatinphosphat *n*
2124	creatinine	Kreatinin *n*
2125	creeper, creeping plant	Kriechpflanze *f*
2126	crenate	gekerbt
2127	Cretaceous (period)	Kreide *f geol.*
2128	cricoid cartilage	Ringknorpel *m*
2129	criss-crossing	Wechselkreuzung *f*, Überkreuzzüchtung *f*, alternierende Rückkreuzung *f*
2130	criss-cross inheritance	Überkreuzvererbung *f*
2131	crop *zool.*	Kropf *m zool.*
2132	crop milk, pigeon's ~	Kropfmilch *f*
2133	cross	Kreuzung *f*
2134	to cross, to crossbreed	kreuzen
2135	to cross back	rückkreuzen
2136	cross breeding	Kreuzungszüchtung *f*
2137	cross experiment	Kreuzungsversuch *m*
2138	cross fertilization → allogamy, xenogamy	—
2139	cross pollination → allogamy	—
2140	cross reactivation	Kreuzungsreaktivierung *f*

2082	—	—
2083	cortine *f*	cortina *f*
2084	—	—
2085	cortisone *f*	cortisona *f*
2086	corymbe *m*	corimbo *m*
2087	rayons *mpl* cosmiques	rayos *mpl* cósmicos
2088	cosmopolite *m*	cosmopolita *m*
2089	cartilage *m* costal	cartílago *m* costal
2090	plèvre *f* costale	pleura *f* costal, ~ parietal
2091	respiration *f* costale	respiración *f* costal, ~ torácica
2092	côtelé, nervuré	nervado
2093	cotyle *f*, cavité *f* cotyloïde	cotilo *m*, cotila *f*, cótila *f*, cavidad *f* cotiloidea
2094	cotylédon *m*, feuille *f* séminale	cotiledón *m*, hoja *f* seminal
2095	—	—
2096	comptage *m*, numération *f*, dénombrement *m*	conteo *m*, recuento *m*
2097	cellule *f* compte-globules, appareil *m* ~	célula *f* de recuento, cámara *f* ~ ~, célula *f* cuentaglóbulos
2098	phase *f* d'accouplement, ~ d'attraction	fase *f* de acoplamiento, ~ de atracción
2099	mode *m* de transmission	curso *m* de herencia, vía *f* hereditaria
2100	parader, courtiser, faire la cour à la femelle	cortejar, galantear
2101	parade *f* sexuelle, ~ nuptiale	cortejo *m*, parada *f* sexual, ~ nupcial
2102	comportement *m* de parade, ~ nuptial, ~ de cour	comportamiento *m* durante el cortejo, conducta *f* de galanteo
2103	parure *f* de noce	adorno *m* vistoso (durante el cortejo)
2104	mouvements *mpl* de cour, activité *f* ~ ~	actitud *f* de galanteo
2105	covariation *f*, covariance *f*	covariación *f*, covarianza *f*
2106	couvre-objet *m*	cubreobjetos *m*
2107	bractée-écaille *f*, écaille *f* couvrante	escama *f* tectriz
2108	—	—
2109	—	—
2110	glande *f* coxale	glándula *f* coxal
2111	crânien	craneal, craneano
2112	os *m* crânien	hueso *m* del cráneo
2113	capacité *f* crânienne	capacidad *f* craneana
2114	boîte *f* crânienne	cápsula *f* craneal
2115	cavité *f* crânienne	fosa *f* craneal
2116	nerf *m* crânien	nervio *m* craneal
2117	—	—
2118	craniologie *f*	craneología *f*
2119	craniométrie *f*	craneometría *f*
2120	crâne *m*	cráneo *m*
2121	mouvement *m* rampant, reptation *f*	movimiento *m* reptante, reptación *f*
2122	créatine *f*	creatina *f*
2123	créatine-phosphate *m*	fosfato *m* de creatina, fosfocreatina *f*
2124	créatinine *f*	creatinina *f*
2125	plante *f* rampante	planta *f* rastrera
2126	crénelé, échancré	crenado
2127	crétacé *m*, crétacique *m*	cretáceo *m*
2128	cartilage *m* cricoïde	cartílago *m* cricoides, ~ anular
2129	croisement *m* alternatif	cruzamiento *m* alternativo, ~ alterno, criss-crossing *m*
2130	hérédité *f* alternative	herencia *f* alternativa, ~ cruzada
2131	jabot *m*	buche *m*
2132	«lait» *m* du jabot	leche *f* de buche, ~ ~ paloma
2133	croisement *m*	cruzamiento *m*, cruce *m*, cruza *f*
2134	croiser	cruzar
2135	croiser en retour, recroiser	retrocruzar, cruzar en retroceso
2136	croisement *m*	cruzamiento *m*
2137	croisement *m* expérimental, essai *m* de croisement	cruzamiento *m* experimental, experimento *m* de cruce
2138	—	—
2139	—	—
2140	réactivation *f* croisée	reactivación *f* cruzada

2141	crossing-over	Crossing-over *n*
2142	crossing-over modifier	Crossing-over-Modifikator *m*
2143	crossing-over percentage → crossover value	—
2144	crossover suppressor	Cross(ing)-over-Suppressor *m*
2145	crossover unit	Cross-over-Einheit *f*
2146	crossover value, crossing-over percentage	Crossing-over-Prozentsatz *m*, ~ -Wert *m*
2147	cross-reacting material, CRM	kreuzreagierendes Protein *n*
2148	cross section, transverse ~, transversal ~	Querschnitt *m*
2149	cross striation → transverse ~	—
2150	cross wall, transverse septum	Querwand *f*
2151	crown (of the tooth)	Zahnkrone *f*
2152	Cruciferae	Kreuzblütler *mpl*
2153	crude protein value	Rohproteinwert *m*
2154	crumb structure	Krümelstruktur *f*
2155	crushed preparation	Quetschpräparat *n*
2156	Crustacea, crustaceans	Krebse *mpl*, Krustazeen *fpl*
2157	crustose lichen	Krustenflechte *f*
2158	cryophil(ic)	kälteliebend
2159	cryophyte	Kältepflanze *f*, Kryophyt *m*
2160	cryoplankton	Kryoplankton *n*
2161	cryoscopy	Kryoskopie *f*
2162	cryostat	Kryostat *m*
2163	cryptic colo(u)ration	Tarnfärbung *f*, Tarntracht *f*
2164	cryptogam(ous)	kryptogam
2165	cryptogams, spore-plants	Kryptogamen *pl*, Sporenpflanzen *fpl*, blütenlose Pflanzen *fpl*
2166	cryptogenic, cryptogenetic	kryptogen(etisch)
2167	cryptomerism	Kryptomerie *f*
2168	cryptomitose	Kryptomitose *f*
2169	cryptophyte	Kryptophyt *m*
2170	crystalline	kristallin
2171	crystalline lens	Linse *f* (Auge)
2172	crystallizable	kristallisierbar
2173	crystallization	Kristallisation *f*
2174	to crystallize	kristallisieren
2175	crystallography	Kristallographie *f*
2176	C^{14}-test → radiocarbon method	—
2177	ctenidium	Ktenidie *f*, Kammkieme *f*
2178	Ctenophora, ctenophores, ctenophorans	Kammquallen *fpl*, Rippenquallen *fpl*
2179	ctetosome	Ktetosom *n*
2180	cubical epithelium, cuboidal ~	Pflasterepithel *n*, kubisches Epithel *n*
2181	cubitus, ulna	Elle *f*, Ulna *f*
2182	culling → elimination	—
2183	culm, ha(u)lm	Halm *m*
2184	cultivar → variety	—
2185	cultivated form	Kulturform *f*
2186	cultivated plant	Kulturpflanze *f*
2187	cultural landscape	Kulturlandschaft *f*
2188	culture experiment	Kulturexperiment *n*
2189	culture flask, ~ vessel	Kulturgefäß *n*
2190	culture fluid	Kulturflüssigkeit *f*
2191	culture medium	Kulturmedium *n*
2192	culture plate	Kulturplatte *f*
2193	culture solution → nutrient ~	—
2194	culture tube	Kulturröhrchen *n*
2195	cupulate, cup-like, cup-shaped	becherförmig
2196	cupule *bot.*	Fruchtbecher *m*, Kupula *f*
2197	curvature movement	Krümmungsbewegung *f*
2198	cushion plants	Polsterpflanzen *fpl*
2199	cut surface	Schnittfläche *f*
2200	cutaneous, dermic, dermal	Haut ..., kutan
2201	cutaneous gland → skin gland	—

2141	crossing-over *m*, «enjambement» *m*	crossing-over *m*, entrecruzamiento *m*
2142	modificateur *m* de crossing-over	modificador *m* de crossing-over
2143	—	—
2144	suppresseur *m* de crossing-over	supresor *m* de crossing-over
2145	unité *f* de crossing-over	unidad *f* de crossing-over
2146	pourcentage *m* de crossing-over, taux *m* ~ ~	porcentaje *m* de crossing-over, proporción *f* ~ ~, tasa *f* ~ ~
2147	cross-reacting material	material *m* de reacción cruzada
2148	section *f* transversale	sección *f* transversal, corte *m* ~
2149	—	—
2150	cloison *f* transversale	tabique *m* transversal
2151	couronne *f* dentaire	corona *f* dental
2152	crucifères *fpl*, cruciféracées *fpl*	crucíferas *fpl*
2153	valeur *f* protéique brute	índice *m* de proteína bruta
2154	structure *f* grumeleuse, ~ granuleuse	estructura *f* grumosa
2155	préparation *f* contuse	preparación *f* aplastada
2156	crustacés *mpl*	crustáceos *mpl*
2157	lichen *m* crustacé	liquen *m* crustáceo
2158	cryophile	criófilo
2159	cryophyte *m*	criófito *m*
2160	cryoplancton *m*	crioplancton *m*
2161	cryoscopie *f*	crioscopia *f*
2162	cryostat *m*	criostato *m*
2163	coloration *f* dissimulatrice, pigmentation *f* cryptique	coloración *f* (pro)críptica, ~ de ocultación
2164	cryptogame	criptógamo
2165	cryptogames *mpl*	criptógamas *fpl*
2166	cryptogène, cryptogén(ét)ique	criptógeno, criptogenético
2167	cryptomérie *f*	criptomerismo *m*, criptomería *f*
2168	cryptomitose *f*	criptomitosis *f*
2169	cryptophyte *m*	criptófito *m*
2170	cristallin	cristalino
2171	cristallin *m*	cristalino *m*
2172	cristallisable	cristalizable
2173	cristallisation *f*	cristalización *f*
2174	cristalliser	cristalizar
2175	cristallographie *f*	cristalografía *f*
2176	—	—
2177	cténidie *f*	ctenidia *f*, branquia *f* ctenidial
2178	cténaires *mpl*, cténophores *mpl*	ctenarios *mpl*, ctenóforos *mpl*
2179	ctétosome *m*	ctetosoma *m*
2180	épithélium *m* cubique	epitelio *m* cúbico
2181	cubitus *m*	cúbito *m*
2182	—	—
2183	chaume *m*	caña *f*
2184	—	—
2185	forme *f* de culture, ~ cultivée	forma *f* de cultivo
2186	plante *f* de culture, ~ cultivée	planta *f* cultivada
2187	paysage *m* transformé par la main de l'homme, ~ «de civilisation»	paisaje *m* transformado por el hombre
2188	culture *f* expérimentale	experimento *m* de cultivo
2189	bouteille *f* à culture, récipient *m* de ~	frasco *m* de cultivo
2190	liquide *m* à culture	líquido *m* de cultivo
2191	milieu *m* de culture	medio *m* de cultivo
2192	plaque *f* à culture	placa *f* de cultivo
2193	—	—
2194	tube *m* à culture	tubo *m* de cultivo
2195	cupulaire, cupuliforme	acopado, cupuliforme
2196	cupule *f* *bot.*	cúpula *f* *bot.*
2197	mouvement *m* de courbure, ~ d'incurvation	movimiento *m* de curvatura
2198	plantes *fpl* pulviniformes	plantas *fpl* pulviniformes
2199	surface *f* de la section (ou: coupe)	superficie *f* cortada, ~ de corte
2200	cutané, dermique	cutáneo, dérmico
2201	—	—

No.	English	German
2202	cutaneous receptor	Hautsinnesorgan *n*
2203	cutaneous respiration	Hautatmung *f*
2204	cuticle, cuticula	Kutikula *f*
2205	cuticular transpiration	kutikuläre Transpiration *f*
2206	cutin	Kutin *n*
2207	cutinization	Kutinisierung *f*
2208	cyanin	Zyanin *n*
2209	cyanocobalamine → cobalamine	—
2210	Cyanophyta, blue-green algae	Spaltalgen *fpl*, blaugrüne Algen *fpl*, Blaualgen *fpl*
2211	cyathium	Cyathium *n* (*pl* Cyathien)
2212	cybernetics	Kybernetik *f*
2213	cyclic, cyclical	zyklisch
2214	cyclic photophosphorylation	zyklische Photophosphorylierung *f*
2215	cyclomorphosis	Zyklomorphose *f*
2216	cyclosis	Zyklosis *f*, Plasmazirkulation *f*
2217	Cyclostomata, cyclostomes	Zyklostomen *mpl*, Rundmäuler *npl*
2218	cyme	Trugdolde *f*
2219	cymose	zymös, trugdoldig
2220	cyst	Zyste *f*
2221	cysteine	Zystein *n*
2222	cystid(ium)	Zystide *f*
2223	cystine	Zystin *n*
2224	cystocarp	Zystokarp *m*, Hüllfrucht *f*
2225	cystolith	Zystolith *m*
2226	cytidylic acid	Zytidylsäure *f*
2227	cyto ... see also cell ...	—
2228	cytoactive	zellaktiv
2229	cytochemism	Zellchemismus *m*
2230	cytochemistry	Zellchemie *f*
2231	cytochrome	Zytochrom *n*
2232	cytode	Zytode *f*
2233	cytodiaeresis	Zytodiärese *f*
2234	cytogene	Zytogen *n*
2235	cytogenesis	Zellbildung *f*, Zytogenese *f*
2236	cytogenetic	zytogenetisch
2237	cytogeneticist	Zytogenetiker *m*
2238	cytogenetics	Zytogenetik *f*
2239	cytogony	Zytogonie *f*
2240	cytokinesis	Zytokinese *f*
2241	cytokinin, phytokinin, kinin	Zytokinin *n*, Phytokinin *n*, Kinin *n*
2242	cytological	zytologisch
2243	cytologist	Zytologe *m*, Zellforscher *m*
2244	cytology	Zytologie *f*, Zellforschung *f*, Zellenlehre *f*
2245	cytolymph	Zytolymphe *f*
2246	cytolysin	Zytolysin *n*
2247	cytolysis	Zytolyse *f*
2248	cytome	Zytom *n*
2249	cytomorphosis	Zytomorphose *f*
2250	cytophysics	Zellphysik *f*
2251	cytoplasm	Zellplasma *n*, Zytoplasma *n*
2252	cytoplasm(at)ic	zytoplasmatisch
2253	cytoplasmic inclusion	Plasmaeinschluß *m*
2254	(cyto)plasmic inheritance	Plasmavererbung *f*, (zyto)plasmatische Vererbung *f*, extrachromosomale ~
2255	cytoplasmic matrix → hyaloplasm	—
2256	cytopyge	Zytopyge *n*, Zellafter *m*
2257	cytosine	Zytosin *n*
2258	cytosome	Zytosom *n*
2259	cytostome, cell-mouth	Zytostom *n*, Zellmund *m*
2260	cytotaxonomy	Zytotaxonomie *f*
2261	cytotoxin	Zytotoxin *n*
2262	cytotrophoblast, layer of Langhans	Zytotrophoblast *m*
2263	cytotropism	Zytotropismus *m*

2202	récepteur *m* cutané	receptor *m* cutáneo
2203	respiration *f* cutanée	respiración *f* cutánea
2204	cuticule *f*	cutícula *f*
2205	transpiration *f* cuticulaire	transpiración *f* cuticular
2206	cutine *f*	cutina *f*
2207	cutinisation *f*	cutinización *f*
2208	cyanine *f*	cianina *f*
2209	—	—
2210	cyanophytes *fpl*, cyanophycées *fpl*, algues *fpl* bleues	cianofitos *mpl*, cianofíceas *fpl*, algas *fpl* azules, ~ verdeazules
2211	cyathe *f*	ciatio *m*
2212	cybernétique *f*	cibernética *f*
2213	cyclique	cíclico
2214	photophosphorylation *f* cyclique	fotofosforilación *f* cíclica
2215	cyclomorphose *f*	ciclomorfosis *f*
2216	cyclose *f*	ciclosis *f*
2217	cyclostomes *mpl*	ciclóstomos *mpl*
2218	cyme *f*	cima *f*
2219	cymeux	cimoso
2220	cyste *m*, kyste *m*	ciste *m*
2221	cystéine *f*	cisteína *f*
2222	cystide *m*	cistidio *m*
2223	cystine *f*	cistina *f*
2224	cystocarpe *m*	cistocarpo *m*
2225	cystolithe *m*	cistolito *m*
2226	acide *m* cytidylique	ácido *m* citidílico
2227	—	—
2228	cytoactif	citoactivo
2229	cytochimisme *m*	citoquimismo *m*
2230	cytochimie *f*	citoquímica *f*
2231	cytochrome *m*	citocroma *m*, citocromo *m*
2232	cytode *m/f*	citodo *m*
2233	cytodiérèse *f*	citodiéresis *f*
2234	cytogène *m*	citogen *m*
2235	cytogenèse *f*	citogénesis *f*
2236	cytogénétique	citogenético
2237	cytogénéticien *m*	citogeneticista *m*
2238	cytogénétique *f*	citogenética *f*
2239	cytogonie *f*	citogonia *f*
2240	cytocinèse *f*, cytokinèse *f*	citocinesis *f*, citoquinesis *f*
2241	cytokinine *f*, kinine *f*	citoquinina *f*, fitoquinina *f*, quinina *f*
2242	cytologique	citológico
2243	cytologiste *m*	citólogo *m*
2244	cytologie *f*	citología *f*
2245	cytolymphe *f*	citolinfa *f*
2246	cytolysine *f*	citolisina *f*
2247	cytolyse *f*	citólisis *f*
2248	cytome *m*	citoma *m*
2249	cytomorphose *f*	citomorfosis *f*
2250	cytophysique *f*	citofísica *f*
2251	cytoplasme *m*	citoplasma *m*
2252	cytoplasm(at)ique	citoplasmático, citoplásmico
2253	enclave *f* cytoplasmique	inclusión *f* del plasma
2254	hérédité *f* (cyto)plasmique, ~ extra-chromosomique, ~ non-chromosomique	herencia *f* (cito)plásmica
2255	—	—
2256	cytopyge *m*	citopigio *m*
2257	cytosine *f*	citosina *f*
2258	cytosome *m*	citosoma *m*
2259	cytostome *m*	citostoma *m*
2260	cytotaxonomie *f*, caryosystématique *f*	citotaxonomía *f*
2261	cytotoxine *f*	citotoxina *f*
2262	cytotrophoblaste *m*	citotrofoblasto *m*
2263	cytotropisme *m*	citotropismo *m*

2264	dacryocyst → lacrimal sac	—
2265	daily rhythmicity → diurnal ~	—
2266	daltonism → colour blindness	—
2267	dam	Muttertier *n*
2268	dam-daughter comparison	Mutter-Tochter-Vergleich *m*
2269	dammar resin	Dammarharz *n*
2270	danger threshold	Schädlichkeitsschwelle *f*
2271	dark field	Dunkelfeld *n*
2272	dark-field microscopy, dark-ground ~	Dunkelfeldmikroskopie *f*
2273	dark reaction *bot.*	Dunkelreaktion *f* *bot.*
2274	dark room	Dunkelkammer *f*
2275	Darwinism	Darwinismus *m*
2276	dauermodification, persistent modification	Dauermodifikation *f*
2277	daughter cell	Tochterzelle *f*
2278	daughter chromatid	Tochterchromatide *f*
2279	daughter chromosome	Tochterchromosom *n*
2280	daughter nucleus	Tochterkern *m*
2281	daughter star → diaster	—
2282	day-neutral plant	tagneutrale Pflanze *f*
2283	deacidification	Entsäuerung *f*
2284	to deacidify, to disacidify	entsäuern
2285	deadaptation	Deadaptation *f*, Anpassungsverlust *m*
2286	deafferentiation	Desafferentierung *f*
2287	deaminase	Desaminase *f*
2288	to deaminate	desaminieren
2289	deamination	Desamination *f*, Desaminierung *f*
2290	death rate, mortality ~	Sterbeziffer *f*, Todesrate *f*
2291	decalcification	Entkalkung *f*
2292	to decalcify	entkalken
2293	decapitation	Dekapitation *f*, Dekapitieren *n*
2294	decarboxylase	Dekarboxylase *f*
2295	decarboxylation	Dekarboxylierung *f*
2296	decay → decomposition	—
2297	deceptive behavio(u)r, protean ~	Ablenkungsverhalten *n*
2298	deceration	Entparaffinierung *f*
2299	decidua	Dezidua *f*
2300	deciduous	laubabwerfend
2301	deciduous forest, broad-leaved ~, hardwood ~	Laubwald *m*
2302	deciduous tooth → milk ~	—
2303	deciduous tree, broad-leaved ~	Laubbaum *m*
2304	to decode	entschlüsseln
2305	decoding, decodification	Entschlüsselung *f* (des genetischen Kodes)
2306	decoloration, decolorization, discoloration	Entfärbung *f*, Farbverlust *m*
2307	to decolorize, to discolo(u)r	entfärben
2308	to decompose, to decay	zersetzen, verwesen
2309	decomposer *ecol.*	Zersetzer *m*, Zerleger *m*, Reduzent *m* *ecol.*
2310	decomposition, decay	Zersetzung *f*, Verwesung *f*
2311	decomposition product	Zersetzungsprodukt *n*
2312	deconjugation	Dekonjugation *f*
2313	to decontaminate	entseuchen
2314	decontaminating agent	Entseuchungsmittel *n*
2315	decontamination	Entseuchung *f*
2316	decontamination plant	Entseuchungsanlage *f*
2317	decrement	Dekrement *n*
2318	decussate	dekussiert, kreuzgegenständig
2319	decussation of the pyramids	Pyramiden(bahn)kreuzung *f*
2320	dedifferentiation	Dedifferenzierung *f*, Entdifferenzierung *f*
2321	deep-rooting plant	Tiefwurzler *m*
2322	deep-sea fauna → abyssal ~	—

2264	—	—
2265	—	—
2266	—	—
2267	mère *f*	madre *f*
2268	comparaison *f* mère-fille	comparación *f* madre-hija
2269	résine *f* dammar	resina *f* damar
2270	seuil *m* de nocivité	umbral *m* de nocividad
2271	fond *m* obscur, ~ noir	campo *m* oscuro, fondo *m* ~
2272	microscopie *f* à fond noir	microscopia *f* de campo oscuro
2273	réaction *f* sombre *bot.*	reacción *f* "oscura" *bot.*
2274	chambre *f* obscure	cámara *f* oscura
2275	darwinisme *m*	darwinismo *m*, darvinismo *m*
2276	modification *f* durable, dauermodification *f*	modificación *f* duradera, ~ persistente, dauermodificación *f*
2277	cellule *f* fille	célula *f* hija
2278	chromatide *f* fille	cromátida *f* hija
2279	chromosome *m* fils	cromosoma *m* hijo
2280	noyau *m* fils	núcleo *m* hijo
2281	—	—
2282	plante *f* indifférente, ~ photoapériodique	planta *f* de día neutro, ~ fotoperiódicamente neutra, mesohémera *f*
2283	désacidification *f*	desacidificación *f*
2284	désacidifier	desacidificar
2285	désadaptation *f*	desadaptación *f*
2286	désafférentation *f*	desaferentización *f*, desaferenciación *f*
2287	désaminase *f*	desaminasa *f*
2288	désaminer	desaminar
2289	désamination *f*	desaminación *f*
2290	taux *m* de mortalité	índice *m* de mortalidad, tasa *f* ~ ~
2291	décalcification *f*	descalcificación *f*
2292	décalcifier	descalcificar
2293	décapitation *f*	decapitación *f*
2294	décarboxylase *f*	descarboxilasa *f*
2295	décarboxylation *f*	descarboxilación *f*
2296	—	—
2297	comportement *m* de distraction	actitud *f* de distracción
2298	décération *f*, déparaffinage *m*	de(s)ceración *f*, desparafinado *m*
2299	décidue *f*, caduque *f*, membrane *f* déciduale	decidua *f*, caduca *f*
2300	à feuilles caduques, caducifolié	deciduo, caducifolio, de hoja caduca, de hojas caedizas
2301	forêt *f* caducifoliée, ~ à feuilles caduques, ~ feuillue	bosque *m* caducifolio
2302	—	—
2303	arbre *m* feuillu, ~ à feuilles caduques	árbol *m* deciduo, ~ de hojas caducas
2304	déchiffrer (le code génétique)	descifrar (el código genético)
2305	déchiffrement *m* (du code génétique), décodage *m*, décodification *f*	desciframiento *m* (del código genético)
2306	décoloration *f*, décolorisation *f*	descoloración *f*
2307	décolorer	descolorar
2308	se décomposer	descomponerse
2309	décomposeur *m*, réducteur *m* *ecol.*	disgregante *m*, desintegrador *m*, degradador *m*, descomponedor *m* *ecol.*
2310	décomposition *f*	descomposición *f*
2311	produit *m* de décomposition	producto *m* de descomposición
2312	déconjugaison *f*	deconjugación *f*
2313	décontaminer	de(s)contaminar
2314	produit *m* décontaminant, agent *m* ~	producto *m* de de(s)contaminación, agente *m* de(s)contaminador
2315	décontamination	de(s)contaminación *f*
2316	installation *f* de décontamination	planta *f* de(s)contaminadora
2317	décrément *m*	decremento *m*
2318	décussé	decusado
2319	décussation *f* des pyramides	decusación *f* de las pirámides
2320	dédifférenciation *f*	dediferenciación *f*
2321	plante *f* à enracinement profond	planta *f* de enraizamiento profundo
2322	—	—

2323	def(a)ecation	Defäkation *f*, Stuhlentleerung *f*
2324	defect experiment, lesion ~	Beschädigungsversuch *m*, Verletzungs~ *m*
2325	defensive posture	Abwehrstellung *f*
2326	defensive reaction	Verteidigungsreaktion *f*
2327	defensive substance	Abwehrstoff *m*
2328	deferent duct	Samenleiter *m*
2329	deficiency *gen.* → deletion	—
2330	deficiency disease	Mangelkrankheit *f*
2331	deficiency symptom	Mangelsymptom *n*
2332	definite inflorescence	begrenzter Blütenstand *m*
2333	definitive host, final ~	Endwirt *m*, Hauptwirt *m*
2334	definitive tissue → permanent ~	—
2335	defoliant	Entlaubungsmittel *n*
2336	defoliated	entlaubt, entblättert
2337	defoliation	Entlaubung *f*, Entblätterung *f*
2338	to degenerate	degenerieren, entarten
2339	degenerated code	degenerierter Kode *m*
2340	degeneration, degeneracy	Degeneration *f*, Entartung *f*
2341	deglutition, swallowing	Schlucken *n*
2342	degradation → breakdown	—
2343	degree of acidity	Säuregrad *m*
2344	degree of aggregation	Aggregationsgrad *m*
2345	degree of coverage, amount ~ ~	Deckungsgrad *m*
2346	degree of freedom	Freiheitsgrad *m*
2347	degree of hardness → hardness value	—
2348	degree of maturity, ~ ~ ripeness	Reifegrad *m*
2349	degree of pollution	Verunreinigungsgrad *m*, Verschmutzungsgrad *m*
2350	degree of relationship	Verwandtschaftsgrad *m*
2351	degree of saturation	Sättigungsgrad *m*
2352	dehiscence	Dehiszenz *f*, Kapselöffnung *f*
2353	dehiscent capsule	Spaltkapsel *f*
2354	dehiscent fruit	Streufrucht *f*, Springfrucht *f*
2355	dehydration	Dehydra(ta)tion *f*, Entwässerung *f*, Wasserentzug *m*
2356	dehydrocorticosterone	Dehydrokortikosteron *n*
2357	dehydroepiandrosterone	Dehydroepiandrosteron *n*, Androstenolon *n*
2358	dehydrogenase	Dehydrogenase *f*, Dehydrase *f*
2359	dehydrogenation, dehydrogenization	Dehydrierung *f*, Wasserstoffentzug *m*
2360	delamination	Delamination *f*
2361	delayed development, retarded ~	Entwicklungsverzögerung *f*
2362	delayed germination	Keimverzug *m*
2363	delayed reaction, ~ response	aufgeschobene Reaktion *f*
2364	deleterious → noxious	—
2365	deletion, deficiency *gen.*	Deletion *f*
2366	demarcation potential	Demarkationspotential *n*
2367	deme	Dem *n* (*pl* Deme)
2368	to demonstrate experimentally	experimentell nachweisen
2369	denaturation	Denaturierung *f*
2370	to denature	denaturieren
2371	dendrite	Dendrit *m*
2372	dendrochronology	Dendrochronologie *f*
2373	dendrology	Dendrologie *f*, Gehölzkunde *f*
2374	to denervate	denervieren, entnervieren
2375	denervation	Denervierung *f*, Entnervierung *f*
2376	denitrification	Denitrifizierung *f*, Denitrifikation *f*
2377	to denitrify	denitrifizieren
2378	denitrifying bacteria	denitrifizierende Bakterien *fpl*, Denitrifikanten *mpl*
2379	denitrogenation	Stickstoffentzug *m*
2380	density dependent *ecol.*	dichteabhängig *ecol.*
2381	density gradient	Dichtegradient *m*, Dichtegefälle *n*
2382	density gradient centrifugation	Dichtegradient-Zentrifugation *f*

2323	défécation *f*, évacuation *f* fécale	defecación *f*
2324	lésion *f* expérimentale, dommage *m* expérimental	experimento *m* por defecto
2325	attitude *f* de défense	posición *f* defensiva
2326	réaction *f* de défense, ~ défensive	reacción *f* defensiva
2327	substance *f* de défense	sustancia *f* defensiva
2328	canal *m* déférent	conducto *m* deferente
2329	—	—
2330	maladie *f* de carence, ~ carentielle	enfermedad *f* por carencia, ~ carencial
2331	symptôme *m* de carence, ~ ~ déficience	síntoma *m* de deficiencia
2332	inflorescence *f* définie	inflorescencia *f* definida
2333	hôte *m* définitif, ~ principal	huésped *m* definitivo
2334	—	—
2335	défoliant *m*	defoliante *m*
2336	défeuillé	defoliado
2337	défoliation *f*, défeuillaison *f*, défoliaison *f*	defoliación *f*
2338	dégénérer	degenerar
2339	code *m* dégénéré	código *m* degenerado
2340	dégénérescence *f*, dégénération *f*	degeneración *f*
2341	déglutition *f*	deglución *f*
2342	—	—
2343	degré *m* d'acidité	grado *m* de acidez
2344	coefficient *m* d'agrégation	coeficiente *m* de agregación
2345	degré *m* de recouvrement	grado *m* de cobertura
2346	degré *m* de liberté	grado *m* de libertad
2347	—	—
2348	degré *m* de maturité	grado *m* de madurez
2349	degré *m* de pollution	índice *m* de polución, nivel *m* ~ ~, grado *m* de contaminación
2350	degré *m* de parenté	grado *m* de parentesco
2351	degré *m* de saturation	grado *m* de saturación
2352	déhiscence *f*	dehiscencia *f*
2353	capsule *f* déhiscente	cápsula *f* dehiscente
2354	fruit *m* déhiscent	fruto *m* dehiscente
2355	déshydration *f*	deshidratación *f*
2356	dé(s)hydrocorticostérone *f*	de(s)hidrocorticosterona *f*
2357	dé(s)hydroépiandrostérone *f*	de(s)hidroepiandrosterona *f*
2358	déshydrogénase *f*	de(s)hidrogenasa *f*, de(s)hidrasa *f*
2359	déshydrogénation *f*, déshydrogénisation *f*	de(s)hidrogenación *f*
2360	délamination *f*	deslaminación *f*
2361	retard *m* du développement	retraso *m* del desarrollo
2362	germination *f* retardée, ~ ralentie	germinación *f* retardada
2363	réponse *f* retardée, délai *m* de la réaction	reacción *f* retardada
2364	—	—
2365	délétion *f*, déficience *f* *gen.*	deleción *f*, deficiencia *f* *gen.*
2366	potentiel *m* de démarcation	potencial *m* de demarcación
2367	dème *m*	demo *m*
2368	mettre expérimentalement en évidence, démontrer ~	demostrar experimentalmente
2369	dénaturation *f*, dénaturalisation *f*	desnaturalización *f*
2370	dénaturer	desnaturalizar
2371	dendrite *m*	dendrita *f*
2372	dendrochronologie *f*	dendrocronología *f*
2373	dendrologie *f*	dendrología *f*
2374	dénerver, énerver	denervar
2375	dénervation *f*, énervation *f*	denervación *f*
2376	dénitrification *f*	desnitrificación *f*
2377	dénitrifier	desnitrificar
2378	bactéries *f pl* dénitrifiantes, dénitrificateurs *m pl*	bacterias *f pl* desnitrificantes
2379	dénitrogénation *f*	desnitrogenación *f*
2380	dépendant de la densité *ecol.*	dependiente de la densidad *ecol.*
2381	gradient *m* de densité	gradiente *m* de densidad
2382	centrifugation *f* en gradients de densité	centrifugación *f* sobre gradiente de densidades

No.	English	German
2383	density independent *ecol.*	dichteunabhängig *ecol.*
2384	density of species	Artdichte *f*
2385	dental	Zahn ...
2386	dental alveolus, tooth socket	Zahnfach *n*
2387	dental formula	Zahnformel *f*
2388	dental pulp, ~ papilla	Zahnmark *n*
2389	dentate, toothed	gezähnt
2390	denticle, dermal ~, placoid scale	Plakoidschuppe *f*, Hautzahn *m*
2391	dentification	Zahnbildung *f*
2392	dentine	Dentin *n*, Zahnbein *n*
2393	dentition	Dentition *f*, Zahnen *n*, Bezahnung *f*
2394	denture	Gebiß *n anat.*
2395	denucleination	Denukleinisierung *f*
2396	deoxycorticosterone	Desoxykortikosteron *n*
2397	deoxydation, deoxygenation, deoxygenization	Sauerstoffentzug *m*
2398	deoxyribonuclease	Desoxyribonuklease *f*
2399	deoxyribonucleic acid, DNA	Desoxyribonukleinsäure *f*, DNS
2400	deoxyribose	Desoxyribose *f*
2401	dephosphorylation	Dephosphorylierung *f*
2402	depigmentation	Depigmentierung *f*, Depigmentation *f*
2403	deplasmolysis	Deplasmolyse *f*
2404	depolarization	Depolarisation *f*
2405	to depollute	entschmutzen
2406	depollution	Entschmutzung *f*, Beseitigung *f* von Umweltverschmutzungen
2407	depot fat	Depotfett *n*
2408	derepression	Derepression *f*
2409	derivation, leading off, recording *neur.*	Ableitung *f*, Abgriff *m neur.*
2410	derivative hybrid	Doppelhybrid *m*
2411	derm → dermis	—
2412	dermal → cutaneous	—
2413	dermal bone → membrane ~	—
2414	dermal gland → skin ~	—
2415	dermal papilla	Hautpapille *f*
2416	dermatogen	Dermatogen *n*
2417	dermic → cutaneous	—
2418	derm(is), corium	Lederhaut *f*, Corium *n*
2419	dermoskeleton → exoskeleton	—
2420	to desalt, to desalinate	entsalzen
2421	desalting (or: desalination) of sea water	Entsalzung *f* von Meerwasser, Meerwasserentsalzung *f*
2422	desalting (or: desalination) plant	Entsalzungsanlage *f*
2423	to descend of	abstammen von
2424	descendant	Nachkomme *m*
2425	descending colon	absteigendes Kolon *n*, absteigender Grimmdarm *m*
2426	descent; filiation	Abstammung *f*
2427	descent theory	Abstammungslehre *f*, ~theorie *f*, Deszendenzlehre *f*, ~theorie *f*
2428	descriptive botany → phytography	—
2429	desensitization	Desensibilisierung *f*
2430	to desensitize	desensibilisieren
2431	desert plant, deserticolous ~	Wüstenpflanze *f*
2432	desmocyte → fibrocyte	—
2433	desmolase	Desmolase *f*
2434	desmone	Desmon *n*
2435	desmosome	Desmosom *n*
2436	desquamation	Abschuppung *f*
2437	dessication, drying	Trocknen *n*, Austrocknung *f*, Vertrocknung *f*
2438	destructive lumbering	Raubbau *m* (Forstw.)
2439	destructive metabolism → catabolism	—
2440	desulphurization; desulfurization *US*	Desulfurikation *f*, Entschwefelung *f*
2441	to desulphurate; to desulfurize *US*	entschwefeln
2442	desynapsis	Desynapsis *f*
2443	detergent	Detergens *n*, (*pl* Detergentia, Detergenzien)

2383	indépendant de la densité *ecol.*	independiente de la densidad *ecol.*
2384	densité *f* des espèces	densidad *f* de las especies
2385	dentaire	dental, dentario
2386	alvéole *f* dentaire	alvéolo *m* dentario
2387	formule *f* dentaire	fórmula *f* dental
2388	pulpe *f* dentaire	pulpa *f* dentaria
2389	denté, dentelé	dentado
2390	denticule *m* cutané, écaille *f* placoïde	dentículo *m*, escama *f* placoidea
2391	dentification *f*	dentificación *f*
2392	dentine *f*	dentina *f*
2393	dentition *f*	dentición *f*
2394	denture *f* *anat.*	dentadura *f* *anat.*
2395	dénucléinisation *f*	desnucleinización *f*
2396	déoxycorticostérone *f*	desoxicorticosterona *f*
2397	désoxydation *f*	desoxidación *f*, desoxigenación *f*
2398	déoxyribonucléase *f*	desoxirribonucleasa *f*
2399	acide *m* désoxyribonucléique, ADN	ácido *m* desoxirribonucleico, ADN
2400	désoxyribose *m*	desoxirribosa *f*
2401	déphosphorylation *f*	defosforilación *f*
2402	dépigmentation *f*	depigmentación *f*
2403	déplasmolyse *f*	deplasmólisis *f*
2404	dépolarisation *f*	despolarización *f*
2405	dépolluer	de(s)polucionar
2406	dépollution *f*	de(s)polución *f*
2407	graisse *f* de dépôt	grasa *f* de depósito
2408	dérépression *f*	derrepresión *f*
2409	dérivation *f* *neur.*	derivación *f* *neur.*
2410	hybride *m* double	híbrido *m* doble, ~ derivado
2411	—	—
2412	—	—
2413	—	—
2414	—	—
2415	papille *f* dermique	papila *f* dérmica
2416	dermatogène *m*	dermatógeno *m*
2417	—	—
2418	derme *m*	dermis *f*, corion *m*
2419	—	—
2420	dessaler	desalar, desalinar, desalinizar
2421	dessalement *m* de l'eau de la mer	desalación *f* (o: desalinización *f*) del agua del mar
2422	usine *f* de dessalement	planta *f* de desalación
2423	descendre de	descender de
2424	descendant *m*	descendiente *m*
2425	côlon *m* descendant	colon *m* descendiente
2426	descendance *f*; filiation *f*	descendencia *f*; filiación *f*
2427	théorie *f* de la descendance	teoría *f* de la descendencia
2428	—	—
2429	désensibilisation *f*	desensibilización *f*
2430	désensibiliser	desensibilizar
2431	plante *f* désertique	planta *f* desértica, ~ desertícola
2432	—	—
2433	desmolase *f*	desmolasa *f*
2434	desmone *f*	desmona *f*
2435	desmosome *m*	desmosoma *m*
2436	desquamation *f*	descamación *f*
2437	dessication *f*, dessèchement *m*, assèchement *m*	desecación *f*
2438	coupes *f pl* abusives, exploitation *f* abusive	tala *f* incontrolada, explotación *f* abusiva
2439	—	—
2440	désulfuration *f*	desulfur(ic)ación *f*
2441	désulfurer	desulfurar
2442	désynapsis *f*	desinapsis *f*
2443	détergent *m*, produit *m* ~	detergente *m*

2444	deterioration of environment → environmental degradation	—
2445	deterioration of strains	Sortenabbau *m*
2446	determinant	determinierend; Determinante *f*
2447	to detoxicate, to detoxify	entgiften
2448	detoxication	Entgiftung *f*
2449	detritus	Detritus *m*
2450	detritus-feeding animal	Detritusfresser *m*
2451	deuterocoele → coelom	—
2452	deuteromycetes → fungi imperfectae	—
2453	Deuterostomia	Deuterostomier *pl*, Zweitmünder *mpl*, Neumünder *mpl*, Rückenmarkstiere *npl*
2454	deuterotoky, deuterotocia	Deuterotokie *f*
2455	deutoplasm, deuteroplasm, food yolk, nutritive ~	Deutoplasma *n*, Deuteroplasma *n*, Nährdotter *m*
2456	developable, capable of development	entwicklungsfähig
2457	development	Entwicklung *f*
2458	development laboratory	Versuchslabor(atorium) *n*
2459	developmental biology, evolutionary ~	Entwicklungsbiologie *f*
2460	developmental capacity, evolutivity	Entwicklungsfähigkeit *f*
2461	developmental failure	Entwicklungsstörung *f*
2462	developmental genetics	Entwicklungsgenetik *f*
2463	developmental physiology, ~ mechanics, experimental embryology	Entwicklungsphysiologie *f*, Entwicklungsmechanik *f*
2464	developmental process	Entwicklungsprozeß *m*
2465	developmental rate	Entwicklungsrate *f*, ~geschwindigkeit *f*
2466	developmental stage	Entwicklungsstadium *n*
2467	devernalization	Devernalisation *f*
2468	deviation	Abweichung *f*, Deviation *f*
2469	to devitalize	abtöten
2470	Devonian (period)	Devon *n*
2471	dew point	Taupunkt *m*
2472	dextran(e)	Dextran *n*
2473	dextrin	Dextrin *n*
2474	dextrorota(to)ry, dextrogyral, dextrogyrate	rechtsdrehend
2475	dextrorse	rechtswindend
2476	dextrose → glucose	—
2477	diachene	Doppelachäne *f*
2478	diageotropism	Diageotropismus *m*
2479	diakinesis	Diakinese *f*
2480	dialypetalous, choripetalous	frei(kron)blättrig, choripetal
2481	dialysate	Dialysat *n*
2482	dialysis	Dialyse *f*
2483	to dialyze	dialysieren
2484	diamine	Diamin *n*
2485	diamino acid	Diaminosäure *f*
2486	diaminopimelic acid	Diaminopimelinsäure *f*
2487	diapause	Diapause *f*
2488	diapedesis	Diapedese *f*
2489	diaphragm	Zwerchfell *n*
2490	diaphragmatic respiration	Zwerchfellatmung *f*
2491	diaphysis	Diaphyse *f*
2492	diapophysis, transverse process	Diapophyse *f*, Querfortsatz *m*
2493	diarthrosis, ball-and-socket joint	Diarthrose *f*, Kugelgelenk *n*
2494	diaspore, disseminule, dispersal unit	Diaspore *f*, Verbreitungseinheit *f*
2495	diastase → amylase	—
2496	diaster, daughter star	Diaster *m*
2497	diastole	Diastole *f*
2498	diatomaceous earth → infusorial ~	—
2499	diatoms, Bacillariophyceae	Diatomeen *fpl*, Kieselalgen *fpl*
2500	Dibranchiata, dibranchiates	Zweikiemer *mpl*, Dibranchier *mpl*
2501	dicaryon → dikaryon	—

2444	—	—
2445	dégradation *f* des variétés	degradación *f* de variedades
2446	déterminant (*m*)	determinante (*m*)
2447	désintoxiquer	des(in)toxicar, destoxificar
2448	détoxi(fi)cation *f*, désintoxication *f*	des(in)toxicación *f*, destoxificación *f*
2449	détritus *m*	detrito *m*
2450	détritiphage *m*, détritophage *m*, détritivore *m*	detritóvoro *m*
2451	—	—
2452	—	—
2453	deutérostomiens *m pl*	deuterostomas *m pl*
2454	deutérotocie *f*, deutérotoquie *f*	deuterotocia *f*
2455	deutoplasme *m*, deutéroplasme *m*	deutoplasma *m*, deuteroplasma *m*, vitelo *m* nutritivo, deutolecito *m*
2456	capable d'évolution, apte à évoluer	desarrollable
2457	développement *m*	desarrollo *m*
2458	laboratoire *m* d'expérience	laboratorio *m* de ensayos, ~ de experimentación
2459	biologie *f* du développement	biología *f* del desarrollo
2460	capacité *f* évolutive, évolutivité *f*	capacidad *f* de desarrollo
2461	trouble *m* de développement	trastorno *m* de desarrollo
2462	génétique *f* évolutive	genética *f* del desarrollo
2463	physiologie *f* du développement, mécanique *f* ~ ~, ~ embryonnaire, embryologie *f* causale, ~ expérimentale	fisiología *f* del desarrollo, embriología *f* experimental
2464	processus *m* de développement	proceso *m* de desarrollo
2465	vitesse *f* d'évolution	velocidad *f* de evolución, rapidez *f* ~ ~
2466	stade *m* de développement	estado *m* de desarrollo, etapa *f* ~ ~
2467	dévernalisation *f*	desvernalización *f*
2468	déviation *f*	desviación *f*
2469	dévitaliser	desvitalizar
2470	dévonien *m*	devónico *m*
2471	point *m* de rosée	punto *m* de rocío
2472	dextrane *f*	dextrana *f*
2473	dextrine *f*	dextrina *f*
2474	dextrogyre	dextrógiro, dextrorrotatorio
2475	dextrorsum	dextrorso, dextrovoluble
2476	—	—
2477	diakène *m*, double akène *m*	diaquenio *m*
2478	diagéotropisme *m*	diageotropismo *m*
2479	diacinèse *f*	diacinesis *f*, diaquinesis *f*
2480	dialypétale	dialipétalo, coripétalo
2481	dialysat *m*	dializado *m*
2482	dialyse *f*	diálisis *f*
2483	dialyser	dializar
2484	diamine *f*	diamina *f*
2485	acide *m* diaminé	acido *m* diamínico
2486	acide *m* diaminopimélique	ácido *m* diaminopimélico
2487	diapause *f*	diapausia *f*
2488	diapédèse *f*	diapédesis *f*
2489	diaphragme *m*	diafragma *m*
2490	respiration *f* diaphragmatique	respiración *f* diafragmática
2491	diaphyse *f*	diáfisis *f*
2492	diapophyse *f*, apophyse *f* transverse	diapófisis *f*, apófisis *f* transversa
2493	diarthrose *f*	diartrosis *f*
2494	diaspore *f*	diáspora *f*, diseminulo *m*, unidad *f* de dispersión
2495	—	—
2496	diaster *m*, sphère-fille *f*	diaster *m*
2497	diastole *f*	diástole *f*
2498	—	—
2499	diatomées *f pl*, bacillariophycées *f pl*	diatomeas *f pl*, bacilariofíceas *f pl*
2500	dibranches *m pl*	dibranquiados *m pl*
2501	—	—

2502	dicentric	dizentrisch
2503	dichasium	Dichasium *n*
2504	dichlamydeous	dichlamyd(eisch)
2505	dichogamic, dichogamous	dichogam
2506	dichogamy	Dichogamie *f*
2507	dichotomous, dichotomic	dichotom, gabelig, gabelteilig
2508	dichotomous ramification, ~ branching	dichotome Verzweigung *f*, gabelige ~
2509	dichotomy	Dichotomie *f*, Gabelung *f*, Gabelteilung *f*
2510	diclinous	diklin
2511	Dicotyledoneae, dicotyledons	Dikotyle(done)n *f pl*, zweikeimblättrige Pflanzen *f pl*
2512	dicotyledonous	zweikeimblättrig, dikotyl
2513	dictyokinesis	Diktyokinese *f*
2514	dictyosome	Diktyosom *n*
2515	to die back, to ~ down, to perish	absterben, eingehen *(Pflanzen)*
2516	die-back, dying, perishing	Absterben *n*, Eingehen *n (Pflanzen)*
2517	diencephalon, between-brain, inter-brain	Zwischenhirn *n*, Dienzephalon *n*
2518	differential centrifugation	Differential-Zentrifugation *f*
2519	differential distance	Differentialabstand *m*
2520	differential growth	Differenzierungswachstum *n*
2521	differential sex gene	Geschlechtsrealisator *m*
2522	differential threshold	Unterschiedsschwelle *f*
2523	to differentiate	differenzieren
2524	differentiation	Differenzierung *f*
2525	to diffuse	diffundieren
2526	diffusible	diffusionsfähig
2527	diffusion	Diffusion *f*
2528	diffusion barrier	Diffusionsbarriere *f*
2529	diffusion coefficient	Diffusionskoeffizient *m*
2530	diffusion constant	Diffusionskonstante *f*
2531	diffusion equilibrium	Diffusionsgleichgewicht *n*
2532	diffusion potential	Diffusionspotential *n*
2533	diffusion pressure deficit, DPD	Diffusionsdruckdefizit *n*
2534	diffusion speed	Diffusionsgeschwindigkeit *f*
2535	diffusive respiration	Diffusionsatmung *f*
2536	digametic → heterogametic	—
2537	digenic	digen
2538	to digest	verdauen
2539	digestibility	Verdaulichkeit *f*
2540	digestible	verdaulich
2541	digestion	Verdauung *f*
2542	digestive ferment, ~ enzyme	Verdauungsferment *n*, ~enzym *n*
2543	digestive gland	Verdauungsdrüse *f*
2544	digestive juice	Verdauungssaft *m*
2545	digestive organ	Verdauungsorgan *n*
2546	digestive system, ~ tract, alimentary ~	Verdauungssystem *n*, ~apparat *m*
2547	digitate	gefingert
2548	digitigrade	Zehengänger *m*
2549	dihybrid	dihybrid
2550	dihybridism	Dihybridie *f*
2551	dikaryon, dicaryon	Dikaryon *n*
2552	dikaryophase	Dikaryophase *f*, Zweikernphase *f*, Paarkernphase *f*
2553	dilatability	Dehnbarkeit *f*
2554	dilatable	dehnbar
2555	diluent	Verdünnungsmittel *n*
2556	to dilute	verdünnen
2557	dilution	Verdünnung *f*
2558	dimerization	Dimerisation *f*
2559	dimerous, dimeric(al)	dimer, zweigliederig
2560	dimery	Dimerie *f*
2561	dimorphic, dimorphous	dimorph
2562	dimorphism	Dimorphismus *m*, Dimorphie *f*
2563	dinitrophenol	Dinitrophenol *n*

2502	dicentrique	dicéntrico
2503	cyme *f* bipare, dichasium *m*	dicasio *m*, cima *f* bípara
2504	dichlamyde	diclamídeo
2505	dichogame	dicógamo
2506	dichogamie *f*	dicogamia *f*
2507	dichotome, dichotomique	dicótomo, dicotómico
2508	ramification *f* dichotomique	ramificación *f* dicotómica
2509	dichotomie *f*	dicotomía *f*
2510	dicline	diclino
2511	dicotylédones *f pl*	dicotiledóneas *f pl*
2512	dicotylédone	dicotiledóneo, dicotiledónico
2513	dictyocinèse *f*	dictiocinesis *f*
2514	dictyosome *m*	dictiosoma *m*
2515	dépérir	perecer, morir
2516	dépérissement *m*	muerte *f* lenta
2517	diencéphale *m*, cerveau *m* intermédiaire	diencéfalo *m*, cerebro *m* (inter)medio
2518	centrifugation *f* différentielle	centrifugación *f* diferencial
2519	distance *f* différentielle	distancia *f* diferencial
2520	croissance *f* différentielle	crecimiento *m* diferencial
2521	gène *m* de différenciation sexuelle, réalisateur *m* du sexe	gen *m* sexual diferencial
2522	seuil *m* de différenciation	umbral *m* diferencial
2523	différencier	diferenciar
2524	différenciation *f*	diferenciación *f*
2525	diffuser	difundir(se)
2526	diffusible, diffusif	difundible, difusivo
2527	diffusion *f*	difusión *f*
2528	barrière *f* de diffusion	barrera *f* de difusión
2529	coefficient *m* de diffusion	coeficiente *m* de difusión
2530	constante *f* de diffusion	constante *f* de difusión
2531	équilibre *m* de diffusion	equilibrio *m* de difusión
2532	potentiel *m* de diffusion	potencial *m* de difusión
2533	déficit *m* de pression de diffusion, DPD	déficit *m* de la presión de difusión, DPD
2534	vitesse *f* de diffusion	velocidad *f* de difusión
2535	respiration *f* diffusive	respiración *f* por difusión
2536	—	—
2537	digénique	digénico
2538	digérer	digerir
2539	digestibilité *f*	digestibilidad *f*
2540	digestible, digérable	digestible, digerible
2541	digestion *f*	digestión *f*
2542	ferment *m* digestif	fermento *m* digestivo, enzima *f* digestiva
2543	glande *f* digestive	glándula *f* digestiva
2544	suc *m* digestif	jugo *m* digestivo
2545	organe *m* de la digestion	órgano *m* digestivo, ~ de la digestión
2546	appareil *m* digestif	sistema *m* digestivo
2547	digité	digitado
2548	digitigrade *m*	digitígrado *m*
2549	dihybride	dihíbrido
2550	dihybridisme *m*	dihibridismo *m*
2551	dicaryon *m*, bicaryon *m*	dicario(n) *m*, dicarionte *m*
2552	dicaryophase *f*	dicariófase *f*, fase *f* dicariótica
2553	dilatabilité *f*	dilatabilidad *f*, extensibilidad *f*
2554	dilatable	dilatable
2555	diluant *m*	diluente *m*
2556	diluer	diluir
2557	dilution *f*	dilución *f*
2558	dimérisation *f*	dimerización *f*
2559	dimère, dimérique	dímero, dimérico
2560	dimérie *f*	dimería *f*
2561	dimorphe, dimorphique	dimorfo, dimórfico
2562	dimorphisme *m*	dimorfismo *m*
2563	dinitrophénol *m*	dinitrofenol *m*

2564	Dinoflagellata, Dinophyceae	Dinoflagellaten *pl*, Peridineen *pl*
2565	dioecious	zweihäusig, diözisch
2566	dioecism, dioecy	Zweihäusigkeit *f*, Diözie *f*
2567	di(o)estrus	Diöstrus *m*
2568	dipeptide	Dipeptid *n*
2569	diphosphopyridine nucleotide → nicotinamide adenine dinucleotide	—
2570	diphyletic	diphyletisch
2571	diplobiont	Diplobiont *m*
2572	diplochromosome	Diplochromosom *n*
2573	diplohaplont	Diplohaplont *m*
2574	diploid	diploid
2575	diploidization	Diploidisation *f*
2576	diploidy	Diploidie *f*
2577	diplokaryon	Diplokaryon *n*
2578	diplont	Diplont *m*
2579	diplophase, diploid phase	Diplophase *f*
2580	Diplopoda, diplopods	Doppelfüßer *mpl*, Diplopoden *pl*
2581	diplosis	Diplosis *f*
2582	diplosome	Diplosom *n*
2583	diplostemonous	diplostemon
2584	diplotene	Diplotän *n*
2585	Dipneusti, Dipnoi, lung-fish	Lungenfische *mpl*
2586	Diptera, dipterans	Zweiflügler *mpl*, Dipteren *pl*
2587	direct observation	Direktbeobachtung *f*
2588	directed mutation → induced ~	—
2589	directed selection, directional ~	gerichtete Selektion *f*
2590	disaccharide	Disa(c)charid *n*
2591	disassimilation → catabolism	—
2592	disc floret	Scheibenblüte *f*
2593	discharge	Entladung *f*
2594	discharge frequency	Entladungsfrequenz *f*
2595	disclimax, disturbance climax	Disklimax *f*
2596	to discolour → to decolorize	—
2597	discontinuity → disjunction *ecol.*	—
2598	discontinuity layer → thermocline	—
2599	discontinuous area	disjunktes Areal *n*, diskontinuierliches ~
2600	discontinuous variability, qualitative ~	diskontinuierliche Variabilität *f*, alternative ~, qualitative ~
2601	disease germ	Krankheitskeim *m*
2602	to disinfect	desinfizieren
2603	disinfectant	Desinfektionsmittel *n*, Desinfiziens *n*
2604	disinfection	Desinfektion *f*
2605	to disinhibit	enthemmen
2606	disinhibition	Enthemmung *f*
2607	to disintegrate	zerfallen
2608	disintegration	Zerfall *m*
2609	disjunction *gen.*	Disjunktion *f*, Chromosomentrennung *f*
2610	disjunction *ecol.*, discontinuity	Disjunktion *f*, Teilareal *n*
2611	dislocation *gen.*	Dislokation *f* *gen.*
2612	disomy	Disomie *f*
2613	dispermous, dispermic	zweisamig, disperm
2614	dispermy, dispermic fertilization	Dispermie *f*, disperme Befruchtung *f*
2615	dispersal by man → anthropochory	—
2616	dispersal unit → diaspore	—
2617	dispersed	dispers
2618	dispersion	Streuung *f*, Dispersion *f*
2619	dispersion agent	Dispersionsmittel *n*
2620	displacement *gen.* → shift	—
2621	displacement activity	Übersprungbewegung *f*, ~handlung *f*

2564	dinoflagellés *mpl*, péridiniens *mpl*	dinoflagelados *mpl*, dinofíceas *fpl*, peridineas *fpl*
2565	dioïque, diécique	dioico
2566	di(o)écie *f*, dioïcité *f*	di(o)ecia *f*, dioecismo *m*
2567	dioestrus *m*	diestro *m*
2568	dipeptide *m*	dipéptido *m*
2569	—	—
2570	diphylétique	difilético
2571	diplobionte *m*	diplobionte *m*
2572	diplochromosome *m*	diplocromosoma *m*
2573	diplohaplonte *m*	diplohaplonte *m*
2574	diploïde	diploide
2575	diploïdisation *f*	diploidización *f*
2576	diploïdie *f*	diploidía *f*
2577	diplocaryon *m*	diplocario(n) *m*
2578	diplonte *m*	diplonte *m*
2579	diplophase *f*	diplofase *f*, fase *f* diploide
2580	diplopodes *mpl*	diplópodos *mpl*
2581	diplosie *f*	diplosis *f*
2582	diplosome *m*	diplosoma *m*
2583	diplostémone	diplostémono
2584	diplotène *m*	diplóteno *m*, diplótena *f*
2585	dipneustes *mpl*	dipneustos *mpl*, dipnoos *mpl*, peces *mpl* pulmonados
2586	diptères *mpl*	dípteros *mpl*
2587	observation *f* directe	observación *f* directa
2588	—	—
2589	sélection *f* dirigée, ~ directionnelle, ~ dynamique	selección *f* direccional, ~ dinámica
2590	disaccharide *m*, dioside *m*	disacárido *m*
2591	—	—
2592	fleuron *m*	flósculo *m* discoide
2593	décharge *f*	descarga *f*
2594	fréquence *f* de décharge	frecuencia *f* de descarga
2595	disclimax *m*, dysclimax *m*	disclímax *m*
2596	—	—
2597	—	—
2598	—	—
2599	aire *f* disjointe, ~ discontinue	área *f* disyunta, ~ discontinua
2600	variabilité *f* discontinue, ~ qualitative	variabilidad *f* discontinua, ~ cualitativa
2601	germe *m* pathogène	gérmen *m* patógeno
2602	désinfecter	desinfectar
2603	désinfectant *m*	desinfectante *m*
2604	désinfection *f*	desinfección *f*
2605	désinhiber	desinhibir
2606	désinhibition *f*	desinhibición *f*
2607	se désintégrer	desintegrarse
2608	désintégration *f*	desintegración *f*
2609	disjonction *f* *gen.*	disyunción *f* *gen.*, separación *f* de cromosomas
2610	disjonction *f* *ecol.*, zone *f* partielle de répartition	disyunción *f* *ecol.*
2611	dislocation *f* *gen.*	dislocación *f* *gen.*
2612	disomie *f*	disomia *f*
2613	dispermique	dispermo, dispérmico
2614	dispermie *f*, fertilisation *f* dispermique	dispermia *f*, fertilización *f* dispérmica
2615	—	—
2616	—	—
2617	dispersé	esparcido
2618	dispersion *f*	dispersión *f*
2619	agent *m* de dispersion	medio *m* dispersante, ~ dispersivo, ~ de dispersión
2620	—	—
2621	mouvements *mpl* de déplacement	actividad *f* de desplazamiento, comportamiento *m* desplazado

2622	display	Imponiergehaben *n*
2623	disruptive selection	disruptive Selektion *f*
2624	to dissect	sezieren
2625	dissecting microscope	Präpariermikroskop *n*
2626	dissecting needle → microneedle	—
2627	dissection	Sezieren *n*
2628	dissemination → seed dispersal	—
2629	disseminule → diaspore	—
2630	dissepiment	Dissepiment *n*
2631	dissimilation	Dissimilation *f*
2632	dissociation	Dissoziation *f*
2633	dissociation constant	Dissoziationskonstante *f*
2634	dissociation curve	Dissoziationskurve *f*
2635	dissolvent → solvent	—
2636	distance orientation	Fernorientierung *f*
2637	distance receptor, tele(re)ceptor	Distanzrezeptor *m*
2638	distichous	distich, zweizeilig
2639	distillation	Destillation *f*
2640	distinctive character	Unterscheidungsmerkmal *n*
2641	distress call	Notruf *m*
2642	distribution area → range of distribution	—
2643	distribution map	Arealkarte *f*
2644	disturbance climax → disclimax	—
2645	disturbance of growth	Wachstumsstörung *f*
2646	disymmetrical	disymmetrisch
2647	diurnal animal	Tagtier *n*
2648	diurnal rhythm → circadian ~	—
2649	diurnal rhythmicity, daily ~	Tagesperiodik *f*, ~rhythmik *f*
2650	divergency	Divergenz *f*
2651	divergent evolution	aufspaltende Evolution *f*, divergente ~
2652	division *tax.*	Abteilung *f* *tax.*
2653	division of labo(u)r, repartition ~ ~	Arbeitsteilung *f*
2654	division plane	Teilungsebene *f*
2655	division zone	Teilungszone *f*
2656	dizygotic, binovular	dizygotisch, zweieiig
2657	dizygotic twins, binovular ~, fraternal ~	zweieiige Zwillinge *m pl*, dizygotische ~, Geschwisterzwillinge *m pl*
2658	DNA → deoxyribonucleic acid	—
2659	DNA-duplication, DNA-replication	DNS-Replikation *f*
2660	DNA-sequence	DNS-Sequenz *f*
2661	doctrine of panspermatism	Panspermie-Theorie *f*
2662	domestic animal	Haustier *n*
2663	domestic refuse, ~ waste, household refuse	Müll *m*
2664	domestic refuse disposal	Müllbeseitigung *f*
2665	domestication	Domestikation *f*, Haustierwerdung *f*
2666	dominance	Dominanz *f*
2667	dominance hierarchy → rank order	—
2668	dominance-recessiveness relation	Dominanz-Rezessivitäts-Verhältnis *n*
2669	dominant	dominant
2670	dominant species	dominante Art *f*, Dominante *f*
2671	dominigene	Dominanzmodifikator *m*, Dominigen *n*
2672	donor, donator	Donator *m*, Donor *m*, Spender *m*
2673	donor bacterium	Spenderbakterie *f*
2674	donor cell	Spenderzelle *f*, Donorzelle *f*
2675	dormancy	Keimruhe *f*, Wachstumsruhe *f*
2676	dormant bud	schlafendes Auge *n*
2677	dormant cell → resting cell	—
2678	dormant egg → winter egg	—
2679	dormant stage → resting stage	—
2680	dormin	Dormin *n*, Abszisin II *n*
2681	dorsal	Rücken ..., dorsal
2682	dorsal fin, back-fin	Rückenflosse *f*
2683	dorsal horn *neur.*	Hinterhorn *n* *neur.*
2684	dorsal root, sensory ~, posterior ~	dorsale Rückenmarkswurzel *f*, sensible ~

2622	parade *f*, «display» *m*	gesto *m* de destacar, ceremonia *f* de instigación
2623	sélection *f* disruptive	selección *f* disruptiva
2624	disséquer	disecar
2625	microscope *m* à dissection	microscopio *m* de disección
2626	—	—
2627	dissection *f*	disección *f*
2628	—	—
2629	—	—
2630	dissépiment *m*	disepimento *m*
2631	dissimilation *f*	disimilación *f*
2632	dissociation *f*	disociación *f*
2633	constante *f* de dissociation	constante *f* de disociación
2634	courbe *f* de dissociation	curva *f* de disociación
2635	—	—
2636	orientation *f* lointaine	orientación *f* a distancia
2637	récepteur *m* de distance	receptor *m* de distancia, teleceptor *m*
2638	distique	dístico
2639	distillation *f*	destilación *f*
2640	caractère *m* distinctif	rasgo *m* distintivo, carácter *m* ~
2641	cri *m* de détresse	llamada *f* de apuro
2642	—	—
2643	carte *f* de répartition	mapa *m* de área
2644	—	—
2645	trouble *m* de croissance	trastorno *m* de crecimiento
2646	dissymétrique	disimétrico
2647	animal *m* diurne	animal *m* diurno
2648	—	—
2649	périodicité *f* journalière, ~ diurne	periodicidad *f* diaria, ~ diurna
2650	divergence *f*	divergencia *f*
2651	évolution *f* divergente	evolución *f* divergente
2652	embranchement *m*, division *f* *tax.*	división *f* *tax.*
2653	répartition *f* du travail	división *f* del trabajo
2654	plan *m* de division	plano *m* de división
2655	zone *f* de division	zona *f* de división
2656	dizygotique, biovulaire	dizigótico, biovular
2657	jumeaux *mpl* dizygotiques, ~ biovulaires, faux jumeaux *mpl*	gemelos *mpl* dizigóticos, ~ biovulares, ~ bivitelinos, ~ fraternos
2658	—	—
2659	réplication *f* de l'ADN	replicación *f* del ADN, duplicación *f* ~ ~
2660	séquence *f* d'ADN	secuencia *f* del ADN
2661	hypothèse *f* de la panspermie	teoría *f* de la panspermia
2662	animal *m* domestique	animal *m* doméstico
2663	ordures *fpl* ménagères	basuras *fpl* domésticas, ~ domiciliarias
2664	évacuation *f* des ordures ménagères	eliminación *f* de las basuras
2665	domestication *f*	domesticación *f*
2666	dominance *f*	dominancia *f*
2667	—	—
2668	rapport *m* de dominance-récessivité	relación *f* predominio-recesividad
2669	dominant	dominante
2670	espèce *f* dominante	especie *f* dominante
2671	dominigène *m*	dominigen *m*
2672	donneur *m*, donateur *m*	donador *m*, dador *m*, donante *m*
2673	bactérie *f* donatrice, ~ donneuse	bacteria *f* donadora
2674	cellule *f* donatrice	célula *f* donadora, ~ dadora, ~ donante
2675	dormance *f*, repos *m* germinatif	letargo *m*, "dormancia" *f*
2676	œil *m* dormant	yema *f* durmiente, ~ latente
2677	—	—
2678	—	—
2679	—	—
2680	dormine *f*, abscissine II *f*, acide *m* abscissique	dormina *f*
2681	dorsal	dorsal
2682	nageoire *f* dorsale	aleta *f* dorsal
2683	corne *f* postérieure *neur.*	cuerno *m* dorsal *neur.*
2684	racine *f* dorsale, ~ sensitive	raíz *f* dorsal, ~ sensitiva, ~ posterior

2685	dorsal shield	Rückenschild *m*
2686	dorsiventrality	Dorsiventralität *f*
2687	dosage compensation	Dosiskompensation *f*
2688	dose effect	Dosiseffekt *m*
2689	dose effect curve	Dosis-Effekt-Kurve *f*
2690	to double → to duplicate	—
2691	double bond	Doppelbindung *f*
2692	double cross, four-way ~	Doppelkreuzung *f*
2693	double crossing-over	Doppel-Crossing-over *n*, Doppelaustausch *m*
2694	double dominant	Doppeldominant *m*
2695	double fertilization	Doppelbefruchtung *f*
2696	double helix	Doppelhelix *f*, Doppelspirale *f*
2697	double refraction → birefringence	—
2698	doubling → duplication	—
2699	doubling time	Verdoppelungszeit *f*
2700	down	Flaum *m*
2701	down feather, plumule	Flaumfeder *f*, Daune *f*
2702	DPN → nicotinamide adenine dinucleotide	—
2703	drepanium	Sichel *f* *bot*
2704	drinking water, potable ~	Trinkwasser *n*
2705	drinking-water supply	Trinkwasserversorgung *f*
2706	drive	Trieb *m*, Drang *m*
2707	drive to fight	Kampftrieb *m*
2708	drone	Drohne *f*
2709	drop method	Tröpfchenkultur *f*
2710	drought, dryness	Trockenheit *f*, Dürre *f*
2711	drought period	Trockenperiode *f*
2712	drought resistance	Dürrefestigkeit *f*, ~resistenz *f*, ~härte *f*, Trockenresistenz *f*
2713	drought resistant	dürrefest, dürreresistent
2714	drupe → stone fruit	—
2715	dry culture medium	Trockennährboden *m*
2716	dry extract	Trockenextrakt *m*
2717	dry forest	Trockenwald *m*
2718	dry fruit	Trockenfrucht *f*
2719	dry mount, ~ preparation	Trockenpräparat *n*
2720	dry season	Trockenzeit *f*
2721	dry substance, ~ matter	Trockensubstanz *f*, ~masse *f*
2722	dry weight	Trockengewicht *n*
2723	drying → dessication	—
2724	dryness → drought	—
2725	ductless gland → endocrine gland	—
2726	dug → teat	—
2727	dump(ing) site	(Müll)deponie *f*
2728	duodenum	Zwölffingerdarm *m*, Duodenum *n*
2729	duplex	duplex
2730	duplicate	Duplikat *n*
2731	to duplicate, to double	(sich) verdoppeln, duplizieren
2732	duplication, doubling	Verdoppelung *f*, Duplikation *f*
2733	dura mater	harte Gehirnhaut *f*
2734	duramen, heart-wood	Kernholz *n*, Hartholz *n*
2735	duration of germination	Keimdauer *f*
2736	dwarf mutant	Zwergmutante *f*
2737	dwarf plant	Zwergpflanze *f*
2738	dwarfism → nanism	—
2739	dyad	Dyade *f*
2740	dye → stain	—
2741	dysphotic zone	dysphotische Zone *f*
2742	dysploidy	Dysploidie *f*
2743	dystrophic	dystroph
2744	ear *bot.* → spike	—
2745	ear drum → tympanic membrane	—
2746	ear ossicle, auditory ~	Gehörknöchelchen *n*
2747	early wood → primary ~	—

2685	bouclier *m* (dorsal), carapace *f* dorsale, dossière *f*	escudo *m* dorsal, caparazón *m* ~, espaldar *m*
2686	dorsiventralité *f*	dorsiventralidad *f*
2687	compensation *f* de dosage	compensación *f* de dosificación
2688	effet *m* de dose	efecto *m* de la dosis
2689	courbe *f* de l'effet de dose, ~ dose-effet	curva *f* del efecto de la dosis
2690	—	—
2691	double liaison *f*	doble enlace *m*
2692	croisement *m* double	cruzamiento *m* doble, ~ de cuatro líneas
2693	double crossing-over *m*	entrecruzamiento *m* doble
2694	double dominant *m*	doble dominante *m*
2695	double fertilisation *f*, ~ fécondation	fertilización *f* doble, fecundación *f* ~
2696	double hélice *f*, ~ spirale *f*	doble hélice *f*
2697	—	—
2698	—	—
2699	période *f* de duplication	tiempo *m* de duplicación
2700	duvet *m*	plumón *m*
2701	plumule *f*	plúmula *f*
2702	—	—
2703	cyme *f* unipare scorpioïde	drepanio *m*
2704	eau *f* potable	agua *f* potable
2705	approvisionnement *m* en eau potable	abastecimiento *m* de agua potable
2706	pulsion *f*, impulsion *f*	impulso *m*, pulsión *f*
2707	instinct *m* combatif, impulsion *f* combative	instinto *m* combativo, ~ de lucha
2708	faux-bourdon *m*, abeille *f* mâle	zángano *m*
2709	méthode *f* de goutte	método *m* de las gotitas
2710	sécheresse *f*	sequía *f*, sequedad *f*
2711	période *f* de sécheresse	período *m* de sequía
2712	résistance *f* à la sécheresse	resistencia *f* a la sequía
2713	résistant à la sécheresse	resistente a la sequía
2714	—	—
2715	milieu *m* de culture sec	medio *m* de cultivo seco
2716	extrait *m* sec	extracto *m* seco
2717	forêt *f* sèche, ~ xérophile	bosque *m* seco, ~ xerofítico
2718	fruit *m* sec	fruto *m* seco
2719	préparation *f* sèche	preparación *f* en seco
2720	saison *f* sèche	estación *f* de sequía
2721	substance *f* sèche, matière *f* ~	materia *f* seca
2722	poids *m* sec	peso *m* seco
2723	—	—
2724	—	—
2725	—	—
2726	—	—
2727	dépôt *m* d'ordures	vertedero *m* de basuras
2728	duodénum *m*	duodeno *m*
2729	duplex	duplexo
2730	duplicata *m*	duplicado *m*
2731	doubler	duplicarse, desdoblarse
2732	doublement *m*, duplication *f*	desdoblamiento *m*, duplicación *f*
2733	dure-mère *f*	duramadre *f*
2734	duramen *m*, bois *m* de cœur, ~ parfait	duramen *m*, corazón *m* de la madera
2735	durée *f* germinative	duración *f* de la germinación
2736	mutant *m* nain	mutante *m* enano
2737	plante *f* naine	planta *f* enana
2738	—	—
2739	dyade *f*	díada *f*
2740	—	—
2741	zone *f* dysphotique, ~ crépusculaire	zona *f* disfótica
2742	dysploïdie *f*	disploidia *f*
2743	dystrophique	distrófico
2744	—	—
2745	—	—
2746	osselet *m* de l'oreille	huesecillo *m* del oído
2747	—	—

2748	eccrine gland	ekkrine Drüse *f*
2749	ecdysial... → moulting...	—
2750	ecdysis → moult(ing)	—
2751	ecdysone, moulting hormone	Häutungshormon *n*, Ekdyson *n*
2752	echinate → prickly	—
2753	Echinodermata, echinoderms	Stachelhäuter *mpl*, Echinodermen *pl*
2754	echolocation	Echolotung *f*, Echoortung *f*
2755	ecobiotic	ökobiotisch
2756	eco-catastrophe → environmental disaster	—
2757	ecoclimatic	ökoklimatisch
2758	ecogeographic	ökogeographisch
2759	ecological	ökologisch
2760	ecological balance, ~ equilibrium	ökologisches Gleichgewicht *n*
2761	ecological dominant	ökologische Dominante *f*
2762	ecological indicator	ökologischer Indikator *m*
2763	ecological isolation	ökologische Isolation *f*
2764	ecological niche	ökologische Nische *f*
2765	ecological race	ökologische Rasse *f*
2766	ecological succession	ökologische Sukzession *f*
2767	ecological system → ecosystem	—
2768	ecological valence	ökologische Valenz *f*
2769	ecologist	Ökologe *m*
2770	ecology	Ökologie *f*
2771	ecomorphosis	Ökomorphose *f*
2772	ecophysiology, physiological ecology	Ökophysiologie *f*
2773	ecospecies	Ökospezies *f*
2774	ecosphere	Ökosphäre *f*
2775	ecosystem, ecological system	Ökosystem *n*, ökologisches System *n*
2776	ecotone	Ökoton *n*
2777	ecotype, habitat type	Ökotyp(us) *m*
2778	ecto... see also exo...	—
2779	ectoderm, ectoblast, epiblast	Ektoderm *n*, Ektoblast *m*, Epiblast *m*, Exoderm *n*, äußeres Keimblatt *n*
2780	ectoenzyme, exoenzyme	Ektoenzym *n*, Exoenzym *n*
2781	ectogenesis	Ektogenese *f*
2782	ectolecithal	ektolezithal
2783	ectoparasite, ectosite, ectozoon, external parasite	Außenschmarotzer *m*, Ektoparasit *m*
2784	ectoplasm, exoplasm	Ektoplasma *n*
2785	Ectoprocta	Ectoprocta *pl*
2786	ectotrophic	ektotroph
2787	ectozoon → ectoparasite	—
2788	edaphic	edaphisch
2789	edaphic climax	Bodenklimax *f*, edaphische Klimax *f*
2790	edaphic conditions → soil ~	—
2791	edaphic factor	edaphischer Faktor *m*
2792	edaphology → soil science	—
2793	edaphon, soil organisms	Bodenorganismen *mpl*, Bodenlebewelt *f*, Edaphon *n*
2794	effectiveness	Wirksamkeit *f*
2795	effector (organ)	Effektor *m*, Erfolgsorgan *n*
2796	efference	Efferenz *f*
2797	efferent	efferent
2798	efficiency	Leistungsfähigkeit *f*
2799	egg	Ei *n*, Ovum *n*
2800	egg albumin → ovalbumin	—
2801	egg-apparatus	Eiapparat *m*
2802	egg-cell → ovum	—
2803	egg-laying → oviposition	—
2804	egg-membrane, egg-envelope	Eihaut *f*, Eihülle *f*
2805	egg-nucleus	Eikern *m*
2806	egg-rolling movement	Eirollbewegung *f*
2807	egg-shell	Ei(er)schale *f*
2808	egg-tube → ovariole	—
2809	egg-white → albumen[1]	—

2748	glande *f* eccrine	glándula *f* ecrina
2749	—	—
2750	—	—
2751	ecdysone *f*, hormone *f* de la mue	ecdisona *f*, hormona *f* de la muda
2752	—	—
2753	échinodermes *mpl*	equinodermos *mpl*
2754	écholocation *f*	ecolocalización *f*
2755	écobiotique	ecobiótico
2756	—	—
2757	écoclimatique	ecoclimático
2758	écogéographique	ecogeográfico
2759	écologique	ecológico
2760	équilibre *m* écologique	equilibrio *m* ecológico
2761	dominant *m* écologique	dominante *m* ecológico
2762	indicateur *m* écologique	indicador *m* ecológico
2763	isolement *m* écologique	aislamiento *m* ecológico
2764	niche *f* écologique	nicho *m* ecológico
2765	race *f* écologique	raza *f* ecológica
2766	succession *f* écologique	sucesión *f* ecológica
2767	—	—
2768	valence *f* écologique	valencia *f* ecológica
2769	écologiste *m*, écologue *m*	ecólogo *m*
2770	écologie *f*	ecología *f*
2771	écomorphose *f*	ecomorfosis *f*
2772	écophysiologie *f*, écologie *f* physiologique	ecofisiología *f*, ecología *f* fisiológica
2773	écoespèce *f*	ecospecie *f*
2774	écosphère *f*	ecosfera *f*
2775	écosystème *m*, système *m* écologique	ecosistema *m*, sistema *m* ecológico
2776	écotone *m*	ecotono *m*
2777	écotype *m*	ecotipo *m*
2778	—	—
2779	ectoderme *m*, ectoblaste *m*, feuillet *m* externe	ectodermo *m*, ectoblasto *m*, epiblasto *m*, exodermo *m*, hoja *f* germinal externa
2780	ectoenzyme *m*	ectoenzima *f*, exoenzima *f*
2781	ectogenèse *f*	ectogenia *f*
2782	ectolécithe, ectolécithique	ectolecito, ectolecítico
2783	ectoparasite *m*	ectoparásito *m*, ectozoo *m*
2784	ectoplasme *m*	ectoplasma *m*, exoplasma *m*
2785	ectoproctes *mpl*, bryozoaires *mpl* vrais	ectoproctos *mpl*
2786	ectotrophe	ectótrofo
2787	—	—
2788	édaphique	edáfico
2789	climax *m* édaphique, pédoclimax *m*	clímax *m* edáfico
2790	—	—
2791	facteur *m* édaphique	factor *m* edáfico
2792	—	—
2793	édaphon *m*	edafón *m*
2794	efficacité *f*	eficacidad *f*
2795	(organe *m*) effecteur *m*	(órgano *m*) efector *m*
2796	efférence *f*	sistema *m* eferente
2797	efférent	eferente
2798	efficience *f*	eficiencia *f*
2799	œuf *m*	huevo *m*
2800	—	—
2801	appareil *m* de ponte	aparato *m* ovular
2802	—	—
2803	—	—
2804	enveloppe *f* de l'œuf, ~ ovarienne	membrana *f* del huevo, ~ ovular
2805	noyau *m* ovulaire	núcleo *m* ovular, ovonúcleo *m*
2806	rattrapage *m* des œufs	movimiento *m* de rotación del huevo, ~ de rodar el huevo
2807	coquille *f* (de l'œuf)	cáscara *f* del huevo
2808	—	—
2809	—	—

2810	ejaculate	Ejakulat *n*
2811	ejaculation	Ejakulation *f*, Samenerguß *m*
2812	ejaculatory duct	Samengang *m*
2813	elaioplast, elaiosome	Ölkörperchen *n*, Elaioplast *m*, Elaiosom *n*
2814	Elasmobranchii, elasmobranchs	Elasmobranchier *mpl*, Haiartige *mpl*
2815	elastin	Elastin *n*
2816	elater	Elatere *f*
2817	electric organ	elektrisches Organ *n*
2818	electric stimulus	elektrischer Reiz *m*, Elektroreiz *m*
2819	electric titration	Elektrotitration *f*
2820	electrical brain stimulation	elektrische Hirnreizung *f*
2821	electrical potential	elektrisches Potential *n*
2822	electrocardiogram, ECG	Elektrokardiogramm *n*, EKG
2823	electrocardiography	Elektrokardiographie *f*
2824	electrode	Elektrode *f*
2825	electrodialyse	Elektrodialyse *f*
2826	electroencephalogram, EEG	Elektroenzephalogramm *n*, EEG
2827	electroencephalography	Elektroenzephalographie *f*
2828	electrolyt	Elektrolyt *m*
2829	electromyogram	Elektromyogramm *n*, EMG
2830	electromyography	Elektromyographie *f*
2831	electron	Elektron *n*
2832	electron carrier	Elektronenträger *m*
2833	electron image	Elektronenbild *n*
2834	electronmicrograph	elektronenmikrographische Aufnahme *f*
2835	electron microscope	Elektronenmikroskop *n*
2836	electron microscopy	Elektronenmikroskopie *f*
2837	electron radiation	Elektronenstrahlung *f*
2838	electron ray	Elektronenstrahl *m*
2839	electron transport, ~ transfer	Elektronentransport *m*
2840	electron transport chain	Elektronentransportkette *f*
2841	electron transport particle, ETP	Elektronen-Transport-Partikel *f*, ETP
2842	electron transport system, ETS	Elektronentransportsystem *n*
2843	electronics	Elektronik *f*
2844	electro-osmosis, electric osmosis	Elektroosmose *f*
2845	electrophoresis	Elektrophorese *f*
2846	electrophysiology	Elektrophysiologie *f*
2847	electroplate, electroplax *zool.*	Elektroplatte *f*, ~plaxe *f* *zool.*
2848	electrotonus	Elektrotonus *m*
2849	electrotropism → galvanotropism	—
2850	elementary body	Elementarkörperchen *n*
2851	elementary particle, fundamental ~	Elementarteilchen *n*
2852	to eliminate, to cull	ausmerzen
2853	elimination; culling	Ausmerzen *n*; Elimination *f* *gen.*
2854	ellipsoid joint	Ellipsoidgelenk *n*
2855	elongation	Streckungswachstum *n*
2856	elongation zone	Streckungszone *f*
2857	eluate	Eluat *n*
2858	elution	Elution *f*
2859	eluvial horizon → A-horizon	—
2860	elytron (*pl* elytra)	Deckflügel *m*
2861	emasculation → castration	—
2862	to embed	einbetten
2863	embedding	Einbettung *f*
2864	embedding medium	Einbettungsmasse *f*, ~mittel *n*
2865	embole, emboly *embr.*	Embolie *f* *embr.*
2866	embryo	Embryo *m*; Keimling *m* *bot.*
2867	embryo cell, embryonic ~	Embryonalzelle *f*
2868	embryo-sac	Embryosack *m*
2869	embryo-sac mother-cell, megasporocyte, megaspore mother-cell	Embryosackmutterzelle *f*
2870	embryogenesis, embryogeny	Embryogenese *f*, Embryogenie *f*, Keimesentwicklung *f*
2871	embryoid	Embryoid *m*

2810	sperme *m* éjaculé	eyaculado *m*
2811	éjaculation *f*	eyaculación *f*
2812	canal *m* éjaculateur	conducto *m* eyaculador
2813	élaïosome *m*, corpuscule *m* huileux	elaioplasto *m*, eleoplasto *m*, cuerpo *m* oleoso
2814	élasmobranches *mpl*	elasmobranquios *mpl*
2815	élastine *f*	elastina *f*
2816	élatère *f*	elater *m*
2817	organe *m* électrique, ~ électrogène	órgano *m* eléctrico
2818	stimulus *m* électrique	estímulo *m* eléctrico, electroestímulo *m*
2819	électrotitration *f*	titulación *f* eléctrica
2820	stimulation *f* électrique du cerveau	estimulación *f* eléctrica del cerebro
2821	potentiel *m* électrique	potencial *m* eléctrico
2822	électrocardiogramme *m*, ECG	electrocardiograma *m*, ECG
2823	électrocardiographie *f*	electrocardiografía *f*
2824	électrode *f*	electrodo *m*
2825	électrodialyse *f*	electrodiálisis *f*
2826	électroencéphalogramme *m*, EEG	electroencefalograma *m*, EEG
2827	électroencéphalographie *f*	electroencefalografía *f*
2828	électrolyte *m*	electrolito *m*, electrólito *m*
2829	électromyogramme *m*	electromiograma *m*
2830	électromyographie *f*	electromiografía *f*
2831	électron *m*	electrón *m*
2832	transporteur *m* d'électrons	(trans)portador *m* de electrones
2833	électronographie *f*	electronografía *f*
2834	micrographie *f* électronique	micro(foto)grafía *f* electrónica
2835	microscope *m* électronique	microscopio *m* electrónico
2836	microscopie *f* électronique	microscopia *f* electrónica, electronomicroscopia *f*
2837	radiation *f* électronique	radiación *f* electrónica
2838	rayon *m* électronique	rayo *m* de electrones
2839	transport *m* d'électrons, transfert *m* ~	transporte *m* de electrones
2840	chaîne *f* de transport d'électrons	cadena *f* transportadora de electrones
2841	electron transport particle	partícula *f* transportadora de electrones, PTE
2842	système *m* transporteur d'électrons	sistema *m* transportador de electrones
2843	électronique *f*	electrónica *f*
2844	électro-osmose *f*	electroosmosis *f*
2845	électrophorèse *f*	electroforesis *f*
2846	électrophysiologie *f*	electrofisiología *f*
2847	électroplaque *f* *zool.*	electroplaca *f* *zool.*
2848	électrotonus *m*	electrotono *m*
2849	—	—
2850	corps *m* élémentaire	corpúsculo *m* elemental
2851	particule *f* élémentaire	partícula *f* elemental, ~ fundamental
2852	éliminer	eliminar
2853	élimination *f*	eliminación *f*
2854	articulation *f* ellipsoïde	articulación *f* elipsoidea, ~ condílea
2855	croissance *f* par allongement	crecimiento *m* por elongación
2856	zone *f* d'élongation, ~ d'allongement	zona *f* de alargamiento
2857	éluat *m*	eluato *m*
2858	élution *f*	elución *f*
2859	—	—
2860	élytre *m/f*	élitro *m*
2861	—	—
2862	inclure	incluir
2863	inclusion *f*	inclusión *f*
2864	masse *f* d'inclusion, milieu *m* ~	medio *m* de inclusión
2865	embolie *f* *embr.*	embolia *f* *embr.*
2866	embryon *m*	embrión *m*
2867	cellule *f* embryonnaire	célula *f* embrional
2868	sac *m* embryonnaire	saco *m* embrional, ~ embrionario
2869	cellule *f* mère du sac embryonnaire, mégasporocyte *m*	célula *f* madre del saco embrional, megasporocito *m*
2870	embryogénie *f*, embryogenèse *f*	embriogénesis *f*, embriogenia *f*
2871	embryoïde *m*	embrioide *m*

2872	embryologist	Embryologe *m*
2873	embryology	Embryologie *f*
2874	embryonal knot	Embryonalknoten *m*
2875	embryonic	embryonal
2876	embryonic determination	embryonale Determination *f*
2877	embryonic development	Embryonalentwicklung *f*
2878	embryonic induction	embryonale Induktion *f*
2879	embryonic membrane, extraembryonic ~, foetal ~	Embryonalhülle *f*, Keimhülle *f*, Fruchthülle *f*
2880	embryonic shield	Embryonalschild *m*
2881	embryonic stage	Embryonalstadium *n*
2882	embryonic tissue	Embryonalgewebe *n*
2883	embryophytes	Embryophyten *mpl*
2884	embryotrophe	Embryotrophe *f*, Keimlingsnahrung *f*
2885	emergence	Emergenz *f*
2886	emission *env.*	Emission *f env.*
2887	emulsification	Emulgierung *f*
2888	to emulsify	emulgieren
2889	emulsifyer	Emulgator *m*
2890	emulsin	Emulsin *n*
2891	emulsion	Emulsion *f*
2892	emulsoid	Emulsoid *n*
2893	enamel	Zahnschmelz *m*
2894	encephalic → cerebral	—
2895	encephalon → brain	—
2896	encephalization	Enzephalisierung *f*
2897	to encyst	sich einkapseln
2898	encystation, encystment	Einkapseln *n*, Enzystierung *f*, Zystenbildung *f*
2899	endbrain → telencephalon	—
2900	end cell, terminal ~	Endzelle *f*
2901	endemic	endemisch
2902	endemic species	Endemit *m*
2903	endemism	Endemismus *m*
2904	endergonic	endergonisch, energieverbrauchend
2905	endobiosis	Endobiose *f*
2906	endoblast → endoderm	—
2907	endocardium	Herzinnenhaut *f*, Endokard(ium) *n*
2908	endocarp	Endokarp *m*
2909	endocranium	Endokranium *n*
2910	endocrine gland, ductless ~, incretory ~	endokrine Drüse *f*, inkretorische ~, innersekretorische ~
2911	endocrine secretion → incretion	—
2912	endocrinologist	Endokrinologe *m*
2913	endocrinology	Endokrinologie *f*
2914	endoderm, entoderm, endoblast, entoblast, hypoblast	Entoderm *n*, Entoblast *m*, Hypoblast *m*, inneres Keimblatt *n*
2915	endodermis *bot.*	Endodermis *f bot.*
2916	endoenzyme	Endoenzym *n*, Endoferment *n*
2917	endogamic, endogamous	endogam
2918	endogamy	Endogamie *f*
2919	endogenic, endogenous	endogen
2920	endogenous rhythm	endogener Rhythmus *m*
2921	endogeny	Endogenese *f*
2922	endolymph	Endolymphe *f*
2923	endometrium	Endometrium *n*
2924	endomitosis	Endomitose *f*
2925	endomixis	Endomixis *f*
2926	endomysium	Endomysium *n*
2927	endoneurium	Endoneurium *n*
2928	endoparasite, entoparasite, endosite, internal parasite	Endoparasit *m*, Entoparasit *m*, Innenschmarotzer *m*
2929	endopeptidase	Endopeptidase *f*
2930	endophyte	Endophyt *m*
2931	endoplasm, entoplasm	Endoplasma *n*, Entoplasma *n*
2932	endoplasmic	endoplasmatisch
2933	endoplasmic reticulum	endoplasmatisches Retikulum *n*

2872	embryologiste *m*, embryologue *m*	embriólogo *m*
2873	embryologie *f*	embriología *f*
2874	bouton *m* embryonnaire	nódulo *m* embrionario
2875	embryonnaire	embrional, embrionario
2876	détermination *f* embryonnaire	determinación *f* embrionaria
2877	développement *m* embryonnaire	desarrollo *m* embrionario
2878	induction *f* embryonnaire	inducción *f* embrionaria
2879	enveloppe *f* embryonnaire, ~ foetale	membrana *f* (o: envoltura *f*) (extra)-embrionaria, ~ fetal
2880	écusson *m* embryonnaire	escudo *m* embrionario, rodete *m* ~
2881	stade *m* embryonnaire	estado *m* embrionario
2882	tissu *m* embryonnaire	tejido *m* embrionario
2883	embryophytes *mpl*	embriofitas *fpl*, embriófitos *mpl*
2884	embryotrophe *m*	embriótrofo *m*
2885	émergence *f*	emergencia *f*
2886	émission *f* *env.*	emisión *f* *env.*
2887	émulsionnement *m*, émulsification *f*	emulsificación *f*
2888	émulsionner	emulsionar, emulsificar
2889	émulsionneur *m*, agent *m* émulsionnant	emulsificador *m*
2890	émulsine *f*	emulsina *f*
2891	émulsion *f*	emulsión *f*
2892	émulsoïde *m*	emulsoide *m*
2893	émail *m*	esmalte *m* dental
2894	—	—
2895	—	—
2896	encéphalisation *f*	encefalización *f*
2897	s'enkyster	enquistarse
2898	enkystement *m*	enquistamiento *m*
2899	—	—
2900	cellule *f* terminale	célula *f* terminal
2901	endémique	endémico
2902	endémique *m*	especie *f* endémica, endémico *m*
2903	endémisme *m*	endemismo *m*
2904	endergonique	endergónico
2905	endobiose *f*	endobiosis *f*
2906	—	—
2907	endocarde *m*	endocardio *m*
2908	endocarpe *m*	endocarpo *m*
2909	endocrâne *m*	endocránco *m*
2910	glande *f* endocrine, ~ incrétoire, ~ à sécrétion interne	glándula *f* endocrina, ~ cerrada, ~ de secreción interna
2911	—	—
2912	endocrinologiste *m*	endocrinólogo *m*
2913	endocrinologie *f*	endocrinología *f*
2914	endoderme *m*, endoblaste *m*, hypoblaste *m*, feuillet *m* interne	endodermo *m*, entodermo *m*, endoblasto *m*, hipoblasto *m*
2915	endoderme *m* *bot.*	endodermis *f* *bot.*
2916	endoenzyme *m*	endoenzima *f*
2917	endogamique	endogámico
2918	endogamie *f*	endogamia *f*
2919	endogène	endógeno
2920	rythme *m* endogène	ritmo *m* endógeno
2921	endogenèse *f*	endogénesis *f*
2922	endolymphe *f*	endolinfa *f*
2923	endomètre *m*	endometrio *m*
2924	endomitose *f*	endomitosis *f*
2925	endomixie *f*	endomixis *f*
2926	endomysium *m*	endomisio *m*
2927	endonèvre *m*	endoneur(i)o *m*
2928	endoparasite *m*	endoparásito *m*, endosito *m*
2929	endopeptidase *f*	endopeptidasa *f*
2930	endophyte *m*	endófito *m*
2931	endoplasme *m*	endoplasma *m*
2932	endoplasm(at)ique	endoplásmico, endoplasmático
2933	réticulum *m* endoplasm(at)ique	retículo *m* endoplásmico, ~ endoplasmático

2934	endopolyploidy	Endopolyploidie *f*
2935	Endoprocta, endoprocts	Endoprocta *pl*, Kelchwürmer *mpl*
2936	Endopterygota → Holometabola	—
2937	end-organ, terminal organ	Endorgan *n*
2938	endosite → endoparasite	—
2939	endoskeleton, internal skeleton	Innenskelett *n*, Entoskelett *n*
2940	endosmosis	Endosmose *f*
2941	endosome	Endosom *n*
2942	endosperm	Endosperm *n*
2943	endospore	Endospore *f*
2944	endosporium	Endospor(ium) *n*
2945	endosteum	Endost *n*
2946	endostome	Endostom *n*
2947	endostyle	Endostyl *n*
2948	endosymbiosis	Endosymbiose *f*
2949	endothecium	Endothezium *n*
2950	endothelial cell	Endothelzelle *f*
2951	endothelium	Endothel(ium) *n*
2952	endothermic	endotherm
2953	endotoxin	Endotoxin *n*
2954	endotrophic	endotroph
2955	end-plate *neur.*	Endplatte *f neur.*
2956	end-plate potential	Endplattenpotential *n*
2957	end product, final ~	Endprodukt *n*
2958	energic nucleus → metabolic ~	—
2959	energid	Energide *f*
2960	energy consumption	Energieverbrauch *m*
2961	energy demand, ~ requirement, energetic ~	Energiebedarf *m*
2962	energy flow	Energiefluß *m*
2963	energy level	Energieniveau *n*
2964	energy loss	Energieverlust *m*
2965	energy metabolism	Energiestoffwechsel *m*, Betriebs~ *m*
2966	energy potential	Energiepotential *n*
2967	energy production	Energieerzeugung *f*
2968	energy requirement → energy demand	—
2969	energy-rich bond	energiereiche Bindung *f*
2970	energy source	Energiequelle *f*
2971	energy storage, accumulation of energy	Energiespeicherung *f*
2972	energy transfer	Energieübertragung *f*
2973	energy transformation	Energieumformung *f*
2974	energy-yielding → exergonic	—
2975	engram	Engramm *n*, Gedächtnisinhalt *m*
2976	to enrich	anreichern
2977	enriched horizon → B-horizon	—
2978	enrichment	Anreicherung *f*
2979	enrichment culture, concentrating ~	Anreicherungskultur *f*
2980	enrichment experiment	Anreicherungsversuch *m*
2981	entelechy	Entelechie *f*
2982	enteramine → serotonine	—
2983	enteric → intestinal	—
2984	enteric cavity → gut ~	—
2985	enterocoel	Enterozöl *n*
2986	enterocrinin	Enterokrinin *n*
2987	enterogastrone	Enterogastron *n*
2988	enteropeptidase, enterokinase *obs.*	Enteropeptidase *f*, Enterokinase *f obs.*
2989	enterozoon, enterosite, intestinal parasite	Darmparasit *m*, ~schmarotzer *m*, Enterozoon *n*
2990	enthalpy	Enthalpie *f*
2991	entire (leaf)	glattrandig, ganzrandig (Blatt)
2992	ento ... see also endo ...	—
2993	entomogamy → entomophily	—
2994	entomologist	Entomologe *m*, Insektenforscher *m*
2995	entomology	Entomologie *f*, Insektenkunde *f*
2996	entomophage → insectivorous	—

2934	endopolyploïdie *f*	endopoliploidia *f*
2935	endoproctes *mpl*, kamptozoaires *mpl*	endoproctos *mpl*, camptozoos *mpl*
2936	—	—
2937	endorgane *m*	órgano *m* terminal
2938	—	—
2939	endosquelette *m*	endo(e)squeleto *m*, esqueleto *m* interno
2940	endosmose *f*	endosmosis *f*
2941	endosome *m*	endosoma *m*
2942	endosperme *m*	endosperma *m*, endospermo *m*
2943	endospore *f*	endóspora *f*
2944	endosporium *m*	endosporio *m*
2945	endoste *m*	endostio *m*
2946	endostome *f*	endóstoma *m*
2947	endostyle *m*	endostilo *m*
2948	endosymbiose *f*	endosimbiosis *f*
2949	endothécium *m*	endotecio *m*
2950	cellule *f* endothéliale	célula *f* endotelial
2951	endothélium *m*	endotelio *m*
2952	endothermique	endotérmico
2953	endotoxine *f*	endotoxina *f*
2954	endotrophe, endotrophique	endótrofo
2955	plaque *f* terminale *neur.*	placa *f* terminal *neur.*
2956	potentiel *m* de plaque terminale	potencial *m* de placa terminal
2957	produit *m* final, ~ terminal	producto *m* final
2958	—	—
2959	énergide *f*	enérgida *f*
2960	consommation *f* d'énergie, ~ énergétique	consumo *m* de energía
2961	demande *f* d'énergie, besoins *mpl* énergétiques, exigences *fpl* ~	necesidades *fpl* energéticas, requerimientos *mpl* energéticos
2962	flux *m* d'énergie	flujo *m* de energía, ~ energético
2963	niveau *m* d'énergie, ~ énergétique	nivel *m* de energía, ~ energético
2964	perte *f* d'énergie, ~ énergétique	pérdida *f* de energía
2965	métabolisme *m* énergétique	metabolismo *m* energético
2966	potentiel *m* énergétique	potencial *m* energético
2967	production *f* d'énergie, ~ énergétique	producción *f* de energía
2968	—	—
2969	liaison *f* riche en énergie	enlace *m* rico en energía
2970	source *f* d'énergie, ~ énergétique	fuente *f* de energía, ~ energética
2971	accumulation *f* d'énergie, mise *f* en réserve ~	almacenamiento *m* de energía, ~ energético
2972	transmission *f* d'énergie, transfert *m* ~	transmisión *f* de energía, transferencia *f* ~ ~
2973	transformation *f* d'énergie, ~ énergétique	transformación *f* de energía, ~ energética
2974	—	—
2975	engramme *m*	engrama *m*
2976	enrichir	enriquecer
2977	—	—
2978	enrichissement *m*	enriquecimiento *m*
2979	culture *f* sur milieu concentrant	cultivo *m* de enriquecimiento
2980	essai *m* d'enrichissement	experimento *m* de enriquecimiento
2981	entéléchie *f*	entelequia *f*
2982	—	—
2983	—	—
2984	—	—
2985	entérocoele *m*	enterocele *m*
2986	entérocrinine *f*	enterocrinina *f*
2987	entérogastrone *f*	enterogastrona *f*
2988	entéropeptidase *f*, entérokinase *f*	enteropeptidasa *f*, enteroquinasa *f* *obs.*
2989	entérozoaire *m*, parasite *m* intestinal	enterozoo *m*, enterosito *m*, parásito *m* intestinal
2990	enthalpie *f*	entalpía *f*
2991	(à bord) entier	(de borde) entero
2992	—	—
2993	—	—
2994	entomologiste *m*, entomologue *m*	entomólogo *m*
2995	entomologie *f*	entomología *f*
2996	—	—

2997	entomophilous	insektenblütig, entomogam
2998	entomophilous flower	Insektenblume *f*
2999	entomophily, entomogamy, insect pollination	Insektenblütigkeit *f*, ~bestäubung *f*, Entomophilie *f*, Entomogamie *f*
3000	entozoon (*pl* entozoa)	Entozoon *n*
3001	entropy	Entropie *f*
3002	to enucleate	entkernen (Zellen etc.)
3003	enucleation	Entkernung *f*
3004	environment	Umwelt *f*
3005	environmental	Umwelt ...
3006	environmental adaptation, adaptation to environment	Umweltanpassung *f*
3007	environmental alteration, ~ modification	Umweltveränderung *f*
3008	environmental biology	Umweltbiologie *f*
3009	environmental change	Umweltänderung *f*
3010	environmental conditions	Umweltbedingungen *fpl*
3011	environmental contamination, ~ pollution	Umweltverseuchung *f*, ~verschmutzung *f*
3012	environmental degradation, deterioration of environment	Verschlechterung *f* der Umweltverhältnisse
3013	environmental disaster, eco-catastrophe	Umweltkatastrophe *f*, Ökokatastrophe *f*
3014	environmental disturbance	Umweltstörung *f*
3015	environmental factor	Umweltfaktor *m*
3016	environmental hygiene	Umwelthygiene *f*
3017	environmental influence	Umwelteinfluß *m*
3018	environmental legislation	Umweltgesetzgebung *f*
3019	environmental noise → ambient ~	—
3020	environmental planning	Umweltplanung *f*
3021	environmental policy	Umweltpolitik *f*
3022	environmental pollution → ~ contamination	—
3023	environmental protection, protection of environment	Umweltschutz *m*
3024	environmental quality	Umweltqualität *f*
3025	environmental resistance	Umweltwiderstand *m*
3026	environmental stimulus	Umweltreiz *m*
3027	environmentalist, environmental scientist	Umweltspezialist *m*, ~forscher *m*, Fachmann *m* für Umweltfragen
3028	enzymatic	enzymatisch
3029	enzyme, ferment	Enzym *n*, Ferment *n*
3030	enzyme activity	Enzymaktivität *f*
3031	enzyme catalysis	Enzymkatalyse *f*
3032	enzyme-catalyzed reaction	enzymkatalisierte Reaktion *f*
3033	enzyme induction	Enzyminduktion *f*
3034	enzyme inhibition	Enzymhemmung *f*
3035	enzyme inhibitor	Enzymhemmer *m*, Enzyminhibitor *m*
3036	enzyme precursor → zymogen	—
3037	enzyme protein	Enzymprotein *n*, Fermentprotein *n*
3038	enzyme reaction	Enzymreaktion *f*
3039	enzyme repression	Enzymrepression *f*
3040	enzyme specifity	Enzymspezifität *f*
3041	enzyme-substrate complex	Enzym-Substrat-Komplex *m*
3042	enzyme synthesis, enzymatic ~	Enzymsynthese *f*
3043	enzymology	Enzymologie *f*
3044	Eocene	Eozän *n*
3045	eosin	Eosin *n*
3046	eosinocyte, eosinophil leucocyte	eosinophiler Leukozyt *m*
3047	eosinophil	eosinophil
3048	ependyma	Ependym *n*
3049	ephapse	Ephapse *f*
3050	epharmosis, cpharmony	Epharmonie *f*, Epharmose *f*

2997	entomophile	entomófilo, entomógamo
2998	fleur *f* entomophile	flor *f* entomófila
2999	entomophilie *f*, pollinisation *f* entomophile	entomofilia *f*, polinización *f* entomófila
3000	entozoaire *m*	entozoo *m*, entozoario *m*
3001	entropie *f*	entropía *f*
3002	énucléer	enuclear, denuclear
3003	énucléation *f*	enucleación *f*, denucleación *f*
3004	environnement *m*, milieu *m* environnant, ~ ambiant	medio *m* ambiente
3005	ambiant	ambiental
3006	adaptation *f* au milieu, ~ à l'environnement	adaptación *f* al (medio) ambiente
3007	altération *f* du milieu, modification *f* ~ ~	alteración *f* del medio ambiente, ~ ecológica
3008	biologie *f* de l'environnement	biología *f* del medio ambiente
3009	changement *m* de milieu	cambio *m* ambiental
3010	conditions *fpl* du milieu, ~ ambiantes, ~ environnantes	condiciones *fpl* del medio, ~ ambientales
3011	pollution *f* du milieu, ~ de l'environnement	polución *f* del medio ambiente, contaminación *f* ambiental
3012	détérioration *f* de l'environnement, dégradation *f* ~ ~	deterioro *m* del medio ambiente, degradación *f* ~ ~ ~, deterioración *f* ambiental
3013	catastrophe *f* écologique	catástrofe *f* ecológica, desastre *m* ecológico
3014	perturbation *f* du milieu	perturbación *f* del medio ambiente, trastorno *m* ecológico
3015	facteur *m* du milieu, ~ ambiant	factor *m* ambiental
3016	hygiène *f* de l'environnement	higiene *f* del medio ambiente, ~ ambiental
3017	influence *f* du milieu	influencia *f* del medio ambiente, ~ ambiental
3018	législation *f* relative à l'environnement	legislación *f* sobre el medio ambiente
3019	—	—
3020	planification *f* de l'environnement, aménagement *m* ~ ~	planificación *f* del medio ambiente, ordenación *f* ~ ~ ~
3021	politique *f* de l'environnement	política *f* ambiental, ~ del medio ambiente
3022	—	—
3023	protection *f* de l'environnement, défense *f* ~ ~	protección *f* del medio ambiente, defensa *f* ~ ~ ~, protección *f* ambiental
3024	qualité *f* de l'environnement	calidad *f* del medio (ambiente)
3025	résistance *f* du milieu, ~ de l'environnement	resistencia *f* ambiental
3026	stimulus *m* environnant	estímulo *m* ambiental
3027	expert *m* en matière d'environnement, environnementaliste *m*	experto *m* en (problemas del) medio ambiente
3028	enzymatique	enzimático
3029	enzyme *m/f*, ferment *m*	enzima *m/f*, fermento *m*
3030	activité *f* enzymatique	actividad *f* enzimática
3031	catalyse *f* enzymatique	catálisis *f* enzimática
3032	réaction *f* catalysée par des enzymes	reacción *f* enzimáticamente catalizada
3033	induction *f* enzymatique	inducción *f* enzimática
3034	inhibition *f* enzymatique	inhibición *f* enzimática
3035	inhibiteur *m* d'enzymes, poison *m* ~	inhibidor *m* enzimático
3036	—	—
3037	protéine *f* enzymatique	proteína *f* enzimática, enzimoproteína *f*
3038	réaction *f* enzymatique	reacción *f* enzimática
3039	répression *f* enzymatique, ~ fermentative	represión *f* enzimática
3040	spécificité *f* enzymatique	especificidad *f* enzimática
3041	complexe *m* enzyme-substrat	complejo *m* enzima-substrato
3042	synthèse *f* enzymatique	síntesis *f* enzimática
3043	enzymologie *f*	enzimología *f*
3044	éocène *m*	eoceno *m*
3045	éosine *f*	eosina *f*
3046	éosinocyte *m*	eosinocito *m*, leucocito *m* eosinófilo
3047	éosinophile	eosinófilo
3048	épendyme *m*	epéndimo *m*
3049	éphapse *f*	contacto *m* efáptico
3050	épharmonie *f*	efarmonía *f*

No.	English	German
3051	ephedrine	Ephedrin *n*
3052	ephemeral, ephemer	Ephemere *f*
3053	epiblast → ectoderm	—
3054	epiblem → rhizodermis	—
3055	epibole, epiboly	Epibolie *f*
3056	epicardium	Epikard(ium) *n*
3057	epicarp → exocarp	—
3058	epicotyl	Epikotyl *n*
3059	epidemiology	Epidemiologie *f*, Seuchenlehre *f*
3060	epidermal cell	Oberhautzelle *f*, Epidermiszelle *f*
3061	epidermis *anat., bot.*	Epidermis *f*, Oberhaut *f* *anat., bot.*
3062	epididymis	Nebenhoden *m*
3063	epigastrium	Epigastrium *n*, Oberbauch *m*
3064	epigeal	epigäisch, oberirdisch
3065	epigenesis	Epigenese *f*
3066	epigenetics	Epigenetik *f*
3067	epiglottis	Kehldeckel *m*
3068	epigynous	epigyn
3069	epilimnion	Epilimnion *n*
3070	epilithic plant → rupicolous ~	—
3071	epimorphosis	Epimorphose *f*
3072	epimysium	Epimysium *n*
3073	epinasty	Epinastie *f*
3074	epinephrine → adrenaline	—
3075	epineurium	Epineurium *n*
3076	epipelagic zone	Epipelagial *n*
3077	epiphyllous	epiphyll
3078	epiphyseal cartilage, epiphysial ~	Epiphysenknorpel *m*
3079	epiphyseal eye → parietal ~	—
3080	epiphysis^{1}	Epiphyse *f* (Knochen)
3081	epiphysis^{2}, pineal gland, ~ body	Epiphyse *f*, Zirbeldrüse *f*
3082	epiphytes	Epiphyten *mpl*, Überpflanzen *fpl*
3083	epiphytic	epiphytisch
3084	epiploon → omentum	—
3085	episome	Episom *n*
3086	episperm → seed coat	—
3087	epistasis, epistasy	Epistasie *f*, Epistasis *f*
3088	epithecium	Epithezium *n*
3089	epithelial body → parathyroid	—
3090	epithelial cell	Epithelzelle *f*
3091	epithelial tissue	Epithelgewebe *n*
3092	epithelium	Epithel *n*
3093	epithem	Epithem *n*
3094	epizoic	epizoisch
3095	epizoon (*pl* epizoa)	Epizoon *n*
3096	equation division, equational ~	Äquationsteilung *f*
3097	equatorial body	Äquatorialkörper *m*
3098	equatorial plane	Äquatorialebene *f*
3099	equatorial plate, nuclear ~	Äquatorialplatte *f*, Kernplatte *f*
3100	erectile tissue → cavernous tissue	—
3101	erepsin, ereptase	Erepsin *n*
3102	ergastoplasm	Ergastoplasma *n*
3103	ergine, ergone	Ergin *n*, Ergon *n*
3104	ergosome → polyribosome	—
3105	ergosterin, ergosterol	Ergosterin *n*
3106	ergotinine	Ergotinin *n*
3107	erosion	Erosion *f*
3108	erythroblast	Erythroblast *m*
3109	erythrocyte, red blood corpuscle	Erythrozyt *m*, rotes Blutkörperchen *n*
3110	escape behavio(u)r	Fluchtverhalten *n*
3111	escape drive	Fluchttrieb *m*
3112	escape reaction, fugitive ~	Fluchtreaktion *f*
3113	esophagus → oesophagus	—
3114	essential amino acid	essentielle Aminosäure *f*
3115	essential oil, volatile ~, ethereal ~	ätherisches Öl *n*

3051	ephédrine *f*	efedrina *f*
3052	éphémère *f*	efímera *f*
3053	—	—
3054	—	—
3055	épibolie *f*	epibolia *f*
3056	épicardium *m*	epicardio *m*
3057	—	—
3058	épicotyle *m*	epicotilo, epicótilo *m*
3059	épidémiologie *f*	epidemiología *f*
3060	cellule *f* épidermique	célula *f* epidérmica
3061	épiderme *m* *anat.*, *bot.*	epidermis *f* *anat.*, *bot.*
3062	épididyme *m*	epidídimo *m*
3063	épigastre *m*	epigastrio *m*
3064	épigé	epigeo
3065	épigenèse *f*	epigénesis *f*
3066	épigénétique *f*	epigenética *f*
3067	épiglotte *f*	epiglotis *f*
3068	épigyne	epígino
3069	épilimnion *m*	epilimneo *m*
3070	—	—
3071	épimorphose *f*	epimorfosis *f*
3072	épimysium *m*	epimisio *m*
3073	épinastie *f*	epinastia *f*
3074	—	—
3075	épinèvre *m*	epineur(i)o *m*
3076	zone *f* épipélagique	zona *f* epipelágica
3077	épiphylle	epifilo
3078	cartilage *m* épiphysaire, ~ de conjugaison	cartílago *m* epifisario, ~ de conjunción
3079	—	—
3080	épiphyse *f*	epífisis *f*
3081	épiphyse *f*, glande *f* pinéale, corps *m* pinéal	epífisis *f*, glándula *f* pineal, cuerpo *m* ~
3082	épiphytes *mpl*	epífitos *mpl*, epífitas *fpl*
3083	épiphyte, épiphytique	epífito, epifítico
3084	—	—
3085	épisome *m*	episoma *m*
3086	—	—
3087	épistasie *f*	epistasia *f*, epistasis *f*
3088	épithécium *m*	epitecio *m*
3089	—	—
3090	cellule *f* épithéliale	célula *f* epitelial
3091	tissu *m* épithélial	tejido *m* epitelial
3092	épithélium *m*	epitelio *m*
3093	épithème *m*	epitema *m*
3094	épizoïque	epizoico
3095	épizoaire *m*	epizoo *m*, epizoario *m*
3096	division *f* équationnelle	división *f* ecuacional
3097	cellule *f* équatoriale	cuerpo *m* ecuatorial
3098	plan *m* équatorial	plano *m* ecuatorial
3099	plaque *f* équatoriale	placa *f* ecuatorial, ~ nuclear
3100	—	—
3101	érepsine *f*, éreptase *f*	erepsina *f*, ereptasa *f*
3102	ergastoplasme *m*	ergastoplasma *m*
3103	ergine *f*	ergina *f*
3104	—	—
3105	ergostérine *f*, ergostérol *m*	ergosterina *f*, ergosterol *m*
3106	ergotinine *f*	ergotinina *f*
3107	érosion *f*	erosión *f*
3108	érythroblaste *m*	eritroblasto *m*
3109	érythrocyte *m*, globule *m* rouge (du sang), hématie *f*	eritrocito *m*, glóbulo *m* rojo (de la sangre), hematíe *m*
3110	comportement *m* de fuite	comportamiento *m* de huida
3111	instinct *m* de fuite	instinto *m* de huida
3112	réaction *f* de fuite	reacción *f* de huida
3113	—	—
3114	amino-acide *m* essentiel	aminoácido *m* esencial
3115	huile *f* essentielle, ~ volatile, ~ éthérée	aceite *m* esencial, ~ etéreo, ~ volátil

3116	ester	Ester *m*
3117	esterase	Esterase *f*
3118	esterification	Veresterung *f*
3119	to esterify, to esterize	verestern
3120	estivation → aestivation	—
3121	estr … → oestr …	—
3122	ethanol → ethyl alcohol	—
3123	etheogenesis, ethiogenesis	Etheogenese *f*
3124	ethereal oil → essential oil	—
3125	ethmoid(al) bone	Siebbein *n*
3126	ethogram, behavioral inventory	Ethogramm *n*, Aktionskatalog *m*, Verhaltensinventar *n*
3127	ethological	ethologisch
3128	ethologist	Verhaltensforscher *m*
3129	ethology, study of behavio(u)r	Ethologie *f*, Verhaltensforschung *f*
3130	ethyl alcohol, ethanol	Äthylalkohol *m*, Äthanol *n*
3131	ethylene	Äthylen *n*
3132	etiolation	Vergeilung *f*, Etiolement *n*
3133	euchromatic	euchromatisch
3134	euchromatin	Euchromatin *n*
3135	euchromosome	Euchromosom *n*
3136	eugenic	eugenisch
3137	eugenics	Eugenik *f*, Erbhygiene *f*
3138	eukaryotes, eucaryotic cells	Eukaryoten *f pl*
3139	Eulamellibranchia	Blattkiemer *m pl*, Eulamellibranchier *pl*
3140	Eumycophyta → true fungi	—
3141	euphotic zone	euphotische Zone *f*
3142	euploid	euploid
3143	euploidy	Euploidie *f*
3144	euryhaline	euryhalin
3145	euryphagous	euryphag
3146	eurythermous, eurythermic	eurytherm
3147	Eustachian tube, auditory ~	Eustachische Röhre *f*, Ohrtrompete *f*
3148	eutely → cell constancy	—
3149	eutrophic	eutroph
3150	eutrophication	Eutrophierung *f*
3151	evagination, outfolding	Ausstülpung *f*
3152	to evaporate	verdunsten
3153	evaporation	Verdunstung *f*
3154	evaporation loss	Verdunstungsverlust *m*
3155	evapotranspiration	Verdunstung *f* und Transpiration *f*
3156	even-toed ungulates → Artiodactyla	—
3157	evergreen, everlasting	immergrün
3158	evocation *embr.*	Evokation *f embr.*
3159	evocator	Evokator *m*
3160	evoked potential	Reaktionspotential *n*, „evoked potential"
3161	evolution	Evolution *f*
3162	evolution cycle	Entwicklungszyklus *m*
3163	evolutionary factor	Evolutionsfaktor *m*
3164	evolutionary forces	Entwicklungskräfte *f pl*
3165	evolutionary line(age)	Evolutionslinie *f*
3166	evolutionary potential	Evolutionspotential *n*
3167	evolutionary process	Evolutionsprozeß *m*, ~vorgang *m*
3168	evolutionary series	Evolutionsreihe *n*, Entwicklungsreihe *f*
3169	evolutionary stage	Evolutionsstadium *n*
3170	evolutionary success	Evolutionserfolg *m*
3171	evolutionary trend	Evolutionsrichtung *f*
3172	evolutionism → theory of evolution	—
3173	evolutionist	Evolutionsforscher *m*
3174	evolutive	evolutiv
3175	evolutivity → developmental capacity	—
3176	examination of living tissue	Lebendbeobachtung *f*
3177	to examine microscopically	mikroskopisch untersuchen
3178	exarate pupa	freie Puppe *f*, gemeißelte ~
3179	exchange of substances, ~ ~ materials	Stoffaustausch *m*
3180	exciple, excipulum	Excipulum *n*

3116	ester *m*, éther-sel *m*	éster *m*
3117	estérase *f*	esterasa *f*
3118	estérification *f*	esterificación *f*
3119	estérifier	esterificar, esterizar
3120	—	—
3121	—	—
3122	—	—
3123	éthéogenèse *f*	eteogénesis *f*
3124	—	—
3125	(os *m*) ethmoïde *m*	(hueso *m*) etmoides *m*
3126	éthogramme *m*	repertorio *m* de esquemas de conducta
3127	éthologique	etológico
3128	éthologiste *m*	etólogo *m*
3129	éthologie *f*, étude *f* du comportement	etología *f*, estudio *m* del comportamiento
3130	alcool *m* éthylique, éthanol *m*	alcohol *m* etílico, etanol *m*
3131	éthylène *m*	etileno *m*
3132	étiolement *m*	ahilamiento *m*, etiolación *f*
3133	euchromatique	eucromático
3134	euchromatine *f*	eucromatina *f*
3135	euchromosome *m*	eucromosoma *m*
3136	eugénique	eugenético
3137	eugénique *f*, eugénisme *m*	eugenesia *f*
3138	eucaryotes *f pl*	eucariotas *f pl*, células *f pl* eucarióticas
3139	eulamellibranches *m pl*	eulamelibranquios *m pl*
3140	—	—
3141	zone *f* euphotique	zona *f* eufótica
3142	euploïde	euploide
3143	euploïdie *f*	euploidia *f*
3144	euryhalin, halotolérant	eurihalino
3145	euryphage	eurifágico
3146	eurytherme, eurythermique	euritermo, euritérmico
3147	trompe *f* d'Eustache, ~ auditive	trompa *f* de Eustaquio, ~ auditiva
3148	—	—
3149	eutrophe, eutrophique	eutrófico
3150	eutrophisation *f*	eutroficación *f*
3151	évagination *f*	evaginación *f*
3152	évaporer	evaporar
3153	évaporation *f*	evaporación *f*
3154	perte *f* par évaporation	pérdida *f* por evaporación
3155	evapotranspiration *f*	evapotranspiración *f*
3156	—	—
3157	toujours vert, sempervirent	siempreverde, sempervirente
3158	évocation *f* *embr.*	evocación *f* *embr.*
3159	évocateur *m*	evocador *m*
3160	potentiel *m* évoqué	potencial *m* de reacción, ~ evocado
3161	évolution *f*	evolución *f*
3162	cycle *m* évolutif, ~ d'évolution	ciclo *m* evolutivo
3163	facteur *m* d'évolution	factor *m* de evolución, ~ evolutivo
3164	forces *f pl* évolutives	fuerzas *f pl* evolutivas
3165	lignée *f* évolutive	linaje *m* evolutivo, línea *f* evolutiva
3166	potentiel *m* évolutif	potencial *m* evolutivo
3167	processus *m* évolutif	proceso *m* evolutivo, ~ de evolución
3168	série *f* évolutive	serie *f* evolutiva, ~ filogenética
3169	stade *m* évolutif, ~ d'évolution	etapa *f* evolutiva
3170	succès *m* évolutif	éxito *m* evolutivo
3171	orientation *f* évolutive, sens *m* d'évolution	dirección *f* evolutiva
3172	—	—
3173	évolutionniste *m*	evolucionista *m*
3174	évolutif	evolutivo
3175	—	—
3176	observation *f* vitale	observación *f* sobre material vivo
3177	examiner au microscope	examinar al microscopio
3178	chrysalide *f* libre	pupa *f* libre, ~ exarada
3179	échange *m* de substances	intercambio *m* de sustancias, ~ ~ materia
3180	excipulum *m*	excípulo *m*

3181	excitability	Erregbarkeit *f*, Erregungsfähigkeit *f*
3182	excitable	erregbar, erregungsfähig
3183	excitation	Erregung *f*
3184	excitation pattern	Erregungsmuster *n*
3185	excitation threshold	Erregungsschwelle *f*
3186	excitatory current	Erregungsstrom *m*
3187	excitatory postsynaptic potential, EPSP	exzitatorisches postsynaptisches Potential *n*, EPSP
3188	excitatory substance	Erregungssubstanz *f*
3189	excitatory value	Erregungshöhe *f*, ~niveau *n*
3190	exciting electrode → stimulation ~	—
3191	exconjugant	Exkonjugant *m*
3192	excrements	Exkremente *npl*
3193	excrescence, outgrowth	Auswuchs *m*
3194	excreta	Exkrete *npl*
3195	to excrete	ausscheiden, abscheiden
3196	excretion	Ausscheidung *f*, Exkretion *f*
3197	excretory	exkretorisch, Ausscheidungs ...
3198	excretory canal	Ausscheidungsgang *m*, Ausführungsgang *m*, Exkretionskanal *m*
3199	excretory organ	Ausscheidungsorgan *n*, Exkretionsorgan *n*
3200	excretory product, excretion ~	Ausscheidungsprodukt *n*, Exkretionsprodukt *n*
3201	excretory system	Exkretionssystem *n*
3202	exergonic, energy-yielding	energieliefernd, exergonisch
3203	exine, extine	Exine *f*
3204	exo ... see also ecto ...	—
3205	exocarp, epicarp	Exokarp *n*, Epikarp *n*
3206	exocoelom	Exozölom *n*
3207	exocranium	Kopfskelett *n*
3208	exocrine gland	exokrine Drüse *f*
3209	exocrine secretion	äußere Sekretion *f*, exokrine ~
3210	exodermis *bot.*	Exodermis *f* *bot.*
3211	exogamy	Exogamie *f*
3212	exogastrulation	Exogastrulation *f*
3213	exogenous	exogen
3214	exopeptidase	Exopeptidase *f*
3215	Exopterygota → Hemimetabola	—
3216	exoskeleton, ectoskeleton, dermoskeleton	Außenskelett *n*, Ektoskelett *n*, Hautskelett *n*
3217	exosmosis	Exosmose *f*
3218	exospore, ectospore	Exospore *f*, Ektospore *f*
3219	exosporium	Exospor(ium) *n*
3220	exostome	Exostom *n*
3221	exothecium	Exothezium *n*
3222	exothermic, exothermal	exotherm
3223	exotic	Exot *m*
3224	exotoxin, ectotoxin	Exotoxin *n*, Ektotoxin *n*
3225	experiment	Experiment *n*, Versuch *m*
3226	to experiment	experimentieren
3227	experimental	experimentell
3228	experimental animal	Versuchstier *n*
3229	experimental arrangement → ~ setup	—
3230	experimental breeding	Versuchszüchtung *f*
3231	experimental conditions, test ~	Versuchsbedingungen *fpl*
3232	experimental embryology → developmental physiology	—
3233	experimental evidence, ~ proof	experimenteller Beweis *m*, ~ Nachweis *m*
3234	experimental method	Versuchsmethode *f*
3235	experimental plant	Versuchspflanze *f*
3236	experimental result, result of experiments	Versuchsergebnis *n*

3181	excitabilité *f*	excitabilidad *f*
3182	excitable	excitable
3183	excitation *f*	excitación *f*
3184	schème *m* d'excitation	modelo *m* de excitación
3185	seuil *m* d'excitation, ~ excitatif	umbral *m* de excitación
3186	courant *m* de stimulation	corriente *m* estimulante, ~ excitadora
3187	potentiel *m* post-synaptique excitateur, PPSE	potencial *m* postsináptico excitatorio
3188	substance *f* excitante, excitant *m*	sustancia *f* excitante
3189	valeur *f* excitatoire	nivel *m* de excitación
3190	—	—
3191	ex-conjugant *m*	exconjugante *m*
3192	excréments *mpl*	excrementos *mpl*, materias *fpl* excrementicias
3193	excroissance *f*	excrecencia *f*
3194	excreta *mpl*	excreta *mpl*
3195	excréter	excretar
3196	excrétion *f*	excreción *f*
3197	excréteur; excrétoire *(peu usité)*	excretor(io)
3198	canal *m* excréteur, conduit *m* ~	conducto *m* excretor(io)
3199	organe *m* excréteur	órgano *m* excretor(io)
3200	produit *m* d'excrétion	producto *m* de excreción
3201	système *m* excréteur, appareil *m* ~	sistema *m* de excreción, ~ excretor
3202	exergonique	exergónico
3203	exine *f*	exina *f*
3204	—	—
3205	exocarpe *m*, épicarpe *m*	exocarpo *m*, epicarpo *m*
3206	exocoelome *m*, coelome *m* externe, ~ extra-embryonnaire	exoceloma *m*
3207	squelette *m* céphalique	exocráneo *m*
3208	glande *f* exocrine, ~ à sécrétion externe	glándula *f* exocrina, ~ de secreción externa
3209	sécrétion *f* exocrine	secreción *f* exocrina
3210	exoderme *m* *bot.*	exodermis *f* *bot.*
3211	exogamie *f*	exogamia *f*
3212	exogastrulation *f*	exogastrulación *f*
3213	exogène	exógeno
3214	exopeptidase *f*	exopeptidasa *f*
3215	—	—
3216	exosquelette *m*, squelette *m* dermique	exo(e)squeleto *m*, esqueleto *m* externo, dermosqueleto *m*
3217	exosmose *f*	exósmosis *f*
3218	ectospore *f*	exóspora *f*
3219	exosporium *m*	exosporio *m*
3220	exostome *m*	exóstoma *m*
3221	épiderme *m* de l'anthère	exotecio *m*
3222	exothermique	exotérmico
3223	plante *f* (ou animal *m*) exotique	planta *f* exótica; animal *m* exótico
3224	exotoxine *f*, ectotoxine *f*	exotoxina *f*
3225	expérience *f*	experimento *m*, experiencia *f*
3226	expérimenter	experimentar
3227	expérimental	experimental
3228	animal *m* d'expérience, ~ expérimental	animal *m* de experimentación, ~ experimental
3229	—	—
3230	élevage *m* expérimental	cría *f* experimental
3231	conditions *fpl* expérimentales	condiciones *fpl* experimentales
3232	—	—
3233	preuve *f* expérimentale, mise *f* en évidence ~, vérification *f* ~	prueba *f* experimental, comprobación *f* ~
3234	méthode *f* expérimentale	método *m* experimental, ~ de ensayo
3235	plante *f* d'expérience	planta *f* experimental
3236	résultat *m* expérimental, ~ des expériences	resultado *m* experimental

3237	experimental setup, ~ arrangement	Versuchsanordnung *f*
3238	experimental stage, test ~	Versuchsstadium *n*
3239	experimental study	experimentelle Untersuchung *f*
3240	experimental subject	Versuchsperson *f*
3241	experimentation on animals, animal experiment	Tierversuch *m*
3242	experimenter, experimentalist	Experimentator *m*
3243	expiration	Ausatmung *f*
3244	expiratory reserve → reserve air	—
3245	expired air	Ausatmungsluft *f*
3246	explant	Explantat *n*
3247	explantation	Explantation *f*
3248	exploratory behavio(u)r	Erkundungsverhalten *n*
3249	exponential growth, logarithmic ~	Exponentialwachstum *n*, exponentielles Wachstum *n*, logarithmisches ~
3250	exponential phase, logarithmic ~, logphase	Exponentialphase *f*, logarithmische Phase *f*, „Log"-Phase *f*
3251	expressive movement *eth.*	Ausdrucksbewegung *f* *eth.*
3252	expressivity (gene)	Ausprägungsgrad *m* (Gen), Expressivität *f*
3253	expulsion reaction *eth.*	Ausstoß(ungs)reaktion *f* *eth.*
3254	extensor (muscle)	Strecker *m*, Extensor *m*
3255	extensor reflex	Streckreflex *m*
3256	extermination	Ausrottung *f*
3257	external ear, outer ~	äußeres Ohr *n*
3258	external factor	Außenfaktor *m*
3259	external fertilization	äußere Befruchtung *f*
3260	external medium, ~ environment	Außenmedium *n*
3261	external parasite → ectoparasite	—
3262	external stimulus	Außenreiz *m*
3263	exteroceptive	exterozeptiv
3264	extero(re)ceptor	Extero(re)zeptor *m*
3265	extinct	ausgestorben
3266	to extinct, to become extinct	aussterben
3267	extinction	Aussterben *n*; Auslöschen *n* (eines Reizes)
3268	extine → exine	—
3269	extracellular	extrazellulär
3270	extrachromosomic	extrachromosomal
3271	extract	Extrakt *m*
3272	to extract	extrahieren
3273	extrapolation	Extrapolieren *n*
3274	extrapyramidal	extrapyramidal
3275	extremities, limbs, members	Gliedmaßen *f pl*, Extremitäten *f pl*
3276	extrinsic factor → cobalamine	—
3277	extrorse	extrors
3278	exudate	Exsudat *n*
3279	exudation	Exsudation *f*, Ausschwitzung *f*
3280	exuvia	Exuvie *f*
3281	exuvial gland → moulting gland	—
3282	exuviation	Häutung *f*
3283	eyeball	Augapfel *m*
3284	eye cup → optic cup	—
3285	eyelid, palpebra	Augenlid *n*
3286	eyelid reflex	Lidschlußreflex *m*
3287	eye-piece *micr.*	Okular *n* *micr.*
3288	eye-spot, stigma	Augenfleck *m*, Flachauge *n*, Stigma *n*
3289	eye-stalk	Augenstiel *m*
3290	facial expressions, mimic ~	Mimik *f*
3291	facial nerve	Gesichtsnerv *m*, Fazialis *m*
3292	facies	Fazies *f*
3293	facilitation *neur.*	Bahnung *f* *neur.*
3294	factor map → genetic map	—
3295	factor pair → pair of genes	—

3237	montage *m* expérimental, disposition *f* des expériences	disposición *f* del ensayo, ordenación *f* del experimento
3238	stade *m* expérimental, période *f* d'essai	período *m* experimental, fase *f* de ensayo
3239	étude *f* expérimentale	estudio *m* experimental
3240	sujet *m* (expérimental)	sujeto *m* (de experimentación)
3241	expérimentation *f* sur des animaux	experimento *m* en animales, experimentación *f* animal
3242	expérimentateur *m*	experimentador *m*
3243	expiration *f*	espiración *f*
3244	—	—
3245	air *m* expiratoire	aire *m* espiratorio
3246	explant *m*	explante *m*
3247	explantation *f*	explantación *f*
3248	comportement *m* explorateur, ~ exploratoire	conducta *f* exploratoria
3249	croissance *f* exponentielle	crecimiento *m* exponencial, ~ logarítmico
3250	phase *f* exponentielle, ~ logarithmique	fase *f* exponencial, ~ logarítmica
3251	mouvement *m* expressif *eth.*	movimiento *m* expresivo *eth.*
3252	expressivité *f* (du gène), degré *m* d'expression, ~ de réalisation	expresividad *f* (génica)
3253	réaction *f* d'expulsion *eth.*	reacción *f* de expulsión *eth.*
3254	(muscle *m*) extenseur *m*	(músculo *m*) extensor *m*
3255	réflexe *m* d'extension	reflejo *m* de extensión
3256	extermination *f*	exterminación *f*, exterminio *m*
3257	oreille *f* externe	oído *m* externo
3258	facteur *m* externe	factor *m* externo
3259	fertilisation *f* externe, fécondation *f* ~	fertilización *f* externa
3260	milieu *m* extérieur	medio *m* externo
3261	—	—
3262	stimulus *m* externe	estímulo *m* externo
3263	extéroceptif	exteroceptivo
3264	extéro(ré)cepteur *m*	exteroceptor *m*
3265	éteint	extinguido, extinto
3266	s'éteindre	extinguirse
3267	extinction *f*	extinción *f*
3268	—	—
3269	extracellulaire	extracelular
3270	extrachromosomique	extracromosómico
3271	extrait *m*	extracto *m*
3272	extraire	extraer
3273	extrapolation *f*	extrapolación *f*
3274	extra-pyramidal	extrapiramidal
3275	extrémités *f pl*, membres *m pl*	extremidades *f pl*, miembros *m pl*
3276	—	—
3277	extrorse	extrorso
3278	exsudat *m*	exudado *m*
3279	exsudation *f*	exudación *f*
3280	exuvie *f*	exuvia *f*
3281	—	—
3282	exuviation *f*	exuviación *f*
3283	globe *m* oculaire	globo *m* ocular
3284	—	—
3285	paupière *f*	párpado *m*, pálpebra *f*
3286	réflexe *m* palpébral	reflejo *m* palpebral
3287	oculaire *m* *micr.*	ocular *m* *micr.*
3288	tache *f* oculaire, ~ oculiforme, stigma *m*	mancha *f* ocular, punto *m* oculiforme, estigma *m* ocular
3289	pédoncule *m* oculaire, ~ optique	pedúnculo *m* ocular, tallo *m* óptico
3290	mimique *f*, expressions *f pl* faciales	mímica *f*
3291	nerf *m* facial	nervio *m* facial
3292	faciès *m*	facies *f*
3293	facilitation *f* *neur.*	facilitación *f* *neur.*
3294	—	—
3295	—	—

3296	factorial analysis	Faktorenanalyse *f*
3297	facultative parasite	fakultativer Parasit *m*
3298	FAD → flavin adenine dinucleotide	—
3299	faeces, feces	Fäkalien *pl*, Faeces *pl*
3300	falciform	sichelförmig
3301	Fallopian tube, uterine ~	Eileiter *m (Säugetiere)*
3302	false axis → sympodium	—
3303	false fruit → pseudocarp	—
3304	false rib, asternal ~	falsche Rippe *f*
3305	family *tax.*	Familie *f tax.*
3306	family selection	Familienzüchtung *f*, ~auslese *f*, Sippenzüchtung *f*
3307	fang	Fangzahn *m*
3308	fan-shaped → flabellate	—
3309	fascia	Faszie *f*
3310	fasciation	Fasziation *f*, Verbänderung *f*
3311	fascicular cambium	faszikuläres Kambium *n*, Bündelkambium *n*
3312	fasciculated root	Büschelwurzel *f*
3313	fast growing	schnellwachsend, schnellwüchsig
3314	fat	Fett *n*
3315	fat body	Fettkörper *m*
3316	fat deposition	Fettablagerung *f*
3317	fat granule, ~ droplet	Fett-Tröpfchen *n*
3318	fat layer	Fettschicht *f*
3319	fat metabolism, lipid ~, lipometabolism	Fettstoffwechsel *m*
3320	fat reserve → lipid reserve	—
3321	fat solubility	Fettlöslichkeit *f*
3322	fat soluble, lipo-soluble	fettlöslich
3323	fat-splitting → lipoclastic	—
3324	fatigue	Ermüdung *f*
3325	fatigueability	Ermüdbarkeit *f*
3326	fatty acid	Fettsäure *f*
3327	fatty inclusion	Fetteinschluß *m*
3328	fatty substance	Fettstoff *m*
3329	fatty tissue → adipose tissue	—
3330	fauna	Fauna *f*
3331	faunal element	Faunenelement *n*
3332	faunal region	Faunengebiet *n*
3333	feather	Feder *f*
3334	feather shaft → rachis[3]	—
3335	feces → faeces	—
3336	fecundation; fecundity → fertilization; fertility	—
3337	feedback	Rückkoppelung *f*, Feedback *n*
3338	feedback control	Rückregulierung *f*
3339	feedback control system	Regelsystem *n*
3340	feedback inhibition	Rückkoppelungshemmung *f*
3341	feeding behavio(u)r	Freßverhalten *n*
3342	feeding drive	Nahrungstrieb *m*, Freßtrieb *m*
3343	feeding interrelations	Nahrungsbeziehungen *f pl*
3344	feeding place, ~ ground	Futterplatz *m*, Nahrungsplatz *m*
3345	feeding tube → sucking tube	—
3346	feeding of youngs	Jungenfüttern *n*
3347	feeler → antenna	—
3348	female, gynic	weiblich
3349	female	Weibchen *n*
3350	female line (of breeding)	weibliche Linie *f*, mütterliche ~
3351	female parent	weiblicher Elter *m*, mütterlicher ~
3352	femini(ni)ty, femaleness	Weiblichkeit *f*
3353	femur, thigh bone	Oberschenkelknochen *m*, Femur *m*

3296	analyse *f* factorielle, ~ des facteurs	análisis *m* factorial, ~ de factores
3297	parasite *m* facultatif	parásito *m* facultativo
3298	—	—
3299	fèces *f pl*, matières *f pl* fécales	heces *f pl*, materias *f pl* fecales
3300	falciforme	falciforme
3301	trompe *f* de Fallope, ~ utérine	trompa *f* de Falopio, tubo *m* uterino
3302	—	—
3303	—	—
3304	fausse côte *f*, côte *f* asternale	costilla *f* falsa, ~ asternal
3305	famille *f tax.*	familia *f tax.*
3306	sélection *f* familiale, ~ de sippes	selección *f* familiar
3307	crochet *m*, croc *m*	colmillo *m*
3308	—	—
3309	fascia *m*	fascia *f*
3310	fasciation *f* (rubanée)	fasciación *f*
3311	cambium *m* fasciculaire	cámbium *m* fascicular
3312	racine *f* fasciculée	raíz *f* fasciculada
3313	à croissance rapide	de crecimiento rápido
3314	graisse *f*	grasa *f*
3315	corps *m* adipeux, ~ gras, ~ graisseux	cuerpo *m* adiposo, ~ graso
3316	dépôt *m* adipeux	deposición *f* de grasas
3317	gouttelette *f* lipidique	gotita *f* de grasa, ~ lipídica
3318	couche *f* adipeuse, ~ lipidique	capa *f* adiposa
3319	métabolisme *m* lipidique, ~ de lipides	metabolismo *m* graso, ~ lipídico, lipometabolismo *m*
3320	—	—
3321	liposolubilité *f*	liposolubilidad *f*
3322	liposoluble	liposoluble
3323	—	—
3324	fatigue *f*	fatiga *f*
3325	fatigabilité *f*	fatigabilidad *f*
3326	acide *m* gras	ácido *m* graso
3327	enclave *f* graisseuse, ~ lipidique	inclusión *f* de grasa, ~ lipídica
3328	matière *f* grasse	materia *f* grasa
3329	—	—
3330	faune *f*	fauna *f*
3331	élément *m* faunistique	elemento *m* faunístico
3332	région *f* faunistique	región *f* faunística, zona *f* faunal
3333	plume *f*	pluma *f*
3334	—	—
3335	—	—
3336	—	—
3337	rétroaction *f*, action *f* rétrograde, «feed-back» *m*	retroacción *f*, retroalimentación *f*, "feedback" *m*
3338	rétrocontrôle *m*, rétrorégulation *f*, régulation *f* par retour	control *m* por retroacción, ~ retroactivo
3339	système *m* de rétrocontrôle, ~ à régulation	sistema *m* de retroacción, ~ de regulación, ~ regulador
3340	inhibition *f* par rétrocontrôle, rétroinhibition *f*	inhibición *f* por retroacción, retroinhibición *f*
3341	comportement *m* alimentaire, ~ d'alimentation	comportamiento *m* de alimentación, ~ nutricio
3342	instinct *m* alimentaire	instinto *m* de alimentación
3343	relations *f pl* alimentaires, rapports *m pl* ~	relaciones *f pl* alimenticias
3344	région *f* alimentaire, terrain *m* d'alimentation	zona *f* de alimentación, terreno *m* alimentario
3345	—	—
3346	nourrissage *m* des jeunes	alimentación *f* de las crías
3347	—	—
3348	femelle	femenino
3349	femelle *f*	hembra *f*
3350	lignée *f* femelle, ~ maternelle	línea *f* femenina, ~ materna
3351	parent *m* femelle	(pro)genitor *m* hembra, ~ femenino
3352	féminité *f*	feminidad *f*, femineidad *f*
3353	fémur *m*	fémur *m*

3354	ferment → enzyme	—
3355	to ferment	fermentieren, gären, vergären
3356	fermentable	gärfähig, vergärbar
3357	fermentation	Fermentation *f*, Gärung *f*, Vergärung *f*
3358	fermentation tube	Gärröhrchen *n*
3359	fermentative bacteria	Gärungsbakterien *fpl*
3360	ferns → Pteropsida	—
3361	ferredoxin	Ferredoxin *n*
3362	fertile, fecund	fruchtbar
3363	fertility, fecundity	Fruchtbarkeit *f*, Fertilität *f*
3364	fertility coefficient	Fruchtbarkeitsziffer *f*
3365	fertility factor, sex ~, F-factor	F-Faktor *m*, Fertilitätsfaktor *m*
3366	fertilizable, fecundable	befruchtungsfähig
3367	fertilization, fecundation	Befruchtung *f*
3368	fertilization cone, attraction ~	Befruchtungshügel *m*, Empfängnishügel *m*
3369	fertilization membrane	Befruchtungsmembran *f*
3370	fertilization tube, copulation ~	Befruchtungsschlauch *m*, Kopulationsschlauch *m*
3371	to fertilize, to fecundate	befruchten
3372	fertilizin	Fertilizin *n*
3373	fertilizing capacity	Befruchtungsfähigkeit *f*, ~vermögen *n*
3374	fetus → foetus	—
3375	Feulgen reaction, ~ method	Feulgenreaktion *f*, Feulgensche Nuklearreaktion *f*
3376	F-factor → fertility factor	—
3377	F_1 generation, first filial ~	F_1-Generation *f*, erste Filialgeneration *f*, erste Tochtergeneration *f*
3378	F_2 generation, second filial ~	F_2-Generation *f*, zweite Filialgeneration *f*, zweite Tochtergeneration *f*
3379	fibre (US: fiber)	Faser *f*
3380	fibre plant	Faserpflanze *f*
3381	fibre tracheid	Fasertracheide *f*
3382	fibril	Fibrille *f*
3383	fibrillar	fibrillär
3384	fibrin	Fibrin *n*, Blutfaserstoff *m*
3385	fibrinogen	Fibrinogen *n*
3386	fibrinolysine	Fibrinolysin *n*
3387	fibroblast	Fibroblast *m*
3388	fibrocartilage, fibrous cartilage	Faserknorpel *m*, Bindegewebsknorpel *m*
3389	fibrocyte, desmocyte, connective tissue cell	Fibrozyt *m*, Desmozyt *m*, Bindegewebszelle *f*
3390	fibroin	Fibroin *n*
3391	fibrous	faserig
3392	fibrous protein → scleroprotein	—
3393	fibrous root	Faserwurzel *f*
3394	fibrous tissue	Fasergewebe *n*
3395	fibula	Wadenbein *n*, Fibula *f*
3396	fidelity	Gesellschaftstreue *f*
3397	field capacity	Feldkapazität *f*
3398	field experiment(ation)	Feldversuch *m*, Freilandversuch *m*
3399	field of investigation → research field	—
3400	field observation	Feldbeobachtung *f*
3401	field strain	Wildstamm *m*
3402	fighting behavio(u)r *eth.*	Kampfverhalten *n eth.*
3403	filament *bot.*	Staubfaden *m*, Filament *n*
3404	filamentous, filiform, thread-like	fadenförmig, fadenartig
3405	filamentous algae	Fadenalgen *fpl*
3406	filial generation	Filialgeneration *f*, Tochter~ *f*, Spaltungs~ *f*
3407	filiation → descent	—
3408	filoplume, hair feather	Fadenfeder *f*, Haarfeder *f*
3409	filopodia *pl*	Filopodien *pl*, Fadenfüßchen *npl*
3410	to filter, to filtrate	filtern, filtrieren
3411	filter feeder	Filtrierer *m zool.*
3412	filter-paper	Filterpapier *n*, Filtrierpapier *n*

3354	—	—
3355	fermenter	fermentar
3356	fermentescible	fermentable, fermentescible
3357	fermentation *f*	fermentación *f*
3358	tube *m* à fermentation	tubo *m* de fermentación
3359	bactéries *f pl* de fermentation	bacterias *f pl* fermentativas
3360	—	—
3361	ferrédoxine *f*	ferredoxina *f*
3362	fertile, fécond	fértil, fecundo
3363	fertilité *f*, fécondité *f*	fertilidad *f*, fecundidad *f*
3364	coefficient *m* de fécondité	tasa *f* de fertilidad, cifra *f* ~ ~
3365	facteur *m* F	factor *m* de fertilidad, ~ sexual, ~ F
3366	fécondable, fertilisable	fertilizable, apto para la fecundación
3367	fertilisation *f*, fécondation *f*	fertilización *f*, fecundación *f*
3368	cône *m* d'attraction	cono *m* de fertilización
3369	membrane *f* de fécondation	membrana *f* de fecundación
3370	tube *m* de copulation, ~ copulateur	tubo *m* de copulación, ~ copulador, ~ fecundante
3371	fertiliser, féconder	fertilizar, fecundar
3372	fertilisine *f*	fertilizina *f*
3373	capacité *f* fécondante, pouvoir *m* fécondant, fécondabilité *f*	capacidad *f* de fertilización, poder *m* fecundante, fecundabilidad *f*
3374	—	—
3375	réaction *f* (nucléale) de Feulgen	reacción *f* (nuclear) de Feulgen
3376	—	—
3377	génération *f* F_1, première génération *f* filiale	generación *f* F_1, primera generación *f* filial
3378	génération *f* F_2, deuxième génération *f* filiale	generación *f* F_2, segunda generación *f* filial
3379	fibre *f*	fibra *f*
3380	plante *f* à fibre	planta *f* fibrosa
3381	fibro-trachéide *f*	traqueida *f* fibriforme
3382	fibrille *f*	fibrilla *f*
3383	fibrillaire, fibrilleux	fibrilar
3384	fibrine *f*	fibrina *f*
3385	fibrinogène *m*	fibrinógeno *m*
3386	fibrinolysine *f*	fibrinolisina *f*
3387	fibroblaste *m*	fibroblasto *m*
3388	fibrocartilage *m*, cartilage *m* fibreux	fibrocartílago *m*, cartílago *m* fibroso
3389	fibrocyte *m*, cellule *f* conjonctive	fibrocito *m*
3390	fibroïne *f*	fibroína *f*
3391	fibreux	fibroso
3392	—	—
3393	racine *f* fibreuse	raíz *f* fibrosa
3394	tissu *m* fibreux	tejido *m* fibroso
3395	péroné *m*	peroné *m*
3396	fidélité *f* sociale	fidelidad *f* asociativa
3397	capacité *f* (de rétention) au champs	capacidad *f* de campo
3398	expérimentation *f* sur le terrain, expérience *f* de plein champ	experimento *m* en el campo
3399	—	—
3400	observation *f* dans le champ	observación *f* en campo abierto
3401	souche *f* sauvage	cepa *f* salvaje
3402	comportement *m* de combat	comportamiento *m* de lucha, conducta *f* combativa
3403	filet *m* (staminal)	filamento *m* *bot.*
3404	filamenteux, filiforme	filamentoso, filiforme
3405	algues *f pl* filamenteuses	algas *f pl* filamentosas
3406	génération *f* filiale	generación *f* filial
3407	—	—
3408	filoplume *f*	filopluma *f*
3409	filopodes *m pl*	filopodios *m pl*
3410	filtrer	filtrar
3411	filtreur *m* *zool.*	comedor *m* por filtración
3412	papier-filtre *m*, ~ filtrant	papel-filtro *m*, ~ filtrante

3413	filtrate	Filtrat *n*
3414	filtration	Filterung *f*, Filtration
3415	fin	Flosse *f*
3416	fin-ray	Flossenstrahl *m*
3417	final host → definite ~	—
3418	fine structure	Feinstruktur *f*, Feinbau *m*
3419	fire algae → Pyrrophyta	—
3420	first filial generation → F_1 generation	—
3421	fish fry	Fischbrut *f*
3422	fissibility, fissionability	Spaltbarkeit *f*
3423	fissile, fissible, scissile, cleavable	spaltbar
3424	fission, scission, splitting	Spaltung *f*, Aufspaltung *f*
3425	fission product	Spaltprodukt *n*
3426	fissiparity → scissiparity	—
3427	fitness	Eignung *f*
3428	fitted → adapted	—
3429	to fix	fixieren
3430	fixation, fixing	Fixierung *f*
3431	fixed action pattern	Erbkoordination *f*, Bewegungsnorm *f*
3432	fixing agent, fixer, fixative	Fixier(ungs)mittel *n*
3433	fixing fluid	Fixier(ungs)flüssigkeit *f*
3434	fixity of species → constancy ~ ~	—
3435	flabellate, flabelliform, fan-shaped	fächerförmig
3436	flagellar movement	Geißelbewegung *f*
3437	Flagellata, flagellates, Mastigophora, mastigophores	Flagellaten *pl*, Mastigophoren *pl*, Geißeltierchen *npl*, ~infusorien *pl*
3438	flagellate, mastigote	begeißelt
3439	flagellated cell	Geißelzelle *f*
3440	flagellation	Begeißelung *f*
3441	flagelliform	geißelförmig
3442	flagellum (*pl* flagella)	Geißel *f*, Flagellum *n*
3443	flaky → flocculent	—
3444	flame cell	Wimperflamme *f*, Flammenzelle *f*
3445	flat bone	platter Knochen *m*, breiter ~
3446	flavin, lyochrome	Flavin *n*, Lyochrom *n*
3447	flavin adenine dinucleotide, FAD	Flavinadenindinukleotid *n*, FAD
3448	flavin mononucleotide, FMN	Flavinmononukleotid *n*, FMN
3449	flavone	Flavon *n*
3450	flavonoid	Flavonoid *n*
3451	flavoprotein, yellow enzyme	Flavoprotein *n*, Flavinenzym *n*, gelbes Ferment *n*
3452	fleshy	fleischig *(Frucht)*
3453	flexor (muscle)	Beuger *m*, Beugemuskel *m*, Flexor *m*
3454	flexor reflex	Beugereflex *m*
3455	flexural strength → bending ~	—
3456	flight capacity	Flugfähigkeit *f*
3457	flight feather	Flugfeder *f*
3458	flight muscle	Flugmuskel *m*
3459	flightless	flugunfähig
3460	floating leaf	Schwimmblatt *n*
3461	floating plant	Schwebepflanze *f*
3462	floating rib	freie Rippe *f*
3463	to flocculate	ausflocken
3464	flocculation	Ausflockung *f*, Flockenbildung *f*
3465	flocculent, flocky, flaky	flockig
3466	flock	Schar *f*, Herde *f*
3467	flora	Flora *f*
3468	floral axis	Blütenachse *f*
3469	floral diagram	Blütendiagramm *n*
3470	floral element	Florenelement *n*
3471	floral formula	Blütenformel *f*
3472	floral kingdom	Florenreich *n*
3473	floral leaf	Blütenblatt *n*
3474	floral organ, ~ part	Blütenorgan *n*
3475	floral primordium, flower ~	Blütenanlage *f*
3476	floral region	Florengebiet *n*

3413	filtrat *m*	filtrado *m*
3414	filtrage *m*, filtration *f*	filtración *f*
3415	nageoire *f*	aleta *f*
3416	rayon *m* des nageoires	rayo *m* de la aleta
3417	—	—
3418	structure *f* fine	estructura *f* fina
3419	—	—
3420	—	—
3421	frai *m*, alevin *m*	alevín *m*
3422	fissilité *f*	fisibilidad *f*
3423	fissile, fissible	fisible, escindible
3424	fission *f*, scission *f*	fisión *f*, escisión *f*
3425	produit *m* de fission, ~ de scission	producto *m* de fisión
3426	—	—
3427	«fitness»	eficacia *f* reproductora
3428	—	—
3429	fixer	fijar
3430	fixation *f*	fijación *f*
3431	coordination *f* héréditaire	modelo *m* fijo de movimiento
3432	fixateur *m*, fixatif *m*, substance *f* fixatrice	sustancia *f* fijadora, fijador *m*
3433	solution *f* fixante	líquido *m* fijador
3434	—	—
3435	flabellé, flabelliforme	flabelado, flabeliforme
3436	mouvement *m* flagellaire	movimiento *m* flagelar
3437	flagellés *mpl*, mastigophores *mpl*	flagelados *mpl*, mastigóforos *mpl*
3438	flagellé	flagelado
3439	cellule *f* à flagelles, ~ flagellée	célula *f* flagelada
3440	flagellation *f*	flagelación *f*
3441	flagelliforme	flageliforme
3442	flagelle *m*	flagelo *m*
3443	—	—
3444	cellule-flamme *f*, flamme *f* vibratile	célula *f* llama, ~ flamígera
3445	os *m* plat	hueso *m* plano
3446	flavine *f*, lyochrome *m*	flavina *f*, liocromo *m*
3447	flavine-adénine-dinucléotide *m*, FAD	flavin-adenin-dinucleótido *m*, FAD
3448	flavine-mononucléotide *m*, FMN	flavin-mononucleótido *m*, FMN
3449	flavone *m*	flavona *f*
3450	flavonoïde *m*	flavonoide *m*
3451	flavoprotéine *f*, flavine-enzyme *f*, ferment *m* jaune	flavoproteína *f*, fermento *m* amarillo
3452	charnu	carnoso
3453	(muscle *m*) fléchisseur *m*	(músculo *m*) flexor *m*
3454	réflexe *m* de flexion	reflejo *m* de flexión, ~ flexor
3455	—	—
3456	capacité *f* de vol	capacidad *f* de vuelo
3457	plume *f* voilière *f*, penne *f*	pena *f*
3458	muscle *m* du vol	músculo *m* de vuelo, ~ volador
3459	incapable de voler, non voilier	no volador
3460	feuille *f* nageante, ~ flottante	hoja *f* flotante
3461	plante *f* flottante	planta *f* flotante
3462	côte *f* flottante	costilla *f* flotante
3463	floculer	flocular
3464	floculation *f*	floculación *f*
3465	floconneux	coposo
3466	bande *f*; troupeau *m*	banda *f*; rebaño *m*
3467	flore *f*	flora *f*
3468	axe *m* floral	eje *m* floral
3469	diagramme *m* floral	diagrama *m* floral
3470	élément *m* floristique	elemento *m* florístico, ~ floral
3471	formule *f* florale	fórmula *f* floral
3472	empire *m* floral	reino *m* floral
3473	feuille *f* florale, anthophylle *f*	hoja *f* floral, antofilo *m*
3474	organe *m* floral, pièce *f* florale	órgano *m* floral, pieza *f* ~
3475	ébauche *f* florale	primordio *m* floral
3476	région *f* florale, zone *f* floristique	región *f* floral, ~ florística

3477	floriferous	blütentragend
3478	florigen → flowering hormone	—
3479	floristics, floristic plant geography, ~ geobotany	Florenkunde *f*, Floristik *f*, floristische Pflanzengeographie *f*
3480	flow of electrons	Elektronenfluß *m*
3481	to flower, to bloom, to blossom	blühen
3482	flower, blossom	Blüte *f* *(Pflanzenteil)*
3483	flower bud	Blütenknospe *f*
3484	flower cluster → glomerule	—
3485	flower envelope → perianth	—
3486	flower formation	Blütenbildung *f*
3487	flower head → capitulum	—
3488	flower stalk → peduncle	—
3489	flowering	Blühen *n*, Blüte *f*
3490	flowering glume → lemma	—
3491	flowering hormone, florigen	Blühhormon *n*, Florigen *n*
3492	flowering period, blossom-time	Blütezeit *f*
3493	flowering shoot	Blütensproß *m*
3494	flowing water	Fließgewässer *n*, fließendes Wasser *n*
3495	flowing-water ecosystem	Fließend-Wasser-Ökosystem *n*
3496	fluorescence	Fluoreszenz *f*
3497	fluorescence microscopy	Fluoreszenzmikroskopie *f*
3498	fluvial ecosystem	Flußökosystem *n*
3499	FMN → flavin mononucleotide	—
3500	foetal, fetal	fötal, fetal
3501	foetal circulation, placental ~	fötaler Kreislauf *m*, plazentärer ~
3502	foetal membrane → embryonic ~	—
3503	foetalization	Fötalisation *f*
3504	foetus, fetus	Fötus *m*, Fetus *m*
3505	foliaceous	blattartig
3506	foliage, leafage	Laub(werk) *n*
3507	foliage leaf	Laubblatt *n*
3508	foliage plant	Blattpflanze *f*
3509	foliar	Blatt ...
3510	foliation, leafing	Blattbildung *f*; Belaubung *f*
3511	folic acid	Folsäure *f*
3512	foliole, leaflet	Teilblättchen *n* (eines zusammengesetzten Blattes)
3513	foliose, leafy	beblättert, belaubt
3514	foliose lichen, leaf-lichen	Laubflechte *f*
3515	follicle[1] *anat.*	Follikel *m* *anat.*
3516	follicle[2] *bot.*, follicular fruit	Balg *m*, Balgfrucht *f*
3517	follicle cell	Follikelzelle *f*
3518	follicle-stimulating hormone, prolan A, FSH	follikelstimulierendes Hormon *n*, Follikelreifungshormon *n*, Gonadotropin *n* A, Prolan *n* A, FSH
3519	follicular hormone → oestrone	—
3520	folliculin → oestrone	—
3521	following reaction *eth.*	Nachfolgereaktion *f* *eth*
3522	fontanelle	Fontanelle *f*
3523	food-chain	Nahrungskette *f*
3524	food cycle, alimentary ~	Nahrungskreislauf *m*
3525	food deprivation	Nahrungsentzug *m*
3526	food energy	Nahrungsenergie *f*
3527	food factor → nutritive ~	—
3528	food-getting, procurement of food	Nahrungserwerb *m*
3529	food habit	Ernährungsgewohnheit *f*
3530	food intake, ingestion of food	Nahrungsaufnahme *f*
3531	food material, ~ substance	Nahrungsstoff *m*
3532	food particle	Nahrungsteilchen *n*
3533	food plant	Nährpflanze *f*, Nahrungspflanze *f*
3534	food requirements, nutritional ~	Nahrungsbedürfnisse *n pl*, ~bedarf *m*

3477	florifère	florífero
3478	—	—
3479	floristique *f*, phytogéographie *f* floristique	florística *f*, fitogeografía *f* florística
3480	flux *m* électronique, ~ d'électrons	flujo *m* de electrones
3481	fleurir, être en fleurs	florecer
3482	fleur *f*	flor *f*
3483	bouton *m* floral, ~ à fleurs, bourgeon *m* floral	yema *f* floral, ~ florífera, botón *m* floral
3484	—	—
3485	—	—
3486	formation *f* de fleurs, mise *f* à fleurs	formación *f* de flores
3487	—	—
3488	—	—
3489	floraison *f*, fleuraison *f*	floración *f*
3490	—	—
3491	hormone *f* de floraison, florigène *m*, hormone *f* ~	hormona *f* floral, ~ florígena, florígeno *m*
3492	période *f* de floraison, saison *f* ~ ~	florescencia *f*, época *f* de floración
3493	pousse *f* florale	brote *m* floral
3494	eau *f* courante	agua *f* corriente, ~ fluente
3495	écosystème *m* d'eau courante	ecosistema *m* de agua fluente
3496	fluorescence *f*	fluorescencia *f*
3497	microscopie *f* fluorescente, ~ à fluorescence	microscopia *f* fluorescente
3498	écosystème *m* fluvial	ecosistema *m* fluvial
3499	—	—
3500	foetal	fetal
3501	circulation *f* foetale, ~ placentaire	circulación *f* fetal, ~ placentaria
3502	—	—
3503	foetalisation *f*	fetalización *f*
3504	foetus *m*	feto *m*
3505	foliacé	foliáceo
3506	feuillage *m*	follaje *m*
3507	feuille *f*	hoja *f* vegetativa, ~ foliácea
3508	plante *f* à feuilles	planta *f* de hoja
3509	foliaire	foliar
3510	foliation *f*, feuillaison *f*	foliación *f*
3511	acide *m* folique	ácido *m* fólico
3512	foliole *f*	folíolo *m*
3513	folié, feuillu	folioso
3514	lichen *m* foliacé	liquen *m* foliáceo
3515	follicule *m* *anat.*	folículo *m* *anat.*
3516	follicule *m* *bot.*	folículo *m* *bot.*
3517	cellule *f* folliculaire	célula *f* folicular
3518	hormone *f* folliculostimulante, ~ gonado-stimulante, gonadotrop(h)ine *f* A, FSH	hormona *f* foliculoestimulante, gonado-trofina *f* A, prolan *m* A, FSH
3519	—	—
3520	—	—
3521	réaction *f* de suite *eth.*	reacción *f* de seguir (a la madre), ~ de seguimiento *eth.*
3522	fontanelle *f*	fontanela *f*
3523	chaîne *f* alimentaire, ~ trophique	cadena *f* alimentaria, ~ alimenticia
3524	cycle *m* alimentaire	ciclo *m* alimentario, ~ alimenticio
3525	privation *f* de nourriture	privación *f* de alimentos
3526	énergie *f* des aliments	energía *f* alimenticia, ~ nutritiva, ~ trófica
3527	—	—
3528	capture *f* de nourriture, collecte *f* de la ~	captura *f* del alimento, recolección *f* de los alimentos
3529	habitude *f* alimentaire	hábito *m* alimenticio
3530	ingestion *f* des aliments, prise *f* ~ ~	ingestión *f* de alimentos, toma *f* ~ ~
3531	matière *f* alimentaire, substance *f* ~	sustancia *f* alimenticia
3532	particule *f* alimentaire	partícula *f* alimenticia
3533	plante *f* nourricière, ~ nutritive	planta *f* nutriz, ~ alimenticia
3534	besoins *mpl* alimentaires, exigences *fpl* ~, ~ nutritionnelles	requerimientos *mpl* alimenticios, nece-sidades *fpl* alimenticias, exigencias *fpl* ~

3535	food reserve, nutritive ~, reserve food material	Nahrungsreserve *f*
3536	food residues	Nahrungsrückstände *mpl*
3537	food seeking, search for food	Nahrungssuche *f*
3538	food selection	Nahrungswahl *f*
3539	food shortage	Nahrungsmangel *m*
3540	food source	Nahrungsquelle *f*
3541	food storage, ~ storing	Nahrungsspeicherung *f*
3542	food substance → food material	—
3543	food supply	Nahrungsangebot *n*
3544	food vacuole	Nahrungsvakuole *f*
3545	food yolk → deutoplasm	—
3546	foramen	Foramen *n*
3547	fore-brain → prosencephalon	—
3548	foregut	Vorderdarm *m ins.*
3549	foreign body	Fremdkörper *m*
3550	foreign protein	körperfremdes Eiweiß *n*
3551	foreign substance	Fremdstoff *m*
3552	fore-kidney → pronephros	—
3553	foremilk → colostrum	—
3554	fore-skin → prepuce	—
3555	forest botany	Forstbotanik *f*
3556	forest community	Waldgesellschaft *f*
3557	forest limit	Waldgrenze *f*
3558	forestation → afforestation	—
3559	forked → bifurcate	—
3560	form constancy, shape ~	Formkonstanz *f*
3561	form sense	Formensinn *m*
3562	form vision	Formsehen *n*
3563	formaldehyde	Formaldehyd *n*
3564	formation, plant ~	Formation *f (Pfl. soz.)*
3565	formic acid	Ameisensäure *f*
3566	fossil	fossil; Fossil *n* (*pl* Fossilien)
3567	fossiliferous, fossil-bearing	fossilführend
3568	fossilization	Fossilisation *f*
3569	to fossilize	fossilisieren
3570	foundation animal	Stammtier *n*, Linienbegründer *m*
3571	four-way cross → double cross	—
3572	fovea	Fovea *f*, Sehgrube *f*
3573	fraction collector	Fraktionskollektor *m*
3574	fractionated centrifugation	fraktionierte Zentrifugation *f*
3575	fraternal twins → dizygotic twins	—
3576	free-martin	Zwicke *f*
3577	freeze-drying → lyophilization	—
3578	freezing microtome	Gefriermikrotom *n*
3579	freezing point	Gefrierpunkt *m*
3580	frequency distribution	Frequenzverteilung *f*
3581	frequency modulation	Frequenzmodulation *f*
3582	frequency of the heart	Herzfrequenz *f*
3583	frequency of respiration	Atemfrequenz *f*
3584	fresh preparation	Frischpräparat *n*
3585	fresh water	Süßwasser *n*
3586	freshwater animal	Süßwassertier *n*
3587	freshwater plankton → limnoplankton	—
3588	freshwater plant	Süßwasserpflanze *f*
3589	fresh weight	Frischgewicht *n*
3590	fright reaction	Schreckreaktion *f*
3591	frond	(Blatt)wedel *m*, Farnwedel *m*
3592	frontal (bone)	Stirnbein *n*
3593	frontal lobe	Stirnlappen *m*
3594	frontal sinus	Stirnhöhle *f*
3595	frost damage, ~ injury	Frostschaden *m*

3535	réserve *f* alimentaire, ~ nutritive, ~ nourricière	reserva *f* alimenticia, ~ nutritiva, materiales *mpl* nutritivos de reserva
3536	résidus *mpl* alimentaires	residuos *mpl* alimenticios
3537	recherche *f* de nourriture, quête *f* ~ ~	busca *f* de alimento, búsqueda *f* ~ ~
3538	choix *m* de la nourriture	elección *f* del alimento
3539	manque *m* de nourriture, carence *f* alimentaire	escasez *f* alimentaria
3540	source *f* d'aliments, ~ de nourriture	fuente *f* de alimentación, ~ alimenticia
3541	accumulation *f* d'aliments, ~ de nourriture	almacenamiento *m* de alimento, acumulación *f* de materiales alimenticios
3542	—	—
3543	apport *m* nutritif, offre *f* en nourriture	aporte *m* alimenticio, disponibilidad *f* de alimento
3544	vacuole *f* alimentaire, ~ digestive	vacuola *f* alimenticia, ~ digestiva
3545	—	—
3546	foramen *m*	foramen *m*
3547	—	—
3548	intestin *m* antérieur	intestino *m* anterior
3549	corps *m* étranger	cuerpo *m* extraño
3550	protéine *f* étrangère	proteína *f* extraña
3551	substance *f* étrangère	sustancia *f* extraña, materia *f* ~
3552	—	—
3553	—	—
3554	—	
3555	botanique *f* forestière	botánica *f* forestal
3556	communauté *f* forestière	comunidad *f* nemoral
3557	limite *f* forestière	límite *m* del bosque, línea *f* de bosques
3558	—	—
3559	—	—
3560	constance *f* de forme	constancia *f* de forma, ~ formal
3561	sens *m* des formes	sentido *m* de formas
3562	vision *f* des formes	visión *f* de formas
3563	formaldéhyde *m*, aldéhyde *m* formique	formaldehído *m*, aldehído *m* fórmico
3564	formation *f* (végétale)	formación *f* (vegetal)
3565	acide *m* formique	ácido *m* fórmico
3566	fossile (*m*)	fósil (*m*)
3567	fossilifère	fosilífero
3568	fossilisation *f*	fosilización *f*
3569	(se) fossiliser	fosilizar(se)
3570	animal *m* fondateur, ~ tête de ligne, individu *m* souche	cabeza *f* de línea
3571	—	—
3572	fovéa *f*	fóvea *f*
3573	collecteur *m* de fractions	colector *m* de fracciones
3574	centrifugation *f* fractionnée	centrifugación *f* fraccionada
3575	—	—
3576	free-martin *m*	freemartin *m*
3577	—	—
3578	microtome *m* à congélation	micrótomo *m* de congelación
3579	point *m* de congélation	punto *m* de congelación
3580	distribution *f* des fréquences	distribución *f* de frecuencias
3581	modulation *f* de fréquence	modulación *f* de frecuencia
3582	fréquence *f* du cœur	frecuencia *f* cardíaca
3583	fréquence *f* respiratoire	frecuencia *f* respiratoria
3584	préparation *f* non fixée	preparación *f* en fresco
3585	eau *f* douce	agua *f* dulce
3586	animal *m* dulçaquicole	animal *m* dulciacuícola
3587	—	—
3588	plante *f* d'eau douce	planta *f* de agua dulce
3589	poids *m* frais	peso *m* fresco
3590	réaction *f* d'alarme	reacción *f* de alarma
3591	fronde *f*	fronda *f*
3592	(os *m*) frontal *m*	(hueso *m*) frontal *m*
3593	lobe *m* frontal	lóbulo *m* frontal
3594	sinus *m* frontal	seno *m* frontal
3595	dégât *m* de gel, ~ par la gelée	daño *m* causado por las heladas

3596	frost resistance, ~ hardiness	Frostresistenz *f*, Frosthärte *f*
3597	frozen section	Gefrierschnitt *m*
3598	fructiferous, fruit-bearing	fruchttragend
3599	fructification	Fruchtbildung *f*, Fruktifikation *f*
3600	to fructify	Frucht tragen, ~ bilden, fruktifizieren
3601	fructose, l(a)evulose, fruit sugar	Fruktose *f*, Lävulose *f*, Fruchtzucker *m*
3602	frugivorous	früchtefressend
3603	fruit	Frucht *f*
3604	fruit-bearing → fructiferous	—
3605	fruit-body, fruiting body	Fruchtkörper *m*
3606	fruit bud	Tragknospe *f*, Fruchtauge *n*
3607	fruit drop	Fruchtfall *m*, Fruchtabwurf *m*
3608	fruit setting	Fruchtansatz *m*
3609	fruit-stalk	Fruchtstiel *m*
3610	fruit sugar → fructose	—
3611	fruiting lateral	Fruchtholz *n*
3612	frustule	Frustel *f*
3613	frutex → shrub	—
3614	fruticose, shrubby	strauchförmig, strauchartig
3615	fruticose lichen	Strauchflechte *f*
3616	FSH → follicle-stimulating hormone	—
3617	fucoxanthin	Fukoxanthin *n*
3618	fumarase	Fumarase *f*
3619	fumaric acid	Fumarsäure *f*
3620	function(al) unit	Funktionseinheit *f*
3621	fundamental number → basic number	—
3622	fundamental particle → elementary particle	—
3623	fundamental structure	Grundstruktur *f*
3624	fungi (*sing*. fungus), Mycophyta	Pilze *mpl*, Fungi *pl*
3625	fungi imperfectae, imperfect fungi, deuteromycetes	unvollständige Pilze *mpl*, Fungi imperfecti
3626	fungicide	Fungizid *n*, pilztötendes Mittel *n*
3627	funicle *bot*., seed-stalk	Funikulus *m bot*., Samenstiel *m*
3628	funnel-shaped → infundibular	—
3629	to fuse	verschmelzen
3630	fusiform	spindelförmig
3631	fusion	Verschmelzung *f*
3632	fusion nucleus	Fusionskern *m*
3633	galactin → lactogenic hormone	—
3634	galactolipid, galactolipin(e)	Galaktolipid *n*
3635	galactophorous duct → lactiferous duct	—
3636	galactose	Galaktose *f*
3637	galbulus	Beerenzapfen *m*
3638	gall-bladder, cholecyst	Gallenblase *f*
3639	galvanotaxis	Galvanotaxis *f*
3640	galvanotropism, electrotropism	Galvanotropismus *m*, Elektrotropismus *m*
3641	gametangiogamy	Gametangiogamie *f*
3642	gametangium	Gametangium *n*
3643	gamete, sexual cell	Gamet *m*, Geschlechtszelle *f*, Sexualzelle *f*
3644	gametic	gametisch, Gameten...
3645	gametic nucleus	Gametenkern *m*
3646	gametocyte	Gametozyt *m*
3647	gametogamy	Gametogamie *f*
3648	gametogenesis, gametogeny	Gametogenese *f*, Gametogenie *f*, Keimzellenbildung *f*
3649	gametogenic, gametogenous	gametogen
3650	gametogony → gamogony	—
3651	gametophore	Gametophor *m*
3652	gametophyte	Gametophyt *m*
3653	gammaglobulin	Gammaglobulin *n*
3654	gamma rays	Gammastrahlen *mpl*
3655	gamobium	Gamobium *n*
3656	gamogony, gametogony	Gamogonie *f*, Gametogonie *f*
3657	gamone	Gamon *n*, Befruchtungsstoff *m*

3596	résistance *f* au gel	resistencia *f* a la(s) helada(s)
3597	coupe *f* à la congélation, ~ de tissu congelé	corte *m* por congelación, sección *f* ~ ~
3598	fructifère, fructifiant	fructífero
3599	fructification *f*	fructificación *f*
3600	fructifier	fructificar
3601	fructose *m*, lévulose *m*, sucre *m* de fruit	fructosa *f*, levulosa *f*, azúcar *m* de fruto
3602	frugivore	frugívoro
3603	fruit *m*	fruto *m*
3604	—	—
3605	corpuscule *m* reproducteur	cuerpo *m* fructífero
3606	bourgeon *m* à fruits	yema *f* de fruta
3607	chute *f* des fruits, abscission *f* ~ ~	desprendimiento *m* de los frutos, abscisión *f* ~ ~ ~
3608	mise *f* à fruit	cuajado *m* del fruto
3609	pédoncule *m* fructifère	pedúnculo *m* fructífero
3610	—	—
3611	branche *f* fruitière	rama *f* fructífera
3612	frustule *m*	frústulo *m*
3613	—	—
3614	frutescent, buissonnant, arbustif	fruticoso, arbustivo
3615	lichen *m* fruticuleux	liquen *m* fruticuloso
3616	—	—
3617	fucoxanthine *f*, phykoxanthine *f*	fucoxantina *f*, ficoxantina *f*
3618	fumarase *f*	fumarasa *f*
3619	acide *m* fumarique	ácido *m* fumárico
3620	unité *f* fonctionnelle	unidad *f* funcional
3621	—	—
3622	—	—
3623	structure *f* fondamentale	estructura *f* fundamental
3624	champignons *mpl*, mycophytes *mpl*	micofitos *mpl*, hongos *mpl*
3625	champignons *mpl* imparfaits, deutéromycètes *mpl*	hongos *mpl* imperfectos, deuteromicetos *mpl*
3626	fongicide *m*	fungicida *m*
3627	funicule *m* *bot.*	funículo *m* *bot.*
3628	—	—
3629	fusionner	fusionar(se)
3630	fusiforme	fusiforme
3631	fusion *f*	fusión *f*
3632	noyau *m* de fusion	núcleo *m* de fusión
3633	—	—
3634	galactolipide *m*	galactolípido *m*, galactolipina *f*
3635	—	—
3636	galactose *m*	galactosa *f*
3637	galbule *m*	gálbulo *m*, estróbilo *m* abayado
3638	vésicule *f* biliaire, cholécyste *m*	vesícula *f* biliar, colecisto *m*
3639	galvanotaxie *f*, galvanotactisme *m*	galvanotaxis *f*
3640	galvanotropisme *m*	galvanotropismo *m*, electrotropismo *m*
3641	gamétangie *f*	gametangiogamia *f*
3642	gamétange *m*	gametangio *m*
3643	gamète *m*, cellule *f* sexuelle	gameto *m*, gámeta *m*, célula *f* sexual
3644	gamétique	gamético
3645	noyau *m* gamétique	núcleo *m* gamético
3646	gamétocyte *m*	gametocito *m*
3647	gamétogamie *f*	gametogamia *f*
3648	gamétogenèse *f*, gamétogénie *f*	gametogénesis *f*
3649	gamétogénique	gametogénico
3650	—	—
3651	gamétophore *m*	gametóforo *m*
3652	gamétophyte *m*	gametofito *m*, gametófito *m*
3653	gammaglobuline *f*	gammaglobulina *f*, globulina *f* gamma
3654	rayons *mpl* gamma	rayos *mpl* gamma
3655	gamobium *m*	gamobio *m*
3656	gamogonie *f*, gamétogonie *f*	gamogonia *f*
3657	gamone *f*, substance *f* de fécondation	gamona *f*, sustancia *f* de fecundación

3658	gamont	Gamont *m*
3659	gamopetalous, sympetalous	gamopetal, sympetal, verwachsenblättrig
3660	gamophase	Gamophase *f*
3661	gamosepalous	gamosepal
3662	gamotropism	Gamotropismus *m*
3663	ganglion (*pl* ganglia)	Ganglion *n*
3664	ganglion cell, gangliocyte	Ganglienzelle *f*
3665	ganoid scale	Ganoidschuppe *f*, Schmelzschuppe *f*
3666	gaping *eth.*	Sperren *n* *eth.*
3667	gaping reaction	Sperreaktion *f*
3668	gas chromatography	Gaschromatographie *f*
3669	gaseous interchange, gas exchange	Gasaustausch *m*
3670	gastral cavity	Gastralhöhle *f*
3671	gastric, gastral, stomachic	Magen..., Gastral..., gastrisch
3672	gastric gland, stomach ~	Magendrüse *f*
3673	gastric juice	Magensaft *m*
3674	gastrocoel	Gastrozöl *n*
3675	gastrointestinal	Magendarm...
3676	Gastropoda, gastropods	Bauchfüßer *m pl*, Gastropoden *pl*, Schnecken *f pl*
3677	gastrovascular cavity	Gastrovaskularraum *m*, ~system *n*
3678	gastrula	Gastrula *f*, Becherkeim *m*
3679	gastrulation	Gastrulation *f*
3680	Geiger(-Müller) counter	Geigerzähler *m*
3681	geitonocarpy	Geitonokarpie *f*
3682	geitonogamy	Geitonogamie *f*, Nachbarbefruchtung *f*, ~bestäubung *f*
3683	gel	Gel *n*
3684	gelatin(e)	Gelatine *f*, Gallert *n*, Gallerte *f*
3685	gelatin(iz)ation, gelation, gelatification	Gelatinierung *f*, Gelieren *n*
3686	to gelatinize, to gelate	gelatinieren, gelieren
3687	gelatinizing agent	Gelatinierungsmittel *n*
3688	gelatinous, jelly-like	gelartig, gelatinös, gallertig
3689	gelose → agar	—
3690	geminate species → twin species	—
3691	geminus → bivalent	—
3692	gemmation	Knospung *f* (*Fortpflanzung*)
3693	gemmiferous, bud-bearing	knospentragend
3694	gemmiparity	Vermehrung *f* durch Knospenbildung
3695	gemmule	Gemmula *f*, Brutknospe *f*
3696	gene, hereditary factor	Gen *n* (*pl* Gene), Erbfaktor *m*
3697	gene action, genic ~	Genwirkung *f*
3698	gene activation	Genaktivierung *f*
3699	gene activity	Gentätigkeit *f*
3700	gene balance, genic ~	Genbalance *f*, genetische Balance *f*
3701	gene centre (*US:* center)	Genzentrum *n*
3702	gene combination	Genkombination *f*
3703	gene complex	Genkomplex *m*
3704	gene-cytoplasm isolation	Gen-Zytoplasma-Isolation *f*
3705	gene dosage, genic ~	Gendosis *f*
3706	gene-enzyme relationship	Gen-Enzym-Beziehung *f*
3707	gene exchange, genetic ~	Genaustausch *m*
3708	gene expression	Genmanifestierung *f*
3709	gene flow	Genfluß *m*, Gen-„flow“ *m*
3710	gene frequency	Genhäufigkeit *f*, Genfrequenz *f*
3711	gene function	Genfunktion *f*
3712	gene interaction	Genwechselwirkung *f*
3713	gene linkage, genetic coupling	Genkopplung *f*, Faktorenkopplung *f*
3714	gene mutation, point ~, genovariation; transgenation *obs.*	Genmutation *f*, Punktmutation *f*, Genovariation *f*; Transgenation *f* *obs.*
3715	gene pool	Gen-Pool *m*
3716	gene product	Genprodukt *n*
3717	gene recombination	Genrekombination *f*

3658	gamonte *m*	gamonte *m*
3659	gamopétale, à pétales concrescents	gamopétalo, simpétalo, de pétalos concrescentes
3660	gamophase *f*	gamófase *f*
3661	gamosépale, à sépales concrescents	gamosépalo, de sépalos concrescentes
3662	gamotropisme *m*	gamotropismo *m*
3663	ganglion *m*	ganglio *m*
3664	cellule *f* ganglionnaire, gangliocyte *m*	célula *f* ganglionar, gangliocito *m*
3665	écaille *f* ganoïde	escama *f* ganoidea
3666	ouverture *f* du bec *eth.*	apertura *f* del pico *eth.*
3667	réaction *f* d'ouverture du bec	reacción *f* en abrir el pico
3668	chromatographie *f* gazeuse	cromatografía *f* de gases
3669	échange *m* gazeux	intercambio *m* gaseoso
3670	cavité *f* gastrale, ~ gastrique	cavidad *f* gástrica
3671	gastrique, stomacal	gástrico, estomacal
3672	glande *f* gastrique	glándula *f* gástrica
3673	suc *m* gastrique	jugo *m* gástrico
3674	gastrocèle *m*	gastrocele *m*
3675	gastro-intestinal	gastrointestinal
3676	gastéropodes *mpl*	gast(e)rópodos *mpl*
3677	système *m* gastro-vasculaire	cavidad *f* gastrovascular
3678	gastrula *f*	gástrula *f*
3679	gastrulation *f*	gastrulación *f*
3680	compteur *m* (de) Geiger(-Müller)	contador *m* (de) Geiger
3681	géitonocarpie *f*	geitonocarpia *f*
3682	géitonogamie *f*	geitonogamia *f*
3683	gel *m*	gel *m*
3684	gélatine *f*	gelatina *f*
3685	gélatin(is)ation *f*, gélation *f*, gélification *f*	gelatinización *f*, gelación *f*, gelificación *f*, gelatificación *f*
3686	gélatiniser, (se) gélifier	gelatinizar, gelificar, gelatinificar
3687	gélifiant *m*, gélatinisant *m*	agente *m* gelificante, ~ gelatinizante
3688	gélatineux	gelatinoso
3689	—	—
3690	—	—
3691		—
3692	gemmation *f*	gemación *f*
3693	gemmifère	gemífero
3694	gemmiparité *f*	gemiparidad *f*
3695	gemmule *f*	gémula *f*
3696	gène *m*, facteur *m* héréditaire	gen(e) *m*, factor *m* hereditario
3697	action *f* des gènes, ~ génique	acción *f* génica
3698	activation *f* du gène	activación *f* del gen
3699	activité *f* génique	actividad *f* de los genes
3700	balance *f* génique, équilibre *m* ~	equilibrio *m* génico
3701	centre *m* génétique, génocentre *m*	centro *m* genético, genocentro *m*
3702	combinaison *f* de gènes	combinación *f* de genes
3703	complexe *m* de gènes	complejo *m* de genes
3704	séparation *f* du gène du cytoplasme	aislamiento *m* genocitoplásmico
3705	dosage *m* des gènes, ~ génique	dosis *f* génica
3706	relation *f* enzyme-gène	relación *f* enzima-gen
3707	échange *m* de gènes, ~ génétique	intercambio *m* genético
3708	expression *f* génétique	expresión *f* genética, ~ génica
3709	écoulement *m* de gènes	flujo *m* de genes
3710	fréquence *f* des gènes, ~ génique	frecuencia *f* de genes, ~ génica
3711	fonction *f* génétique	función *f* génica
3712	interaction *f* des gènes	interacción *f* de los genes, ~ factorial
3713	liaison *f* génétique, accouplement *m* ~	ligamiento *m* genético, "linkaje" *m* ~
3714	mutation *f* génique, ~ au point, ~ ponctuelle, ~ factorielle, génovariation *f*; transgénation *f* *obs.*	mutación *f* génica, ~ puntual; transgenación *f* *obs.*
3715	fonds *m* gén(ét)ique, pool *m* de gènes	fondo *m* genético, "pool" *m* de genes
3716	produit *m* du gène	producto *m* del gen, ~ génico
3717	recombinaison *f* des gènes	recombinación *f* de genes, ~ génica

3718	gene reduplication	Genreduplikation *f*
3719	gene replica hypothesis	Gen-Replica-Hypothese *f*
3720	gene transfer	Genübertragung *f*
3721	genealogical table	Ahnentafel *f*
3722	genecology	Genökologie *f*
3723	generation	Generation *f*
3724	generation index	Generationsindex *m*
3725	generation interval	Generationsintervall *n*
3726	generation time	Generationszeit *f*
3727	generative	generativ
3728	generative nucleus	generativer Kern *m*
3729	generator potential	Generatorpotential *n*
3730	generic	Gattungs ..., generisch
3731	generic hybrid → intergeneric ~	—
3732	generitype	Gattungstyp *m*
3733	genesis	Genese *f*, Genesis *f*
3734	genetic	genetisch
3735	genetic analysis	Erbanalyse *f*
3736	genetic assimilation	genetische Assimilation *f*
3737	genetic blocking	genetische Blockierung *f*
3738	genetic code	genetischer Kode *m*
3739	genetic complement	Genbestand *m*
3740	genetic drift, gene ~	Gendrift *f*, Allelendrift *f*
3741	genetic equilibrium	genetisches Gleichgewicht *n*
3742	genetic information	genetische Information *f*
3743	genetic load	genetische Ladung *f*
3744	genetic map, factor ~	Genkarte *f*
3745	genetic material	Genmaterial *n*
3746	genetic tolerance dose, permissible dose	genetische Toleranzdosis *f*
3747	geneticist, genetician	Genetiker *m*
3748	genetics	Genetik *f*, Vererbungslehre *f*, Erblehre *f*
3749	genic	Gen ...
3750	genital aperture, reproductive ~	Geschlechtsöffnung *f*, Genitalöffnung *f*
3751	genital duct	Genitalkanal *m*
3752	genital gland → gonad	—
3753	genital organ → sex organ	—
3754	genital pouch	Begattungstasche *f*
3755	genital ridge, germinal ~	Genitalleiste *f*, Keimleiste *f*
3756	genital tract, ~ system	Geschlechtsapparat *m*, Genitalsystem *n*
3757	genitals, genitalia, genital parts	Genitalien *npl*, Geschlechtsteile *npl*
3758	genocopy	Genokopie *f*
3759	genom(e)	Genom *n*
3760	genomal mutation, genome ~	Genommutation *f*
3761	genomere	Genomere *f*
3762	genomic	Genom ...
3763	genonema (*pl* genonemata)	Genonema *n* (*pl* Genonemata)
3764	genosome	Genosom *n*
3765	genotype	Genotyp(us) *m*, Erbtyp(us) *m*, Erbbild *n*
3766	genotypic(al)	genotypisch
3767	genotypic character	genotypisches Merkmal *n*
3768	genovariation → gene mutation	—
3769	genus (*pl* genera)	Gattung *f*, Genus *n* (*pl* Genera)
3770	genus name	Gattungsname *m*
3771	geobiology	Geobiologie *f*
3772	geobotany → phytogeography	—
3773	geocarpy	Erdfrüchtigkeit *f*, Geokarpie *f*
3774	geographic isolation	geographische Isolation *f*
3775	geographic(al) race	geographische Rasse *f*
3776	geological epoch	Erdzeitalter *n*
3777	geophytes	Geophyten *mpl*, Erdpflanzen *fpl*
3778	geotaxis	Geotaxis *f*
3779	geotropism	Geotropismus *m*, Erdwendigkeit *f*
3780	germ	Keim *m*
3781	to germ	keimen

3718 réduplication *f* des gènes — reduplicación *f* de genes
3719 hypothèse *f* sur la réplication du gène — hipótesis *f* de la genocopia
3720 transfert *m* de gènes — transferencia *f* de genes
3721 tableau *m* généalogique — tabla *f* genealógica
3722 écologie *f* génétique — genecología *f*
3723 génération *f* — generación *f*
3724 indice *m* de générations — número *m* de generaciones
3725 intervalle *m* entre les générations — intervalo *m* de generaciones, ~ entre las generaciones
3726 temps *m* de génération — período *m* de generación
3727 génératif, générateur — generativo, generador
3728 noyau *m* génératif, ~ reproducteur — núcleo *m* generativo
3729 potentiel *m* générateur — potencial *m* generador
3730 générique — genérico
3731 — — —
3732 génértype *m* — generotipo *m*
3733 genèse *f* — génesis *f*
3734 génétique — genético
3735 analyse *f* génétique — análisis *m* genético
3736 assimilation *f* génétique — asimilación *f* genética
3737 blocage *m* génétique — bloqueo *m* genético
3738 code *m* génétique — código *m* genético
3739 stock *m* de gènes, patrimoine *m* génétique — equipo *m* de genes, patrimonio *m* genético, dotación *f* génica, ~ genética
3740 dérive *f* génique, «genetic drift» — deriva *f* genética
3741 équilibre *m* génétique — equilibrio *m* genético
3742 information *f* génétique — información *f* genética
3743 charge *f* génétique — carga *f* genética
3744 carte *f* génétique, ~ génique, ~ factorielle — mapa *m* genético, ~ génico, ~ factorial
3745 matériel *m* génique — material *m* genético
3746 dose *f* de tolérance génétique, ~ tolérée — dosis *f* permisible
3747 généticien *m*, génétiste *m* — geneticista *m*, genetista *m*
3748 génétique *f* — genética *f*
3749 génique — génico
3750 orifice *m* génital — abertura *f* genital, orificio *m* ~
3751 conduit *m* génital, canal *m* ~ — conducto *m* genital
3752 — — —
3753 — — —
3754 bourse *f* copulatrice — bolsa *f* copuladora, ~ copulatriz
3755 crête *f* génitale — pliegue *m* gonadal, ~ germinativo
3756 appareil *m* génital, tractus *m* ~ — sistema *m* genital, aparato *m* ~, tracto *m* ~
3757 génitalias *fpl*, parties *fpl* génitales — genitales *mpl*, partes *fpl* sexuales
3758 génocopie *f* — genocopia *f*
3759 génome *m* — genoma *m*
3760 mutation *f* de génomes — mutación *f* genómica
3761 génomère *m* — genómero *m*
3762 génomique — genómico
3763 génonème *m* — genonema *m*
3764 génosome *m* — genósoma *m*
3765 génotype *m* — genotipo *m*
3766 génotypique — genotípico
3767 caractère *m* génotypique — característica *f* genotípica
3768 — — —
3769 genre *m* — género *m*
3770 nom *m* générique — nombre *m* genérico
3771 géobiologie *f* — geobiología *f*
3772 — — —
3773 géocarpie *f* — geocarpia *f*
3774 isolement *m* géographique — aislamiento *m* geográfico
3775 race *f* géographique — raza *f* geográfica
3776 ère *f* géologique, période *f* ~ — época *f* geológica, período *m* geológico
3777 géophytes *mpl* — geófitos *mpl*
3778 géotaxie *f*, géotactisme *m* — geotaxis *f*, geotactismo *m*
3779 géotropisme *m* — geotropismo *m*
3780 germe *m* — germen *m*
3781 germer — germinar

3782	germ band, germinal streak	Keimstreifen *m*
3783	germ cell, germinative ~	Keimzelle *f*
3784	germ centre (US: center)	Keimzentrum *n*
3785	germfree	keimfrei
3786	germ layer	Keimblatt *n* *embr.*
3787	germ plasm, idioplasm	Keimplasma *n*, Erbplasma *n*, Idioplasma *n*
3788	germ-proof filter → bacteria-proof ~	—
3789	germ track	Keimbahn *f*
3790	germ tube	Keimschlauch *m*
3791	germarium	Germarium *n*, Keimfach *n*
3792	germicidal	keimtötend
3793	germinal	Keim ...
3794	germinal area	Keimzone *f*
3795	germinal continuity → continuity of the germ plasm	—
3796	germinal disc → blastodisc	—
3797	germinal epithelium	Keimepithel *n*
3798	germinal pole	Keimpol *m*
3799	germinal ridge → genital ~	—
3800	germinal selection	Germinalselektion *f*
3801	germinal spot	Keimfleck *m*
3802	germinal streak → germ band	—
3803	germinal vesicle	Keimbläschen *n*
3804	germinating power	Keimkraft *f*
3805	germination	Keimung *f*
3806	germination capacity, ~ ability	Keimfähigkeit *f*
3807	germination inhibition	Keim(ungs)hemmung *f*
3808	germination inhibitor	Keim(ungs)hemmer *m*
3809	germinative, capable of germination	keimfähig
3810	germinative layer → Malpighian layer	—
3811	gerontic stage	Altersstadium *n*
3812	gerontology	Gerontologie *f*, Alternsforschung *f*
3813	gestagene	Gestagen *n*
3814	gestalt psychologist	Gestaltpsychologe *m*
3815	gestalt psychology	Gestaltpsychologie *f*
3816	gestaltism	Gestalttheorie *f*
3817	gestation	Trächtigkeit *f*, Schwangerschaft *f*
3818	gestation time, ~ period	Trächtigkeitsdauer *f*, Tragezeit *f*
3819	giant axon	Riesenaxon *n*
3820	giant cell, gigantocyte	Riesenzelle *f*
3821	giant chromosome	Riesenchromosom *n*
3822	giant fibre (US: fiber)	Riesen(nerven)faser *f*, Kolossalfaser *f*
3823	gibberellic acid	Gibberellinsäure *f*
3824	gibberellin	Gibberellin *n*
3825	gigantism	Riesenwuchs *m*, Gigantismus *m*
3826	gigantocyte → giant cell	—
3827	gill, branchia	Kieme *f*, Branchie *f*
3828	gill arch, branchial ~	Kiemenbogen *m*, Branchialbogen *m*
3829	gill basket	Kiemenkorb *m*
3830	gill book	Kiemenblatt *n*, Kiemenlamelle *f*
3831	gill cavity, branchial ~	Kiemenhöhle *f*
3832	gill chamber	Kiemenkammer *f*
3833	gill circulation, branchial ~	Kiemenkreislauf *m*
3834	gill cleft, branchial ~, gill slit	Kiemenspalte *f*
3835	gill cover, ~ operculum	Kiemendeckel *m*
3836	gill filament	Kiemenfaden *m*
3837	gill opening, branchial ~	Kiemenöffnung *f*
3838	gill pouch, branchial ~	Kiementasche *f*
3839	gill raker	Kiemensieb *n*
3840	gill respiration, branchial ~	Kiemenatmung *f*
3841	gill slit → gill cleft	—

3782	bandelette *f* germinative	banda *f* germinal
3783	cellule *f* germinale, ~ germinative	célula *f* germinal, ~ germinativa
3784	centre *m* germinal	centro *m* germinal
3785	sans germes	libre de gérmenes, exento ~ ~
3786	feuillet *m* germinatif, ~ embryonnaire	hoja *f* embrionaria, ~ germinal
3787	plasme *m* germinatif, ~ germinal, idioplasme *m*, germen *m*	plasma *m* germinal, idioplasma *m*
3788	—	—
3789	lignée *f* germinale	linaje *m* germinal
3790	tube *m* germinatif	tubo *m* germinal
3791	germarium *m*	germario *m*, germígeno *m*
3792	germicide	germicida
3793	germinal	germinal
3794	aire *f* germinative, zone *f* ~	área *f* germinativa, ~ embrionaria, zona *f* germinal
3795	—	—
3796	—	—
3797	épithélium *m* germinal	epitelio *m* germinal
3798	pôle *m* germinal	polo *m* germinativo
3799	—	—
3800	sélection *f* germinale	selección *f* germinal
3801	tache *f* germinale	mancha *f* germinal
3802	—	—
3803	vésicule *f* germinative	vesícula *f* germinativa, ~ germinal
3804	pouvoir *m* de germination, ~ germinatif	poder *m* germinativo
3805	germination *f*	germinación *f*
3806	faculté *f* germinative, capacité *f* ~	facultad *f* germinativa, capacidad *f* ~, germinabilidad *f*
3807	inhibition *f* de la germination	inhibición *f* de la germinación
3808	inhibiteur *m* de la germination	inhibidor *m* de la germinación
3809	germinatif, capable de germer, apte à la germination	germinativo, capaz de germinar
3810	—	—
3811	stade *m* gérontique, ~ de vieillesse	fase *f* gerónticа
3812	gérontologie *f*	gerontología *f*
3813	gestagène *f*	gestágeno *m*
3814	gestaltiste *m*	psicólogo *m* gestaltista
3815	gestaltpsychologie *f*, psychologie *f* de la forme	psicología *f* de la gestalt
3816	gestaltisme *m*, gestalt-théorie *f*, théorie *f* de la forme	gestaltismo *m*, teoría *f* de la gestalt
3817	gestation *f*	gestación *f*
3818	durée *f* de la gestation, période *f* de ~	duración *f* de la gestación, período *m* de ~
3819	axone *m* géant	axón *m* gigante
3820	cellule *f* géante, gigantocyte *m*	célula *f* gigante, gigantocito *m*
3821	chromosome *m* géant	cromosoma *m* gigante
3822	fibre *f* géante	fibra *f* (nerviosa) gigante
3823	acide *m* gibbérellique	ácido *m* giberélico
3824	gibbérelline *f*	giberelina *f*
3825	gigantisme *m*	gigantismo *m*
3826	—	—
3827	branchie *f*; ouïe *f* *pop.*	branquia *f*; agalla *f* *pop.*
3828	arc *m* branchial	arco *m* branquial
3829	corbeille *f* branchiale	canastilla *f* branquial, cesta *f* ~
3830	feuillet *m* branchial, lame *f* branchiale	lámina *f* branquial
3831	cavité *f* branchiale	cavidad *f* branquial
3832	chambre *f* branchiale	cámara *f* branquial
3833	circulation *f* branchiale	circulación *f* branquial
3834	fente *f* branchiale	hendidura *f* branquial
3835	opercule *m* branchial	opérculo *m* branquial
3836	filament *m* branchial	filamento *m* branquial
3837	orifice *m* branchial	abertura *f* branquial
3838	poche *f* branchiale	bolsa *f* branquial
3839	filtre *m* branchial	criba *f* branquial
3840	respiration *f* branchiale	respiración *f* branquial
3841	—	—

No.	English	German
3842	gingival	Zahnfleisch ...
3843	ginglymus → hinge-joint	—
3844	girdling *bot.*	Ringelung *f bot.*
3845	gizzard, masticatory stomach	Kaumagen *m*, Muskelmagen *m*
3846	glacial epoch, ice age, glacial ~	Eiszeit *f*
3847	glacial refuge, ~ refugium	Glazialrefugium *n*
3848	glacial relict	Glazialrelikt *n*, Eiszeitrelikt *n*
3849	glaciation	Vereisung *f*, Vergletscherung *f*
3850	gland	Drüse *f*
3851	glandular	Drüsen ...
3852	glandular cell, adenocyte	Drüsenzelle *f*
3853	glandular epithelium	Drüsenepithel *n*
3854	glandular hair	Drüsenhaar *n*
3855	glandular scale	Drüsenschuppe *f*
3856	glandular stomach → proventriculus	—
3857	glandular tissue	Drüsengewebe *n*
3858	glandulous, adenose	drüsig
3859	glass-slide → microscope slide	—
3860	glia, neuroglia	Glia *f*, Neuroglia *f*
3861	gliacyte, neuroglia cell, glial ~	Gliazelle *f*
3862	gliadin	Gliadin *n*
3863	gliding movement	Gleitbewegung *f*
3864	globigerina ooze	Globigerinenschlamm *m*
3865	globin	Globin *n*
3866	globoid	Globoid *n*
3867	globulin	Globulin *n*
3868	glomerule, flower cluster	Blütenknäuel *n*
3869	glossal → lingual	—
3870	glossopharyngeal nerve	Schlundnerv *m*, Glossopharyngeus *m*
3871	glottis	Stimmritze *f*, Glottis *f*
3872	glucagon	Glukagon *n*
3873	glucide → carbohydrate	—
3874	glucocorticoid	Glukokortikoid *n*
3875	gluconic acid	Glukonsäure *f*
3876	glucoprotein, glycoprotein, mucoprotein	Glukoprotein *n*, Glykoprotein *n*, Mukoprotein *n*
3877	glucosamine	Glukosamin *n*
3878	glucose, dextrose, grape sugar	Glukose *f*, Dextrose *f*, Traubenzucker *m*
3879	glucose phosphate	Glukosephosphat *n*
3880	glucoside → glycoside	—
3881	glume	Spelze *f*, Hüllspelze *f*
3882	glutamic acid	Glutaminsäure *f*
3883	glutamine	Glutamin *n*
3884	glutaric acid	Glutarsäure *f*
3885	glutathione	Glutathion *n*
3886	glutelin	Glutelin *n*
3887	gluten	Kleber *m*, Gluten *n*
3888	glyc(a)emia	Blutzuckergehalt *m*, Glykämie *f*
3889	glyceric acid	Glyzerinsäure *f*
3890	glycerin, glycerol	Glyzerin *n*
3891	glycine, glycocoll, aminoacetic acid	Glyzin *n*, Glykokoll *n*, Aminoessigsäure *f*
3892	glycogen, animal starch	Glykogen *n*, tierische Stärke *f*
3893	glycogenesis, glycogeny	Glykogenie *f*, Glykogenaufbau *m*
3894	glycogenolysis	Glykogenolyse *f*
3895	glycolic acid	Glykolsäure *f*
3896	glycolysis	Glykolyse *f*
3897	glycometabolism, sugar metabolism	Zuckerstoffwechsel *m*
3898	glycoprotein → glucoprotein	—
3899	glycoside, glucoside	Glykosid *n*, Glukosid *n*
3900	glyoxylic cycle, glyoxalate ~	Glyoxylsäurezyklus *m*, Glyoxalatzyklus *m*
3901	gnathic → maxillary	—
3902	Gnathostomata	Gnathostomen *pl*
3903	goblet cell, chalice ~, caliciform ~	Becherzelle *f*

3842	gingival	gingival
3843	—	—
3844	décortication *f* annulaire, annélation *f*	decorticación *f* anular
3845	gésier *m*, estomac *m* musculaire, ~ masticateur	molleja *f*
3846	période *f* glaciaire, époque *f* ~	período *m* glacial, época *f* ~
3847	refuge *m* glaciaire	refugio *m* glacial
3848	relique *f* glaciaire	reliquia *f* glacial
3849	glaciation *f*	glaciación *f*
3850	glande *f*	glándula *f*
3851	glandulaire	glandular
3852	cellule *f* glandulaire, adénocyte *m*	célula *f* glandular, adenocito *m*
3853	épithélium *m* glandulaire	epitelio *m* glandular
3854	poil *m* glanduleux	pelo *m* glandular
3855	écaille *f* glanduleuse	escama *f* glandular
3856	—	—
3857	tissu *m* glandulaire	tejido *m* glandular
3858	glanduleux	glanduloso, adenoso
3859	—	—
3860	névroglie *f*	glía *f*, neuroglía *f*
3861	cellule *f* névroglique, ~ gliale	célula *f* (neuro)glial, gliacito *m*
3862	gliadine *f*	gliadina *f*
3863	mouvement *m* de glissement	movimiento *m* deslizante
3864	boues *f pl* à globigérines	barro *m* de globigerina, fango *m* ~ ~
3865	globine *f*	globina *f*
3866	globoïde *m*	globoide *m*
3867	globuline *f*	globulina *f*
3868	glomérule *m* floral	glomérulo *m* floral
3869	—	—
3870	nerf *m* glosso-pharyngien	nervio *m* glosofaríngeo
3871	glotte *f*	glotis *f*
3872	glucagon *m*	glucagón *m*
3873	—	—
3874	glucocorticoïde *m*	glucocorticoide *m*
3875	acide *m* gluconique	ácido *m* glucónico
3876	glucoprotéine *f*, glycoprotéine *f*	glucoproteína *f*, glicoproteína *f*, mucoproteína *f*
3877	glucosamine *f*, dextrosamine *f*	glucosamina *f*, glicosamina *f*, dextrosamina *f*
3878	glucose *m*, dextrose *m*	glucosa *f*, dextrosa *f*
3879	glucose-phosphate *m*	glucosafosfato *m*
3880	—	—
3881	glume *f*	gluma *f*
3882	acide *m* glutamique	ácido *m* glutámico
3883	glutamine *f*	glutamina *f*
3884	acide *m* glutarique	ácido *m* glutárico
3885	glutathion *m*	glutation *m*, glutationa *f*
3886	glutéline *f*	glutelina *f*
3887	gluten *m*	gluten *m*
3888	glycémie *f*	glicemia *f*
3889	acide *m* glycérique	ácido *m* glicérico
3890	glycérine *f*, glycérol *m*	glicerina *f*, glicerol *m*
3891	glycine *f*, glycocolle *m*, acide *m* amino-acétique	glicina *f*, glicocola *f*, ácido *m* aminoacético
3892	glycogène *m*, amidon *m* animal	glicógeno *m*, glucógeno *m*, almidón *m* animal
3893	glycogenèse *f*	glicogénesis *f*, glicogenia *f*
3894	glycogénolyse *f*	glicogenólisis *f*, glucogenólisis *f*
3895	acide *m* glycolique	ácido *m* glicólico
3896	glycolyse *f*	glicólisis *f*
3897	glycométabolisme *m*	glucometabolismo *m*, metabolismo *m* de los azúcares
3898	—	—
3899	glucoside *m*, glycoside *m*	glucósido *m*, glicósido *m*
3900	cycle *m* glyoxylique	ciclo *m* glioxílico, ~ del glioxalato
3901	—	—
3902	gnathostomes *m pl*	gnatóstomos *m pl*
3903	cellule *f* caliciforme	célula *f* caliciforme

No.	English	German
3904	golden brown algae → Chrysophyta	—
3905	Golgi apparatus, ~ complex	Golgi-Apparat *m*, Golgi-Feld *n*
3906	Golgi body	Golgikörperchen *n*
3907	golgiokinesis	Golgiokinese *f*
3908	gonad, genital gland, sex(ual) ~	Keimdrüse *f*, Geschlechtsdrüse *f*, Gonade *f*
3909	gonad hormone → sex hormone	—
3910	gonadotrop(h)ic	gonadotrop
3911	gonadotrop(h)ic hormone, gonadotrop(h)in	gonadotropes Hormon *n*, Gonadotropin *n*
3912	gone	Gone *f*
3913	gonid(ium) (*pl* gonidia)	Gonidium *n*, Gonidie *f*
3914	gonimoblast	Gonimoblast *m*
3915	gonochorism	Gonochorismus *m*
3916	gonocyte	Gonozyte *f*
3917	gonoduct, gonaduct	Gonodukt *m*
3918	gonomery	Gonomerie *f*
3919	Graafian follicle, Graffian ~, ovarian ~	Graafscher Follikel *m*, Bläschenfollikel *m*
3920	gradation → mass propagation	—
3921	grading up, substitution crossing, absorption ~	Verdrängungskreuzung *f*
3922	graft[1] → transplant	—
3923	graft[2], scion; cion *US*	Pfropfreis *n*, Edelreis *n*
3924	graft hybrid, ~ chim(a)era	Pfropfbastard *m*, Pfropfchimäre *f*
3925	grafting → transplantation	—
3926	grain	Samenkorn *n*
3927	grain size, size of particles	Korngröße *f*
3928	gram-calorie	Kleinkalorie *f*, Grammkalorie *f*
3929	gram-negative	gramnegativ
3930	gram-positive	grampositiv
3931	Gram reaction	Gramreaktion *f*
3932	Gram stain, Gram's ~	Gramfärbung *f*
3933	grana *pl*	Grana *pl*
3934	granivorous, seed-eating	körnerfressend
3935	granular	körnig
3936	to granulate	granulieren
3937	granulation	Granulation *f*, Granulierung *f*
3938	granulocyte, polymorph, polymorphonuclear leucozyte	Granulozyt *m*, polymorphkerniger Leukozyt *m*
3939	grape sugar → glucose	—
3940	grasping hand	Greifhand *f*
3941	gravidity → pregnancy	—
3942	graviperception	Schwerewahrnehmung *f*
3943	gravitational field	Schwerefeld *n*, Gravitationsfeld *n*
3944	gravitational force, force of gravity	Schwerkraft *f*
3945	gravitational organ, gravity receptor	Schweresinnesorgan *n*
3946	gravitational stimulus, stimulus of gravity	Schwerereiz *m*, Schwerkraftreiz *m*
3947	gravitational water	Senkwasser *n*
3948	gray → grey	—
3949	greater circulation → systemic ~	—
3950	green algae → Chlorophyta	—
3951	green belt	Grünfläche *f*
3952	gregarious	gesellig, in Herden lebend
3953	gregariousness → herd instinct	—
3954	grey crescent	grauer Halbmond *m*
3955	grey matter	graue Substanz *f*
3956	groin	Leiste *f*
3957	grooming behavio(u)r *eth.*	Hautpflegehandlungen *fpl*, Putzbewegungen *fpl* *eth.*
3958	gross production *ecol.*	Bruttoproduktion *f* *ecol.*
3959	ground animal → subterranean ~	—
3960	ground layer → herbaceous ~	—
3961	ground-nester	Bodenbrüter *m*, Bodennister *m*
3962	ground substance	Grundsubstanz *f*
3963	ground tissue, basic ~	Grundgewebe *n*
3964	ground water	Grundwasser *n*

3904	—	—
3905	appareil *m* de Golgi	aparato *m* de Golgi, complejo *m* ~ ~
3906	corps *m* de Golgi	cuerpo *m* de Golgi
3907	golgiocinèse *f*	golgiocinesis *f*
3908	gonade *f*, glande *f* génitale, ~ sexuelle	gónada *f*, gonada *f*, glándula *f* sexual
3909	—	—
3910	gonadotrope, gonadotrophique	gonadótropo, gonadotrópico
3911	hormone *f* gonadotrope, gonadotrophine *f*, gonadostimuline *f*	hormona *f* gonadotrópica, ~ gonadotrófica, gonadotropina *f*, gonadotrofina *f*
3912	gone *m*, gonie *f*	gonio *m*
3913	gonidie *f*	gonidio *m*, gonidia *f*
3914	gonimoblaste *m*	gonimoblasto *m*
3915	gonochorisme *m*	gonocorismo *m*
3916	gonocyte *m*	gonócito *m*
3917	gonoducte *m*	gonaducto *m*
3918	gonomérie *f*	gonomería *f*
3919	follicule *m* de Graaf, ~ ovarien	folículo *m* de Graaf, ~ ovárico
3920	—	—
3921	croisement *m* d'absorption, ~ d'implantation, ~ de substitution	cruzamiento *m* de absorción, ~ por sustitución
3922	—	—
3923	greffon *m*	injerto *m*
3924	hybride *m* de greffe, chimère *f* ~ ~	híbrido *m* de injerto
3925	—	—
3926	graine *f*	grano *m*
3927	dimension *f* des particules (du sol)	tamaño *m* de las partículas, ~ de los granos
3928	gramme-calorie *f*, petite calorie *f*	caloría *f* pequeña, ~ gramo
3929	gramnégatif	gramnegativo
3930	grampositif	grampositivo
3931	réaction de Gram, méthode *f* ~ ~	reacción *f* de Gram, método *m* ~ ~
3932	coloration *f* de Gram	coloración *f* de Gram
3933	grana *mpl*	grana *mpl*
3934	granivore	granívoro
3935	granuleux, granulé	granular, granulado
3936	granuler	granular
3937	granulation *f*	granulación *f*
3938	granulocyte *m*, poly(morpho)nucléaire *m*	granulocito *m*, polimorfo *m*, leucocito *m* polimorfonuclear
3939	—	—
3940	main *f* préhensile	mano *f* prensil
3941	—	—
3942	perception *f* de la gravité	percepción *f* de la gravedad
3943	champ *m* de gravitation, ~ gravitationnel	campo *m* gravitacional, ~ gravitatorio
3944	gravitation *f*, force *f* de pesanteur	gravitación *f*, fuerza *f* de la gravedad
3945	gravicepteur *m*	receptor *m* gravitatorio
3946	stimulus *m* de la gravité, ~ gravitique	estímulo *m* de la gravedad, ~ gravitacional
3947	eau *f* de gravitation	agua *f* de gravitación, ~ gravitacional
3948	—	—
3949	—	—
3950	—	—
3951	espace *m* vert, zone *f* verte	espacio *m* verde, zona *f* ~
3952	grégaire	gregario
3953	—	—
3954	croissant *m* gris	lúnula *f* gris, media luna *f* gris
3955	substance *f* grise	sustancia *f* gris
3956	aine *f*	ingle *f*
3957	comportement *m* de toilette *eth.*	movimientos *mpl* de limpieza *eth.*
3958	production *f* brute *ecol.*	producción *f* bruta *ecol.*
3959	—	—
3960	—	—
3961	oiseau *m* qui niche à terre	ave *f* que anida en el suelo
3962	substance *f* fondamentale	sustancia *f* fundamental
3963	tissu *m* fondamental	tejido *m* fundamental
3964	eaux *fpl* souterraines, ~ phréatiques	aguas *fpl* subterráneas, ~ freáticas

No.	English	German
3965	ground water level	Grundwasserspiegel *m*
3966	group cohesion	Gruppenbindung *f*
3967	group effect	Gruppeneffekt *m*
3968	group formation	Gruppenbildung *f*
3969	growing bud	treibendes Auge *n*
3970	growing point → vegetative ~	—
3971	growing region, ~ zone, growth ~	Wachstumszone *f*
3972	growing season, ~ period, growth ~	Wachstumsperiode *f*
3973	growth	Wachstum *n*
3974	growth control → growth regulation	—
3975	growth curvature	Wachstumskrümmung *f*
3976	growth curve	Wachstumskurve *f*
3977	growth factor	Wachstumsfaktor *m*
3978	growth form, form of growth	Wuchsform *f*
3979	growth hormone → somatotrophic ~	—
3980	growth in breadth, increase in girth	Breitenwachstum *n*
3981	growth in height, height growth	Höhenwachstum *n*, Höhenzuwachs *m*
3982	growth in length, longitudinal growth	Längenwachstum *n*
3983	growth in thickness, increase in diameter	Dickenwachstum *n*, Stärkenzuwachs *m*
3984	growth inhibition	Wachstumshemmung *f*
3985	growth inhibitor	Wachstumshemmstoff *m*
3986	growth promotion, ~ stimulation	Wachstumsförderung *f*
3987	growth promotor, ~ stimulant	Wachstumsförderer *m*
3988	growth rate	Wachstumsrate *f*, Wachstums-geschwindigkeit *f*
3989	growth regulation, ~ control	Wachstumsregulierung *f*, Wachstums-steuerung *f*
3990	growth regulator	Wachstumsregulator *m*
3991	growth retardation, retarded growth	Wachstumsverzögerung *f*
3992	growth ring	Wachstumsring *m*
3993	growth stage, stage of growth	Wachstumsphase *f*, Wachstumsstadium *n*
3994	growth stimulation → growth promotion	—
3995	growth substance	Wuchsstoff *m*
3996	growth zone → growing region	—
3997	guanidine	Guanidin *n*
3998	guanine	Guanin *n*
3999	guanylic acid	Guanylsäure *f*
4000	guard cell	Schließzelle *f*
4001	gum(s)	Zahnfleisch *n*
4002	gum duct	Gummigang *m*
4003	gustation	Schmecken *n*
4004	gustatory	Geschmacks...
4005	gustatory cell, taste ~	Geschmackszelle *f*
4006	gustatory organ, taste ~, organ of taste	Geschmacksorgan *n*
4007	gustatory pore, taste ~	Geschmacksporus *m*
4008	gustatory sensation	Geschmacksempfindung *f*
4009	gustatory sense, sense of taste	Geschmack(s)sinn *m*
4010	gustatory stimulus, taste ~	Geschmacksreiz *m*
4011	gut → intestine	—
4012	gut cavity	Darmhöhle *f*
4013	guttation	Guttation *f*
4014	gymnocarpic, gymnocarpous	nacktfrüchtig, gymnokarp
4015	gymnospermous, naked-seeded	nacktsamig
4016	gymnosperms, naked-seeded plants	Gymnospermen *pl*, Nacktsamer *mpl*, nacktsamige Pflanzen *fpl*
4017	gynaeceum, gyn(o)ecium	Gynäzeum *n*
4018	gynandrism	Gynandrie *f*
4019	gynandromorph	gynandromorph
4020	gynandromorphism	Gynandromorphismus *m*
4021	gynic → female	—
4022	gynodioecious	gynodiözisch

3965	nappe *f* phréatique, ~ souterraine	capa *f* freática, nivel *m* freático
3966	cohésion *f* du groupe	cohesión *f* de grupo
3967	effet *m* de groupe	efecto *m* de grupo
3968	groupement *m*, formation *f* de groupes	agrupación *f*, formación *f* del grupo
3969	œil *m* poussant	ojo *m* velado
3970	—	—
3971	zone *f* de croissance	zona *f* de crecimiento, región *f* ~ ~
3972	période *f* de croissance, saison *f* ~ ~	período *m* de crecimiento
3973	croissance *f*	crecimiento *m*
3974	—	—
3975	courbure *f* de croissance	curvatura *f* de crecimiento, encorvadura *f* ~ ~
3976	courbe *f* de croissance	curva *f* de crecimiento
3977	facteur *m* de croissance	factor *m* de crecimiento
3978	forme *f* de croissance	forma *f* de crecimiento
3979	—	—
3980	croissance *f* en largeur	crecimiento *m* en anchura, ~ en latitud
3981	croissance *f* en hauteur	crecimiento *m* en altura
3982	croissance *f* en longueur, ~ longitudinale	crecimiento *m* en longitud, ~ longitudinal
3983	croissance *f* en épaisseur, accroissement *m* diamétral	crecimiento *m* en grosor, ~ diametral
3984	inhibition *f* de la croissance	inhibición *f* del crecimiento
3985	inhibiteur *m* de croissance	inhibidor *m* del crecimiento, sustancia *f* inhibidora del crecimiento
3986	stimulation *f* de croissance	estimulación *f* del crecimiento
3987	stimulant *m* de croissance	estimulante *m* del crecimiento
3988	taux *m* de croissance, vitesse *f* ~ ~	índice *m* de crecimiento, tasa *f* ~ ~
3989	régulation *f* de croissance	regulación *f* del crecimiento, control *m* ~ ~
3990	régulateur *m* de croissance	regulador *m* de crecimiento, fitorregulador *m*
3991	retard *m* de croissance	retraso *m* del crecimiento, retardo *m* ~ ~
3992	anneau *m* de croissance	anillo *m* de crecimiento
3993	phase *f* de croissance, stade *m* ~ ~	fase *f* de crecimiento, etapa *f* ~ ~
3994	—	—
3995	substance *f* de croissance	sustancia *f* de crecimiento
3996	—	—
3997	guanidine *f*	guanidina *f*
3998	guanine *f*	guanina *f*
3999	acide *m* guanylique	ácido *m* guanílico
4000	cellule *f* stomatique	célula *f* de guarda, ~ de cierre, ~ oclusiva, ~ estomática
4001	gencive *f*	encía(s) *f* (*pl*)
4002	canal *m* gommifère	conducto *m* gumífero
4003	gustation *f*	gustación *f*
4004	gustatif	gustativo, gustatorio
4005	cellule *f* gustative, ~ du goût	célula *f* gustativa
4006	organe *m* gustatif, ~ du goût	órgano *m* del gusto
4007	pore *m* gustatif	poro *m* gustativo
4008	sensation *f* gustative	sensación *f* gustativa
4009	sens *m* du goût	sentido *m* del gusto
4010	stimulus *m* gustatif	estímulo *m* gustativo
4011	—	—
4012	cavité *f* intestinale	cavidad *f* intestinal
4013	guttation *f*; sudation *f* (*peu usité*)	gutación *f*
4014	gymnocarpe	gimnocarpo
4015	gymnospermique	gimnospermo, de semillas desnudas
4016	gymnospermes *f pl*	gimnospermas *f pl*, gimnospermos *m pl*, plantas *f pl* de semilla desnuda
4017	gynécée *m*	gineceo *m*
4018	gynandrie *f*	ginandrismo *m*
4019	gynandromorphe	ginandromorfo
4020	gynandromorphisme *m*	ginandromorfismo *m*
4021	—	—
4022	gynodioïque	ginodioico

4023	gynogenesis	Gynogenese *f*
4024	gynomerogony	Gynomerogonie *f*
4025	gynomonoecious	gynomonözisch
4026	gynophore	Gynophor *m*
4027	gynospore → macrospore	—
4028	gynostegium	Gynostegium *n*
4029	gynostemium	Gynostemium *n*
4030	gyratory shaker	Schüttelbecher *m*
4031	gyrus, cerebral convolution, brain ~	(Ge)hirnwindung *f*, Gyrus *m*
4032	habitat	Habitat *n*, Standort *m*
4033	habitat selection	Habitatselektion *f*
4034	habituation	Habituation *f*
4035	habitus	Habitus *m*
4036	Haeckel's law → biogenetic law	—
4037	haem, heme	Häm *n*
4038	haemagglutination	Hämagglutination *f*
4039	haemagglutinin	Hämagglutinin *n*
4040	haemal arch	Hämalbogen *m*
4041	haematin	Hämatin *n*
4042	haemato... see also haemo...	—
4043	haematocryal → cold-blooded	—
4044	haematogenesis	Hämatogenese *f*
4045	haematology	Hämatologie *f*
4046	haematophagous	blutsaugend, hämatophag
4047	haematothermal → warm-blooded	—
4048	haematoxylin	Hämatoxylin *n*
4049	haemerythrin	Hämerythrin *n*
4050	haemic, haemal	Blut...
4051	haemin	Hämin *n*
4052	haemoblast	Hämoblast *m*, Blutstammzelle *f*
4053	haemocoel(e)	Hämozöl *n*
4054	haemocyanin, haematocyanin	Hämozyanin *n*
4055	haemocyte, blood cell	Blutzelle *f*, Hämozyt *m*
4056	haemodynamics	Hämodynamik *f*
4057	haemoglobin	Hämoglobin *n*
4058	haemogram → blood picture	—
4059	haemolymph	Hämolymphe *f*
4060	haemolysin	Hämolysin *n*
4061	haemolysis	Hämolyse *f*
4062	haemopoiesis, haematopoiesis	Hämatopoese *f*, Blutbildung *f*
4063	haemopoietic, haematopoietic	blutbildend
4064	hair	Haar *n*
4065	hair bulb	Haarzwiebel *f*
4066	hair feather → filoplume	—
4067	hair follicle	Haarbalg *m*, Haarfollikel *m*
4068	hair root	Haarwurzel *f*
4069	hair shaft	Haarschaft *m*
4070	hairy → pilose	—
4071	half-antigen → hapten(e)	—
4072	half-chiasma	Halbchiasma *n*
4073	half-chromatid	Halbchromatide *f*
4074	half-chromosome	Halbchromosom *n*
4075	half-life (period), half-value period	Halbwertzeit *f*
4076	half-mutant	Halbmutante *f*
4077	half-race	Halbrasse *f*
4078	half-shrub → suffrutex	—
4079	half-sider	Halbseitenzwitter *m*
4080	halophilous, halophilic, salt-loving	halophil, salzliebend
4081	halophytes	Salzpflanzen *f pl*, Halophyten *m pl*
4082	haloplankton, haliplankton, marine plankton	Meeresplankton *n*, Salzwasserplankton *n*, Haloplankton *n*, Haliplankton *n*
4083	haltere, balancer, poiser	Haltere *f*, Schwingkölbchen *n*
4084	hanging-drop culture	Hängetropfenkultur *f*, Adhäsionskultur *f*
4085	haplobiont	Haplobiont *m*

4023	gynogenèse *f*	ginogénesis *f*
4024	gynomérogonie *f*	ginomerogonia *f*
4025	gynomonoïque	ginomonoico
4026	gynéphore *m*	ginóforo *m*
4027	—	—
4028	gynostège *m*	ginostegio *m*
4029	gynostème *m*	ginostemo *m*
4030	agitateur *m*	agitador *m* giratorio
4031	circonvolution *f* cérébrale	circunvolución *f* cerebral
4032	habitat *m*, résidence *f* écologique, station *f*	hábitat *m*, residencia *f* ecológica, estación *f*
4033	choix *f* de l'habitat	selección *f* del hábitat
4034	habituation *f*, accoutumance *f*	habituación *f*
4035	habitus *m*	hábito *m*
4036	—	—
4037	hème *m*	hemo *m*
4038	hémagglutination *f*	hem(o)aglutinación *f*
4039	hémagglutinine *f*	hem(o)aglutinina *f*
4040	arc *m* hémal	arco *m* hemal
4041	hématine *f*, hématosine *f*	hematina *f*
4042	—	—
4043	—	—
4044	hématogenèse *f*	hematogénesis *f*
4045	hématologie *f*	hematología *f*
4046	hématophage	hematófago
4047	—	—
4048	hématoxyline *f*	hematoxilina *f*
4049	hémérythrine *f*	hemeritrina *f*
4050	sanguin, hémal	sanguíneo, hemático
4051	hémine *f*	hemina *f*
4052	hémoblaste *m*	hemoblasto *m*
4053	hémocoele *m*	hemocele *m*
4054	hémocyanine *f*	hemocianina *f*, hematocianina *f*
4055	hémocyte *m*, cellule *f* sanguine	hemocito *m*, célula *f* sanguínea
4056	hémodynamique *f*	hemodinámica *f*
4057	hémoglobine *f*	hemoglobina *f*
4058	—	—
4059	hémolymphe *f*	hemolinfa *f*
4060	hémolysine *f*	hemolisina *f*
4061	hémolyse *f*	hemólisis *f*
4062	hématopoïèse *f*, hémopoïèse *f*	hematopoyesis *f*, hemopoyesis *f*
4063	hématopoïétique, hémopoïétique	hematopoyético
4064	cheveu *m*; poil *m*	cabello *m*; pelo *m*
4065	bulbe *m* pileux	bulbo *m* piloso
4066	—	—
4067	follicule *m* pileux	folículo *m* piloso
4068	racine *f* pileuse, ~ capillaire	raíz *f* capilar
4069	tige *f* pileuse	tallo *m* del pelo
4070	—	—
4071	—	—
4072	demi-chiasma *m*	semiquiasma *m*
4073	demi-chromatide *f*	hemicromatida *f*
4074	demi-chromosome *m*	hemicromosoma *m*
4075	période *f* de demi-valeur, demi-période *f*, demi-vie *f*	período *m* de vida media, semiperíodo *m* de vida
4076	demi-mutant *m*	semi-mutante *m*
4077	demi-race *f*	media raza *f*
4078	—	—
4079	gynandromorphe *m* biparti	semilateral *m*
4080	halophile	halófilo
4081	halophytes *mpl*, plantes *fpl* halophiles	halófitos *mpl*, halofitas *fpl*
4082	plancton *m* marin	plancton *m* marino, haloplancton *m*
4083	haltère *f*, balancier *m*	halterio *m*, balancín *m*
4084	culture *f* de la goutte pendante	cultivo *m* en gota pendiente (o: colgante)
4085	haplobionte *m*, haplobiontique *m*	haplobionte *m*

4086	haplochlamydeous	haplochlamyd(eisch)
4087	haploid	haploid
4088	haploidy	Haploidie *f*
4089	haplont	Haplont *m*
4090	haplophase, haploid phase	Haplophase *f*
4091	haplosis	Haplosis *f*
4092	haplostemonous	haplostemon
4093	hapten(e), half-antigen	Hapten *n*, Halbantigen *n*
4094	haptera (*sing*. hapteron)	Hapteren *f pl*
4095	haptic stimulus → contact ~	—
4096	haptonasty	Haptonastie *f*, Thigmonastie *f*
4097	haptotropism	Haptotropismus *m*
4098	hard-leaved → sclerophyllous	—
4099	hard palate	harter Gaumen *m*
4100	hard water	hartes Wasser *n*
4101	to harden	härten
4102	hardening	Härtung *f*, Verhärtung *f*
4103	hardness value, degree of hardness	Härtegrad *m*
4104	hardwood forest → deciduous ~	—
4105	harmful → noxious	—
4106	harmful insect → insect pest	—
4107	harmfulness → nocuity	—
4108	harmless → innocuous	—
4109	to hatch	(aus)schlüpfen
4110	hatching, eclosion	(Aus)schlüpfen *n*
4111	haustorium (*pl* haustoria)	Haustorie *f*, Saugfortsatz *m*
4112	hay infusion	Heuaufguß *m*
4113	haze canopy, smoggy bowl *env*.	Dunstglocke *f env*.
4114	head gut	Kopfdarm *m*
4115	head shield	Kopfschild *m*
4116	hearing	Gehör *n*
4117	hearing organ → auditory ~	—
4118	heart-beat	Herzschlag *m*
4119	heart minute volume	Herzminutenvolumen *n*
4120	heart valve	Herzklappe *f*
4121	heartwood → duramen	—
4122	heat → oestrus	—
4123	heat balance	Wärmebilanz *f*
4124	heat centre, thermogenic ~, calorific ~	Wärmezentrum *n*
4125	heat conductivity, thermal ~	Wärmeleitfähigkeit *f*
4126	heat cycle → oestrous ~	—
4127	heat emission	Wärmeabgabe *f*
4128	heat energy, thermal ~	Wärmeenergie *f*
4129	heat exchange	Wärmeaustausch *m*
4130	heat injury, ~ lesion	Wärmeschaden *m*
4131	heat loss	Wärmeverlust *m*
4132	heat period → oestrous ~	—
4133	heat radiation	Wärmestrahlung *f*
4134	heat resistance, thermal ~	Hitzeresistenz *f*
4135	heat resistant	hitzeresistent
4136	heat rigour → thermal rigidity	—
4137	heat stability	Hitzestabilität *f*
4138	heat treatment	Wärmebehandlung *f*
4139	hectocotylus	Hektokotyl *m*, Geschlechtstentakel *m/n*
4140	helical structure	Helix-Struktur *f*
4141	helical thickening → spiral ~	—
4142	helical tracheid → spiral ~	—
4143	helical vessel	Schraubengefäß *n*
4144	helicoid	schraubenförmig, schraubig
4145	helicoid cyme → bostryx	—
4146	heliophil(ic), heliophilous	heliophil, sonnenliebend

4086	haplochlamydé	haploclamídeo
4087	haploïde	haploide
4088	haploïdie *f*	haploidia *f*
4089	haplonte *m*	haplonte *m*
4090	haplophase *f*	haplofase *f*, fase *f* haploide
4091	haplose *f*	haplosis *f*
4092	haplostémone	haplostémono
4093	haptène *m*, semi-antigène *m*	hapteno *m*
4094	haptères *f pl*	hapterios *m pl*
4095	—	—
4096	haptonastie *f*, thigmonastie *f*	haptonastia *f*, tigmonastia *f*
4097	haptotropisme *m*	haptotropismo *m*
4098	—	—
4099	palais *m* dur	paladar *m* duro
4100	eau *f* dure	agua *f* dura
4101	durcir	endurecer
4102	durcissement *m*	endurecimiento *m*
4103	degré *m* de dureté	grado *m* de dureza
4104	—	—
4105	—	—
4106	—	—
4107	—	—
4108	—	—
4109	éclore	eclosionar, hacer eclosión
4110	éclosion *f*	eclosión *f*
4111	haustorium *m*, haustorie *f*	haustorio *m*
4112	infusion *f* de foin	infusión *f* de heno
4113	dôme *m* de brume, cloche *f* ~ ~	capa *f* flotante de calina, cúpula *f* de gases y humos
4114	intestin *m* céphalique	intestino *m* cefálico
4115	bouclier *m* céphalique	escudo *m* cefálico
4116	ouïe *f*	oído *m*
4117	—	—
4118	battement *m* du cœur, ~ cardiaque	latido *m* del corazón, ~ cardíaco
4119	volume-minute *m* du cœur	volumen *m* minuto circulatorio
4120	valvule *f* cardiaque	válvula *f* cardíaca
4121	—	—
4122	—	—
4123	bilan *m* de chaleur	balance *m* calorífico
4124	centre *m* thermique, ~ calorifiant	centro *m* termogénico, ~ calorífico
4125	conductibilité *f* thermique, ~ calorifique	conductibilidad *f* calorífica
4126	—	—
4127	dégagement *m* de chaleur	desprendimiento *m* de calor
4128	énergie *f* calorique, ~ calorifique, ~ thermique	energía *f* calórica, ~ calorífica
4129	échange *m* de chaleur	intercambio *m* de calor
4130	dégât *m* causé par la chaleur, lésion *f* par chaleur	daño *m* causado por calor
4131	perte *f* calorique, déperdition *f* de chaleur, ~ thermique	pérdida *f* de calor
4132	—	—
4133	rayonnement *m* thermique, ~ calorifique	radiación *f* calorífica
4134	thermorésistance *f*, résistance *f* à la chaleur	termorresistencia *f*, resistencia *f* térmica, ~ al calor
4135	thermorésistant, résistant à la chaleur	termorresistente, resistente al calor
4136	—	—
4137	stabilité *f* à la chaleur, thermostabilité *f*	estabilidad *f* al calor
4138	traitement *m* par chaleur	tratamiento *m* por el calor
4139	hectocotyle *m*	hectocótilo *m*
4140	structure *f* en hélice, ~ hélicoïdale	estructura *f* en hélice, ~ helicoidal
4141	—	—
4142	—	—
4143	vaisseau *m* spiralé	vaso *m* helicoidal, ~ helicado
4144	hélicoïde	helicoidal
4145	—	—
4146	héliophile	heliófilo

4147	heliophobic	heliophob
4148	heliophyll → sun leaf	—
4149	heliophyte, sun plant, heliophilous ~	Sonnenpflanze *f*, Starklichtpflanze *f*, Heliophyt *m*
4150	heliotropism → phototropism	—
4151	helix (*pl* helices), spiral	Helix *f*
4152	helminthology	Helminthologie *f*
4153	helophyte, marsh plant	Helophyt *m*, Sumpfpflanze *f*
4154	helotism	Helotismus *m*
4155	help-cell → synergid	—
4156	hem..., hemato → haem..., haemato...	—
4157	hemeranthous, hemeranthic	tagblühend
4158	hemicellulose	Hemizellulose *f*
4159	Hemichordata, hemichordates	Hemichordaten *pl*, Stomochordaten *pl*
4160	hemicryptophyte	Hemikryptophyt *m*, Erdschürfepflanze *f*
4161	hemikaryon	Hemikaryon *n*
4162	Hemimetabola, Heterometabola, Exopterygota	Hemimetabolen *pl*, Heterometabolen *pl*
4163	hemimetabolism, incomplete metamorphosis	Hemimetabolie *f*, Heterometabolie *f*, unvollkommene Metamorphose *f*, unvollständige ~
4164	hemiparasite, semiparasite	Halbschmarotzer *m*, Halbparasit *m*, Hemiparasit *m*
4165	Hemiptera, Rhynchota, hemipterans	Hemipteren *pl*, Schnabelkerfe *mpl*, Halbflügler *mpl*
4166	hemitropous	hemitrop
4167	hemizygous	hemizygot
4168	hemo... → haemo...	—
4169	Hensen's node → primitive ~	—
4170	heparin	Heparin *n*
4171	hepatic	Leber...
4172	Hepaticae, liverworts	Lebermoose *npl*
4173	herb	Kraut *n*
4174	herbaceous	krautig
4175	herbaceous layer, ground ~	Krautschicht *f*
4176	herbarium	Herbar(ium) *n*
4177	herbicide	Herbizid *n*, Unkrautvernichtungsmittel *n*
4178	herbivore	Krautfresser *m*, Herbivore *m*
4179	herbivorous	krautfressend
4180	hercogamy, herkogamy	Herkogamie *f*
4181	herd	Herde *f*
4182	herd instinct, gregariousness	Herdentrieb *m*
4183	hereditary, heritable	erblich, erbbedingt
4184	heredity → inheritance	—
4185	hereditary change, heritable ~	Erbänderung *f*
4186	hereditary character, ~ trait	Erbanlage *f*, Erbmerkmal *n*, Vererbungsmerkmal *n*
4187	hereditary constitution	Erbgefüge *n*
4188	hereditary defect	Erbfehler *m*, Erbschaden *m*
4189	hereditary factor → gene	—
4190	hereditary information	Erbinformation *f*
4191	hereditary material, ~ endowment	Erbgut *n*, Erbmaterial *n*
4192	hereditary property	Erbeigenschaft *f*
4193	hereditary unit	Erbeinheit *f*
4194	heritability, hereditability	Erblichkeit *f*; Erblichkeitsgrad *m*, Heritabilität *f*
4195	heritable → hereditary	—
4196	hermaphrodite	Zwitter *m*, Hermaphrodit *m*
4197	hermaphroditic	zwitterig, hermaphroditisch
4198	hermaphroditic flower	Zwitterblüte *f*
4199	hermaphroditism	Zwittrigkeit *f*, Zwittertum *n*, Hermaphroditismus *m*
4200	herpetology	Herpetologie *f*, Lehre *f* von den Kriechtieren
4201	heteroallele	Heteroallel *n*
4202	heteroauxin → indole-3-acetic acid	—

4147	héliophobe	heliófobo
4148	—	—
4149	plante *f* héliophile, ~ de soleil, ~ de lumière	planta *f* heliófila
4150	—	—
4151	hélix *m*, spirale *f*	hélice *f*, espiral *f*
4152	helminthologie *f*	helmintología *f*
4153	hélophyte *m*, plante *f* palustre, ~ des marais	helófito *m*, planta *f* palustre, ~ paludícola
4154	hélotisme *m*, ilotisme *m*	helotismo *m*
4155	—	—
4156	—	—
4157	à floraison diurne	de floración diurna
4158	hémi-cellulose *f*	hemicelulosa *f*
4159	stomoc(h)ordés *mpl*	hemicordados *mpl*, hemicordios *mpl*, estomocordados *mpl*, estomocordios *mpl*
4160	hémicryptophyte *m*	hemicriptófito *m*
4161	hémicaryon *m*	hemicarión *m*
4162	hémimétaboles *mpl*, hétérométaboles *mpl*, exoptérygotes *mpl*	hemimetábolos *mpl*, heterometábolos *mpl*, exopterigotos *mpl*
4163	métamorphose *f* directe, ~ incomplète	hemimetabolia *f*, heterometabolia *f*, metamorfosis *f* directa, ~ incompleta
4164	hémiparasite *m*	hemiparásito *m*, semiparásito *m*
4165	hémiptères *mpl*	hemípteros *mpl*
4166	hémitrope	hemítropo
4167	hémizygote	hemicigoto, hemizigótico
4168	—	—
4169	—	—
4170	héparine *f*	heparina *f*
4171	hépatique	hepático
4172	hépatiques *fpl*	hepáticas *fpl*
4173	herbe *f*	hierba *f*
4174	herbacé	herbáceo
4175	strate *f* herbacée	estrato *m* herbáceo
4176	herbier *m*	herbario *m*
4177	herbicide *m*	herbicida *m*
4178	herbivore *m*	herbívoro *m*
4179	herbivore	herbívoro
4180	hercogamie *f*	hercogamia *f*
4181	troupeau *m*, troupe *f*	rebaño *m*, tropel *m*
4182	instinct *m* grégaire, grégarisme *m*	instinto *m* gregario, gregarismo *m*
4183	héréditaire	hereditario, heredable
4184	—	—
4185	modification *f* héréditaire	modificación *f* hereditaria
4186	caractère *m* héréditaire	carácter *m* hereditario
4187	constitution *f* héréditaire	constitución *f* hereditaria
4188	défaut *m* héréditaire, tare *f* ~	defecto *m* hereditario, tara *f* hereditaria
4189	—	—
4190	information *f* héréditaire	información *f* hereditaria
4191	patrimoine *m* héréditaire	patrimonio *m* hereditario, material *m* ~, dotación *f* hereditaria
4192	propriété *f* héréditaire	propiedad *f* hereditaria
4193	unité *f* héréditaire	unidad *f* hereditaria
4194	héritabilité *f*	heredabilidad *f*
4195	—	—
4196	hermaphrodite *m*	hermafrodita *m*
4197	hermaphrodite	hermafrodita
4198	fleur *f* hermaphrodite	flor *f* hermafrodita
4199	hermaphrodi(ti)sme *m*	hermafroditismo *m*
4200	(h)erpétologie *f*	herpetología *f*
4201	hétéroallèle *m*	heteroalelo *m*
4202	—	—

No.	English	German
4203	heterocarpous	heterokarp
4204	heterocentric	heterozentrisch
4205	heterochlamydeous	heterochlamyd(eisch)
4206	heterochromatic	heterochromatisch
4207	heterochromatin	Heterochromatin *n*
4208	heterochromatism	Heterochromatie *f*
4209	heterochromosome, sex chromosome, allosome	Heterochromosom *n*, Geschlechtschromosom *n*, Allosom *n*
4210	heterocyst, boundary cell	Grenzzelle *f*, Heterozyste *f*
4211	heterodont	heterodont
4212	heteroecious, heteroicous	getrenntgeschlechtig, heterözisch
4213	heteroecism	Getrenntgeschlechtigkeit *f*, Heterözie *f*
4214	heterogamete → anisogamete	—
4215	heterogametic, digametic	heterogametisch, digametisch
4216	heterogamy → anisogamy	—
4217	heterogeneity	Heterogenität *f*, Verschiedenartigkeit *f*
4218	heterogenetic	heterogenetisch
4219	heterogenote	Heterogenote *f*
4220	heterogenous, heterogenic	heterogen(isch)
4221	heterogeny	Heterogenie *f*
4222	heterogony	Heterogonie *f*
4223	heterograft, heteroplastic graft	Heteroplantat *n*
4224	heterogynous, heterogynic	heterogyn
4225	heterokaryon, heterocaryon	Heterokaryon *n*
4226	heterokaryosis	Heterokaryose *f*
4227	heterokinesis	Heterokinese *f*
4228	heterokont	heterokont
4229	heterologous	heterolog
4230	heterology	Heterologie *f*
4231	heterolysis	Heterolyse *f*
4232	heteromerous	heteromer
4233	Heterometabola → Hemimetabola	—
4234	heteromorphic, heteromorphous	heteromorph, verschiedengestaltig
4235	heteromorphism	Heteromorphismus *m*, Heteromorphie *f*
4236	heteromorphosis	Heteromorphose *f*
4237	heteronomous	heteronom
4238	heterophylly	Heterophyllie *f*
4239	heterophytic	heterophytisch
4240	heteroplasmony	Heteroplasmonie *f*
4241	heteroplasty	Heteroplastie *f*
4242	heteroploid	heteroploid
4243	heteroploidy	Heteroploidie *f*
4244	heteropycnosis, heteropyknosis	Heteropyknose *f*
4245	heterosis, hybrid vigo(u)r	Heterosis *f*, Bastardwüchsigkeit *f*
4246	heterosomal	heterosomal
4247	heterosporous	heterospor
4248	heterospory	Heterosporie *f*
4249	heterostyled, heterostylic	heterostyl, verschiedengriff(e)lig
4250	heterostyly, heterostylism	Heterostylie *f*, Verschiedengriff(e)ligkeit *f*
4251	heterosynapsis	Heterosynapsis *f*
4252	heterosyndesis	Heterosyndese *f*
4253	heterothallic	heterothallisch
4254	heterothally, heterothallism	Heterothallie *f*
4255	heterotrichous	heterotrich
4256	heterotrophia	Heterotrophie *f*
4257	heterotrophic	heterotroph
4258	heterotype	Heterotyp *m*
4259	heterozygosis, heterozygo(si)ty	Heterozygotie *f*, Mischerbigkeit *f*, Ungleicherbigkeit *f*
4260	heterozygote	Heterozygote *f*
4261	heterozygotic, heterozygous	heterozygot, mischerbig, ungleicherbig
4262	heterozygotic breeding	Heterosiszüchtung *f*
4263	hexaploid	hexaploid
4264	hexaploidy	Hexaploidie *f*
4265	Hexapoda → Insecta	—

4203	hétérocarpe	heterocarpo, heterocárpico
4204	hétérocentrique	heterocéntrico
4205	hétérochlamydé	heteroclamídeo
4206	hétérochromatique	heterocromático
4207	hétérochromatine *f*	heterocromatina *f*
4208	hétérochromatisme *m*	heterocromatismo *m*
4209	hétérochromosome *m*, chromosome *m* sexuel, allosome *m*	heterocromosoma *m*, cromosoma *m* sexual, alosoma *m*
4210	hétérocyste *m*	célula *f* límite, heterocisto *m*
4211	hétérodonte	heterodonto
4212	hétéroécique, hétéroïque	heterecio, heteroico
4213	hétéroécie *f*	heter(o)ecismo *m*
4214	—	—
4215	hétérogamétique, digamétique	heterogamético, digamético
4216	—	—
4217	hétérogénéité *f*	heterogeneidad *f*
4218	hétérogénétique	heterogenético
4219	hétérogénote *m*	heterogenote *m*
4220	hétérogène, hétérogénique	heterogéneo, heterógeno, heterogénico
4221	hétérogénie *f*	heterogenia *f*
4222	hétérogonie *f*	heterogonia *f*
4223	hétérogreffe *f*, greffe *f* hétérologue, ~ hétéroplastique	heteroinjerto *m*
4224	hétérogyne	heterogino
4225	hétérocaryon *m*	heterocarión *m*, heterocarionte *m*
4226	hétérocaryose *f*	heterocariosis *f*
4227	hétérocinésie *f*, hétérocinèse *f*	heterocinesis *f*
4228	hétéroconté	heteroconto
4229	hétérologue	heterólogo
4230	hétérologie *f*	heterología *f*
4231	hétérolyse *f*	heterólisis *f*
4232	hétéromère	heterómero
4233	—	—
4234	hétéromorphe	heteromorfo, heteromórfico
4235	hétéromorphisme *m*, hétéromorphie *f*	heteromorfismo *m*, heteromorfia *f*
4236	hétéromorphose *f*	heteromorfosis *f*
4237	hétéronome	heterónomo
4238	hétérophyllie *f*	heterofilia *f*
4239	hétérophytique	heterofítico
4240	hétéroplasmonie *f*	heteroplasmonia *f*
4241	hétéroplastie *f*	heteroplastia *f*
4242	hétéroploïde	heteroploide
4243	hétéroploïdie *f*	heteroploidia *f*
4244	hétéropycnose *f*	heteropicnosis *f*
4245	hétérosis *f*, vigueur *f* hybride	heterosis *f*, vigor *m* híbrido
4246	hétérosomal	heterosomal
4247	hétérosporé	heteróspoto, heterospóreo
4248	hétérosporie *f*	heterosporia *f*
4249	hétérostylé	heterostilo, heterostílico
4250	hétérostylie *f*	heterostilia *f*, heteroestilismo *m*
4251	hétérosynapse *f*	heterosinapsis *f*
4252	hétérosyndèse *f*	heterosíndesis *f*
4253	hétérothallique	heterotálico
4254	hétérothallie *f*, hétérothallisme *m*	heterotalia *f*, heterotalismo *m*
4255	hétérotriche	heterótrico
4256	hétérotrophie *f*	heterotrofia *f*
4257	hétérotrophe, hétérotrophique	heterótrofo, heterotrófico
4258	hétérotype *m*	heterotipo *m*
4259	hétérozygose *f*, hétérozygotie *f*	heterocigosis *f*, heterozigosis *f*, heterozigotismo *m*
4260	hétérozygote *m*	heterocigoto *m*, heterozigoto *m*
4261	hétérozygote	heterocigótico, heterozigótico
4262	sélection *f* hétérosique	selección *f* heterósica
4263	hexaploïde	hexaploide
4264	hexaploïdie *f*	hexaploidia *f*
4265	—	—

4266	hexokinase	Hexokinase *f*
4267	hexose	Hexose *f*
4268	hibernaculum, hibernacle, winter bud	Hibernakel *n*, Überwinterungsknospe *f*
4269	to hibernate	Winterschlaf halten
4270	hibernating organ	Überwinterungsorgan *n*
4271	hibernation	Winterschlaf *m*
4272	hibernator	Winterschläfer *m*
4273	hierarchy of instincts	Instinkthierarchie *f*
4274	high moor	Hochmoor *n*
4275	higher animal	höheres Tier *n*
4276	higher plant	höhere Pflanze *f*
4277	hilum *bot.*	Hilum *n*, Nabel *m* *bot.*
4278	hind-brain → metencephalon	—
4279	hind-gut	Enddarm *m*
4280	hind-kidney → metanephros	—
4281	hinge-joint, ginglymus	Scharniergelenk *n*
4282	hip, coxa	Hüfte *f*
4283	hip-bone, iliac bone	Hüftbein *n*
4284	hip-girdle → pelvic girdle	—
4285	hip-joint	Hüftgelenk *n*
4286	hirudin	Hirudin *n*
4287	Hirudinea, leeches	Egel *m pl*
4288	histamine	Histamin *n*
4289	histidine	Histidin *n*
4290	histiocyte	Histiozyt *m*
4291	histochemical	histochemisch
4292	histochemistry	Histochemie *f*
4293	histocompatibility	Gewebeverträglichkeit *f*
4294	histogen	Histogen *n*
4295	histogenesis, histogeny	Histogenese *f*, Gewebebildung *f*
4296	histogram	Histogramm *n*
4297	histological	histologisch
4298	histological section → tissue ~	—
4299	histologist	Histologe *m*
4300	histology	Histologie *f*, Gewebelehre *f*
4301	histolysis	Histolyse *f*
4302	histomorphology	Gewebemorphologie *f*
4303	histone	Histon *n*
4304	histophysiology	Gewebephysiologie *f*
4305	historadiography	Historadiographie *f*
4306	hit theory → target theory	—
4307	holandric	holandrisch
4308	holarctic region	Holarktis *f*
4309	holdfast → adhesive organ	—
4310	hole-nester → cave-nesting animal	—
4311	hollow organ	Hohlorgan *n*
4312	hollow spindle	Hohlspindel *f*
4313	holobasidium	Holobasidie *f*
4314	holoblastic	holoblastisch
4315	Holocene (epoch), Recent ~	Holozän *n*, Alluvium *n*
4316	holocrine gland	holokrine Drüse *f*
4317	holoenzyme	Holoenzym *n*
4318	hologamete	Hologamet *m*
4319	hologamous	hologam
4320	hologamy, macrogamy	Hologamie *f*, Makrogamie *f*
4321	hologynic	hologyn
4322	Holometabola, Endopterygota	Holometabolen *pl*
4323	holometabolism, complete metamorphosis	Holometabolie *f*, vollkommene Metamorphose *f*, vollständige ~
4324	holoparasite	Vollschmarotzer, Vollparasit *m*, Holoparasit *m*, Ganzschmarotzer *m*
4325	holotrichous	holotrich
4326	holotype, type specimen	Holostandard *m*
4327	home range	Heimrevier *n*

4266	hexokinase *f*	hexoquinasa *f*
4267	hexose *m*	hexosa *f*
4268	hibernacle *m*	hibernáculo *m*, yema *f* invernal, ~ persistente, ~ invernante
4269	hiberner	hibernar
4270	organe *m* de résistance	órgano *m* invernante
4271	hibernation *f*	hibernación *f*
4272	animal *m* hibernant	(animal *m*) hibernante *m*
4273	organisation *f* hiérarchique des instincts	organización *f* jerárquica de los instintos
4274	tourbière *f* haute	turbera *f* alta
4275	animal *m* supérieur	animal *m* superior
4276	plante *f* supérieure, végétal *m* supérieur	planta *f* superior
4277	hile *m* *bot.*	hilo *m* *bot.*
4278	—	—
4279	intestin *m* postérieur	intestino *m* posterior
4280	—	—
4281	ginglyme *m*	gínglimo *m*, articulación *f* en charnela
4282	hanche *f*	cadera *f*, coxa *f*
4283	os *m* coxal, ~ iliaque	hueso *m* coxal, ~ ilíaco
4284	—	—
4285	articulation *f* de la hanche, ~ coxofémorale	articulación *f* de la cadera, ~ coxofemoral
4286	hirudine *f*	hirudina *f*
4287	hirudinées *f pl*	hirudíneos *m pl*
4288	histamine *f*	histamina *f*
4289	histidine *f*	histidina *f*
4290	histiocyte *m*	histiocito *m*
4291	histochimique	histoquímico
4292	histochimie *f*	histoquímica *f*
4293	histocompatibilité *f*	histocompatibilidad *f*
4294	histogène *m*	histógeno *m*
4295	histogenèse *f*, histogénie *f*	histogénesis *f*, histogenia *f*
4296	histogramme *m*	histograma *m*
4297	histologique	histológico
4298	—	—
4299	histologiste *m*	histólogo *m*
4300	histologie *f*	histología *f*
4301	histolyse *f*	histólisis *f*
4302	histomorphologie *f*, morphologie *f* histologique	histomorfología *f*
4303	histone *f*	histona *f*
4304	histophysiologie *f*	histofisiología *f*
4305	historadiographie *f*	historradiografía *f*
4306	—	—
4307	holandrique	holándrico
4308	région *f* holarctique	región *f* holártica
4309	—	—
4310	—	—
4311	organe *m* creux	órgano *m* hueco
4312	fuseau *m* creux	huso *m* cóncavo
4313	holobaside *f*	holobasidio *m*
4314	holoblastique	holoblástico
4315	holocène *m*, alluvium *m*	holoceno *m*, reciente *m*
4316	glande *f* holocrine	glándula *f* holocrina
4317	holoenzyme *m*	holoenzima *f*
4318	hologamète *m*	hologameto *m*
4319	hologame	hológamo, hologámico
4320	hologamie *f*	hologamia *f*
4321	hologynique	hologínico
4322	holométaboles *m pl*, endoptérygotes *m pl*	holometábolos *m pl*, endopterigotos *m pl*
4323	holométabolie *f*, métamorphose *f* complète, ~ indirecte	metamorfosis *f* completa, ~ indirecta
4324	holoparasite *m*	holoparásito *m*
4325	holotriche	holotrico
4326	holotype *m*, spécimen-type *m*	holótipo *m*, ejemplar *m* tipo
4327	zone *f* d'habitat, domaine *m* individuel	zona *f* batida, "ámbito *m* del hogar"

4328	homeokinesis, homoeokinesis	Homöokinese *f*
4329	homeologous, homoeologous	homöolog, homoiolog
4330	homeosis, homoeosis	Homöosis *f*
4331	homeostasis, homoeostasis	Homöostasis *f*, Homöostasie *f*
4332	homeosynapsis, homosynapsis	Homosynapsis *f*
4333	homeothermic → homoiothermic	—
4334	homeotypic	homöotypisch
4335	homing	Heimfinden *n*, Heimkehrvermögen *n* *eth.*
4336	hominids	Hominiden *pl*
4337	homoallele	Homoallel *n*
4338	homocentric	homozentrisch
4339	homochromy	Homochromie *f*
4340	homodont	homodont
4341	homoeo ... → homeo ...	—
4342	homogametic	homogametisch
4343	homogamous	homogam
4344	homogamy	Homogamie *f*
4345	homogenate	Homogenat *n*
4346	homogeneity, homogeny	Homogenität *f*, Gleichartigkeit *f*
4347	homogeneous	homogen, gleichartig
4348	homogenetic	homogenetisch
4349	homogentisic acid	Homogentisinsäure *f*
4350	homograft, homoplastic graft	Homoplantat *n*
4351	homoiotherm, homeotherm	dauerwarmes Tier *n*
4352	homoiothermic, homoiothermal, homeothermic	homöotherm, dauerwarm
4353	homokaryon	Homokaryon *n*
4354	homokaryosis	Homokaryose *f*
4355	homolecithal	homolezithal
4356	homologous chromosome	homologes Chromosom *n*
4357	homology	Homologie *f*
4358	homomery	Homomerie *f*
4359	homomorphous, homomorphic	homomorph, gleichgestaltig
4360	homophytic	homophytisch
4361	homoplastic	homoplastisch
4362	homoploidy	Homoploidie *f*
4363	homopolar	homopolar
4364	Homoptera, homopterans	Homopteren *pl*, Gleichflügler *mpl*
4365	homosomal	homosomal
4366	homosporous, isosporous	isospor
4367	homospory, isospory	Isosporie *f*
4368	homostyled, homostilic	homostyl, gleichgriff(e)lig
4369	homostyly	Homostylie *f*, Gleichgriff(e)ligkeit *f*
4370	homothallic	homothallisch
4371	homothallism	Homothallie *f*
4372	homotype	Homotyp *m*
4373	homotypy	Homotypie *f*
4374	homozygote	Homozygote *f*
4375	homozygosis, homozygo(si)ty	Homozygotie *f*, Reinerbigkeit *f*
4376	homozygous	homozygot, reinerbig
4377	honeycomb bag → reticulum	—
4378	hoof	Huf *m*
4379	hoofed animals → Ungulata	—
4380	hooked → uncinate	—
4381	horde	Horde *f*
4382	hormogonium, hormogone	Hormogonium *n*
4383	hormonal	hormonal, hormonell
4384	hormonal action	Hormontätigkeit *f*
4385	hormonal control, ~ regulation	hormonale Steuerung *f*
4386	hormone	Hormon *n*
4387	hormonogenic	hormonbildend
4388	horn	Horn *n*
4389	horny plate	Hornplatte *f*
4390	host	Wirt *m*

4328	homéocinèse *f*	homeocinesis *f*
4329	homéologue	homeólogo
4330	homéose *f*	homeosis *f*
4331	homéostasie *f*	homeostasis *f*
4332	homosynapse *f*	homeosinapsis *f*
4333	—	—
4334	homéotypique	homeotípico
4335	retour *m* au gîte, ~ au nid, ~ à l'habitat	retorno *m* al nido, ~ al hogar
4336	hominiens *m pl*	homínidos *m pl*
4337	homoallèle *m*	homoalelo *m*
4338	homocentrique	homocéntrico
4339	homochromie *f*	homocromía *f*
4340	homodonte	homodonto
4341	—	—
4342	homogamétique	homogamético
4343	homogame	homógamo
4344	homogamie *f*	homogamia *f*
4345	homogénat *m*	homogeneizado *m*, homogenato *m*
4346	homogénéité *f*	homogeneidad *f*
4347	homogène	homógeno
4348	homogénétique	homogenético
4349	acide *m* homogentisique	ácido *m* homogentísico, ~ homogentisínico
4350	homogreffe *f*, greffe *f* homologue, ~ homoplastique	homoinjerto *m*, isoinjerto *m*, injerto *m* homoplástico
4351	animal *m* homoeotherme, ~ homéotherme	hom(e)otermo *m*
4352	homoeotherme, homéothermal	hom(e)otermo, homotérmico
4353	homocaryon *m*	homocarión *m*, homocario *m*
4354	homocaryose *f*	homocariosis *f*
4355	homolécithe, homolécithique	homolecito, homolecítico
4356	chromosome *m* homologue	cromosoma *m* homólogo
4357	homologie *f*	homología *f*
4358	homomérie *f*	homomeria *f*
4359	homomorphe	homomórfico
4360	homophytique	homofítico
4361	homoplastique	homoplástico
4362	homoploïdie *f*	homoploidia *f*
4363	homopolaire	homopolar
4364	homoptères *m pl*	homópteros *m pl*
4365	homosomal	homosomal
4366	homosporé, isosporé	homósporo, isósporo
4367	homosporie *f*, isosporie *f*	homosporia *f*, isosporia *f*
4368	homostylé	homostilo, homostílico
4369	homostylie *f*	homostilia *f*
4370	homothallique	homotálico
4371	homothallie *f*, homothallisme *m*	homotalia *f*, homotalismo *m*
4372	homotype *m*	homotipo *m*
4373	homotypie *f*	homotipia *f*
4374	homozygote *m*	homocigoto *m*, homozigoto *m*
4375	homozygose *f*, homozygotie *f*	homocigosis *f*, homozigosis *f*, homozigotismo *m*
4376	homozygote	homocigoto, homozigótico
4377	—	—
4378	sabot *m*, onglon *m*	casco *m*, uña *f*
4379	—	—
4380	—	—
4381	horde *f*	horda *f*
4382	hormogonie *f*	hormogonio *m*
4383	hormonal	hormonal
4384	activité *f* hormonale	acción *f* hormonal, actividad *f* ~
4385	régulation *f* hormonale	regulación *f* hormonal
4386	hormone *f*	hormona *f*
4387	hormonogène	hormonogénico
4388	corne *f*	cuerno *m*, asta *f*
4389	plaque *f* cornée	placa *f* córnea
4390	hôte *m*	huésped *m*, hospedante *m*

4391	host alternation	Wirtswechsel *m*
4392	host bacterium, bacterial host	Wirtsbakterium *n*, Bakterienwirt *m*
4393	host cell	Wirtszelle *f*
4394	host-parasite interaction, ~ relationship	Wirt-Parasit-Wechselwirkung *f*
4395	host plant	Wirtspflanze *f*
4396	host species	Wirtsart *f*
4397	host specificity	Wirtsspezifität *f*
4398	human biology	Humanbiologie *f*
4399	human ecology	Humanökologie *f*, Anthropo-Ökologie *f*
4400	human genetics	Humangenetik *f*
4401	human palaeontology → palaeoanthropology	—
4402	humerus	Oberarmknochen *m*, Humerus *m*
4403	humic acid	Huminsäure *f*
4404	humid preparation	Naßpräparat *n*
4405	humid weight → wet weight	—
4406	humification	Humifizierung *f*, Humusbildung *f*
4407	humus	Humus *m*
4408	hyaline	Hyalin *n*
4409	hyaline cartilage	Glasknorpel *m*, hyaliner Knorpel *m*
4410	hyaloplasm, cytoplasmic matrix	Hyaloplasma *n*, Grundzytoplasma *n*, zytoplasmatische Matrix *f*
4411	hyaluronic acid	Hyaluronsäure *f*
4412	hybrid *adj.*	hybrid
4413	hybrid *sub.*	Hybride *f*, Hybrid *m*, Bastard *m*
4414	hybrid lethality	Bastardletalität *f*
4415	hybrid plant	Bastardpflanze *f*
4416	hybrid sterility, ~ incapacitation	Bastardsterilität *f*, Hybridensterilität *f*
4417	hybrid swarm, ~ population	Hybrid(en)schwarm *m*, Bastardpopulation *f*
4418	hybrid vigour → heterosis	—
4419	hybrid zone	Hybridenzone *f*
4420	hybridism, hybridity	Hybridität *f*, Hybridismus *m*, Bastardnatur *f*
4421	hybridization	Hybridisierung *f*, Bastardierung *f*
4422	to hybridize	bastardieren
4423	hydathode	Hydathode *f*
4424	hydratation	Hydratation *f*, Hydratisierung *f*
4425	to hydrate	hydratisieren
4426	hydrobiological	hydrobiologisch
4427	hydrobiology	Hydrobiologie *f*
4428	hydrocarbon	Kohlenwasserstoff *m*
4429	hydrocarbon chain	Kohlenwasserstoffkette *f*
4430	hydrochloric acid	Salzsäure *f*
4431	hydrochory, water dispersal	Hydrochorie *f*, Wasserverbreitung *f* (des Samens)
4432	hydrocoel	Hydrozöl *n*
4433	hydrocortisone, cortisol	Hydrokortison *n*, Kortisol *n*
4434	hydrogel	Hydrogel *n*
4435	hydrogen	Wasserstoff *m*
4436	hydrogen acceptor	Wasserstoffakzeptor *m*
4437	hydrogen bond	Wasserstoffbrücke *f*
4438	hydrogen donor	Wasserstoffdonator *m*
4439	hydrogen exponent	Wasserstoffexponent *m*
4440	hydrogen ion	Wasserstoffion *n*
4441	hydrogen-ion concentration	Wasserstoffionenkonzentration *f*
4442	hydrogenase	Hydrogenase *f*
4443	to hydrogenate, to hydrogenize	hydrieren
4444	hydrogenation	Hydrierung *f*
4445	hydrolase	Hydrolase *f*
4446	hydrolysate	Hydrolysat *n*
4447	hydrolysis	Hydrolyse *f*
4448	hydrolytic	hydrolytisch
4449	to hydrolyze	hydrolysieren
4450	hydronasty	Hydronastie *f*
4451	hydrophilous	hydrophil, wasserblütig
4452	hydrophily	Hydrophilie *f*, Hydrogamie *f*, Wasserblütigkeit *f*, Wasserbestäubung *f*

4391	changement *m* d'hôte, alternance *f* ~	cambio *m* de huésped
4392	bactérie *f* hôte	huésped *m* bacteriano
4393	cellule *f* hôte	célula *f* huésped
4394	rapports *mpl* hôte-parasite	interacción *f* huésped-parásito
4395	plante-hôte *f*	planta *f* huésped, ~ hospedante
4396	espèce *f* hôte	especie *f* huésped
4397	spécificité *f* de l'hôte, ~ parasitaire	especificidad *f* al huésped
4398	biologie *f* humaine	biología *f* humana
4399	écologie *f* humaine	ecología *f* humana
4400	génétique *f* humaine	genética *f* humana
4401	—	—
4402	humérus *m*	húmero *m*
4403	acide *m* humique	ácido *m* húmico
4404	préparation *f* humide	preparación *f* húmeda
4405	—	—
4406	humification *f*	humificación *f*
4407	humus *m*	humus *m*, mantillo *m*
4408	hyaline *f*	hialina *f*
4409	cartilage *m* hyalin	cartílago *m* hialino
4410	hyaloplasme *m*, cytoplasme *m* fondamental, ~ hyalin	hialoplasma *m*, citoplasma *m* fundamental, matriz *f* citoplasmática
4411	acide *m* hyaluronique	ácido *m* hialurónico
4412	hybride	híbrido
4413	hybride *m*	híbrido *m*
4414	létalité *f* de l'hybride	letalidad *f* del híbrido
4415	plante *f* hybride	planta *f* híbrida
4416	stérilité *f* (des) hybride(s)	esterilidad *f* híbrida, incapacitación *f* ~
4417	population *f* hybride	población *f* híbrida
4418	—	—
4419	zone *f* hybride	zona *f* híbrida
4420	hybridité *f*, hybridisme *m*	hibridismo *m*, hibridez *f*
4421	hybridation *f*	hibridación *f*
4422	hybrider	hibridar
4423	hydathode *m*	hidátodo *m*
4424	hydratation *f*	hidratación *f*
4425	hydrater	hidratar
4426	hydrobiologique	hidrobiológico
4427	hydrobiologie *f*	hidrobiología *f*
4428	hydrocarbure *m*, carbure *m* d'hydrogène	hidrocarburo *m*
4429	chaîne *f* d'hydrocarbures	cadena *f* de hidrocarburos
4430	acide *m* chlorhydrique	ácido *m* clorhídrico
4431	hydrochorie *f*	hidrocoria *f*
4432	hydrocoele *m*	hidrocele *m*
4433	hydrocortisone *f*, cortisol *m*	hidrocortisona *f*, cortisol *m*
4434	hydrogel *m*	hidrogel *m*
4435	hydrogène *m*	hidrógeno *m*
4436	accepteur *m* (d') hydrogène	acept(ad)or *m* de hidrógeno
4437	pont *m* (d') hydrogène	puente *m* de hidrógeno, enlace *m* ~ ~
4438	donateur *m* d'hydrogène, donneur *m* ~	donador *m* de hidrógeno
4439	exposant *m* d'hydrogène	exponente *m* de hidrógeno
4440	ion *m* (d') hydrogène	hidrogenión *m*, ion *m* hidrógeno
4441	concentration *f* des ions (d')hydrogène	concentración *f* de hidrogeniones, ~ de iones hidrógeno
4442	hydrogénase *f*	hidrogenasa *f*
4443	hydrogéner	hidrogenar
4444	hydrogénation *f*	hidrogenación *f*
4445	hydrolase *f*	hidrolasa *f*
4446	hydrolysat *m*	hidrolizado *m*
4447	hydrolyse *f*	hidrólisis *f*
4448	hydrolytique	hidrolítico
4449	hydrolyser	hidrolizar
4450	hydronastie *f*	hidronastia *f*
4451	hydrophile	hidrófilo, hidrofílico
4452	hydrophilie *f*, hydrogamie *f*, pollinisation *f* aquatique, ~ par l'eau	hidrofilia *f*, hidrogamia *f*

4453	hydrophobe	hydrophob, wassermeidend
4454	hydrophoric canal → stone canal	—
4455	hydrophytes, aquatic plants	Hydrophyten *mpl*, Wasserpflanzen *fpl*
4456	hydroponics	Hydroponik *f*, Hydrokultur *f*
4457	hydrosere	Hydroserie *f*
4458	hydrosol	Hydrosol *n*
4459	hydrosoluble → water-soluble	—
4460	hydrosphere	Hydrosphäre *f*
4461	hydrotaxis	Hydrotaxis *f*
4462	hydrotropism	Hydrotropismus *m*
4463	hydroxide	Hydroxyd *n*
4464	hydroxyl group	Hydroxylgruppe *f*
4465	hydroxylamine	Hydroxylamin *n*
4466	hydroxylation	Hydroxylierung *f*
4467	Hydrozoa, hydrozoans	Hydrozoen *pl*
4468	hygrometer	Feuchtigkeitsmesser *m*, Hygrometer *n*
4469	hygronasty	Hygronastie *f*
4470	hygrophilous	hygrophil, feuchtigkeitsliebend
4471	hygrophytes, hygrophilous plants	Hygrophyten *mpl*, Feuchtpflanzen *fpl*
4472	hygroscopic(al)	hygroskopisch, wasserziehend
4473	hygroscopicity	Hygroskopizität *f*
4474	hymenium	Hymenium *n*, Sporenlager *n*
4475	Hymenoptera, hymenopterans	Hymenopteren *pl*, Hautflügler *mpl*
4476	hyoid arch	Hyoidbogen *m*, Zungenbeinbogen *m*
4477	hyoid (bone)	Zungenbein *n*
4478	hypanthium	Hypanthium *n*, Blütenbecher *m*
4479	hyperchim(a)era	Hyperchimäre *f*
4480	hyperchromatosis, hyperchromatism	Hyperchromatose *f*, Hyperchromatizität *f*
4481	hypermetamorphosis	Hypermetamorphose *f*, Hypermetabolie *f*
4482	hypermorphic	hypermorph
4483	hypermorphosis	Hypermorphose *f*
4484	hyperparasite, superparasite	Hyperparasit *m*, Überparasit *m*
4485	hyperplasia	Hyperplasie *f*
4486	hyperploid	hyperploid
4487	hyperploidy	Hyperploidie *f*
4488	hyperpolarization	Hyperpolarisation *f*
4489	hypertely, hypertelia, overdevelopment	Hypertelie *f*
4490	hypertonic	hypertonisch
4491	hypertonicity, hypertonia	Hypertonie *f*
4492	hypha (*pl* hyphae)	Hyphe *f*, Pilzfaden *m*
4493	hyphomycetes	Fadenpilze *mpl*, Hyphomyzeten *pl*
4494	hypnospore → resting spore	—
4495	hypoblast → endoblast	—
4496	hypocotyl	Hypokotyl *n*, Keimachse *f*
4497	hypoderm → subcutaneous tissue	—
4498	hypodermis, hypoderm(a) *bot.*	Hypoderm *n*, Hypodermis *f* *bot.*
4499	hypogastrium	Hypogastrium *n*, Unterbauch *m*
4500	hypogeal, hypogean	hypogäisch
4501	hypogenesis	Hypogenese *f*, Hypogenesie *f*
4502	hypoglossal nerve	Hypoglossus *m*, Zungen(muskel)nerv *m*
4503	hypogynous	hypogyn
4504	hypolimnion	Hypolimnion *n*
4505	hypomicron → submicron	—
4506	hypomorph(ic)	hypomorph
4507	hyponasty	Hyponastie *f*
4508	hypophyllous	hypophyll
4509	hypophyseal hormone, hypophysial ~, pituitary ~	Hypophysenhormon *n*
4510	hypophyseal stalk, hypophysial ~	Hypophysenstiel *m*
4511	hypophysis, pituitary (body)	Hypophyse *f*, Hirnanhangdrüse *f*
4512	hypoplasia, hypoplastic development	Hypoplasie *f*, Unterentwicklung *f*
4513	hypoploid	hypoploid
4514	hypoploidy	Hypoploidie *f*
4515	hypostasis, hypostasy *gen.*	Hypostasis *f*, Hypostase *f* *gen.*
4516	hypostomatic chamber, ~ cavity, respiratory ~ *bot.*	Atemhöhle *f* *bot.*

4453	hydrophobe	hidrofóbico
4454	—	—
4455	hydrophytes *mpl*, plantes *fpl* aquatiques	hidrófitos *mpl*, plantas *fpl* acuáticas
4456	culture *f* hydroponique	cultivo *m* hidropónico, hidrocultivo *m*
4457	série *f* hygrophile	hidroserie *f*
4458	hydrosol *m*	hidrosol *m*
4459	—	—
4460	hydrosphère *f*	hidrosfera *f*
4461	hydrotaxie *f*, hydrotactisme *m*	hidrotaxis *f*, hidrotactismo *m*
4462	hydrotropisme *m*	hidrotropismo *m*
4463	hydroxyde *m*	hidróxido *m*
4464	groupe(ment) *m* hydroxyle	grupo *m* hidroxilo
4465	hydroxylamine *f*	hidroxilamina *f*
4466	hydroxylation *f*	hidroxilación *f*
4467	hydrozoaires *mpl*	hidrozoos *mpl*
4468	hygromètre *m*	higrómetro *m*
4469	hygronastie *f*	higronastia *f*
4470	hygrophile	higrófilo
4471	hygrophytes *mpl*	higrófitos *mpl*, higrófilas *fpl*
4472	hygroscopique	higroscópico
4473	hygroscopicité *f*	higroscopicidad *f*
4474	hyménium *m*	himenio *m*
4475	hyménoptères *mpl*	himenópteros *mpl*
4476	arc *m* hyoïdien	arco *m* hioideo
4477	(os *m*) hyoïde *m*	hioides *m*
4478	hypanthium *m*	hipanto *m*
4479	hyperchimère *f*	hiperquimera *f*
4480	hyperchromatose *f*	hipercromatosis, hipercromatismo *m*
4481	hypermétamorphose *f*	hipermetamorfosis *f*
4482	hypermorphe	hipermórfico
4483	hypermorphose *f*	hipermorfosis *f*
4484	hyperparasite *m*	hiperparásito *m*, superparásito *m*
4485	hyperplasie *f*, hyperplastie *f*	hiperplasia *f*
4486	hyperploïde	hiperploide
4487	hyperploïdie *f*	hiperploidia *f*
4488	hyperpolarisation *f*	hiperpolarización *f*
4489	hypertélie *f*, surévolution *f*	hipertelia *f*, sobreevolución *f*
4490	hypertonique	hipertónico
4491	hypertonie *f*, hypertonicite *f*	hipertonía *f*, hipertonicidad *f*
4492	hyphe *f*	hifa *f*
4493	champignons *mpl* filamenteux	hifomicetos *mpl*, hongos *mpl* filamentosos
4494	—	—
4495	—	—
4496	hypocotyle *m*	hipocótilo *m*
4497	—	—
4498	hypoderme *m* *bot.*	hipodermis *f*, hipodermo *m* *bot.*
4499	hypogastre *m*	hipogastrio *m*
4500	hypogé	hipogeo
4501	hypogenèse *f*, hypogénésie *f*	hipogénesis *f*
4502	nerf *m* hypoglosse	nervio *m* hipogloso
4503	hypogyne	hipógino
4504	hypolimnion *m*	hipolimneo *m*
4505	—	—
4506	hypomorphe	hipomorfo, hipomórfico
4507	hyponastie *f*	hiponastia *f*
4508	hypophylle	hipofilo
4509	hormone *f* hypophysaire, ~ pituitaire	hormona *f* hipofisaria, ~ pituitaria
4510	tige *f* pituitaire	tallo *m* hipofisario, pedúnculo *m* ~
4511	hypophyse *f*, glande *f* pituitaire	hipófisis *f*, (glándula *f*) pituitaria *f*
4512	hypoplasie *f*	hipoplasia *f*
4513	hypoploïde	hipoploide
4514	hypoploïdie *f*	hipoploidia *f*
4515	hypostasie *f*, hypostase *f* *gen.*	hipostasia *f*, hipostasis *f* *gen.*
4516	chambre *f* sous-stomatique	cámara *f* substomática, ~ respiratoria *bot.*

4517	hypothalamus	Hypothalamus *m*
4518	hypothallus	Hypothallus *m*
4519	hypothecium	Hypothezium *n*, Fruchtboden *m*
4520	hypotonic	hypotonisch
4521	hypoxanthine	Hypoxanthin *n*
4522	hypsophyll	Hochblatt *n*
4523	hysteresis	Hysterese *f*, Hysteresis *f*
4524	IAA → indole-3-acetic acid	—
4525	ichthyology	Ichthyologie *f*, Fischkunde *f*
4526	ICSH → luteinizing hormone	—
4527	id	Id *n*
4528	idant	Idant *m*
4529	identical twins → monozygotic twins	—
4530	identification key	Bestimmungsschlüssel *m*
4531	idioblast	Idioblast *m*
4532	idiochromatin	Idiochromatin *n*
4533	idiogram, karyogram	Idiogramm *n*, Karyogramm *n*
4534	idioplasm → germ plasm	—
4535	idiotype	Idiotyp *m*
4536	idiovariation	Idiovariation *f*
4537	idiozome	Idiozom *n*
4538	ileum	Krummdarm *m*, Ileum *n*
4539	iliac bone → hip-bone	—
4540	ilium	Darmbein *n*
4541	illuvial horizon → B-horizon	—
4542	imaginal disc	Imaginalscheibe *f*
4543	imaginal mo(u)lt → adult ~	—
4544	imago, perfect insect	Imago *f*, Vollkerf *m*, Vollinsekt *n*
4545	imbibition *bot.*	Quellung *f* *bot.*
4546	imbibition pressure *bot.*	Quellungsdruck *m* *bot.*
4547	imbricate	imbrikat, dachziegelig
4548	immature	unreif
4549	immaturity	Unreife *f*
4550	immersion lens	Immersionsobjektiv *n*
4551	immigration pressure	Wanderungsdruck *m*
4552	immiscible	unmischbar
4553	immission *env.*	Immission *f* *env.*
4554	immune	immun
4555	immune body, immunizator	Immunkörper *m*
4556	immune reaction, immunological ~, immunoreaction	Immun(o)reaktion *f*, immunologische Reaktion *f*
4557	immunity	Immunität *f*
4558	immunity breeding	Immunitätszüchtung *f*
4559	immunization	Immunisierung *f*
4560	immunizator → immune body	—
4561	to immunize	immunisieren
4562	immunobiology	Immunbiologie *f*
4563	immunochemistry	Immun(o)chemie *f*
4564	immuno-electrophoresis	Immunoelektrophorese *f*
4565	immunoglobulin	Immunoglobulin *n*
4566	immunologic(al)	immunologisch
4567	immunologist	Immunologe *m*
4568	immunology	Immunologie *f*
4569	immunomechanism, immune mechanism	Immunmechanismus *m*
4570	immunoreaction → immune reaction	—
4571	immunosuppression, immunological suppression	Immunkörperunterdrückung *f*
4572	immunotolerance, immunological tolerance	Immun(o)toleranz *f*
4573	impar, unpaired	unpaarig
4574	imparidigitate	unpaarzehig
4575	imparipinnate	unpaarig gefiedert
4576	imperfect fungi → fungi imperfectae	—
4577	impermeability, imperviousness	Undurchlässigkeit *f*, Impermeabilität *f*
4578	impermeable, impervious	undurchlässig, impermeabel

4517	hypothalamus *m*	hipotálamo *m*
4518	hypothalle *m*	hipotalo *m*
4519	hypothécie *f*	hipotecio *m*
4520	hypotonique	hipotónico
4521	hypoxanthine *f*	hipoxantina *f*
4522	feuille *f* bractéale	hipsofilo *m*
4523	hystérésis *f*	histeresis *f*, histéresis *f*
4524	—	—
4525	ichthyologie *f*	ictiología *f*
4526	—	—
4527	ide *m*	id *m*, ido *m*
4528	idante *m*	idante *m*
4529	—	—
4530	clé *f* de détermination	clave *f* de identificación
4531	idioblaste *m*	idioblasto *m*
4532	idiochromatine *f*	idiocromatina *f*
4533	idiogramme *m*, caryogramme *m*	idiograma *m*, cariograma *m*
4534	—	—
4535	idiotype *m*	idiótipo *m*
4536	idiovariation *f*	idiovariación *f*
4537	idiozome *m*	idiosoma *m*
4538	iléon *m*	íleon *m*
4539	—	—
4540	ilion *m*	ilion *m*
4541	—	—
4542	disque *m* imaginal	disco *m* imaginal
4543	—	—
4544	imago *f*, insecte *m* parfait	imago *m*, insecto *m* adulto, ~ perfecto
4545	imbibition *f* *bot.*	imbibición *f* *bot.*
4546	pression *f* d'imbibition	presión *f* de imbibición
4547	imbriqué	imbricado
4548	immature	inmaturo, inmaduro
4549	immaturité *f*	inmadurez *f*
4550	objectif *m* à immersion	objetivo *m* de inmersión
4551	pression *f* d'immigration	presión *f* de (in)migración
4552	immiscible	inmiscible, no miscible
4553	immission *f* *env.*	inmisión *f* *env.*
4554	immunisé	inmune
4555	immunisateur *m*	cuerpo *m* inmune
4556	immunoréaction *f*, réaction *f* immunitaire	inmunoreacción *f*, reacción *f* inmunitaria, ~ inmunológica
4557	immunité *f*	inmunidad *f*
4558	sélection *f* d'espèces immunisées	selección *f* de especies inmunizadas
4559	immunisation *f*	inmunización *f*
4560	—	—
4561	immuniser	inmunizar
4562	immunobiologie *f*	inmunobiología *f*
4563	immunochimie *f*	inmunoquímica *f*
4564	immunoélectrophorèse *f*	inmunoelectroforesis *f*
4565	immunoglobuline *f*	inmunoglobulina *f*
4566	immunologique	inmunológico
4567	immunologiste *m*	inmunólogo *m*
4568	immunologie *f*	inmunología *f*
4569	immunomécanisme *m*	mecanismo *m* inmunológico, ~ inmunitario
4570	—	—
4571	immunosuppression *f*	inmunosupresión *f*
4572	immunotolérance *f*, tolérance *f* immunologique	tolerancia *f* inmunológica
4573	impair	impar, no apareado
4574	imparidigité	imparidigitado
4575	imparipenné	imparipinnado
4576	—	—
4577	imperméabilité *f*	impermeabilidad *f*
4578	imperméable	impermeable

4579	implant	Implantat *n*
4580	implantation → nidation	—
4581	impression preparation	(Ab)klatschpräparat *n*
4582	imprinting *eth.*	Prägung *f* *eth.*
4583	improving (quality) of environment	Umweltverbesserung *f*
4584	impulse frequency	Impulsfrequenz *f*
4585	impurity	Verunreinigung *f*
4586	to inactivate	inaktivieren
4587	inactivation	Inaktivierung *f*
4588	inanimate → lifeless	—
4589	inbreeding, close breeding	Inzucht *f*, Verwandtschaftszucht *f*
4590	inbreeding coefficient	Inzuchtkoeffizient *m*
4591	inbreeding depression	Inzuchtdepression *f*, Inzuchtdegeneration *f*
4592	inbreeding population	Inzuchtpopulation *f*
4593	incineration	Veraschung *f*
4594	incipient plasmolysis	Grenzplasmolyse *f*
4595	incipient species	Prospezies *f*
4596	incisor (tooth)	Schneidezahn *m*
4597	inclusion body	Einschlußkörper *m*
4598	incompatibility	Unverträglichkeit *f*, Inkompatibilität *f*
4599	imcompatibility factor	Inkompatibilitätsfaktor *m*
4600	incompatible	unverträglich, inkompatibel
4601	incomplete dominance → semi-dominance	—
4602	incomplete penetrance	unvollständige Penetranz *f*
4603	incongruent crossing, ~ hybridization	inkongruente Kreuzung *f*
4604	incorporation of radioisotopes	Einbau *m* von Radioisotopen
4605	increase in diameter → growth in thickness	—
4606	increase in girth → growth in breath	—
4607	increta	Inkret *n*, Inkretionsstoff *m*
4608	incretion, internal secretion, endocrine ~	Inkretion *f*, innere Sekretion *f*
4609	incretory	inkretorisch
4610	incretory gland → endocrine ~	—
4611	incrustation	Inkrustation *f*
4612	to incubate, to brood	brüten, bebrüten, ausbrüten
4613	incubation	Brüten *n*, Brut *f*, Bebrüten *n*
4614	incubation behavio(u)r, breeding ~	Brutverhalten *n*
4615	incubation drive	Bruttrieb *m*
4616	incubation patch, brood ~	Brutfleck *m*
4617	incubation period	Inkubationszeit *f*
4618	incubation time	Brutdauer *f*
4619	incubator	Brutschrank *m*
4620	incurvation	Inkurvation *f*, Einkrümmung *f*
4621	incus	Amboß *m* *anat.*
4622	indeciduous → nondeciduous	—
4623	indefinite inflorescence	unbegrenzter Blütenstand *m*
4624	indehiscent fruit	Schließfrucht *f*
4625	independent assortment (of genes)	unabhängige Genverteilung *f*
4626	independent character	unabhängiges Merkmal *n*
4627	index fossil	Leitfossil *n*
4628	index of refraction → refractive index	—
4629	index species → characteristic species	—
4630	indicator plant	Indikatorpflanze *f*, Zeigerpflanze *f*, Leitpflanze *f*, Bodenzeiger *m*
4631	indicator species → characteristic species	—
4632	indigenous plant, native ~	einheimische Pflanze *f*
4633	indigestibility	Unverdaulichkeit *f*
4634	indigestible, indigerible	unverdaulich
4635	individual distance	Individualdistanz *f*
4636	individual selection	Individualauslese *f*
4637	indoleacetic acid, indolylacetic ~	Indol(yl)essigsäure *f*
4638	indole-3-acetic acid, heteroauxin, IAA	Indolyl-3-Essigsäure *f*, Heteroauxin *n*, IES
4639	induced mutation, directed ~	induzierte Mutation *f*
4640	inducer → inductor	—

4579	implant *m*	implante *m*
4580	—	—
4581	préparation *f* par impression	preparación *f* por impresión
4582	empreinte *f*, «imprinting» *m* *eth.*	impresión *f*, "imprinting" *m* *eth.*
4583	amélioration *f* de l'environnement	mejora *f* del medio ambiente
4584	fréquence *f* des influx	frecuencia *f* de los impulsos
4585	impureté *f*	impureza *f*, impuridad *f*
4586	inactiver	inactivar
4587	inactivation *f*	inactivación *f*
4588	—	—
4589	sélection *f* consanguine, consanguinité *f*, inbreeding *m*, croisement *m* consanguin	cruzamiento *m* consanguíneo, consanguinidad *f*, inbreeding *m*
4590	coefficient *m* de consanguinité	coeficiente *m* de consanguinidad, ~ de inbreeding
4591	dégénérescence *f* consanguine	degeneración *f* consanguínea, depresión *f* ~
4592	population *f* consanguine	población *f* consanguínea
4593	incinération *f*	incineración *f*
4594	plasmolyse *f* limite, ~ commençante	plasmólisis *f* límite, ~ incipiente
4595	espèce *f* naissante	especie *f* naciente
4596	(dent *f*) incisive *f*	(diente *m*) incisivo *m*
4597	corpuscule *m* intracellulaire	cuerpo *m* de inclusión, corpúsculo *m* ~ ~
4598	incompatibilité *f*	incompatibilidad *f*
4599	facteur *m* d'incompatibilité	factor *m* de incompatibilidad
4600	incompatible	incompatible
4601	—	—
4602	pénétrance *f* incomplète	penetración *f* incompleta
4603	croisement *m* incongru	cruzamiento *m* incongruente, hibridación *f* ~
4604	incorporation *f* de radioisotopes	incorporación *f* de radioisótopos
4605	—	—
4606	—	—
4607	incrément *m*	incremento *m*
4608	incrétion *f*, sécrétion *f* interne, ~ endocrine	increción *f*, secreción *f* interna
4609	incrétoire	incretor(io)
4610	—	—
4611	incrustation *f*	incrustación *f*
4612	incuber; couver	incubar; empollar
4613	incubation *f*; couvaison *f*	incubación *f*
4614	comportement d'incubation	comportamiento *m* incubador
4615	impulsion *f* de couver, instinct *m* ~ ~	instinto *m* de incubación
4616	plaque *f* incubatrice	placa *f* incubatriz, mancha *f* incubadora
4617	période *f* d'incubation	período *m* de incubación
4618	durée *f* de l'incubation, ~ de la couvaison	duración *f* de la incubación
4619	incubateur *m*	incubadora *f*
4620	incurvation *f*	encorvadura *f*
4621	enclume *f* *anat.*	yunque *m* *anat.*
4622	—	—
4623	inflorescence *f* indéfinie	inflorescencia *f* indefinida
4624	fruit *m* indéhiscent	fruto *m* indehiscente
4625	assortiment *m* indépendant (des gènes)	distribución *f* independiente (de los genes)
4626	caractère *m* indépendant	carácter *m* independiente
4627	fossile *m* caractéristique	fósil *m* característico, ~ indicador
4628	—	—
4629	—	—
4630	plante *f* indicatrice	planta *f* indicadora
4631	—	—
4632	plante *f* indigène	planta *f* indígena, ~ nativa
4633	indigestibilité *f*	indigestibilidad *f*
4634	indigestible, indigérable	indigerible, indigestible
4635	distance *f* individuelle, ~ entre individus	distancia *f* entre los individuos
4636	sélection *f* individuelle	selección *f* individual
4637	acide *m* indol(yl)acétique	ácido *m* indol(il)acético
4638	acide *m* 3-indolyl-acétique, hétéroauxine *f*, AIA	ácido *m* indol-3-acético, heteroauxina *f*, AIA
4639	mutation *f* provoquée	mutación *f* inducida, ~ provocada
4640	—	—

4641	inducible enzyme → adaptive enzyme	—
4642	induction of flowering	Blühinduktion *f*
4643	induction of mutation	Mutationsauslösung *f*
4644	inductive capacity	Induktionsfähigkeit *f*
4645	inductor, inducer, inductive agent	Induktor *m*, Induktionsstoff *m*
4646	indusium	Indusium *n*
4647	industrial melanism	Industriemelanismus *m*
4648	industrial waste	Industrieabfälle *mpl*
4649	inexcitability	Unerregbarkeit *f*
4650	inexcitable	unerregbar
4651	infecundity → infertility	—
4652	inferior *bot.*	unterständig
4653	inferior palea → lemma	—
4654	infertile	unfruchtbar
4655	infertility, infecundity	Unfruchtbarkeit *f*
4656	inflorescence	Blütenstand *m*, Infloreszenz *f*
4657	inflorescence axis	Blütenstandachse *f*
4658	information theory	Informationstheorie *f*
4659	infrared light	Infrarotlicht *n*
4660	infrared rays	Infrarotstrahlen *mpl*
4661	infrasound	Infraschall *m*
4662	infructescence	Fruchtstand *m*
4663	infundibular, infundibuliform, funnel-shaped	trichterförmig
4664	infundibulum	Infundibulum *n*
4665	to infuse	aufgießen
4666	Infusoria, infusorians	Infusorien *pl* (*sing*. Infusorium *n*), Infusionstierchen *npl*, Aufgußtierchen *npl*
4667	infusorial earth, diatomaceous ~, siliceous ~, kieselguhr	Kieselgur *m*, Diatomeenerde *f*, Infusorienerde *f*
4668	ingestion	Stoffaufnahme *f*
4669	inguinal	Leisten ...
4670	to inherit from	erben von
4671	inheritable	vererbbar, vererblich
4672	inheritance, heredity	Vererbung *f*
4673	inheritance of acquired characters	Vererbung *f* erworbener Eigenschaften
4674	inhibiting gene	Inhibitorgen *n*
4675	inhibitor	Hemmer *m*, Inhibitor *m*
4676	inhibitory, inhibitive	inhibitorisch, hemmend
4677	inhibitory action, ~ effect	Hemmwirkung *f*
4678	inhibitory factor	Hemm(ungs)faktor *m*
4679	inhibitory postsynaptic potentiel, IPSP	hemmendes postsynaptisches Potential *n*
4680	inhibitory reaction	Hemmungsreaktion *f*
4681	inhibitory substance, inhibiting ~	Hemmstoff *m*, Hemmsubstanz *f*
4682	inhibitory synapse	Hemmsynapse *f*
4683	initial	Initialzelle *f*, Initiale *f*
4686	initial population → original ~	—
4685	initial stage	Anfangsstadium *n*
4686	injurious insect → insect pest	—
4687	ink gland	Tintendrüse *f*
4688	ink sac	Tintensack *m*
4689	innate → congenital	—
4690	innate releasing mechanism, IRM	angeborener Auslösemechanismus *m*, AAM
4691	inner ear	Innenohr *n*, inneres Ohr *n*
4692	to innervate, to innerve	innervieren
4693	innervation	Innervation *f*, Innervierung *f*
4694	innocuous, harmless	unschädlich
4695	innocuousness, harmlessness	Unschädlichkeit *f*
4696	innovation shoot	Erneuerungssproß *m*, Innovationstrieb *m*
4697	to inoculate	impfen, beimpfen, überimpfen
4698	inoculation	Impfung *f* *microb.*
4699	inoculum	Inokulum *n*, Impfkultur *f*
4700	inorganic	anorganisch
4701	inquiline	Einmieter *m*, Inquilin *m*
4702	to insalivate	einspeicheln

4641	—	—
4642	induction *f* de floraison	inducción *f* de la floración
4643	induction *f* de mutations	inducción *f* de mutaciones
4644	capacité *f* inductive	capacidad *f* inductiva
4645	inducteur *m*	inductor *m*, agente *m* inductivo
4646	indusium *m*, indusie *f*	indusio *m*
4647	mélanisme *m* industriel	melanismo *m* industrial
4648	déchets *mpl* industriels	desechos *mpl* industriales, desperdicios *mpl* ~
4649	inexcitabilité *f*	inexcitabilidad *f*
4650	inexcitable	inexcitable
4651	—	—
4652	infère	ínfero
4653	—	—
4654	infertile, infécond	infértil, infecundo
4655	infertilité *f*, infécondité *f*	infertilidad *f*, infecundidad *f*
4656	inflorescence *f*	inflorescencia *f*
4657	axe *m* de l'inflorescence	eje *m* de la inflorescencia
4658	théorie *f* de l'information	teoría *f* de la información
4659	lumière *f* infrarouge	luz *f* infrarroja
4660	rayons *mpl* infrarouges	rayos *mpl* infrarrojos
4661	infrason(s) *m(pl)*	infrasonido *m*
4662	infrutescence *f*	infrutescencia *f*
4663	infundibuliforme	infundibiliforme
4664	infundibulum *m*	infundíbulo *m*
4665	infuser	infundir
4666	infusoires *mpl*	infusorios *mpl*
4667	terre *f* d'infusoires	tierra *f* de diatomeas, ~ silícea, "kieselgur" *m*
4668	ingestion *f*	ingestión *f*
4669	inguinal	inguinal
4670	hériter de	heredar de
4671	transmissible héréditairement	transmisible hereditariamente, ~ por herencia
4672	hérédité *f*, transmission *f* héréditaire	herencia *f*, transmisión *f* hereditaria
4673	hérédité *f* des caractères acquis	herencia *f* de los caracteres adquiridos
4674	gène *m* inhibiteur	gen *m* inhibidor
4675	inhibiteur *m*	inhibidor *m*
4676	inhibiteur, inhibitif	inhibitorio, inhibidor
4677	effet *m* inhibiteur, action *f* inhibitrice	efecto *m* inhibidor, acción *f* inhibidora
4678	facteur *m* inhibiteur	factor *m* inhibitorio
4679	potentiel *m* postsynaptique inhibiteur	potencial *m* postsináptico inhibitorio
4680	réaction *f* inhibitrice	reacción *f* de inhibición
4681	substance *f* inhibitrice, ~ inhibitive	sustancia *f* inhibidora
4682	synapse *f* inhibitrice	sinapsis *f* inhibidora
4683	initiale *f*	(célula *f*) inicial *f*
4684	—	—
4685	stade *m* initial, phase *f* initiale	fase *f* inicial
4686	—	—
4687	glande *f* du noir	glándula *f* secretora de la tinta
4688	poche *f* du noir, ~ d'encre	bolsa *f* negra, ~ de la tinta
4689	—	—
4690	mécanisme *m* inné de déclenchement, MID	mecanismo *m* desencadenante innato
4691	oreille *f* interne	oído *m* interno
4692	innerver	inervar
4693	innervation *f*	inervación *f*
4694	inoffensif	in(n)ocuo, inofensivo
4695	innocuité *f*	inocuidad *f*
4696	pousse *f* de régénérescence	renuevo *m* vegetativo
4697	inoculer	inocular
4698	inoculation *f*	inoculación *f*
4699	inoculum *m*	inóculo *m*
4700	inorganique	inorgánico
4701	inquilin *m*, locataire *m*	inquilino *m*
4702	insaliver	insalivar

4703	insect pest, noxious insect, harmful ~, injurious ~	Schadinsekt *n*, Insektenschädling *m*
4704	insect pollination → entomophily	—
4705	insect society, ~ state	Insektenstaat *m*
4706	Insecta, Hexapoda, insects	Insekten *npl*, Kerbtiere *npl*, Kerfe *mpl*
4707	insecticide	Insektizid *n*
4708	insectivore	Insektenfresser *m*
4709	insectivorous, entomophagous	insektenfressend
4710	insectivorous plant, carnivorous ~	insektenfressende Pflanze *f*, fleischfressende ~
4711	to inseminate	besamen
4712	insemination	Besamung *f*
4713	insertion breakage *gen.*	Insertionsbruch *m gen.*
4714	insertion region	Insertionsstelle *f*, Ansatzstelle *f*
4715	insertional translocation	insertionale Translokation *f*
4716	insight behavio(u)r	Einsichtverhalten *n*
4717	insolubility	Unlöslichkeit *f*
4718	insoluble	unlöslich
4719	inspiration	Einatmung *f*
4720	inspiratory reserve → complemental air	—
4721	inspired air	Einatmungsluft *f*
4722	instinct	Instinkt *m*
4723	instinct of self-preservation	Selbsterhaltungstrieb *m*
4724	instinctive action	Instinkthandlung *f*, Triebhandlung *f*
4725	instinctive behavio(u)r	Instinktverhalten *n*
4726	instinctive movement	Instinktbewegung *f*
4727	instrumental conditioning	instrumentelle Konditionierung *f*
4728	insulin	Insulin *n*
4729	integument, tegument	Integument *n*
4730	intemperate phage, lytic ~, virulent ~	nichttemperierter Phage *m*, lytischer ~, virulenter ~
4731	intensifying factor	Intensitätsfaktor *m*, Intensivierungsgen *n*
4732	intention movement	Intentionsbewegung *f*
4733	interaction	Wechselwirkung *f*, Interaktion *f*
4734	interband *gen.*	Zwischenscheibe *f gen.*
4735	inter-brain → diencephalon	—
4736	intercalary growth	interkalares Wachstum *n*
4737	intercalary meristem	Restmeristem *n*
4738	intercellular air space system	Interzellular(en)system *n*
4739	intercellular fluid → tissue fluid	—
4740	intercellular space	Interzellularraum *m*, Zwischenzellraum *m*
4741	intercellular substance	Interzellularstoff *m*, ~substanz *f*
4742	intercostal	Zwischenrippen...
4743	interfascicular cambium	interfaszikuläres Kambium *n*
4744	interference *gen.*	Interferenz *f gen.*
4745	interference distance	Interferenzabstand *m*
4746	interference microscopy	Interferenzmikroskopie *f*
4747	interferon	Interferon *n*
4748	intergeneric hybrid, (bi)generic ~	Gattungsbastard *m*
4749	intergeneric hybridization, bigeneric cross	Gattungskreuzung *f*
4750	interglacial period	Zwischeneiszeit *f*
4751	intergrana lamella	Intergranalamelle *f*
4752	interkinesis	Interkinese *f*
4753	interlocking *gen.*	Interlocking *n gen.*
4754	intermediary metabolism	Zwischenstoffwechsel *m*, Intermediärstoffwechsel *m*, intermediärer Stoffwechsel *m*
4755	intermediate host	Zwischenwirt *m*
4756	intermediate product	Zwischenprodukt *n*, Intermediärprodukt *n*
4757	intermediate stage	Zwischenstadium *n*
4758	intermediate streak	Intermedialstreifen *m*
4759	intermedin, melanocyte-stimulating hormone, melanophore ~, melanotropin, MSH	Intermedin *n*, Melanotropin *n*, melanophorenstimulierendes Hormon *n*, Pigmenthormon *n*, MSH

4703	insecte *m* nuisible	insecto *m* nocivo, ~ dañino
4704	—	—
4705	société *f* d'insectes	sociedad *f* de insectos
4706	insectes *mpl*, hexapodes *mpl*	insectos *mpl*, hexápodos *mpl*
4707	insecticide *m*	insecticida *m*
4708	insectivore *m*	insectívoro *m*
4709	insectivore, entomophage	insectívoro, entomófago
4710	plante *f* insectivore, ~ carnivore	planta *f* insectívora, ~ carnívora
4711	inséminer	inseminar
4712	insémination *f*	inseminación *f*
4713	cassure *f* d'insertion *gen.*	ruptura *f* de inserción *gen.*
4714	région *f* d'insertion, point *m* ~	zona *f* de inserción, punto *m* ~ ~
4715	translocation *f* insertionnelle	tra(n)slocación *f* de inserción
4716	compréhension *f* brusque, «insight»	discernimiento *m*, "insight"
4717	insolubilité *f*	insolubilidad *f*
4718	insoluble	insoluble
4719	inspiration *f*	inspiración *f*
4720	—	—
4721	air *m* inspiratoire	aire *m* inspiratorio, ~ inspirado
4722	instinct *m*	instinto *m*
4723	instinct *m* de la conservation	instinto *m* de conservación
4724	acte *m* instinctif	acto *m* instintivo
4725	comportement *m* instinctif, conduite *f* instinctive	comportamiento *m* instintivo, conducta *f* instintiva
4726	mouvement *m* instinctif	movimiento *m* instintivo
4727	conditionnement *m* instrumentel	condicionamiento *m* instrumental
4728	insuline *f*	insulina *f*
4729	intégument *m*, tégument *m*	integumento *m*, tegumento *m*
4730	phage *m* intempéré, ~ virulent	fago *m* intemperado, ~ virulento, ~ lítico
4731	facteur *m* d'intensification	factor *m* de intensificación
4732	mouvement *m* intentionnel	movimiento *m* intencional, ~ de intención
4733	interaction *f*	interacción *f*
4734	interbande *f* *gen.*	entrebanda *f* *gen.*
4735	—	—
4736	croissance *f* intercalaire	crecimiento *m* intercalar
4737	méristème *m* intercalaire	meristema *m* intercalar
4738	système *m* d'espaces intercellulaires	sistema *m* de espacios intercelulares, ~ intercelular de espacios de aire
4739	—	—
4740	espace *m* intercellulaire, interstice *m* ~	espacio *m* intercelular, meato *m* ~
4741	substance *f* intercellulaire	sustancia *f* intercelular
4742	intercostal	intercostal
4743	cambium *m* interfasciculaire	cámbium *m* interfascicular
4744	interférence *f* *gen.*	interferencia *f* *gen.*
4745	distance *f* d'interférence	distancia *f* de interferencia
4746	microscopie *f* interférante	microscopia *f* interferencial, ~ de interferencia
4747	interférone *f*	interferón *m*
4748	hybride *m* (inter)générique	híbrido *m* (inter)genérico
4749	croisement *m* intergénérique, ~ bigénérique	hibridación *f* intergenérica, cruza *f* ~
4750	période *f* interglaciaire	período *m* interglacial, interglaciación *f*
4751	lamelle *f* intergranaire	laminilla *f* intergranular
4752	intercinèse *f*	intercinesis *f*
4753	interlocking *m* *gen.*	intercierre *m* *gen.*
4754	métabolisme *m* intermédiaire	metabolismo *m* intermediario
4755	hôte *m* intermédiaire	hospedante *m* intermedio, huésped *m* intermediario
4756	produit *m* intermédiaire	producto *m* intermedio
4757	stade *m* intermédiaire	estad(i)o *m* intermedio, etapa *f* intermedia
4758	bande *f* médiane	línea *f* intermedia
4759	intermédine *f*, mélanotropine *f*, hormone *f* mélano(cyto)stimulante, ~ mélanotrope, mélanostimuline *f*	intermedina *f*, melanotrofina *f*, hormona *f* melanocito-estimulante

4760	intermicellar space	Intermizellarraum *m*
4761	intermittent	intermittierend
4762	internal clock → biological clock	—
4763	internal fertilization	innere Befruchtung *f*
4764	internal milieu, ~ environment	inneres Milieu *n*
4765	internal nare → choana	—
4766	internal parasite → endoparasite	—
4767	internal secretion → incretion	—
4768	internal skeleton → endoskeleton	—
4769	International Rules of Nomenclature	internationale Nomenklaturregeln *f pl*
4770	interneuron(e), relay cell, internuncial ~, association ~	Interneuron *n*, Zwischenneuron *n*, Schaltneuron *n*, Schaltzelle *f*
4771	internodal cell	Internodalzelle *f*, Zwischenknotenzelle *f*
4772	internode	Internodium *n*, Stengelglied *n*
4773	intero(re)ceptor	Intero(re)zeptor *m*
4774	interphase	Interphase *f*
4775	interphase nucleus, interphasic ~	Interphasekern *m*
4776	intersex	Intersex *n*
4777	intersexuality	Intersexualität *f*
4778	interspecific	zwischenartlich, interspezifisch
4779	interspecific competition, interspecies ~	zwischenartliche Konkurrenz *f*, interspezifische ~
4780	interspecific hybrid, species ~	Artbastard *m*
4781	interspecific hybridization, species cross	Artkreuzung *f*, interspezifische Kreuzung *f*
4782	intersterility	Intersterilität *f*
4783	interstice	Zwischenraum *m*, Interstitium *n*
4784	interstitial	interstitiell
4785	interstitial cell	interstitielle Zelle *f*, Zwischenzelle *f*
4786	interstitial cell stimulating hormone → luteinizing hormone	—
4787	interstitial fluid → tissue fluid	—
4788	interstitial water	Porenwasser *n*
4789	intertidal zone, tidal ~	Gezeitenzone *f*
4790	intervertebral disc	Zwischenwirbelscheibe *f*, Bandscheibe *f*
4791	intestinal, enteric	Darm...
4792	intestinal canal, enteric ~	Darmkanal *m*
4793	intestinal flora	Darmflora *f*
4794	intestinal juice	Darmsaft *m*
4795	intestinal parasite → enterozoon	—
4796	intestinal tract	Darmtrakt *m*
4797	intestinal villus (*pl* villi)	Darmzotte *f*
4798	intestinal wall	Darmwand *f*
4799	intestine, gut	Darm *m*
4800	intine	Intine *f*
4801	intracellular	intrazellular
4802	intraspecific	innerartlich, intraspezifisch
4803	intraspecific competition, intraspecies ~	innerartliche Konkurrenz *f*, intraspezifische ~
4804	intravital staining, intra vitam ~	Intravitalfärbung *f*
4805	intrinsic factor	Intrinsic-Faktor *m*, innerer Faktor *m*
4806	introgression *gen.*	Introgression *f* *gen.*
4807	introgressive hybridization	introgressive Hybridisierung *f*, ~ Bastardierung *f*
4808	introrse	intrors
4809	intussusception	Intussuszeption *f*
4810	inulin	Inulin *n*
4811	invagination, infolding	Einstülpung *f*, Invagination *f*
4812	inverse size-metabolism law	Größe-Stoffwechsel-Gesetz *n*
4813	inversion *gen.*	Inversion *f* *gen.*
4814	inversion heterozygosis	Inversionsheterozygotie *f*
4815	invertase, invertin, saccharase, sucrase	Invertase *f*, Invertin *n*, Sa(c)charase *f*
4816	Invertebrata, invertebrates	Wirbellose *m pl*
4817	to investigate, investigation, investigator → to research, research, research worker	—

4760	espace *m* intermicellaire	espacio *m* intermicelar
4761	intermittent	intermitente
4762	—	—
4763	fécondation *f* interne, fertilisation *f* ~	fertilización *f* interna
4764	milieu *m* intérieur	medio *m* interno
4765	—	—
4766	—	—
4767	—	—
4768	—	—
4769	Règles *f pl* Internationales de Nomenclature	Reglas *f pl* Internacionales de Nomenclatura
4770	interneurone *m*, neurone *m* internoncial, ~ intermédiaire, ~ d'association, ~ intercalaire	interneurona *f*, neurona *f* intermedia, ~ de asociación, ~ internuncial, célula *f* intercalar
4771	cellule *f* internodale	célula *f* internodal
4772	entre-nœud *m*	entrenudo *m*, internodio *m*
4773	intéro(ré)cepteur *m*	interoceptor *m*
4774	interphase *f*	interfase *f*
4775	noyau *m* interphasique	núcleo *m* (de) interfase
4776	intersexe *m*, intersexué *m*	intersexo *m*, intersexuado *m*
4777	intersexualité *f*	intersexualidad *f*
4778	interspécifique	interespecífico
4779	concurrence *f* interspécifique, compétition *f* ~	competencia *f* interespecífica
4780	hybride *m* (inter)spécifique	híbrido *m* (inter)específico, ~ interspecie
4781	hybridation *f* interspécifique, croisement *m* ~	hibridación *f* interespecífica, ~ interspecie
4782	interstérilité *f*	interesterilidad *f*
4783	interstice *m*	intersticio *m*
4784	interstitiel	intersticial
4785	cellule *f* interstitielle	célula *f* intersticial
4786	—	—
4787	—	—
4788	eau *f* interstitielle	agua *f* intersticial
4789	zone *f* des marées, ~ inter(co)tidale	zona *f* de (entre)marea
4790	disque *m* intervertébral	disco *m* intervertebral
4791	intestinal, entérique	intestinal, entérico
4792	canal *m* intestinal	canal *m* intestinal, ~ entérico
4793	flore *f* intestinale	flora *f* intestinal
4794	suc *m* intestinal	jugo *m* intestinal, ~ entérico
4795	—	—
4796	tractus *m* intestinal	tracto *m* intestinal
4797	villosité *f* intestinale	vellosidad *f* intestinal
4798	paroi *f* intestinale	pared *f* intestinal
4799	intestin *m*	intestino *m*
4800	intine *f*	intina *f*
4801	intracellulaire	intracelular
4802	intraspécifique	intraespecífico
4803	concurrence *f* intraspécifique, compétition *f* ~	competencia *f* intraespecífica
4804	coloration *f* intravitale	coloración *f* intravital, tinción *f* ~
4805	facteur *m* intrinsèque	factor *m* intrínseco
4806	introgression *f gen.*	introgresión *f gen.*
4807	hybridation *f* introgressive	hibridación *f* introgresiva
4808	introrse	introrso
4809	intussusception *f*	intususcepción *f*
4810	inuline *f*	inulina *f*
4811	invagination *f*	invaginación *f*
4812	loi *f* du rapport inverse entre la masse corporelle de l'organisme et son métabolisme	ley *f* del metabolismo inverso al tamaño
4813	inversion *f gen.*	inversión *f gen.*
4814	hétérozygose *f* d'inversion	heterocigosis *f* de inversión
4815	invertase *f*, invertine *f*, sucrase *f*, saccharase *f*	invertasa *f*, invertina *f*, sucrasa *f*, sacarasa *f*
4816	invertébrés *m pl*	invertebrados *m pl*
4817	—	—

4818	inviability	Lebensunfähigkeit *f*
4819	inviable → nonviable	—
4820	in vitro	in vitro, im Reagenzglas
4821	in vivo	in vivo, im lebenden Organismus
4822	involucel(lum)	Hüllchen *n* *bot.*
4823	involucre	Hüllkelch *m*
4824	involuntary muscle → smooth muscle	—
4825	involute	involutiv
4826	involution	Involution *f*, Rückbildung *f*
4827	iodiferous	jodhaltig
4828	iodine	Jod *n*
4829	iodine solution	Jodlösung *f*
4830	ion	Ion *n*
4831	ion antagonism	Ionenantagonismus *m*
4832	ion exchange	Ionenaustausch *m*
4833	ion exchange chromatography	Ionenaustausch-Chromatographie *f*
4834	ion exchange resin	Ionenaustauscherharz *n*
4835	ion exchanger	Ionenaustauscher *m*
4836	ion transport	Ionentransport *m*
4837	ionic balance	Ionengleichgewicht *n*
4838	ionic concentration	Ionenkonzentration *f*
4839	ionic current, ~ flow	Ionenfluß *m*, Ionenstrom *m*
4840	ionic pump	Ionenpumpe *f*
4841	ionic theory	Ionentheorie *f* (der Erregung)
4842	ionization	Ionisierung *f*, Ionisation *f*
4843	to ionize	ionisieren
4844	ionizing radiation	ionisierende Strahlen *mpl*
4845	iris	Regenbogenhaut *f*, Iris *f*
4846	IRM → innate releasing mechanism	—
4847	iron bacteria	Eisenbakterien *fpl*
4848	to irradiate	bestrahlen
4849	irradiation	Bestrahlung *f*, Ausstrahlung *f*
4850	irradiation dose	Bestrahlungsmenge *f*
4851	irreversibility of evolution	Irreversibilität *f* der Evolution, Nichtumkehrbarkeit *f* ~ ~
4852	irritability	Reizbarkeit *f*
4853	irritable	reizbar
4854	ischium	Sitzbein *n*
4855	isidium (*pl* isidia)	Isidie *f*
4856	islets of Langerhans	Langerhanssche Inseln *fpl*
4857	isoagglutinin	Isoagglutinin *n*
4858	isoallele	Isoallel *n*
4859	isoantibody, isobody	Isoantikörper *m*
4860	isoantigen	Isoantigen *n*
4861	isocaloric, isoenergetic	isokalorisch
4862	isochromosome	Isochromosom *n*
4863	iso-citric acid	Isozitronensäure *f*
4864	isodiametric	isodiametrisch
4865	isoelectric point	isoelektrischer Punkt *m*
4866	isoenzyme, isozyme	Iso(en)zym *n*
4867	isogamete	Isogamet *m*
4868	isogamous	isogam
4869	isogamy	Isogamie *f*
4870	isogenic, isogenous	isogen(isch)
4871	isogenomatic, isogenomic	isogenomatisch
4872	isogeny	Isogenie *f*
4873	isoiony	Isoionie *f*
4874	isokont	isokont
4875	isolate	Isolat *n*, Heiratskreis *m*
4876	to isolate	isolieren
4877	isolating mechanism	Isolationsmechanismus *m*
4878	isolation	Isolation *f*, Isolierung *f*
4879	isolation experiment	Isolationsexperiment *n*
4880	isolation gene	Isolationsgen *n*
4881	isolation index	Isolationsindex *m*
4882	isolecithal	isolezithal

4818	inviabilité *f*	inviabilidad *f*
4819	—	—
4820	in vitro	in vitro
4821	in vivo	in vivo
4822	involucelle *m*	involucelo *m*, involucela *f*
4823	involucre *m*	involucro *m*
4824	—	—
4825	involuté	involuto
4826	involution *f*	involución *f*
4827	iodifère, iodé	yodífero, yodado
4828	iode *m*	yodo *m*
4829	solution *f* d'iode	solución *f* de yodo, ~ yodada
4830	ion *m*	ion *m*
4831	antagonisme *m* ionique	antagonismo *m* iónico, ~ de iones
4832	échange *m* ionique, ~ d'ions	intercambio *m* iónico, ~ de iones
4833	chromatographie *f* par échange d'ions	cromatografía *f* de intercambio iónico
4834	résine *f* échangeante d'ions	resina *f* intercambiadora (de iones)
4835	échangeur *m* d'ions	intercambiador *m* iónico, ~ de iones
4836	transport *m* d'ions	transporte *m* de iones
4837	équilibre *m* ionique	equilibrio *m* iónico
4838	concentration *f* en ions	concentración *f* iónica
4839	flux *m* ionique	flujo *m* de iones
4840	pompe *f* ionique	bomba *f* iónica
4841	théorie *f* ionique	teoría *f* iónica
4842	ionisation *f*	ionización *f*
4843	ioniser	ionizar
4844	rayonnement *m* ionisant, radiation *f* ionisante	radiaciones *f pl* ionizantes
4845	iris *m*	iris *m*
4846	—	—
4847	ferrobactéries *f pl*, bactéries *f pl* du fer	ferrobacterias *f pl*
4848	irradier	irradiar
4849	irradiation *f*	irradiación *f*
4850	dose *f* d'irradiation	dosis *f* de irradiación
4851	irréversibilité *f* de l'évolution	irreversibilidad *f* de la evolución
4852	irritabilité *f*	irritabilidad *f*
4853	irritable	irritable
4854	ischion *m*	ísquion *m*
4855	isidie *f*, isidium *m*	isidio *m*
4856	îlots *m pl* de Langerhans	islotes *m pl* de Langerhans
4857	isoagglutinine *f*	isoaglutinina *f*
4858	isoallèle *m*	isoalelo *m*
4859	isoanticorps *m*	isocuerpo *m*
4860	isoantigène *m*	isoantígeno *m*
4861	isoénergétique	isoenergético
4862	isochromosome *m*	isocromosoma *m*
4863	acide *m* isocitrique	ácido *m* isocítrico
4864	isodiamétrique	isodiamétrico
4865	point *m* is(o)électrique	punto *m* isoeléctrico
4866	isoenzyme *m*, isozyme *m*	isoenzima *m*
4867	isogamète *m*	isogameto *m*
4868	isogame	isógamo
4869	isogamie *f*	isogamia *f*
4870	isogène, isogénique	isogénico
4871	isogénomique	isogenómico
4872	isogénie *f*	isogenia *f*
4873	iso-ionie *f*	isoionía *f*
4874	isokonté	isoconto
4875	isolat *m*	—
4876	isoler	aislar
4877	mécanisme *m* d'isolement	mecanismo *m* aislante, ~ de aislamiento
4878	isolement *m*, isolation *f*	aislamiento *m*
4879	expérience *f* d'isolement	experimento *m* de aislamiento
4880	gène *m* d'isolation	gen *m* de aislamiento
4881	indice *m* d'isolation	índice *m* de aislamiento
4882	isolécithe, isolécithique	isolecito, isolecítico

4883	isoleucine	Isoleuzin *n*
4884	isomar → isophene	—
4885	isomer	Isomer *n*
4886	isomerase	Isomerase *f*
4887	isomeric(al)	isomer
4888	isomery, isomerism	Isomerie *f*
4889	isometric	isometrisch
4890	isometry	Isometrie *f*
4891	isomorphic, isomorphous	isomorph
4892	isomorphism	Isomorphie *f*
4893	iso-osmotic → isotonic	—
4894	isophene, isophane, isomar	Isophäne *f*, Isomare *f*
4895	isophenic, isophenous, isophan	isophän
4896	isophenogamy	Isophänogamie *f*
4897	isopolyploid	isopolyploid
4898	isoprene	Isopren *n*
4899	isoprenoid	Isoprenoid *n*
4900	isosporous, isospory → homosporous, homospory	—
4901	isotonia	Isotonie *f*
4902	isotonic, iso-osmotic	isotonisch, is(o)osmotisch
4903	isotope	Isotop *n*
4904	isotope labelling	Isotopenmarkierung *f*
4905	isotopic tracer → radioactive tracer	—
4906	isotropic(al), isotropous	isotrop
4907	isotype	Isotyp *m*
4908	isotypy	Isotypie *f*
4909	isozyme → isoenzyme	—
4910	jarovization → vernalization	—
4911	jaw	Kiefer *m*
4912	jaw bone → maxillary (bone)	—
4913	jaw foot → maxilliped	—
4914	jejunum	Leerdarm *m*, Jejunum *n*
4915	joint *bot.* → node	—
4916	joint *anat.* → articulation	—
4917	jugal → zygomatic bone	—
4918	Jurassic (period)	Jura *m*
4919	juvenile form	Jugendform *f*
4920	juvenile hormone, neotenin	Juvenilhormon *n*, Larvalhormon *n*, Neotenin *n*
4921	juvenile plumage	Jugendgefieder *n*
4922	juvenile stage, young ~	Jugendstadium *n*
4923	kainogenesis → caenogenesis	—
4924	kakogenesis → cacogenesis	—
4925	kappa-factor	Kappa-Faktor *m*
4926	karyenchyma → nuclear sap	—
4927	karyogamy, caryogamy	Karyogamie *f*
4928	karyogene	Karyogen *n*
4929	karyogenesis	Karyogenese *f*, Zellkernbildung *f*
4930	karyogramme → idiogramme	—
4931	karyokinesis → mitosis	—
4932	karyology, nuclear cytology	Karyologie *f*, Zellkernlehre *f*
4933	karyolymph → nuclear sap	—
4934	karyolysis	Karyolyse *f*, Zellkernauflösung *f*
4935	karyomere, karyomerite	Karyomer *n*, Karyomerit *m*
4936	karyon → nucleus	—
4937	karyoplasm → nucleoplasm	—
4938	karyoplasmic ratio → nucleo-(cyto)plasmic ~	—
4939	karyorrhexis	Karyorrhexis *f*, Zellkernzerfall *m*
4940	karyosome, nucleosome	Karyosom *n*, Nukleosom *n*
4941	karyostasis	Kernruhe *f*
4942	karyotheca → nuclear membrane	—
4943	karyotin	Karyotin *n*
4944	karyotype	Karyotyp *m*

4883	isoleucine *f*	isoleucina *f*
4884	—	—
4885	isomère *f*	isómero *m*
4886	isomérase *f*	isomerasa *f*, enzima *f* isomerizante
4887	isomère	isómero
4888	isomérie *f*, isomérisme *m*	isomerismo *m*
4889	isométrique	isométrico
4890	isométrie *f*	isometría *f*
4891	isomorphe	isomorfo, isomórfico
4892	isomorphisme *m*	isomorfismo *m*
4893	—	—
4894	isophane *m*, isomar *m*	isófano *m*, isófeno *m*
4895	isophène, isophénique	isófeno, isofénico
4896	isophénogamie *f*	isofenogamia *f*
4897	isopolyploïde	isopoliploide
4898	isoprène *m*	isopreno *m*
4899	isoprénoïde *m*	isoprenoide *m*
4900	—	—
4901	isotonie *f*, isotonicité *f*	isotonia *f*
4902	isotonique, isosmotique	isotónico, is(o)osmótico
4903	isotope *m*	isótopo *m*
4904	marquage *m* isotopique	marcado *m* isotópico, ~ con isótopos
4905	—	—
4906	isotrope	isotropo, isotrópico
4907	isotype *m*	isotipo *m*
4908	isotypie *f*	isotipia *f*
4909	—	—
4910	—	—
4911	maxillaire *m*, mâchoire *f*	quijada *f*, maxilar *m*
4912	—	—
4913	—	—
4914	jéjunum *m*	yeyuno *m*
4915	—	—
4916	—	—
4917	—	—
4918	jurassique *m*	jurásico *m*
4919	forme *f* de jeunesse	forma *f* juvenil
4920	hormone *f* juvénile, ~ juvénilisante	hormona *f* juvenil, neotenina *f*
4921	plumage *m* juvénil	plumaje *m* juvenil
4922	période *f* juvénile, phase *f* de jeunesse	fase *f* juvenil
4923	—	—
4924	—	—
4925	facteur *m* kappa	factor *m* kappa
4926	—	—
4927	caryogamie *f*	cariogamia *f*
4928	caryogène *m*	cariogen *m*
4929	caryogenèse *f*	cariogénesis *f*
4930	—	—
4931	—	—
4932	caryologie *f*	cariología *f*
4933	—	—
4934	caryolyse *f*	cariólisis *f*
4935	caryomère *m*	cariómero *m*
4936	—	—
4937	—	—
4938	—	—
4939	caryorrhexie *f*, caryorrhexis *f*	cariorrexis *f*
4940	caryosome *m*, nucléosome *m*	cariosoma *m*
4941	caryostasie *f*	cariostasis *f*
4942	—	—
4943	caryotine *f*	cariotina *f*
4944	caryotype *m*	cariotipo *m*

4945	katabolism → catabolism	—
4946	kataphase, cataphase	Kataphase *f*
4947	kathepsin → cathepsin	—
4998	kation → cation	—
4949	kephalin → cephalin	—
4950	keratin	Keratin *n*, Hornsubstanz *f*
4951	keratinization	Keratinisierung *f*
4952	to keratinize	keratinisieren
4953	kernel	Kern *m bot.*
4954	ketone	Keton *n*
4955	ketonic acid, ketoacid, ketone acid	Keto(n)säure *f*
4956	ketose	Ketose *f*
4957	key gene → oligogene	—
4958	key stimulus → sign stimulus	—
4959	kidney, nephros	Niere *f*
4960	kieselguhr → infusorial earth	—
4961	kilocalorie, large calorie	Kilokalorie *f*, Großkalorie *f*
4962	kin(a)esthesis, kin(a)esthesia	Kinästhesie *f*, Bewegungsgefühl *n*
4963	kin(a)esthetic	kinästhetisch, bewegungsempfindlich
4964	kinase	Kinase *f*
4965	kinesis	Kinesis *f*
4966	kinetic	kinetisch, Bewegungs ...
4967	kinetic energy	kinetische Energie *f*, Bewegungsenergie *f*
4968	kinetics	Kinetik *f*
4969	kinetin	Kinetin *n*
4970	kinetochore → centromere	—
4971	kinetogene	Kinetogen *n*
4972	kinetogenesis	Kinetogenese *f*
4973	kinetoplasm	Kinetoplasma *n*
4974	kinetoplast	Kinetoplast *m*
4975	kinetosome → basal granule	—
4976	kinin → cytokinin	—
4977	klinotaxis	Klinotaxis *f*
4978	klon → clon	—
4979	knee-cap, patella, rotula	Kniescheibe *f*
4980	knee reflex, patellar ~	Kniereflex *m*
4981	knot	Astknoten *m*
4982	Krebs cycle, citric acid cycle, tricarboxylic acid cycle	Krebszyklus *m*, Zitronensäurezyklus *m*, Trikarbonsäurezyklus *m*
4983	kynurenine	Kynurenin *n*
4984	to label (by), to mark	markieren (mit)
4985	labellum *bot.*	Labellum *n*, Lippe *f bot.*
4986	labial	Lippen...
4987	labial palp, labipalp(us)	Lippentaster *m*, Labialtaster *m*
4988	labium	Labium *n*; Unterlippe *f ins.*
4989	laboratory animal	Laboratoriumstier *n*
4990	lab(oratory) culture	Laboratoriumskultur *f*
4991	laboratory experiment, ~ test	Labor(atoriums)versuch *m*
4992	laboratory population	Laborpopulation *f*
4993	labrocyte → mast cell	—
4994	labrum	Labrum *n*, Oberlippe *f ins.*
4995	lachrymal → lacrimal	—
4996	laciniate	gefranst, geschlitztblättrig
4997	lacrimal bone, lachrymal ~	Tränenbein *n*
4998	lacrimal duct, lachrymal ~	Tränenkanal *m*
4999	lacrimal gland, lachrymal ~, tear ~	Tränendrüse *f*
5000	lacrimal sac, lachrymal ~, tear ~, dacryocyst	Tränensack *m*
5001	lacrimonasal duct, nasolacrimal ~	Tränennasengang *m*
5002	lactalbumin	Laktalbumin *n*, Milchalbumin *n*
5003	lactase	Laktase *f*
5004	lactation	Laktation *f*
5005	lacteal → chyliferous vessel	—
5006	lactic acid	Milchsäure *f*
5007	lactic acid bacteria	Milchsäurebakterien *fpl*
5008	lactic (acid) fermentation	Milchsäuregärung *f*

4945	—	—
4946	cataphase *f*	catafase *f*
4947	—	—
4948	—	—
4949	—	—
4950	kératine *f*	queratina *f*
4951	kératinisation *f*	queratinización *f*
4952	kératiniser	queratinizar
4953	pépin *m*	pepita *f*, pepa *f* (LA)
4954	cétone *f*	quetona *f*, cetona *f*
4955	acide *m* cétonique, cétoacide *m*	ácido *m* cetónico, cetoácido *m*
4956	cétose *m*	quetosa *f*, cetosa *f*
4957	—	—
4958	—	—
4959	rein *m*	riñón *m*
4960	—	—
4961	kilocalorie *f*, grande calorie *f*	kilocaloría *f*, caloría *f* grande
4962	cinesthésie *f*, kinesthésie *f*	cinestesia *f*
4963	cinesthésique, kinesthésique	cinestético
4964	kinase *f*	quinasa *f*
4965	cinésie *f*, cinèse *f*	cinesis *f*
4966	cinétique	cinético
4967	énergie *f* cinétique, ~ de mouvement	energía *f* cinética
4968	cinétique *f*	cinética *f*
4969	kinétine *f*	cinetina *f*, quinetina *f*
4970	—	—
4971	cinétogène *m*	cinetogen *m*
4972	cinétogenèse *f*, kinétogenèse *f*	cinetogénesis *f*
4973	cinétoplasme *m*	cinetoplasma *m*
4974	cinétoplaste *m*, kinétoplaste *m*	cinetoplasto *m*
4975	—	—
4976	—	—
4977	clinotaxie *f*	klinotaxia *f*
4978	—	—
4979	rotule *f*	rótula *f*, patela *f*
4980	réflexe *m* rotulien	reflejo *m* rotuliano, ~ patelar
4981	nœud *m* *bot.*	nudo *m* *bot.*
4982	cycle *m* de Krebs, ~ citrique, ~ d'acide tricarbonique	ciclo *m* de Krebs, ~ (del ácido) cítrico, ~ del ácido tricarboxílico
4983	cynurénine *f*	cinurenina *f*
4984	marquer (à, par)	marcar (con)
4985	labelle *m*, lèvre *f* *bot.*	labelo *m*
4986	labial	labial
4987	palpe *m* labial	palpo *m* labial
4988	labium *m*	labio *m* *ins.*
4989	animal *m* de laboratoire	animal *m* de laboratorio
4990	culture *f* de laboratoire	cultivo *m* de laboratorio
4991	expérience *f* de laboratoire, essai *m* ~ ~	experimento *m* de laboratorio, prueba *f* ~ ~
4992	population *f* de laboratoire	población *f* de laboratorio
4993	—	—
4994	labre *m*	labro *m*
4995	—	—
4996	lacinié	laciniado, de hoja laciniada
4997	os *m* lacrymal	hueso *m* lagrimal
4998	conduit *m* lacrymal, canal *m* ~	conducto *m* lagrimal
4999	glande *f* lacrymale	glándula *f* lagrimal
5000	sac *m* lacrymal	saco *m* lagrimal, dacriocisto *m*
5001	canal *m* naso-lacrymal, ~ lacrymo-nasal	conducto *m* nasolacrimal
5002	lactalbumine *f*	lactalbúmina *f*
5003	lactase *f*	lactasa *f*
5004	lactation *f*	lactación *f*
5005	—	—
5006	acide *m* lactique	ácido *m* láctico
5007	bactéries *f pl* lactiques	bacterias *f pl* (ácido)lácticas
5008	fermentation *f* lactique	fermentación *f* láctica

5009	lactiferous duct, galactophorous ~	Milchgang *m*
5010	lactoflavin → riboflavin	—
5011	lactogenic hormone, luteotrop(h)ic ~, luteotropin, prolactin, galactin, LTH	Prolaktin *n*, Laktotropin *n*, Laktationshormon *n*, Luteotrophin *n*, laktotropes Hormon *n*, luteotrop(h)es ~, LT
5012	lactoglobulin	Laktoglobulin *n*
5013	lactose, milk sugar	Laktose *f*, Laktobiose *f*, Milchzucker *m*
5014	lacunar system	Lakunensystem *n*
5015	laevulose → fructose	—
5016	lag-phase, lag-period	Lag-Phase *f*, Lag-Periode *f*, Verzögerungsphase *f*
5017	Lamarckism	Lamarckismus *m*
5018	lamella	Lamelle *f*
5019	lamellar structure	Lamellenstruktur *f*
5020	Lamellibranchiata → Bivalvia	—
5021	laminal *bot.*	flächenständig, laminal *bot.*
5022	lampbrush chromosome	Lampenbürstenchromosom *n*
5023	lamp shells → Brachiopoda	—
5024	lanceolate	lanzettförmig
5025	land-dwelling → terrestrial	—
5026	land plant → terrestrial plant	—
5027	land race → native breed	—
5028	landscape husbandry	Landschaftspflege *f*
5029	landscape protection	Landschaftsschutz *m*
5030	lanuginous, lanuginose	flaumig
5031	lanugo	Lanugo *f*, Wollhaar *n*
5032	lapping *ins.*	leckend *ins.*
5033	large calorie → kilocalorie	—
5034	large intestine	Dickdarm *m*
5035	larmier → tear pit	—
5036	larva	Larve *f*
5037	larval form	Larvenform *f*
5038	larval mo(u)lt	Larvenhäutung *f*
5039	larval skin, ~ cuticle	Larvenhaut *f*
5040	larval stage	Larvenstadium *n*
5041	larynx	Kehlkopf *m*, Larynx *m*
5042	late wood, autumn ~	Spätholz *n*, Sommerholz *n*, Engholz *n*
5043	latency	Latenz *f*
5044	latent	latent
5045	latent learning	latentes Lernen *n*
5046	latent period	Latenzzeit *f*, Latenzperiode *f*
5047	lateral bud → axillary bud	—
5048	lateral line organ	Seitenlinienorgan *n*
5049	lateral ramification	Seitenverzweigung *f*
5050	lateral root, secondary ~	Nebenwurzel *f*, Seitenwurzel *f*
5051	latex	Milchsaft *m*, Latex *m*
5052	laticifer, laticiferous vessel, milk-tube	Milchröhre *f*, Milchsaftgefäß *n*
5053	laticiferous cell	Milchsaftzelle *f*
5054	latifoliate	breitblättrig
5055	law of heterogenous summation	Reizsummengesetz *n*
5056	law of independent assortment → Mendel's third law	—
5057	law of mass action	Massenwirkungsgesetz *n*
5058	law of segregation → Mendel's second law	—
5059	law of the minimum, Liebig's law	Gesetz *n* vom Minimum, Minimumgesetz *n*
5060	law of thermodynamics	thermodynamisches Gesetz *n*
5061	law of uniformity → Mendel's first law	—
5062	laws of inheritance	Vererbungsgesetze *n pl*
5063	layer, layered → stratum, stratified	—
5064	leader *zool.*	Leittier *n*, Führer *m* (Vogelzug)
5065	leader *bot.*, leading shoot	Leittrieb *m*, Haupttrieb *m*
5066	leaf	Blatt *n*

5009	canal *m* lactifère, ~ galactophore	conducto *m* lactífero, ~ galactóforo
5010	—	—
5011	hormone *f* lactogène, ~ galactogène, ~ lutéotrope, prolactine *f*, galactine *f*, lacthormone *f*, LTH	hormona *f* lactógena, ~ luteótropa, ~ luteotrófica, ~ lactotropa, prolactina *f*, luteotrofina *f*, LTH
5012	lactoglobuline *f*	lactoglobulina *f*
5013	lactose *m*, sucre *m* de lait	lactosa *f*, lactobiosa *f*, azúcar *m* de leche
5014	système *m* lacunaire	sistema *m* lagunar
5015	—	—
5016	phase *f* de latence, «lag phase» *f*	fase *f* de retardo, ~ retrasada, ~ retardada
5017	lamarckisme *m*	lamarckismo *m*, lamarquismo *m*
5018	lamelle *f*	laminilla *f*
5019	structure *f* lamellaire	estructura *f* lamelar
5020	—	—
5021	laminaire *bot.*	laminal *bot.*
5022	chromosome *m* plumeux, ~ en goupillon, ~ en écouvillon, ~ en rince-bouteille	cromosoma *m* plumoso, ~ plumulado
5023	—	—
5024	lancéolé	lanceolado
5025	—	—
5026	—	—
5027	—	—
5028	aménagement *m* paysagiste	ordenación *f* paisajística
5029	protection *f* des paisages (naturels)	protección *f* (o: defensa *f*) del paisaje (natural), ~ de espacios naturales
5030	lanugineux, duveteux	lanuginoso
5031	lanugo *m*	lanugo *m*
5032	lécheur *ins.*	lamedor *ins.*
5033	—	—
5034	gros intestin *m*	intestino *m* grueso
5035	—	—
5036	larve *f*	larva *f*
5037	forme *f* larvaire	forma *f* larval
5038	mue *f* larvaire	muda *f* larval, ~ larvaria
5039	dépouille *f* larvaire, exuvie *f* ~	envoltura *f* larval, tegumento *m* larvario
5040	stade *m* larvaire	estad(i)o *m* larvario, fase *f* larval
5041	larynx *m*	laringe *f*
5042	bois *m* tardif, ~ d'automne	leño *m* tardío, ~ de otoño, ~ estrecho
5043	latence *f*, état *m* de vie latent	latencia *f*, estado *m* de vida latente
5044	latent	latente
5045	apprentissage *m* latent	aprendizaje *m* latente
5046	période *f* de latence, temps *m* ~ ~	período *m* de latencia, tiempo *m* ~ ~
5047	—	—
5048	organe *m* de la ligne latérale	órgano *m* de la línea lateral
5049	ramification *f* latérale	ramificación *f* lateral
5050	racine *f* latérale, ~ secondaire	raíz *f* lateral, ~ secundaria
5051	latex *m*	látex *m*
5052	laticifère *m*, tube *m* ~, vaisseau *m* ~	laticífero *m*, vaso *m* ~, tubo *m* ~
5053	cellule *f* laticifère	célula *f* laticífera
5054	latifolié	latifolio
5055	loi *f* de la sommation hétérogène des stimuli	ley *f* (o: regla *f*) de la suma heterogénea de estímulos
5056	—	—
5057	loi *f* d'action de masse	ley *f* de acción de masas
5058	—	—
5059	loi *f* du minimum, ~ des minima, ~ de Liebig	ley *f* del mínimo
5060	principe *m* de la thermodynamique	ley *f* (de la) termodinámica
5061	—	—
5062	lois *fpl* de l'hérédité, règles *fpl* ~ ~	leyes *fpl* de la herencia, ~ hereditarias
5063	—	—
5064	chef *m* de file, conducteur *m*, guide *m*, «leader» *m*	animal *m* conductor, caudillo *m*, guía *m*
5065	pousse *f* principale, flèche *f*	tallo *m* principal, guía *m* (del árbol), flecha *f*
5066	feuille *f*	hoja *f*

5067	leaf abscission, ~ shed	Blattabwurf *m*
5068	leaf absorption	Blattabsorption *f*
5069	leaf axil	Blattachsel *f*
5070	leaf base	Blattgrund *m*, Blattbasis *f*
5071	leaf blade	Blattspreite *f*
5072	leaf bud	Blattknospe *f*
5073	leaf bundle	blatteigenes Leitbündel *n*
5074	leaf cushion, pulvinus	Blattpolster *n*
5075	leaf cutting	Blattsteckling *m*
5076	leaf fall	Blattfall *m*, Laubfall *m*
5077	leaf gap, ~ lacuna	Blattlücke *f*
5078	leaf insertion, ~ attachment	Blattansatz *m*
5079	leaf-lichen → foliose lichen	—
5080	leaf primordium, embryonic leaf	Blattanlage *f*
5081	leaf rib	Blattrippe *f*
5082	leaf scar	Blattnarbe *f*
5083	leaf sheath	Blattscheide *f*
5084	leaf spine	Blattdorn *m*
5085	leaf surface	Blattfläche *f*
5086	leaf-stalk → petiole	—
5087	leaf tendril	Blattranke *f*
5088	leaf tip, apex of a leaf	Blattspitze *f*
5089	leaf trace	Blattspur *f*
5090	leaf vein, nervure	Blattader *f*, Blattnerv *m*
5091	leaf venation	Blattaderung *f*
5092	leafage → foliage	—
5093	leafing → foliation	—
5094	leafless → aphyllous	—
5095	leaflet → foliole	—
5096	leafy → foliose	—
5097	learning *eth.*	Lernen *n eth.*
5098	learning capacity, ~ ability	Lernfähigkeit *f*, Lernvermögen *n*
5099	learning curve	Lernkurve *f*
5100	lecithin	Lezithin *n*
5101	lectotype	Lektotyp *m*
5102	leeches → Hirudinea	—
5103	legume, pod, pulse	Hülse *f*, Legumen *n*
5104	Leguminosae	Hülsenfrüchtler *mpl*, Leguminosen *fpl*
5105	leisure area, recreational ~ *env.*	Erholungsraum *m*, Freizeitgebiet *n env.*
5106	lemma, lower palea, inferior ~, flowering glume	Deckspelze *f*
5107	lens → crystalline lens	—
5108	lens placode	Linsenanlage *f*
5109	lenticel, cortical pore	Lentizelle *f*, Korkpore *f*
5110	lenticular, lentiform, lentil-shaped	linsenförmig, lentikulär
5111	Lepidoptera, lepidopterans	Lepidopteren *pl*, Schmetterlinge *mpl*
5112	leptonema	Leptonema *n*
5113	leptosporangiate	leptosporangiat
5114	leptotene	Leptotän *n*
5115	lesion experiment → defect experiment	—
5116	lesser circulation → pulmonary circulation	—
5117	lethal	letal
5118	lethal dose	Letaldosis *f*
5119	lethal factor, ~ gene	Letalfaktor *m*
5120	lethal mutation	Letalmutation *f*
5121	lethality	Letalität *f*
5122	leucine	Leuzin *n*
5123	leucocyte, white blood corpuscle	Leukozyt *m*, weißes Blutkörperchen *n*
5124	leucoplast	Leukoplast *m*
5125	levogyrate, levogyric, laevorota(to)ry	linksdrehend
5126	LH → luteinizing hormone	—
5127	liane	Liane *f*

5067	abscis(s)ion *f* des feuilles	abscisión *f* de las hojas, desprendimiento *m* ~ ~ ~
5068	absorption *f* foliaire	absorción *f* foliar
5069	aisselle *f* de la feuille	axila *f* de la hoja
5070	base *f* foliaire	base *f* foliar
5071	limbe *m* (foliaire)	limbo *m* (foliar)
5072	bourgeon *m* foliaire, œil *m* à feuilles	yema *f* foliar
5073	faisceau *m* foliaire	haz *m* foliar
5074	coussinet *m* foliaire	pulvínulo *m* foliar, cojinete *m* ~
5075	bouture *f* de feuille	esqueje *m* de hoja
5076	chute *f* des feuilles, effeuillaison *f*, défeuillaison *f*	caída *f* de las hojas
5077	fenêtre *f* foliaire, lacune *f* ~	espacio *m* foliar, intersticio *m* ~
5078	insertion *f* foliaire	inserción *f* foliar
5079	—	—
5080	ébauche *f* foliaire, primordium *m* ~	primordio *m* foliar, hoja *f* embrionaria
5081	côte *f* de feuille	costilla *f* foliar
5082	cicatrice *f* foliaire	cicatriz *f* foliar
5083	gaine *f* foliaire	vaina *f* foliar
5084	épine *f* foliaire	espina *f* foliar
5085	surface *f* foliaire	superficie *f* foliar
5086	—	—
5087	vrille *f* foliaire	zarcillo *f* foliar
5088	sommet *m* du limbe	ápice *m* foliar
5089	trace *f* foliaire	rastro *m* foliar, traza *f* ~
5090	nervure *f* foliaire	nervio *m* foliar
5091	nervation *f* foliaire, vénation *f* ~	nervadura *f* foliar, nerv(i)ación *f* ~, venación *f* ~
5092	—	—
5093	—	—
5094	—	—
5095	—	—
5096	—	—
5097	apprentissage *m* *eth.*	aprendizaje *m* *eth.*
5098	capacité *f* d'apprentissage, faculté *f* ~	capacidad *f* de aprendizaje, facultad *f* de aprender
5099	courbe *f* d'apprentissage	curva *f* del aprendizaje
5100	lécithine *f*	lecitina *f*
5101	lectotype *m*	lectotipo *m*
5102	—	—
5103	gousse *f*, légume *m*, cosse *f*	vaina *f*, legumbre *f*
5104	légumineuses *f pl*	leguminosas *f pl*
5105	zone *f* de loisir(s), ~ de détente *env.*	zona *f* de recreo *env.*
5106	glumelle *f* inférieure, lemma *f*	lema *f*, palea *f* inferior
5107	—	—
5108	placode *f* cristallinienne	placoda *f* del cristalino
5109	lenticelle *f*	lenticela *f*
5110	lenticulé, lenticulaire	lenticular
5111	lépidoptères *m pl*	lepidópteros *m pl*
5112	leptonème *m*	leptonema *m*
5113	leptosporangié	leptosporangiado
5114	leptotène *m*	leptóteno *m*
5115	—	—
5116	—	—
5117	lét(h)al	letal
5118	dose *f* lét(h)ale	dosis *f* letal
5119	facteur *m* lét(h)al, gène *m* ~	factor *m* letal, gen *m* ~
5120	mutation *f* lét(h)ale	mutación *f* letal
5121	lét(h)alité *f*	letalidad *f*
5122	leucine *f*	leucina *f*
5123	leucocyte *m*, globule *m* blanc (du sang)	leucocito *m*, glóbulo *m* blanco (de la sangre)
5124	leucoplaste *m*	leucoplasto *m*
5125	lévogyre, sénestrogyre	levógiro
5126	—	—
5127	liane *f*	bejuco *m*, liana *f*

No.	English	German
5128	Lias	Lias *m/f*, schwarzer Jura *m*
5129	liber, bast	Bast *m*
5130	to liberate (energy etc.)	freisetzen (Energie etc.)
5131	lichen	Flechte *f*
5132	lichenin(e), lichen starch, moss ~	Lichenin *n*, Flechtenstärke *f*, Moosstärke *f*
5133	lichenology	Lichenologie *f*, Flechtenkunde *f*
5134	Liebig's law → law of the minimum	—
5135	lien, lienal → spleen, splenic	—
5136	life conditions, conditions of existence	Lebensbedingungen *f pl*
5137	life cycle	Lebenszyklus *m*
5138	life expectation, ~ expectancy	Lebenserwartung *f*
5139	life form	Lebensform *f*
5140	life process	Lebensvorgang *m*
5141	life quality	Lebensqualität *f*
5142	life span, length of life	Lebensdauer *f*
5143	life table	Sterbetafel *f*
5144	life zone	Lebensgebiet *n*
5145	lifeless, inanimate	leblos, unbelebt
5146	ligament	Band *n anat.*, Ligament *n*
5147	ligase, synthetase	Ligase *f*, Synthetase *f*
5148	light energy	Lichtenergie *f*
5149	light intensity	Lichtintensität *f*
5150	light microscope, optical ~	Lichtmikroskop *n*
5151	light quantum → photon	—
5152	light reaction *bot.*	Lichtreaktion *f bot.*
5153	light receptor → photoreceptor	—
5154	light scattering	Lichtstreuung *f*
5155	light-sensitive, photosensitive, sensitive to light	lichtempfindlich, photosensibel
5156	light sensitivity, photosensitivity, sensitivity to light	Lichtempfindlichkeit *f*
5157	light source	Lichtquelle *f*
5158	light stimulus	Lichtreiz *m*
5159	ligneous, woody	holzig
5160	ligneous plant, woody ~	Holzpflanze *f*
5161	ligneous tissue → woody ~	—
5162	lignification	Verholzung *f*
5163	lignified fibre (US: fiber)	Holzfaser *f*
5164	to lignify	verholzen
5165	lignin	Lignin *n*, Holzstoff *m*
5166	ligulate flower	Zungenblüte *f*
5167	ligule	Ligula *f*, Blatthäutchen *n*
5168	limb bud	Gliedmaßenknospe *f*
5169	limbic system	limbisches System *n*
5170	limbs → extremities	—
5171	liminal stimulus, threshold ~	Schwellenreiz *m*
5172	limit of distribution	Verbreitungsgrenze *f*
5173	limiting factor	Begrenzungsfaktor *m*, limitierender Faktor *m*
5174	limiting membrane	Grenzmembran *f*
5175	limivorous	schlammfressend
5176	limnetic	limnetisch
5177	limnetic zone, limnion	limnische Zone *f*, Limnion *n*
5178	limnology	Limnologie *f*, Seenkunde *f*, Süßwasserkunde *f*
5179	limnoplankton, freshwater plankton	Limnoplankton *n*, Süßwasserplankton *n*
5180	line breeding	Linienzucht *f*
5181	line of breeding	Zuchtlinie *f*
5182	line of descent	Abstammungslinie *f*
5183	line of inbreeding	Inzuchtlinie *f*
5184	linear arrangement of genes	lineare Genanordnung *f*
5185	linear regression	lineare Regression *f*
5186	lingual, glossal	Zungen ...
5187	lingual gland	Zungendrüse *f*
5188	lingual papilla, tongue ~	Zungenpapille *f*
5189	linin → achromatin	—

5128	lias *m*	liásico *m*
5129	liber *m*	líber *m*
5130	libérer (énergie etc.)	liberar (energía etc.)
5131	lichen *m*	liquen *m*
5132	lichénine *f*	liquenina *f*, fécula *f* de liquen
5133	lichénologie *f*	liquenología *f*
5134	—	—
5135	—	—
5136	conditions *fpl* vitales, ~ d'existence	condiciones *fpl* de vida, ~ de existencia
5137	cycle *m* vital	ciclo *m* de vida, ~ vital
5138	espérance *f* de (sur)vie, expectance *f* ~ ~	expectativa *f* de vida
5139	forme *f* biologique	forma *f* biológica
5140	processus *m* vital	proceso *m* vital
5141	qualité *f* de la vie	calidad *f* de (la) vida
5142	durée *f* de vie	duración *f* de (la) vida, lapso *m* vital
5143	table *f* de survie, ~ de mortalité	tabla *f* de vida
5144	zone *f* biologique	zona *f* biológica, biozona *f*
5145	inanimé	inanimado, no viviente
5146	ligament *m*	ligamento *m*
5147	ligase *f*, synthétase *f*	ligasa *f*, sintetasa *f*
5148	énergie *f* lumineuse	energía *f* luminosa, ~ lumínica
5149	intensité *f* lumineuse	intensidad *f* luminosa, ~ lumínica
5150	microscope *m* photonique, ~ optique	microscopio *m* de luz, ~ óptico, ~luminoso
5151	—	—
5152	réaction *f* lumineuse *bot.*	reacción *f* lumínica *bot.*
5153	—	—
5154	dispersion *f* de la lumière	dispersión *f* de la luz, ~ luminosa
5155	sensible à la lumière, photosensible	sensible a la luz, fotosensible
5156	photosensibilité *f*	fotosensibilidad *f*
5157	source *f* lumineuse	foco *m* luminoso
5158	stimulus *m* lumineux	estímulo *m* luminoso, ~ lumínico
5159	ligneux	leñoso
5160	plante *f* ligneuse	planta *f* leñosa
5161	—	—
5162	lignification *f*	lignificación *f*
5163	fibre *f* ligneuse	fibra *f* leñosa
5164	lignifier	lignificarse
5165	lignine *f*	lignina *f*
5166	fleur *f* ligulée	flor *f* ligulada
5167	ligule *f*	lígula *f*
5168	bourgeon *m* de membres	brote *m* de miembro
5169	système *m* limbique	sistema *m* límbico
5170	—	—
5171	stimulus *m* liminaire	estímulo *m* liminar, ~ umbral, ~ mínimo
5172	limite *f* de distribution	límite *m* de distribución
5173	facteur *m* limitant, ~ limitatif	factor *m* limitante, ~ limitador
5174	membrane *f* limitante	membrana *f* limitante
5175	lim(n)ivore	lim(n)ívoro
5176	limnétique	limnético
5177	zone *f* limnétique	zona *f* limnética
5178	limnologie *f*	limnología *f*
5179	limnoplancton *m*, plancton *m* dulçaquicole	limnoplancton *m*
5180	line-breeding *m*, élevage *m* selon les lignes	crianza *f* de líneas
5181	lignée *f* généalogique	línea *f* genealógica
5182	filiation *f* généalogique	filiación *f* genealógica, línea *f* de descendencia
5183	lignée *f* consanguine, filiation *f* ~	línea *f* consanguínea, filiación *f* ~
5184	disposition *f* (ou: arrangement *m*) linéaire des gènes	disposición *f* lineal de los genes
5185	régression *f* linéaire	regresión *f* lineal
5186	lingual, glossien	lingual
5187	glande *f* linguale	glándula *f* lingual
5188	papille *f* linguale	papila *f* lingual
5189	—	—

5190	linkage	Kopp(e)lung *f*
5191	linkage group	Kopp(e)lungsgruppe *f*
5192	linoleic acid	Linolsäure *f*, Leinölsäure *f*
5193	linolenic acid	Linolensäure *f*
5194	lipase	Lipase *f*
5195	lipid(e)	Lipid *n*
5196	lipid metabolism → fat metabolism	—
5197	lipid reserve, fat ~	Fettreserve *f*
5198	lipochrome	Lipochrom *n*
5199	lipoclastic, fat-splitting, lipolytic	fettspaltend, fettlösend
5200	lipocyte	Fettzelle *f*
5201	lipoid	Lipoid *n*
5202	lipolyse	Lipolyse *f*
5203	lipolytic → lipoclastic	—
5204	liponic acid → thioctic acid	—
5205	lipopolysaccharide	Lipopolysa(c)charid *n*
5206	lipoprotein	Lipoprotein *n*
5207	lipo-soluble → fat-soluble	—
5208	liquefaction, liquation	Verflüssigung *f*
5209	to liquefy	verflüssigen
5210	liquid culture	Flüssigkeitskultur *f*
5211	liquid medium, fluid ~	flüssiges Medium *n*
5212	lithophyte	Steinpflanze *f*, Lithophyt *m*
5213	lithosere	Lithoserie *f*
5214	lithosphere	Lithosphäre *f*
5215	lithotroph	lithotroph
5216	lit(t)oral zone	Litoralzone *f*, Litoral *n*
5217	liver	Leber *f*
5218	liverworts → Hepaticae	—
5219	living organism	Lebewesen *n*
5220	living space	Lebensraum *m*
5221	living weight	Lebendgewicht *n*
5222	lobulate, lobose, lobate	gelappt
5223	local circuit theory	Strömchentheorie *f*
5224	local race	Lokalrasse *f*
5225	localization of genes	Genlokalisation *f*
5226	locomotion	Fortbewegung *f*
5227	locomotor organ, organ of locomotion	Fortbewegungsorgan *n*, Bewegungsorgan *n*
5228	locomotory apparatus	Bewegungsapparat *m*
5229	locomotory capacity, locomotor ability	Fortbewegungsfähigkeit *f*
5230	locomotory movement	lokomotorische Bewegung *f*
5231	loculicidal	lokulizid, fachspaltig, rückenspaltig
5232	locus (*pl* loci)	Locus *m* (*pl* Loci), Genlocus *m*, Genort *m*
5233	logarithmic phase → exponential phase	—
5234	loins	Lende *f*
5235	loment(um)	Bruchfrucht *f*, Gliederfrucht *f*
5236	long bone	Röhrenknochen *m*
5237	long-day plant	Langtagpflanze *f*
5238	long-lived → longeval	—
5239	long-styled → macrostylous	—
5240	long-tailed	langschwänzig
5241	longeval, long-lived, macrobiotic	langlebig
5242	longevity	Langlebigkeit *f*
5243	longitudinal axis	Längsachse *f*
5244	longitudinal dehiscence	Längsspaltung *f* *(Frucht)*
5245	longitudinal division	Längsteilung *f*
5246	longitudinal growth → growth in length	—
5247	longitudinal muscle	Längsmuskel *m*
5248	longitudinal section, longisection	Längsschnitt *m*
5249	longitudinal separation	Längstrennung *f*
5250	lophophore	Lophophor *n*
5251	low moor	Nieder(ungs)moor *n*, Flachmoor *n*
5252	lower eyelid	Unterlid *n*
5253	lower jaw → mandible	—
5254	lower lip	Unterlippe *f*

5190	liaison *f*, «linkage» *m*	ligamiento *m*, "linkage" *m*
5191	groupe *m* de liaison	grupo *m* de ligamiento
5192	acide *m* linoléique	ácido *m* linoleico
5193	acide *m* linolénique	ácido *m* linolénico
5194	lipase *f*	lipasa *f*
5195	lipide *m*	lípido *m*
5196	—	—
5197	réserve *f* lipidique, ~ grasse, ~ adipeuse	reserva *f* grasa
5198	lipochrome *m*	lipocromo *m*
5199	lipolytique, adipolytique	lipoclástico, lipolítico
5200	lipocyte *m*, cellule *f* adipeuse	lipocito *m*, célula *f* adiposa, ~ grasa
5201	lipoïde *m*	lipoide *m*
5202	lipolyse *f*	lipólisis *f*
5203	—	—
5204	—	—
5205	lipopolyoside *m*, lipopolysaccharide *m*	lipopolisacárido *m*
5206	lipoprotéine *f*	lipoproteína *f*
5207	—	—
5208	liquéfaction *f*	licuefacción *f*, licuación *f*
5209	liquéfier	licuar, licuefacer
5210	culture *f* liquide	cultivo *m* líquido
5211	milieu *m* liquide	medio *m* líquido
5212	lithophyte *m*	litófito *m*
5213	série *f* qui commence par du rocher nu	litoserie *f*, litosere *f*
5214	lithosphère *f*	litosfera *f*
5215	lithotrophe	litótrofo
5216	littoral *m*	zona *f* litoral, litoral *m*
5217	foie *m*	hígado *m*
5218	—	—
5219	être *m* vivant, organisme *m* ~	ser *m* vivo, ~ viviente, organismo *m* ~
5220	espace *m* vital	espacio *m* vital
5221	poids *m* vif	peso *m* vivo
5222	lobé, lobulaire	lobulado, lobular
5223	théorie *f* des courants locaux	teoría *f* del circuito local
5224	race *f* locale	raza *f* local
5225	localisation *f* génique, ~ des gènes	localización *f* génica
5226	locomotion *f*	locomoción *f*
5227	organe *m* locomoteur	órgano *m* locomotor
5228	appareil *m* locomoteur	aparato *m* locomotor
5229	aptitude *f* locomotrice	aptitud *f* locomotora
5230	mouvement *m* locomoteur	movimiento *m* locomotor, ~ de locomoción
5231	loculicide	loculicida
5232	locus *m* (*pl* loci)	locus *m* (génico) (*pl* loci)
5233	—	—
5234	lombes *m pl*	lomo *m*
5235	fruit *m* lomentacé	lomento *m*
5236	os *m* long	hueso *m* largo
5237	plante *f* de journée longue, ~ héméro-périodique	planta *f* macrohémera, ~ de día largo
5238	—	—
5239	—	—
5240	longicaude, à queue longue	rabilargo, de cola larga
5241	macrobite, à vie longue	longevo
5242	longévité *f*	longevidad *f*
5243	axe *m* longitudinal	eje *m* longitudinal
5244	déhiscence *f* longitudinale	dehiscencia *f* longitudinal
5245	division *f* longitudinale	división *f* longitudinal
5246	—	—
5247	muscle *m* longitudinal	músculo *m* longitudinal
5248	section *f* longitudinale, coupe *f* ~	sección *f* longitudinal
5249	séparation *f* longitudinale	separación *f* longitudinal
5250	lophophore *m*, cercle *m* tentaculaire	lofóforo *m*
5251	tourbière *f* basse	turbera *f* baja
5252	paupière *f* inférieure	párpado *m* inferior, pálpebra *f* ~
5253	—	—
5254	lèvre *f* inférieure	labio *m* inferior

5255	lower palea → lemma	—
5256	lower plant	niedere Pflanze *f*
5257	LTH → lactogenic hormone	—
5258	luciferin	Luziferin *n*
5259	lumbar	lumbal, Lenden...
5260	lumbar vertebra	Lendenwirbel *m*
5261	lumen (*pl* lumina)	Lumen *n*
5262	luminescence	Lumineszenz *f*
5263	luminescent bacteria, luminous ~	Leuchtbakterien *f pl*
5264	luminescent organ, photophore	Leuchtorgan *n*, Photophor *n*
5265	lung	Lunge *f*
5266	lung-fish → Dipnoi	—
5267	luteal hormone → progesterone	—
5268	lutein	Lutein *n*
5269	luteinizing hormone, prolan B, interstitial cell stimulating hormone, LH, ICSH	luteinisierendes Hormon *n*, zwischenzellenstimulierendes ~, Prolan *n* B, Gonadotropin *n* B, LH, ICSH
5270	luteotrop(h)ic hormone → lactogenic ~	—
5271	lyase	Lyase *f*
5272	lycopene, lycopin	Lykopin *n*
5273	Lycopsida, lycopods, club mosses	Bärlappartige *pl*, Bärlappgewächse *n pl*
5274	lymph	Lymphe *f*
5275	lymph circulation	Lymphkreislauf *m*
5276	lymph heart	Lymphherz *n*
5277	lymph(atic) node, ~ gland	Lymphknoten *m*, Lymphdrüse *f* (*fälschlich*)
5278	lymph space	Lymphraum *m*
5279	lymph(atic) system	Lymph(gefäß)system *n*
5280	lymph(atic) vessel	Lymphgefäß *n*
5281	lymphocyte	Lymphozyt *m*
5282	lymphoid tissue	Lymphgewebe *n*
5283	lyochrome → flavin	—
5284	lyo-enzyme	Lyoferment *n*
5285	lyophilization, freeze-drying	Gefriertrocknung *f*, Lyophilisierung *f*
5286	lysate	Lysat *n*, Lösungsprodukt *n*
5287	lysergic acid	Lysergsäure *f*
5288	lysine	Lysin *n*
5289	lysis	Lysis *f*, Lyse *f*
5290	lysogenous, lysigenous, lysigenic	lysogen, lysigen
5291	lysogeny, lysogenicity	Lysogenie *f*
5292	lysosome	Lysosom *n*
5293	lysotype	Lysotyp *m*, Phagentyp *m*
5294	lysozyme	Lysozym *n*
5295	lytic	lytisch
5296	lytic enzyme	Lösungsenzym *n*
5297	lytic phage → intemperate phage	—
5298	to lyze	lysieren
5299	maceration	Mazeration *f*
5300	macrobiotic → longeval	—
5301	macroblast	Langtrieb *m*
5302	macrocarpous	großfrüchtig
5303	macroclimate	Makroklima *n*, Großklima *n*
5304	macroconsumer	Makrokonsument *m*, Großkonsument *m*
5305	macro-element → macronutrient	—
5306	macroevolution	Makroevolution *f*
5307	macrofauna	Makrofauna *f*
5308	macroflora	Makroflora *f*
5309	macrogamete, megagamete	Makrogamet *m*, Gynogamet *m*
5310	macroglia	Makroglia *f*
5311	macromere	Makromer *n*
5312	macromethod	Makromethode *f*
5313	macromolecule	Makromolekül *n*, Großmolekül *n*
5314	macromutation, major mutation	Makromutation *f*, Großmutation *f*

5255	—	—
5256	plante *f* inférieure, végétal *m* inférieur	planta *f* inferior
5257	—	—
5258	luciférine *f*	luciferina *f*
5259	lombaire	lumbar
5260	vertèbre *f* lombaire	vértebra *f* lumbar
5261	lumière *f*, lumen *m*	luz *f*, lumen *m*
5262	luminescence *f*	luminescencia *f*
5263	bactéries *f pl* lumineuses, ~ photogènes, photobactéries *f pl*	bacterias *f pl* luminosas, ~ fotógenas
5264	organe *m* luminescent, ~ lumineux, photophore *m*	órgano *m* luminescente, fotóforo *m*
5265	poumon *m*	pulmón *m*
5266	—	—
5267	—	—
5268	lutéine *f*	luteína *f*
5269	hormone *f* lutéinisante, gonadotrop(h)ine *f* B, luté(in)ostimuline *f*, LH	hormona *f* luteinizante, ~ estimulante de las células intersticiales, prolan *m* B, gonadotrofina *f* B, LH, ICSH
5270	—	—
5271	lyase *f*	liasa *f*, sintasa *f*
5272	lycopène *m*	licopeno *m*, licopina *f*
5273	lycopodinées *f pl*	licópsidos *m pl*
5274	lymphe *f*	linfa *f*
5275	circulation *f* lymphatique	circulación *f* linfática
5276	cœur *m* lymphatique	corazón *m* linfático
5277	ganglion *m* lymphatique, nœud *m* ~	ganglio *m* linfático, nódulo *m* ~
5278	espace *m* lymphatique	espacio *m* linfático
5279	système *m* lymphatique	sistema *m* linfático
5280	vaisseau *m* lymphatique	vaso *m* linfático
5281	lymphocyte *m*	linfocito *m*
5282	tissu *m* lymphoïde	tejido *m* linfoide
5283	—	—
5284	lyodiastase *f*	lioenzima *f*
5285	lyophilisation *f*, cryodessication *f*	liofilización *f*, desecación-congelación *f* criodesecación *f*
5286	lysat *m*	lisado *m*
5287	acide *m* lysergique	ácido *m* lisérgico
5288	lysine *f*	lisina *f*
5289	lyse *f*	lisis *f*
5290	lysogène, lysogénique, lysigène	lisógeno, lisogénico, lisígeno
5291	lysogénie *f*, lysogénicité *f*	lisogenia *f*
5292	lysosome *m*	lisosoma *m*
5293	lysotype *m*	lisótipo *m*
5294	lysozyme *m*	lisozima *f*
5295	lytique	lítico
5296	enzyme *m* lytique	enzima *f* lítica
5297	—	—
5298	lyser	lisar
5299	macération	maceración *f*
5300	—	—
5301	macroblaste *m*, auxiblaste *m*	macroblasto *m*
5302	macrocarpe	macrocarpo
5303	macroclimat *m*	macroclima *m*
5304	macroconsommateur *m*	macroconsumidor *m*
5305	—	—
5306	macroévolution *f*	macroevolución *f*
5307	macrofaune *f*	macrofauna *f*
5308	macroflore *f*	macroflora *f*
5309	macrogamète *m*, gynogamète *m*	macrogameto *m*, macrogámeta *m*
5310	macroglie *f*	macroglia *f*
5311	macromère *m*	macrómero *m*
5312	macrométhode *f*	macrométodo *m*
5313	macromolécule *f*, grosse molécule *f*	macromolécula *f*
5314	macromutation *f*, mutation *f* majeure	macromutación *f*, mutación *f* mayor

5315	macronucleus, meganucleus	Makronukleus *m*, Großkern *m*
5316	macronutrient, macro-element, major element	Makronährstoff *m*, Hauptnährstoff *m*
5317	macrophage	Makrophage *m*
5318	macrophyll, megaphyll	Makrophyll *n*
5319	macrophylogenesis	Makrophylogenese *f*
5320	macroscopic	makroskopisch
5321	macrosmatic animal	Makrosmat *m*
5322	macrosporangium, megasporangium, gynosporangium	Makrosporangium *n*, Megasporangium *n*
5323	macrospore, megaspore, gynospore	Makrospore *f* Megaspore *f*, Gynospore *f*
5324	macrosporogenesis, megasporogenesis	Makrosporogenese *f*, Megasporogenese *f*
5325	macrosporophyll, megasporophyll	Makrosporophyll *n*, Megasporophyll *n*
5326	macrostylous, long-styled	langgriff(e)lig
5327	macrotome	Makrotom *n*
5328	macula lutea → yellow spot	—
5329	madreporic canal → stone canal	—
5330	madreporic plate, madreporite	Madreporenplatte *f*
5331	magnifying glass	Lupe *f*
5332	main root	Hauptwurzel *f*
5333	major element → macronutrient	—
5334	major gene → oligogene	—
5335	major mutation → macromutation	—
5336	malacology	Malako(zoo)logie *f*, Weichtierkunde *f*
5337	malar bone → zygomatic bone	—
5338	male, masculine	männlich
5339	male	Männchen *n*
5340	male hormone → androgen	—
5341	male line (of breeding)	männliche Linie *f*, väterliche ~
5342	male parent	männlicher Elter *m*, väterlicher ~
5343	maleic acid	Maleinsäure *f*
5344	maleness → masculinity	—
5345	malformation	Mißbildung *f*
5346	malic acid	Apfelsäure *f*
5347	malleus	Hammer *m* *anat.*
5348	malonic acid	Malonsäure *f*
5349	Malpighian body, ~ corpuscle	Malpighisches Körperchen *n*
5350	Malpighian layer, germinative ~	Keimschicht *f* (Haut)
5351	Malpighian tubules	Malpighische Gefäße *n pl*
5352	maltase	Maltase *f*
5353	maltose, malt sugar	Malzzucker *m*, Maltose *f*, Maltobiose *f*
5354	mamilla, mammilla, nipple	Brustwarze *f*, Mamille *f*
5355	Mammalia, mammals	Säugetiere *n pl*
5356	mammalogy	Säugetierkunde *f*, Mammalogie *f*
5357	mammary gland, milk ~, mamma	Brustdrüse *f*, Milchdrüse *f*, Mamma *f*
5358	mandible, lower jaw	Unterkiefer *m*; Mandibel *f* *ins.*
5359	mandibular arch	Kieferbogen *m*
5360	Mandibulata	Mandibulata *pl*
5361	mannose	Mannose *f*
5362	mantle, pallium	Mantel *m* *zool.*
5363	mantle cavity, pallial ~	Mantelhöhle *f*
5364	manubrium	Manubrium *n*
5365	manyplies → omasum	—
5366	many-seeded → polyspermous	—
5367	to map *(chromosomes etc.)*	kartieren *(Chromosomen etc.)*
5368	map distance	Genabstand *m*
5369	mapping	Kartierung *f*
5370	mapping of chromosomes	Chromosomenkartierung *f*
5371	mapping of vegetation	Vegetationskartierung *f*
5372	marginal *bot.*	randständig
5373	marine animal	Meerestier *n*
5374	marine biology	Meeresbiologie *f*
5375	marine ecology	Meeresökologie *f*

5315	macronucléus *m*, macronoyau *m*, grand noyau *m*	macronúcleo *m*, meganúcleo *m*
5316	macro-élément *m*, élément *m* majeur	macronutrim(i)ento *m*, macronutriente *m*
5317	macrophage *m*	macrófago *m*
5318	mégaphylle *f*	megafilo *m*, macrofilo *m*
5319	macrophylogenèse *f*	macrofilogénesis *f*
5320	macroscopique	macroscópico
5321	(animal *m*) macrosmatique *m*	animal *m* macrosmático
5322	macrosporange *m*, mégasporange *m*, gynosporange *m*	macrosporangio *m*, megasporangio *m*
5323	macrospore *f*, mégaspore *f*, gynospore *f*	macrospora *f*, megaspora *f*, ginóspora *f*
5324	macrosporogenèse *f*, megasporogenèse *f*	macrosporogénesis *f*, megasporogénesis *f*
5325	mégasporophylle *f*	macrosporofilo *m*, megasporofilo *m*
5326	longistylé, à style long	macrostilo, longistilo
5327	macrotome *m*	macrótomo *m*
5328	—	—
5329	—	
5330	plaque *f* madréporique	placa *f* madrepórica
5331	loupe *f*	lupa *f*
5332	racine *f* principale	raíz *f* principal
5333	—	—
5334	—	—
5335	—	
5336	malacologie *f*	malacología *f*
5337	—	—
5338	mâle, masculin	masculino
5339	mâle *m*	macho *m*
5340	—	—
5341	lignée *f* mâle, ~ paternelle	línea *f* masculina, ~ paterna
5342	parent *m* mâle	(pro)genitor *m* macho
5343	acide *m* maléique	ácido *m* maleico
5344	—	—
5345	malformation *f*	malformación *f*
5346	acide *m* malique	ácido *m* málico
5347	marteau *m* *anat.*	martillo *m* *anat.*
5348	acide *m* malonique	ácido *m* malónico
5349	corpuscule *m* de Malpighi	corpúsculo *m* de Malpighi(o), cuerpo *m* ~ ~
5350	couche *f* germinative	estrato *m* de Malpighi(o), ~ germinativo
5351	tubes *m pl* de Malpighi	tubos *m pl* de Malpighi(o)
5352	maltase *f*	maltasa *f*
5353	maltose *m*, sucre *m* de malt	maltosa *f*, azúcar *m* de malta
5354	mamelon *m*, tétin *m*	mamila *f*, pezón *m*
5355	mammifères *m pl*	mamíferos *m pl*
5356	mammalogie *f*	mamalogía *f*
5357	glande *f* mammaire, ~ lactéale	glándula *f* mamaria, ~ láctea
5358	mâchoire *f* inférieure, maxillaire *m* inférieur, mandibule *f*	mandíbula *f* (inferior), maxilar *m* inferior
5359	arc *m* mandibulaire	arco *m* mandibular
5360	mandibulates *m pl*	mandibulados *m pl*
5361	mannose *m*	manosa *f*
5362	manteau *m* *zool.*, pallium *m*	manto *m* *zool.*, palio *m*
5363	cavité *f* palléale	cavidad *f* paleal, ~ del manto
5364	manubrium *m*	manubrio *m*
5365	—	—
5366	—	—
5367	dresser des cartes *(chromosomiques etc.)*	trazar mapas *(cromosómicos etc.)*
5368	distance *f* sur la carte	distancia *f* de mapa, ~ en el mapa
5369	établissement *m* de cartes	confección *f* de mapas, trazado *m* ~ ~
5370	établissement *m* de cartes chromosomiques	confección *f* de mapas cromosómicos
5371	cartographie *f* de la végétation, relevé *m* cartographique de la ~	levantamiento *m* fitocartográfico
5372	marginal *bot.*	marginal *bot.*
5373	animal *m* marin	animal *m* marino
5374	biologie *f* marine	biología *f* marina
5375	écologie *f* marine	ecología *f* marina

5376	marine fauna	Meeresfauna *f*
5377	marine flora	Meeresflora *f*
5378	marine plankton → haloplankton	—
5379	to mark → to label	—
5380	mark and recapture method	Markierungs- und Wiederfangmethode *f*
5381	marker, label	Markierungsstoff *m*
5382	marker gene	Markierungsgen *n*
5383	marking, labelling	Markierung *f*
5384	marking method, labelling ~	Markierungsmethode *f*
5385	marrow → medulla, medullary	—
5386	marsh plant → helophyte	—
5387	Marsupiala, marsupials	Beuteltiere *npl*, Marsupialier *mpl*
5388	marsupium	Brutbeutel *m*, Marsupium *n*
5389	masculine → male	—
5390	masculinity, maleness	Männlichkeit *f*
5391	mass action	Massenwirkung *f*
5392	mass culture	Massenkultur *f*
5393	mass effect	Masseneffekt *m*
5394	mass movement	Massenwanderung *f*
5395	mass mutation	Massenmutation *f*
5396	mass propagation, gradation	Massenvermehrung *f*, Gradation *f*
5397	mass selection	Massenauslese *f*, Massenselektion *f*
5398	mass spectrograph	Massenspektrograph *m*
5399	masseter	Kaumuskel *m*
5400	mast cell, mastocyte, labrocyte	Mastzelle *f*
5401	mastax	Mastax *m*, Kaumagen *m* (*Rädertiere*)
5402	mastication, chewing	Kauen *n*
5403	masticatory stomach → gizzard	—
5404	Mastigophora → Flagellata	—
5405	mastigote → flagellate	—
5406	mastoid process	Warzenfortsatz *m*
5407	mate, mating partner, sexual ~	Paarungspartner *m*, Geschlechtspartner *m*
5408	to mate	paaren, anpaaren
5409	maternal	mütterlich
5410	maternal instinct	Mutterinstinkt *m*
5411	mating	Paarung *f*
5412	mating behavio(u)r	Paarungsverhalten *n*
5413	mating call	Paarungsruf *m*, Balzruf *m*
5414	mating flight → nuptial flight	—
5415	mating partner → mate	—
5416	mating time, ~ season	Paarungszeit *f*
5417	mating type	Paarungstyp *m*
5418	matrix[1] *gen.*	Matrix *f* *gen.*
5419	matrix[2] → uterus	—
5420	matrix bridge	Matrixbrücke *f*
5421	matroclinous, matriclinous, matroclinal	matroklin
5422	matrocliny	Matroklinie *f*
5423	maturation, ripening	Reifung *f*
5424	maturation division	Reifungsteilung *f*, Reifeteilung *f*
5425	maturation period	Reifezeit *f*
5426	maturation phase	Reifungsphase *f*
5427	mature, ripe	reif
5428	to mature, to ripe	reifen
5429	maturity, ripeness	Reife *f*
5430	maxilla, upper jaw	Oberkiefer *m*; Maxille *f* *Ins.*
5431	maxillary, gnathic	Kiefer ...
5432	maxillary bone, jaw ~	Kieferknochen *m*
5433	maxillary gland	Maxillardrüse *f*, Kieferdrüse *f*, Schalendrüse *f*
5434	maxillary palp(us)	Kiefertaster *m*, Maxillartaster *m*
5435	maxillary sinus, ~ cavity	Kieferhöhle *f*
5436	maxilliped(e), jaw foot	Kieferfuß *m*
5437	maze experiment	Labyrinthversuch *m*

5376	faune *f* marine	fauna *f* marina, ~ marítima
5377	flore *f* marine	flora *f* marina, ~ marítima
5378	—	—
5379	—	—
5380	technique *f* du marquage-et-recapture	método *m* de marcado y recaptura
5381	marqueur *m*	marcador *m*
5382	gène *m* marqueur	gen *m* marcador
5383	marquage *m*	marcado *m*, marcación *f*
5384	méthode *f* de marquage	método *m* de marcado
5385	—	—
5386	—	—
5387	marsupiaux *m pl*	marsupiales *m pl*
5388	marsupie *f*, poche *f* marsupiale	marsupio *m*, bolsa *f* marsupial
5389	—	—
5390	masculinité *f*	masculinidad *f*
5391	action *f* de masse	acción *f* de masas
5392	culture *f* en masse	cultivo *m* en masa
5393	effet *m* de masse	efecto *m* de masa
5394	migration *f* massive	movimiento *m* en masa, migración *f* ~ ~, desplazamiento *m* ~ ~
5395	mutation *f* massale	mutación *f* masal
5396	multiplication *f* massive, pullulation *f*, pullulement *m*, gradation *f*	multiplicación *f* en masa, reproducción *f* masiva
5397	sélection *f* massale, ~ en masse	selección *f* masal, ~ masiva, ~ de masa
5398	spectrographe *m* de masse	espectrógrafo *m* de masas
5399	muscle *m* masticateur, (~) masséter *m*	(músculo *m*) masticador *m*, masetero *m*
5400	mastocyte *m*, labrocyte *m*, cellule-engrais *f*, mastzelle *f*	mastocito *m*, labrocito *m*, célula *f* cebada, „mastzelle" *f*
5401	mastax *m*	mástax *m*
5402	mastication *f*	masticación *f*
5403	—	—
5404	—	—
5405	—	—
5406	apophyse *f* mastoïdienne, (~) mastoïde *f*	(apófisis *f*) mastoides *f*
5407	conjoint *m*, partenaire *m* sexuel	compañero *m* sexual, cónyuge *m*
5408	apparier, accoupler	aparear, acoplar
5409	maternel	materno, maternal
5410	instinct *m* maternel	instinto *m* materno, ~ maternal
5411	appariement *m*, accouplement *m*	apareamiento *m*, acoplamiento *m*
5412	comportement *m* d'appariement	comportamiento *m* de apareamiento
5413	appel *m* sexuel	llamada *f* sexual, ~ de apareamiento
5414	—	—
5415	—	—
5416	saison *f* d'accouplement, ~ d'appariement	época *f* de apareamiento, estación *f* ~ ~
5417	type *m* d'appariement	tipo *m* de apareamiento, ~ copulante
5418	matrice *f* *gen.*	matriz *f* *gen.*
5419	—	—
5420	pont *m* de matrice	puente *m* de matriz
5421	matrocline, matroclinal	matroclino, matroclinal
5422	matroclinie *f*	matroclinia *f*
5423	maturation *f*	maduración *f*
5424	division *f* de maturation	división *f* de maduración
5425	période *f* de maturité	período *m* de maduración
5426	phase *f* de maturation	fase *f* de maduración
5427	mûr	maduro
5428	mûrir	madurar
5429	maturité *f*	maturidad *f*, madurez *f*
5430	mâchoire *f* supérieure, maxillaire *m* supérieur; maxille *f* *ins.*	maxilar *m* superior, mandíbula *f* ~; maxila *f* *ins.*
5431	maxillaire, gnathique	maxilar, gnático
5432	os *m* maxillaire	hueso *m* maxilar
5433	glande *f* maxillaire	glándula *f* maxilar
5434	palpe *m* maxillaire	palpo *m* maxilar
5435	sinus *m* maxillaire	seno *m* maxilar
5436	patte-mâchoire *f*, maxillipède *m*	maxilípedo *m*, pata *f* maxila
5437	expérience *f* de labyrinthe	prueba *f* del laberinto

No.	English	German
5438	mean life-span	durchschnittliche Lebensdauer *f*
5439	measurement, measuring, mensuration	Messung *f*
5440	mechanical tissue → supporting tissue	—
5441	mechanoreceptor	Mechanorezeptor *m*
5442	medial, median	mittelständig
5443	median plane	Medianebene *f*, Mediane *f*
5444	mediastinal pleura	Mittelfell *n*
5445	mediastinum	Mediastinum *n*, Mittelfellraum *m*
5446	medulla, marrow; pith *bot.*	Mark *n*, Medulla *f*
5447	medulla oblongata, spinal bulb	verlängertes Mark *n*
5448	medullary canal	Markkanal *m*
5449	medullary cavity, marrow ~	Markraum *m*, Markhöhle *f*
5450	medullary cell, marrow ~	Markzelle *f*
5451	medullary plate	Medullarplatte *f*
5452	medullary ray, pith ~	Markstrahl *m*
5453	medullary ray cell	Markstrahlzelle *f*
5454	medullary sheath → myelin sheath	—
5455	medullated → myelinated	—
5456	medusa	Meduse *f*, Qualle *f*
5457	mega ... see also macro ...	—
5458	megaevolution	Megaevolution *f*
5459	meiocyte	Meiozyte *f*
5460	meiosis, meiotic division	Meiose *f*, meiotische Teilung *f*
5461	meiospore	Meiospore *f*
5462	meiotic	meiotisch
5463	meiotic product	Meioseprodukt *n*
5464	melanin	Melanin *n*
5465	melanism	Melanismus *m*
5466	melanocyte-stimulating hormone → intermedin	—
5467	melanophore	Melanophor *n*
5468	melanotropin → intermedin	—
5469	melatonin, melanotonin	Mela(no)tonin *n*
5470	melitose → raffinose	—
5471	melting point	Schmelzpunkt *m*
5472	membrane	Membran *f*
5473	membrane bone, dermal ~	Hautknochen *m*, Deckknochen *m*, Belegknochen *m*
5474	membrane capacity	Membrankapazität *f*
5475	membrane current	Membranstrom *m*
5476	membrane permeability	Membranpermeabilität *f*
5477	membrane potential	Membranpotential *n*
5478	membrane resistance	Membranresistenz *f*
5479	membranella	Membranelle *f*
5480	membranous	membranartig, häutig
5481	membranous filter	Membranfilter *m*
5482	membranous labyrinth	häutiges Labyrinth *n*
5483	Mendel's laws, Mendelian ~	Mendelsche Gesetze *n pl*, Mendelgesetze *n pl*
5484	Mendel's first law: law of dominance or of uniformity (of hybrids)	1. Mendelsches Gesetz: Uniformitäts- oder Reziprozitätsgesetz
5485	Mendel's second law: law of segregation	2. Mendelsches Gesetz: Spaltungsgesetz
5486	Mendel's third law: law of independent assortment	3. Mendelsches Gesetz: Unabhängigkeitsgesetz, Gesetz der Neukombination der Gene
5487	Mendelian factor	Mendelscher Faktor *m*
5488	Mendelian population	Mendelpopulation *f*
5489	Mendelian ratio, segregation ~	Aufspaltungsverhältnis *n*
5490	mendelism	Mendeln *n*, Mendelismus *m*
5491	meninges (*sing.* meninx)	Hirnhäute *f pl*, Meningen *f pl* (*sing.* Meninx *f*)
5492	menotaxis	Menotaxis *f*
5493	mericarp	Teilfrucht *f*, Merikarp *n*
5494	mericlinal chim(a)era	Meriklinalchimäre *f*
5495	meristem, meristematic tissue	Meristem *n*, Bildungsgewebe *n*
5496	meroblastic	meroblastisch

5438	vie *f* moyenne	vida *f* media, promedio *m* vital, período *m* medio de vida
5439	mesurage *m*, mensuration *f*	medición *f*
5440	—	—
5441	méchano-récepteur *m*	mecanorreceptor *m*
5442	médian	medio, mediano
5443	plan *m* médian	plano *m* mediano
5444	plèvre *f* médiastinale	pleura *f* mediastínica
5445	médiastin *m*	mediastino *m*
5446	moelle *f*, médulle *f*	medula *f*, médula *f*; tuétano *m* (hueso)
5447	bulbe *m* rachidien, moelle *f* allongée	bulbo *m* raquídeo, medula *f* oblonga(da)
5448	canal *m* médullaire	canal *m* medular
5449	cavité *f* médullaire	cavidad *f* medular
5450	cellule *f* médullaire	célula *f* medular
5451	plaque *f* médullaire	placa *f* medular
5452	rayon *m* médullaire	radio *m* medular, rayo *m* ~
5453	cellule *f* des rayons médullaires	célula *f* de los radios medulares
5454	—	—
5455	—	—
5456	méduse *f*	medusa *f*
5457	—	—
5458	mégaévolution *f*	megaevolución *f*
5459	méiocyte *m*	meiocito *m*
5460	méiose *f*	meiosis *f*, meyosis *f*, división *f* meiótica
5461	méiospore *f*	meiospora *f*
5462	méiotique	meiótico, meyótico
5463	produit *m* méiotique	producto *m* meiótico
5464	mélanine *f*	melanina *f*
5465	mélanisme *m*	melanismo *m*
5466	—	—
5467	mélanophore *m*	melanóforo *m*
5468	—	—
5469	mélatonine *f*	melatonina *f*
5470	—	—
5471	point *m* de fusion	punto *m* de fusión
5472	membrane *f*	membrana *f*
5473	os *m* dermique, ~ de membrane	hueso *m* dérmico, ~ cutáneo, ~ membranoso
5474	capacité *f* de membrane	capacidad *f* de membrana
5475	courant *m* de membrane	corriente *f* de membrana
5476	perméabilité *f* membranaire	permeabilidad *f* de la membrana
5477	potentiel *m* de membrane	potencial *m* de membrana
5478	résistance *f* membranaire	resistencia *f* de membrana
5479	membranelle *f*	membranela *f*
5480	membraneux	membranoso
5481	filtre *m* membraneux	filtro *m* de membrana
5482	labyrinthe *m* membraneux	laberinto *m* membranoso
5483	lois *f pl* de Mendel, ~ mendéliennes	leyes *f pl* de Mendel
5484	première loi de Mendel: loi de la dominance ou de l'uniformité (des hybrides)	primera ley de Mendel: ley de la dominancia o uniformidad (de los híbridos)
5485	seconde loi de Mendel: loi de (la) ségrégation	segunda ley de Mendel: ley de la segregación
5486	troisième loi de Mendel: loi de l'assortiment indépendant, ~ de la disjonction indépendante, ~ de pureté des caractères, ~ de l'indépendance ~ ~	tercera ley de Mendel: ley de la distribución independiente
5487	facteur *m* mendélien	factor *m* mendeliano
5488	population *f* mendélienne	población *f* mendeliana
5489	proportion *f* mendélienne	proporción *f* mendeliana
5490	mendélisme *m*	mendelismo *m*
5491	méninges *f pl*	meninges *f pl*
5492	ménotaxie *f*	menotaxia *f*
5493	méricarpe *m*	mericarpo *m*
5494	chimère *f* mériclinale	quimera *f* mericlinal
5495	méristème *m*, tissu *m* méristématique	meristema *m*, meristemo *m*
5496	méroblastique	meroblástico

5497	merogamete	Merogamet *m*
5498	merogamy	Merogamie *f*
5499	merogony	Merogonie *f*
5500	meromixis	Meromixis *f*
5501	merozygote	Merozygote *f*
5502	mesencephalon, mid-brain	Mittelhirn *n*, Mesenzephalon *n*
5503	mesenchyme	Mesenchym *n*
5504	mesenterium, mesentery	Mesenterium *n*, Gekröse *n*
5505	mesenteron, mid-gut	Mitteldarm *m*, Mesenteron *n*
5506	mesoblast → mesoderm	—
5507	mesocarp	Mesokarp *n*
5508	mesocoel	Mesozöl *n*
5509	mesocotyl	Mesokotyl *n*
5510	mesoderm, mesoblast	Mesoderm *n*, Mesoblast *m*, mittleres Keimblatt *n*
5511	mesogloea	Mesogloea *f*
5512	mesomitosis	Mesomitose *f*
5513	mesonephric duct, Wolffian ~	Urnierengang *m*, Wolffscher Gang *m*
5514	mesonephros, Wolffian body	Urniere *f*, Mesonephros *m*, Wolffscher Körper *m*
5515	mesophil(ic), mesophilous	mesophil
5516	mesophyll	Mesophyll *n*
5517	mesophyte	Mesophyt *m*
5518	mesothelium	Mesothel *n*
5519	mesotrophic	mesotroph
5520	mesozoan	Mesozoon *n*, Morulatier *n*
5521	Mesozoic (era)	Mesozoikum *n*, Erdmittelalter *n*
5522	messenger RNA, m-RNA	Boten-RNS *f*, Messenger-RNS *f*, Matrizen-RNS *f*, m-RNS
5523	metabiosis	Metabiose *f*
5524	Metabola → Pterygota	—
5525	metabolic	Stoffwechsel ...
5526	metabolic activity	Stoffwechseltätigkeit *f*
5527	metabolic antagonist	Stoffwechselantagonist *m*
5528	metabolic disorder, ~ disturbance	Stoffwechselstörung *f*
5529	metabolic end product	Stoffwechselendprodukt *n*
5530	metabolic energy	Stoffwechselenergie *f*
5531	metabolic inhibitor	Stoffwechselgift *n*, -inhibitor *m*
5532	metabolic intermediate	Stoffwechselzwischenprodukt *n*
5533	metabolic nucleus, energic ~	Arbeitskern *m*
5534	metabolic pathway	Stoffwechselweg *m*
5535	metabolic process	Stoffwechselprozeß *m*, -vorgang *m*
5536	metabolic product → metabolite	—
5537	metabolic rate	Stoffwechselrate *f*, ~umsatz *m*
5538	metabolic reaction	Stoffwechselreaktion *f*
5539	metabolic wastes	Stoffwechselschlacken *fpl*
5540	metabolically active	stoffwechselaktiv
5541	metabolism	Stoffwechsel *m*, Metabolismus *m*
5542	metabolite, metabolic product	Stoffwechselprodukt *n*, Metabolit *m*
5543	metabolizable	umsetzbar
5544	to metabolize	umsetzen, metabolisieren
5545	metacarpal bone	Mittelhandknochen *m*
5546	metacarpus	Mittelhand *f*
5547	metacentric	metazentrisch
5548	metachromatic	metachromatisch
5549	metacoel	Metazöl *n*
5550	metafemale → superfemale	—
5551	metagenesis	Metagenese *f*, Ammenzeugung *f*
5552	metakinesis	Metakinese *f*
5553	metalimnion → thermocline	—
5554	metamere	Metamer *n*
5555	metamerism, metameric segmentation	Metamerie *f*

5497	mérogamète *m*	merogameto *m*
5498	mérogamie *f*	merogamia *f*
5499	mérogonie *f*	merogonia *f*
5500	méromixie *f*	meromixia *f*, meromixis *f*
5501	mérozygote *m*	merocigoto *m*
5502	mésencéphale *m*, cerveau *m* moyen	mesocéfalo *m*, mesencéfalo *m*, cerebro *m* medio
5503	mésenchyme *m*	mesénquima *m*
5504	mésentère *m*	mesenterio *m*
5505	intestin *m* moyen	mesenterón *m*, intestino *m* medio
5506	—	—
5507	mésocarpe *m*	mesocarpo *m*
5508	mésocoele *m*	mesocele *m*
5509	mésocotyle *m*	mesocótilo *m*
5510	mésoderme *m*, mésoblaste *m*, feuillet *m* moyen	mesodermo *m*, mesoblasto *m*
5511	mésoglée *f*	mesoglea *f*
5512	mésomitose *f*	mesomitosis *f*
5513	conduit *m* mésonéphrique, canal *m* de Wolff	conducto *m* mesonéfrico, ~ de Wolff
5514	mésonéphros *m*, corps *m* de Wolff	mesonefros *m*, cuerpo *m* de Wolff
5515	mésophile	mesofílico
5516	mésophylle *m*	mesofilo *m*
5517	mésophyte *m*	mesófito *m*, mesofita *m*
5518	mésothélium *m*	mesotelio *m*
5519	mésotrophe	mesótrofo
5520	mésozoaire *m*	mesozoo *m*
5521	mésozoïque *m*	mesozoico *m*
5522	ARN-messager *m*, ARN-matrice *m*, ARN m	ARN-mensajero, ARN-matriz, ARN m
5523	métabiose *f*	metabiosis *f*
5524	—	—
5525	métabolique	metabólico
5526	activité *f* métabolique	actividad *f* metabólica
5527	antagoniste *m* métabolique	antagonista *m* metabólico
5528	trouble *m* du métabolisme, perturbation *f* ~ ~, ~ métabolique	desorden *m* metabólico, trastorno *m* ~
5529	produit *m* terminal du métabolisme	producto *m* final del matabolismo
5530	énergie *f* métabolique	energía *f* metabólica
5531	inhibiteur *m* métabolique	inhibidor *m* metabólico
5532	métabolite *m* intermédiaire	producto *m* metabólico intermedio
5533	noyau *m* métabolique, ~ en activité	núcleo *m* metabólico, ~ enérgico
5534	voie *f* métabolique	ruta *f* metabólica, vía *f* ~
5535	processus *m* métabolique	proceso *m* metabólico
5536	—	—
5537	taux *m* métabolique, rapidité *f* du métabolisme	proporción *f* metabólica, velocidad *f* ~
5538	réaction *f* métabolique	reacción *f* metabólica
5539	déchets *m pl* métaboliques	desechos *m pl* del metabolismo, subproductos *m pl* metabólicos
5540	métaboliquement actif	metabólicamente activo
5541	métabolisme *m*	metabolismo *m*
5542	métabolite *m*, produit *m* métabolique	metabolito *m*, producto *m* metabólico
5543	métabolisable	metabolizable
5544	métaboliser	metabolizar
5545	os *m* métacarpien	hueso *m* metacarpiano
5546	métacarpe *m*	metacarpo *m*
5547	métacentrique	metacéntrico
5548	métachromatique	metacromático
5549	métacoele *m*	metacele *m*
5550	—	—
5551	métagenèse *f*	metagénesis *f*
5552	métacinésie *f*	metacinesis *f*
5553	—	—
5554	métamère *m*	metámero *m*
5555	métamérisme *m*, métamérie *f*	metamerismo *m*, segmentación *f* metamérica

5556	metamitosis	Metamitose *f*
5557	to metamorphose	metamorphosieren
5558	metamorphosis	Metamorphose *f*
5559	metanephros, hind-kidney	Nachniere *f*, Metanephros *m*
5560	metaphase	Metaphase *f*
5561	metaphase pairing index	Metaphasepaarungsindex *m*
5562	metaphytes	Metaphyten *mpl*
5563	metaplasia, metaplasy	Metaplasie *f*
5564	metatarsal bone	Mittelfußknochen *m*
5565	metatarsus	Mittelfuß *m*
5566	metaxenia	Metaxenie *f*
5567	metaxylem	Metaxylem *n*
5568	metazoon, metazoan	Metazoon *n*, Gewebetier *n*
5569	metencephalon, hind-brain	Hinterhirn *n*, Metenzephalon *n*
5570	methionine	Methionin *n*
5571	method of locomotion	Fortbewegungsweise *f*
5572	method of preparation	Präparationstechnik *f*
5573	method of reproduction, mode ~ ~	Fortpflanzungsweise *f*
5574	methyl group	Methylgruppe *f*
5575	methylation	Methylierung *f*
5576	met(o)estrus, met(o)estrum	Nachbrunst *f*, Metöstrus *m*
5577	mevalonic acid	Mevalonsäure *f*
5578	micella	Mizelle *f*
5579	microanalysis	Mikroanalyse *f*
5580	microautoradiography	Mikroautoradiographie *f*
5581	microbalance	Mikrowaage *f*
5582	microbe, microbion	Mikrobe *f*, Mikrobion *n*
5583	microbial, microbic, microbian	mikrobisch, mikrobiell
5584	microbial content	Keimzahl *f*
5585	microbicidal	mikrobizid
5586	microbiological	mikrobiologisch
5587	microbiologist	Mikrobiologe *m*
5588	microbiology	Mikrobiologie *f*
5589	microcentrum	Mikrozentrum *n*
5590	microchemistry	Mikrochemie *f*
5591	microchromosome	Mikrochromosom *n*
5592	microcinematography	Mikrokinematographie *f*
5593	microclimate	Mikroklima *n*, Kleinklima *n*
5594	microconsumer	Mikrokonsument *m*
5595	microcyst	Mikrozyste *f*
5596	microdissection	Mikrodissektion *f*
5597	microelectrode	Mikroelektrode *f*
5598	micro-element → trace-element	—
5599	microenvironment, micromilieu	Mikromilieu *n*
5600	microevolution	Mikroevolution *f*
5601	microfauna	Mikrofauna *f*
5602	microfibril	Mikrofibrille *f*
5603	microflora	Mikroflora *f*
5604	microfossil	Mikrofossil *n*
5605	microgamete	Mikrogamet *m*, Androgamet *m*
5606	microgene	Mikrogen *n*
5607	microglia	Mikroglia *f*
5608	microhabitat	Kleinstandort *m*
5609	microincineration	Mikroveraschung *f*, mikroskopische Aschenanalyse *f*
5610	microinjection	Mikroinjektion *f*
5611	micromanipulation	Mikromanipulation *f*
5612	micromanipulator	Mikromanipulator *m*
5613	micromere	Mikromer *n*
5614	micrometer	Mikrometer *n*
5615	micromethod	Mikromethode *f*
5616	micrometric screw	Mikrometerschraube *f*
5617	micromilieu → microenvironment	—
5618	micromutation	Mikromutation *f*, Kleinmutation *f*
5619	micron	Mikron *n*
5620	microneedle, dissecting needle	Präpariernadel *f*

5556	métamitose *f*	metamitosis *f*
5557	métamorphoser	metamorfosear
5558	métamorphose *f*	metamorfosis *f*
5559	métanéphros *m*	metanefros *m*
5560	métaphase *f*	metafase *f*
5561	indice *m* d'appariement métaphasique	índice *m* de apareamiento metafásico
5562	métaphytes *mpl*	metafitas *fpl*, metáfitos *mpl*
5563	métaplasie *f*	metaplasia *f*
5564	os *m* métatarsien	hueso *m* metatarsiano
5565	métatarse *m*	metatarso *m*
5566	métaxénie *f*	metaxenia *f*
5567	métaxylème *m*	metaxilema *m*
5568	métazoaire *m*	metazoo *m*, metazoario *m*
5569	métencéphale *m*, cerveau *m* postérieur	metencéfalo *m*, cerebro *m* posterior
5570	méthionine *f*	metionina *f*
5571	mode *m* de locomotion	modo *m* de locomoción
5572	technique *f* de préparation	técnica *f* de preparación
5573	mode *m* de reproduction	modo *m* de reproducción
5574	groupe(ment) *m* méthyle	grupo *m* metilo
5575	méthylation *f*	metilación *f*
5576	métoestrus *m*	metaestro *m*, postcelo *m*
5577	acide *m* mévalonique	ácido *m* mevalónico
5578	micelle *f*	micela *f*
5579	micro-analyse *f*	microanálisis *m*
5580	microautoradiographie *f*	microautorradiografía *f*
5581	microbalance *f*	microbalanza *f*
5582	microbe *m*	microbio *m*
5583	microbien	microbiano, micróbico
5584	nombre *m* de germes	número *m* de gérmenes
5585	microbicide	microbicida
5586	microbiologique	microbiológico
5587	microbiologiste *m*	microbiólogo *m*
5588	microbiologie *f*	microbiología *f*
5589	microcentre *m*	microcentro *m*
5590	microchimie *f*	microquímica *f*
5591	microchromosome *m*	microcromosoma *m*
5592	microcinématographie *f*	microcinematografía *f*
5593	microclimat *m*	microclima *m*
5594	microconsommateur *m*	microconsumidor *m*
5595	microcyste *m*	microcisto *m*
5596	microdissection *f*	microdisección *f*
5597	micro-électrode *f*	microelectrodo *m*
5598	—	—
5599	microenvironnement *m*	micromedio *m*, microambiente *m*
5600	micro-évolution *f*	microevolución *f*
5601	microfaune *f*	microfauna *f*
5602	microfibrille *f*	microfibrilla *f*
5603	microflore *f*	microflora *f*
5604	microfossile *m*	microfósil *m*
5605	microgamète *m*, androgamète *m*	microgameto *m*
5606	microgène *m*	microgen *m*
5607	microglie *f*	microglía *f*
5608	microhabitat *m*	microhábitat *m*, microrresidencia *f*
5609	micro-incinération *f*	microincineración *f*
5610	micro-injection *f*	microinyección *f*
5611	micromanipulation *f*	micromanipulación *f*
5612	micromanipulateur *m*	micromanipulador *m*
5613	micromère *m*	micrómero *m*
5614	micromètre *m*	micrómetro *m*
5615	microméthode *f*	micrométodo *m*
5616	vis *f* micrométrique	tornillo *m* micrométrico
5617	—	—
5618	micromutation *f*	micromutación *f*
5619	micron *m*	micra *f*, micrón *m*
5620	micro-aiguille *f*	microaguja *f*

5621	micronucleus	Mikronukleus *m*, Kleinkern *m*
5622	micro-nutrient	Mikronährstoff *m*
5623	microorganism	Mikroorganismus *m*, Kleinlebewesen *n*
5624	micropal(a)eontology	Mikropaläontologie *f*
5625	microphage	Mikrophage *m*
5626	microphotography, photomicrography	Mikrophotographie *f*
5627	microphyll	Mikrophyll *n*
5628	microphylogenesis	Mikrophylogenese *f*
5629	micropipette, micropipet (US)	Mikropipette *f*
5630	microprotein	Mikroprotein *n*, Bakterieneiweiß *n*
5631	micropyle	Mikropyle *f*, Keimmund *m*
5632	microradiography	Mikroradiographie *f*
5633	microscope	Mikroskop *n*
5634	microscope-slide, glass-slide, slide	Objektträger *m*
5635	microscope-stage	Objekttisch *m*
5636	microscopic(al)	mikroskopisch
5637	microscopic(al) examination, ~ study	mikroskopische Untersuchung *f*
5638	microscopic(al) preparation	mikroskopisches Präparat *n*
5639	microscopist	Mikroskopiker *m*
5640	microscopy	Mikroskopie *f*
5641	microsmatic animal	Mikrosmat *m*
5642	microsome	Mikrosom *n*
5643	microspecies	Mikrospezies *f*, Kleinart *f*
5644	microspectrophotometer	Mikrospektrophotometer *n*
5645	microspectrophotometry	Mikrospektrophotometrie *f*
5646	microsphere	Mikrosphäre *f*
5647	microsporangium, androsporangium	Mikrosporangium *n*
5648	microspore, androspore	Mikrospore *f*, Androspore *f*
5649	microsporocyte → pollen mother-cell	—
5650	microsporophyll	Mikrosporophyll *n*
5651	microstylous, short-styled	kurzgriff(e)lig
5652	microsubspecies	Mikrosubspezies *f*
5653	microsurgery	Mikrochirurgie *f*
5654	microsurgical	mikrooperativ, mikrochirurgisch
5655	microtechnique	Mikrotechnik *f*
5656	microtome, section cutter	Mikrotom *n*
5657	microtomy	Mikrotomie *f*
5658	microtubules	Mikrotubuli *pl* (*sing.* ~tubulus *m*), Zytotubuli *pl*
5659	microvilli (*sing.* ~villus)	Mikrovilli *pl* (*sing.* ~villus *m*), Mikrozotten *f pl*
5660	mid-brain → mesencephalon	—
5661	mid-gut → mesenteron	—
5662	mid-rib, central nervure	Mittelnerv *m*, Hauptnerv *m* *bot.*, Mittelrippe *f*
5663	middle ear	Mittelohr *n*
5664	middle lamella	Mittellamelle *f*
5665	to migrate	wandern
5666	migrating fish	Wanderfisch *m*
5667	migrating flock	Wanderherde *f*
5668	migration	Wanderung *f*, Migration *f*, Zug *m* *(Vögel)*
5669	migration instinct	Wandertrieb *m*, Zugtrieb *m*
5670	migratory behavio(u)r	Migrationsverhalten *n*, Zugverhalten *n*
5671	migratory bird	Zugvogel *m*
5672	migratory cell → wandering cell	—
5673	migratory community, ~ society	Wandergesellschaft *f*
5674	migratory direction	Zugrichtung *f*, Wanderrichtung *f*
5675	migratory group	Wanderschwarm *m*
5676	migratory nucleus	Wanderkern *m*
5677	migratory restlessness	Zugunruhe *f*
5678	migratory route	Zugstraße *f*, Zugweg *m*
5679	mild humus, mull	milder Humus *m*, Mull *m*
5680	milk gland → mammary gland	—
5681	milk secretion	Milchabsonderung *f*, ~sekretion *f*
5682	milk sugar → lactose	—

5621	micronucléus *m*, micronoyau *m*	micronúcleo *m*
5622	substance *f* micronutritive	micronutrim(i)ento *m*, micronutriente *m*
5623	microorganisme *m*	microorganismo *m*
5624	micropaléontologie *f*	micropaleontología *f*
5625	microphage *m*	micrófago *m*
5626	microphotographie *f*	microfotografía *f*, fotomicrografía *f*
5627	microphylle *f*	microfilo *m*
5628	microphylogenèse *f*	microfilogénesis *f*
5629	micropipette *f*	micropipeta *f*
5630	microprotéine *f*, protéine *f* bactérienne	microproteína *f*
5631	micropyle *m*	micrópilo *m*
5632	microradiographie *f*	microrradiografía *f*
5633	microscope *m*	microscopio *m*
5634	porte-objet *m*, lame *f* ~	portaobjetos *m*
5635	platine *f* (porte-objet)	platina *f* (microscópica)
5636	microscopique	microscópico
5637	examen *m* microscopique	examen *m* microscópico
5638	préparation *f* microscopique	preparación *f* microscópica
5639	microscopiste *m*	microscopista *m*
5640	microscopie *f*	microscopia *f*
5641	microsmatique *m*, animal *m* ~	animal *m* microsmático
5642	microsome *m*	microsoma *m*
5643	micro-espèce *f*	microespecie *f*, especie *f* menor
5644	microspectrophotomètre *m*	microespectrofotómetro *m*
5645	microspectrophotométrie *f*	microespectrofotometría *f*
5646	microsphère *f*	micrósfera *f*
5647	microsporange *m*, androsporange *m*	microsporangio *m*, androsporangio *m*
5648	microspore *f*, androspore *f*	micróspora *f*, andróspora *f*
5649	—	—
5650	microsporophylle *f*	microsporófilo *m*
5651	brévistylé, à style court	microstilo, brevistilo
5652	micro-sous-espèce *f*	microsubespecie *f*
5653	microchirurgie *f*	microcirugía *f*
5654	microchirurgical	microquirúrgico
5655	microtechnique *f*	microtécnica *f*
5656	microtome *m*	micrótomo *m*
5657	microtomie *f*	microtomía *f*
5658	microtubules *m pl*	microtúbulos *m pl*
5659	microvillosités *f pl*	microvellos *m pl*, microvellosidades *f pl*
5660	—	—
5661	—	—
5662	nervure *f* médiane, ~ centrale	nervio *m* medio, ~ medial, ~ central
5663	oreille *f* moyenne	oído *m* medio
5664	lamelle *f* mitoyenne, ~ moyenne	lámina *f* media, laminilla *f* ~
5665	migrer	migrar
5666	poisson *m* migrateur	pez *m* migrante
5667	troupeau *m* migrateur	rebaño *m* migrante
5668	migration *f*	migración *f*
5669	instinct *m* migrateur, (im)pulsion *f* migratoire	impulso *m* migratorio
5670	comportement *m* migrateur	comportamiento *m* migratorio
5671	oiseau *m* migrateur	ave *f* migratoria, ~ migrante
5672	—	—
5673	association *f* migratrice	asociación *f* migratoria
5674	direction *f* migratoire, ~ de migration	dirección *f* de la migración
5675	bande *f* migratrice, essaim *m* migrateur	bando *m* migratorio, enjambre *m* ~
5676	noyau *m* migrateur	núcleo *m* emigrante
5677	agitation *f* migratoire, «inquiétude» *f* migratrice	inquietud *f* motora en las aves migratorias
5678	route *f* de migration	ruta *f* migratoria, vía *f* de migración
5679	humus *m* doux, mull *m*	humus *m* suave, mantillo *m* ~
5680	—	—
5681	sécrétion *f* lactée	secreción *f* láctea
5682	—	—

No.	English	German
5683	milk-tooth, deciduous ~	Milchzahn *m*
5684	milk tube → laticifer	—
5685	milt[1], soft roe	Milch *f* *zool.*
5686	milt[2] → spleen	—
5687	milter	Milcher *m*, Milchner *m*
5688	mimesis	Mimese *f*
5689	mimetic	mimetisch
5690	mimic	Nachahmer *m*
5691	to mimic	nachahmen
5692	mimic expressions → facial ~	—
5693	mimicry	Mimikry *f*
5694	mineral element	Mineralstoff *m*
5695	mineral metabolism	Mineralstoffwechsel *m*
5696	mineral nutrient	Mineralnährstoff *m*
5697	mineral nutrition	Mineralernährung *f*
5698	mineral salt	Mineralsalz *n*
5699	mineralization	Mineralisation *f*, Mineralisierung *f*
5700	mineralocorticoid	Mineral(o)kortikoid *n*
5701	minimal medium	Minimalmedium *n*
5702	minimum area	Minimalareal *n*
5703	minor element → trace-element	—
5704	minor gene → polygene	—
5705	Miocene	Miozän *n*
5706	miscibility	Mischbarkeit *f*
5707	miscible	mischbar
5708	misdivision	Mißteilung *f*
5709	mitochondrion (*pl* mitochondria), chondriosome, plastosome	Mitochondrium (*pl* Mitochondrien), Chondriosom *n*, Plastosom *n*
5710	mitogenetic rays, ~ radiation	mitogenetische Strahlungen *f pl*
5711	mitosis, mitotic division, karyokinesis	Mitose *f*, mitotische Teilung *f*, Karyokinese *f*
5712	mitosome	Mitosom *n*
5713	mitotic	mitotisch
5714	mitotic apparatus	Mitoseapparat *m*
5715	mitotic cycle	Mitosezyklus *m*
5716	mitotic index	Mitoseindex *m*
5717	mitotic inhibition	Mitosehemmung *f*
5718	mitotic poison	Mitosegift *n*
5719	mitotic spindle	Mitosespindel *f*
5720	mitral valve	Mitralklappe *f*
5721	mixed culture	Mischkultur *f*
5722	mixed population	Mischpopulation *f*
5723	mixoploidy	Mixoploidie *f*
5724	mode of action	Wirkungsweise *f*
5725	mode of life	Lebensweise *f*
5726	mode of nutrition	Ernährungsweise *f*
5727	mode of reproduction → method ~ ~	—
5728	mode of transmission	Übertragungsweise *f*
5729	model experiment	Attrappenversuch *m*, Modellversuch *m*
5730	modification *gen.*	Modifikation *f* *gen.*
5731	modificatory	modifizierend
5732	modifier *gen.*, modifying factor	Modifikationsgen *n*, ~faktor *m*, Modifikator *m*
5733	modulation *gen.*	Modulation *f* *gen.*
5734	modulator	Modulator *m*
5735	moist chamber	feuchte Kammer *f*
5736	moist forest	Feuchtwald *m*
5737	moisture content	Feuchtigkeitsgehalt *m*
5738	moisture degree	Feuchtigkeitsgrad *m*
5739	molar (tooth)	Backenzahn *m*, Molar *m*
5740	molecular	molekular

No.	Français	Español
5683	dent *f* de lait, ~ lactéale	diente *m* de leche, ~ deciduo, ~ temporal
5684	—	—
5685	laitance *f*, laite *f*	lecha *f*, lechaza *f*
5686	—	—
5687	poisson *m* laité	pez *m* lechal
5688	mimèse *f*	mimesis *f*
5689	mimétique	mimético
5690	imitateur *m*, mime *m*	mimético *m*
5691	mimer	mimetizar
5692	—	—
5693	mimétisme *m*	mimetismo *m*
5694	élément *m* minéral, substance *f* minérale	elemento *m* mineral
5695	métabolisme *m* minéral	metabolismo *m* mineral
5696	substance *f* nutritive minérale	sustancia *f* nutritiva mineral, mineral *m* nutritivo
5697	nutrition *f* minérale	nutrición *f* mineral
5698	sel *m* minéral	sal *f* mineral
5699	minéralisation *f*	mineralización *f*
5700	minéralocorticoïde *m*	mineralocorticoide *m*
5701	milieu *m* minimal	medio *m* mínimo
5702	aire *f* minima	área *f* minima
5703	—	—
5704	—	—
5705	miocène *m*	mioceno *m*
5706	miscibilité *f*	miscibilidad *f*
5707	miscible	miscible
5708	misdivision *f*	misdivisión *f*
5709	mitochondrie *f*, chondriosome *m*, plastosome *m*	mitocondrio *m*, mitocondria *f*, condriosoma *m*, plastosoma *m*
5710	rayons *m pl* mitogènes	rayos *m pl* mitogenéticos, ~ mitógenos
5711	mitose *f*, caryocinèse *f*	mitosis *f*, división *f* mitótica, cariocinesis *f*
5712	mitosome *m*	mitosoma *m*
5713	mitotique	mitótico
5714	appareil *m* mitotique	aparato *m* mitótico
5715	cycle *m* mitotique	ciclo *m* mitótico
5716	indice *m* mitotique	índice *m* mitótico
5717	inhibition *f* mitotique, mito-inhibition *f*	inhibición *f* mitótica
5718	poison *m* mitotique, inhibiteur *m* ~, mito-inhibiteur *m*	tóxico *m* mitótico, veneno *m* ~
5719	fuseau *m* mitotique	huso *m* mitótico
5720	valvule *f* mitrale	válvula *f* mitral
5721	culture *f* mixte	cultivo *m* mezclado, ~ mixto
5722	peuplement *m* mélangé	población *f* mezclada
5723	mixoploïdie *f*	mixoploidia *f*
5724	mode *m* d'action, mécanisme *m* ~	modo *m* de acción, mecanismo *m* ~ ~
5725	mode *m* de vie, façon *f* de vivre	modo *m* de vida
5726	mode *m* de nutrition, ~ nutritionnel, régime *m* alimentaire	modo *m* de nutrición, régimen *m* alimenticio
5727	—	—
5728	mode *m* de transmission	modo *m* de transmisión
5729	essai *m* de modèle	experiencia *f* con modelos, prueba *f* con simulacros
5730	modification *f* *gen.*	modificación *f* *gen.*
5731	modificateur	modificativo, modificatorio
5732	(gène *m*) modifieur *m*, ~ modificateur	(gen *m*) modificador *m*, factor *m* ~
5733	modulation *f* *gen.*	modulación *f* *gen.*
5734	modulateur *m*	modulador *m*
5735	chambre *f* humide	cámara *f* húmeda
5736	forêt *f* humide	bosque *m* húmedo
5737	teneur *f* en humidité	contenido *m* en humedad
5738	degré *m* d'humidité, ~ hygrométrique	grado *m* de humedad, ~ higrométrico
5739	molaire *f*	(diente *m*) molar *m*
5740	moléculaire	molecular

5741	molecular biology	Molekularbiologie *f*
5742	molecular genetics	Molekulargenetik *f*
5743	molecular model	Molekülmodell *n*
5744	molecular structure	Molekulargefüge *n*, ~struktur *f*
5745	molecular weight	Molekulargewicht *n*
5746	molecule	Molekül *n*
5747	Mollusca, molluscs, mollusks	Mollusken *fpl*, Weichtiere *npl*
5748	molt → moult	—
5749	monad	Monade *f*
5750	monaster, mother star	Monaster *m*
5751	monaxial, uniaxial	einachsig
5752	Monera, monerans	Moneren *fpl*, niedere Protisten *mpl*
5753	monid	Monide *f*
5754	monocarpic	einfrüchtig, monokarp
5755	monocellular → unicellular	—
5756	monocentric	monozentrisch
5757	monochasium	Monochasium *n*
5758	monochlamydeous	monochlamyd
5759	monoclinous, monoclinic	monoklin
5760	Monocotyledoneae, monocotyledons	Monokotyle(done)n *fpl*, einkeimblättrige Pflanzen *fpl*
5761	monocotyledonous	einkeimblättrig, monokotyl
5762	monocyclic	monozyklisch
5763	monocyte	Monozyt *m*
5764	monoecious	einhäusig, monözisch
5765	monoecism, monoecy	Einhäusigkeit *f*, Monözie *f*
5766	monoestrous	monoöstrisch
5767	monofactorial, unifactorial	monofaktoriell, unifaktoriell
5768	monogamous	monogam, einehig
5769	monogamy	Monogamie *f*, Einehe *f*
5770	monogenesis	Monogenese *f*
5771	monogenic	monogen
5772	monogeny	Monogenie *f*
5773	monogerminal → unigerminal	—
5774	monogony	Monogonie *f*
5775	monohaploidy	Monohaploidie *f*
5776	monohybrid	monohybrid
5777	monohybridism	Monohybridie *f*
5778	monokaryon, monocaryon	Monokaryon *n*
5779	monolocular → unilocular	—
5780	monomastigote → uniflagellate	—
5781	monomeric(al)	monomer
5782	monomolecular layer	monomolekulare Schicht *f*
5783	monomorphic, monomorphous	monomorph, eingestaltig
5784	mononuclear, uninuclear, uninucleate, monokaric	einkernig
5785	monophagous	monophag
5786	monophyletic	monophyletisch
5787	monophyletism, monophyletic evolution	Monophylie *f*
5788	monophyllous, unifoliate	einblättrig
5789	monoploidy	Monoploidie *f*
5790	monopodial	monopodial
5791	monopodium	Monopodium *n*
5792	monopolar → unipolar	—
5793	monosaccharide, monose	Monosa(c)charid *n*, Monose *f*
5794	monosome	Monosom *n*
5795	monosomy	Monosomie *f*
5796	monospermous, monospermic, one-seeded	einsamig
5797	monosymmetrical → zygomorphic	—
5798	Monotremata, monotremes	Monotremen *pl*, Kloakentiere *npl*
5799	monotrichous, monotrichic	monotrich
5800	monozygotic, monozygous, uniovular	eineiig, monozygot
5801	monozygotic twins, uniovular ~, identical ~	eineiige Zwillinge *mpl*, monozygotische ~, echte ~
5802	monsoon(al) forest	Monsunwald *m*
5803	mood *eth.*	Gestimmtheit *f*, Stimmung *f* *eth.*

5741	biologie *f* moléculaire	biología *f* molecular
5742	génétique *f* moléculaire	genética *f* molecular
5743	modèle *m* moléculaire	modelo *m* molecular
5744	structure *f* moléculaire	estructura *f* molecular
5745	poids *m* moléculaire	peso *m* molecular
5746	molécule *f*	molécula *f*
5747	mollusques *m pl*	moluscos *m pl*
5748	—	—
5749	monade *f*	mónada *f*
5750	monaster *m*, sphère-mère *f*	monáster *m*
5751	monaxial, uniaxial	uniaxial, uniáxico
5752	monères *f pl*	monerados *m pl*, móneras *f pl*
5753	monide *f*	mónida *f*
5754	monocarpique	monocárpico
5755	—	—
5756	monocentrique	monocéntrico
5757	cyme *f* unipare	monocasio *m*, cima *f* unípara
5758	monochlamydé	monoclamídeo
5759	monocline	monoclino
5760	monocotylédones *f pl*	monocotiledóneas *f pl*
5761	monocotylédone	monocotiledóneo, monocotiledónico
5762	monocyclique	monocíclico
5763	monocyte *m*	monocito *m*
5764	monoïque, monoecique	monoico
5765	mon(o)écie *f*	monoecia *f*
5766	monoestrien	monoéstrico, monoestral
5767	monofactoriel	monofactorial
5768	monogame, monogamique	monógamo, monogámico
5769	monogamie *f*	monogamia *f*
5770	monogenèse *f*	monogénesis *f*
5771	monogénique	monogénico
5772	monogénie *f*	monogenia *f*
5773	—	—
5774	monogonie *f*	monogonia *f*
5775	monohaploïdie *f*	monohaploidia *f*
5776	monohybride	monohíbrido
5777	monohybridisme *m*	monohibridismo *m*
5778	monocaryon *m*	monocario *m*
5779	—	—
5780	—	—
5781	monomérique	monomérico
5782	couche *f* monomoléculaire	capa *f* monomolecular
5783	monomorphe	monomorfo
5784	mononucléaire, uninucléé	mononuclear, mononucleado, uninuclear
5785	monophage	monófago
5786	monophylétique	monofilético
5787	monophylie *f*	monofiletismo *m*
5788	monophylle, unifolié	monofilo, unifoliado
5789	monoploïdie *f*	monoploidia *f*
5790	monopodique, monopodial	monopódico, monopodial
5791	monopode *m*	monopodio *m*
5792	—	—
5793	monosaccharide *m*, monose *m*, ose *m*	monosacárido *m*, monosa *f*, osa *f*
5794	monosome *m*	monosoma *m*
5795	monosomie *f*	monosomia *f*
5796	monosperme, unisperme	monospermo
5797	—	—
5798	monotrèmes *m pl*	monotremas *m pl*
5799	monotriche	monotrico
5800	monozygote, uniovulaire	monocigótico, uniovular
5801	jumeaux *m pl* monozygotes, ~ uniovulaires, ~ univitellins, ~ vrais	gemelos *m pl* monocigóticos, ~ uniovulares, ~ univitelinos, ~ idénticos
5802	forêt *f* de moussons	bosque *m* monzónico
5803	disposition *f eth.*	disposición *f eth.*

5804	mor → raw humus	—
5805	mordant	Beizmittel *n*
5806	morgan	Morgan-Einheit *f*
5807	morphallaxis	Morphallaxis *f*, Morpholaxis *f*
5808	morphine	Morphin *n*, Morphium *n*
5809	morphogenesis, morphogeny	Morphogenese *f*, Morphogenie *f*
5810	morphogenetic	morphogenetisch, gestaltbildend
5811	morphogenetic field	morphogenetisches Feld *n*
5812	morphogenetic movement	morphogenetische Bewegung *f*
5813	morphologic(al)	morphologisch
5814	morphology	Morphologie *f*, Gestaltlehre *f*
5815	morphoplasm	Morphoplasma *n*
5816	morphosis	Morphose *f*
5817	mortality	Sterblichkeit *f*, Mortalität *f*
5818	mortality rate → death rate	—
5819	morula	Morula *f*, Maulbeerkeim *m*
5820	morulation	Morulabildung *f*
5821	mosaic development, mosaicism	Mosaikentwicklung *f*
5822	mosaic egg	Mosaikei *n*
5823	moss starch → lichenin	—
5824	mosses, Musci	Laubmoose *npl*
5825	mother cell	Mutterzelle *f*
5826	mother plant → parent plant	—
5827	mother star → monaster	—
5828	motile	beweglich
5829	motility	Beweglichkeit *f*, Motilität *f*
5830	motivating factor, motivational ~	motivierender Faktor *m*
5831	motoneuron → motor neurone	—
5832	motor cortex	Motorcortex *m*, motorischer Cortex *m*, motorische Großhirnrinde *f*
5833	motor end-organ, ~ end-plate	motorische Endplatte *f*
5834	motor nerve	motorischer Nerv *m*
5835	motor neurone, motoneuron	Motoneuron *n*, motorische Nervenzelle *f*
5836	motor root → ventral root	—
5837	motoricity	Motorik *f*
5838	to mo(u)lt	mausern, sich häuten
5839	mo(u)lt, mo(u)lting, ecdysis	Mauser *f*, Häutung *f*, Haarwechsel *m*, Federwechsel *m*, Ekdyse *f*
5840	mo(u)lting fluid, ecdysial ~	Häutungsflüssigkeit *f*
5841	mo(u)lting gland, exuvial ~, ecdysial ~	Häutungsdrüse *f*, Exuvialdrüse *f*
5842	mo(u)lting hormone → ecdysone	—
5843	mouth breeder, ~ brooder	Maulbrüter *m*
5844	mouth cavity → buccal cavity	—
5845	mouth parts	Mundwerkzeuge *npl*, Mundgliedmaßen *fpl*
5846	movement perception	Bewegungswahrnehmung *f*
5847	movement of substances	Stoffwanderung *f*
5848	m-RNA → messenger RNA	—
5849	MSH → intermedin	—
5850	mucilage	Pflanzenschleim *m*
5851	mucilaginous	schleimig, schleimartig
5852	mucilaginous substance	Schleimstoff *m*
5853	mucin	Muzin *n*
5854	muciparous, mucus-secreting	schleimabsondernd
5855	mucopolysaccharide	Mukopolysa(c)charid *n*
5856	mucoprotein → glucoprotein	—
5857	mucosa → mucous membrane	—
5858	mucous, slimy	schleimig, schleimartig
5859	mucous cell	Schleimzelle *f*
5860	mucous gland, slime ~	Schleimdrüse *f*
5861	mucous membrane, mucosa	Schleimhaut *f*
5862	mucous sheath, mucilaginous ~	Schleimscheide *f*
5863	mucus, slime	Schleim *m*
5864	mucus-secreting → muciparous	—
5865	mucus secretion	Schleimabsonderung *f*
5866	mull → mild humus	—

5804	—	—
5805	mordant *m*	mordiente *m*
5806	morgan *m*	morgan *m*
5807	morphallaxie *f*	morfalasis *f*
5808	morphine *f*	morfina *f*
5809	morphogenèse *f*	morfogénesis *f*, morfogenia *f*
5810	morphogénétique, morphogène	morfogenético, morfógeno, morfogénico
5811	champ *m* morphogénétique	campo *m* morfogenético
5812	mouvement *m* morphogène	movimiento *m* morfogenético
5813	morphologique	morfológico
5814	morphologie *f*	morfología *f*
5815	morphoplasme *m*	morfoplasma *m*
5816	morphose *f*	morfosis *f*
5817	mortalité *f*	mortalidad *f*
5818	—	—
5819	morula *f*	mórula *f*
5820	morulation *f*	morulación *f*
5821	développement *m* en mosaïque	desarrollo *m* en mosaico
5822	œuf *m* en mosaïque	huevo *m* en mosaico
5823	—	—
5824	mousses *f pl*	musgos *m pl*
5825	cellule *f* mère	célula *f* madre
5826	—	—
5827	—	—
5828	mobile	móvil
5829	motilité *f*	motilidad *f*
5830	facteur *m* motivationnel, ~ de motivation	factor *m* motivante, ~ de motivación
5831	—	—
5832	cortex *m* moteur	corteza *f* motora
5833	plaque *f* terminale motrice	placa *f* terminal motora
5834	nerf *m* moteur	nervio *m* motor
5835	motoneurone *m*, cellule *f* motrice, neurone *m* moteur	neurona *f* motora, motoneurona *f*
5836	—	—
5837	motricité *f*	motricidad *f*
5838	muer	mudar
5839	mue *f*, ecdysie *f*	muda *f*, ecdisis *f*
5840	liquide *m* exuvial	líquido *m* exuvial, fluido *m* de la muda
5841	glande *f* de mue, ~ prothoracique	glándula *f* protorácica
5842	—	—
5843	incubateur *m* buccal, ~ oral	incubador *m* bucal
5844	—	—
5845	pièces *f pl* buccales	piezas *f pl* bucales
5846	perception *f* du mouvement	percepción *f* del movimiento
5847	transit *m* des substances	movimiento *m* de sustancias
5848	—	—
5849	—	—
5850	mucilage *m*	mucílago *m*
5851	mucilagineux	mucilaginoso
5852	substance *f* mucilagineuse	sustancia *f* mucilaginosa
5853	mucine *f*	mucina *f*
5854	mucipare	mucíparo, mucífero
5855	mucopolysaccharide *m*, mucopolyoside *m*	mucopolisacárido *m*
5856	—	—
5857	—	—
5858	muqueux	mucoso
5859	cellule *f* muqueuse, ~ à mucus	célula *f* mucífera
5860	glande *f* muqueuse, ~ à mucus	glándula *f* mucosa, ~ mucípara
5861	(membrane *f*) muqueuse *f*	(membrana *f*) mucosa *f*
5862	gaine *f* mucilagineuse	vaina *f* mucilaginosa
5863	mucus *m*, mucosité *f*	mucus *m*, mucosidad *f*, moco *m*
5864	—	—
5865	sécrétion *f* muqueuse	secreción *f* mucosa
5866	—	—

5867	Müllerian duct, paramesonephric ~	Müllerscher Gang *m*
5868	Müllerian mimicry	Müllersche Mimikry *f*
5869	multicamerate, multilocular	vielkammerig, mehrfächerig
5870	multicellular, many-celled	vielzellig, mehrzellig
5871	multicellular organism	Vielzeller *m*
5872	multicellularity	Vielzelligkeit *f*
5873	multienzyme complex	Multienzymkomplex *m*, Multienzymsystem *n*
5874	multifactorial → polyfactorial	—
5875	multiflagellate, polymastigote, polykont	vielgeiß(e)lig
5876	multigene	Multigen *n*
5877	multilayered	vielschichtig, mehrschichtig
5878	multilobar, multilobate, multilobular	viellappig
5879	multilocular → multicamerate	—
5880	multinucleate, multinuclear, plurinucleate, polykaric	vielkernig, mehrkernig
5881	multipartial, pluripartial	vielteilig, mehrteilig
5882	multiple allelles, ~ allelomorphs	multiple Allele *n pl*
5883	multiple allelism	multiple Allelie *f*
5884	multiple factor → polygene	—
5885	multiple fission, ~ division	Vielteilung *f*, multiple Zellteilung *f*
5886	multiple fruit, collective ~	Sammelfrucht *f*
5887	multiple reactivation	Mehrfachreaktivierung *f*
5888	multiplication capacity	Vermehrungsfähigkeit *f*
5889	multiplication rate	Vermehrungsgeschwindigkeit *f*, ~rate *f*
5890	to multiply	sich vermehren
5891	multipolar	multipolar
5892	multivalent	Multivalent *n*
5893	muramic acid	Muraminsäure *f*
5894	Musci → mosses	—
5895	muscle	Muskel *m*
5896	muscle action potential	Muskelaktionspotential *n*
5897	muscle cell, myocyte	Muskelzelle *f*
5898	muscle fibril → myofibril	—
5899	muscle insertion, ~ attachment	Muskelansatz *m*
5900	muscle protein, myoprotein	Muskelprotein *n*
5901	muscle spindle	Muskelspindel *f*
5902	muscology → bryology	—
5903	muscular	muskulär, Muskel ...
5904	muscular activity, muscle action	Muskeltätigkeit *f*
5905	muscular contraction	Muskelkontraktion *f*
5906	muscular fibre, muscle ~	Muskelfaser *f*
5907	muscular relaxation	Muskelerschlaffung *f*
5908	muscular system	Muskelsystem *n*
5909	muscular tone, ~ tonicity, muscle tonus	Muskeltonus *m*
5910	musculature	Muskulatur *f*
5911	mushroom bodies, corpora pedunculata, pedunculate bodies	Pilzkörper *m pl*, Corpora pedunculata
5912	mutability	Mutabilität *f*, Mutierbarkeit *f*
5913	mutable	mutabel
5914	mutagene	Mutagen *n*
5915	mutagene specificity	Mutagenspezifität *f*
5916	mutagene stability	Mutagenstabilität *f*
5917	mutagenesis	Mutagenese *f*
5918	mutagenic, mutagenous	mutagen, mutationsauslösend
5919	mutagenicity	Mutationsfähigkeit *f*
5920	mutant	Mutant *m*, Mutante *f*
5921	mutant phenotype	Mutantenphänotyp *m*
5922	mutant strain	Mutantenstamm *m*
5923	to mutate	mutieren
5924	mutation	Mutation *f*
5925	mutation breeding, selection by mutation	Mutationszüchtung *f*
5926	mutation delay, mutational lag	Mutationsverzögerung *f*

5867	canal *m* de Müller, conduit *m* paraméso-néphritique	conducto *m* de Müller
5868	mimétisme *m* müllérien	mimetismo *m* mülleriano
5869	pluriloculaire	plurilocular
5870	multicellulaire, pluricellulaire	pluricelular, multicelular
5871	organisme *m* multicellulaire	organismo *m* multicelular
5872	multicellularité *f*	pluricelularidad *f*, multicelularidad *f*
5873	système *m* multienzymatique	complejo *m* multienzimático
5874	—	—
5875	muni de plusieurs flagelles	multiflagelado
5876	multigène *m*	multigen *m*
5877	multistratifié, pluristratifié	pluristratificado
5878	multilobaire, multilobulaire	multilobular
5879	—	—
5880	multinucléaire, multinucléé, polynucléaire, plurinucléé	multinuclear, multinucleado, plurinucleado, policario
5881	multipartiel, pluripartiel	multiparcial
5882	allèles *mpl* multiples	alelos *mpl* múltiples
5883	allélisme *m* multiple, polyallélie *f*	alelismo *m* múltiple, polialelia *f*
5884	—	—
5885	schizogamie *f* multiple	escisión *f* múltiple, multipartición *f*, pluripartición *f*
5886	fruit *m* multiple	fruto *m* múltiplo, ~ colectivo
5887	réactivation *f* multiple	reactivación *f* múltiple
5888	pouvoir *m* multiplicateur, ~ de multiplication	capacidad *f* de multiplicación
5889	vitesse *f* de multiplication, taux *m* ~ ~	velocidad *f* de multiplicación
5890	se multiplier	multiplicarse
5891	multipolaire	multipolar
5892	multivalent *m*	multivalente *m*
5893	acide *m* muramique	ácido *m* murámico
5894	—	—
5895	muscle *m*	músculo *m*
5896	potentiel *m* d'action musculaire	potencial *m* de acción muscular
5897	cellule *f* musculaire	célula *f* muscular, miocito *m*
5898	—	—
5899	insertion *f* musculaire, ~ du muscle	inserción *f* del músculo
5900	protéine *f* musculaire	proteína *f* muscular, mioproteína *f*
5901	fuseau *m* musculaire	huso *m* muscular
5902	—	—
5903	musculaire	muscular
5904	fonction *f* musculaire	acción *f* muscular
5905	contraction *f* musculaire	contracción *f* muscular
5906	fibre *f* musculaire	fibra *f* muscular
5907	relaxation *f* musculaire	relajación *f* muscular, relajamiento *m* ~
5908	système *m* musculaire	sistema *m* muscular
5909	tonus *m* musculaire	tono *m* muscular
5910	musculature *f*	musculatura *f*
5911	corps *mpl* pédonculés	cuerpos *mpl* pedunculados
5912	mutabilité *f*	mutabilidad *f*
5913	mutable	mutable
5914	(agent *m*) mutagène *m*	mutágeno *m*, sustancia *f* mutágena, agente *m* mutagénico
5915	spécificité *f* mutagénique	especificidad *f* mutagénica
5916	stabilité *f* mutagénique	estabilidad *f* mutagénica
5917	mutagenèse *f*	mutagénesis *f*
5918	mutagénique, mutagène	mutagénico, mutágeno
5919	mutagénicité *f*	capacidad *f* mutante
5920	mutant *m*, mutante *f*	mutante *m/f*
5921	phénotype *m* mutant	fenotipo *m* mutante
5922	souche *f* mutante	cepa *f* mutante
5923	muter	mudar, mutar
5924	mutation *f*	mutación *f*
5925	sélection *f* par mutation	selección *f* por mutación
5926	délai *m* de mutation	retardo *m* de mutación

5927	mutation factor	Mutationsfaktor *m*
5928	mutation pressure	Mutationsdruck *m*
5929	mutation rate	Mutationsrate *f*
5930	mutation theory, mutationist ~	Mutationstheorie *f*
5931	mutational	Mutations ...
5932	mutational equilibrium	Mutationsgleichgewicht *n*
5933	mutational lag → mutation delay	—
5934	mutator gene	Mutatorgen *n*
5935	muton	Muton *n*
5936	mutualism	Mutualismus *m*
5937	mycelium	Myzel(ium) *n*
5938	mycetology → mycology	—
5939	mycetoma	Myzetom *n*
5940	mycobacterium	Mykobakterie *f*
5941	mycology, mycetology	Mykologie *f*, Pilzkunde *f*
5942	Mycophyta → Fungi	—
5943	mycor(r)hiza	Mykorrhiza *f*, Pilzwurzel *f*
5944	mycotrophy	Mykotrophie *f*
5945	myelencephalon, after-brain, marrow-brain	Nachhirn *n*, Markhirn *n*, Myelenzephalon *n*
5946	myelin	Myelin *n*, Nervenmark *n*
5947	myelin sheath, medullary ~	Myelinscheide *f*, Markscheide *f*
5948	myelinated, medullated	markhaltig, myelinisiert
5949	myoblast	Myoblast *m*
5950	myocard(ium), cardiac muscle	Herzmuskel *m*, Myokard *m*
5951	myocyte → muscle cell	—
5952	myofibril(la), muscle fibril	Muskelfibrille *f*, Myofibrille *f*
5953	myogenic, myogenous	myogen
5954	myoglobin, myohaemoglobin	Myoglobin *n*, Myohämoglobin *n*
5955	myolemma, sarcolemma	Myolemm *n*, Sarkolemm *n*
5956	myology	Myologie *f*, Muskellehre *f*
5957	myomere	Myomer *n*
5958	myometrium	Myometrium *n*
5959	myoplasm	Myoplasma *n*
5960	myoprotein → muscle protein	—
5961	myosin	Myosin *n*
5962	myotatic reflex → stretch reflex	—
5963	myotome	Myotom *n*, Ursegment *n*
5964	Myriapoda, myriapods	Myriapoden *mpl*, Myriopoden *mpl*, Vielfüßer *mpl*
5965	myrmecophilous	myrmekophil
5966	myrmecophyte, myrmecophilous plant	Myrmekophyt *m*, Ameisenpflanze *f*
5967	myxam(o)eba	Myxamöbe *f*
5968	myxoflagellates	Myxoflagellaten *pl*
5969	Myxomycophyta, Myxomycetes, Mycetozoa, slime mo(u)lds	Schleimpilze *mpl*, Myxomyzeten *mpl*, Myxophyta *pl*
5970	NAD → nicotinamide adenine dinucleotide	—
5971	NADP → nicotinamide adenine dinucleotide phosphate	—
5972	nanism, dwarfishness, dwarfism	Zwergwuchs *m*, Nanismus *m*
5973	nan(n)oplankton	Nanoplankton *n*, Zwergplankton *n*
5974	nare, nasal aperture	Nasenloch *n*, Nasenöffnung *f*
5975	narrow-leaved → angustifoliate	—
5976	nasal, rhinal	Nasen ..., nasal
5977	nasal bone	Nasenbein *n*
5978	nasal cavity	Nasenhöhle *f*
5979	nasal meatus	Nasengang *m*
5980	nasal septum, internasal ~	Nasenscheidewand *f*
5981	nasolacrimal duct → lacrimonasal duct	—
5982	nasopharynx, rhinopharynx	Nasenrachen *m*
5983	nastic	nastisch
5984	nasty, nastic movement	Nastie *f*, nastische Bewegung *f*
5985	natality	Geburtenziffer *f*
5986	native breed, land race	Landrasse *f*, Primitivrasse *f*, Naturrasse *f*
5987	native plant → indigenous plant	—

5927	facteur *m* de mutation	factor *m* de mutación
5928	pression *f* de mutation	presión *f* de mutación
5929	taux *m* de mutation	tasa *f* de mutación, frecuencia *f* ~ ~
5930	mutationnisme *m*	mutacionismo *m*, teoría *f* mutacionista
5931	mutationnel	mutacional
5932	équilibre *m* mutationnel	equilibrio *m* mutacional
5933	—	—
5934	gène-mutateur *m*	gen *m* mutador, ~ mutante
5935	muton *m*, site *m* mutable	mutón *m*
5936	mutualisme *m*, symbiose *f* mutualiste	mutualismo *m*, simbiosis *f* mutualista
5937	mycélium *m*	micelio *m*
5938	—	—
5939	mycétome *m*	micetoma *m*
5940	mycobactérie *f*	micobacteria *f*
5941	mycologie *f*, mycétologie *f*	micología *f*, micetología *f*
5942	—	—
5943	mycor(r)hize *f*	micorriza *f*
5944	mycotrophie *f*	micotrofia *f*
5945	myélencéphale *m*, arrière-cerveau *m*	mielencéfalo *m*
5946	myéline *f*	mielina *f*
5947	gaine *f* myélinique, ~ à myéline	vaina *f* mielínica, ~ medulada, ~ de mielina
5948	myélinique, myélinisé	mielínico, mielinizado, medulado
5949	myoblaste *m*	mioblasto *m*
5950	myocarde *m*, muscle *m* cardiaque	miocardio *m*, músculo *m* cardíaco
5951	—	—
5952	myofibrille *f*, fibrille *f* musculaire	miofibrilla *f*, fibrilla *f* muscular
5953	myogène	miogénico, miógeno
5954	myoglobine *f*	mioglobina *f*, miohemoglobina *f*
5955	myolemme *m*, sarcolemme *m*	miolema *m*, sarcolema *m*
5956	myologie *f*	miología *f*
5957	myomère *m*	miomero *m*
5958	myomètre *m*	miometrio *m*
5959	myoplasme *m*, plasma *m* musculaire	mioplasma *m*
5960	—	—
5961	myosine *f*	miosina *f*
5962	—	—
5963	myotome *m*	miótomo *m*
5964	myriapodes *mpl*	miriápodos *mpl*
5965	myrmécophile	mirmecófilo
5966	myrmécophyte *m*	mirmecófito *m*, planta *f* mirmecófila
5967	myxamibe *f*	mixameba *f*
5968	myxoflagellés *mpl*	mixoflagelados *mpl*
5969	myxomycètes *mpl*, mycétozoaires *mpl*	mixomicetos *mpl*, mixomicofitos *mpl*, micetozoos *mpl*, mohos *mpl* mucosos
5970	—	—
5971	—	—
5972	nanisme *m*	nanismo *m*
5973	nannoplancton *m*	nanoplancton *m*
5974	narine *f*	nar *f*
5975	—	—
5976	nasal	nasal
5977	os *m* nasal	hueso *m* nasal
5978	fosse *f* nasale, cavité *f* ~	fosa *f* nasal, cavidad *f* ~
5979	conduit *m* nasal	conducto *m* nasal
5980	cloison *f* nasale	tabique *m* nasal
5981	—	—
5982	nasopharynx *m*, rhinopharynx *m*	nasofaringe *f*, faringe *f* nasal, rinofaringe *f*
5983	nastique	nástico
5984	nastisme *m*, nastie *f*	nastia *f*, movimiento *m* nástico
5985	natalité *f*	natalidad *f*
5986	race *f* primitive	raza *f* original
5987	—	—

5988	natural landscape	Naturlandschaft *f*
5989	natural selection	natürliche Auslese *f*, ~ Selektion *f*
5990	natural substance	Naturstoff *m*
5991	nature leaf	Folgeblatt *n*, Altersblatt *n*
5992	nature park	Natur(schutz)park *m*, Nationalpark *m*
5993	nature reserve	Naturschutzgebiet *n*, Reservat *n*
5994	navel, umbilicus	Nabel *m*
5995	navel cord → umbilical cord	—
5996	nearctic	nearktisch
5997	nebenkern → paranucleus	—
5998	necrobiosis	Nekrobiose *f*
5999	necrocytosis → cell death	—
6000	necrohormone → wound hormone	—
6001	necrophage	Nekrophage *m*
6002	necrosis	Nekrose *f*
6003	nectar	Nektar *m*
6004	nectary, nectar gland	Nektarium *n*, Nektardrüse *f*
6005	necton → nekton	—
6006	needle culture → stab culture	—
6007	needle-leaved forest → conifer(ous) forest	—
6008	negative staining	Negativfärbung *f*
6009	nekton, necton	Nekton *n*
6010	nemathelminthes, aschelminthes	Schlauchwürmer *mpl*, Aschelminthen *pl*, Nemathelminthen *pl obs.*
6011	nematicide	Nematizid *n*, Nematozid *n*
6012	nematocyst, cnidocyst, cnida, stinging capsule	Nesselkapsel *f*, Nematozyste *f*, Knide *f*
6013	nematodes, threadworms	Fadenwürmer *mpl*, Nematoden *pl*
6014	Nemertea, nemerteans, ribbon worms	Schnurwürmer *mpl*
6015	neoblaste	Neoblast *m*
6016	neocortex	Neocortex *m*
6017	Neo-Darwinism	Neodarwinismus *m*
6018	neoformation, new formation	Neubildung *f*
6019	Neogaea	Neogäa *f*
6020	Neogene (period)	Neogen *n*, Jungtertiär *n*
6021	Neo-Lamarckism	Neolamarckismus *m*
6022	Neolithic (age), newer stone age	Jungsteinzeit *f*, Neolithikum *n*
6023	neomorph(ic)	neomorph
6024	neopallium	Neopallium *n*, Großhirnmantel *m*
6025	neotenin → juvenile hormone	—
6026	neoteny, neoteinia	Neotenie *f*
6027	neotropical region	Neotropis *f*
6028	neotype	Neostandard *m*
6029	Neozoic → Cenozoic	—
6030	nephric → renal	—
6031	nephridium	Nephridium *n*
6032	nephrocoel	Nephrozöl *n*
6033	nephron	Nephron *n*
6034	nephropore, nephridiopore	Nephroporus *m*
6035	nephrostome	Nephrostom *n*, Wimpertrichter *m*, Flimmertrichter *m*
6036	nephrotome	Nephrotom *n*, Ursegmentstiel *m*
6037	neritic	neritisch
6038	nervation, nervature, venation	Nervatur *f*, Aderung *f*, Geäder *n (Blatt, Flügel)*
6039	nerve	Nerv *m*
6040	nerve cell, neuron(e), neurocyte	Nervenzelle *f*, Neuron *n*, Neurozyt *m*
6041	nerve centre, nervous ~ (US: center)	Nervenzentrum *n*
6042	nerve conduction, nervous ~	Nervenleitung *f*
6043	nerve cord, nervous ~	Nervenstrang *m*
6044	nerve ending	Nervenendigung *f*
6045	nerve fibre (US: fiber)	Nervenfaser *f*
6046	nerve impulse	Nervenimpuls *m*
6047	nerve net	Nervennetz *n*

5988	paysage *m* naturel	paisaje *m* natural
5989	sélection *f* naturelle	selección *f* natural
5990	substance *f* naturelle	sustancia *f* natural
5991	feuille *f* adulte, ~ secondaire	hoja *f* adulta, ~ siguiente
5992	parc *m* national	parque *m* nacional
5993	réserve *f* naturelle, ~ biologique	reserva *f* natural, ~ nacional, ~ ecológica, ~ biológica
5994	nombril *m*, ombilic *m*	ombligo *m*
5995	—	—
5996	néarctique	neártico
5997	—	—
5998	nécrobiose *f*	necrobiosis *f*
5999	—	—
6000	—	—
6001	nécrophage *m*	necrófago *m*
6002	nécrose *f*	necrosis *f*
6003	nectar *m*	néctar *m*
6004	nectaire *m*, glande *f* nectarifère, ~ nectaire	nectario *m*, glándula *f* nectarífera
6005	—	—
6006	—	—
6007	—	—
6008	coloration *f* négative	coloración *f* negativa, tinción *f* ~
6009	necton *m*	necton *m*
6010	némathelminthes *mpl*	nematelmintos *mpl*, asquelmintos *mpl*, gusanos *mpl* redondos
6011	nématicide *m*	nematicida *m*
6012	nématocyste *m*, cnidocyste *m*	nematocisto *m*, (c)nidocisto *m*, cápsula *f* urticante
6013	nématodes *mpl*	nematodos *mpl*, nemátodos *mpl*
6014	némertes *mpl*, némertiens *mpl*	nemertinos *mpl*, gusanos *mpl* en cinta
6015	néoblaste *m*	neoblasto *m*
6016	néocortex *m*	neocórtex *f*
6017	néo-darwinisme *m*	neodarwinismo *m*, neodarvinismo *m*
6018	néoformation *f*	neoformación *f*
6019	néogée *f*	neogeo *m*
6020	néogène *m*	neogeno *m*
6021	néo-lamarckisme *m*	neolamarckismo *m*, neolamarquismo *m*
6022	néolithique *m*	neolítico *m*
6023	néomorphe	neomorfo
6024	néopallium *m*	neopalio *m*
6025		—
6026	néoténie *f*	neotenia *f*
6027	région *f* néotropicale	región *f* neotrópica
6028	néotype *m*	neotipo *m*
6029	—	—
6030	—	—
6031	néphridie *f*	nefridio *m*
6032	néphrocoele *m*	nefrocele *m*
6033	néphron *m*	nefrón *m*, nefrona *f*
6034	néphridiopore *m*, pore *m* néphridien	nefridióporo *m*
6035	néphrostome *m*, entonnoir *m* cilié	nefrostoma *m*, embudo *m* ciliado
6036	néphrotome *m*	nefrotomo *m*
6037	néritique	nerítico
6038	nervation *f*, vénation *f*	nerv(i)ación *f*, nervadura *f*, venación *f*
6039	nerf *m*	nervio *m*
6040	cellule *f* nerveuse, neurone *m*, neurocyte *m*	célula *f* nerviosa, neurona *f*, neurocito *m*
6041	centre *m* nerveux, neurocentre *m*	centro *m* nervioso, neurocentro *m*
6042	conduction *f* nerveuse	conducción *f* nerviosa
6043	chaîne *f* nerveuse	cordón *m* nervioso, cuerda *f* nerviosa
6044	terminaison *f* nerveuse	terminación *f* nerviosa
6045	fibre *f* nerveuse	fibra *f* nerviosa
6046	influx *m* nerveux	impulso *m* nervioso
6047	réseau *m* nerveux	red *f* nerviosa, retículo *m* nervioso

6048	nerve node, neuroganglion	Nervenknoten *m*
6049	nerve plexus	Nervenplexus *m*, Nervengeflecht *n*
6050	nerve ring	Nervenring *m*
6051	nerve root	Nervenwurzel *f*
6052	nerve stem	Nervenstamm *m*
6053	nerve tract, nervous path(way)	Nervenbahn *f*
6054	nervous	Nerven ...
6055	nervous activity	Nerventätigkeit *f*
6056	nervous excitation	Nervenerregung *f*
6057	nervous ramification	Nervenverzweigung *f*
6058	nervous system	Nervensystem *n*
6059	nervous tissue	Nervengewebe *n*
6060	nervure → leaf vein	—
6061	to nest, to nidify, to nidificate	nisten
6062	nest building	Nestbau *m*
6063	nest-building behavio(u)r	Nestbauverhalten *n*
6064	nest-building drive	Nestbautrieb *m*
6065	nest defense	Nestverteidigung *f*
6066	nest parasitism → brood parasitism	—
6067	nest(ing) site, ~ area	Nistplatz *m*
6068	nesting	Nisten *n*
6069	nesting material	Nestmaterial *n*, Nistmaterial *n*
6070	nestling *(bird)*	Nestling *m*
6071	net production *ecol.*	Nettoproduktion *f* *ecol.*
6072	net(ted)-veined, retinerved, of reticulate nervation	netznervig
6073	neural	neural
6074	neural arch	Neuralbogen *m*
6075	neural canal	Neuralkanal *m*
6076	neural crest	Neuralleiste *f*
6077	neural fold	Neuralfalte *f*
6078	neural groove	Neuralrinne *f*
6079	neural plate	Neuralplatte *f*
6080	neural tube	Neuralrohr *n*
6081	neurapophysis, spinous process	Dornfortsatz *m*
6082	neuraxis	Neuralachse *f*
6083	neuric	nervlich
6084	neurilemma → Schwann's sheath	—
6085	neurite → axon	—
6086	neuroblast	Neuroblast *m*
6087	neurocranium	Hirnschädel *m*, Neurokranium *n*
6088	neurocrinia	Neurokrinie *f*
6089	neurocyte → nerve cell	—
6090	neuroendocrine	neuroendokrin
6091	neuro-epithelium, nerve-epithelium	Neuroepithel *n*
6092	neurofibril	Neurofibrille *f*
6093	neuroganglion → nerve node	—
6094	neurogenic, neurogenous	neurogen
6095	neuroglia, neuroglia cell → glia, gliacyte	—
6096	neurohormone	Neurohormon *n*
6097	neurohypophysis	Neurohypophyse *f*
6098	neurologist	Neurologe *m*
6099	neurology	Neurologie *f*, Nervenlehre *f*
6100	neuron(e) → nerve cell	—
6101	neuron theory	Neuronentheorie *f*
6102	neurophysiologist	Neurophysiologe *m*
6103	neurophysiology, nerve physiology	Neurophysiologie *f*, Nervenphysiologie *f*
6104	neuropile	Neuropil *n*
6105	Neuroptera, neuropterans	Netzflügler *mpl*, Neuropteren *pl*
6106	neurosecretion	Neurosekretion *f*; Neurosekret *n*
6107	neurosecretory	neurosekretorisch
6108	neurosensory	neurosensorisch
6109	neurotransmitter → transmitter	—
6110	neurovegetative	neurovegetativ

6048	nœud *m* nerveux, ~ ganglionnaire	ganglio *m* nervioso, neuroganglio *m*
6049	plexus *m* nerveux	plexo *m* nervioso
6050	anneau *m* nerveux	anillo *m* nervioso
6051	racine *f* nerveuse	raíz *f* nerviosa
6052	tronc *m* nerveux	tronco *m* nervioso
6053	voie *f* nerveuse	vía *f* nerviosa, tracto *m* nervioso
6054	nerveux	nervioso
6055	activité *f* nerveuse	actividad *f* nerviosa
6056	excitation *f* nerveuse	excitación *f* nerviosa
6057	ramification *f* nerveuse	ramificación *f* nerviosa
6058	système *m* nerveux	sistema *m* nervioso
6059	tissu *m* nerveux	tejido *m* nervioso
6060	—	—
6061	nidifier, nicher	nidificar, anidar
6062	construction *f* du nid	construcción *f* del nido
6063	comportement *m* nidificateur	comportamiento *m* de construcción del nido, actividad *f* nidificadora
6064	instinct *m* nidificateur	instinto *m* nidificador
6065	défense *f* du nid	defensa *f* del nido
6066	—	—
6067	lieu *m* de nidification, aire *f* ~ ~, site *m* du nid	área *f* de nidificación, lugar *m* de anidamiento, emplazamiento *m* del nido
6068	nidification *f*	nidificación *f*, anidación *f*
6069	matériaux *mpl* du nid	material *m* del nido
6070	jeune oiseau *m* au nid	ave *f* mientras está en el nido
6071	production *f* nette *ecol.*	producción *f* neta *ecol.*
6072	à nervation réticulée, rétinerve	de nerviación (o: nervadura) reticulada
6073	neural	neural
6074	arc *m* neural	arco *m* neural
6075	canal *m* neural	canal *m* neural, conducto *m* ~
6076	crête *f* neurale	cresta *f* neural
6077	pli *m* neural	pliegue *m* neural
6078	gouttière *f* neurale	surco *m* neural
6079	plaque *f* neurale	placa *f* neural
6080	tube *m* neural	tubo *m* neural
6081	apophyse *f* épineuse	apófisis *f* espinosa, neurapófisis *f*
6082	axe *m* neural, névraxe *m*	neuraxis *m*, neuroeje *m*
6083	névrique	néurico
6084	—	—
6085	—	—
6086	neuroblaste *m*	neuroblasto *m*
6087	neurocrâne *m*	neurocráneo *m*
6088	neurocrinie *f*	neurocrinia *f*
6089	—	—
6090	neuroendocrine	neuroendocrino
6091	neuro-épithélium *m*	neuroepitelio *m*, epitelio *m* neural
6092	neuro-fibrille *f*	neurofibrilla *f*
6093	—	—
6094	neurogène	neurogénico
6095	—	—
6096	neurohormone *f*	neurohormona *f*
6097	neurohypophyse *f*	neurohipófisis *f*
6098	neurologiste *m*, neurologue *m*	neurólogo *m*
6099	neurologie *f*	neurología *f*
6100	—	—
6101	doctrine *f* du neurone	teoría *f* de la neurona, doctrina *f* ~ ~ ~
6102	neurophysiologiste *m*	neurofisiólogo *m*
6103	neurophysiologie *f*, physiologie *f* nerveuse	neurofisiología *f*
6104	neuropile *m*	neuropilo *m*, neurópilo *m*
6105	neuroptères *mpl*	neurópteros *mpl*
6106	neurosécrétion *f*	neurosecreción *f*
6107	neurosécréteur, neurosécrétoire	neurosecretor
6108	neurosensoriel	neurosensorial
6109	—	—
6110	neurovégétatif	neurovegetativo

6111	neurula	Neurula *f*
6112	neurulation	Neurulation *f*
6113	neuston	Neuston *n*
6114	neutron	Neutron *n*
6115	neutrophil(ic)	neutrophil
6116	niacin → nicotinic acid	—
6117	nicotinamide, niacinamide, nicotinic acid amide, antipellagrous vitamin, pellegra-preventive factor, P.P. factor	Nikotin(säure)amid *n*, Niazinamid *n*, Antipellagravitamin *n*
6118	nicotinamide adenine dinucleotide, NAD, diphosphopyridine nucleotide *obs.*, DPN	Nikotinamid-Adenin-Dinukleotid *n*, NAD, Diphosphopyridin-Nukleotid *n obs.*, DPN
6119	nicotinamide adenine dinucleotide phosphate, NADP, triphosphopyridine nucleotide *obs.*, TPN	Nikotinamid-Adenin-Dinukleotidphosphat *n*, NADP, Triphosphopyridinnukleotid *n obs.*, TPN
6120	nicotine	Nikotin *n*
6121	nicotinic acid, niacin	Nikotinsäure *f*, Niazin *n*
6122	nictitating membrane	Blinzhaut *f*, Nickhaut *f*
6123	nidamental gland	Nidamentaldrüse *f*
6124	nidation, implantation	Nidation *f*, Einnistung *f*
6125	nidicolous bird, altricial ~	Nesthocker *m*
6126	nidifugous bird, precocial ~	Nestflüchter *m*
6127	to nidify → to nest	—
6128	night flowering plant	Nachtblüher *m*
6129	nipple → mamilla	—
6130	nitrification	Nitrifikation *f*, Nitrifizierung *f*
6131	to nitrify	nitrifizieren
6132	nitrifying bacteria	nitrifizierende Bakterien *fpl*, stickstoffaufbauende ~, Nitrifikanten *mpl*
6133	nitrogen balance	Stickstoffbilanz *f*
6134	nitrogen base, nitrogenous ~	Stickstoffbase *f*
6135	nitrogen compound	Stickstoffverbindung *f*
6136	nitrogen cycle	Stickstoffzyklus *m*, ~kreislauf *m*
6137	nitrogen equilibrium	Stickstoffgleichgewicht *n*
6138	nitrogen fixation	Stickstoffbindung *f*, ~fixierung *f*
6139	nitrogen-fixing bacteria, nitrogen fixer	stickstoffbindende Bakterien *fpl*, Stickstoffbinder *mpl*
6140	nitrogen metabolism	Stickstoff-Stoffwechsel *m*
6141	nitrogenous	stickstoffhaltig
6142	nitrogenous wastes	Stickstoffabfälle *mpl*
6143	nitrous acid	salpetrige Säure *f*
6144	nociceptive stimulus	Schmerzreiz *m*
6145	nociceptor	Nozizeptor *m*
6146	nocturnal animal	Nachttier *n*
6147	nocuity, harmfulness	Schädlichkeit *f*
6148	node, nodus, joint *bot.*	Knoten *m bot.*, Stengelknoten *m*
6149	node of Ranvier → Ranvier's node	—
6150	nodule bacteria	Knöllchenbakterien *fpl*
6151	noise control, anti-noise campaign	Lärmbekämpfung *f*
6152	noise trouble → nuisance noise	—
6153	nonallelic	nicht allel
6154	noncellular → acellular	—
6155	noncompetitive inhibition	nichtkompetitive Hemmung *f*
6156	noncyclic photophosphorylation	nichtzyklische Photophosphorylierung *f*
6157	nondeciduous, indeciduous	nichtlaubabwerfend
6158	non-disjunction	Non-Disjunktion *f*
6159	nonessential amino acid	nichtessentielle Aminosäure *f*
6160	non-linear regression	nichtlineare Regression *f*
6161	non-medullated → amyelinate	—
6162	non-mendelian inheritance	nichtmendelnde Vererbung *f*
6163	nonmotile	unbeweglich
6164	non-nucleated, anuclear, akaryotic	kernlos
6165	nonoverlapping code	nichtüberlappender Kode *m*
6166	nonparental ditype tetrad, NPD	Nonparental-Dityp-Tetrade *f*
6167	nonpoisonous → nontoxic	—
6168	nonprotein nitrogen → rest nitrogen	—

6111	neurula *f*	neurula *f*, néurula *f*
6112	neurulation *f*	neurulación *f*
6113	neuston *m*	neuston *m*
6114	neutron *m*	neutrón *m*
6115	neutrophile	neutrófilo
6116	—	—
6117	nicotinamide *m*, niacinamide *m*, amide *m* nicotinique, vitamine *f* antipellagreuse, ~ PP	nicotinamida *f*, niacinamida *f*, vitamina *f* PP, factor *m* antipelagroso, ~ PP
6118	nicotinamide-adénine-dinucléotide *m*, NAD, diphosphopyridine-nucléotide *m obs.*, DPN	nicotinamida-adenina-dinucleótido *m*, NAD, difosfopiridina-nucleótido *m obs.*, DPN
6119	nicotinamide-adénine-dinucléotide-phosphate *m*, NADP, triphosphopyridine-nucléotide *m obs.*, TPN	nicotinamida-adenina-dinucleótido-fosfato *m*, NADP, trifosfopiridina-nucleótido *m obs.*, TPN
6120	nicotine *f*	nicotina *f*
6121	acide *m* nicotinique, niacine *f*	ácido *m* nicotínico, niacina *f*
6122	paupière *f* nictitante, membrane *f* ~	membrana *f* nictitante
6123	glande *f* nidamentaire	glándula *f* nidamental
6124	nidation *f*, implantation *f*	nidación *f*, implantación *f*
6125	nidicole *m*	ave *f* nidícola, ~ altricial
6126	nidifuge *m*	ave *f* nidífuga
6127	—	—
6128	plante *f* à floraison nocturne	planta *f* noctíflora, ~ de floración nocturna
6129	—	—
6130	nitrification *f*	nitrificación *f*
6131	nitrifier	nitrificar
6132	bactéries *fpl* nitrifiantes	bacterias *fpl* nitrificantes
6133	bilan *m* azoté, balance *f* azotée	balance *m* nitrogenado
6134	base *f* azotée	base *f* nitrogenada
6135	combiné *m* azoté	compuesto *m* nitrogenado
6136	cycle *m* de l'azote	ciclo *m* del nitrógeno
6137	équilibre *m* azoté	equilibrio *m* nitrogenado
6138	fixation *f* de l'azote	fijación *f* de nitrógeno
6139	bactéries *fpl* fixatrices d'azote	bacterias *fpl* fijadoras de nitrógeno
6140	métabolisme *m* azoté	metabolismo *m* nitrogenado, ~ del nitrógeno
6141	azoté	nitrogenado
6142	déchets *mpl* azotés	desechos *mpl* nitrogenados, residuos *mpl* ~
6143	acide *m* azoteux	ácido *m* nitroso
6144	stimulus *m* nociceptif, ~ nocicepteur	estímulo *m* nociceptivo
6145	nocicepteur *m*, récepteur *m* nociceptif	nociceptor *m*
6146	animal *m* nocturne	animal *m* nocturno
6147	nocivité *f*, nuisibilité *f*	nocividad *f*, perniciosidad *f*
6148	nœud *m* (caulinaire)	nudo *m* (caulinar)
6149	—	—
6150	bactéries *fpl* des nodosités, ~ des nodules	bacterias *fpl* nodulares, ~ de las nudosidades
6151	lutte *f* contre le bruit	lucha *f* contra el ruido, control *m* de los ruidos
6152	—	—
6153	non-allélique	no alélico
6154	—	—
6155	inhibition *f* non compétitive	inhibición *f* no competitiva
6156	photophosphorylation *f* non cyclique	fotofosforilación *f* no cíclica
6157	à feuillage persistant	no deciduo, perennifolio, de hoja persistente
6158	non-disjonction *f*	no disyunción *f*
6159	amino-acide *m* nonessentiel	aminoácido *m* no esencial
6160	régression *f* non-linéaire	regresión *f* no lineal
6161	—	—
6162	hérédité *f* non-mendélienne	herencia *f* no mendeliana
6163	nonmotile	inmóvil, no móvil
6164	anucléé, anucléaire, acaryote	anucleado, no nucleado, acario
6165	code *m* non chevauchant	código *m* no sobrepuesto
6166	tétrade *f* ditype non-parentale	tétrade *f* dítipo no progenitor
6167	—	—
6168	—	—

6169	non-reduction	Nichtreduktion *f*
6170	nonseptate, aseptate	nichtseptiert, unseptiert
6171	non-sexual → asexual	—
6172	non-sister chromatid	Nichtschwester-Chromatide *f*
6173	nonspecies member	Artfremder *m*
6174	nonspecific	unspezifisch
6175	nontoxic, nonpoisonous, atoxic, poisonless	ungiftig
6176	nontoxicity	Ungiftigkeit *f*
6177	nonvascular	gefäßlos
6178	nonviable, inviable	lebensunfähig
6179	noradrenaline, norepinephrine, arterenol	Noradrenalin *n*, Arterenol *n*
6180	norm of reaction	Reaktionsnorm *f*
6181	nostrils	Nüstern *f pl*
6182	notochord, chorda dorsalis	Rückensaite *f*, Chorda dorsalis
6183	Notogaea	Notogäa *f*
6184	notomorph	notomorph
6185	notthreshold substance	Nichtschwellensubstanz *f*, ~stoff *m*
6186	nourishment → nutrition	—
6187	noxious, deleterious, harmful	schädlich
6188	noxious insect → insect pest	—
6189	noxious substance	Schadstoff *m*
6190	nucellus	Nuzellus *m*, Nucellus *m*
6191	nuclear cap	Kernkappe *f*
6192	nuclear cytology → karyology	—
6193	nuclear dimorphism	Kerndimorphismus *m*
6194	nuclear division	Kernteilung *f*
6195	nuclear energy → atomic energy	—
6196	nuclear envelope	Kernhaut *f*
6197	nuclear explosion	Kernexplosion *f*
6198	nuclear fission	Kernspaltung *f*
6199	nuclear fragmentation	Kernfragmentation *f*
6200	nuclear fusion	Kernverschmelzung *f*
6201	nuclear gene	Kerngen *n*
6202	nuclear membrane, karyotheca	Kernmembran *f*, Karyotheka *f*
6203	nuclear plate → equatorial plate	—
6204	nuclear pore	Kernpore *f*
6205	nuclear radiation	Kernstrahlung *f*
6206	nuclear reactor	Kernreaktor *m*
6207	nuclear sap, karyolymph, karyenchyma	Kernsaft *m*, Karyolymphe *f*, Karyoenchym *n*
6208	nuclear spindle	Kernspindel *f*
6209	nuclear stain, ~ dye	Kernfarbstoff *m*
6210	nuclear staining	Kernfärbung *f*
6211	nuclear transplantation	Kernüberpflanzung *f*, ~transplantation *f*
6212	nuclease	Nuklease *f*
6213	nucleated	kernhaltig
6214	nucleic acid	Nukleinsäure *f*
6215	nuclein	Nuklein *n*
6216	nucleination	Nukleinisierung *f*
6217	nucleo-(cyto)plasmic ratio, karyoplasmic ~	Kern-Plasma-Relation *f*, ~ ~ Verhältnis *n*
6218	nucleohistone	Nukleohiston *n*
6219	nucleoid	Nukleoid *n*
6220	nucleolar constriction, secondary ~	Nukleolareinschnürung *f*, Sekundär~
6221	nucleolar zone, SAT-zone *obs.*	Nukleolarzone *f*, SAT-Zone *f obs.*
6222	nucleolonema	Nukleolonema *n*
6223	nucleolus	Nukleolus *m*, Kernkörperchen *n*
6224	nucleolus organizer region, NOR	Nukleolus-Organisator-Region *f*, NOR
6225	nucleoplasm, karyoplasm	Nukleoplasma *n*, Kernplasma *n*, Karyoplasma *n*
6226	nucleoplasm(at)ic, karyoplasmic	kernplasmatisch
6227	nucleoprotein	Nukleoprotein *n*, Kerneiweißkörper *m*
6228	nucleosid(e)	Nukleosid *n*
6229	nucleosome → karyosome	—
6230	nucleotid(e)	Nukleotid *n*

6169	non-réduction *f*	no reducción *f*
6170	non cloisonné	no tabicado, no septado
6171	—	—
6172	chromatide *f* non-sœur	cromátida *f* no hermana
6173	étranger *m* à l'espèce, membre *m* d'une espèce étrangère	extraño *m* a la especie
6174	nonspécifique	no específico
6175	atoxique, nontoxique	atóxico, no tóxico
6176	non-toxicité *f*	atoxicidad *f*
6177	non vasculaire	no vascular
6178	nonviable, inviable	no viable, inviable
6179	noradrénaline *f*, norépinéphrine *f*	noradrenalina *f*, arterenol *m*
6180	norme *f* de réaction	norma *f* de reacción
6181	naseaux *mpl*	ollares *mpl*
6182	notoc(h)orde *f*, c(h)orde *f* dorsale	notocord(i)o *m*, notocorda *f*
6183	notogée *f*	notogeo *m*
6184	notomorphe	notomorfo
6185	substance *f* sans seuil	sustancia *f* no liminar
6186	—	—
6187	nocif, nuisible, délétère	nocivo, dañino, deleterio, perjudicial
6188	—	—
6189	substance *f* nocive, ~ nuisible	sustancia *f* nociva
6190	nucelle *m*	nucelo *m*, nucela *f*
6191	calotte *f* nucléaire	casquete *m* nuclear
6192	—	—
6193	dimorphisme *m* nucléaire	dimorfismo *m* nuclear
6194	division *f* nucléaire	división *f* nuclear
6195	—	—
6196	enveloppe *f* nucléaire	envoltura *f* nuclear
6197	explosion *f* nucléaire	explosión *f* nuclear
6198	fission *f* nucléaire	fisión *f* nuclear
6199	fragmentation *f* nucléaire	fragmentación *f* nuclear
6200	fusion *f* nucléaire	fusión *f* nuclear
6201	gène *m* nucléaire	gen *m* nuclear
6202	membrane *f* nucléaire, caryothèque *f*	membrana *f* nuclear, carioteca *f*
6203	—	
6204	pore *m* nucléaire	poro *m* nuclear
6205	rayonnement *m* nucléaire	radiación *f* nuclear
6206	réacteur *m* nucléaire	reactor *m* nuclear
6207	suc *m* nucléaire, caryolymphe *f*	jugo *m* nuclear, cariolinfa *f*
6208	fuseau *m* nucléaire	huso *m* nuclear
6209	colorant *m* nucléaire	colorante *m* nuclear
6210	coloration *f* nucléaire	coloración *f* nuclear
6211	transplantation *f* nucléaire, greffage *m* ~	trasplante *m* nuclear
6212	nucléase *f*	nucleasa *f*
6213	nucléé	nucleado
6214	acide *m* nucléique	ácido *m* nucleico
6215	nucléine *f*	nucleína *f*
6216	nucléisation *f*	nucleinación *f*
6217	relation *f* nucléo-plasm(at)ique, ~ caryoplasmique, rapport *m* N/P	relación *f* nucleoplásmica, ~ nucleoplasmática
6218	nucléohistone *f*	nucleohistona *f*
6219	nucléoïde *m*	nucleoide *m*
6220	constriction *f* nucléolaire, ~ secondaire	constricción *f* nucleolar, ~ secundaria
6221	zone *f* nucléolaire, olisthérozone *f*, zone *f* SAT *obs*	zona *f* nucleolar, ~ SAT *obs.*
6222	nucléolonème *m*	nucleolonema *m*
6223	nucléole *m*	nucléolo *m*
6224	région *f* de l'organisateur nucléolaire, RON	región *f* del organizador nucleolar, RON
6225	nucléoplasme *m*, caryoplasme *m*	nucleoplasma *m*, carioplasma *m*
6226	nucléo-plasm(at)ique, caryoplasmique	nucleoplásmico, nucleoplasmático
6227	nucléoprotéine *f*, protéine *f* nucléaire	nucleoproteína *f*
6228	nucléoside *m*	nucleósido *m*
6229	—	—
6230	nucléotide *m*	nucleótido *m*, nucleótida *f*

6231	nucleus (*pl* nuclei), cell ~, karyon	Zellkern *m*, Nukleus *m*, Karyon *n*
6232	nuisance noise, noise trouble, ~ pollution	Lärmbelästigung *f*
6233	nulliplex	nulliplex
6234	nullisomic, nullosomic	nullisom, nullosom
6235	nullisomy, nullosomy	Nullisomie *f*, Nullosomie *f*
6236	number of individuals	Individuenzahl *f*
6237	numeric constancy of chromosomes	Chromosomenzahlenkonstanz *f*
6238	numerical hybrid	numerische Hybride *f*, Chromosomenzahlenbastard *m*
6239	nuptial dress, breeding plumage, reproductive colo(u)r	Hochzeitskleid *n*, Brutkleid *n*
6240	nuptial flight, mating ~	Hochzeitsflug *m*
6241	nurse cell → nutritive cell	—
6242	nut	Nuß *f*, Nußfrucht *f*
6243	nutation	Nutation *f*
6244	nutricism	Nutritismus *m*
6245	nutrient, nutritive substance	Nährstoff *m*
6246	nutrient agar	Nähragar *m/n*
6247	nutrient bouillon, broth	Nährbouillon *f*
6248	nutrient content	Nährstoffgehalt *m*
6249	nutrient deficiency, nutritive ~	Nährstoffmangel *m*
6250	nutrient medium	Nährboden *m*, Nährmedium *n*
6251	nutrient needs, ~ requirements, nutritive ~	Nährstoffbedarf *m*
6252	nutrient solution, culture ~	Nährlösung *f*
6253	nutrition, nourishment	Ernährung *f*
6254	nutritional mutant	Mangelmutante *f*
6255	nutritive cell, nurse ~, trophocyte	Nährzelle *f*, Trophozyt *m*
6256	nutritive factor, nutritional ~, food ~	Ernährungsfaktor *m*, Nahrungsfaktor *m*, trophischer Faktor *m*
6257	nutritive rate, nutrient ratio	Nährstoffverhältnis *n*
6258	nutritive reserve → food ~	—
6259	nutritive salt	Nährsalz *n*
6260	nutritive substrate	Nährsubstrat *n*
6261	nutritive tissue	Nährgewebe *n*
6262	nutritive value	Nährwert *m*
6263	nutritive yolk → deutoplasm	—
6264	nyctanthous, night-blooming	nachtblühend
6265	nyctinastic	nyktinastisch
6266	nyctinasty, nyctinastic movement, sleep ~	Nyktinastie *f*, nyktinastische Bewegung *f*, Schlafbewegung *f*
6267	nymph	Nymphe *f*
6268	nymphal mo(u)lt	Nymphenhäutung *f*
6269	nymphal stage	Nymphenstadium *n*
6270	nymphosis → pupation	—
6271	obdiplostemonous	obdiplostemon
6272	objective *micr.*	Objektiv *n* *micr.*
6273	obligate parasite	obligater Parasit *m*
6274	obtect pupa	Mumienpuppe *f*, bedeckte Puppe *f*
6275	occipital bone	Hinterhauptsbein *n*
6276	occipital condyle	Hinterhauptshöcker *m*
6277	occipital lobe	Hinterhauptslappen *m*
6278	occiput	Hinterhaupt *n*
6279	oceanography	Ozeanographie *f*, Meereskunde *f*
6280	ocellus (*pl* ocelli)	Ozelle *f*, Punktauge *n*, Nebenauge *n*, Stirnauge *f*
6281	oculomotor nerve	Augenmuskelnerv *m*, Okulomotorius *m*
6282	odontoblast	Odontoblast *m*
6283	odoriferous gland, scent ~	Duftdrüse *f*
6284	odoriferous substance, odorous ~	Duftstoff *m*, Riechstoff *m*
6285	odo(u)r mark, scent ~	Duftmarke *f*
6286	odo(u)r trail, scent ~	Duftspur *f*

6231	nucléus *m*, noyau *m* cellulaire	núcleo *m* (celular)
6232	nuisance *f* acoustique, ~ sonore	contaminación *f* sonora, ~ sónica
6233	nulliplex	nuliplexo
6234	nullisomique, nullosomique	nulisómico
6235	nullisomie *f*, nullosomie *f*	nulisomia *f*
6236	nombre *m* des individus	número *m* de individuos
6237	constance *f* numérique des chromosomes	constancia *f* numérica de los cromosomas
6238	hybride *m* numérique	híbrido *m* numérico
6239	habit *m* nuptial, livrée *f* nuptiale, robe *f* de noce	traje *m* nupcial, librea *f* ~, plumaje *m* ~
6240	vol *m* nuptial	vuelo *m* nupcial
6241	—	—
6242	noix *f*	nuez *f*
6243	nutation *f*	nutación *f*
6244	nutricisme *m*	nutricismo *m*
6245	substance *f* nutritive, élément *m* nutritif	sustancia *f* nutritiva, nutriente *m*
6246	gélose *f* nutritive	agar *m* nutritivo, ~ de cultivo
6247	bouillon *m* de culture	caldo *m* de cultivo, ~ nutritivo
6248	teneur *f* nutritive, ~ en substances nutritives	contenido *m* nutritivo, ~ en sustancias nutritivas
6249	carence *f* en substances nutritives	carencia *f* en sustancias nutritivas
6250	milieu *m* nutritif	medio *m* nutritivo
6251	besoin *m* nutritif, ~ en substances nutritives	exigencias *f pl* (de sustancias) nutritivas, necesidades *f pl* (~ ~) ~
6252	solution *f* nutritive, liquide *m* de culture	solución *f* nutritiva, ~ de cultivo
6253	nutrition *f*, alimentation *f*	nutrición *f*, alimentación *f*
6254	mutant *m* nutritionnel, ~ auxotrophe	mutante *f* nutricional, ~ auxotrófica
6255	cellule *f* nourricière, ~ nutritive, trophocyte *m*	célula *f* nutritiva, trofocito *m*
6256	facteur *m* nutritif, ~ alimentaire	factor *m* alimentario, ~ nutricional
6257	relation *f* nutritive	relación *f* nutritiva
6258	—	—
6259	sel *m* nutritif	sal *f* nutritiva, ~ alimenticia
6260	substrat *m* nutritif	substrato *m* nutritivo
6261	tissu *m* nourricier	tejido *m* nutritivo, ~ nutricio
6262	valeur *f* nutritive	valor *m* nutritivo
6263	—	—
6264	à floraison nocturne	nictanto, de floración nocturna
6265	nyctinastique	nictinástico
6266	nyctinastie *f*, mouvement *m* nyctinastique	nictinastia *f*, movimiento *m* nictinástico, ~ de sueño
6267	nymphe *f*	ninfa *f*
6268	mue *f* nymphale	muda *f* ninfal
6269	stade *m* nymphal	estadio *m* ninfal
6270	—	—
6271	obdiplostémone	obdiplostémono
6272	objectif *m micr.*	objetivo *m micr.*
6273	parasite *m* obligé, ~ obligatoire	parásito *m* obligado, ~ obligatorio, ~ forzado
6274	chrysalide *f* obtectée	pupa *f* obtecta
6275	os *m* occipital	hueso *m* occipital
6276	condyle *m* occipital	cóndilo *m* occipital
6277	lobe *m* occipital	lóbulo *m* occipital
6278	occiput *m*	occipucio *m*
6279	océanographie *f*	oceanografía *f*
6280	ocelle *m*	ocelo *m*
6281	nerf *m* moteur oculaire (commun)	nervio *m* oculomotor
6282	odontoblaste *m*	odontoblasto *m*
6283	glande *f* odorante, ~ odoriférante	glándula *f* odorífica, ~ odorípara
6284	substance *f* odorante, ~ odoriférante	sustancia *f* olorosa, ~ odorífera
6285	marque *f* odorante	marca *f* olorosa
6286	piste *f* odorante	pista *f* olorosa

6287	oenocyte	Önozyte *f*
6288	(o)esophagus	Speiseröhre *f*, Ösophagus *m*
6289	(o)estradiol	Östradiol *n*
6290	(o)estrane	Östran *n*
6291	(o)estrin → (o)estrone	—
6292	(o)estriol	Östriol *n*
6293	(o)estrogen, (o)estrogenic hormone	Östrogen *n*
6294	(o)estrone, (o)estrin, folliculin, follicular hormone	Östron *n*, Follikelhormon *n*
6295	(o)estrous cycle, heat ~	Brunstzyklus *m*, Östralzyklus *m*
6296	(o)estrous period, heat ~, rutting season	Brunstzeit *f*
6297	(o)estrus, (o)estrum, heat, rut	Brunst *f*, Östrus *m*; Brunft *f* (*Wild*)
6298	to be in (o)estrus, to be in heat, to rut	brünstig (sein), brunsten; brunften (*Wild*)
6299	offspring → progeny	—
6300	oil duct	Ölgang *m*
6301	oil gland → uropygial ~	—
6302	oil immersion	Ölimmersion *f*
6303	oil pollution	Ölverschmutzung *f*; Ölpest *f* *pop.*
6304	oil slick	Ölteppich *m*
6305	olecranon	Ellbogenhöcker *m*
6306	oleic acid	Oleinsäure *f*, Ölsäure *f*
6307	olein	Olein *n*
6308	olfaction	Riechen *n*
6309	olfactory	Riech ..., Geruchs ...
6310	olfactory brain, ~ bulb → rhinencephalon	—
6311	olfactory lobe	Riechlappen *m*
6312	olfactory nerve	Riechnerv *m*, Olfaktorius *m*
6313	olfactory organ	Riechorgan *n*, Geruchsorgan *n*
6314	olfactory perception	Geruchswahrnehmung *f*
6315	olfactory receptor, smell ~	Geruchsrezeptor *m*
6316	olfactory sensation	Geruchsempfindung *f*
6317	olfactory sense, sense of smell	Geruch(s)sinn *m*
6318	olfactory stimulus	Geruchsreiz *m*
6319	Oligocene	Oligozän *n*
6320	Oligochaeta	Wenigborster *mpl*
6321	oligodendroglia	Oligodendroglia *f*
6322	oligogene, key gene, major ~	Oligogen *n*, Hauptgen *n*
6323	oligolecithal	oligolezithal, dotterarm
6324	oligosaccharide	Oligosac(c)harid *n*
6325	oligospermous, few-seeded	wenigsamig
6326	oligotrophic	oligotroph
6327	omasum, psalterium, manyplies	Blättermagen *m*
6328	ombrophil(ous), rain-loving	ombrophil, regenliebend
6329	ombrophil(ous) plant	Regenpflanze *f*, ombrophile Pflanze *f*
6330	omentum, epiploon	Netz *n* *anat.*
6331	ommatidium (*pl* ommatidia)	Ommatidium *n*, Augenkeil *m*, Sehkeil *m*
6332	ommochrome, ommatochrome	Ommochrom *n*
6333	omnipotency → totipotency	—
6334	omnivore	Allesfresser *m*, Omnivore *m*
6335	omnivorous	allesfressend, omnivor
6336	omphalic → umbilical	—
6337	one gene one enzyme hypothesis	Ein-Gen-ein-Enzym-Hypothese *f*
6338	ontogenesis, ontogeny	Ontogenese *f*, Ontogenie *f*, Entwicklungsgeschichte *f*
6339	ontogenetic	ontogenetisch, entwicklungsgeschichtlich
6340	Onychophora, onychophores, onychophorans	Stummelfüßer *mpl*
6341	oocyte, ovocyte, oöcyte	Oozyte *f*
6342	oogamous	oogam
6343	oogamy	Oogamie *f*, Eibefruchtung *f*
6344	oogenesis, ovogenesis, oögenesis	Oogenese *f*, Ovogenese *f*, Eizellenbildung *f*
6345	oogonium, ovogonium	Oogonium *n*, Ureizelle *f*
6346	ookinesis	Ookinesis *f*
6347	oolemma → vitelline membrane	—
6348	ooplasm, ovoplasm	Ooplasma *n*, Bildungsdotter *m*

6287	oenocyte *m*	enocito *m*
6288	oesophage *m*	esófago *m*
6289	(o)estradiol *m*	estradiol *m*
6290	(o)estrane *f*	estrano *m*
6291	—	—
6292	(o)estriol *m*	estriol *m*
6293	oestrogène *m*, hormone *f* ~	estrógeno *m*, hormona *f* estrógena
6294	(o)estrone *f*, folliculine *f*, hormone *f* folliculaire	estrona *f*, estrina *f*, foliculina *f*, hormona *f* folicular
6295	cycle *m* œstral, ~ œstrien	ciclo *m* estral, ~ del estro, ~ del celo
6296	période *f* des chaleurs, ~ du rut	período *m* del celo, ~ de calores
6297	œstrus *m*, rut *m*, chaleur(s) *f* (*pl*)	estro *m*, celo *m*, calores *m pl*
6298	être en chaleur, ~ ~ rut	estar en celo, marcar el ~, encelarse
6299	—	—
6300	canal *m* oléifère	conducto *m* oleífero, canal *m* oleoso
6301	—	—
6302	immersion *f* d'huile	inmersión *f* en aceite
6303	pollution *f* par les hydrocarbures, peste *f* huileuse *pop.*	polución *f* por hidrocarburos, contaminación *f* petrolífera
6304	nappe *f* d'huile, film *m* ~	capa *f* de aceite
6305	olécrâne *m*	olécranon *m*
6306	acide *m* oléique	ácido *m* oleico
6307	oléine *f*	oleína *f*
6308	olfaction *f*	olfacción *f*
6309	olfactif	olfativo, olfatorio
6310	—	—
6311	lobe *m* olfactif	lóbulo *m* olfativo
6312	nerf *m* olfactif	nervio *m* olfativo
6313	organe *m* olfactif, ~ de l'olfaction	órgano *m* olfatorio
6314	perception *f* olfactive	percepción *f* olfativa
6315	récepteur *m* olfactif	receptor *m* olfativo, olfatorreceptor *m*
6316	sensation *f* olfactive	sensación *f* olfativa
6317	odorat *m*, sens *m* de l'odorat, ~ olfactif	olfato *m*, sentido *m* del olfato
6318	stimulus *m* olfactif	estímulo *m* olfativo
6319	oligocène *m*	oligoceno *m*
6320	oligochètes *m pl*	oligoquetos *m pl*
6321	oligodendroglie *f*	oligodendroglía *f*
6322	gène-clé *m*, oligogène *m*, gène *m* majeur	oligogen *m*, gen *m* mayor
6323	oligolécithe, oligolécithique	oligolecito, oligolecítico
6324	oligosaccharide *m*, oligoside *m*	oligosacárido *m*, oligósido *m*
6325	oligosperme	oligospermo
6326	oligotrophe, oligotrophique	oligotrófico
6327	feuillet *m*, omasum *m*	omaso *m*, libro *m*, librillo *m*
6328	ombrophile	ombrófilo
6329	plante *f* ombrophile	ombrófito *m*, planta *f* ombrófila
6330	épiploon *m*	epiplón *m*, omento *m*
6331	ommatidie *f*	omatidio *m*
6332	ommochrome *m*	ommocromo *m*
6333	—	—
6334	omnivore *m*	omnívoro *m*
6335	omnivore	omnívoro
6336	—	—
6337	hypothèse *f* «un gène, un enzyme», ~ d'un gène pour un enzyme	hipótesis *f* de un gen por una enzima
6338	ontogenèse *f*, ontogénie *f*	ontogénesis *f*, ontogenia *f*
6339	ontogénétique, ontogénique	ontogenético, ontogénico
6340	onychophores *m pl*, péripates *m pl*	onicóforos *m pl*
6341	oocyte *m*, ovocyte *m*	oocito *m*, oócito *m*, ovocito *m*
6342	oogame	oógamo
6343	oogamie *f*	oogamia *f*
6344	oogenèse *f*, ovogenèse *f*, ovogénie *f*	oogénesis *f*, ovogénesis *f*
6345	oogone *m*, oogonie *f*, ovogonie *f*, ovigerme *m*	oogonio *m*, ovogonio *m*, oogonia *f*
6346	oocinèse *f*	oocinesis *f*
6347	—	—
6348	ovoplasme *m*	ooplasma *m*, ovoplasma *m*, vitelo *m* formativo

6349 oosome — Oosom *n*
6350 oosphere — Oosphäre *f*
6351 oospore — Oospore *f*
6352 ootid — Ootide *f*, Reifei *n*
6353 opaque — lichtundurchlässig
6354 opaque area → area opaca — —
6355 operator gene — Operatorgen *n*, Operator *m*
6356 operculum (*pl* opercula) — Operkulum *n*
6357 operon — Operon *n*
6358 opposite *bot.* — gegenständig
6359 opsonin, bacteriotropin — Opsonin *n*, Bakteriotropin *n*
6360 optic chiasma — optisches Chiasma *n*
6361 optic cup, eye ~ — Augenbecher *m*
6362 optic nerve — Sehnerv *m*, Optikus *m*
6363 optic stalk → eye-stalk — —
6364 optic vesicle, ophthalmic ~ — Augenbläschen *n*, Blasenauge *n*, Bläschenauge *n*
6365 optical density — optische Dichte *f*
6366 optical microscope → light microscope — —
6367 optical rod → rhabdom — —
6368 optical stimulus → visual stimulus — —
6369 optimum temperature, optimal ~ — Optimaltemperatur *f*
6370 oral → buccal — —
6371 orbit — Augenhöhle *f*
6372 order *tax.* — Ordnung *f* *tax.*
6373 Ordovician (period) — Ordovizium *n*
6374 organ culture — Organkultur *f*
6375 organ of sight, ~ ~ vision — Sehorgan *n*, Gesichtsorgan *n*, Lichtsinnesorgan *n*
6376 organ system — Organsystem *n*
6377 organelle, organella — Organell *n*, Organelle *f*
6378 organic — organisch
6379 organism — Organismus *m*
6380 organization centre, organizing ~ — Organisationszentrum *n*
6381 organizational level, level of organization — Organisationshöhe *f*, Organisationsebene *f*
6382 organizer *embr.* — Organisator *m* *embr.*
6383 organogenesis, organogeny — Organogenese *f*, Organbildung *f*
6384 organogenetic, organogenic — organogen, organbildend
6385 organography — Organographie *f*
6386 organoid — organoid, organähnlich
6387 organology — Organologie *f*, Organlehre *f*
6388 organotroph(ic) — organotroph
6389 orientation — Orientierung *f*
6390 orientation movement, orienting ~ — Orientierungsbewegung *f*
6391 orientation response — Orientierungsreaktion *f*
6392 original population, initial ~ — Ausgangspopulation *f*, Ursprungspopulation *f*
6393 ornithine — Ornithin *n*
6394 ornithine cycle, urea ~ — Ornithinzyklus *m*, Harnstoffzyklus *m*
6395 ornithogamy → ornithophily — —
6396 ornithologist — Ornithologe *m*
6397 ornithology — Ornithologie *f*, Vogelkunde *f*
6398 ornithophilous — vogelblütig, ornithophil
6399 ornithophilous flower — Vogelblume *f*
6400 ornithophily, ornithogamy, bird pollination — Ornithogamie *f*, Ornithophilie *f*, Vogelblütigkeit *f*, Vogelbestäubung *f*
6401 orotic acid — Orotsäure *f*
6402 orthodromic — orthodrom
6403 orthogenesis — Orthogenese *f*
6404 orthogenic evolution — Orthoevolution *f*
6405 orthoploidy — Orthoploidie *f*
6406 Orthoptera, orthopterans — Geradflügler *mpl*, Orthopteren *pl*
6407 orthotropic, orthotropous — orthotrop
6408 oscillograph — Oszillograph *m*, Schwingungsschreiber *m*

6349	oosome *m*	oosoma *m*
6350	oosphère *f*	oosfera *f*
6351	oospore *f*	oospora *f*
6352	ovotide *f*	oótida *f*, ovótida *f*
6353	opaque	opaco
6354	—	—
6355	gène *m* opérateur	gen *m* operador
6356	opercule *m*	opérculo *m*
6357	opéron *m*	operón *m*
6358	opposé *bot.*	opuesto *bot.*
6359	opsonine *f*, bactériotropine *f*	opsonina *f*, bacteriotropina *f*
6360	chiasma *m* optique	quiasma *m* óptico
6361	cupule *f* optique	cúpula *f* óptica, cáliz *m* óptico
6362	nerf *m* optique	nervio *m* óptico
6363	—	—
6364	vésicule *f* optique, ~ ophtalmique	vesícula *f* óptica
6365	densité *f* optique	densidad *f* óptica
6366	—	—
6367	—	—
6368	—	—
6369	température *f* optimale, ~ optimum	temperatura *f* óptima
6370	—	—
6371	orbite *f*, cavité *f* orbitaire	órbita *f*
6372	ordre *m* *tax.*	orden *m* *tax.*
6373	ordovicien *m*	ordoviciense *m*, ordovícico *m*
6374	culture *f* d'organes	cultivo *m* de órganos
6375	organe *m* de la vue	órgano *m* visual, ~ de la visión
6376	système *m* organe	sistema *m* orgánico
6377	organite *m*	orgánulo *m*, organela *f*
6378	organique	orgánico
6379	organisme *m*	organismo *m*
6380	centre *m* d'organisation, ~ organisateur	centro *m* de organización
6381	niveau *m* d'organisation	nivel *m* de organización
6382	organisateur *m* *embr.*	organizador *m* *embr.*
6383	organogenèse *f*	organogénesis *f*, organogenia *f*
6384	organogène, organoformateur, organoformatif	organógeno, organogenético, organoformativo
6385	organographie *f*	organografía *f*
6386	organoïde	organoide
6387	organologie *f*	organología *f*
6388	organotrophique	organótrofo, organotrófico
6389	orientation *f*	orientación *f*
6390	mouvement *m* d'orientation	movimiento *m* de orientación
6391	réaction *f* d'orientation	reacción *f* de orientación, respuesta *f* orientativa
6392	population *f* originelle, ~ initiale, ~ d'origine	población *f* original, ~ inicial, ~ de origen
6393	ornithine *f*	ornitina *f*
6394	cycle *m* de l'ornithine, ~ de l'urée	ciclo *m* de la ornitina, ~ de la urea
6395	—	—
6396	ornithologue *m*, ornithologiste *m*	ornitólogo *m*
6397	ornithologie *f*	ornitología *f*
6398	ornithogame, ornithophile	ornitógamo, ornitófilo
6399	fleur *f* ornithogame	flor *f* ornitófila
6400	ornithogamie *f*, ornithophilie *f*, pollinisation *f* ornithophile	ornitogamia *f*, ornitofilia *f*
6401	acide *m* orotique	ácido *m* orótico
6402	orthodromique	ortodrómico
6403	orthogenèse *f*	ortogénesis *f*
6404	évolution *f* orthogénétique, ~ en ligne droite	evolución *f* ortogenética, ~ en línea directa
6405	orthoploïdie *f*	ortoploidía *f*
6406	orthoptères *mpl*	ortópteros *mpl*
6407	orthotrope	ortótropo, ortotrópico
6408	oscillographe *m*	oscilógrafo *m*

6409	osmic acid	Osmiumsäure *f*
6410	osmiophil(ic)	osmiophil
6411	osmium tetroxide	Osmiumtetroxid *n*
6412	osmolarity	Osmolarität *f*
6413	osmometer	Osmometer *n*
6414	osmo(re)ceptor	Osmorezeptor *m*
6415	osmoregulation, osmotic regulation	Osmoregulation *f*
6416	osmosis	Osmose *f*
6417	osmotaxis	Osmotaxis *f*
6418	osmotic	osmotisch
6419	osmotic balance	osmotisches Gleichgewicht *n*
6420	osmotic pressure	osmotischer Druck *m*
6421	osmotic regulation → osmoregulation	—
6422	osmotic value	osmotischer Wert *m*, Saugwert *m*
6423	osphradium	Osphradium *n*
6424	ossein	Ossein *n*
6425	osseous, bony	knöchern, knochig
6426	ossification	Verknöcherung *f*, Ossifikation *f*
6427	to ossify	verknöchern
6428	Osteichthyes, bony fish	Knochenfische *mpl*
6429	osteoblast	Osteoblast *m*, Knochenbildner *m*
6430	osteoclast	Osteoklast *m*, Knochenzerstörer *m*
6431	osteocranium, bony skull	Knochenschädel *m*, Osteokranium *n*
6432	osteocyte, bony cell	Knochenzelle *f*, Osteozyt *m*
6433	osteogenesis	Knochenbildung *f*, Osteogenese *f*
6434	osteology	Knochenlehre *f*, Osteologie *f*
6435	osteometry	Osteometrie *f*
6436	ostiole	Ostiolum *n*
6437	ostium (*pl* ostia)	Ostium *n*
6438	otic	Ohr ...
6439	otic capsule → auditory capsule	—
6440	otocyst → auditory vesicle	—
6441	otolith	Otolith *m*, Gehörstein *m*
6442	outbreeding, outcrossing	Fremdzucht *f*, Outbreeding *n*
6443	outer ear → external ear	—
6444	outgrowth → excrescence	—
6445	oval window *anat.*	ovales Fenster *n* *anat.*
6446	ovalbumin, egg albumin	Eieralbumin *n*, Ovalbumin *n*
6447	ovarian cycle	Ovarialzyklus *m*
6448	ovarian follicle → Graafian follicle	—
6449	ovarian hormone	Ovarialhormon *n*
6450	ovariole, egg tube	Ovariole *f*, Eiröhre *f*, Eischlauch *m*
6451	ovariotestis, ovotestis	Zwitterdrüse *f*, Ovariotestis *m*
6452	ovary *anat.*, *bot.*	a) Eierstock *m*, Ovar(ium) *n*, b) Fruchtknoten *m* *bot.*
6453	overdevelopment → hypertely	—
6454	overdominance, superdominance	Superdominanz *f*, Überdominanz *f*
6455	overheating	Überhitzung *f*
6456	overlapping code	überlappender Kode *m*
6457	overpopulation	Übervölkerung *f*
6458	to overwinter, to winter	überwintern
6459	overwintering, wintering	Überwinterung *f*
6460	oviduct	Eileiter *m*, Ovidukt *m*
6461	ovification	Eibildung *f*
6462	oviparity	Oviparie *f*
6463	oviparous	ovipar, eierlegend
6464	to oviposit, to lay eggs	Eier legen
6465	oviposition, egg-laying	Eiablage *f*
6466	oviposition site	Eiablageplatz *m*
6467	ovipositor, oviscapte	Legeröhre *f*, Eilegeapparat *m*
6468	ovo ... see also oo ...	—
6469	ovocentre, oocentre (US: ... center)	Oozentrum *n*, Ovozentrum *n*
6470	ovoid	eiförmig
6471	ovotestis → ovariotestis	—
6472	ovovitellin	Ovovitellin *n*
6473	ovoviviparity	Ovoviviparie *f*

6409	acide *m* osmique	ácido *m* ósmico
6410	osmiophile	osmiófilo, osmiofílico
6411	tétroxyde *m* d'osmium	tetróxido *m* de osmio
6412	osmolarité *f*	osmolaridad *f*
6413	osmomètre *m*	osmómetro *m*
6414	osmo(ré)cepteur *m*	osmorreceptor *m*
6415	osmorégulation *f*, régulation *f* osmotique	osmorregulación *f*, regulación *f* osmótica
6416	osmose *f*	ósmosis *f*, osmosis *f*
6417	osmotaxie *f*, osmotactisme *m*	osmotaxis *f*, osmotactismo *m*
6418	osmotique	osmótico
6419	équilibre *m* osmotique	equilibrio *m* osmótico
6420	pression *f* osmotique	presión *f* osmótica
6421	—	—
6422	valeur *f* osmotique	valor *m* osmótico
6423	osphradie *f*	osfradio *m*
6424	osséine *f*	oseína *f*
6425	osseux	óseo
6426	ossification *f*	osificación *f*
6427	ossifier	osificar
6428	ostéichthyens *mpl*, poissons *mpl* osseux	osteicties *mpl*, peces *mpl* óseos
6429	ostéoblaste *m*	osteoblasto *m*
6430	ostéoclaste *m*	osteoclasto *m*
6431	crâne *m* osseux	osteocráneo *m*
6432	ostéocyte *m*, cellule *f* osseuse	osteocito *m*, célula *f* ósea
6433	ostéogenèse *f*, ostéogénie *f*	osteogénesis *f*
6434	ostéologie *f*	osteología *f*
6435	ostéométrie *f*	osteometría *f*
6436	ostiole *m*	ostíolo *m*
6437	orifice *m*	óstium *m*, orificio *m*
6438	otique	ótico
6439	—	—
6440	—	—
6441	otolithe *m*	otolito *m*
6442	outbreeding *m*	outbreeding *m*
6443	—	—
6444	—	—
6445	fenêtre *f* ovale *anat.*	ventana *f* oval *anat.*
6446	ov(o)albumine *f*	ovoalbúmina *f*, albúmina *f* de huevo
6447	cycle *m* ovigène	ciclo *m* ovárico
6448	—	—
6449	hormone *f* ovarienne	hormona *f* ovárica
6450	ovariole *f*	ovariolo *m*
6451	ovariotestis *m*, ovotestis *m*	ovariotestis *m*, ovotestis *m*
6452	ovaire *m* *anat.*, *bot.*	ovario *m* *anat.*, *bot.*
6453	—	—
6454	overdominance *f*, superdominance *f*	sobredominancia *f*, superdominancia *f*
6455	échauffement *m* excessif, surchauffe *f*	sobrecalentamiento *m*
6456	code *m* chevauchant	código *m* sobrepuesto
6457	surpopulation *f*, surpeuplement *m*	superpoblación *f*
6458	hiverner	invernar
6459	hivernage *m*	invernación *f*
6460	oviducte *m*	oviducto *m*
6461	ovification *f*	ovificación *f*
6462	oviparité *f*, oviparisme *m*	oviparidad *f*
6463	ovipare	ovíparo
6464	pondre (des œufs)	poner huevos, aovar, desovar
6465	oviposition *f*, ponte *f*	ovoposición *f*, desove *m*, puesta *f*, postura *f*
6466	lieu *m* de ponte	sitio *m* de puesta, lugar *m* ~ ~
6467	oviscapte *m*, tarière *f*, ovipositeur *m*	oviscapto *m*, ovopositor *m*
6468	—	—
6469	ovocentre *m*	ovocentro *m*
6470	ovoïde	ovoide(o), ovado
6471	—	—
6472	ovovitelline *f*	ovovitelina *f*
6473	ovoviviparité *f*, ovoviviparisme *m*	ovoviviparidad *f*, ovoviviparismo *m*

6474	ovoviviparous	ovovivipar
6475	ovulation	Ovulation *f*, Follikelsprung *m*
6476	ovule *bot.*	Samenanlage *f* *bot.*
6477	ovuliferous scale → seed-bearing ~	—
6478	ovum, egg-cell	Eizelle *f*
6479	oxalacetic acid	Oxalessigsäure *f*, Oxalazetat *n*
6480	oxalic acid	Oxalsäure *f*, Kleesäure *f*
6481	oxidable	oxydierbar
6482	oxidant, oxidizing agent, oxidator	Oxydationsmittel *n*, Oxydant *m*, Oxydator *m*
6483	oxidase	Oxydase *f*
6484	to oxidate, to oxidize	oxydieren
6485	oxidation	Oxydation *f*, Oxydierung *f*
6486	oxidation pond	Oxydationsteich *m*
6487	oxidation-reduction potential, redox ~	Oxydations-Reduktionspotential *n*, Redoxpotential *n*
6488	oxidation-reduction reaction, redox ~	Redoxreaktion *f*
6489	oxidation-reduction system, redox ~	Redoxsystem *n*
6490	oxidative metabolism	oxydativer Stoffwechsel *m*
6491	oxidative phosphorylation, respiratory chain ~	oxydative Phosphorylierung *f*, Atmungskettenphosphorylierung *f*
6492	oxidoreductase	Oxydoreduktase *f*, Redoxase *f*
6493	oxidoreduction, redox	Oxydoreduktion *f*, Redox *n*
6494	oxybiotic → aerobic	—
6495	oxygen absorption	Sauerstoffaufnahme *f*
6496	oxygen carrier	Sauerstoff(über)träger *m*
6497	oxygen consumption	Sauerstoffverbrauch *m*
6498	oxygen content	Sauerstoffgehalt *m*
6499	oxygen debt	Sauerstoffschuld *f*
6500	oxygen demand, ~ requirement	Sauerstoffbedarf *m*
6501	oxygen lack	Sauerstoffmangel *m*
6502	oxygen supply	Sauerstoffversorgung *f*, ~zufuhr *f*
6503	oxygenase	Oxygenase *f*
6504	oxygenation	Oxygenierung *f*, Oxygen(is)ation *f*
6505	oxyh(a)emoglobin	Oxyhämoglobin *n*
6506	oxytocin, pitocin	Oxytozin *n*
6507	pacemaker potential	Schrittmacherpotential *n*
6508	pachytene	Pachytän *n*
6509	pack	Rudel *n*
6510	p(a)edogamy	Pädogamie *f*
6511	p(a)edogenesis	Pädogenese *f*
6512	pain receptor	Schmerzrezeptor *m*
6513	pain sensation	Schmerzempfindung *f*
6514	to pair *(chromosomes)*	paaren *(Chromosomen)*
6515	pair formation	Paarbildung *f*
6516	pair of genes, factor-pair	Genpaar *n*
6517	paired	paarig
6518	pairing of chromosomes → synapsis	—
6519	Palade's granula → ribosome	—
6520	palaearctic	paläarktisch, paläoarktisch
6521	pal(a)eoanthropology, human pal(a)eontology	Palä(o)anthropologie *f*
6522	pal(a)eobiology	Paläobiologie *f*
6523	pal(a)eobotany, pal(a)eophytology	Paläobotanik *f*, Paläophytologie *f*, Phytopaläontologie *f*
6524	Pal(a)eocene (epoch)	Paläozän *n*
6525	pal(a)eoclimatology	Paläoklimatologie *f*
6526	palaeo-ecology, paleocology	Paläoökologie *f*
6527	palaeo-encephalon → palencephalon	—
6528	Pal(a)eogene (period), early Tertiary period	Paläogen *n*, Alttertiär *n*
6529	pal(a)eogeography	Paläogeographie *f*
6530	Pal(a)eolithic (age), older stone age	Altsteinzeit *f*, Paläolithikum *n*
6531	pal(a)eontologist	Paläontologe *m*
6532	pal(a)eontology	Paläontologie *f*, Versteinerungskunde *f*
6533	pal(a)eophytology → palaeobotany	—
6534	pal(a)eotropical region	Paläotropis *f*

6474	ovovivipare	ovovivíparo
6475	ovulation *f*, ponte *f* ovulaire, ~ ovarique	ovulación *f*, puesta *f* ovular
6476	ovule *m* *bot.*	óvulo *m* *bot.*
6477	—	—
6478	cellule-œuf, ovule *m*	célula *f* huevo, ovocélula *f*, óvulo *m*
6479	acide *m* oxalacétique	ácido *m* oxalacético
6480	acide *m* oxalique	ácido *m* oxálico
6481	oxydable	oxidable
6482	(agent *m*) oxydant *m*, oxydateur *m*	(agente *m*) oxidante *m*
6483	oxydase *f*	oxidasa *f*
6484	oxyder	oxidar
6485	oxydation *f*	oxidación *f*
6486	étang *m* d'oxydation	estanque *m* de oxidación
6487	potentiel *m* d'oxydoréduction	potencial *m* redox, ~ de oxidorreducción
6488	réaction *f* d'oxydoréduction	reacción *f* de oxidorreducción, ~ redox
6489	système *m* oxydo-réducteur, ~ redox	sistema *m* de oxidorreducción, ~ redox, ~ oxidorreductor
6490	métabolisme *m* oxydatif	metabolismo *m* oxidativo
6491	phosphorylation *f* oxydative, ~ de la chaîne respiratoire	fosforilación *f* oxidativa, ~ de la cadena respiratoria
6492	oxydoréductase *f*	oxidorreductasa *f*
6493	oxydoréduction *f*, redox *m*	oxidorreducción *f*, redox *m*
6494	—	—
6495	absorption *f* d'oxygène	absorción *f* de oxígeno
6496	transporteur *m* d'oxygène	transportador *m* de oxígeno
6497	consommation *f* d'oxygène	consumo *m* de oxígeno
6498	teneur *f* en oxygène	contenido *m* en oxígeno
6499	débet *m* d'oxygène	débito *m* de oxígeno
6500	besoin *m* d'oxygène	demanda *f* de oxígeno
6501	manque *m* d'oxygène, pénurie *f* ~	falta *f* de oxígeno, déficit *m* ~ ~
6502	apport *m* d'oxygène	aporte *m* de oxígeno, suministro *m* ~ ~
6503	oxygénase *f*	oxigenasa *f*
6504	oxygénation *f*	oxigenación *f*
6505	oxyhémoglobine *f*	oxihemoglobina *f*
6506	oxytocine *f*, ocytocine *f*, pitocine *f*	oxitocina *f*, ocitocina *f*, pitocina *f*
6507	potentiel *m* de stimulateur cardiaque	potencial *m* marcapaso
6508	pachytène *m*	paquíteno *m*
6509	bande *f*	manada *f*
6510	paedogamie *f*	paidogamia *f*
6511	paedogenèse *f*, pédogenèse *f*	pedogénesis *f*, paidogénesis *f*
6512	algocepteur *m*	receptor *m* del dolor, algesiorreceptor *m*
6513	sensation *f* de douleur	sensación *f* dolorosa
6514	s'accoupler *(chromosomes)*	aparearse, emparejarse *(cromosomas)*
6515	formation *f* du couple	emparejamiento *m*, formación *f* de pareja
6516	couple *m* de gènes	par *m* de genes
6517	pair	par
6518	—	—
6519	—	—
6520	paléarctique	paleártico
6521	paléoanthropologie *f*, paléontologie *f* humaine	paleoantropología *f*, paleontología *f* humana
6522	paléobiologie *f*	paleobiología *f*
6523	paléobotanique *f*, paléophytologie *f*, phyto-paléontologie *f*, paléontologie *f* végétale	paleobotánica *f*, fitopaleontología *f*, paleofitología *f*
6524	paléocène *m*	paleoceno *m*
6525	paléoclimatologie *f*	paleoclimatología *f*
6526	paléoécologie *f*	paleoecología *f*
6527	—	—
6528	paléogène *m*	paleógeno *m*
6529	paléogéographie *f*	paleogeografía *f*
6530	paléolithique *m*	paleolítico *m*
6531	paléontologiste *m*	paleontólogo *m*
6532	paléontologie *f*	paleontología *f*
6533	—	—
6534	région *f* paléotropicale	región *f* paleotropical, ~ paleotrópica

6535	Pal(a)eozoic (era)	Paläozoikum *n*, Erdaltertum *n*
6536	pal(a)eozoology	Paläozoologie *f*
6537	palatal, palatine	Gaumen ...
6538	palatal bone, palatine ~	Gaumenbein *n*
6539	palatal reflex → swallowing reflex	—
6540	palatal tonsil, amygdala	Gaumenmandel *f*
6541	palate	Gaumen *m*
6542	palatine velum	Gaumensegel *n*
6543	palea	Spreublatt *n*
6544	palencephalon, palaeo-encephalon	Althirn *n*, Palenzephalon *n*
6545	paleo ... → palaeo ...	—
6546	palingenesis, palingeny	Palingenese *f*
6547	palisade cell	Palisadenzelle *f*
6548	palisade tissue, ~ parenchyma	Palisadengewebe *n*, ~parenchym *n*
6549	pallial cavity → mantle cavity	—
6550	pallium[1] → mantle	—
6551	pallium[2], brain mantle	Hirnmantel *m*, Pallium *n*
6552	palmate	handförmig *(Blatt)*
6553	palmatilobate	handförmig gelappt
6554	palmatipartite	handförmig geteilt
6555	palmigrade → plantigrade	—
6556	palmiped	Schwimmfüßler *m*, Schwimmvogel *m*
6557	palmitic acid	Palmitinsäure *f*
6558	palpus (*pl* palpi), palp	Taster *m*, Palpus *m* (*pl* Palpen)
6559	palynology	Palynologie *f*
6560	pancreas	Bauchspeicheldrüse *f*, Pankreas *m*
6561	pancreatic juice	Bauchspeichel *m*, Pankreassaft *m*
6562	pancreatin	Pankreatin *n*
6563	pancreozymin	Pankreozymin *n*
6564	pangen(e)	Pangen *n*
6565	pangenesis	Pangenesis *f*
6566	panicle	Rispe *f*
6567	panmixis, panmixia	Panmixie *f*, Panmixis *f*
6568	panspermy, panspermism	Panspermie *f*
6569	pantothenic acid	Pantothensäure *f*
6570	papain	Papain *n*
6571	paper chromatography	Papierchromatographie *f*
6572	paper electrophoresis	Papierelektrophorese *f*
6573	papilla	Papille *f*
6574	pappus	Pappus *m*, Haarkrone *f*
6575	parabasal apparatus	Parabasalapparat *m*
6576	parabiosis	Parabiose *f*
6577	paracarpous	parakarp
6578	paracasein → casein	—
6579	paracentric	parazentrisch
6580	paraffin block	Paraffinblock *m*
6581	paraffin embedding	Paraffineinbettung *f*
6582	paraffin section	Paraffinschnitt *m*
6583	parallel mutation	Parallelmutation *f*
6584	parallelinervate	parallelnervig, streifennervig
6585	paramesonephric duct → Müllerian duct	—
6586	parameter	Parameter *m*
6587	paramitosis	Paramitose *f*
6588	paramylon, paramylum	Paramylon *n*
6589	paranasal sinus	Nasennebenhöhle *f*
6590	paranucleus, nebenkern	Nebenkern *m*, Paranukleus *m*
6591	paraphysis	Paraphyse *f*
6592	parapineal organ → parietal organ	—
6593	paraplasm	Paraplasma *n*
6594	parapodium	Parapodium *n*
6595	parapophysis	Parapophyse *f*
6596	parasexual	parasexuell
6597	parasexual cycle	parasexueller Zyklus *m*
6598	parasexuality	Parasexualität *f*
6599	parasite	Parasit *m*, Schmarotzer *m*

6535	paléozoïque *m*	paleozóico *m*
6536	paléozoologie *f*, paléontologie *f* animale	paleozoología *f*
6537	palatal, palatin	palatal, palatino
6538	(os *m*) palatin *m*	hueso *m* palatino
6539	—	—
6540	amygdale *f* palatine	tonsila *f* palatina, amígdala *f* ~
6541	palais *m*	paladar *m*
6542	voile *m* palatin, ~ du palais	velo *m* palatino, ~ del paladar
6543	glumelle *f*, palea *f*	pálea *f*
6544	pal(é)encéphale *m*	paleoencéfalo *m*, paleoncéfalo *m*
6545	—	—
6546	palingenèse *f* palingénésie *f*	palingenesia *f*, palingenia *f*
6547	cellule *f* palissadique	célula *f* en empalizada
6548	tissu *m* palissadique, parenchyme *m* ~, tissu *m* en palissade	tejido *m* en empalizada
6549	—	—
6550	—	—
6551	pallium *m*, manteau *m* des hémisphères	palio *m*
6552	palmé	palmado, palmeado
6553	palmatilobé	palmatilobulado
6554	palmatipartite	palmatipartido
6555	—	—
6556	palmipède *m*	palmípeda *f*
6557	acide *m* palmitique	ácido *m* palmítico
6558	palpe *m*	palpo *m*
6559	palynologie *f*	palinología *f*
6560	pancréas *m*	páncreas *m*
6561	suc *m* pancréatique	jugo *m* pancreático
6562	pancréatine *f*	pancreatina *f*
6563	pancréozymine *f*, pancréozyme *m*	pancreozimina *f*
6564	pangène *m*	pángen *m*
6565	pangenèse *f*	pangénesis *f*
6566	panicule *m*	panícula *f*
6567	panmixie *f*, reproduction *f* panmictique	panmixis *f*, panmixia *f*
6568	panspermie *f*	panspermia *f*
6569	acide *m* pantothénique	ácido *m* pantoténico
6570	papaïne *f*	papaína *f*
6571	chromatographie *f* en papier, ~ sur papier	cromatografía *f* en papel, ~ sobre ~
6572	électrophorèse *f* en papier	electroforesis *f* sobre papel
6573	papille *f*	papila *f*
6574	pappus *m*	papo *m*, vilano *m*
6575	appareil *m* parabasal	aparato *m* parabasal
6576	parabiose *f*	parabiosis *f*
6577	paracarpe	paracárpico
6578	—	—
6579	paracentrique	paracéntrico
6580	bloc *m* de paraffine	bloque *m* de parafina
6581	inclusion *f* dans la paraffine	inclusión *f* en parafina
6582	coupe *f* à la paraffine	corte *m* en parafina
6583	mutation *f* parallèle	mutación *f* paralela
6584	parallélinerve, à nervures parallèles	paralelinervio, de nervadura paralela
6585	—	—
6586	paramètre *m*	parámetro *m*
6587	paramitose *f*	paramitosis *f*
6588	paramylon *m*	paramilo *m*
6589	sinus *m* paranasal	seno *m* paranasal
6590	nebenkern *m*, paranucléus *m*	paranúcleo *m*
6591	paraphyse *f*	paráfisis *f*
6592	—	—
6593	paraplasme *m*	paraplasma *m*
6594	parapode *m*, parapodie *f*	parapodio *m*, parápodo *m*
6595	parapophyse *f*	parapófisis *f*
6596	parasexuel	parasexual
6597	cycle *m* parasexuel	ciclo *m* parasexual
6598	parasexualité *f*	parasexualidad *f*
6599	parasite *m*	parásito *m*

6600	parasitic	parasitisch, parasitär, schmarotzend
6601	parasitic plant, phytoparasite	Schmarotzerpflanze *f*, Phytoparasit *m*, pflanzlicher Schmarotzer *m*
6602	parasitism	Parasitismus *m*, Parasitie *f*, Schmarotzertum *n*
6603	parasitization	Parasitenbefall *m*
6604	to parasitize	parasitieren, schmarotzen
6605	parasitized	befallen (von Parasiten)
6606	parasitologist	Parasitologe *m*
6607	parasitology	Parasitologie *f*
6608	parasympathetic (nervous) system	Parasympathikus *m*, parasympathisches Nervensystem *n*
6609	parasynapsis, parasyndesis	Parasynapsis *f*, Parasyndese *f*
6610	parathormone, parathyrin, parathyroid hormone	Parathormon *n*, Nebenschilddrüsenhormon *n*
6611	parathyroid, epithelial body	Nebenschilddrüse *f*, Beischilddrüse *f*, Epithelkörperchen *n*
6612	paratype	Paratyp(us) *m*
6613	paravariation	Paravariation *f*
6614	parazoon (*pl* parazoa)	Parazoon *n* (*pl* Parazoa)
6615	parenchyma	Parenchym *n*
6616	parenchymatous cell	Parenchymzelle *f*
6617	parent	Elter *m* (*pl* Eltern)
6618	parent cell	Elternzelle *f*
6619	parent material, ~ rock	Urgestein *n*, Muttergestein *n*
6620	parent plant, mother ~, stock ~	Mutterpflanze *f*
6621	parental	elterlich, parental
6622	parental care	Brutfürsorge *f*
6623	parental care behavio(u)r	Brutpflegeverhalten *n*
6624	parental ditype, PD	Parental-Dityp *m*, PD
6625	parental generation, P_1 generation	Elterngeneration *f*, Parentalgeneration *f*, P-Generation *f*
6626	parental type	Parentaltyp *m*
6627	parietal	wandständig, parietal
6628	parietal bone	Scheitelbein *n*
6629	parietal eye, epiphyseal ~	Scheitelauge *n*, Parietalauge *n*
6630	parietal lobe	Scheitellappen *m*
6631	parietal organ, parapineal ~	Parietalorgan *n*, Parapinealorgan *n*
6632	paripinnate	paarig gefiedert
6633	paroecious, paroicous	parökisch
6634	parotid (gland)	Ohrspeicheldrüse *f*
6635	parthenocarpy	Parthenokarpie *f*, Jungfernfrüchtigkeit *f*
6636	parthenogenesis	Parthenogenese *f*, Jungfernzeugung *f*
6637	partial dominance → semi-dominance	—
6638	partial pressure	Partialdruck *m*
6639	partial regression	partielle Regression *f*
6640	particulate inheritance	Mendelvererbung *f*, mendelnde Vererbung *f*
6641	partition chromatography	Verteilungschromatographie *f*
6642	partly decomposed vegetable mould	Moder *m*
6643	parturition → birth	—
6644	passage cell	Durchlaßzelle *f*
6645	passive immunity	passive Immunität *f*
6646	pasteurization	Pasteurisation *f*, Pasteurisierung *f*
6647	to pasteurize	pasteurisieren
6648	patagium	Flughaut *f*
6649	patella → knee-cap	—
6650	patellar reflex → knee ~	—
6651	paternal	väterlich
6652	path of conduction *neur.*	Leitungsbahn *f* *neur.*
6653	pathogenic	pathogen, krankheitserregend
6654	pathogenic agent, pathogen	Krankheitserreger *m*
6655	pathogenicity	Pathogenität *f*
6656	patroclinal, patriclinous	patroklin
6657	patrocliny	Patroklinie *f*
6658	paunch → rumen	—

6600	parasitaire	parasitario
6601	plante *f* parasite, phytoparasite *m*	planta *f* parásita, ~ parasitaria, fitoparásito *m*
6602	parasitisme *m*	parasitismo *m*
6603	parasitisation *f*	invasión por parásitos
6604	parasiter	parasitar
6605	parasité	parasitado
6606	parasitologue *m*, parasitologiste *m*	parasitólogo *m*
6607	parasitologie *f*	parasitología *f*
6608	(système *m*) parasympathique *m*	sistema *m* (nervioso) parasimpático
6609	parasyndèse *f*	parasíndesis *f*
6610	parathormone *f*	parathormona *f*, hormona *f* paratiroides
6611	parathyroïde *f*, corpuscule *m* épithélial	paratiroides *f*, glándula *f* ~
6612	paratype *m*	paratipo *m*
6613	paravariation *f*	paravariación *f*
6614	parazoaire *m*	parazoo *m*, parazoario *m*
6615	parenchyme *m*	parénquima *m*
6616	cellule *f* parenchymateuse	célula *f* parenquimática
6617	parent *m*, géniteur *m*	genitor *m*, progenitor *m*
6618	cellule *f* parentale	célula *f* (pro)genitora
6619	roche-mère *f*	roca *f* madre
6620	plante-mère *f*	planta *f* madre
6621	parental, géniteur	parental, genitor
6622	soins *m pl* paternels	cuidado *m* paternal
6623	comportement *m* parental	comportamiento *m* epimelético
6624	ditype *m* parental, DP	dítipo *m* progenitor, DP
6625	génération *f* P_1, ~ des parents	generación *f* P_1
6626	type *m* parental	tipo *m* genitor
6627	pariétal	parietal
6628	(os *m*) pariétal *m*	(hueso *m*) parietal *m*
6629	œil *m* pariétal	ojo *m* parietal, ~ epifisario
6630	lobe *m* pariétal	lóbulo *m* parietal
6631	organe *m* pariétal	órgano *m* parietal
6632	paripenné	paripinnado
6633	paroecique	paroico
6634	parotide *f*	(glándula *f*) parótida *f*
6635	parthénocarpie *f*	partenocarpia *f*
6636	parthénogenèse *f*	partenogénesis *f*
6637	—	—
6638	pression *f* partielle, tension *f* ~	presión *f* parcial
6639	régression *f* partielle	regresión *f* parcial
6640	hérédité *f* mendélienne	herencia *f* mendeliana
6641	chromatographic *f* de partage, ~ de répartition	cromatografía *f* de partición
6642	moder *m*, amphi-mull *m*	sustancia *f* vegetal en descomposición
6643	—	—
6644	cellule *f* traversée	célula *f* de paso
6645	immunité *f* passive	inmunidad *f* pasiva
6646	pasteurisation *f*	paste(u)rización *f*
6647	pasteuriser	paste(u)rizar
6648	patagium *m*, membrane *f* alaire	patagio *m*
6649	—	—
6650	—	—
6651	paternel	paternal, paterno
6652	voie *f* de conduction *neur.*	vía *f* de conducción *neur.*
6653	pathogène	patógeno
6654	agent *m* pathogène	agente *m* patógeno
6655	pathogénicité *f*, pouvoir *m* pathogène	patogenicidad *f*, poder *m* patógeno
6656	patrocline	patroclino
6657	patroclinie *f*	patroclinia *f*
6658	—	—

6659	paurometabolism, paurometabolic metamorphosis	Paurometabolie *f*
6660	pavement epithelium, tesselated ~	Plattenepithel *n*
6661	peat	Torf *m*
6662	peat moss, sphagnum ~, bog ~	Torfmoos *n*, Sphagnum *n*
6663	peck order, peck-right hierarchy	Hackordnung *f*
6664	pectase	Pektase *f*, Pektinesterase *f*
6665	pectic acid	Pektinsäure *f*
6666	pectin	Pektin *n*
6667	pectoral	Brust ...
6668	pectoral fin	Brustflosse *f*
6669	pectoral girdle → shoulder girdle	—
6670	pedicel	Blütenstiel *m* (der Einzelblüte)
6671	pedigree	Stammbaum *m* (*Tiere*)
6672	pedigree breeding	Stammbaumzüchtung *f*, ~selektion *f*
6673	pedigree certificate	Abstammungsnachweis *m*
6674	pedipalp	Beintaster *m*, Pedipalpus *m*
6675	pedo ... see also paedo ...	—
6676	pedogenesis → soil formation	—
6677	pedology → soil science	—
6678	peduncle, flower stalk	Blütenstiel *m*
6679	pedunculate, stalked	gestielt
6680	pedunculate bodies → mushroom ~	—
6681	pelage, coat	Fell *n*, Haarkleid *n*
6682	pelagic	pelagisch
6683	pelagic zone	Pelagial *n*
6684	Pelecypoda → Bivalvia	—
6685	pellucide area → area pellucida	—
6686	peloria	Pelorie *f*
6687	peltate → scutiform	—
6688	peltate leaf	Schildblatt *n*
6689	pelvic	Becken ...
6690	pelvic cavity	Beckenhöhle *f*
6691	pelvic fin → ventral fin	—
6692	pelvic girdle, hip girdle	Beckengürtel *m*
6693	pelvis	Becken *n*
6694	pendular movement	Pendelbewegung *f*
6695	penetrance *gen.*	Penetranz *f* *gen.*
6696	penicillic acid	Penizill(in)säure *f*
6697	penicillin	Penizillin *n*
6698	penis	Penis *m*
6699	pentosan	Pentosan *n*
6700	pentose	Pentose *f*
6701	pepsin	Pepsin *n*
6702	pepsinogen	Pepsinogen *n*
6703	peptidase	Peptidase *f*
6704	peptide	Peptid *n*
6705	peptide bond	Peptidbindung *f*
6706	peptide chain	Peptidkette *f*
6707	peptide hormone	Peptidhormon *n*
6708	peptone	Pepton *n*
6709	peptonization	Peptonisierung *f*
6710	perception	Wahrnehmung *f*
6711	perceptiveness	Wahrnehmungsfähigkeit *f*
6712	perennation	Mehrjährigkeit *f*
6713	perennial, perennating	perennierend, ausdauernd
6714	perennial (plant)	Perenne *f*, perenne Pflanze *f*, ausdauernde ~, Staude *f*
6715	perfect insect → imago	—
6716	perforatorium → acrosome	—
6717	perianth, flower envelope	Blütenhülle *f*, Perianth(ium) *n*
6718	periblast	Periblast *m*
6719	periblem	Periblem *n*
6720	pericardial cavity, ~ sac	Herzbeutelhöhle *f*, Perikardialhöhle *f*
6721	pericardium	Herzbeutel *m*, Perikard(ium) *n*

6659	métamorphose *f* paurométabole	paurometabolia *f*
6660	épithélium *m* pavimenteux	epitelio *m* pavimentoso, ~ plano
6661	tourbe *f*	turba *f*
6662	tourbière *f* à sphaignes, sphaigne *f*	turbera *f* de esfagnales, esfagno *m*
6663	ordre *m* du becquetage, hiérarchie *f* ~ ~	orden *m* de picoteo, jerarquía *f* ~ ~
6664	pectase *f*	pectasa *f*
6665	acide *m* pectique	ácido *m* péctico
6666	pectine *f*	pectina *f*
6667	pectoral	pectoral
6668	nageoire *f* pectorale	aleta *f* pectoral
6669	—	—
6670	pédicelle *m*	pedicelo *m* (floral)
6671	pedigree *m*	pedigree *m*, registro *m* genealógico
6672	sélection *f* sur pedigree, ~ généalogique	selección *f* genealógica
6673	certificat *m* d'ascendance	certificado *m* de ascendencia
6674	pédipalpe *m*	pedipalpo *m*
6675	—	—
6676	—	—
6677	—	—
6678	pédoncule *m*, tige *f* florale	pedúnculo *m* (floral)
6679	pédonculé	pedunculado
6680	—	—
6681	pelage *m*	pelaje *m*
6682	pélagique	pelágico
6683	zone *f* pélagique	zona *f* pelágica
6684	—	—
6685	—	—
6686	pélorie *f*	peloria *f*
6687	—	—
6688	feuille *f* peltée	hoja *f* peltada
6689	pelvien	pélvico, pelviano
6690	cavité *f* pelvienne	cavidad *f* pelviana
6691	—	—
6692	ceinture *f* pelvienne	cinturón *m* pélvico, ~ pelviano
6693	pelvis *m*, bassin *m*	pelvis *f*
6694	mouvement *m* pendulaire	movimiento *m* pendular
6695	pénétrance *f*, fréquence *f* de manifestation *gen.*	penetración *f* *gen.*
6696	acide *m* pénicillique	ácido *m* penicilínico
6697	pénicilline *f*	penicilina *f*
6698	pénis *m*	pene *m*
6699	pentosanne *m*	pentosana *f*
6700	pentose *m*	pentosa *f*
6701	pepsine *f*	pepsina *f*
6702	pepsinogène *m*	pepsinógeno *m*
6703	peptidase *f*	peptidasa *f*
6704	peptide *m*	péptido *m*
6705	liaison *f* peptidique	enlace *m* peptídico, ~ péptido
6706	chaîne *f* peptidique	cadena *f* peptídica, ~ péptida
6707	hormone *f* peptidique	hormona *f* peptídica
6708	peptone *f*	peptona *f*
6709	peptonisation *f*	peptonización *f*
6710	perception *f*	percepción *f*
6711	perceptivité *f*	perceptibilidad *f*
6712	pérennance *f*, pérennité *f*	perennidad *f*
6713	pérenne, pérennant	perenne
6714	plante *f* pérenne, ~ vivace	planta *f* perenne, ~ vivaz
6715	—	—
6716	—	—
6717	périanthe *m*, enveloppe *f* florale	perianto *m*, envoltura *f* floral
6718	périblaste *m*	periblasto *m*
6719	périblème *m*	periblema *m*
6720	cavité *f* péricardique	cavidad *f* pericardíaca, saco *m* pericardíaco
6721	péricarde *m*	pericardio *m*

6722	pericarp	Fruchtwand *f*, Perikarp *n*
6723	pericentric	perizentrisch
6724	perichondrium	Knorpelhaut *f*, Perichondrium *n*
6725	periclinal	periklin
6726	periclinal chim(a)era	Periklinalchimäre *f*, Mantelchimäre *f*
6727	pericycle	Perizykel *m*
6728	periderm	Periderm *n*
6729	peridium	Peridie *f*
6730	perigone, perigonium	Perigon(ium) *n*
6731	perigynous	perigyn
6732	perikaryon	Perikaryon *n*
6733	perilymph	Perilymphe *f*
6734	perimetrium	Perimetrium *n*
6735	perimysium	Perimysium *n*
6736	perineum	Perineum *n*, Damm *m* *anat.*
6737	perineurium	Perineurium *n*
6738	period of development	Entwicklungsphase *f*
6739	periodic-acid-Schiff-reaction	PAS-Reaktion *f*, Perjodsäure-Schiff-Reaktion *f*
6740	periodicity, rhythmicity	Periodizität *f*
6741	periosteum	Knochenhaut *f*, Periost *n*
6742	peripheral nervous system	peripheres Nervensystem *n*
6743	peripheric movement	Peripheriewanderung *f*
6744	periphyton	Aufwuchs *m*
6745	periplasm	Periplasma *n*
6746	to perish, perishing → to die back, die-back	—
6747	perisperm	Perisperm *n*
6748	perispore, perisporium	Perispor *n*
6749	Perissodactyla, odd-toed ungulates	Unpaarhufer *mpl*, Unpaarzeher *mpl*
6750	peristalsis, peristaltic movement	Peristaltik *f*
6751	peristasis	Peristase *f*
6752	peristome	Peristom *n*
6753	perithecium	Perithezium *n*
6754	peritoneal cavity → abdominal cavity	—
6755	peritoneum	Bauchfell *n*, Peritoneum *n*
6756	peritrichous, peritrichic	peritrich
6757	permanent association	Dauergesellschaft *f*
6758	permanent parasite	ständiger Parasit *m*
6759	permanent preparation	Dauerpräparat *n*
6760	permanent tissue, definitive ~	Dauergewebe *n*
6761	permanent tooth	bleibender Zahn *m*, Dauerzahn *m*
6762	permanent wilting	Dauerwelke *f*
6763	permanent wilting point	permanenter Welkepunkt *m*
6764	permeability, perviousness	Durchlässigkeit *f*, Permeabilität *f*
6765	permeable, pervious	durchlässig, permeabel
6766	permease	Permease *f*
6767	permeation	Permeation *f*
6768	Permian (period)	Perm *n*, Dyas *n*
6769	permissible dose → genetic tolerance dose	—
6770	permutation	Permutation *f*
6771	peroxidase	Peroxydase *f*
6772	persistence *gen.*	Persistenz *f* *gen.*
6773	persistent	persistent
6774	persistent modification → dauermodification	—
6775	perspiration	Ausdünstung *f*, Perspiration *f*
6776	pervious, perviousness → permeable, permeability	—
6777	pest	Schädling *m*
6778	pesticide	Schädlingsbekämpfungsmittel *n*, Pestizid *n*
6779	pesticide residues	Rückstände *mpl* von Schädlingsbekämpfungsmitteln
6780	petal	Kron(en)blatt *n*
6781	petiole, leaf-stalk	Blattstiel *m*

6722	péricarpe *m*	pericarpo *m*, pericarpio *m*
6723	péricentrique	pericéntrico
6724	périchondre *m*	pericondrio *m*
6725	périclinal	periclinal
6726	chimère *f* périclinale	quimera *f* periclinal
6727	péricycle *m*	periciclo *m*
6728	périderme *m*	peridermo *m*, peridermis *f*
6729	péridium *m*	peridio *m*
6730	périgone *m*	perigonio *m*
6731	périgyne	perígino
6732	péricaryon *m*	pericarión *m*, pericario *m*
6733	périlymphe *f*	perilinfa *f*
6734	périmétrium *m*	perimetrio *m*
6735	périmysium *m*	perimisio *m*
6736	périnée *m*	perineo *m*
6737	périnèvre *m*	perineur(i)o *m*
6738	phase *f* de développement	fase *f* de desarrollo, período *m* ~ ~
6739	réaction *f* PAS, ~ à l'acide périodique-Schiff	reacción *f* PAS, ~ del ácido periódico de Schiff
6740	périodicité *f*, rythmicité *f*	periodicidad *f*
6741	périoste *m*	periostio *m*
6742	système *m* nerveux périphérique	sistema *m* nervioso periférico
6743	mouvement *m* périphérique	movimiento *m* periférico
6744	periphyton *m*, «Aufwuchs» *m*	perifiton *m*, perífiton *m*
6745	périplasme *m*	periplasma *m*
6746	—	—
6747	périsperme *m*	perisperma *m*, perispermo *m*
6748	périspore *f*	perisporio *m*
6749	périssodactyles *mpl*	perisodáctilos *mpl*
6750	péristaltisme *m*, mouvements *mpl* péristaltiques	peristaltismo *m*, peristalsis *f*, movimiento *m* peristáltico
6751	péristase *f*	peristasis *f*
6752	péristome *m*	peristoma *m*
6753	périthécium *m*, périthèce *m*	peritecio *m*
6754	—	—
6755	péritoine *m*	peritoneo *m*
6756	péritriche	peritrico, perítrico
6757	association *f* permanente	comunidad *f* permanente
6758	parasite *m* permanent	parásito *m* permanente
6759	préparation *f* permanente	preparación *f* permanente
6760	tissu *m* permanent	tejido *m* permanente, ~ adulto
6761	dent *f* permanente, ~ définitive	diente *m* permanente
6762	fanaison *f* permanente, flétrissement *m* permanent	marchitez *f* permanente
6763	point *m* de fanaison permanente, ~ de flétrissement permanent	punto *m* de marchitez permanente
6764	perméabilité *f*	permeabilidad *f*
6765	perméable	permeable
6766	perméase *f*, enzyme *m* de perméation	permeasa *f*
6767	perméation *f*	permeación *f*
6768	permien *m*	pérmico *m*
6769	—	—
6770	permutation *f*	permutación *f*
6771	peroxydase *f*	peroxidasa *f*
6772	persistance *f gen.*	persistencia *f gen.*
6773	persistant	persistente
6774	—	—
6775	perspiration *f*	perspiración *f*
6776	—	—
6777	animal *m* nuisible	animal *m* nocivo, ~ dañino, plaga *f* animal
6778	pesticide *m*	pesticida *m*, plaguicida *m*
6779	résidus *mpl* des pesticides	residuos *mpl* de pesticidas
6780	pétale *m*	pétalo *m*
6781	pétiole *m*	pecíolo *m* (foliar)

6782	petiolule	Fiederblattstiel *m*
6783	Petri dish, ~ plate	Petrischale *f*
6784	petrifaction	Versteinerung *f*
6785	to petrify	versteinern
6786	petrophyte → rupicolous plant	—
6787	P_1 generation → parental generation	—
6788	pH, pH value	pH-Wert *m*
6789	Phaeophyta, brown algae	Braunalgen *f pl*, Phäophyzeen *f pl*
6790	phage → bacteriophage	—
6791	phagocyte	Phagozyt *m*, Freßzelle *f*
6792	to phagocytize	phagozytieren
6793	phagocytosis	Phagozytose *f*
6794	phanerogams → spermatophyta	—
6795	phanerophyte	Phanerophyt *m*
6796	pharyngeal	Rachen ..., Schlund ...
6797	pharyngeal canal	Schlundkanal *m*
6798	pharyngeal cavity	Rachenhöhle *f*
6799	pharyngeal tonsil, ~ amygdala	Rachenmandel *f*
6800	pharynx *anat.*	Rachen *m*, Schlund *m*, Pharynx *m*
6801	pharynx *zool.*	Kiemendarm *m*
6802	phase-contrast microscope	Phasenkontrastmikroskop *n*
6803	phase specificity, phasic ~	Phasenspezifität *f*
6804	phellem → cork tissue	—
6805	phelloderm	Phelloderm *n*
6806	phellogen, cork cambium	Phellogen *n*, Korkkambium *n*, Korkbildungsgewebe *n*
6807	phen	Phän *n*
6808	phenocopy	Phänokopie *f*
6809	phenocytology	Phänozytologie *f*
6810	phenogenetics	Phänogenetik *f*
6811	phenolase, phenoloxidase	Phenolase *f*, Phenoloxydase *f*
6812	phenological isolation	phänologische Isolation *f*
6813	phenology	Phänologie *f*
6814	phenotype	Phänotyp(us) *m*, Erscheinungstyp *m*, Erscheinungsbild *n*
6815	phenotypic	phänotypisch
6816	phenotypic character	phänotypisches Merkmal *n*
6817	phenotypic expression	phänotypische Ausprägung *f*
6818	phenylalanine	Phenylalanin *n*
6819	pheromone	Pheromon *n*
6820	philopatry, site-attachment	Ortstreue *f*
6821	phloem	Phloem *n*, Bastteil *m*
6822	phobotaxis	Phobotaxis *f*
6823	phonation, sound-production	Lauterzeugung *f*, Lautbildung *f*
6824	phonoreceptor	Phonorezeptor *m*
6825	phoresis, phoresy	Phoresie *f*
6826	Phoronidea, phoronids	Hufeisenwürmer *m pl*
6827	phosphatase, phosphomonoesterase	Phosphatase *f*, Phosphomonoesterase *f*
6828	phosphatide, phospholipid	Phosphatid *n*, Phospholip(o)id *n*
6829	phosphatidic acid	Phosphatidsäure *f*
6830	phosphoarginine → arginine phosphate	—
6831	phosphocreatine → creatine phosphate	—
6832	phosphoglyceric acid	Phosphoglyzerinsäure *f*
6833	phospholipid → phosphatide	—
6834	phosphoprotein	Phosphoproteid *n*
6835	phosphorescence	Phosphoreszenz *f*
6836	phosphoric acid	Phosphorsäure *f*
6837	phosphorolysis	Phosphorolyse *f*
6838	phosphorylase	Phosphorylase *f*
6839	to phosphorylate	phosphorylieren
6840	phosphorylation	Phosphorylierung *f*
6841	photic zone	photische Zone *f*
6842	photobiology	Photobiologie *f*
6843	photochemical	photochemisch

6782	pétiolule *m*	peciólulo *m*
6783	plaque *f* de Petri, boîte *f* ~ ~	placa *f* de Petri, caja *f* ~ ~
6784	pétrification *f*	petrificación *f*
6785	se pétrifier	petrificarse
6786	—	—
6787	—	—
6788	pH *m*, valeur *f* (du) pH	pH *m*, valor *m* del pH
6789	phéophytes *fpl*, algues *fpl* brunes, phéophycées *fpl*	feofitos *mpl*, algas *fpl* pardas, feofíceas *fpl*
6790	—	—
6791	phagocyte *m*	fagocito *m*
6792	phagocyter	fagocitar
6793	phagocytose *f*	fagocitosis *f*
6794	—	—
6795	phanérophyte *m*	fanerófito *m*, fanerófita *f*
6796	pharyngien	faríngeo
6797	canal *m* pharyngien	canal *m* faríngeo
6798	cavité *f* pharyngée	cavidad *f* faríngea
6799	amygdale *f* pharyngienne	amígdala *f* faríngea, tonsila *f* ~
6800	pharynx *m*, arrière-gorge *f*	faringe *f*
6801	pharynx *m* branchial	faringe *f* branquial, intestino *m* ~
6802	microscope *m* à contraste de phase	microscopio *m* de contraste de fase
6803	spécificité *f* phasique	especificidad *f* fásica
6804	—	—
6805	phelloderme *m*	felodermo *m*, felodermis *f*
6806	phellogène, assise *f* subéro-phellodermique	felógeno *m*, cámbium *m* suberígeno
6807	phène *m*	fen *m*
6808	phénocopie *f*	fenocopia *f*
6809	phénocytologie *f*	fenocitología *f*
6810	phénogénétique *f*	fenogenética *f*
6811	phénoloxydase *f*	fenolasa *f*
6812	isolement *m* phénologique	aislamiento *m* fenológico
6813	phénologie *f*	fenología *f*
6814	phénotype *m*	fenotipo *m*, fenótipo *m*
6815	phénotypique	fenotípico
6816	caractère *m* phénotypique	característica *f* fenotípica
6817	expression *f* phénotypique	expresión *f* fenotípica
6818	phénylalanine *f*	fenilalanina *f*
6819	phéromone *f*, phérormone *f*	feromona *f*
6820	fidélité *f* au domaine, ~ au lieu	fidelidad *f* al lugar, apego *m* ~ ~
6821	phloème *m*	floema *m*
6822	phobotaxie *f*, phobotactisme *m*	fobotaxia *f*, fobotactismo *m*
6823	phonation *f*	fonación *f*
6824	phonorécepteur *m*	fonorreceptor *m*, receptor *m* del sonido
6825	phorésie *f*	foresis *f*, foresia *f*
6826	phoronidiens *mpl*	foronídeos *mpl*
6827	phosphatase *f*, phosphomonoestérase *f*	fosfatasa *f*, fosfomonoesterasa *f*
6828	phosphatide *m*, phospholipide *m*	fosfátido *m*, fosfolípido *m*, fosfolipoide *m*, fosfolipina *f*
6829	acide *m* phosphatidique	ácido *m* fosfatídico
6830	—	—
6831	—	—
6832	acide *m* phosphoglycérique	ácido *m* fosfoglicérico
6833	—	—
6834	phosphoprotéide *m*	fosfoproteína *f*
6835	phosphorescence *f*	fosforescencia *f*
6836	acide *m* phosphorique	ácido *m* fosfórico
6837	phosphorolyse *f*	fosforólisis *f*
6838	phosphorylase *f*	fosforilasa *f*
6839	phosphoryler	fosforilar
6840	phosphorylation *f*	fosforilación *f*
6841	zone *f* photique	zona *f* fótica
6842	photo(bio)logie *f*	fotobiología *f*
6843	photochimique	fotoquímico

6844	photochemistry	Photochemie *f*
6845	photokinesis	Photokinese *f*
6846	photolysis	Photolyse *f*
6847	photometry	Photometrie *f*
6848	photomicrography → microphotography	—
6849	photomorphism, photomorphosis	Photomorphose *f*, Photomorphismus *m*
6850	photon, light quantum	Photon *n*, Lichtquant *m*
6851	photonasty	Photonastie *f*
6852	photoperception	Photoperzeption *f*, Lichtwahrnehmung *f*
6853	photoperiod	Photoperiode *f*
6854	photoperiodism, photoperiodicity	Photoperiodismus *m*
6855	photophilous, photophil(ic)	photophil
6856	photophore → luminescent organ	—
6857	photophosphorylation, photosynthetic phosphorylation	Photophosphorylierung *f*
6858	photoreactivation	Photoreaktivierung *f*
6859	photoreceptor, light receptor	Photorezeptor *m*, Lichtrezeptor *m*
6860	photosensitive, photosensitivity → light sensitive, light sensitivity	—
6861	photosynthesis	Photosynthese *f*
6862	photosynthetic	photosynthetisch
6863	photosynthetic rate	Photosyntheserate *f*, Photosyntheseverhältnis *n*
6864	phototaxis	Phototaxis *f*
6865	phototonus	Phototonus *m*
6866	phototrophic	phototroph
6867	phototropism, heliotropism	Phototropismus *m*, Heliotropismus *m*
6868	phragmoplast	Phragmoplast *m*
6869	phycobilin	Phykobilin *n*
6870	phycocyanin	Phykozyan(in) *n*, Algenblau *n*
6871	phycoerythrin	Phykoerythrin *n*, Algenrot *n*
6872	phycology → algology	—
6873	phycomycetes, algal fungi	Algenpilze *mpl*, Phykomyzeten *mpl*
6874	phycophaein	Phykophäin *n*
6875	phyletic	phyletisch
6876	phyletic series	Stammesreihe *f*
6877	phylloclade, cladophyll	Phyllokladium *n*
6878	phyllode	Phyllodium *n*, Blattstielblatt *n*
6879	phylloquinone, antih(a)emorrhagic vitamin, vitamin K	Phyllochinon *n*, antihämorrhagisches Vitamin *n*, Koagulationsvitamin *n*, Vitamin *n* K
6880	phyllotaxis, phyllotaxy	Blattstellung *f*
6881	phylogenesis, phylogeny	Phylogenese *f*, Phylogenie *f*, Stammesentwicklung *f*, Stammesgeschichte *f*
6882	phylogenetic	phylogenetisch, stammesgeschichtlich
6883	phylogenetic tree	Stammbaum *m*
6884	phylogenetics	Phylogenetik *f*
6885	phylum (pl. phyla)	Stamm *m*, Phylum *n* (*pl.* Phyla)
6886	physiologic(al)	physiologisch
6887	physiologic(al) race	physiologische Rasse *f*
6888	physiological ecology → eco-physiology	—
6889	physiological equilibrium	Stoffwechselgleichgewicht *n*
6890	physiological saline	physiologische Kochsalzlösung *f*, Blutersatzlösung *f*
6891	physiologist	Physiologe *m*
6892	physiology	Physiologie *f*
6893	physiology of nutrition	Ernährungsphysiologie *f*
6894	phytin	Phytin *n*
6895	phyto-albumin	Pflanzenalbumin *n*
6896	phytobenthos	Phytobenthos *n*
6897	phytobiology, plant biology	Phytobiologie *f*, Pflanzenbiologie *f*
6898	phytochemical	pflanzenchemisch, phytochemisch
6899	phytochemistry	Pflanzenchemie *f*, Phytochemie *f*

6844	photochimie *f*	fotoquímica *f*
6845	photocinèse *f*	fotocinesis *f*
6846	photolyse *f*	fotólisis *f*
6847	photométrie *f*	fotometría *f*
6848	—	—
6849	photomorphose *f*	fotomorfosis *f*
6850	photon *m*, quantum *m* de lumière	fotón *m*, cuanto *m* de luz
6851	photonastie *f*	fotonastia *f*
6852	photoperception *f*	fotopercepción *f*
6853	photopériode *f*	fotoperíodo *m*
6854	photopériodisme *m*	fotoperiodicidad *f*, fotoperiodismo *m*
6855	photophile	fotófilo
6856	—	—
6857	photophosphorylation *f*	fotofosforilación *f*, fosforilación *f* fotosintética
6858	photoréactivation *f*	fotorreactivación *f*
6859	photorécepteur *m*	fotorreceptor *m*, receptor *m* luminoso
6860	—	—
6861	photosynthèse *f*	fotosíntesis *f*
6862	photosynthétique	fotosintético
6863	indice *m* photosynthétique, rapport *m* ~	proporción *f* fotosintética
6864	phototaxie *f*, phototactisme *m*	fototaxis *f*, fototaxia *f*, fototactismo *m*
6865	phototonus *m*	fototono *m*
6866	phototrophique	fotótrofo, fototrófico
6867	phototropisme *m*, héliotropisme *m*	fototropismo *m*, heliotropismo *m*
6868	phragmoplaste *m*	fragmoplasto *m*
6869	phycobiline *f*	ficobilina *f*
6870	phycocyanine *f*, cyanophycine *f*	ficocianina *f*
6871	phycoérythrine *f*	ficoeritrina *f*
6872	—	—
6873	phycomycètes *mpl*	ficomicetos *mpl*, hongos *mpl* algoides
6874	phycophéine *f*	ficofeína *f*
6875	phylétique	filético
6876	lignée *f* phylétique, série *f* ~	serie *f* filética
6877	phylloclade *f*	filóclado *m*, filocladio *m*, cladofilo *m*
6878	phyllode *m*	filodio *m*
6879	phylloquinone *f*, vitamine *f* antihémorragique, ~ K	filoquinona *f*, vitamina *f* antihemorrágica, ~ de la coagulación, ~ K
6880	phyllotaxie *f*, disposition *f* phyllotaxique, ~ des feuilles	filotaxis *f*, filotaxia *f*, disposición *f* de las hojas
6881	phylogenèse *f*, phylogénie *f*	filogénesis *f*, filogenia *f*
6882	phylogénétique, phylogénique	filogenético, filogénico
6883	arbre *m* phylogénétique, ~ généalogique	árbol *m* filogenético, ~ genealógico
6884	phylogénétique *f*	filogenética *f*
6885	phylum *m*, clade *m*, embranchement *m*	filum *m* (pl. fila o filums)
6886	physiologique	fisiológico
6887	race *f* physiologique	raza *f* fisiológica
6888	—	—
6889	équilibre *m* métabolique, ~ physiologique	equilibrio *m* metabólico
6890	solution *f* saline physiologique	solución *f* salina fisiológica
6891	physiologiste *m*	fisiólogo *m*
6892	physiologie *f*	fisiología *f*
6893	physiologie *f* de la nutrition	fisiología *f* de la nutrición
6894	phytine *f*	fitina *f*
6895	phyto-albumine *f*	fitoalbúmina *f*, albúmina *f* vegetal
6896	phytobenthos *m*	fitobentos *m*
6897	phytobiologie *f*, biologie *f* végétale	fitobiología *f*, biología *f* vegetal, ~ botánica, biobotánica *f*
6898	phytochimique	fitoquímico
6899	phytochimie *f*, chimie *f* végétale	fitoquímica *f*

6900 phytochrome | Phytochrom *n*
6901 phytocoenosis → plant community | —
6902 phytoflagellates, Phytomastigophora | Geißelalgen *fpl*, Phytoflagellaten *mpl*
6903 phytogenesis | Phytogenese *f*
6904 phytogenetics, plant genetics | Phytogenetik *f*, Pflanzengenetik *f*
6905 phytogenous, phytogenic | phytogen
6906 phytogeographer | Pflanzengeograph *m*
6907 phytogeography, plant geography, geobotany, geographic botany | Phytogeographie *f*, Pflanzengeographie *f*, Geobotanik *f*
6908 phyto-globulin | Pflanzenglobulin *n*
6909 phytography, descriptive botany | Phytographie *f*, Pflanzenbeschreibung *f*
6910 phytohormone, plant hormone | Phytohormon *n*, Pflanzenhormon *n*, pflanzlicher Wuchsstoff *m*
6911 phytokinin → cytokinin | —
6912 phytol | Phytol *n*
6913 phytology → botany | —
6914 phytomastigophora → phytoflagellates | —
6915 phytoparasite → parasitic plant | —
6916 phytopathology | Phytopathologie *f*
6917 phytophage, plant-eater | Pflanzenfresser *m*, Phytophage *m*
6918 phytophagous, plant-eating | pflanzenfressend, phytophag
6919 phytophysiology, plant physiology | Pflanzenphysiologie *f*
6920 phytoplankton, plant plankton | Phytoplankton *n*
6921 phytosociology, plant sociology | Phytosoziologie *f*, Pflanzensoziologie *f*, soziologische Pflanzengeographie *f*, Phytozönologie *f*, Vegetationskunde *f*
6922 phytostérol, plant sterol | Phytosterin *n*
6923 phytotomy → plant anatomy | —
6924 phytotoxic | phytotoxisch, pflanzenschädlich
6925 phytotoxicity | Phytotoxizität *f*, Pflanzenschädlichkeit *f*
6926 phytotoxin, plant poison, vegetable ~ | Pflanzengift *n*, Phytotoxin *n*
6927 phytotron | Phytotron *n*
6928 phytozoon | Phytozoon *n*
6929 pia mater | weiche Gehirnhaut *f*
6930 pigeon's milk → crop milk | —
6931 pigment | Pigment *n*, Farbkörper *m*
6932 to pigment | pigmentieren
6933 pigment cell, chromocyte | Pigmentzelle *f*, Chromozyt *m*
6934 pigment cup | Pigmentbecher *m*
6935 pigment layer, pigmentary ~ | Pigmentschicht *f*
6936 pigment migration | Pigmentwanderung *f*
6937 pigment spot | Pigmentfleck *m*
6938 pigmentary colo(u)r | Pigmentfarbe *f*
6939 pigmentation | Pigmentierung *f*
6940 pileorhiza → root-cap | —
6941 pileus | Pilzhut *m*
6942 piliferous layer, root-hair zone | Wurzelhaarzone *f*
6943 pilose, hairy | haarig, behaart
6944 pilosity | Behaarung *f*
6945 pincer → chela | —
6946 pincers | Pinzette *f*
6947 pineal body → epiphyse[2] | —
6948 pineal organ, ~ eye | Pinealorgan *n*
6949 pinna *bot.* | Fieder *f*
6950 pinnate | gefiedert
6951 pinnate leaf | Fiederblatt *n*
6952 pinniform | fiederförmig
6953 pinninervate, pinnate-veined, feather-veined | fiedernervig
6954 pinocytic vesicle | Pinozytosebläschen *n*, pinozytäres Bläschen *n*
6955 pinocytosis | Pinozytose *f*
6956 pioneer plant | Pionierpflanze *f*
6957 pipette, pipet (US) | Pipette *f*
6958 pistil | Stempel *m*, Pistillum *n*

6900	phytochrome *m*	fitocroma *m*
6901	—	—
6902	phytoflagellés *mpl*	fitoflagelados *mpl*
6903	phytogenèse *f*, phytogénie *f*	fitogénesis *f*, fitogenia *f*
6904	phytogénétique *f*, génétique *f* végétale	fitogenética *f*, genética *f* vegetal
6905	phytogène, phytogénique	fitógeno
6906	phytogéographe *m*	fitogeógrafo *m*
6907	phytogéographie *f*, géographie *f* botanique, biogéographie *f* végétale	fitogeografía *f*, geografía *f* botánica, ~ vegetal, geobotánica *f*
6908	phyto-globuline *f*	fitoglobulina *f*, globulina *f* vegetal
6909	phytographie *f*	fitografía *f*
6910	phytohormone *f*	fitohormona *f*, hormona *f* vegetal
6911	—	—
6912	phytol *m*	fitol *m*
6913	—	—
6914	—	—
6915	—	—
6916	phytopathologie *f*, pathologie *f* végétale	fitopatología *f*
6917	phytophage *m*	fitófago *m*
6918	phytophage	fitófago
6919	physiologie *f* végétale	fitofisiología *f*, fisiología *f* vegetal
6920	phytoplancton *m*, plancton *m* végétal	fitoplancton *m*, plancton *m* vegetal
6921	phytosociologie *f*	fitosociología *f*, sociología *f* vegetal, ~ botánica, fitocenología *f*
6922	phytostérol *m*	fitosterol *m*, fitosterina *f*
6923	—	—
6924	phytotoxique	fitotóxico
6925	phytotoxicité *f*	fitotoxicidad *f*
6926	phytotoxine *f*, toxine *f* végétale, poison *m* végétal	fitotoxina *f*, veneno *m* vegetal, tóxico *m* ~
6927	phytotron *m*	fitotrón *m*
6928	phytozoaire *m*	fitozoo *m*
6929	pie-mère *f*	piamadre *f*
6930	—	—
6931	pigment *m*	pigmento *m*
6932	pigmenter	pigmentar
6933	cellule *f* pigmentaire, chromocyte *m*	célula *f* pigmentaria, cromocito *m*
6934	cupule *f* pigmentaire, ~ ocellaire	cúpula *f* pigmentaria
6935	couche *f* pigmentaire	capa *f* pigmentaria
6936	migration *f* pigmentaire, mouvement *m* ~	movilización *f* del pigmento
6937	tache *f* pigmentaire	mancha *f* de pigmento
6938	pigment *m* colorant	color *m* pigmentario
6939	pigmentation *f*	pigmentación *f*
6940	—	—
6941	chapeau *m*	píleo *m*
6942	assise *f* pilifère, zone *f* ~	capa *f* pilífera, zona *f* ~
6943	pileux, poilu	piloso, peludo
6944	pilosité *f*	pilosidad *f*
6945	—	—
6946	pincette *f*	pinzas *fpl*
6947	—	—
6948	œil *m* pinéal	ojo *m* pineal
6949	penne *f*	pinna *f*
6950	pinné, penné	pinnado
6951	feuille *f* pinnée, ~ pennée	hoja *f* pinnada
6952	penniforme	penniforme
6953	penninervé	penninervio, penninervado, pinnatinervio
6954	vésicule *f* de pinocytose	vesícula *f* pinocítica
6955	pinocytose *f*	pinocitosis *f*
6956	plante *f* pionnière	planta *f* pionera
6957	pipette *f*	pipeta *f*
6958	pistil *m*	pistilo *m*

6959	pistillate flower	Stempelblüte *f*, weibliche Blüte *f*
6960	pit *bot.*	Tüpfel *m bot.*
6961	pith → medulla	—
6962	pitocin → oxytocin	—
6963	pitressin → vasopressin	—
6964	pitted vessel	Tüpfelgefäß *n*
6965	pituicyte	Pituizyt *m*
6966	pituitary → hypophyse	—
6967	pituitary hormone → hypophyseal hormone	—
6968	placenta *anat.*, *bot.*	Plazenta *f*, a) Mutterkuchen *m anat.*, b) Samenleiste *f bot.*
6969	placental circulation → foetal ~	—
6970	Placentalia, placental mammals, Eutheria	Plazentatiere *npl*, Plazentalier *mpl*
6971	placentation *bot.*	Plazentation *f bot.*
6972	placoid scale → denticle	—
6973	plagiogeotropism	Plagiogeotropismus *m*, Transversalgeotropismus *m*
6974	plagiotropic	plagiotrop
6975	plagiotropism	Plagiotropismus *m*
6976	plain muscle → smooth muscle	—
6977	planation	Planation *f*, Verflächung *f*
6978	plankton	Plankton *n*
6979	planktonic	planktonisch
6980	planktonic organism, planktont, plankter	Planktonorganismus *m*, Planktont *m*, Plankter *m*
6981	planktonophage, plankton eater, ~ feeder	Planktonfresser *m*
6982	planocyte → wandering cell	—
6983	planogamete → zoogamete	—
6984	plant	Pflanze *f*
6985	plant . . . see also phyto . . .	—
6986	plant anatomy, phytotomy	Pflanzenanatomie *f*, Phytotomie *f*
6987	plant breeder	Pflanzenzüchter *m*
6988	plant breeding	Pflanzenzüchtung *f*
6989	plant cell	Pflanzenzelle *f*
6990	plant community, phytocoenosis	Pflanzengesellschaft *f*, -gemeinschaft *f*, Phytozönose *f*
6991	plant cover, vegetation(al) ~	Pflanzendecke *f*
6992	plant eater, plant-eating → phytophage, phytophagous	—
6993	plant ecology	Pflanzenökologie *f*, Phyto-Ökologie *f*
6994	plant gall → cecidium	—
6995	plant hair → trichome	—
6996	plant indicating lime soil	Kalkzeiger *m*
6997	plant kingdom	Pflanzenreich *n*
6998	plant morphology	Pflanzenmorphologie *f*
6999	plant nutrient	Pflanzennährstoff *m*
7000	plant nutrition	Pflanzenernährung *f*
7001	plant parasite	Pflanzenparasit *m*, Pflanzenschmarotzer *m*
7002	plant physiologist	Pflanzenphysiologe *m*
7003	plant pigment	Pflanzenpigment *n*
7004	plant poison → phytotoxin	—
7005	plant protection	Pflanzenschutz *m*
7006	plant protein	Pflanzeneiweiß *n*
7007	plant species	Pflanzenart *f*
7008	plant taxonomy, ~ systematics	Pflanzentaxonomie *f*, -systematik *f*, botanische Systematik *f*
7009	plant tissue, vegetable ~	Pflanzengewebe *n*
7010	plant world	Pflanzenwelt *f*
7011	plantigrade, palmigrade	Sohlengänger *m*
7012	plantlet, plantling	Keimpflanze *f*
7013	planula	Planula *f*
7014	plaque	Plaque *f*, Phagenloch *n*
7015	plasm(a)	Plasma *n*

6959	fleur *f* pistillée, ~ femelle	flor *f* pistilada, ~ femenina
6960	ponctuation *f bot.*	punteadura *f bot.*
6961	—	—
6962	—	—
6963	—	—
6964	vaisseau *m* ponctué	vaso *m* punteado
6965	pituicyte *m*	pituicito *m*
6966	—	—
6967	—	—
6968	placenta *m anat., bot.*	placenta *f anat., bot.*
6969	—	—
6970	(mammifères *mpl*) placentaires *mpl*, euthériens *mpl*	(mamíferos *mpl*) placentarios *mpl*, euterios *m*
6971	placentation *f bot.*	placentación *f bot.*
6972	—	—
6973	plagiogéotropisme *m*	geotropismo *m* transversal
6974	plagiotrope	plagiótropo
6975	plagiotropisme *m*	plagiotropismo *m*
6976	—	—
6977	planation *f*	planación *f*
6978	plancton *m*	plancton *m*
6979	planctonique	planctónico
6980	organisme *m* planctonique	organismo *m* planctónico, planctonte *m*
6981	planctonophage *m*, mangeur *m* de plancton	planctófago *m*, planctonófago *m*
6982	—	—
6983	—	—
6984	plante *f*, végétal *m*	planta *f*, vegetal *m*
6985	—	—
6986	anatomie *f* végétale, phytotomie *f*	anatomía *f* vegetal, fitotomía *f*
6987	phytogénéticien *m*, sélectionneur *m* (de plantes)	fitogenetista *m*, fitogeneticista *m*, criador *m* de plantas
6988	sélection *f* des plantes	selección *f* de plantas
6989	cellule *f* végétale	célula *f* vegetal
6990	communauté *f* végétale, phytocénose *f*	comunidad *f* vegetal, fitocenosis *f*
6991	couverture *f* végétale	capa *f* vegetal
6992	—	—
6993	écologie *f* végétale, phytoécologie *f*	ecología *f* vegetal, fitoecología *f*
6994	—	—
6995	—	—
6996	indicateur *m* de chaux	indicador *m* de cal
6997	règne *m* végétal	reino *m* vegetal
6998	morphologie *f* végétale	morfología *f* vegetal
6999	élément *m* nutritif des plantes	nutriente *m* vegetal, ~ de las plantas
7000	nutrition *f* végétale, ~ des plantes	nutrición *f* vegetal, ~ de las plantas
7001	parasite *m* végétal	parásito *m* vegetal
7002	physiologiste *m* végétal	fitofisiólogo *m*
7003	pigment *m* végétal	pigmento *m* vegetal
7004	—	—
7005	protection *f* des plantes	protección *f* de las plantas
7006	protéine *f* végétale	proteína *f* vegetal
7007	espèce *f* végétale	especie *f* vegetal
7008	systématique *f* botanique, botanique *f* systématique	taxonomía *f* botánica, ~ vegetal, botánica *f* sistemática, fitotaxonomía *f*
7009	tissu *m* végétal	tejido *m* vegetal
7010	monde *m* végétal	mundo *m* vegetal
7011	plantigrade *m*	plantígrado *m*
7012	plantule *f*	plántula *f*
7013	planule *f*	plánula *f*
7014	plaque *f*	placa *f*
7015	plasma *m*	plasma *m*

7016 plasma cell, plasmocyte — Plasmazelle *f*

7017 plasmagene — Plasmagen *n*
7018 plasmalemma — Plasmalemma *n*, äußere Plasmahaut *f*

7019 plasma membrane, (cyto)plasmic ~ — Plasmahaut *f*, Plasmamembran *f*
7020 plasma protein — Plasmaprotein *n*, Plasmaeiweiß *n*
7021 plasmatic, plasmic — plasmatisch
7022 plasmatic inheritance → cytoplasmic ~ — —
7023 plasmid — Plasmid *n*
7024 plasmin(e) — Plasmin *n*
7025 plasmodesm(a) (pl. plasmodesmata) — Plasmodesma *n*
7026 plasmodium — Plasmodium *n*
7027 plasmogamy — Plasmogamie *f*, Plasmaverschmelzung *f*
7028 plasmolysis — Plasmolyse *f*
7029 plasmon, plasmotype — Plasmon *n*, Plasmotyp *m*
7030 plasmosome — Plasmosom *n*
7031 plasmotomy — Plasmotomie *f*
7032 plast — Plast *m*
7033 plastid — Plastide *f*, Plastid *m*
7034 plastid inheritance — Plastidenvererbung *f*
7035 plastid mutation — Plastidenmutation *f*
7036 plastidome, plastome, plastidotype — Plastidom *n*, Plastom *n*, Plastidotyp *m*

7037 plastogamy, plasmatogamy — Plastogamie *f*
7038 plastogene — Plastogen *n*
7039 plastomere → chondriomere — —
7040 plastoquinone — Plastochinon *n*
7041 plastosome → mitochondrion — —
7042 plastron, ventral shield — Bauchschild *m*
7043 to plate — plattieren
7044 plate culture, platiculture — Plattenkultur *f*
7045 Plat(y)helminthes, flatworms — Plattwürmer *mpl*
7046 plectenchyma — Plektenchym *n*, Flechtgewebe *n*
7047 pleiochasium — Pleiochasium *n*
7048 pleiomorphism → pleomorphism — —
7049 pleiotropic, polyphene — pleiotrop, polyphän
7050 pleiotropism, pleiotropy, polypheny — Pleiotropie *f*, Polyphänie *f*
7051 Pleistocene — Pleistozän *n*, Diluvium *n*
7052 pleomorphic, pleiomorphous — pleomorph
7053 pleomorphism, pleiomorphism — Pleomorphismus *m*
7054 pleopod, abdominal leg — Bauchfuß *m*, Pleopode *f*
7055 plerome — Plerom *n*
7056 pleura — Brustfell *n*, Pleura *f*
7057 pleural cavity → thoracic cavity — —
7058 pleuston — Pleuston *n*
7059 plicate *(vernation)* — plikativ, gefaltet *(Knospenlage)*
7060 Pliocene — Pliozän *n*
7061 ploughshare bone → vomer — —
7062 plumage — Gefieder *n*, Federkleid *n*
7063 plumule[1] *bot.* — Sproßknospe *f*, Plumula *f*
7064 plumule[2] → down feather — —
7065 plurannual — mehrjährig
7066 plurinucleate → multinucleate — —
7067 pneumatophore, respiratory root, breathing ~ — Atemwurzel *f*
7068 pneumogastric nerve → vagus ~ — —
7069 pod → legume — —
7070 Pogonophora — Bartwürmer *mpl*
7071 poikilotherm — wechselwarmes Tier *n*
7072 poikilothermal, poikilothermic, heterothermal — wechselwarm, poikilotherm
7073 point mutation → gene mutation — —
7074 poiser → haltere — —
7075 poison, venom — Gift *n*
7076 poison gas → toxic gas — —

7016	plasmacyte *m*, plasmocyte *m*, cellule *f* plasmatique, plasmazelle *f*	plasmacito *m*, célula *f* plasmática, plasmazelle *f*
7017	plasmagène *m*	plasmagen(e) *m*
7018	plasmalemme *m*	plasmalema *m*, membrana *f* plasmática externa
7019	membrane *f* plasmatique	membrana *f* plasmática
7020	plasma-protéine *f*	proteína *f* plasmática
7021	plasmatique, plasmique	plasmático, plásmico
7022	—	—
7023	plasmide *m*	plasmidio *m*
7024	plasmine *f*	plasmina *f*
7025	plasmodesme *m*	plasmodesma *m*
7026	plasmode *m*	plasmodio *m*
7027	plasmogamie *f*	plasmogamia *f*
7028	plasmolyse *f*	plasmólisis *f*
7029	plasmon(e) *m*, plasmotype *m*	plasmón *m*, plasmótipo *m*
7030	plasmosome *m*	plasmosoma *m*
7031	plasmotomie *f*	plasmotomía *f*
7032	plaste *m*	plasto *m*
7033	plastide *m*	plastidio *m*
7034	hérédité *f* plastidique	herencia *f* plastídica
7035	mutation *f* plastidique, ~ plastidiale	mutación *f* plastídica, ~ plastidial
7036	plastidome *m*, plastome *m*, plastidotype *m*	plastidoma *m*, plastoma *m*, plastidiótipo *m*
7037	plastogamie *f*	plastogamia *f*, plasmatogamia *f*
7038	plastogène *m*	plastogen(e) *m*
7039	—	—
7040	plastoquinone *f*	plastoquinona *f*
7041	—	—
7042	plastron *m*, bouclier *m* ventral	peto *m*, caparazón *m* ventral
7043	plaquer	poner en placa
7044	culture *f* sur plaques	cultivo *m* en place, platicultivo *m*
7045	plathelminthes *m pl*, vers *m pl* plats	platelmintos *m pl*, gusanos *m pl* planos
7046	plectenchyme *m*	plecténquima *m*
7047	cyme *f* multipare	pleocasio *m*, pleyocasio *m*, policasio *m*
7048	—	—
7049	pléiotropique, polyphène	ple(i)otropo, pleiotrópico
7050	pléiotropie *f*, pléotropie *f*, polyphénie *f*	ple(i)otropismo *m*, ple(i)otropía *f*, polifenia *f*
7051	pléistocène *m*, diluvium *m*	pleistoceno *m*
7052	pléomorphe	ple(i)omórfico
7053	pléomorphisme *m*	ple(i)omorfismo *m*
7054	pléopode *m*, patte *f* abdominale	pleópodo *m*
7055	plérome *m*	pleroma *m*
7056	plèvre *f*	pleura *f*
7057	—	—
7058	pleuston *m*	pleuston *m*
7059	plissé *(vernation)*	plegado *(vernación)*
7060	pliocène *m*	plioceno *m*
7061	—	—
7062	plumage *m*	plumaje *m*
7063	plumule *f*	plúmula *f*, plumilla *f*
7064	—	—
7065	pluriannuel	plurienal
7066	—	—
7067	pneumatophore *m*, racine *f* respiratoire	neumatóforo *m*, raíz *f* respiratoria
7068	—	—
7069	—	—
7070	pogonophores *m pl*	pogonóforos *m pl*
7071	animal *m* poïkilotherme	poiquilotermo *m*
7072	poïkilotherme, poecilotherme, hétérotherme	poiquilotermo, poiquilotérmico, heterotérmico
7073	—	—
7074	—	—
7075	poison *m*, toxique *m*; venin *m (animaux)*	veneno *m*, tóxico *m*
7076	—	—

7077	poison gland, venom ~	Giftdrüse *f*
7078	poison tooth, venomous ~, poison fang	Giftzahn *m*
7079	poisonous → toxic	—
7080	poisonous animal, venomous ~	Gifttier *n*
7081	poisonous plant	Giftpflanze *f*
7082	poisonous substance → toxin	—
7083	polar body, polocyte	Polkörperchen *n*, Richtungskörper *m*, Polozyt *m*
7084	polar cap	Polkappe *f*
7085	polar capsule	Polkapsel *f*
7086	polar nucleus	Polkern *m*
7087	polar plate	Polplatte *f*
7088	polar rays, astral ~	Polstrahlen *mpl*, Sternstrahlen *mpl*
7089	polarity, polar behavio(u)r	Polarität *f*
7090	polarization	Polarisation *f*, Polarisierung *f*
7091	polarization filter, polarizing ~	Polarisationsfilter *m*
7092	polarization microscope, polarizing ~	Polarisationsmikroskop *n*
7093	to polarize	polarisieren
7094	polarized light	polarisiertes Licht *n*
7095	pole field	Polfeld *n*
7096	pollen	Blütenstaub *m*, Pollen *m*
7097	pollen analysis	Pollenanalyse *f*
7098	pollen chamber	Pollenkammer *f*
7099	pollen diagram	Pollendiagramm *n*
7100	pollen grain	Pollenkorn *n*
7101	pollen mother cell, microsporocyte	Pollenmutterzelle *f*, Mikrosporozyt *m*
7102	pollen sac	Pollensack *m*
7103	pollen sterility	Pollensterilität *f*
7104	pollen transfer	Pollenübertragung *f*
7105	pollen tube	Pollenschlauch *m*
7106	pollinarium	Pollinarium *n*
7107	to pollinate	bestäuben
7108	pollinating agent, pollinator	Bestäuber *m*
7109	pollinating insect, insect pollinator	Bestäuberinsekt *n*
7110	pollination	Bestäubung *f*
7111	pollination drop	Bestäubungstropfen *m*
7112	polliniferous, pollen-bearing	pollentragend, pollenführend
7113	pollinium	Pollinium *n*
7114	pollutant	Schmutzstoff *m*, Verschmutzungsstoff *m*, Umweltverschmutzer *m (Anlage)*
7115	to pollute	verschmutzen, verunreinigen
7116	polluter	Umweltverschmutzer *m (Person)*
7117	pollution	Verschmutzung *f*, Verunreinigung *f*
7118	pollution of the sea	Meerwasserverschmutzung *f*
7119	pollution source	Verschmutzungsquelle *f*
7120	polocyte → polar body	—
7121	polyandrous	polyandrisch
7122	polyandry	Polyandrie *f*
7123	polycarpic, polycarpous	polykarp
7124	polycentric	polyzentrisch
7125	Polychaeta, bristle worms	Borstenwürmer *mpl*, Vielborster *mpl*
7126	polycross	Massenkreuzung *f*, Polycrossmethode *f*
7127	polycyclic	polyzyklisch
7128	polyembryony	Polyembryonie *f*
7129	polyenergid	Polyenergide *f*
7130	polyfactorial, multifactorial	plurifaktoriell, multifaktoriell
7131	polygamous	polygam
7132	polygamy	Polygamie *f*
7133	polygene, minor gene, multiple factor	Polygen *n*
7134	polygenesis	Polygenese *f*
7135	polygenic	polygen(isch)
7136	polygeny	Polygenie *f*
7137	polygynous	polygyn

7077	glande *f* à venin, ~ venimeuse	glándula *f* venenosa, ~ tóxica
7078	crochet *m* à venin, dent *f* venimeuse	diente *m* venenoso, colmillo *m* ~
7079	—	—
7080	animal *m* venimeux	animal *m* venenoso
7081	plante *f* vénéneuse, ~ toxique	planta *f* venenosa, ~ tóxica
7082	—	—
7083	corpuscule *m* polaire, polocyte *m*	cuerpo *m* polar, polocito *m*
7084	calotte *f* polaire	casquete *m* polar
7085	capsule *f* polaire	cápsula *f* polar
7086	noyau *m* polaire	núcleo *m* polar
7087	plaque *f* polaire	placa *f* polar
7088	raies *fpl* astrales	rayos *mpl* del áster
7089	polarité *f*	polaridad *f*
7090	polarisation *f*	polarización *f*
7091	filtre *m* de polarisation	filtro *m* polarizador
7092	microscope *m* polarisant, ~ à polarisation	microscopio *m* polarizante
7093	polariser	polarizar
7094	lumière *f* polarisée	luz *f* polarizada
7095	région *f* polaire	área *f* polar
7096	pollen *m*	polen *m*
7097	analyse *f* pollinique	análisis *m* polínico
7098	chambre *f* pollinique, loge *f* ~	cámara *f* polínica
7099	diagramme *m* pollinique	diagrama *m* polínico
7100	grain *m* de pollen	grano *m* de polen, ~ polínico
7101	cellule *f* mère des grains de pollen, microsporocyte *m*	célula *f* madre del polen, microsporocito *m*
7102	sac *m* pollinique	saco *m* polínico
7103	stérilité *f* du pollen, ~ pollinique	esterilidad *f* del polen, ~ polínica
7104	transport *m* du pollen	transferencia *f* del polen
7105	tube *m* pollinique	tubo *m* polínico
7106	pollinarie *f*	polinario *m*
7107	polliniser	polinizar
7108	(agent *m*) pollinisateur *m*	agente *m* polinizador, polinizante *m*
7109	insecte *m* pollinisateur	insecto *m* polinizador
7110	pollinisation *f*	polinización *f*
7111	goutte *f* de pollinisation	gota *f* polinizante
7112	pollinifère	polinífero
7113	pollinie *f*, pollinide *m*	polinio *m*, polinia *f*
7114	polluant *m*, agent *m* ~, matière *f* polluante, contaminant *m*	agente *m* de polución, poluante *m*, contaminante *m*, agente *m* ~, sustancia *f* ~
7115	polluer	polucionar, poluar
7116	pollueur *m*	contaminador *m*, autor *m* de la polución
7117	pollution *f*	polución *f*
7118	pollution *f* de la mer	polución *f* del (agua del) mar, contaminación *f* marina
7119	source *f* de la pollution, ~ de contamination	fuente *f* contaminante, ~ de contaminación, foco *m* de polución
7120	—	—
7121	polyandrique	poliándrico
7122	polyandrie *f*	poliandria *f*
7123	polycarpique, polycarpien	policárpico
7124	polycentrique	policéntrico
7125	polychètes *mpl*	poliquetos *mpl*, poliquétidos *mpl*
7126	polycroisement *m*	policruzamiento *m*
7127	polycyclique	policíclico
7128	polyembryon(n)ie *f*	poliembrionía *f*
7129	polyénergide *f*	polienérgida *f*
7130	polyfactoriel, plurifactoriel, multifactoriel	polifactorial
7131	polygame	polígamo
7132	polygamie *f*	poligamia *f*
7133	polygène *m*, gène *m* mineur	poligen *m*, gen *m* menor, factor *m* múltiple
7134	polygenèse *f*	poligénesis *f*
7135	polygénique	poligénico
7136	polygénie *f*	poligenia *f*
7137	polygyne	poligínico

No.	English	German
7138	polygyny	Polygynie *f*
7139	polyhaploidy	Polyhaploidie *f*
7140	polyhybrid	polyhybrid
7141	polyhybridism	Polyhybridie *f*
7142	polykaric → multinucleate	—
7143	polykaryocyte	Polykaryozyt *m*
7144	polymerase	Polymerase *f*
7145	polymeric(al), polymerous	polymer
7146	polymeric gene	Polymergen *n*
7147	polymerism, polymery	Polymerie *f*
7148	polymerization	Polymerisation *f*
7149	polymorph → granulocyte	—
7150	polymorphic, polymorphous	polymorph
7151	polymorphism	Polymorphismus *m*
7152	polynucleotide	Polynukleotid *n*
7153	polynucleotide chain	Polynukleotidkette *f*
7154	poly(o)estrous	polyöstrisch
7155	polyp	Polyp *m*
7156	polypeptide	Polypeptid *n*
7157	polypeptide chain	Polypeptidkette *f*
7158	polyphagia	Polyphagie *f*
7159	polyphagous	polyphag
7160	polyphene, polypheny → pleiotropic, pleiotropism	—
7161	polyphyletic	polyphyletisch
7162	polyphyletism, polyphyletic evolution	Polyphylie *f*
7163	polyploid	polyploid
7164	polyploidization	Polyploidisierung *f*
7165	polyploidy	Polyploidie *f*
7166	polyribosome, polysome, ergosome	Polyribosom *n*, Polysom *n*, Ergosom *n*
7167	polysaccharide	Polysa(c)charid *n*
7168	polysomic	polysom
7169	polysomy	Polysomie *f*
7170	polyspermous, many-seeded	mehrsamig, vielsamig
7171	polyspermy	Polyspermie *f*
7172	polystely	Polystelie *f*
7173	polytene	polytän
7174	polyteny	Polytänie *f*
7175	polytopic	polytop
7176	polytopy	Polytopie *f*
7177	polytrophic	polytroph
7178	polytypic	polytypisch
7179	Polyzoa → Bryozoa	—
7180	pome, pomaceous fruit	Kernfrucht *f*
7181	pons Varolii, Varolius' bridge	Brücke *f* *anat.*
7182	poor in species	artenarm
7183	population	Population *f*
7184	population analysis	Populationsanalyse *f*
7185	population biology, biology of populations	Populationsbiologie *f*, Bevölkerungsbiologie *f*
7186	population density	Populationsdichte *f*, Bevölkerungsdichte *f*
7187	population dynamics	Populationsdynamik *f*
7188	population ecology, biodemography	Populationsökologie *f*, Demökologie *f*
7189	population equilibrium	Populationsgleichgewicht *n*
7190	population explosion	Bevölkerungsexplosion *f*
7191	population fluctuation	Massenwechsel *m*, Fluktuation *f*
7192	population genetics, genetics of populations	Populationsgenetik *f*
7193	population growth	Populationswachstum *n*, Bevölkerungs~ *n*
7194	population pressure	Populationsdruck *m*, Bevölkerungsdruck *m*
7195	population pyramid	Alterspyramide *f*, Bevölkerungspyramide *f*
7196	population size	Populationsgröße *f*
7197	population waves	Populationswellen *fpl*
7198	P/O ratio	P/O-Quotient *m*, Phosphorylierungsquotient *m*

7138	polygynie *f*	poliginia *f*
7139	polyhaploïdie *f*	polihaploidia *f*
7140	polyhybride	polihíbrido
7141	polyhybridisme *m*	polihibridismo *m*
7142	—	—
7143	polycaryocyte *m*	policariocito *m*
7144	polymérase *f*	polimerasa *f*
7145	polymère, polymérique	polímero, polimérico
7146	gène *m* polymère	gen *m* polímero
7147	polymérisme *m*, polymérie *f*	polimerismo *m*, polimería *f*
7148	polymérisation *f*	polimerización *f*
7149	—	—
7150	polymorphe	polimorfo
7151	polymorphisme *m*	polimorfismo *m*
7152	polynucléotide *m*	polinucleótido *m*
7153	chaîne *f* polynucléotidique	cadena *f* polinucleotídica
7154	polyoestral, polyoestrien	poliéstrico
7155	polype *m*	pólipo *m*
7156	polypeptide *m*	polipéptido *m*
7157	chaîne *f* polypeptidique	cadena *f* polipeptídica
7158	polyphagie *f*	polifagia *f*
7159	polyphage	polífago
7160	—	—
7161	polyphylétique	polifilético
7162	polyphylétisme *m*	polifiletismo *m*
7163	polyploïde	poliploide
7164	polyploïdisation *f*	poliploidización *f*
7165	polyploïdie *f*	poliploidia *f*
7166	polyribosome *m*, polysome *m*	polirribosoma *m*, polisoma *m*, ergosoma *m*
7167	polysaccharide *m*, polyoside *m*, polyholoside *m*	polisacárido *m*, poliósido *m*, poliholósido *m*
7168	polysomique	polisómico
7169	polysomie *f*	polisomía *f*
7170	polysperme	polispermo
7171	polyspermie *f*	polispermia *f*
7172	polystélie *f*	polistelia *f*
7173	polytène	politénico
7174	polyténie *f*	politenia *f*
7175	polytopique	politópico
7176	polytopie *f*, polytopisme *m*	politopismo *m*
7177	polytrophique	politrófico
7178	polytypique	politípico
7179	—	—
7180	fruit *m* à pépin	pomo *m*, fruto *m* de pepita
7181	pont *m* de Varole, protubérance *f* annulaire	puente *m* de Varolio, protuberancia *f* anular
7182	pauvre en espèces	pobre en especies
7183	population *f*	población *f*
7184	analyse *f* de population	análisis *m* de población
7185	biologie *f* des populations	biología *f* de las poblaciones
7186	densité *f* de (la) population	densidad *f* de (la) población, ~ demográfica
7187	dynamique *f* des populations	dinámica *f* de las poblaciones
7188	écologie *f* des populations	ecología *f* de población
7189	équilibre *m* démographique	equilibrio *m* demográfico
7190	explosion *f* démographique	explosión *f* demográfica
7191	fluctuation *f* des populations	fluctuación *f* de población
7192	génétique *f* des populations	genética *f* de (las) poblaciones
7193	croissance *f* de (la) population, ~ démographique	crecimiento *m* de (las) poblaciones, ~ demográfico
7194	pression *f* de population, ~ démographique	presión *f* de población, ~ demográfica
7195	pyramide *f* d'âge, ~ des âges	pirámide *f* de las edades
7196	grandeur *f* de la population, dimension *f* ~ ~ ~	tamaño *m* de la población
7197	ondes *f pl* de population	ondas *f pl* de población
7198	rapport *m* P/O	relación *f* P/O, cociente *m* P/O

7199	pore	Pore *f*
7200	pore capsule, poricidal ~	Porenkapsel *f*
7201	pore volume	Porenvolumen *n*
7202	Porifera, sponges	Poriferen *pl*, Schwämme *mpl*, Schwammtiere *npl*
7203	porogamy	Porogamie *f*
7204	porphyrin	Porphyrin *n*
7205	portal circulation, ~ system	Pfortaderkreislauf *m*
7206	portal vein	Pfortader *f*, Portalgefäß *n*
7207	position effect	Positionseffekt *m*, Lageeffekt *m*
7208	positional pseudoallele	Positionspseudoallel *n*
7209	post-adaptation	Postadaptation *f*
7210	postembryonic, postembryonal	postembryonal
7211	posterior pituitary, ~ lobe (of hypophysis), posthypophysis	Hypophysenhinterlappen *m*
7212	posterior pituitary hormone	Hypophysenhinterlappenhormon *n*
7213	posterior root → dorsal root	—
7214	postganglionic	postganglionär
7215	postglacial epoch, ~ age	Nacheiszeit *f*, Postglazial *n*
7216	postheterokinesis	Postheterokinese *f*
7217	postnatal development	Postnatalentwicklung *f*, postnatale Entwicklung *f*
7218	post-reduction	Postreduktion *f*, Prääquation *f*
7219	postsynaptic	postsynaptisch
7220	postural reflex	Lagereflex *m*
7221	posture sense	Lagesinn *m*
7222	potency, potence *gen.*	Potenz *f gen.*
7223	potential difference	Potentialdifferenz *f*
7224	potential energy	Lageenergie *f*, potentielle Energie *f*
7225	potentiometer	Potentiometer *n*
7226	potometer	Po(te)tometer *n*
7227	P.P. factor → nicotinamide	—
7228	preadaptation	Präadaptation *f*
7229	Precambrian (period)	Präkambrium *n*
7230	precartilage, cartilaginous precursor	Knorpelanlage *f*
7231	precipitable	(aus)fällbar
7232	precipitant	Fällungsmittel *n*
7233	to precipitate	ausfällen
7234	precipitate	Präzipitat *n*
7235	precipitation	Ausfällung *f*, Präzipitation *f*
7236	precipitin	Präzipitin *n*
7237	precipitin reaction	Präzipitinreaktion *f*
7238	precocial bird → nidifugous bird	—
7239	precocious	frühreif
7240	precocity, prematurity	Frühreife *f*
7241	precocity theory	Präkozitätstheorie *f*
7242	precursor	Vorstufe *f*
7243	predaceous → predatory	—
7244	predation	Episitie *f*, Episitismus *m*, Räubertum *n*
7245	predator	Episit *m*, Räuber *m*
7246	predator-prey relation(ship)	Räuber-Beute-Beziehung *f*
7247	predatory, predaceous, preying, raptorial	räuberisch
7248	predatory bird → bird of prey	—
7249	predatory exploitation	Raubbau *m*, Raubwirtschaft *f*
7250	predatory fish, predaceous ~	Raubfisch *m*
7251	predetermination	Prädetermination *f*
7252	preen gland → uropygial gland	—
7253	preferred temperature, temperature preferendum	Vorzugstemperatur *f*
7254	prefloration → aestivation[2]	—
7255	prefoliation → vernation	—
7256	preformation	Präformation *f*
7257	preformation theory, preformism, preformationism	Präformationstheorie *f*
7258	preganglionic	präganglionär

7199	pore *m*	poro *m*
7200	capsule *f* poricide	cápsula *f* porífera
7201	volume *m* des pores	volumen *m* de (los) poros
7202	porifères *mpl*, spongiaires *mpl*	poríferos *mpl*, espongiarios *mpl*
7203	porogamie *f*	porogamia *f*
7204	porphyrine *f*	porfirina *f*
7205	circulation *f* portale, ~ porte, système *m* ~	circulación *f* portal, sistema *m* ~
7206	veine *f* porte	vena *f* porta
7207	effet *m* de position	efecto *m* de posición
7208	pseudoallèle *m* de position	pseudoalelo *m* de posición
7209	post-adaptation *f*	postadaptación *f*
7210	postembryonnaire	postembrionario
7211	posthypophyse *f*, lobe *m* postérieur (de l'hypophyse)	posthipófisis *f*, pituitaria *f* posterior, lóbulo *m* ~ (de la hipófisis)
7212	hormone *f* posthypophysaire	hormona *f* posthipofisaria
7213	—	—
7214	postganglionnaire	postganglionar
7215	période *f* post-glaciaire	período *m* postglacial
7216	post-hétérocinèse *f*	postheterocinesis *f*
7217	développement *m* postnatal	desarrollo *m* postnatal
7218	post-réduction *f*	postreducción *f*
7219	postsynaptique	postsináptico
7220	réflexe *m* de posture	reflejo *m* postural
7221	sens *m* postural	sentido *m* postural, ~ de la posición
7222	potence *f gen.*	potencia *f gen.*
7223	différence *f* de potentiel	diferencia *f* de potencial
7224	énergie *f* potentielle	energía *f* potencial
7225	potentiomètre *m*	potenciómetro *m*
7226	potomètre *m*	po(te)tómetro *m*
7227	—	—
7228	préadaptation *f*	preadaptación *f*
7229	précambrien *m*, antécambrien *m*	precámbrico *m*
7230	procartilage *m*	precartílago *m*, cartílago *m* precursor
7231	précipitable	precipitable
7232	précipitant *m*	precipitante *m*
7233	précipiter	precipitar
7234	précipité *m*	precipitado *m*
7235	précipitation *f*	precipitación *f*
7236	précipitine *f*	precipitina *f*
7237	réaction *f* de précipitation	reacción *f* de precipitación, ~ de la precipitina
7238	—	—
7239	précoce	precoz
7240	précocité *f*, prématurité *f*	precocidad *f*
7241	théorie *f* de la précocité	teoría *f* de la precocidad
7242	précurseur *m*	precursor *m*
7243	—	—
7244	prédation *f*	predación *f*, predatorismo *m*
7245	prédateur *m*, déprédateur *m*	predador *m*, depredador *m*
7246	relation *f* prédateur-proie	relación *f* (de) predador-presa
7247	rapace, (dé)prédateur	predatorio, rapaz
7248	—	—
7249	exploitation *f* abusive	explotación *f* abusiva
7250	poisson *m* prédateur, ~ carnassier	pez *m* (de)predador, ~ rapaz
7251	prédétermination *f*	predeterminación *f*
7252	—	—
7253	température *f* préférentielle	temperatura *f* preferida
7254	—	—
7255	—	—
7256	préformation *f*	preformación *f*
7257	théorie *f* de la préformation, préformisme *m*	teoría *f* de la preformación, preformismo *m*
7258	préganglionnaire	preganglionar

7259	pregnancy, gravidity	Schwangerschaft *f*, Gravidität *f*
7260	pregnandiol	Pregnandiol *n*
7261	pregnant, gravid	schwanger, trächtig
7262	prehensile tail	Greifschwanz *m*
7263	preheterokinesis	Präheterokinese *f*
7264	prehistory	Vorgeschichte *f*, Urgeschichte *f*
7265	prehypophysis → anterior pituitary	—
7266	prematurity → precocity	—
7267	premolar (tooth)	Vormahlzahn *m*, Prämolar *m*
7268	premunity	Prämunität *f*, Präimmunität *f*
7269	premutation	Prämutation *f*
7270	prenatal development	Pränatalentwicklung *f*, pränatale Entwicklung *f*
7271	preparation *micr.*	Präparat *n*
7272	preparator	Präparator *m*
7273	to prepare	präparieren
7274	prepotence, prepotency	Präpotenz *f*, Individualpotenz *f*
7275	prepuce, foreskin	Vorhaut *f*, Präputium *n*
7276	pre-reduction	Präreduktion *f*
7277	presence-absence hypothesis	Präsenz-Absenz-Hypothese *f*
7278	presentation time	Präsentationszeit *f*
7279	pressure difference, ~ gradient	Druckgefälle *n*, Druckdifferenz *f*
7280	pressure receptor	Druckrezeptor *m*
7281	prestreak → primitive streak	—
7282	presumptive *embr.*	präsumtiv *embr.*
7283	presynaptic	präsynaptisch
7284	previtamin → provitamin	—
7285	prey (animal)	Beute *f*, Beutetier *n*
7286	prey catching, ~ capture	Beutefang *m*, Beuteerwerb *m*
7287	prickle, acantha	Stachel *m*
7288	prickle-cell, spinose cell	Stachelzelle *f*
7289	prickly, echinate	stachelig
7290	primary consumer	Primärkonsument *m*
7291	primary leaf, primordial ~	Jugendblatt *n*, Primärblatt *n*
7292	primary meristem	primäres Bildungsgewebe *n*, ~ Meristem *n*
7293	primary production *ecol.*	Primärproduktion *f* *ecol.*
7294	primary sere → prisere	—
7295	primary structure	Primärstruktur *f*
7296	primary wood, early ~	Frühholz *n*, Frühlingsholz *n*, Weitholz *n*
7297	primates	Primaten *mpl*, Herrentiere *npl*
7298	primeval forest → virgin forest	—
7299	primitive form, basic ~	Stammform *f*, Urform *f*
7300	primitive groove	Primitivrinne *f*
7301	primitive gut → archenteron	—
7302	primitive node, Hensen's ~	Primitivknoten *m*, Hensenscher Knoten *m*
7303	primitive streak, prestreak	Primitivstreifen *m*
7304	primordial leaf → primary ~	—
7305	primordial meristem, promeristem	Urmeristem *n*, Promeristem *n*
7306	primordium, anlage	Primordium *n*, Anlage *f*
7307	prisere, primary sere	Priserie *f*
7308	probability of fertilization	Befruchtungswahrscheinlichkeit *f*
7309	proband	Proband *m*, Propositus *m*
7310	probasidium, hypobasidium	Probasidie *f*, Hypobasidie *f*
7311	Proboscidea	Rüsseltiere *npl*
7312	proboscis, trunk	Rüssel *m*
7313	procambium	Prokambium *n*
7314	prochromosome	Prochromosom *n*
7315	procrypsis → camouflage	—
7316	proctodaeum	Proctodaeum *n*, Enddarm *m* *embr.*
7317	producer *ecol.*	Produzent *m* *ecol.*
7318	production biology	Produktionsbiologie *f*
7319	production rate	Produktionsrate *f*
7320	proembryo	Vorkeim *m*, Proembryo *m*
7321	proenzyme → zymogen	—
7322	proflavin	Proflavin *n*
7323	profundal zone	Profundal *n*, Tiefenwasserzone *f*

7259	grossesse *f*, gravidité *f*	embarazo *m*, preñez *f*, gravidez *f*
7260	pregnandiol *m*	pregnandiol *m*
7261	prégnante, gestante, grosse, enceinte, gravide	preñada, gestante, embarazada, grávida
7262	queue *f* préhensile	cola *f* prensil
7263	pré-hétérocinèse *f*	preheterocinesis *f*
7264	préhistoire *f*	prehistoria *f*
7265	—	—
7266	—	—
7267	prémolaire *f*	(diente *m*) premolar *m*
7268	prémunité *f*, prémunition *f*	premunidad *f*
7269	prémutation *f*	premutación *f*
7270	développement *m* prénatal	desarrollo *m* prenatal
7271	préparation *f* *micr.*	preparación *f*, preparado *m* *micr.*
7272	préparateur *m*; dissecteur *m*	preparador *m*; disector *m*
7273	préparer	preparar
7274	prépotence *f*, pouvoir *m* raceur	prepotencia *f*, poder *m* raceador
7275	prépuce *m*	prepucio *m*
7276	pré-réduction *f*	pre-reducción *f*, prerreducción *f*
7277	théorie *f* de la présence et de l'absence	teoría *f* de la presencia-ausencia
7278	temps *m* de présentation, ~ d'impression	tiempo *m* de presentación, período *m* ~ ~
7279	gradient *m* de pression	diferencia *f* de presión
7280	récepteur *m* de pression	receptor *m* de presión
7281	—	—
7282	présomptif *embr.*	presuntivo *embr.*
7283	présynaptique	presináptico
7284	—	—
7285	proie *f*	presa *f*
7286	capture *f* des proies	captura *f* de la presa, apresamiento *m*
7287	aiguillon *m*, piquant *m*	aguijón *m*
7288	cellule *f* à épine, ~ épineuse	célula *f* espinosa, ~ dentada
7289	aiguillonné	aguijonado
7290	consommateur *m* primaire	consumidor *m* primario
7291	feuille *f* primaire, ~ de jeunesse	hoja *f* primordial, ~ juvenil
7292	méristème *m* primaire	meristema *m* primario
7293	production *f* primaire *ecol.*	producción *f* primaria *ecol.*
7294	—	—
7295	structure *f* primaire	estructura *f* primaria
7296	bois *m* primaire, ~ de printemps	leño *m* temprano, ~ de primavera, ~ amplio
7297	primates *mpl*	primates *mpl*
7298	—	—
7299	forme *f* primitive, ~ originale, ~ de souche	forma *f* primitiva, ~ originaria
7300	gouttière *f* primitive	surco *m* primitivo
7301	—	—
7302	nœud *m* de Hensen	nudo *m* de Hensen
7303	ligne *f* primitive	línea *f* primitiva, prelínea *f*
7304	—	—
7305	méristème *m* primordial	meristema *m* primordial, promeristema *m*
7306	primordium *m*, ébauche *f*	primordio *m*
7307	série *f* primaire	priser(i)e *f*, serie *f* primitiva, presere *f*
7308	probabilité *f* de fertilisation	probabilidad *f* de fertilización
7309	proband *m*, propositus *m*	probando *m*
7310	probasidie *f*	probasidio *m*, hipobasidio *m*
7311	proboscidiens *mpl*	proboscidios *mpl*
7312	proboscis *m*, trompe *f*	probóscide *f*, trompa *f*
7313	procambium *m*	procámbium *m*
7314	prochromosome *m*	procromosoma *m*
7315	—	—
7316	proctodéum *m*, intestin *m* postérieur *embr.*	proctodeo *m*
7317	producteur *m* *ecol.*	productor *m* *ecol.*
7318	biologie *f* de la production	biología *f* de la producción
7319	intensité *f* de production	proporción *f* de producción
7320	proembryon *m*	proembrión *m*
7321	—	—
7322	proflavine *f*	proflavina *f*
7323	zone *f* profonde	zona *f* profunda

7324	progamic, progamous	progam
7325	progenesis	Progenese *f*
7326	progeny, offspring	Nachkommen *mpl*, Nachkommenschaft *f*, Nachzucht *f*, Abkömmlinge *mpl*, Deszendenten *mpl*
7327	progeny-test(ing)	Nachkommenschaftsprüfung *f*, Nachzuchtprüfung *f*
7328	progesterone, progestin, luteal hormone	Progesteron *n*, Progestin *n*
7329	prolactin → lactogenic hormone	—
7330	prolamin(e)	Prolamin *n*
7331	prolan → chorionic gonadotropin	—
7332	prolan A → follicle-stimulating hormone	—
7333	prolan B → luteinizing hormone	—
7334	proliferation	Proliferation *f*, Prolifikation *f*
7335	proline	Prolin *n*
7336	promeristem → primordial meristem	—
7337	prometaphase	Prometaphase *f*
7338	prometaphase stretch	Prometaphasestreckung *f*
7339	promitosis	Promitose *f*
7340	promycelium	Promyzel *n*
7341	pronephros, fore-kidney	Vorniere *f*, Pronephros *m*
7342	pronucleus	Vorkern *m*, Pronukleus *m*
7343	pro-oestrus	Vorbrunst *f*, Proöstrus *m*
7344	propagule, propagulum, brood body, propagative ~	Brutkörper *m*
7345	prophage	Prophage *m*
7346	prophase	Prophase *f*
7347	prophase index	Prophaseindex *m*
7348	prophyll	Vorblatt *n*
7349	propionic acid	Propansäure *f*, Propionsäure *f*
7350	proplastid	Proplastide *f*, Plastidenvorstadium *n*
7351	proprio(re)ceptor, proprioceptive receptor	Proprio(re)zeptor *m*
7352	prop-root	Stützwurzel *f*
7353	prosencephalon, fore-brain	Vorderhirn *n*, Prosenzephalon *n*
7354	prosenchyma	Prosenchym *n*
7355	prosimian	Halbaffe *m*
7356	prostaglandin	Prostaglandin *n*
7357	prostate (gland)	Vorsteherdrüse *f*, Prostata *f*
7358	prosthetic group	prosthetische Gruppe *f*
7359	protamine	Protamin *n*
7360	protandrous, proterandrous	protandrisch, proterandrisch
7361	protandry, proterandry	Protandrie *f*, Proterandrie *f*, Vormännlichkeit *f*
7362	protean behaviour → deceptive ~	—
7363	protease, proteolytic enzyme	Protease *f*, proteolytisches Ferment *n*
7364	protection of nature → conservation ~ ~	—
7365	protection of waters	Gewässerschutz *m*
7366	protective colo(u)ration	Schutzfärbung *f*
7367	protective envelope	Schutzhülle *f*
7368	protective tissue	Schutzgewebe *n*
7369	proteid	Proteid *n*
7370	protein	Protein *n*, Eiweiß *n*
7371	protein break-down, catabolism of protein	Eiweißabbau *m*
7372	protein coat	Eiweißhülle *f*, Proteinhülle *f*
7373	protein compound	Eiweißverbindung *f*
7374	protein content	Eiweißgehalt *m*
7375	protein fibre (US: fiber)	Proteinfaser *f*
7376	protein metabolism, proteometabolism	Eiweißstoffwechsel *m*
7377	protein minimum	Eiweißminimum *n*
7378	protein molecule	Eiweißmolekül *n*, Proteinmolekül *n*
7379	protein nitrogen	Eiweißstickstoff *m*
7380	protein requirement	Eiweißbedarf *m*
7381	protein reserve	Eiweißreserve *f*

7324	progame	progámico
7325	progenèse *f*	progénesis *f*
7326	progéniture *f*, descendance *f*, descendants *mpl*	progenie *f*, prole *f*, desdencencia *f*, descendientes *mpl*
7327	épreuve *f* de la descendance, testage *m* ~ ~ ~	prueba *f* de descendencia, ~ de progenie
7328	progestérone *f*, progestine *f*, hormone *f* lutéale	progesterona *f*, progestina *f*, hormona *f* lútea
7329	—	—
7330	prolamine *f*	prolamina *f*
7331	—	—
7332	—	—
7333	—	—
7334	prolifération *f*	proliferación *f*
7335	proline *f*	prolina *f*
7336	—	—
7337	prométaphase *f*	prometafase *f*
7338	élongation *f* prométaphasique	elongación *f* prometafásica
7339	promitose *f*	promitosis *f*
7340	promycélium *m*	promicelio *m*
7341	pronéphros *m*	pronefros *m*
7342	pronucléus *m*	pronúcleo *m*
7343	prooestrus *m*	proestro *m*, precelo *m*
7344	propagule *m*	propágulo *m*
7345	prophage *m*	prófago *m*
7346	prophase *f*	prófase *f*, profase *f*
7347	indice *m* des prophases	índice *m* de las profases
7348	préfeuille *f*	profilo *m*
7349	acide *m* propionique	ácido *m* propiónico
7350	proplastide *m*	proplastidio *m*, primordio *m* plastídico
7351	proprio(ré)cepteur *m*	proprioceptor *m*, propriorreceptor *m*
7352	racine *f* adventice aérienne	raíz *f* aérea de sostén
7353	prosencéphale *m*, cerveau *m* antérieur	prosencéfalo *m* cerebro *m* anterior
7354	prosenchyme *m*	prosénquima *m*
7355	prosimien *m*	prosimio *m*
7356	prostaglandine *f*	prostaglandina *f*
7357	prostate *f*	próstata *f*, glándula *f* prostática
7358	groupe(ment) *m* prosthétique	grupo *m* prostético
7359	protamine *f*	protamina *f*
7360	protandre, protandrique, protérandre, protérandrique	protandro, proterandro, proterándrico
7361	protandrie *f*, protérandrie *f*	protandria *f*, proterandria *f*
7362	—	—
7363	protéase *f*, enzyme *m* protéolytique	proteasa *f*, enzima *f* proteolítica
7364	—	—
7365	protection *f* des eaux	protección *f* de (las) aguas
7366	coloration *f* protectrice	coloración *f* protectora
7367	enveloppe *f* protectrice, membrane *f* ~	membrana *f* protectora, cubierta *f* ~
7368	tissu *m* protecteur	tejido *m* protector
7369	protéide *m*	proteido *m*
7370	protéine *f*	proteína *f*
7371	catabolisme *m* protéique	catabolismo *m* proteico, desdoblamiento *m* ~
7372	enveloppe *f* de protéine	cubierta *f* de proteína
7373	composé *m* protéique	compuesto *m* proteico
7374	teneur *f* en protéines	contenido *m* proteico, ~ en proteínas
7375	fibre *f* protidique	fibra *f* de proteína
7376	métabolisme *m* protidique, protéométabolisme *m*	metabolismo *m* prote(ín)ico, proteometabolismo *m*
7377	minimum *m* de protéine	mínimo *m* proteico
7378	molécule *f* protéique, protéomolécule *f*	molécula *f* prote(ín)ica
7379	azote *m* protéique	nitrógeno *m* proteico
7380	besoin *m* de protéine	necesidades *fpl* proteicas
7381	réserve *f* protéique, ~ protidique	reserva *f* proteica

7382	protein substance	Eiweißstoff *m*
7383	protein synthesis	Proteinsynthese *f*
7384	proteinase	Proteinase *f*
7385	proteinic, proteic, proteinaceous	eiweißartig, proteinartig, Protein...
7386	proteinoid	Proteinoid *n*
7387	proteinoplast, aleuroplast	Proteinoplast *m*
7388	proteohormone	Proteohormon *n*, Proteinhormon *n*
7389	proteolysis	Proteolyse *f*
7390	proteolytic, proteoclastic	proteinspaltend, proteolytisch
7391	proteolytic enzyme → protease	—
7392	proterandrous, proterandry → protandrous, protandry	—
7393	proterogynous, proterogyny → protogynous, protogyny	—
7394	prothallus, prothallium	Prothallium *n*
7395	prothrombin → thrombogen	—
7396	Protista, protists, protistans	Protisten *mpl*, Einzeller *mpl*
7397	protistology	Protistologie *f*, Protistenkunde *f*
7398	protocephalon	Protozephalon *n*
7399	protochlorophyll	Protochlorophyll *n*
7400	protocooperation → alliance	—
7401	protogynous, proterogynous	protogyn, proterogyn
7402	protogyny, proterogyny	Protogynie *f*, Proterogynie *f*, Vorweiblichkeit *f*
7403	proton	Proton *n*
7404	protonema	Protonema *n*
7405	protonephridium	Protonephridium *n*
7406	protophyte, protophyton	Protophyt *m*, Protophyte *f*, einzellige Pflanze *f*
7407	protoplasm	Protoplasma *n*
7408	protoplasmic, protoplasmatic	protoplasmatisch
7409	protoplasmic streaming	Plasmaströmung *f*, Protoplasmaströmung *f*
7410	protoplast	Protoplast *m*
7411	protostele	Protostele *f*, Urstele *f*
7412	Protostomia	Protostomier *mpl*, Erstmünder *mpl*, Urmünder *mpl*
7413	prototrophic	prototroph
7414	protovertebra → somite	—
7415	protoxylem	Protoxylem *n*
7416	protozoology	Protozoologie *f*, Protozoenkunde *f*
7417	protozoon (*pl* protozoa)	Protozoon *n* (*pl* Protozoen), Urtierchen *n*
7418	proventriculus, glandular stomach	Drüsenmagen *m*, Vormagen *m*
7419	provitamin, previtamin	Provitamin *n*
7420	psalterium → omasum	—
7421	psammon	Psammon *n*
7422	psammophytes, sand plants	Sandpflanzen *fpl*, Psammophyten *mpl*
7423	pseudoalleles, twin genes	Pseudoallele *npl*
7424	pseudoallelism	Pseudoallelie *f*
7425	pseud(o)axis → sympodium	—
7426	pseudobranch	Pseudobranchie *f*
7427	pseudocarp, false fruit	Scheinfrucht *f*
7428	pseudodominance	Pseudodominanz *f*
7429	pseudogamy, pseudomixis	Pseudogamie *f*, Pseudomixis *f*
7430	pseudohaploidy	Pseudohaploidie *f*
7431	pseudohermaphroditism	Pseudohermaphroditismus *m*, Scheinzwittertum *n*
7432	pseudoimago → subimago	—
7433	pseudoparenchyma	Pseudoparenchym *n*, Scheingewebe *n*
7434	pseudopod(ium), (*pl* pseudopodia)	Pseudopodium *n*, Scheinfüßchen *n*
7435	pseudopolyploidy	Pseudopolyploidie *f*
7436	pseudopupa, coarctate larva	Scheinpuppe *f*, Pseudochrysalis *f*
7437	pseudo-reduction	Pseudoreduktion *f*
7438	Psilopsida	Nacktfarne *mpl*
7439	pteridophytes	Farnpflanzen *fpl*, Pteridophyten *pl*
7440	pteridosperms, seed ferns	Samenfarne *mpl*, Pteridospermen *pl*

7382	substance *f* protéique	sustancia *f* proteica
7383	synthèse *f* de(s) protéine(s), ~ protéique	síntesis *f* prote(ín)ica, ~ de proteínas
7384	protéinase *f*	proteinasa *f*
7385	protéique, protéinique, protidique, protéinacé	proteico, proteínico
7386	protéinoïde *m*	proteinoide *m*
7387	protéoplaste *m*	proteoplasto *m*, aleuroplasto *m*
7388	protéohormone *f*	proteohormona *f*, hormona *f* proteínica
7389	protéolyse *f*	proteólisis *f*
7390	protéolytique, protéoclastique	proteolítico, proteoclástico
7391	—	—
7392	—	—
7393	—	—
7394	prothalle *m*	protalo *m*
7395	—	—
7396	protistes *mpl*	protistos *mpl*, protistas *mpl*
7397	protistologie *f*	protistología *f*
7398	protocéphalon *m*	protocéfalo *m*
7399	protochlorophylle *f*	protoclorofila *f*
7400	—	—
7401	protogyne, protérogyne	protógino, proterógino, proterogínico
7402	protogynie *f*, protérogynie *f*	protoginia *f*, proteroginia *f*
7403	proton *m*	protón *m*
7404	protonéma *m*	protonema *m*
7405	protonéphridie *f*	protonefridio *m*
7406	protophyte *m*, végétal *m* unicellulaire	protófito *m*, protófita *f*
7407	protoplasme *m*	protoplasma *m*
7408	protoplasmique, protoplasmatique	protoplásmico, protoplasmático
7409	courant *m* plasm(at)ique	corriente *f* plasmática, flujo *m* plasmático
7410	protoplaste *m*	protoplasto *m*
7411	protostèle *f*	protostela *f*
7412	protostomiens *mpl*	proteróstomos *mpl*
7413	prototrophe	protótrofo
7414	—	—
7415	protoxylème *m*	protoxilema *m*
7416	protozoologie *f*	protozoología *f*
7417	protozoaire *m*	protozoo *m*, protozoario *m*
7418	estomac *m* glandulaire, ventricule *m* succenturié	proventrículo *m*, ventrículo *m* subcenturiado
7419	provitamine *f*	provitamina *f*, previtamina *f*
7420	—	—
7421	psammon *m*	psamon *m*
7422	psammophytes *mpl*	psamófitas *fpl*, psamófilas *fpl*
7423	pseudo-allèles *mpl*, gènes *mpl* jumeaux	(p)seudoalelos *mpl*, genes *mpl* gemelos
7424	pseudo-allélisme *m*	(p)seudoalelismo *m*
7425	—	—
7426	pseudobranchie *f*	(p)seudobranquia *f*
7427	faux-fruit *m*	(p)seudocarpo *m*, falso fruto *m*
7428	pseudo-dominance *f*	(p)seudodominancia *f*
7429	pseudogamie *f*, pseudomixie *f*	(p)seudogamia *f*, (p)seudomixis *f*
7430	pseudo-haploïdie *f*	(p)seudohaploidia *f*
7431	pseudo-hermaphrodi(ti)sme *m*	(p)seudohermafroditismo *m*
7432	—	—
7433	pseudo-parenchyme *m*	(p)seudoparénquima *m*
7434	pseudopode *m*	(p)seudopodio *m*, (p)seudópodo *m*
7435	pseudo-polyploïdie *f*	(p)seudopoliploidia *f*
7436	pseudo-nymphe *f*, pseudo-chrysalide *f*	(p)seudopupa *f*
7437	pseudo-réduction *f*	(p)seudorreducción *f*
7438	psilophytes *mpl*	psilópsidos *mpl*
7439	ptéridophytes *mpl*	pteridófitos *mpl*, pteridofitas *fpl*
7440	ptéridospermées *fpl*, fougères *fpl* à graines	pteridospermas *fpl*, helechos *mpl* de semilla

7441	pterin	Pterin *n*
7442	Pterobranchia, pterobranchs	Flügelkiemer *mpl*
7443	Pteropsida, ferns	Farne *mpl*
7444	Pterygota, Metabola	Pterygoten *pl*, geflügelte Insekten *npl*
7445	ptyalin, salivary amylase	Ptyalin *n*, Speicheldiastase *f*
7446	puberal, pubertal	pubertär, puberal
7447	puberty	Pubertät *f*
7448	pubic symphysis	Schambeinfuge *f*, Schoßfuge *f*
7449	pubis, pubic bone	Schambein *n*
7450	puff	Puff *m* (*pl* Puffs)
7451	puffing	Puffbildung *f*
7452	pull root → contractile ~	—
7453	pulmonary	Lungen..., pulmonar
7454	pulmonary alveolus	Lungenbläschen *n*, Lungenalveole *f*
7455	pulmonary capacity, respiratory minute-volume	Atemminutenvolumen *n*, Ventilationsgröße *f*
7456	pulmonary circulation, lesser ~	Lungenkreislauf *m*, kleiner Kreislauf *m*
7457	pulmonary lobe	Lungenlappen *m*
7458	pulmonary pleura	Lungenfell *n*
7459	pulmonary respiration	Lungenatmung *f*
7460	pulp *bot.*	Fruchtfleisch *n*
7461	pulp cavity	Pulpahöhle *f*
7462	pulsating vacuole → contractile vacuole	—
7463	pulsation	Pulsen *n*, Pulsieren *n*
7464	pulse[1]	Puls *m*
7465	pulse[2] → legume	—
7466	pulverized	pulverisiert
7467	pulvinus → leaf cushion	—
7468	pupa; chrysalis	Puppe *f*; Chrysalis *f*, Chrysalide *f*
7469	pupal case, ~ integument	Puppenhülle *f*
7470	pupal stage	Puppenstadium *n*
7471	puparium	Puparium *n*, Tönnchen *n*
7472	to pupate	verpuppen
7473	pupation; nymphosis	Verpuppung *f*
7474	pupil	Pupille *f*
7475	pupil reflex	Pupillenreflex *m*
7476	pupiparous	pupipar
7477	pure-bred	reinrassig
7478	pure breed	reine Rasse *f*
7479	pure culture, axenic ~	Reinkultur *f*
7480	pure line	reine Linie *f*
7481	pure (line) breeding	Reinzucht *f*
7482	pure preparation	Reinpräparat *n*
7483	purification plant	Kläranlage *f*, Abwasserreinigungsanlage *f*
7484	to purify	reinigen
7485	purine	Purin *n*
7486	purine base	Purinbase *f*
7487	purity of species	Artreinheit *f*
7488	putrefaction, rot	Fäulnis *f*
7489	putrefactive	fäulniserregend
7490	putrefactive bacteria	Fäulnisbakterien *fpl*
7491	to putrefy, to rot	faulen
7492	putrescent, putrefying	faulend
7493	pycnidium	Pyknidium *n*, Pyknidie *f*
7494	pycnosis, pyknosis	Pyknose *f*
7495	pylorus	Pylorus *m*, Pförtner *m*
7496	pyramidal cell	Pyramidenzelle *f*
7497	pyramidal tract	Pyramidenbahn *f*
7498	pyrenoid	Pyrenoid *n*
7499	pyridine	Pyridin *n*
7500	pyridoxine, adermin, vitamin B_6	Pyridoxin *n*, Adermin *n*, Vitamin *n* B_6

7441	ptérine *f*	pterina *f*
7442	ptérobranches *mpl*	pterobranquios *mpl*
7443	fougères *fpl*	pterópsidos *mpl*, helechos *mpl*
7444	ptérygotes *mpl*	pterigotos *mpl*
7445	ptyaline *f*, amylase *f* salivaire	(p)tialina *f*, amilasa *f* salivar, diastasa *f* ~
7446	pubère	púber, púbero
7447	puberté *f*	pubertad *f*
7448	symphyse *f* pubienne	sínfisis *f* púbica
7449	pubis *m*, os *m* pubien	pubis *m*
7450	puff *m*, boursouflure *f* chromosomique	puff *m*
7451	formation *f* de puffs	formación *f* de puffs
7452	—	—
7453	pulmonaire	pulmonar
7454	alvéole *f* pulmonaire, vésicule *f* ~	alvéolo *m* pulmonar
7455	capacité *f* pulmonaire, volume *m* respiratoire par minute	capacidad *f* pulmonar, volumen *m* minuto respiratorio
7456	circulation *f* pulmonaire, petite circulation *f*	circulación *f* pulmonar, ~ menor
7457	lobule *m* pulmonaire	lóbulo *m* pulmonar
7458	plèvre *f* pulmonaire	pleura *f* pulmonar
7459	respiration *f* pulmonaire, ~ pulmonée	respiración *f* pulmonar
7460	pulpe *f*	pulpa *f*
7461	cavité *f* pulpaire	cavidad *f* pulpar
7462	—	—
7463	pulsation *f*	pulsación *f*
7464	pouls *m*	pulso *m*
7465	—	—
7466	pulvérisé	pulverizado
7467	—	—
7468	pupe *f*; chrysalide *f*	pupa *f*; crisálida *f*
7469	tégument *m* pupal, enveloppe *f* pupale, dépouille *f* ~	envoltura *f* pupal, cubierta *f* ~
7470	stade *m* pupal	estad(i)o *m* pupal
7471	puparium *m*, tonnelet *m*	pupario *m*
7472	se nymphoser; se chrysalider; se transformer en pupe	transformarse en pupa, ~ ~ crisálida
7473	pupaison *f*, pupation *f*; nymphose *f*	pupación *f*; ninfosis *f*, ninfalización *f*
7474	pupille *f*	pupila *f*
7475	réflexe *m* pupillaire, ~ photomoteur	reflejo *m* pupilar
7476	pupipare	pupíparo
7477	de race pure	de raza pura
7478	race *f* pure	raza *f* pura
7479	culture *f* pure, ~ axénique	cultivo *m* puro, ~ axénico
7480	lignée *f* pure	línea *f* pura
7481	élevage *m* en lignées pures, sélection *f* en race pure	cría por líneas puras, crianza *f* ~ ~ ~
7482	préparation *f* pure	preparado *m* puro
7483	station *f* d'épuration, ~ de clarification	estación *f* depuradora (de aguas residuales), planta *f* depuradora
7484	purifier	purificar
7485	purine *f*	purina *f*
7486	base *f* purique	base *f* púrica, ~ purínica
7487	pureté *f* des espèces	pureza *f* de (las) especies
7488	putréfaction *f*, pourriture *f*	putrefacción *f*, putrescencia *f*, podredumbre *f*
7489	putréfiant, putréfactif	putrefactor, putrefactivo
7490	bactéries *fpl* de putréfaction	bacterias *fpl* putrificantes, ~ de putrefacción
7491	se putréfier, pourrir	pudrir(se)
7492	putrescent	putrescente, en estado de putrefacción
7493	pycnide *f*	picnidio *m*
7494	pycnose *f*	picnosis *f*
7495	pylore *m*	píloro *m*
7496	cellule *f* pyramidale	célula *f* piramidal
7497	voie *f* pyramidale	vía *f* piramidal
7498	pyrénoïde *m*	pirenoide *m*
7499	pyridine *f*	piridina *f*
7500	pyridoxine *f*, adermine *f*, vitamine *f* B_6	piridoxina *f*, adermina *f*, vitamina *f* B_6

7501	pyrimidine	Pyrimidin *n*
7502	pyrimidine base	Pyrimidinbase *f*
7503	pyrrol(e)	Pyrrol *n*
7504	pyrrole ring	Pyrrolring *m*
7505	pyruvate	Pyruvat *n*
7506	pyruvic acid	Brenztraubensäure *f*
7507	pyxidium	Deckelkapsel *f*
7508	quadriplex, quadruplex	quadriplex
7509	quadrivalent	Quadrivalent *n*
7510	quadruped → tetrapod	—
7511	qualitative character	qualitatives Merkmal *n*
7512	qualitative variability → discontinuous ~	—
7513	quantitative character → blending ~	—
7514	quantitative variability → continuous ~	—
7515	quant(um)	Quant *m*
7516	quantum biology	Quantenbiologie *f*
7517	quantum evolution	Quantenevolution *f*
7518	Quaternary (period)	Quartär *n*
7519	quaternary structure	Quaternärstruktur *f*
7520	queen bee	Bienenkönigin *f*
7521	quercitrin	Querzitrin *n*
7522	quiescent stage → resting stage	—
7523	quill → calamus	—
7524	quinine	Chinin *n*
7525	quinone	Chinon *n*
7526	race	Rasse *f*
7527	raceme	Traube *f* (*Blütenstand*)
7528	racemose	traubig, razemös
7529	rachis[1] → vertebral column	—
7530	rachis[2] *bot.*	Blattspindel *f*, Rhachis *f*
7531	rachis[3], feather shaft	Federschaft *m*
7532	racial crossbreeding	Rassenkreuzung *f*
7533	racial formation, race ~	Rassenbildung *f*
7534	racial hybrid	Rassenbastard *m*
7535	radiability	Strahlendurchlässigkeit *f*
7536	radial	strahlig, radiär
7537	radial symmetry	Radialsymmetrie *f*, radiäre Symmetrie *f*
7538	radially symmetrical, radiosymmetrical	radialsymmetrisch, radiärsymmetrisch
7539	radiant energy	Strahlungsenergie *f*
7540	radiation	Strahlung *f*
7541	radiation damage → radiolesion	—
7542	radiation dose	Strahlungsdosis *f*
7543	radiation ecology → radioecology	—
7544	radiation field	Strahlungsfeld *n*
7545	radiation protection, protection against radiation, radiological protection	Strahlenschutz *m*
7546	radiation source	Strahlungsquelle *f*
7547	radicant plant	Radikante *f*, wurzelnde Pflanze *f*
7548	radicate, rooted	bewurzelt
7549	radication, rooting	Bewurzelung *f*
7550	radicle	Keimwurzel *f*, Radikula *f*
7551	radioactive	radioaktiv
7552	radioactive carbon → radiocarbon	—
7553	radioactive contamination, radiocontamination	radioaktive Verseuchung *f*
7554	radioactive fall-out, atomic ~	radioaktiver Niederschlag *m*, ~ Ausfall *m*
7555	radioactive indicator	Radioindikator *m*
7556	radioactive isotope → radioisotope	—
7557	radioactive radiation	radioaktive Strahlung *f*
7558	radioactive rays	radioaktive Strahlen *mpl*
7559	radioactive tracer, isotopic ~	radioaktiver Tracer *m*
7560	radioactive waste	Atommüll *m*

7501	pyrimidine *f*	pirimidina *f*
7502	base *f* pyrimidique	base *f* pirimidínica
7503	pyrrole *m*	pirrol *m*
7504	anneau *m* pyrrolique	anillo *m* pirrólico
7505	pyruvate *m*	piruvato *m*
7506	acide *m* pyruvique	ácido *m* pirúvico
7507	pyxide *f*	pixidio *m*
7508	quadruplex	cuadruplexo
7509	quadrivalent *m*	cuadrivalente *m*
7510	—	—
7511	caractère *m* qualitatif	carácter *m* cualitativo
7512	—	—
7513	—	—
7514	—	—
7515	quantum *m* (*pl* quanta)	cuanto *m*
7516	biologie *f* des quanta	biología *f* cuántica
7517	évolution *f* du quantum	evolución *f* del cuanto
7518	quaternaire *m*	cuaternario *m*
7519	structure *f* quaternaire	estructura *f* cuaternaria
7520	abeille *f* mère, reine *f* (des abeilles)	abeja *f* reina, ~ madre
7521	quercitrine *f*	quercitrina *f*, cuercitrina *f*
7522	—	—
7523	—	—
7524	quinine *f*	quinina *f*
7525	quinone *f*	quinona *f*
7526	race *f*	raza *f*
7527	grappe *f*, racème *m*	racimo *m*
7528	racémeux	racemoso
7529	—	—
7530	rachis *m* *bot.*	raquis *m* foliar
7531	rachis *m* *zool.*	raquis *m* *zool.*
7532	croisement *m* de races	cruce *m* de razas
7533	formation *f* de races	formación *f* de razas
7534	hybride *m* de races	híbrido *m* de razas
7535	radiabilité *f*	radiabilidad *f*
7536	radié, radial, rayonné, radiaire	radial, radiado
7537	symétrie *f* rayonnée, ~ radiaire	simetría *f* radial, ~ radiada
7538	à symétrie radiaire, ~ ~ rayonnée	radiado simétricamente
7539	énergie *f* rayonnante, ~ des rayonnements	energía *f* radiante
7540	radiation *f*, rayonnement *m*	radiación *f*
7541	—	—
7542	dose *f* de radiation, ~ ~ rayonnement	dosis *f* de radiación
7543	—	—
7544	champ *m* de radiation	campo *m* de radiación
7545	protection *f* contre les radiations, ~ radiologique, radioprotection *f*	protección *f* contra las radiaciones
7546	source *f* de radiation	fuente *f* de radiación
7547	plante *f* radicante, ~ racinée	planta *f* radicante
7548	raciné	radicado
7549	radication *f*	radicación *f*
7550	radicule *f*	radícula *f*, rejo *m*
7551	radioactif	radiactivo
7552	—	—
7553	contamination *f* radioactive, pollution *f* ~, radiocontamination *f*	contaminación *f* radiactiva
7554	retombées *fpl* radioactives	precipitación *f* radiactiva
7555	indicateur *m* radioactif	indicador *m* radiactivo, radioindicador *m*
7556	—	—
7557	radiation *f* radioactive	radiación *f* radiactiva
7558	rayons *mpl* radioactifs	rayos *mpl* radiactivos
7559	traceur *m* radioactif, ~ isotopique, radiotraceur *m*	trazador *m* radiactivo, marcador *m* ~, sustancia *f* trazadora radiactiva
7560	déchets *mpl* radioactifs, ~ atomiques, résidus *mpl* radioactifs	desechos *mpl* radiactivos, ~ atómicos, residuos *mpl* radiactivos

7561	radioactivity	Radioaktivität *f*
7562	radioautography → autoradiography	—
7563	radiobiology, actinobiology	Strahlenbiologie *f*, Radiobiologie *f*
7564	radiocarbon, radioactive carbon, C^{14}	radioaktiver Kohlenstoff *m*, Radiokohlenstoff *m*, ^{14}C
7565	radiocarbon datation, C^{14} dating, carbon-14-dating	Karbondatierung *f*
7566	radiocarbon method, C^{14}-test	Radiokarbonmethode *f*, ^{14}C-Test *m*
7567	radiochemistry	Radiochemie *f*
7568	radiochromatogram	Radiochromatogramm *n*
7569	radiochromatography	Radiochromatographie *f*
7570	radioecology, radiation ecology	Radioökologie *f*, Strahlungsökologie *f*
7571	radioelement	Radioelement *n*
7572	radiogenetics	Strahlengenetik *f*
7573	radioisotope, radioactive isotope	Radioisotop *n*, radioaktives Isotop *n*
7574	Radiolaria, radiolarians	Radiolarien *pl*, Strahlentierchen *npl*
7575	radiolesion, radiation damage, ~ injury	Strahlungsschaden *m*
7576	radiomimetic	radiomimetisch
7577	radiomimetic substances	Radiomimetika *pl*
7578	radionuclid(e)	Radionuklid *n*
7579	radio-opacity, radiopacity	Strahlenundurchlässigkeit *f*
7580	radiopaque	strahlenundurchlässig
7581	radioresistance	Strahlenresistenz *f*
7582	radioresistant	strahlenresistent
7583	radiosensitive, sensitive to radiation	strahlenempfindlich
7584	radiosensitivity, radiosensibility, radiation sensitivity	Strahlenempfindlichkeit *f*
7585	radiosymmetrical → radially symmetrical	—
7586	radiotransparent	strahlendurchlässig
7587	radius *anat.*	Speiche *f*
7588	radula	Reibzunge *f*, Reibplatte *f*, Radula *f*
7589	raffinose, melitose, melitriose	Raffinose *f*, Melitose *f*, Melitriose *f*
7590	rain forest	Regenwald *m*
7591	rainy season	Regenzeit *f*
7592	raising of threshold, increase ~ ~	Schwellenerhöhung *f*, ~anhebung *f*
7593	ramification, branching	Verzweigung *f*
7594	to ramify, to branch	(sich) verzweigen
7595	random distribution	Zufallsverteilung *f*
7596	random drift	Zufallsdrift *f*
7597	random mating	Zufallspaarung *f*
7598	random sample	Stichprobe *f*
7599	randomization	Randomisation *f*
7600	range of distribution, distribution area	Verbreitungsgebiet *n*, Areal *n*
7601	range of variation	Variationsbreite *f*, Variationsbereich *m*
7602	rank order, dominance hierarchy	Rangordnung *f*
7603	Ranvier's node, node of Ranvier	Ranvierscher Schnürring *m*, ~ Knoten *m*
7604	raphe	Raphe *f*, Samennaht *f*
7605	raptorial → predatory	—
7606	raptorial bird → bird of prey	—
7607	rassenkreis → rheogameon	—
7608	rate gene, ~ factor	Reaktionsratengen *n*
7609	rate of increase	Zuwachsrate *f*
7610	raw humus, acid ~, mor	Rohhumus *m*, saurer Humus *m*
7611	ray effect	Strahlenwirkung *f*
7612	ray floret	Strahlenblüte *f*
7613	rayproof	strahlensicher, strahlengeschützt
7614	to react to	reagieren auf
7615	reaction	Reaktion *f*
7616	reaction centre, reactive ~ (US: center)	Reaktionszentrum *n*
7617	reaction chain	Reaktionskette *f*
7618	reaction period	Reaktionsdauer *f*
7619	reaction rate, ~ speed, ~ velocity	Reaktionsrate *f*, Reaktionsgeschwindigkeit *f*

7561	radioactivité *f*	radiactividad *f*
7562	—	—
7563	radiobiologie *f*, biologie *f* de radiation	radiobiología *f*, biología *f* de radiaciones
7564	radiocarbone *m*, carbone *m* radioactif, carbone-14 *m*, ^{14}C	radiocarbono *m*, carbono *m* radiactivo, carbono 14, C^{14}
7565	datation *f* par le radiocarbone	datación *f* radiocarbónico, ~ por radiocarbono
7566	méthode *f* du ^{14}C	método *m* del carbono 14
7567	radiochimie *f*	radioquímica *f*
7568	radiochromatogramme *m*	radiocromatograma *m*
7569	radiochromatographie *f*	radiocromatografía *f*
7570	radioécologie *f*	ecología *f* de la radiación
7571	radio-élément *m*, élément *m* radioactif	radioelemento *m*, elemento *m* radiactivo
7572	radiogénétique *f*	radiogenética *f*
7573	radioisotope *m*, isotope *m* radioactif	radioisótopo *m*, isotópo *m* radiactivo
7574	radiolaires *mpl*	radiolarios *mpl*
7575	radiolésion *f*, dégât *m* dû aux radiations	radiolesión *f*, lesión *f* por radiación, daño *m* debido a la radiación
7576	radiomimétique	radiomimético
7577	substances *fpl* radiomimétiques, agents *mpl* ~	agentes *mpl* radiomiméticos
7578	radionuclide *m*	radionúclido *m*
7579	radiopacité *f*	radiopacidad *f*
7580	radiopaque, radio-opaque, opaque aux radiations	radio(o)paco
7581	radiorésistance *f*	radiorresistencia *f*
7582	radiorésistant	radiorresistente
7583	radiosensible	radiosensible
7584	radiosensibilité *f*	radiosensibilidad *f*
7585	—	—
7586	radiotransparent	radiotransparente
7587	radius *m* *anat.*	radio *m* *anat.*
7588	radula *f*	rádula *f*
7589	raffinose *m*, mélitose *m*, mélitriose *m*	rafinosa *f*, melitosa *f*, melitriosa *f*
7590	forêt *f* ombrophile	pluvi(i)silva *f*, bosque *m* lluvioso, ~ de lluvia
7591	saison *f* des pluies, ~ pluvieuse	época *f* de lluvia, estación *f* lluviosa
7592	élévation *f* du seuil	elevación *f* del umbral
7593	ramification *f*	ramificación *f*
7594	se ramifier	ramificar(se)
7595	répartition *f* au hasard	distribución *f* al azar
7596	dérive *f* fortuite	desviación *f* al azar
7597	accouplement *m* au hasard	apareamiento *m* al azar
7598	échantillon *m* pris au hasard	muestra *f* al azar
7599	randomisation *f*	randomización *f*
7600	aire *f* de distribution, ~ ~ répartition	área *f* de distribución, ~ ~ repartición
7601	étendue *f* de variation	grado *m* de variación
7602	ordre *m* hiérarchique	orden *m* de dominancia
7603	étranglement *m* de Ranvier, nœud *m* ~ ~, étranglement *m* annulaire	nódulo *m* de Ranvier, estrangulación *f* ~ ~
7604	raphé *m*	rafe *f*
7605	—	—
7606	—	—
7607	—	—
7608	gène *m* contrôleur	gen *m* de control
7609	taux *m* d'accroissement	proporción *f* de incremento
7610	humus *m* brut, ~ acide, mor *m*	humus *m* bruto, ~ ácido, mantillo *m* ~
7611	effet *m* de rayons	efecto *m* de la radiación
7612	hémiligule *f* rayonnante	flor *f* radial
7613	protégé contre les radiations	protegido contra rayos
7614	réagir à	reaccionar a
7615	réaction *f*	reacción *f*
7616	centre *m* de réaction	centro *m* de reacción
7617	chaîne *f* de réaction	cadena *f* de reacciones
7618	période *f* de réaction, durée *f* ~ ~	duración *f* de reacción
7619	vitesse *f* de réaction	velocidad *f* de reacción

7620	reaction time	Reaktionszeit *f*
7621	to reactivate	reaktivieren
7622	reactivation	Reaktivierung *f*
7623	reactivity	Reaktionsfähigkeit *f*
7624	ready for mating	paarungsbereit
7625	ready to react	reaktionsbereit
7626	to reafforest, to reforest	wiederaufforsten
7627	reafforestation, reforestation	Wiederaufforstung *f*
7628	reagent	Reagens *n* (*pl* Reagenzien)
7629	recapitulation	Rekapitulation *f*
7630	recapitulation theory	Rekapitulationstheorie *f*
7631	Recent → Holocene	—
7632	receptacle, thalamus	Blütenboden *m*, Rezeptakulum *n*
7633	receptor	Rezeptor *m*
7634	receptor cell, recipient ~	Rezeptorzelle *f*, Empfängerzelle *f*
7635	receptor organ	Rezeptororgan *n*, Rezeptionsorgan *n*
7636	receptor potential	Rezeptorpotential *n*
7637	recessive	rezessiv
7638	recessive allele → allogene	—
7639	recessiveness	Rezessivität *f*
7640	recipient bacterium	Empfängerbakterie *f*, Rezeptorbakterie *f*
7641	recipient cell → receptor cell	—
7642	reciprocal	reziprok
7643	reciprocal cross	reziproke Kreuzung *f*
7644	reciprocal inhibition	reziproke Hemmung *f*
7645	reciprocal translocation	reziproke Translokation *f*
7646	reciprocity law, Bunsen-Roscoe ~	Reizmengengesetz *n*
7647	recombinant	rekombinant
7648	recombination	Rekombination *f*, Neukombination *f*
7649	recombination analysis	Rekombinationsanalyse *f*
7650	recombination frequency	Rekombinationshäufigkeit *f*
7651	recombination rate, ~ percentage, ~ value	Rekombinationswert *m*, ~prozentsatz *m*
7652	recombination unit	Rekombinationseinheit *f*
7653	recon	Recon *n*, Rekon *n*
7654	recording → derivation	—
7655	recreational area → leisure area	—
7656	rectrix → tail feather	—
7657	rectum	Mastdarm *m*, Rektum *n*
7658	recurrent selection	rekurrente Selektion *f*
7659	red algae → Rhodophyta	—
7660	red blood corpuscle → erythrocyte	—
7661	redirection activity, redirected ~	Umorientierung *f*, umorientiertes Verhalten *n*
7662	redox... → oxidation-reduction	—
7663	reducing agent, reducer, reductant	Reduktionsmittel *n*, Reduktant *m*
7664	reducing potential	Reduktionspotential *n*
7665	reductase	Reduktase *f*
7666	reduction division	Reduktionsteilung *f*
7667	reductone	Redukton *n*
7668	reduplication *gen.*	Reduplikation *f* *gen.*
7669	reflex	Reflex *m*
7670	reflex action	Reflexwirkung *f*
7671	reflex arc	Reflexbogen *m*
7672	reflex centre (US: center)	Reflexzentrum *n*
7673	reflex chain → chain reflex	—
7674	reflex inhibition	Reflexhemmung *f*
7675	reflex movement	Reflexbewegung *f*
7676	reflex path	Reflexbahn *f*
7677	reflex response	Reflexantwort *f*
7678	reflexologist	Reflexologe *m*
7679	reflexology	Reflexologie *f*
7680	to reforest, reforestation → to reafforest, reafforestation	—

7620	temps *m* de réaction	tiempo *m* de reacción
7621	réactiver	reactivar
7622	réactivation *f*	reactivación *f*
7623	réactivité *f*, capacité *f* à réagir	reactividad *f*, capacidad *f* reactiva
7624	en humeur de copuler	dispuesto para el apareamiento
7625	prêt à réagir	dispuesto a reaccionar
7626	reboiser	repoblar (bosques); reforestar (LA)
7627	reboisement *m*	repoblación *f* forestal; reforestación *f* (LA)
7628	réactif *m*	reactivo *m*
7629	récapitulation *f*	recapitulación *f*
7630	théorie *f* de la récapitulation	teoría *f* de la recapitulación
7631	—	—
7632	réceptacle *m* (floral), thalamus *m*	receptáculo *m*, tálamo *m*
7633	récepteur *m*	receptor *m*
7634	cellule *f* réceptrice	célula *f* receptora
7635	organe *m* récepteur	órgano *m* receptor
7636	potentiel *m* de réception	potencial *m* receptor, ~ de recepción
7637	récessif	recesivo
7638	—	—
7639	récessivité *f*	recesividad *f*
7640	bactérie *f* réceptrice	bacteria *f* receptora
7641	—	—
7642	réciproque	recíproco
7643	croisement *m* réciproque	cruzamiento *m* recíproco, cruce *m* ~
7644	inhibition *f* réciproque	inhibición *f* recíproca
7645	translocation *f* réciproque	tra(n)slocación *f* recíproca
7646	loi *f* de la quantité de stimulus, ~ de (la) réciprocité	ley *f* de la cantidad de estímulo
7647	recombinant	recombinante
7648	recombinaison *f*	recombinación *f*
7649	analyse *f* de la recombinaison	análisis *m* de la recombinación
7650	fréquence *f* de recombinaison	frecuencia *f* de recombinación
7651	taux *m* de recombinaison, pourcentage *m* ~ ~	porcentaje *m* de recombinación, valor *m* ~ ~
7652	unité *f* de recombinaison	unidad *f* de recombinación, ~ recombinacional
7653	récon *m*	recón *m*
7654	—	—
7655	—	—
7656	—	—
7657	rectum *m*	recto *m*
7658	sélection *f* récurrente	selección *f* recurrente
7659	—	
7660	—	—
7661	redirection *f*	actividad *f* reorientada, movimiento *m* reorientado, reorientación *f*
7662	—	—
7663	réducteur *m*	(agente *m*) reductor *m*
7664	potentiel *m* réducteur	potencial *m* reductor
7665	réductase *f*	reductasa *f*
7666	division *f* de réduction, ~ réductrice, réduction *f* chromatique	división *f* reductora, reducción *f* cromática
7667	réductone *f*	reductona *f*
7668	réduplication *f* *gen.*	reduplicación *f* *gen.*
7669	réflexe *m*	reflejo *m*
7670	action *f* réflexe	acción *f* refleja
7671	arc *m* (de) réflexe	arco *m* reflejo
7672	centre *m* réflexe	centro *m* reflejo
7673	—	—
7674	inhibition *f* (de) réflexe	inhibición *f* del reflejo
7675	mouvement *m* (de) réflexe	movimiento *m* reflejo
7676	voie *f* réflexe	vía *f* refleja
7677	réponse *f* réflexe	respuesta *f* refleja
7678	réflexologiste *m*, réflexologue *m*	reflexólogo *m*
7679	réflexologie *f*	reflexología *f*
7680	—	—

7681	refractility → refringence	—
7682	refraction	Lichtbrechung *f*
7683	refractive → refringent	—
7684	refractive index, index of refraction	Brechungsindex *m*
7685	refractory period, ~ phase	Refraktärzeit *f*, Refraktärphase *f*
7686	refractory stage	Refraktärstadium *n*
7687	refringence, refractility	Lichtbrechungsvermögen *n*
7688	refringent, refractive, refractile	lichtbrechend
7689	refuge, refugium	Refugialgebiet *n*, Refugium *n*, Residualgebiet *n*
7690	refuse → waste	—
7691	refuse incineration	Müllverbrennung *f*
7692	refuse incinerator	Müllverbrennungsanlage *f*
7693	to regenerate	regenerieren, verjüngen
7694	regeneration	Regeneration *f*, Erneuerung *f*, Verjüngung *f*
7695	regeneration capacity	Regenerationsfähigkeit *f*, ~vermögen *n*
7696	regression	Regression *f*
7697	regression coefficient	Regressionskoeffizient *m*
7698	regression equation	Regressionsgleichung *f*
7699	regressive	regressiv
7700	regressive evolution	regressive Entwicklung *f*
7701	regulating circle	Regelkreis *m*
7702	regulation	Regulation *f*
7703	regulation egg, regulative ~	Regulationsei *n*
7704	regulatory factor	Regulationsfaktor *m*
7705	regulatory gene	Regulatorgen *n*
7706	regulatory mechanism	Regulationsmechanismus *m*
7707	reinforcement *eth.*	Bekräftigung *f* *eth.*
7708	rejuvenescence, rejuvenation	Verjüngung *f*
7709	relationship	Verwandtschaft *f*
7710	relay cell → interneuron(e)	—
7711	to release	auslösen *(Reiz, Reflex)*
7712	releaser	Auslöser *m*
7713	releasing factor	Releaserfaktor *m*
7714	releasing stimulus	Auslösereiz *m*, auslösender Reiz *m*
7715	relict endemism	Reliktendemismus *m*
7716	relict fauna	Reliktenfauna *f*
7717	relict flora	Reliktenflora *f*
7718	relict species	Relikt *n*
7719	relief at the nest	Nestablösung *f*
7720	remex (*pl* remiges)	Schwungfeder *f*
7721	renal, nephric	Nieren...
7722	renal calyce	Nierenkelch *m*
7723	renal pelvis	Nierenbecken *n*
7724	renal tubule, nephridial ~	Nierenkanälchen *n*
7725	renewal bud	Erneuerungsknospe *f*
7726	reniform	nierenförmig
7727	rennin, chymosin	Lab(ferment) *n*, Rennin *n*, Chymosin *n*
7728	reoxidation	Reoxydation *f*, Reoxydierung *f*
7729	to reoxidize	reoxydieren
7730	reparation	Reparation *f*
7731	repattering	Chromosomenummusterung *f*
7732	repellent	Repellent *n*, Insektenabwehrmittel *n*
7733	replacing bone → cartilage bone	—
7734	replica plating	Replika-Technik *f*, Stempeltechnik *f*
7735	replication	Replikation *f*
7736	replication unit	Replikationseinheit *f*
7737	repolarization	Repolarisation *f*
7738	repressible enzyme	repressibles Enzym *n*
7739	repression *gen.*	Repression *f* *gen.*
7740	repressor	Repressor *m*

7681	—	—
7682	réfraction *f*	refracción *f*
7683	—	—
7684	indice *m* de réfraction	índice *m* de refracción
7685	période *f* réfractaire, phase *f* ~	período *m* refractario
7686	stade *m* réfractaire	estado *m* refractario
7687	réfringence *f*	refringencia *f*, refractividad *f*
7688	réfringent	refringente, refractario
7689	aire *f* relicte	refugio *m*
7690	—	—
7691	incinération *f* de déchets	incineración *f* de basuras
7692	usine *f* d'incinération de déchets	planta *f* incineradora de basuras
7693	se régénérer	regenerarse
7694	régénération *f*, régénérescence *f*	regeneración *f*
7695	capacité *f* régénérative, pouvoir *m* régénérateur, ~ de régénération	capacidad *f* regenerativa, ~ de regeneración
7696	régression *f*	regresión *f*
7697	cœfficient *m* de régression	coeficiente *m* de regresión
7698	équation *f* de régression	ecuación *f* de regresión
7699	régressif	regresivo
7700	évolution *f* régressive	evolución *f* regresiva
7701	cercle *m* de régulation	círculo *m* regulador
7702	régulation *f*	regulación *f*
7703	œuf *m* à régulation, ~ régulateur	huevo *m* de regulación, ~ regulatorio
7704	facteur *m* régulateur	factor *m* regulador
7705	gène *m* régulateur, ~ de régulation	gen *m* regulador
7706	mécanisme *m* de régulation, ~ régulateur	mecanismo *m* de regulación, ~ regulador
7707	renforcement *m* *eth.*	refuerzo *m* *eth.*
7708	rajeunissement *m*	rejuvenecimiento *m*
7709	parenté *f*	parentesco *m*
7710	—	—
7711	déclencher	desencadenar
7712	déclencheur *m*, évocateur *m*	desencadenante *m*, desencadenador *m*
7713	facteur *m* déchaînant, ~ de libération	factor *m* de liberación
7714	stimulus *m* déclencheur	estímulo *m* desencadenante, ~ desencadenador
7715	endémisme *m* relictuel, ~ conservatif	endemismo *m* conservativo
7716	faune *f* relique	fauna *f* reliquial
7717	flore *f* relique	flora *f* reliquial
7718	espèce-relique *f*	reliquia *f*
7719	relève *f* au nid	relevo *m* en el cuidado del nido
7720	rémige *f*, penne *f*	remera *f*, rémige *f*
7721	rénal, néphrique	renal, néfrico
7722	calice *m* rénal	cáliz *m* renal
7723	bassinet *m*	pelvis *f* renal
7724	tubule *m* néphridien, canalicule *m* rénal	túbulo *m* renal
7725	bourgeon *m* de remplacement	yema *f* de renovación
7726	réniforme	reniforme
7727	rennine *f*, chymosine *f*, présure *f*, labferment *m*	renina *f*, quimosina *f*, fermento *m* lab
7728	réoxydation *f*	reoxidación *f*
7729	réoxyder	reoxidar
7730	réparation *f*	reparación *f*
7731	réarrangement *m* chromosomique, remaniement *m* ~	reajuste *m* cromosómico
7732	répulsif *m*, substance *f* répulsive	repelente *m*
7733	—	—
7734	test *m* des répliques, méthode *f* des réplications	replicación *f* en placas, reproducción *f* en placa
7735	réplication *f*	replicación *f*
7736	unité *f* de réplication	unidad *f* de replicación
7737	répolarisation *f*	repolarización *f*
7738	enzyme *m* répressible	enzima *f* represible
7739	répression *f* *gen.*	represión *f* *gen.*
7740	répresseur *m*	represor *m*

7741	repressor gene	Repressor-Gen *n*
7742	to reproduce	sich fortpflanzen
7743	to reproduce sexually	geschlechtlich fortpflanzen
7744	reproducible, able to reproduce	fortpflanzungsfähig, vermehrungsfähig
7745	reproduction	Fortpflanzung *f*, Reproduktion *f*
7746	reproduction rate, reproductive ~	Fortpflanzungsrate *f*, Reproduktionsrate *f*
7747	reproductive aperture → genital aperture	—
7748	reproductive association	Fortpflanzungsgemeinschaft *f*
7749	reproductive barrier	Fortpflanzungsschranke *f*
7750	reproductive behavio(u)r	Fortpflanzungsverhalten *n*
7751	reproductive biology	Fortpflanzungsbiologie *f*
7752	reproductive body	Fortpflanzungskörper *m*
7753	reproductive capacity, ~ ability	Fortpflanzungsfähigkeit *f*, ~vermögen *n*
7754	reproductive cell	Fortpflanzungszelle *f*
7755	reproductive colo(u)ration → courtship ~	—
7756	reproductive cycle	Fortpflanzungszyklus *m*
7757	reproductive drive	Fortpflanzungstrieb *m*
7758	reproductive isolation	geschlechtliche Isolation *f*, reproduktive ~
7759	reproductive organ	Fortpflanzungsorgan *n*
7760	reproductive period, ~ season	Fortpflanzungszeit *f*
7761	reproductive physiology	Fortpflanzungsphysiologie *f*
7762	reproductive potential	Fortpflanzungspotential *n*
7763	reproductive stage	Fortpflanzungsstadium *n*
7764	reproductive system, ~ tract	Fortpflanzungssystem *n*, Fortpflanzungstrakt *m*
7765	Reptilia, reptiles	Kriechtiere *npl*, Reptilien *npl* (*sing*. Reptil)
7766	repugnatorial gland → stink gland	—
7767	research, investigation	Forschung *f*
7768	to research (on), to investigate	forschen (über)
7769	research field, field of investigation	Forschungsgebiet *n*
7770	research laboratory	Forschungslaboratorium *n*
7771	research programme	Forschungsprogramm *n*
7772	research work	Forschungsarbeit *f*
7773	research worker, researcher, investigator	Forscher *m*
7774	reserve air, supplemental ~, expiratory reserve	Reserveluft *f*, expiratorisches Reservevolumen *n*
7775	reserve protein	Reserveeiweiß *n*
7776	reserve starch, storage ~	Reservestärke *f*
7777	reserve substance, ~ material	Reservestoff *m*, Speicherstoff *m*, Reservematerial *n*
7778	residential region, ~ area *env.*	Wohngebiet *n env.*
7779	residual air	Restluft *f*, Residualluft *f*
7780	residues	Rückstände *mpl*
7781	resin	Harz *n*
7782	resin canal, ~ duct	Harzgang *m*, Harzkanal *m*
7783	resistance	Resistenz *f*, Widerstandsfähigkeit *f*
7784	resistant	resistent, widerstandsfähig
7785	resolving power *micr.*	Auflösungsvermögen *n micr.*
7786	to resorb	resorbieren
7787	resorption	Resorption *f*
7788	respiration	Atmung *f*
7789	respiratory	Atmungs..., Atem..., respiratorisch
7790	respiratory cavity → hypostomatic chamber	—
7791	respiratory centre (US: center)	Atemzentrum *n*
7792	respiratory chain	Atmungskette *f*
7793	respiratory chain phosphorylation → oxidative phosphorylation	—
7794	respiratory control	Atmungsregulation *f*

7741	gène *m* répresseur	gen *m* represor
7742	se reproduire	reproducirse
7743	se reproduire par voie sexuée	reproducirse sexualmente
7744	apte à se reproduire, ~ à la procréation	reproducible, capaz de reproducirse, apto para la reproducción
7745	reproduction *f*, procréation *f*	reproducción *f*, procreación *f*
7746	taux *m* de reproduction	índice *m* de reproducción, tasa *f* ~ ~
7747	—	—
7748	communauté *f* de reproduction	comunidad *f* reproductora
7749	obstacle *m* à la reproduction	barrera *f* reproductiva, ~ a la reproducción
7750	comportement *m* de reproduction, ~ reproducteur, ~ génésique	comportamiento *m* reproductivo, ~ reproductor
7751	biologie *f* de la reproduction	biología *f* reproductora
7752	corps *m* reproducteur	cuerpo *m* reproductor
7753	reproductivité *f*, capacité *f* reproductrice, pouvoir *m* de reproduction	capacidad *f* reproductiva, facultad *f* ~, poder *m* reproductor
7754	cellule *f* reproductrice	célula *f* reproductora
7755	—	—
7756	cycle *m* reproducteur	ciclo *m* reproductor
7757	instinct *m* de reproduction, ~ génésique	instinto *m* reproductivo, ~ genésico, ~ de reproducción, impulso *m* reproductor
7758	isolement *m* reproductif, ~ sexuel	aislamiento *m* reproductivo
7759	organe *m* de reproduction	órgano *m* de reproducción, ~ reproductor
7760	période *f* de reproduction, saison *f* reproductrice	período *m* reproductor, época *f* de reproducción
7761	physiologie *f* de la reproduction	fisiología *f* de la reproducción
7762	potentiel *m* de reproduction	potencial *m* reproductor, ~ reproductivo
7763	stade *m* reproductif	etapa *f* reproductiva
7764	appareil *m* reproducteur	sistema *m* reproductivo, tracto *m* reproductor
7765	reptiles *mpl*	reptiles *mpl*
7766	—	—
7767	recherche *f*, investigation *f*	investigación *f*
7768	rechercher (sur)	investigar (sobre)
7769	champ *m* d'investigation, domaine *m* de recherches	campo *m* de investigación
7770	laboratoire *m* de recherches	laboratorio *m* de investigación
7771	programme *m* de recherche(s)	programa *m* de investigación
7772	travail *m* de recherche	labor *m* de investigación
7773	(re)chercheur *m*, investigateur *m*	investigador *m*
7774	air *m* de réserve, ~ supplémentaire, volume *m* de réserve expiratoire	aire *m* de reserva, ~ suplementario, volumen *m* de reserva espiratoria
7775	protéine *f* de réserve	proteína *f* de reserva
7776	amidon *m* de réserve	almidón *m* de reserva
7777	substance *f* de réserve, matière *f* ~ ~	sustancia *f* de reserva, material *m* ~ ~
7778	zone *f* résidentielle *env.*	zona *f* residencial, área *f* de residencia *env.*
7779	air *m* résiduel	aire *m* residual
7780	résidus *mpl*	residuos *mpl*, sustancias *fpl* residuales
7781	résine *f*	resina *f*
7782	canal *m* résinifère	canal *m* resinífero, conducto *m* resinoso
7783	résistance *f*	resistencia *f*
7784	résistant	resistente
7785	capacité *f* de résolution, pouvoir *m* résolvant *micr.*	poder *m* de resolución, ~ resolutivo, ~ resolvente *micr.*
7786	résorber	re(ab)sorber
7787	résorption *f*	re(ab)sorción *f*
7788	respiration *f*	respiración *f*
7789	respiratoire	respiratorio
7790	—	—
7791	centre *m* respiratoire	centro *m* respiratorio
7792	chaîne *f* respiratoire	cadena *f* respiratoria
7793	—	—
7794	régulation *f* respiratoire	regulación *f* de la respiración, control *m* ~ ~ ~

No.	English	German
7795	respiratory exchange	respiratorischer Gasaustausch *m*, Austausch *m* der Atemgase
7796	respiratory enzyme	Atmungsferment *n*
7797	respiratory gas	Atemgas *n*
7798	respiratory intensity	Atmungsintensität *f*
7799	respiratory loss	Atmungsverlust *m*
7800	respiratory metabolism	Atmungsstoffwechsel *m*
7801	respiratory minute-volume → pulmonary capacity	—
7802	respiratory movement, breathing ~	Atembewegung *f*
7803	respiratory organ, breathing ~	Atmungsorgan *n*, Atemorgan *n*
7804	respiratory pigment	Atmungspigment *n*
7805	respiratory passages, breathing ~, air ~	Atemwege *mpl*
7806	respiratory quotient, R.Q.	Atmungsquotient *m*, respiratorischer Quotient *m*
7807	respiratory root → pneumatophore	—
7808	respiratory tube, breathing ~	Atemröhre *f*
7809	respiratory system, ~ apparatus	Atmungssystem *n*
7810	to respire → to breathe	—
7811	respirometer	Respirometer *n*
7812	to respond (to a stimulus)	reagieren (auf einen Reiz)
7813	response	Reaktion *f* (auf einen Reiz), Antwortreaktion *f*
7814	responsiveness → capacity to respond	—
7815	rest nitrogen, nonprotein ~	Reststickstoff *m*
7816	resting bud	Ruheknospe *f*
7817	resting cell, dormant ~	Ruhezelle *f*
7818	resting egg → winter egg	—
7819	resting form	Dauerform *f*, Überdauerungsform *f*
7820	resting nucleus	Ruhekern *m*
7821	resting potential	Ruhepotential *n*
7822	resting spore, statospore, hypnospore	Ruhespore *f*
7823	resting stage, dormant ~, quiescent ~	Ruhestadium *n*, Ruheperiode *f*
7824	resting state, quiescent ~, quiescence	Ruhezustand *m*
7825	restituability	Restitutionskapazität *f*
7826	restitution	Restitution *f*
7827	resupination	Resupination *f*
7828	resynthesis	Resynthese *f*
7829	reticular, reticulate	netzförmig, netzartig
7830	reticular formation	Retikularformation *f*, retikulare Formation *f*
7831	reticulate thickening	Netzverdickung *f*
7832	reticulate vessel	Netzgefäß *n*
7833	reticulin	Retikulin *n*
7834	reticulin fibre, argyrophil ~	Retikulinfaser *f*, Gitterfaser *f*
7835	reticulocyte	Retikulozyt *m*
7836	reticulo-endothelial system, reticulo-endothelium	retikulo-endotheliales System *n*, RES, Retikuloendothel *n*
7837	reticulum, honey-comb bag	Netzmagen *m*
7838	retina	Netzhaut *f*, Retina *f*
7839	retinal cone	Netzhautzapfen *m*
7840	retinerved → net(ted) veined	—
7841	retractor (muscle)	Rückzieher(muskel) *m*, Retraktor *m*
7842	retranslocation	Retranslokation *f*
7843	reverse mutation, back ~, return ~	Rückmutation *f*
7844	rhabdome, optic rod	Rhabdom *n*, Sehstab *m*
7845	rhabdomere	Rhabdomer *n*
7846	rheobase, rheobasis	Rheobase *f*
7847	rheogameon, rassenkreis, circle of races	Rassenkreis *m*, Rheogameon *n*
7848	rheotaxis	Rheotaxis *f*
7849	rheotropism	Rheotropismus *m*

7795	échange *m* (gazeux) respiratoire	intercambio *m* respiratorio
7796	enzyme *m* respiratoire	enzima *f* respiratoria
7797	gaz *m* respiratoire	gas *m* respiratorio
7798	intensité *f* respiratoire	intensidad *f* respiratoria
7799	perte *f* par respiration	pérdida *f* respiratoria
7800	métabolisme *m* respiratoire	metabolismo *m* respiratorio
7801	—	—
7802	mouvement *m* respiratoire	movimiento *m* respiratorio
7803	organe *m* respiratoire	órgano *m* respiratorio
7804	pigment *m* respiratoire	pigmento *m* respiratorio
7805	voies *fpl* respiratoires, ~ aériennes	vías *fpl* respiratorias
7806	quotient *m* respiratoire, QR	cociente *m* respiratorio, CR
7807	—	—
7808	tube *m* respiratoire	tubo *m* respiratorio
7809	appareil *m* respiratoire	sistema *m* respiratorio
7810	—	—
7811	respiromètre *m*	respirómetro *m*
7812	répondre (à un stimulus)	responder (a un estímulo)
7813	réponse *f*	respuesta *f*, reacción de ~
7814	—	—
7815	azote *m* résiduel	nitrógeno *m* residual, ~ no proteínico
7816	bourgeon *m* dormant, ~ latent, ~ en vie ralentie	yema *f* en reposo
7817	cellule *f* au repos, ~ quiescente	célula *f* en reposo
7818	—	—
7819	stade *m* de durée, ~ résistant, ~ de résistance, forme *f* de durée	estad(i)o *m* de resistencia, forma *f* ~ ~
7820	noyau *m* quiescent, ~ au repos, ~ en repos	núcleo *m* en reposo
7821	potentiel *m* de repos	potencial *m* de reposo
7822	spore *f* dormante, hypnospore *f*	espora *f* en reposo, ~ durmiente, hipnóspora *f*
7823	période *f* de repos, ~ quiescente	período *m* de reposo, fase *f* ~ ~, período *m* de letargo
7824	état *m* quiescent, ~ de repos, quiescence *f*	estado *m* de reposo
7825	restituabilité *f*	capacidad *f* de restitución
7826	restitution *f*	restitución *f*
7827	résupination *f*	resupinación *f*
7828	resynthèse *f*	resíntesis *f*
7829	réticulaire, réticulé	reticular, reticulado
7830	formation *f* réticulaire	formación *f* reticular
7831	épaississement *m* réticulé	engrosamiento *m* reticulado
7832	vaisseau *m* réticulé	vaso *m* reticulado, ~ reticular
7833	réticuline *f*	reticulina *f*
7834	fibre *f* réticulinique, ~ argyrophile	fibra *f* reticulínica, ~ argirófila
7835	réticulocyte *m*	reticulocito *m*
7836	système *m* réticulo-endothélial, réticulo-endothélium *m*	sistema *m* retículoendotelial, retículoendotelio *m*
7837	réseau *m*, bonnet *m*	retículo *m*, redecilla *f*, bonete *m*
7838	rétine *f*	retina *f*
7839	cône *m* rétinien	cono *m* retinal
7840	—	—
7841	rétracteur *m*	(músculo *m*) retractor *m*
7842	retranslocation *f*	retra(n)slocación *f*
7843	mutation *f* reverse, ~ renversée, ~ en retour	mutación *f* reversiva, ~ revertida, ~ invertida, ~ de retroceso, retromutación *f*
7844	rhabdome *m*	rabdoma *m*
7845	rhabdomère *m*	rabdómero *m*, rabdómera *f*
7846	rhéobase *f*, intensité *f* liminaire	reobase *f*
7847	rhéogaméon *m*	círculo *m* de razas
7848	rhéotaxie *f*, rhéotactisme *m*	reotaxis *f*, reotaxia *f*
7849	rhéotropisme *m*	reotropismo *m*

7850	Rhesus factor, Rh factor	Rhesusfaktor *m*
7851	rhinal → nasal	—
7852	rhinencephalon, olfactory brain, ~ bulb	Riechhirn *n*
7853	rhinopharynx → nasopharynx	—
7854	rhipidium	Fächel *m*
7855	rhizodermis, epiblem(a)	Rhizodermis *f*
7856	rhizoid	Rhizoid *n*
7857	rhizome, root-stock, root-stalk	Wurzelstock *m*, Rhizom *n*
7858	rhizomorph	Rhizomorph *n*, Myzelstrang *m*
7859	Rhizopoda, Sarcodina	Wurzelfüßer *mpl*, Rhizopoden *pl*
7860	rhizosphere	Rhizosphäre *f*
7861	Rhodophyta, red algae	Rotalgen *fpl*, Rottange *mpl*, Rhodophyzeen *fpl*
7862	rhodopsin, visual purple	Sehpurpur *m*, Rhodopsin *n*
7863	rhombencephalon	Rautenhirn *n*, Rhombenzephalon *n*
7864	Rhynchota → Hemiptera	—
7865	rhytidome → bark	—
7866	rib	Rippe *f*
7867	ribbed → costate	—
7868	riboflavin, lactoflavin, ovoflavin, vitamin B_2	Riboflavin *n*, Laktoflavin *n*, Vitamin *n* B_2
7869	ribonuclease	Ribonuklease *f*
7870	ribonucleic acid, RNA	Ribonukleinsäure *f*, RNS
7871	ribonucleoprotein	Ribonukleoprotein *n*
7872	ribonucleosid(e)	Ribonukleosid *n*
7873	ribose	Ribose *f*
7874	ribosomal RNA, r-RNA	ribosomale RNS *f*
7875	ribosome, Palade's granules	Ribosom *n*, Palade-Granula *pl*, RNS-Protein-Granula *pl*
7876	ribulose	Ribulose *f*
7877	ribulose diphosphate	Ribulosediphosphat *n*
7878	rigidity	Starre *f*, Starrheit *f*
7879	rind, cortex	Rinde *f*, Kortex *m*
7880	to ring	beringen
7881	ring-bark	Ringelborke *f*
7882	ring canal	Ringkanal *m*
7883	ring chromosome, circular ~	Ringchromosom *n*
7884	ring vessel → annular vessel	—
7885	Ringer, Ringer's fluid, ~ solution	Ringerlösung *f*
7886	ripe, ripeness → mature, maturity	—
7887	ritualization *eth.*	Ritualisation *f*, Ritualisierung *f*
7888	river pollution, stream ~	Flußverschmutzung *f*
7889	RNA → ribonucleic acid	—
7890	rock-plant → rupicolous plant	—
7891	rod *anat.*	Stäbchen *n* (Netzhaut)
7892	rod-shaped → baculiform	—
7893	Rodentia, rodents	Nagetiere *npl*
7894	rodenticide	Rodentizid *n*
7895	roe, hard roe	Rogen *m*
7896	roentgen equivalent	Röntgeneinheit *f*
7897	roentgen dose → X-ray dose	—
7898	roentgen radiation	Röntgenstrahlung *f*
7899	roentgen rays → X-rays	—
7900	root	Wurzel *f*
7901	root-cap, calyptra, pileorhiza	Wurzelhaube *f*, Kalyptra *f*
7902	root collar	Wurzelhals *m*
7903	root formation	Wurzelbildung *f*, Rhizogenese *f*
7904	root hair	Wurzelhaar *n*
7905	root-hair zone → piliferous layer	—
7906	root nodules, ~ tubercles	Wurzelknöllchen *npl*
7907	root pressure	Wurzeldruck *m*
7908	root-stock → rhizome	—
7909	root-sucker	Wurzeltrieb *m*, Wurzelausschlag *m*, Wurzelschößling *m*
7910	root system	Wurzelsystem *n*

7850	facteur *m* Rhésus, ~ Rh	factor *m* Rhesus, ~ Rh
7851	—	—
7852	rhinencéphale *m*, bulbe *m* olfactif	rinencéfalo *m*, bulbo *m* olfativo
7853	—	—
7854	(espèce de cyme unipare)	ripidio *m*, flabelo *m*
7855	épiderme *m* de la racine	rizodermis *f*
7856	rhizoïde *m*	rizoide *m*
7857	rhizome *m*	rizoma *m*
7858	rhizomorphe *m*	rizomorfo *m*
7859	rhizopodes *mpl*	rizópodos *mpl*, sarcodinos *mpl*
7860	rhizosphère *f*	rizosfera *f*
7861	rhodophycées *fpl*, rhodophytes *fpl*, algues *fpl* rouges	rodofíceas *fpl*, rodófitos *mpl*, rodófitas *fpl*, algas *fpl* rojas
7862	rhodopsine *f*, pourpre *m* visuel, ~ rétinien	rodopsina *f*, púrpura *f* visual
7863	rhombencéphale *m*	rombencéfalo *m*
7864	—	—
7865	—	—
7866	côte *f*	costilla *f*
7867	—	—
7868	riboflavine *f*, lactoflavine *f*, vitamine *f* B_2	riboflavina *f*, lactoflavina *f*, vitamina *f* B_2
7869	ribonucléase *f*	ribonucleasa *f*
7870	acide *m* ribonucléique, ARN	ácido *m* ribonucleico, ARN
7871	ribonucléoprotéine *f*	ribonucleoproteína *f*
7872	ribonucléoside *m*	ribonucleósido *m*
7873	ribose *m*	ribosa *f*
7874	ARN *m* ribosomique, ~ ribosomal	ARN *m* ribosomático
7875	ribosome *m*, grains *mpl* de Palade, ~ ribonucléoprotéiques	ribosoma *m*, granos *mpl* de Palade, partículas *fpl* ribonucleoproteicas
7876	ribulose *m*	ribulosa *f*
7877	ribulosediphosphate *m*	ribulosa *f* difosfato
7878	rigidité *f*	rigidez *f*
7879	écorce *f*, cortex *m*	corteza *f*
7880	baguer	anillar
7881	rhytidome *m* annulaire	ritidoma *m* anular
7882	canal *m* annulaire	canal *m* circular, conducto *m* ~
7883	chromosome *m* annulaire, ~ circulaire	cromosoma *m* anular
7884	—	—
7885	solution *f* de Ringer, liquide *m* ~ ~	solución *f* de Ringer
7886	—	—
7887	ritualisation *f* *eth.*	ritualización *f* *eth.*
7888	pollution *f* des fleuves, ~ des cours d'eau	contaminación *f* de (los) ríos
7889	—	—
7890	—	—
7891	bâtonnet *m* *anat.*	bastón *m* *anat.*
7892	—	—
7893	rongeurs *mpl*	roedores *mpl*
7894	rodenticide *m*	rodenticida *m*
7895	rogue *f*	huevas *fpl*
7896	équivalent *m* roentgenien, unité *f* de rayonnement X	unidad *f* roentgen
7897	—	—
7898	radiation *f* roentgen	radiación *f* roentgen, ~ de rayos X
7899	—	—
7900	racine *f*	raíz *f*
7901	coiffe *f*, calyptre *m*, pilorhize *f*	caliptra *f*, pilorriza *f*, casquete *m* radical
7902	collet *m* de la racine	cuello *m* de la raíz
7903	rhizogenèse *f*	formación *f* de raíces
7904	poil *m* absorbant, ~ radiculaire	pelo *m* absorbente, ~ radicular
7905	—	—
7906	nodosités *fpl* de la racine	nódulos *mpl* radiculares, nudosidades *fpl* radicales
7907	poussée *f* radiculaire	presión *f* radical
7908	—	—
7909	drageon *m*, surgeon *m*	retoño *m* de la raíz
7910	système *m* radiculaire, appareil *m* radical	sistema *m* radicular, ~ radical

7911	root tip	Wurzelspitze *f*
7912	root tuber	Wurzelknolle *f*
7913	rooted, rooting → radicate, radication	—
7914	rosette	Blattrosette *f*
7915	rosette plant	Rosettenpflanze *f*
7916	rosular, rosulate, rosellate	rosettenständig
7917	rot, to rot → putrefaction, to putrefy	—
7918	rotational crossbreeding	Rotationskreuzung *f*
7919	Rotifera, Rotatoria, rotifers, rotiferans	Rädertierchen *npl*
7920	rotula → knee-cap	—
7921	round window *anat.*	rundes Fenster *n anat.*
7922	r-RNA → ribosomal RNA	—
7923	ruderal plant	Ruderalpflanze *f*
7924	rudiment	Rudiment *n*
7925	rudimentary	rudimentär
7926	rudimentary organ; vestigial ~	Rudimentärorgan *n*
7927	rumen, paunch	Pansen *m*
7928	ruminant	Wiederkäuer *m*
7929	to ruminate	wiederkäuen
7930	rumination	Wiederkäuen *n*
7931	to run wild, to return to the wild state	verwildern
7932	runner → stolon	—
7933	running wild, return to the wild state	Verwilderung *f*
7934	rupicolous plant, rock-plant, petrophyte, epilithic plant	Felspflanze *f*, Petrophyt *m*, Epilith *m*
7935	rut → oestrus	—
7936	sac fungi → ascomycetes	—
7937	saccharase → invertase	—
7938	saccharide	Sa(c)charid *n*
7939	saccharimeter	Sa(c)charimeter *n*, Zuckermesser *m*
7940	saccharose, sucrose, saccharobiose, cane sugar	Sa(c)charose *f*, Sa(c)charobiose *f*, Rohrzucker *m*
7941	sacciform, saccular	sackförmig
7942	sacral vertebra	Kreuz(bein)wirbel *m*
7943	sacrum	Kreuzbein *n*
7944	saddle joint	Sattelgelenk *n*
7945	sagittal plane	Sagittalebene *f*
7946	sagittal section	Sagittalschnitt *m*
7947	sagittate	pfeilförmig
7948	salicylic acid	Salizylsäure *f*
7949	saline	salzhaltig, salzig
7950	salinity, salt content	Salzgehalt *m*, Salzhaltigkeit *f*
7951	saliva	Speichel *m*
7952	salivary amylase → ptyalin	—
7953	salivary chromosome	Speicheldrüsenchromosom *n*
7954	salivary gland	Speicheldrüse *f*
7955	salivary secretion	Speichelabsonderung *f*
7956	salivation	Speichelfluß *m*
7957	salt balance	Salzhaushalt *m*
7958	salt concentration	Salzkonzentration *f*
7959	salt content → salinity	—
7960	salt gland	Salzdrüse *f*
7961	salt-loving → halophilous	—
7962	salt regulation	Salzregulierung *f*
7963	salt solution	Salzlösung *f*
7964	saltation *bot.*	Saltation *f bot.*
7965	saltatory conduction	Saltation *f neur.*, saltatorische Leitung *f*
7966	samara → winged fruit	—
7967	to sample	Proben entnehmen, ~ ziehen
7968	sample area	Aufnahmefläche *f*, Probefläche *f*
7969	sampling	Probeentnahme *f*
7970	sand plants → psammophytes	—
7971	sap	Pflanzensaft *m*

7911	apex *m* de racine, extrémité *f* radiculaire	punta *f* radicular, ápice *m* radical
7912	tubercule *m* radiculaire, ~ radical	tubérculo *m* radicular
7913	—	—
7914	rosette *f* foliacée, ~ de feuilles	roseta *f* foliar
7915	plante *f* en rosette	planta *f* (en) roseta, ~ arrosetada
7916	en rosette	arrosetado, arrosariado
7917	—	—
7918	croisement *m* en rotation	cruzamiento *m* de rotación, cruce *m* rotativo
7919	rotifères *mpl*	rotíferos *mpl*
7920	—	—
7921	fenêtre *f* ronde *anat.*	ventana *f* redonda *anat.*
7922	—	—
7923	plante *f* rudérale	planta *f* ruderal
7924	rudiment *m*	rudimento *m*
7925	rudimentaire	rudimentario
7926	organe *m* rudimentaire; ~ vestigial	órgano *m* rudimentario; ~ vestigial
7927	panse *f*, rumen *m*, herbier *m*	rumen *m*, panza *f*
7928	ruminant *m*	rumiante *m*
7929	ruminer	rumiar
7930	rumination *f*	rumia *f*
7931	retourner à l'état sauvage	retornar al estado silvestre, asilvestrar
7932	—	—
7933	retour *m* à l'état sauvage	retorno *m* al estado silvestre
7934	plante *f* rupestre, ~ rupicole, ~ de rocher	planta *f* rupestre, ~ rupícola
7935	—	—
7936	—	—
7937	—	—
7938	saccharide *m*	sacárido *m*
7939	saccharimètre *m*	sacarímetro *m*
7940	saccharose *m*, sucrose *m*, sucre *m* de canne	sacarosa *f*, sucrosa *f*, sacarobiosa *f*, azúcar *m* de caña
7941	sacciforme, sacculaire	sacciforme, sacular
7942	vertèbre *f* sacrée	vértebra *f* sacra
7943	sacrum *m*	sacro *m*, hueso *m* ~
7944	articulation *f* en selle	articulación *f* sellar
7945	plan *m* sagittal	plano *m* sagital
7946	section *f* sagittale, coupe *f* ~	sección *f* sagital
7947	sagitté	sagitado
7948	acide *m* salicylique	ácido *m* salicílico
7949	salin	salino
7950	salinité *f*, teneur *f* saline, salure *f*	salinidad *f*, grado *m* salínico, contenido *m* salino
7951	salive *f*	saliva *f*
7952	—	—
7953	chromosome *m* salivaire	cromosoma *m* salival, ~ de las glándulas salivales
7954	glande *f* salivaire	glándula *f* salival
7955	sécrétion *f* salivaire	secreción *f* salival
7956	salivation *f*	salivación *f*
7957	balance *f* des sels	equilibrio *m* salino
7958	concentration *f* saline	concentración *f* salina
7959	—	—
7960	glande *f* à sel	glándula *f* salífera
7961	—	—
7962	régulation *f* saline	regulación *f* salina
7963	solution *f* salée	solución *f* salina
7964	saltation *f* *bot.*	saltación *f* *bot.*
7965	conduction *f* saltatoire	conducción *f* saltatoria
7966	—	—
7967	échantillonner, prélever des échantillons	tomar muestras, sacar ~
7968	aire-échantillon *f*, surface-échantillon *f*	área *m* de muestra
7969	échantillonnage *m*, prise *f* d'essai	muestreo *m*, toma *f* de muestras
7970	—	—
7971	sève *f* (végétale), suc *m* végétal	savia *f*, jugo *m* vegetal

7972	sap-flow, circulation of sap	Saftstrom *m*, Saftströmung *f*
7973	saprobiont	Saprobiont *m*, Saprobie *f*
7974	saprogenic, saprogenous	saprogen, fäulniserregend
7975	sapropel	Sapropel *n*, Faulschlamm *m*
7976	saprophage, saprovore	Saprophage *m*, Fäulnisfresser *m*
7977	saprophyte, saprophytic plant	Saprophyt *m*, Fäulnispflanze *f*
7978	saprophytism	Saprophytismus *m*
7979	saprozoite, saprozoic animal	Saprozoon *n*
7980	sapwood, splint-wood, alburnum	Splint *m*, Splintholz *n*, Weichholz *n*
7981	Sarcodina → Rhizopoda	—
7982	sarcolemma → myolemma	—
7983	sarcomere	Sarkomer *n*
7984	sarcoplasm	Sarkoplasma *n*
7985	sarcoplasmic reticulum	sarkoplasmatisches Retikulum *n*
7986	sarcosome	Sarkosom *n*
7987	sarmentous plant	rankende Pflanze *f*
7988	satellite → trabant	—
7989	satellite chromosome, SAT-chromosome *obs.*	Satellitenchromosom *n*, Trabantenchromosom *n*, SAT-Chromosom *n obs.*
7990	saturated	gesättigt
7991	saturation	Sättigung *f*
7992	saturation effect	Sättigungseffekt *m*
7993	SAT-zone → nucleolar zone	—
7994	Sauropsidia	Sauropsiden *pl*
7995	savanna(h)	Savanne *f*
7996	scalariform tracheid	Leitertracheide *f*, Treppentracheide *f*
7997	scalariform vessel	Leitergefäß *n*, Treppengefäß *n*
7998	scale	Schuppe *f*
7999	scale bark, scaly ~	Schuppenborke *f*
8000	scale leaf, scaly ~	Schuppenblatt *n*
8001	scaly → squamate	—
8002	scapula, shoulder-blade	Schulterblatt *n*
8003	scapular	Schulter...
8004	scapulohumeral articulation → shoulder-joint	—
8005	scent gland → odoriferous gland	—
8006	scent mark, ~ trail → odour mark, ~ trail	—
8007	schizocarp	Spaltfrucht *f*
8008	schizogamy	Schizogamie *f*
8009	schizogenesis	Schizogenese *f*
8010	schizogenous, schizogenetic	schizogen
8011	schizogony	Schizogonie *f*
8012	Schizomycophyta, schizomycetes	Spaltpilze *mpl*, Schizomyzeten *mpl*
8013	schizont	Schizont *m*
8014	schizophyte	Spaltpflanze *f*
8015	school of fish	Fischbank *f*, Fischschwarm *m*
8016	schooling of fish	Schwarmbildung *f (Fische)*
8017	Schwann('s) cell	Schwannsche Zelle *f*
8018	Schwann's sheath, neurolemma, neurilemma	Schwannsche Scheide *f*, Neurolemma *n*, Neurilemma *n*
8019	sciatic(al), coxal	Hüft...
8020	sciatic nerve	Hüftnerv *m*
8021	scintillation counter	Schwingungszähler *m*, Szintillationszähler *m*
8022	scio... → skio...	—
8023	scion → graft	—
8024	scissile → fissile	—
8025	scissiparity, fissiparity	Teilungsvermehrung *f*
8026	sclera, sclerotic	Lederhaut *f (Auge)*
8027	sclereid, stone cell	Steinzelle *f*, Sklereide *f*
8028	sclerenchyma	Sklerenchym *n*
8029	sclerite	Sklerit *n*
8030	sclerophyllous, hard-leaved	hartblättrig
8031	sclerophylls, sclerophyllous plants	Hartlaubgewächse *npl*, Sklerophyllen *pl*

No.	Français	Español
7972	circulation *f* de la sève, courant *m* ~ ~ ~	flujo *m* de savia
7973	saprobionte *m*	saprobio *m*
7974	saprogène	saprógeno
7975	sapropèle *m/f*	saprópelo *m*
7976	saprophage *m*	saprófago *m*, sapróvoro *m*
7977	saprophyte *m*	saprófito *m*, planta *f* saprófita
7978	saprophytisme *m*	saprofitismo *m*
7979	saprozoïte *m*	saprozoo *m*
7980	aubier *m*	albura *f*
7981	—	—
7982	—	—
7983	sarcomère *m*	sarcómera *f*
7984	sarcoplasme *m*	sarcoplasma *m*
7985	réticulum *m* sarcoplasmique	retículo *m* sarcoplasmático
7986	sarcosome *m*	sarcosoma *m*
7987	plante *f* sarmenteuse	planta *f* zarcillosa
7988	—	—
7989	chromosome *m* satellite, SAT-chromosome *m obs.*	cromosoma *m* satélite, ~ SAT *obs.*
7990	saturé	saturado
7991	saturation *f*	saturación *f*
7992	effet *m* de saturation	efecto *m* de saturación
7993	—	—
7994	sauropsidés *mpl*	saurópsidos *mpl*
7995	savane *f*	sabana *f*
7996	trachéide *f* scalariforme	traqueida *f* escalariforme
7997	vaisseau *m* scalariforme	vaso *m* escalariforme
7998	écaille *f*	escama *f*
7999	rhytidome *m* écailleux	ritidoma *m* escamoso
8000	feuille *f* écailleuse, ~ en écaille, ~ squamiforme	hoja *f* escamiforme
8001	—	—
8002	omoplate *f*	escápula *f*, omóplato *m*, paletilla *f*
8003	scapulaire	escapular
8004	—	—
8005	—	—
8006	—	—
8007	schizocarpe *m*	esquizocarpo *m*
8008	schizogamie *f*	esquizogamia *f*
8009	schizogenèse *f*	esquizogénesis *f*
8010	schizogène	esquizógeno
8011	schizogonie *f*	esquizogonia *f*
8012	schizomycètes *mpl*	esquizomicetos *mpl*, esquizomicotitos *mpl*
8013	schizonte *m*	esquizonto *m*
8014	schizophyte *m*	esquizófito *m*
8015	banc *m* de poissons	banco *m* de peces, cardumen *m*
8016	réunion *f* par bancs, formation *f* de bancs, groupement *m* en bancs	reunión *f* en bancos
8017	cellule *f* de Schwann	célula *f* de Schwann
8018	gaine *f* de Schwann, névrilème *m*, neurilemme *m*	vaina *f* de Schwann, neurilema *m*
8019	sciatique, coxal	ciático, coxal
8020	nerf *m* sciatique	nervio *m* ciático
8021	compteur *m* de scintillations	contador *m* de escintilación
8022	—	—
8023	—	—
8024	—	—
8025	scissiparité *f*, fissiparité *f*	escisiparidad *f*, fisiparidad *f*, reproducción *f* esquizógena
8026	sclérotique *f*	esclerótica *f*
8027	cellule *f* pierreuse, ~ scléreuse	célula *f* pétrea, esclereida *f*
8028	sclérenchyme *m*	esclerénquima *m*
8029	sclérite *m*	esclerito *m*
8030	sclérophylle	esclerofilo
8031	sclérophytes *mpl*	esclerofilos *mpl*, (plantas *fpl*) esclerofilas *fpl*

8032	scleroprotein, fibrous protein	Skleroprotein *n*, Gerüsteiweiß *n*
8033	sclerotic → sclera	—
8034	sclerotium	Sklerotium *n*, Dauermyzel *n*
8035	sclerotome	Sklerotom *n*
8036	scolopale	Skolopalorgan *n*, Skoloparium *n*
8037	scorpioid cyme → cincinnus	—
8038	screening	Abschirmung *f*
8039	scrotum	Hodensack *m*, Skrotum *n*
8040	scutiform, peltate	schildförmig
8041	sea-bottom	Meeresboden *m*
8042	sea-bottom water	Meerestiefwasser *n*
8043	sea water	Meerwasser *n*
8044	seasonal dimorphism	Saisondimorphismus *m*
8045	seasonal fluctuations	jahreszeitliche Schwankungen *fpl*
8046	seasonal periodicity, ~ rhythmicity	Saisonperiodik *f*, Saisonrhythmik *f*
8047	sebaceous gland	Talgdrüse *f*
8048	sebum	Hauttalg *m*
8049	second filial generation → F_2 generation	—
8050	secondary constriction → nucleolar constriction	—
8051	secondary consumer	Sekundärkonsument *m*
8052	secondary meristem	Folgemeristem *n*, sekundäres Meristem *n*
8053	secondary root → lateral root	—
8054	secondary sere → subsere	—
8055	secondary sex(ual) character	sekundäres Geschlechtsmerkmal *n*
8056	secondary structure	Sekundärstruktur *f*
8057	to secrete	absondern, sekretieren, sezernieren
8058	secretin	Sekretin *n*
8059	secretion	Sekretion *f*, Absonderung *f*; Sekret *n*
8060	secretory	sekretorisch
8061	secretory action	Drüsentätigkeit *f*
8062	secretory body	Sekretkörperchen *n*
8063	secretory cavity *bot.*	Sekretbehälter *m bot.*
8064	secretory cell	Sekretzelle *f*
8065	secretory product	Sekret *n*
8066	secretory tissue	Sekretionsgewebe *n*, Absonderungsgewebe *n*
8067	sectorial chim(a)era	Sektorialchimäre *f*
8068	sedentary	sitzend
8069	sedimentation rate	Senkungsgeschwindigkeit *f*, Sedimentationsgeschwindigkeit *f*
8070	seed	Samen *m bot.*
8071	seed-bearing → seminiferous	—
8072	seed-bearing scale, ovuliferous ~	Samenschuppe *f*, Fruchtschuppe *f*
8073	seed capsule	Samenkapsel *f*
8074	seed coat, episperm, test(a)	Samenschale *f*, Testa *f*
8075	seed dispersal, dissemination	Samenverbreitung *f*
8076	seed dormancy	Samenruhe *f*
8077	seed ferns → pteridosperms	—
8078	seed leaf → cotyledon	—
8079	seed plants → Spermatophyta	—
8080	seed stalk → funicle	—
8081	segmental allopolyploidy	Segmentallopolyploidie *f*
8082	segmental interchange	Segmentaustausch *m*
8083	segmentation → cleavage	—
8084	segmentation cavity → blastocoel	—
8085	segmentation nucleus → cleavage nucleus	—
8086	to segregate	sich spalten *(Chromosomen etc.)*
8087	segregation	Spaltung *f*, Segregation *f*
8088	segregation ratio → Mendelian ratio	—
8089	seismonasty	Seismonastie *f*
8090	to select	auslesen
8091	selected plants	Elitepflanzen *fpl*
8092	selection	Selektion *f*, Auslese *f*, Zuchtwahl *f*
8093	selection coefficient	Selektionskoeffizient *m*
8094	selection differential	Selektionsdifferential *n*

8032	scléroprotéine *f*, protéine *f* fibrillaire	escleroproteína *f*, proteína *f* fibrosa
8033	—	—
8034	sclérote *m*	esclerocio *m*
8035	sclérotome *m*	esclerótomo *m*
8036	organe *m* scolopal	escolópalo *m*
8037	—	—
8038	blindage *m*	blindaje *m*
8039	scrotum *m*	escroto *m*
8040	scutiforme, pelté	escutiforme, peltado
8041	fond *m* marin	fondo *m* marino, suelo *m* ~
8042	eau *f* du fond de la mer	agua *f* del fondo del mar
8043	eau *f* marine, ~ de mer	agua *f* de mar
8044	dimorphisme *m* saisonnier	dimorfismo *m* estacional
8045	fluctuations *fpl* saisonnières	fluctuaciones *fpl* estacionales
8046	périodicité *f* saisonnière	periodicidad *f* estacional
8047	glande *f* sébacée	glándula *f* sebácea
8048	sébum *m*	sebo *m*
8049	—	—
8050	—	—
8051	consommateur *m* secondaire	consumidor *m* secundario
8052	méristème *m* secondaire	meristema *m* secundario
8053	—	—
8054	—	—
8055	caractère *m* sexuel secondaire	carácter *m* sexual secundario
8056	structure *f* secondaire	estructura *f* secundaria
8057	sécréter, sécerner	secretar, segregar
8058	sécrétine *f*	secretina *f*
8059	sécrétion *f*	secreción *f*
8060	sécréteur, sécrétoire	secretor(io)
8061	action *f* sécrétoire, ~ glandulaire	acción *f* secretoria
8062	corps *m* de sécrétion	cuerpo *m* secretorio
8063	réservoir *m* de sécrétion, appareil *m* sécréteur *bot.*	cavidad *f* secretora, recipiente *m* secretorio *bot.*
8064	cellule *f* sécrétrice	célula *f* secretora
8065	produit *m* sécrété, ~ de sécrétion	producto *m* secretorio
8066	tissu *m* sécréteur	tejido *m* secretor
8067	chimère *f* sectorielle	quimera *f* sectorial
8068	sédentaire	sedentario
8069	vitesse *f* de sédimentation	velocidad *f* de sedimentación
8070	semence *f*	semilla *f*
8071	—	—
8072	écaille *f* fructifère, ~ ovulifère	escama *f* fructífera, ~ seminífera
8073	capsule *f* séminale	cápsula *f* seminal
8074	épisperme *m*, tégument *m* séminal, test *m*	episperma *m*, cubierta *f* seminal, testa *f*
8075	dissémination *f*	diseminación *f*, dispersión *f* de la semilla
8076	dormance *f* des semences	reposo *m* seminal
8077	—	—
8078	—	—
8079	—	—
8080	—	—
8081	allopolyploïdie *f* segmentaire	alopoliploidia *f* segmental
8082	interchangement *m* de segments	intercambio *m* de segmentos
8083	—	—
8084	—	—
8085	—	—
8086	ségréger, se séparer	segregarse
8087	ségrégation *f*	segregación *f*
8088	—	—
8089	s(é)ismonastie *f*, sismonastisme *m*	s(e)ismonastia *f*
8090	sélectionner	seleccionar
8091	lignées *fpl* d'élite	plantas *fpl* élites, ~ selectas
8092	sélection *f*	selección *f*
8093	coefficient *m* de sélection	coeficiente *m* de selección
8094	sélection *f* différentielle	diferencial *m* de selección

No.	English	German
8095	selection method → breeding ~	—
8096	selective	selektiv
8097	selective advantage	Selektionsvorteil *m*
8098	selective breeding	Auslesezüchtung *f*
8099	selective culture medium	Selektivnährboden *m*
8100	selective peak	Selektionsgipfel *m*
8101	selective permeability	selektive Permeabilität *f*
8102	selective pressure, selection ~	Selektionsdruck *m*
8103	selective value	Selektionswert *m*
8104	selectivity	Selektivität *f*
8105	self-compatibility → self-fertility	—
8106	self-differentiation	Selbstdifferenzierung *f*
8107	self-digestion	Selbstverdauung *f*
8108	self-dispersal → autochory	—
8109	self-duplication, self-replication	Auto(re)duplikation *f*, Selbstverdoppelung *f*, identische Reduplikation *f*
8110	self-fertile	selbstfertil, selbstbefruchtend
8111	self-fertility, self-compatibility	Selbstfertilität *f*, Selbstfruchtbarkeit *f*, Selbstkompatibilität *f*
8112	self-fertilization, selfing, autogamy	Selbstbefruchtung *f*, Selbstung *f*, Autogamie *f*
8113	self-fertilizer	Selbstbefruchter *m*
8114	self-incompatibility	Selbstunverträglichkeit *f*, Autoinkompatibilität *f*
8115	self-pollination, autogamy	Selbstbestäubung *f*, Autogamie *f*
8116	self-purification, self-cleaning	Selbstreinigung *f*
8117	self-regulation	Selbstregulierung *f*, Selbststeuerung *f*
8118	self-replication → self-duplication	—
8119	self-reproduction	Selbstreproduktion *f*, Autoreproduktion *f*
8120	self-reproductive	selbstreproduktiv
8121	self-sterile	selbststeril
8122	self-sterility	Selbststerilität *f*
8123	selfing → self-fertilization	—
8124	semen	Samen *m phys.*
8125	semi... see also hemi...	—
8126	semi-arid	semiarid
8127	semicircular canals, ~ ducts	Bogengänge *mpl anat.*
8128	semi-conservative replication	semikonservative Replikation *f*
8129	semidesert	Halbwüste *f*
8130	semidominance, partial dominance, incomplete ~	Semidominanz *f*, unvollständige Dominanz *f*
8131	semifluid	halbflüssig
8132	semilethal → sublethal	—
8133	seminal fluid, spermatic ~	Samenflüssigkeit *f*
8134	seminal receptacle, spermatheca	Samenbehälter *m*, Samentasche *f*
8135	seminal vesicle	Samenbläschen *n*
8136	seminiferous, seed-bearing	samentragend
8137	seminiferous tubule	Samenröhrchen *n*, Samenkanälchen *n*, Hodenkanälchen *n*
8138	semipermeability	Semipermeabilität *f*, Halbdurchlässigkeit *f*
8139	semipermeable	semipermeabel, halbdurchlässig
8140	semipermeable membrane	semipermeable Membran *f*, halbdurchlässige ~
8141	semi-pupa → pseudopupa	—
8142	semisterile	semisteril
8143	semisterility	Semisterilität *f*
8144	senescence, ag(e)ing	Altern *n*, Alterung *f*, Seneszenz *f*
8145	sensation	Empfindung *f*
8146	sense... see also sensory...	—
8147	sense of balance, static sense	Gleichgewichtssinn *m*
8148	sense of hearing → auditory sense	—
8149	sense of orientation	Orientierungssinn *m*
8150	sense of pain	Schmerzsinn *m*
8151	sense of smell → olfactory sense	—

8095	—	—
8096	sélectif	selectivo
8097	avantage *m* sélectif	ventaja *f* selectiva
8098	élevage *m* de races sélectionnées	cría *f* de selección, ~ selectiva
8099	milieu *m* de culture sélectif	medio *m* selectivo
8100	pic *m* de sélection	máximo *m* de selección
8101	perméabilité *f* sélective	permeabilidad *f* selectiva
8102	pression *f* de sélection	presión *f* de selección, ~ selectiva
8103	valeur *f* sélective	valor *m* selectivo
8104	sélectivité *f*	selectividad *f*
8105	—	—
8106	autodifférenciation *f*	autodiferenciación *f*
8107	autodigestion *f*	autodigestión *f*
8108	—	—
8109	autoduplication *f*, auto-réplication *f*	autoduplicación *f*, autorreplicación *f*
8110	autofertile, autofécondant	autofértil, autofecundante
8111	autofertilité *f*, autofécondité *f*	autofertilidad *f*, autocompatibilidad *f*
8112	autofécondation *f*, autogamie *f*	autofecundación *f*, autofertilización *f*, autogamia *f*
8113	individu *m* autofécondant, ~ à autoreproduction	autofecundante *m*, autofertilizador *m*, autofertilizante *m*
8114	auto-incompatibilité *f*	autoincompatibilidad *f*, incompatibilidad *f* propia
8115	autopollinisation *f*, autogamie *f*	autopolinización *f*, autogamia *f*
8116	auto-épuration *f*	autodepuración *f*, autopurificación *f*
8117	autorégulation *f*	autorregulación *f*
8118	—	—
8119	autoreproduction *f*	autorreproducción *f*
8120	autoreproductible	autorreproducible, autorreproductor
8121	autostérile	autostéril
8122	autostérilité *f*	autosterilidad *f*
8123	—	—
8124	semence *f* *phys.*, sperme *m*	semen *m*
8125	—	—
8126	semi-aride	semiárido
8127	canaux *mpl* semi-circulaires	conductos *mpl* semicirculares
8128	réplication *f* semi-conservative	replicación *f* semiconservadora
8129	semi-désert *m*	semidesierto *m*
8130	semi-dominance *f*, dominance *f* incomplète	semidominancia *f*, dominancia *f* incompleta
8131	semi-fluide, semiliquide	semifluido, semilíquido
8132	—	—
8133	liquide *m* séminal, ~ spermatique	líquido *m* seminal
8134	réceptacle *m* séminal, sperma(to)thèque *f*	receptáculo *m* seminífero, espermateca *f*, espermoteca *f*
8135	vésicule *f* séminale	vesícula *f* seminal
8136	séminifère	seminífero
8137	tube *m* séminifère	tubo *m* seminífero
8138	semiperméabilité *f*, hémiperméabilité *f*	semipermeabilidad *f*
8139	semiperméable, hémipermeable	semipermeable
8140	membrane *f* semiperméable	membrana *f* semipermeable
8141	—	—
8142	semi-stérile	semiestéril
8143	semi-stérilité *f*	semiesterilidad *f*
8144	sénescence *f*, vieillissement *m*	senescencia *f*, envejecimiento *m*
8145	sensation *f*	sensación *f*
8146	—	—
8147	sens *m* de l'équilibre, ~ statique	sentido *m* del equilibrio, ~ estático
8148	—	—
8149	sens *m* de l'orientation	sentido *m* de la orientación
8150	sens *m* de la douleur	sentido *m* del dolor
8151	—	—

8152	sense of taste → gustatory sense	—
8153	sense of touch, tactual sense	Tastsinn *m*
8154	sense organ, sensory ~	Sinnesorgan *n*
8155	sensibility, sensitivity	Empfindlichkeit *f*, Sensibilität *f*
8156	sensibilization → sensitization	—
8157	sensillum	Sensillum *n*
8158	sensitive period	sensible Periode *f*
8159	sensitivity to heat → thermal sensitivity	—
8160	sensitivity to stimuli	Reizempfindlichkeit *f*
8161	sensitivity threshold	Empfindungsschwelle *f*
8162	sensitization, sensibilization	Sensibilisierung *f*
8163	to sensitize	sensibilisieren
8164	sensitizing effect	Sensibilisierungswirkung *f*
8165	sensorium	Sensorium *n*
8166	sensory	sensorisch, sensoriell, Sinnes...
8167	sensory bristle	Sinnesborste *f*
8168	sensory cell, sense ~	Sinneszelle *f*
8169	sensory epithelium	Sinnesepithel *n*
8170	sensory hair, sense ~	Sinneshaar *n*
8171	sensory impression, sense ~	Sinneseindruck *m*
8172	sensory nerve	Sinnesnerv *m*, sensibler Nerv *m*
8173	sensory nerve cell, ~ neuron(e)	Sinnesnervenzelle *f*, sensorisches Neuron *n*
8174	sensory physiology, sense ~	Sinnesphysiologie *f*
8175	sensory root → dorsal root	—
8176	sensory stimulus	Sinnesreiz *m*
8177	sepal	Kelchblatt *n*
8178	separation of strains	Formentrennung *f*
8179	septate	septiert
8180	septicidal	septizid, scheidewandspaltig
8181	septifragal	septifrag, scheidewandbrüchig
8182	septum (*pl* septa)	Septum *n*, Scheidewand *f*
8183	sequence of amino acids	Aminosäuresequenz *f*
8184	seral stage, successional ~	Seralstufe *f*
8185	seralbumin → serum albumin	—
8186	sere, successional series	Serie *f*, Sukzessionsserie *f*
8187	serial experiment	Reihenexperiment *n*
8188	serial section	Serienschnitt *m*
8189	series of actions	Handlungskette *f*
8190	series of experiments, set ~ ~	Versuchsserie *f*, Versuchsreihe *f*
8191	serine	Serin *n*
8192	serologist	Serologe *m*
8193	serology	Serologie *f*
8194	serosa, serous membrane	Serosa *f*
8195	serotonin, enteramine	Serotonin *n*, Enteramin *n*
8196	serous	serös
8197	serrate	gesägt
8198	serum	Serum *n*
8199	serum albumin, seralbumin	Serumalbumin *n*
8200	serum diagnosis, serodiagnosis	Serodiagnostik *f*
8201	serum protein	Serumprotein *n*
8202	sesamoid (bone)	Sesambein *n*
8203	sessile	sessil
8204	seston	Seston *n*
8205	set of experiments → series ~ ~	—
8206	set of genes	Gensatz *m*
8207	seta[1] → bristle	—
8208	seta[2] *bot.*	Seta *f*, Kapselstiel *m* *(Moose)*
8209	setaceous → bristly	—
8210	sewage, waste water	Abwasser *n*, Abwässer *n pl*

8152	—	—
8153	(sens *m* du) toucher *m*, sens *m* tactile, tact *m*	(sentido *m* del) tacto *m*
8154	organe *m* des sens, ~ sensoriel	órgano *m* sensorial
8155	sensibilité *f*, sensitivité *f*	sensibilidad *f*
8156	—	—
8157	sensille *f*, sensillum *m*	sensilio *m*
8158	période *f* sensible	período *m* sensible, ~ sensitivo
8159	—	—
8160	sensitivité *f* aux stimuli, sensibilité *f* ~ ~	sensibilidad *f* al estímulo
8161	seuil *m* de sensibilité	umbral *m* de sensación, límite *m* de sensibilidad
8162	sensibilisation *f*	sensibilización *f*
8163	sensibiliser	sensibilizar
8164	effet *m* sensibilisant	efecto *m* sensibilizador
8165	sensorium *m*	sensorio *m*
8166	sensoriel	sensorial
8167	soie *f* sensorielle	cerda *f* sensitiva, queta *f* sensorial
8168	cellule *f* sensorielle	célula *f* sensorial, ~ sensitiva
8169	épithélium *m* sensoriel	epitelio *m* sensorial
8170	poil *m* sensoriel, ~ sensitif	pelo *m* sensorial, ~ sensitivo
8171	impression *f* sensorielle	impresión *f* sensorial
8172	nerf *m* sensitif, ~ sensoriel	nervio *m* sensorial, ~ sensitivo
8173	neurone *m* sensitif	neurona *f* sensorial, ~ sensitiva, célula *f* nerviosa sensorial
8174	physiologie *f* sensorielle	fisiología *f* sensorial
8175	—	—
8176	stimulus *m* sensoriel	estímulo *m* sensorial, ~ sensitivo
8177	sépale *m*	sépalo *m*
8178	isolement *m* des lignées, séparation *f* des souches	aislamiento *m* de líneas
8179	cloisonné	septado, tabicado
8180	septicide	septicida
8181	septifrage	septífrago
8182	septum *m*, septe *m*, cloison *f* séparatrice	septo *m*, tabique *m* separatorio, ~ divisorio, pared *f* divisoria
8183	séquence *f* d'amino-acides	secuencia *f* de aminoácidos, ~ aminoacídica
8184	stade *m* successif, ~ de succession, étape *f* successive	etapa *f* seral, ~ de sucesión
8185		—
8186	série *f* de succession	sere *f*, serie *f* sucesional
8187	expérience *f* en série	experimento *m* seriado
8188	section *f* en série, coupes *f pl* sériées	corte *m* en serie, secciones *f pl* seriadas
8189	séquence *f* d'actes, série *f* d'activité	cadena *f* de actos, ~ ~ actividades
8190	série *f* expérimentale, ~ d'expériences, ~ d'essais	serie *f* experimental, ~ de experimentos, ~ de ensayos
8191	sérine *f*	serina *f*
8192	sérologue *m*, sérologiste *m*	serólogo *m*
8193	sérologie *f*	serología *f*
8194	séreuse *f*	serosa *f*, membrana *f* serosa
8195	sérotonine *f*, entéramine *f*	serotonina *f*, enteramina *f*
8196	séreux	seroso
8197	denticulé	aserrado
8198	sérum *m*	suero *m*
8199	sérum-albumine *f*, séralbumine *f*	seroalbúmina *f*, albúmina *f* sérica
8200	sérodiagnostic *m*	serodiagnóstico *m*
8201	sérum-protéine *f*, protéine *f* sérique	seroproteína *f*, proteína *f* sérica
8202	os *m* sésamoïde	hueso *m* sesamoide
8203	sessile	sésil
8204	seston *m*	seston *m*
8205	—	—
8206	équipement *m* génétique	juego *m* de genes
8207	—	—
8208	soie *f* *bot.*	seta *f* *bot.*
8209	—	—
8210	eaux *f pl* résiduaires, ~ résiduelles, ~ usées, ~ d'égout	aguas *f pl* residuales

8211	sewage disposal, waste water ~	Abwasserbeseitigung *f*
8212	sewage sludge	Klärschlamm *m*
8213	sewage treatment	Abwasseraufbereitung *f*, ~verwertung *f*
8214	sex	Geschlecht *n*
8215	sex chromatin	Geschlechtschromatin *n*
8216	sex chromatin body, Barr body	Geschlechtschromatinkörper *m*, Barrsches Körperchen *n*
8217	sex chromosome → heterochromosome	—
8218	sex cycle, sexual ~	Geschlechtszyklus *m*
8219	sex determination, sexual ~	Geschlechtsbestimmung *f*
8220	sex differentiation	Geschlechtsdifferenzierung *f*
8221	sex dimorphism, sexual ~	Geschlechtsdimorphismus *m*, Sexualdimorphismus *m*
8222	sex duction	Sexduktion *f*
8223	sex ducts, sexual ~	Geschlechtswege *m pl*, Geschlechtsgänge *m pl*
8224	sex expression	Geschlechtsausprägung *f*
8225	sex factor → fertility factor	—
8226	sex gland → gonad	—
8227	sex hormone, gonad(al) ~	Geschlechtshormon *n*, Sexualhormon *n*, Keimdrüsenhormon *n*
8228	sex-influenced	geschlechtsbeeinflußt
8229	sex inheritance, sexual ~	Geschlechtsvererbung *f*
8230	sex limitation	Geschlechtsbegrenzung *f*
8231	sex-limited	geschlechtsbegrenzt
8232	sex linkage	Geschlechtskoppelung *f*
8233	sex-linked	geschlechtsgebunden
8234	sex-linked inheritance	geschlechtsgebundene Vererbung *f*
8235	sex organ, genital ~	Geschlechtsorgan *n*, Genitalorgan *n*, Sexualorgan *n*
8236	sex ratio	Geschlechtsverhältnis *n*, Sexualindex *m*
8237	sex reversal, reversal of sex	Geschlechtsumwandlung *f*, ~umkehr *f*
8238	sexual	sexuell, sexual, geschlechtlich
8239	sexual attractant, ~ attracting substance	Sexuallockstoff *m*
8240	sexual behavio(u)r	Sexualverhalten *n*
8241	sexual cell → gamete	—
8242	sexual character, sex ~	Geschlechtsmerkmal *n*
8243	sexual drive, ~ impulse, ~ urge	Geschlechtstrieb *m*
8244	sexual generation	Geschlechtsgeneration *f*
8245	sexual maturity	Geschlechtsreife *f*
8246	sexual partner → mate	—
8247	sexual reproduction	geschlechtliche Fortpflanzung *f*, sexuelle ~
8248	sexual selection	geschlechtliche Zuchtwahl *f*
8249	sexuality	Sexualität *f*, Geschlechtlichkeit *f*
8250	sexually mature	geschlechtsreif
8251	shade-loving → skiophilous	—
8252	shade-plant → skiophilous plant	—
8253	shallow rooted plant	Flachwurzler *m*
8254	sharing of incubation	Brutablösung *f*
8255	shell → concha	—
8256	shell fruits	Schalenfrüchte *f pl*
8257	shell gland	Schalendrüse *f*
8258	shell membrane	Schalenhaut *f*
8259	shift, displacement *gen.*	Transposition *f*, Shift *n gen.*
8260	shikimic acid	Shikimisäure *f*
8261	shin-bone → tibia	—
8262	to shoot, to sprout	sprießen, sprossen
8263	shoot	Sproß *m*, Schößling *m*, Schoß *m*
8264	short bone	kurzer Knochen *m*
8265	short-day plant	Kurztagpflanze *f*
8266	short-lived	kurzlebig
8267	short-styled → microstylous	—
8268	short-tailed, brachyuric	kurzschwänzig

8211	décharge *f* des eaux résiduaires	eliminación *f* de (las) aguas residuales
8212	vase *m* de filtrage	fango *m* de alcantarrilla, lodo *m* de aguas residuales
8213	traitement *m* des eaux résiduaires	tratamiento *m* de las aguas residuales, aprovechamiento *m* ~ ~ ~ ~
8214	sexe *m*	sexo *m*
8215	chromatine *f* sexuelle	cromatina *f* sexual
8216	corps *m* de chromatine sexuelle, corpuscule *m* de Barr	cuerpo *m* cromatínico sexual, corpúsculo *m* de Barr
8217	—	—
8218	cycle *m* sexuel, ~ sexué	ciclo *m* sexual
8219	détermination *f* du sexe, ~ sexuelle	determinación *f* del sexo, ~ sexual
8220	différenciation *f* sexuelle	diferenciación *f* sexual
8221	dimorphisme *m* sexuel	dimorfismo *m* sexual
8222	sexduction *f*, F-duction *f*	F-ducción *f*
8223	voies *f pl* génitales	conductos *m pl* sexuales, vías *f pl* genitales
8224	expression *f* du sexe	expresión *f* del sexo
8225	—	—
8226	—	—
8227	hormone *f* sexuelle, ~ génitale, ~ gonadique	hormona *f* sexual, ~ gonadal, ~ gonádica
8228	influencé par le sexe	influido por el sexo
8229	hérédité *f* du sexe, ~ sexuelle	herencia *f* del sexo, ~ sexual, transmisión *f* hereditaria del sexo
8230	limitation *f* au sexe	limitación *f* a un sexo
8231	limité à un sexe	limitado por el sexo
8232	linkage *m* au sexe	ligamiento *m* al sexo
8233	lié au sexe	ligado al sexo
8234	hérédité *f* liée au sexe	herencia *f* ligada al sexo
8235	organe *m* génital	órgano *m* sexual, ~ genital
8236	sex-ratio *m*, proportion *f* des sexes	proporción *f* de (los) sexos
8237	réversion *f* sexuelle, inversion *f* du sexe	reversión *f* del sexo, ~ sexual
8238	sexuel, sexué	sexual, sexuado
8239	substance *f* sexuelle	atractivo *m* sexual, atrayente *m* ~, sustancia *f* de atracción sexual
8240	comportement *m* sexuel	comportamiento *m* sexual, conducta *f* ~
8241	—	—
8242	caractère *m* sexuel	carácter *m* sexual
8243	instinct *m* sexuel, pulsion *f* sexuelle	instinto *m* sexual, impulso *m* ~
8244	génération *f* sexuée	generación *f* sexual, ~ sexuada
8245	maturité *f* sexuelle	madurez *f* sexual
8246	—	—
8247	reproduction *f* sexuée	reproducción *f* sexual
8248	sélection *f* sexuelle	selección *f* sexual
8249	sexualité *f*	sexualidad *f*
8250	sexuellement mûr	sexualmente maduro
8251	—	—
8252	—	—
8253	plante *f* à racines traçantes	planta *f* de raíces superficiales
8254	relais *m* entre couveurs, couvaison *f* alternée	cambio *m* de incubadores
8255	—	—
8256	fruits *m pl* à coques	frutos *m pl* de cáscara
8257	glande *f* coquillière	glándula *f* de la concha
8258	membrane *f* coquillière	membrana *f* coclear
8259	transposition *f*, décalage *m* *gen.*	desviación *f* *gen.*
8260	acide *m* shikimique	ácido *m* siquímico
8261	—	—
8262	pousser	brotar
8263	pousse *f*	brote *m*, vástago *m*
8264	os *m* court	hueso *m* corto
8265	plante *f* de journée courte, ~ nyctipériodique	planta *f* microhémera, ~ de día corto
8266	à vie courte	de corta vida
8267	—	—
8268	à queue courte	de cola corta, braquiúrico

8269	shoulder-blade → scapula	—
8270	shoulder girdle, pectoral ~	Schultergürtel *m*
8271	shoulder-joint, scapulohumeral articulation	Schultergelenk *n*
8272	showy structure	Schauapparat *m*
8273	shrub, frutex	Strauch *m*, Busch *m*
8274	shrubby → fruticose	—
8275	side chain	Seitenkette *f*
8276	sieve cell	Siebzelle *f*
8277	sieve plate	Siebplatte *f*
8278	sieve tube	Siebröhre *f*
8279	sieve-tube cell, ~ ~ element, ~ ~ member	Siebröhrenzelle *f*, Siebröhrenglied *n*, Siebröhrenelement *n*
8280	sign stimulus, key ~	Schlüsselreiz *m*, Signalreiz *m*
8281	siliceous, silicious	kieselhaltig
8282	siliceous earth → infusorial earth	—
8283	siliceous skeleton	Kieselskelett *n*
8284	silicic acid	Kieselsäure *f*
8285	silicification	Verkieselung *f*, Silifikation *f*
8286	to silicify	verkieseln, silifizieren
8287	silicle, silicula	Schötchen *n*
8288	silicole	Kieselpflanze *f*
8289	siliqua, silique	Schote *f*
8290	silk gland	Seidendrüse *f*
8291	Silurian (period)	Silur *n*
8292	simple fruit → solitary fruit	—
8293	simple protein	einfaches Eiweiß *n*
8294	simplex	simplex
8295	simultaneous adaptation	Simultanadaptation *f*
8296	simultaneous mutation	Simultanmutation *f*
8297	single cross	Einfachkreuzung *f*
8298	single-grain structure	Einzelkornstruktur *f*
8299	sinistrorse	linkswindend
8300	sinus	Sinus *m*
8301	sinus gland	Sinusdrüse *f*
8302	sinus venosus, venous sinus	Venensinus *m*
8303	siphon	Sipho *m*
8304	siphonogamy	Siphonogamie *f*, Schlauchbefruchtung *f*
8305	Siphonopoda → Cephalopoda	—
8306	siphonostele	Siphonostele *f*
8307	sire	Vatertier *n*
8308	sister chromatid	Schwesterchromatide *f*
8309	site-attachment → philopatry	—
8310	size of particles → grain size	—
8311	skeletal muscle	Skelettmuskel *m*
8312	skeleton	Skelett *n*
8313	skiaphyte → skiophilous plant	—
8314	skin	Haut *f*
8315	skin fold	Hautfalte *f*
8316	skin gland, dermal ~, cutaneous ~	Hautdrüse *f*
8317	skin pigment	Hautpigment *n*
8318	skiophilous, sciophilous, shade-loving	schattenliebend
8319	skiophilous plant, sciophilous ~, shade-loving ~, skiaphyte	Schattenpflanze *f*, Schwachlichtpflanze *f*
8320	skiophyll, sciophyll	Schattenblatt *n*
8321	skull → cranium	—
8322	sleep movement → nyctinasty	—
8323	slide → microscope slide	—
8324	sliding microtome	Schlittenmikrotom *n*
8325	slime, slimy → mucus, mucous	—
8326	slime moulds → Myxomycophyta	—
8327	slow growing	langsamwachsend, langsamwüchsig
8328	small calorie → gram-calorie	—
8329	small intestine	Dünndarm *m*
8330	smear	Abstrich *m*, Ausstrich *m*

8269	—	—
8270	ceinture *f* scapulaire, ~ pectorale, ~ thoracique	cinturón *m* escapular, ~ torácico
8271	articulation *f* scapulo-humérale	articulación *f* escapulohumeral
8272	partie *f* ornementale	estructuras *f pl* vistosas, aparato *m* de reclamo
8273	arbrisseau *m*, arbuste *m*	arbusto *m*, frútice *m*
8274	—	—
8275	chaîne *f* latérale	cadena *f* lateral
8276	cellule *f* criblée	célula *f* cribosa
8277	crible *m*	placa *f* cribosa, criba *f*
8278	tube *m* criblé	tubo *m* criboso
8279	cellule *f* du tube criblé	célula *f* del tubo criboso, elemento *m* tubo-criboso
8280	stimulus *m* signal, stimulus-signe *m*	estímulo *m* clave, ~ llave, ~ signo
8281	siliceux	silíceo
8282	—	—
8283	squelette *m* siliceux	esqueleto *m* silíceo
8284	acide *m* silicique	ácido *m* silícico
8285	silicification *f*	silicificación *f*
8286	silicifier	silicificar
8287	silicule *f*	silícula *f*
8288	plante *f* silicicole	planta *f* silícola
8289	silique *f*, cosse *f*	silicua *f*
8290	glande *f* séricigène	glándula *f* sericígena
8291	silurien *m*	silúrico *m*
8292	—	—
8293	protéine *f* simple, holoprotéine *f*	proteína *f* simple, holoproteido *m*
8294	simplex	simplexo
8295	adaptation *f* simultanée	adaptación *f* simultánea
8296	mutation *f* simultanée	mutación *f* simultánea
8297	croisement simple, ~ unique	cruzamiento *m* simple
8298	structure *f* monogranulaire	estructura *f* monogranular
8299	senestrorsum	levovoluble
8300	sinus *m*	seno *m*
8301	glande *f* du sinus, ~ sinusaire	glándula *f* del seno, ~ sinusal
8302	sinus *m* veineux	seno *m* venoso
8303	siphon *m*	sifón *m*
8304	siphonogamie *f*	sifonogamia *f*
8305	—	—
8306	siphonostèle *f*	sifonostela *f*
8307	père *m*	padre *m*, semental *m*
8308	chromatide *f* sœur	cromátida *f* hermana
8309	—	—
8310	—	—
8311	muscle *m* squelettique	músculo *m* esquelético
8312	squelette *m*	esqueleto *m*
8313	—	—
8314	peau *f*	piel *f*
8315	pli *m* dermique, repli *m* cutané	repliegue *m* cutáneo
8316	glande *f* cutanée, ~ dermique	glándula *f* cutánea, ~ dérmica
8317	pigment *m* cutané	pigmento *m* cutáneo
8318	sciaphile, umbrophile	esciófilo, umbrófilo
8319	plante *f* sciaphile, ~ umbrophile, ~ de sombre, sciaphyte *m*	planta *f* esciófila, ~ de sombra
8320	feuille *f* de sombre	hoja *f* de sombra
8321	—	—
8322	—	—
8323	—	—
8324	microtome *m* à glissière	micrótomo *m* deslizante
8325	—	—
8326	—	—
8327	à croissance lente	de crecimiento lento
8328	—	—
8329	intestin *m* grêle	intestino *m* delgado
8330	frottis *m*	frotis *m*, extensión *f*

8331	to smear	ausstreichen *(Präparat)*
8332	smear culture	Ausstrichkultur *f*
8333	smear preparation	Abstrichpräparat *n*, Ausstrichpräparat *n*
8334	smell receptor → olfactory receptor	—
8335	smoggy bowl → haze canopy	—
8336	smooth muscle, plain ~, involuntary ~	glatter Muskel *m*
8337	snow line, ~ limit	Schneegrenze *f*
8338	social *zool.*	gesellig *zool.*
8339	social attraction	Sozialattraktion *f*
8340	social behavio(u)r	Sozialverhalten *n*
8341	social hierarchy	Gesellschaftsordnung *f*
8342	social insects	staatenbildende Insekten *npl*
8343	social parasitisme	Sozialparasitismus *m*
8344	sociality	Soziabilität *f*, Geselligkeit *f*
8345	sociation	Soziation *f*
8346	society	Sozietät *f*
8347	sodium pump	Natriumpumpe *f*
8348	soft palate	weicher Gaumen *m*
8349	soft roe → milt[1]	—
8350	soft water	weiches Wasser *n*
8351	to soften *(water)*	enthärten *(Wasser)*
8352	softwood forest, ~ tree → conifer forest, ~ tree	—
8353	soil bacteria	Bodenbakterien *fpl*
8354	soil biology	Bodenbiologie *f*
8355	soil conditions, edaphic ~	Bodenbedingungen *fpl*
8356	soil conservation	Bodenerhaltung *f*
8357	soil contamination	Bodenverseuchung *f*
8358	soil exhaustion	Bodenmüdigkeit *f*
8359	soil fauna	Bodenfauna *f*
8360	soil formation, pedogenesis	Bodenbildung *f*, Pedogenese *f*
8361	soil horizon	Bodenhorizont *m*
8362	soil microbiology	Bodenmikrobiologie *f*
8363	soil microfauna	Bodenmikrofauna *f*
8364	soil microflora	Bodenmikroflora *f*
8365	soil microorganisms	Bodenmikroorganismen *mpl*
8366	soil organisms → edaphon	—
8367	soil pollution	Bodenverschmutzung *f*
8368	soil profile	Bodenprofil *n*
8369	soil protection	Bodenschutz *m*
8370	soil reaction	Bodenreaktion *f*
8371	soil sample, ~ specimen	Bodenprobe *f*
8372	soil science, pedology, edaphology	Bodenkunde *f*, Pedologie *f*
8373	soil type	Bodentyp *m*
8374	sol	Sol *n*
8375	solar energy	Sonnenenergie *f*
8376	solar radiation	Sonnenstrahlung *f*
8377	solation	Solbildung *f*
8378	solenocyte	Solenozyt *m*
8379	solid medium	festes Medium *n*
8380	solidification	Erstarrung *f*; Härtung *f*
8381	to solidify	erstarren; härten
8382	solifluction	Solifluktion *f*, Bodenfließen *n*
8383	solitary	solitär
8384	solitary fruit, simple ~	Einzelfrucht *f*
8385	solubility, solvability	Löslichkeit *f*
8386	to solubilize	löslich machen
8387	soluble	löslich
8388	soluble RNA → transfer RNA	—
8389	solute	gelöster Stoff *m*, Solut *n*
8390	solvent, dissolvent	Lösungsmittel *n*
8391	soma	Soma *n*
8392	somatic	somatisch
8393	somatic cell, body ~	Somazelle *f*, Körperzelle *f*
8394	somatogamy	Somatogamie *f*
8395	somatogen(et)ic	somatogen

8331	étaler en frottis	extender
8332	culture *f* par frottis	cultivo *m* por extensión
8333	préparation *f* de frottis	preparación *f* por extensión
8334	—	—
8335	—	—
8336	muscle *m* lisse, ~ involontaire	músculo *m* liso, ~ involuntario
8337	limite *f* des neiges	línea *f* de nieve
8338	social *zool.*	social *zool.*
8339	attraction *f* sociale	atracción *f* social
8340	comportement *m* social	comportamiento *m* social
8341	ordre *m* social	orden *m* social, jerarquía *f* ~
8342	insectes *m pl* sociaux	insectos *m pl* sociales
8343	parasitisme *m* social	parasitismo *m* social
8344	sociabilité *f*	sociabilidad *f*
8345	sociation *f*	sociación *f*
8346	société *f*	sociedad *f*
8347	pompe *f* à sodium	bomba *f* de sodio
8348	palais *m* mou, ~ membraneux	paladar *m* blando
8349	—	—
8350	eau *f* douce	agua *f* blanda
8351	adoucir *(eau)*	ablandar *(agua)*
8352	—	—
8353	bactéries *f pl* du sol	bacterias *f pl* del suelo, ~ edáficas
8354	biologie *f* du sol	biología *f* del suelo
8355	conditions *f pl* édaphiques, ~ pédologiques	condiciones *f pl* edáficas
8356	conservation *f* du sol	conservación *f* del suelo
8357	contamination *f* du sol	contaminación *f* del suelo
8358	fatigue *f* du sol	agotamiento *m* del suelo, cansancio *m* ~ ~
8359	faune *f* du sol	fauna *f* edáfica
8360	pédogenèse *f*	pedogénesis *f*
8361	horizon *m* pédologique	horizonte *m* edáfico
8362	microbiologie *f* du sol	microbiología *f* del suelo
8363	microfaune *f* du sol	microfauna *f* del suelo
8364	microflore *f* du sol, flore *f* microbienne du sol	microflora *f* del suelo
8365	microorganismes *m pl* du sol	microorganismos *m pl* del suelo
8366	—	—
8367	pollution *f* du sol	polución *f* del suelo
8368	profil *m* du sol, ~ pédologique	perfil *m* del suelo
8369	protection *f* du sol	protección *f* del suelo
8370	réaction *f* du sol	reacción *f* del suelo
8371	échantillon *m* du sol	muestra *f* de suelo
8372	science *f* du sol, édaphologie *f*, pédologie *f*	edafología *f*, pedología *f*
8373	type *m* du sol	tipo *m* de suelo
8374	sol *m*	sol *m*
8375	énergie *f* solaire	energía *f* solar
8376	rayonnement *m* solaire	radiación *f* solar
8377	solation *f*	solación *f*
8378	solénocyte *m*	solenocito *m*
8379	milieu *m* solide	medio *m* sólido
8380	solidification *f*	solidificación *f*
8381	(se) solidifier	solidificar(se)
8382	solifluxion *f*, solifluction *f*	soliflucción *f*
8383	solitaire	solitario
8384	fruit *m* solitaire	fruto *m* simple
8385	solubilité *f*	solubilidad *f*
8386	solubiliser	solubilizar
8387	soluble	soluble
8388	—	—
8389	soluté *m*	soluto *m*
8390	solvant *m*, dissolvant *m*	solvente *m*, disolvente *m*
8391	soma *m*	soma *m*
8392	somatique	somático
8393	cellule *f* somatique	célula *f* somática, ~ corporal
8394	somatogamie *f*	somatogamia *f*
8395	somatogénique	somatógeno, somatogénico

8396	somatology	Somatologie *f*
8397	somatopleure	Somatopleura *f*
8398	somatotrophic hormone, somatotrop(h)in, growth hormone, STH	Somatotropin *n*, Wachstumshormon *n*, STH
8399	somatotype → constitutional type	—
8400	somite, protovertebra	Somit *n*, Urwirbel *m*
8401	sonic boom	Überschallknall *m*
8402	soredium, sorede	Soredium *n* (*pl* Soredien)
8403	sorption	Sorption *f*
8404	sorus (*pl* sori)	Sorus *m* (*pl* Sori)
8405	sound barrier, sonic ~	Schallmauer *f*
8406	sound perception	Schallwahrnehmung *f*
8407	sound-producing organ	Lautorgan *n*, lauterzeugendes Organ *n*
8408	sound-production → phonation	—
8409	sound signal → acoustical signal	—
8410	sound source	Schallquelle *f*
8411	sound stimulus	Schallreiz *m*
8412	sound wave	Schallwelle *f*
8413	space orientation, spatial ~, orientation in space	Raumorientierung *f*
8414	space sense	Raumsinn *m*
8415	spadix	Kolben *m*, Spadix *m*
8416	spathe	Spatha *f*
8417	spatial economics	Raumplanung *f*, Raumordnung *f*
8418	spatial facilitation	räumliche Bahnung *f*
8419	spatial isolation	räumliche Isolation *f*
8420	spatial summation	räumliche Summation *f*
8421	spatulate	spatelig, spatelförmig
8422	spawn	Laich *m*
8423	to spawn	laichen, ablaichen
8424	spawner	Rog(e)ner *m*
8425	spawning, deposition of spawn	Laichablage *f*, Laichen *n*, Ablaichen *n*
8426	spawning season, ~ time	Laichzeit *f*
8427	spawning site, ~ ground	Laichplatz *m*
8428	to spay	kastrieren *(weibliches Tier)*
8429	speciation, species formation	Artbildung *f*, Speziation *f*
8430	species	Art *f*, Spezies *f*
8431	species area	Artareal *n*
8432	species composition	Artenzusammensetzung *f*
8433	species cross → interspecific hybridization	—
8434	species diversity	Artenvielfalt *f*
8435	species formation → speciation	—
8436	species group	Artengruppe *f*
8437	species isolation	Artisolierung *f*
8438	species member → conspecific	—
8439	species preservation	Arterhaltung *f*
8440	species-preserving	arterhaltend
8441	species recognition, ~ identification	Arterkennung *f*
8442	species rich, rich in species	artenreich
8443	species-specific	artspezifisch
8444	species specificity, individual ~	Artspezifität *f*
8445	species structure	Artstruktur *f*
8446	species-typical	arttypisch, arteigen
8447	specific	spezifisch, Art...
8448	specific name	Artname *m*
8449	specificity	Spezifität *f*
8450	spectral analysis	Spektralanalyse *f*
8451	spectrophotometer	Spektrophotometer *n*
8452	spectrophotometry	Spektrophotometrie *f*
8453	speech centre (US: center)	Sprachzentrum *n (Gehirn)*
8454	speed of germination	Keimschnelligkeit *f*

8396	somatologie *f*	somatología *f*
8397	somatopleure *f*, somatoplèvre *f*	somatopleura *f*
8398	somatotrop(h)ine *f*, hormone *f* somatotrope, ~ de croissance	somatotrofina *f*, somatotropina *f*, hormona *f* somatotropa, ~ del crecimiento
8399	—	—
8400	somite *m*, protovertèbre *f*	somita *m*, somito *m*, protovértebra *f*
8401	bang *m* supersonique	bang *m* supersónico
8402	sorédie *f*	soredio *m*
8403	sorption *f*	sorción *f*
8404	sore *m*	soro *m*
8405	mur *m* du son	barrera *f* del sonido
8406	perception *f* acoustique, ~ sonore, ~ des sons	percepción *f* acústica, ~ del sonido
8407	organe *m* vocal, ~ phonateur	órgano *m* vocal
8408	—	—
8409	—	—
8410	source *f* sonore	fuente *f* sonora, ~ del sonido
8411	stimulus *m* sonore	estímulo *m* sonoro, ~ acústico
8412	onde *f* sonore	onda *f* sonora, ~ acústica
8413	orientation *f* spatiale	orientación *f* en el espacio
8414	sens *m* de l'espace	sentido *m* del espacio
8415	spadice *m*	espádice *m*
8416	spathe *f*	espata *f*
8417	aménagement *m* du territoire	ordenación *f* territorial, planificación *f* ~, ~ del territorio
8418	facilitation *f* spatiale	facilitación *f* espacial
8419	isolement *m* spatial	aislamiento *m* espacial
8420	sommation *f* spatiale	sumación *f* espacial
8421	spatulé	espatulado
8422	frai *m*	freza *f*
8423	frayer	frezar, desovar
8424	poisson *m* œuvé, ~ rogué	pez *m* ovado
8425	frai *m*	freza *f*, desove *m*
8426	époque *f* du frai, période *f* ~ ~	freza *f*, época *f* de desove
8427	frayère *f*, lieu *m* de frai	lugar *m* de desove, zona *f* de freza
8428	castrer *(femelle)*	castrar *(hembra)*
8429	spéciation *f*, formation *f* des espèces	especiación *f*, formación *f* de las especies
8430	espèce *f*	especie *f*
8431	aire *f* des espèces	área *f* de especies
8432	composition *f* en espèces	composición *f* en especies
8433	—	—
8434	diversité *f* des espèces, ~ spécifique	diversidad *f* de especies, ~ en especies
8435	—	—
8436	groupe *m* d'espèces	grupo *m* de especies
8437	isolement *m* des espèces	aislamiento *m* de (las) especies
8438	—	—
8439	conservation *f* de l'espèce, perpétuation *f* ~ ~, maintien *m* ~ ~	conservación *f* de la(s) especie(s), perpetuación *f* ~ ~ ~, mantenimiento *m* ~ ~ ~
8440	qui perpétue l'espèce	conservador de la especie, que perpetua la especie
8441	reconnaissance *f* spécifique, identification *f* des congénères	reconocimiento *m* entre las especies
8442	riche en espèces	rico en especies
8443	spécifique de l'espèce	específico de la especie
8444	spécificité *f* individuelle	especificidad *f* de especies, ~ individual
8445	structure *f* des espèces	estructura *f* de las especies
8446	propre à l'espèce, typique de ~	propio de la especie, característico ~ ~ ~
8447	spécifique	específico
8448	nom *m* spécifique	nombre *m* específico
8449	spécificité *f*	especificidad *f*
8450	analyse *f* spectrale	análisis *m* espectral
8451	spectrophotomètre *m*	espectrofotómetro *m*
8452	spectrophotométrie *f*	espectrofotometría *f*
8453	centre *m* du langage	centro *m* del lenguaje
8454	rapidité *f* de germination, vitesse *f* ~ ~	rapidez *f* de germinación

No.	English	German
8455	sperm	Sperma *n*
8456	sperm cell	Spermazelle *f*
8457	sperm formation → spermatogenesis	—
8458	sperm nucleus	Spermakern *m*, Spermienkern *m*
8459	sperm-producing → spermatogenetic	—
8460	Spermaphyta → Spermatophyta	—
8461	spermatheca → seminal receptacle	—
8462	spermatic cord, ~ funicle	Samenstrang *m*
8463	spermatic flagellum	Spermienflagellum *n*
8464	spermatic fluid → seminal fluid	—
8465	spermatid	Spermatide *f*
8466	spermatium (*pl* spermatia)	Spermatium *n* (*pl* Spermatien)
8467	spermatocyte	Spermatozyt *m*
8468	spermatogenesis, sperm formation	Spermatogenese *f*, Samenzellenbildung *f*
8469	spermatogen(et)ic, spermatogenous, sperm-producing	spermatogen, spermienbildend
8470	spermatogonium	Spermatogonium *n*, Ursamenzelle *f*
8471	spermatophore	Spermatophore *f*, Spermaträger *m*, Samenträger *m*
8472	Sperma(to)phyta, seed plants, phanerogams, anthophytes	Spermatophyten *mpl*, Samenpflanzen *fpl*, Blütenpflanzen *fpl*, Phanerogamen *fpl*
8473	spermatozoid	Spermatozoid *n*
8474	spermatozoon (*pl* spermatozoa), spermium	Spermatozoon *n* (*pl* ~zoen), Spermium *n* (*pl* Spermien), Samenzelle *f*, Samenfaden *m*
8475	spermin	Spermin *n*
8476	spermiogenesis	Spermiogenese *f*
8477	spermium → spermatozoon	—
8478	sphagnum moss → peat moss	—
8479	sphenoid(bone)	Keilbein *n*
8480	Sphenopsida, Articulatae	Schachtelhalmartige *pl*
8481	spherical	kugelförmig, kugelig
8482	spherobacterium → coccus	—
8483	spherome	Sphärom *n*
8484	spheroprotein	Sphäroprotein *n*
8485	spherosome	Sphärosom *n*
8486	sphincter	Sphinkter *m*, Schließmuskel *m*
8487	sphingomyelin	Sphingomyelin *n*
8488	sphingosin	Sphingosin *n*
8489	spiciform	ährenförmig
8490	spike, ear *bot.*	Ähre *f*
8491	spike *neur.*	Spitzenpotential *n*, Spike
8492	spikelet	Ährchen *n*
8493	spinal bulb → medulla oblongata	—
8494	spinal canal, vertebral ~	Spinalkanal *m*, Wirbelkanal *m*, Rückenmarkskanal *m*
8495	spinal column → vertebral column	—
8496	spinal cord	Rückenmark *n*
8497	spinal ganglion	Spinalganglion *n*
8498	spinal nerve	Spinalnerv *m*, Rückenmarksnerv *m*
8499	spindle attachment → centromere	—
8500	spindle axis	Spindelachse *f*
8501	spindle fibre (US: fiber)	Spindelfaser *f*
8502	spindle poison	Spindelgift *n*
8503	spindle pole	Spindelpol *m*
8504	spine → thorn	—
8505	spinneret	Spinnwarze *f*
8506	spinning gland	Spinndrüse *f*
8507	spinous, spiny	dornig
8508	spinous process → neurapophysis	—
8509	spiracle[1], spiraculum	Spritzloch *n*, Spiraculum *n*
8510	spiracle[2] → stigma[2]	—
8511	spiral cleavage, alternating ~	Spiralfurchung *f*
8512	spiral organ → Corti's organ	—
8513	spiral thickening, helical ~	Schraubenverdickung *f*

8455	sperme *m*	esperma *m*
8456	cellule *f* spermatique	célula *f* espermática
8457	—	—
8458	noyau *m* spermatique	núcleo *m* espermático
8459	—	—
8460	—	—
8461	—	—
8462	cordon *m* spermatique, funicule *m* ~	cordón *m* espermático, funículo *m* ~
8463	flagelle *m* du spermatozoïde	flagelo *m* espermático
8464	—	—
8465	spermatide *f*	espermátida *f*
8466	spermatie *f*	espermacio *m*
8467	spermatocyte *m*	espermatocito *m*
8468	spermatogenèse *f*	espermatogénesis *f*
8469	spermatogén(ét)ique	espermatógeno
8470	spermatogonie *f*	espermatogonio *m*, espermatogonia *f*
8471	spermatophore *m*	espermatóforo *m*
8472	sperma(to)phytes *mpl*, plantes *fpl* à graines, ~ à fleurs, phanérogames *fpl*	esperma(tó)fitos *mpl*, espermáfitas *fpl*, plantas *fpl* de semilla, fanerógamas *fpl*, antófitos *mpl*
8473	spermatozoïde *m*	espermatozoide *m*
8474	spermatozoïde *m*, spermie *f*, spermatozoaire *m*	espermatozoo *m*, espermio *m*
8475	spermine *f*	espermina *f*
8476	spermiogenèse *f*	espermiogénesis *f*
8477	—	—
8478	—	—
8479	(os *m*) sphénoïde *m*	(hueso *m*) esfenoides *m*
8480	sphénopsides *fpl*, articulées *fpl*	esfenópsidos *mpl*
8481	sphérique	esférico
8482	—	—
8483	sphérome *m*	esferoma *m*
8484	sphéroprotéine *f*, protéine *f* globulaire	esferoproteína *f*, proteína *f* globular
8485	sphérosome *m*	esferosoma *m*
8486	sphincter *m*	esfínter *m*
8487	sphingomyéline *f*	esfingomielina *f*
8488	sphingosine *f*	esfingosina *f*
8489	spiciforme	espiciforme
8490	épi *m*	espiga *f*
8491	potentiel *m* de pointe, spike *f*	potencial *m* en aguja, ~ de punta, spike *m*
8492	épillet *m*	espiguilla *f*, espícula *f*
8493		—
8494	canal *m* vertébral, ~ rachidien	conducto *m* espinal, ~ vertebral, ~ raquídeo
8495	—	—
8496	moelle *f* épinière	médula *f* espinal, ~ raquídea, cordón *m* espinal
8497	ganglion *m* spinal	ganglio *m* espinal
8498	nerf *m* spinal, ~ rachidien	nervio *m* espinal, ~ raquídeo
8499	—	—
8500	axe *m* fusorial	eje *m* del huso
8501	fibre *f* fusoriale	fibra *f* del huso
8502	poison *m* fusorial, substance *f* mitoclasique	veneno *m* del huso
8503	pôle *m* du fuseau	polo *m* del huso
8504	—	—
8505	filière *f*	hilera *f*
8506	glande *f* fileuse	glándula *f* hiladora
8507	épineux	espinoso
8508	—	—
8509	spiracle *m*	espiráculo
8510	—	—
8511	segmentation *f* spirale	segmentación *f* espiral
8512	—	—
8513	épaississement *m* spiralé	engrosamiento *m* espiral, ~ helicoidal

8514	spiral tracheid, helical ~	Schraubentracheide *f*, Spiraltracheide *f*
8515	spiralization	Spiralisation *f*, Spiralisierung *f*
8516	spirem(a)	Spirem *n*
8517	spirillum	Spirillum *n*, Spirille *f*
8518	spiroch(a)ete	Spirochäte *f*
8519	spirometer	Spirometer *n*
8520	splanchnic → visceral	—
8521	splanchnic nerve	Eingeweidenerv *m*
8522	splanchnocranium → viscerocranium	—
8523	splanchnology	Splanchnologie *f*, Eingeweidelehre *f*
8524	splanchnopleure	Splanchnopleura *f*
8525	splanchnoskeleton → visceroskeleton	—
8526	spleen, milt	Milz *f*
8527	splenic, lienal	Milz...
8528	splint-wood → sapwood	—
8529	to split (up)	spalten, aufspalten *chem.*
8530	splitting → fission	—
8531	spodogram	Aschenbild *n*, Spodogramm *n*
8532	sponge spicule	Schwammnadel *f*
8533	spongin	Spongin *n*
8534	spongy, spongiose	schwammig
8535	spongy parenchyma, ventilating tissue	Schwammparenchym *n*
8536	spongy tissue	Schwammgewebe *n*
8537	spontaneous generation, abiogenesis	Urzeugung *f*, Abiogenese *f*, Abiogenesis *f*
8538	spontaneous mutation	spontane Mutation *f*
8539	sporangiophore	Sporangienträger *m*
8540	sporangium	Sporangium *n*, Sporenbehälter *m*
8541	spore	Spore *f*
8542	spore-bearing → sporiferous	—
8543	spore capsule	Sporenkapsel *f*
8544	spore dispersal, dissemination of spores	Sporenverbreitung *f*
8545	spore formation → sporulation	—
8546	spore mother-cell, sporocyte	Sporenmutterzelle *f*, Sporozyt *m*
8547	spore-plants → cryptogams	—
8548	sporidium	Sporidie *f*
8549	sporiferous, spore-bearing	sporentragend
8550	sporocarp	Sporokarp *m*
8551	sporocyst	Sporozyste *f*
8552	sporocyte → spore mother-cell	—
8553	sporogenesis, sporogony	Sporogenese *f*, Sporogonie *f*
8554	sporogenous	sporogen, sporenbildend
8555	sporogonium	Sporogon *n*
8556	sporophore	Sporenträger *m*
8557	sporophyll	Sporophyll *n*, Sporenblatt *n*
8558	sporophyte	Sporophyt *m*
8559	sporozoon (*pl* sporozoa)	Sporozoon *n* (*pl* Sporozoen), Sporentierchen *n*
8560	sport, bud mutation	Sport *m*, Knospenmutation *f*, Sproßmutation *f*
8561	to sporulate	Sporen bilden
8562	sporulation	Sporulation *f*, Sporenbildung *f*; Sporenverbreitung *f*
8563	spring-spore → aecidiospore	—
8564	spring water	Quellwasser *n*
8565	sprout	Trieb *m* *bot.*
8566	to sprout → to shoot	—
8567	squalene	Squalen *n*
8568	squamate, squamous, scaly	schuppig, geschuppt
8569	squamiform, scale-like	schuppenartig, schuppenförmig
8570	squamous epithelium	Schuppenepithel *n*
8571	stab culture, needle ~	Stichkultur *f*
8572	stabilizing selection	stabilisierende Selektion *f*
8573	stable humus	Dauerhumus *m*

8514	trachéide *f* spiralée	traqueida *f* espiralada, ~ helicoidal
8515	spiralisation *f*	espiralización *f*
8516	spirème *m*	espirema *m*
8517	spirille *m*	espirilo *m*
8518	spirochète *m*	espiroqueta *f*
8519	spiromètre *m*	espirómetro *m*
8520	—	—
8521	nerf *m* splanchnique	nervio *m* esplácnico, ~ visceral
8522	—	—
8523	splanchnologie *f*	esplacnología *f*
8524	splanchnopleure *f*, splanchnoplèvre *f*	esplacnopleura *f*
8525	—	—
8526	rate *f*	bazo *m*
8527	splénique, liénal, liénique	esplénico, lienal
8528	—	—
8529	scinder	escindir
8530	—	—
8531	spodogramme *m*	espodograma *m*
8532	spicule *m* d'éponge	espícula *f* de esponja
8533	spongine *f*	espongina *f*
8534	spongieux	esponjoso
8535	parenchyme *m* lacuneux, ~ aérifère	parénquima *m* lagunoso, ~ esponjoso, ~ aerífero
8536	tissu *m* spongieux	tejido *m* esponjoso
8537	génération *f* spontanée, abiogenèse *f*	generación *f* espontánea, abiogénesis *f*
8538	mutation *f* spontanée	mutación *f* espontánea
8539	sporangiophore *m*	esporangióforo *m*
8540	sporange *m*	esporangio *m*
8541	spore *f*	espora *f*
8542	—	—
8543	capsule *f* sporifère	cápsula *f* esporífera
8544	dissémination *f* des spores	dispersión *f* de esporas, diseminación *f* ~ ~
8545	—	—
8546	cellule *f* mère des spores, sporocyte *m*	célula *f* madre de esporas, esporocito *m*, espora *f* madre
8547	—	—
8548	sporidie *f*	esporidio *m*
8549	sporifère	esporífero
8550	sporocarpe *m*	esporocarpo *m*
8551	sporocyste *m*	esporocisto *m*
8552	—	—
8553	sporogenèse *f*, sporogonie *f*	esporogénesis *f*, esporogonia *f*
8554	sporogène	esporógeno, esporogénico
8555	sporogone *m*	esporogonio *m*
8556	sporophore *m*	esporóforo *m*
8557	sporophylle *f*	esporófilo *m*
8558	sporophyte *m*	esporófito *m*
8559	sporozoaire *m*	esporozoo *m*, esporozoario *m*
8560	sport *m*, mutation *f* de bourgeons, ~ bourgeonneuse	sport *m*, mutación *f* de brotes, ~ de yema
8561	sporuler	esporular
8562	sporulation *f*	esporulación *f*
8563	—	—
8564	eau *f* de source	agua *f* de manantial
8565	rejet(on) *m*	renuevo *m*
8566	—	—
8567	squalène *m*	escualeno *m*
8568	squamé, squameux	escamoso
8569	squamiforme	escamiforme
8570	épithélium *m* squameux	epitelio *m* escamoso
8571	culture *f* en piqûre	cultivo *m* por picadura, ~ ~ estocada, ~ ~ pinchazo, ~ en aguja
8572	sélection *f* stabilisante, ~ stabilisatrice	selección *f* estabilizadora
8573	humus *m* stable	humus *m* estable, mantillo *m* ~

8574	stage of maturity	Reifestadium *n*
8575	stagnant water	stehendes Wasser *n*
8576	to stain, to dye	färben, anfärben
8577	stain, dye, staining substance, colo(u)ring matter	Farbstoff *m*, Färbemittel *n*
8578	stainability, colorability	Färbbarkeit *f*
8579	stainable	färbbar, anfärbbar
8580	staining	Färbung *f*, Färben *n*
8581	staining procedure, ~ method	Färbemethode *f*, Färbeverfahren *n*
8582	stalk	Stiel *m*
8583	stalk-eye	Stielauge *n*
8584	stalked → pedunculate	—
8585	stamen	Staubgefäß *n*, Staubblatt *n*, Stamen *n*
8586	staminal hair	Staubblatthaar *n*
8587	staminate flower	Staubblüte *f*, männliche Blüte *f*
8588	staminode, staminodium	Staminodium *n*
8589	standard deviation	Standardabweichung *f*, mittlere quadratische Abweichung *f*
8590	standard error	Standardfehler *m*
8591	standard preparation	Standardpräparat *n*
8592	standard solution, standardized ~	Standardlösung *f*
8593	standard type	Standardtyp *m*
8594	standing crop	Bestand *m ecol.*, "standing crop"
8595	stapes, stirrup (bone)	Steigbügel *m anat.*
8596	starch gel	Stärkegel *n*
8597	starch grain	Stärkekorn *n*
8598	starch reserve	Stärkereserve *f*
8599	state of aggregation	Aggregatzustand *m*
8600	state of excitation	Erregungszustand *m*
8601	static sense → sense of balance	—
8602	stationary nucleus	Stationärkern *m*
8603	stationary phase	stationäre Phase *f*
8604	statoblast	Statoblast *m*
8605	statocyst	Statozyste *f*
8606	statolith	Statolith *m*, Schweresteinchen *n*
8607	statospore → resting spore	—
8608	steady state	Fließgleichgewicht *n*
8609	stearic acid	Stearinsäure *f*, Talgsäure *f*
8610	stele	Stele *f*
8611	stem	Stengel *m*, Stiel *m*, Sproß *m*, Stamm *m*
8612	stem apex, ~ tip	Sproßspitze *f*, Sproßscheitel *m*
8613	stem axis → caulome	—
8614	stem-bud	Stengelknospe *f*, Stammknospe *f*
8615	stem bundle, cauline vascular ~	Stengelbündel *n*, stammeigenes Leitbündel *n*
8616	stemless → acauline	—
8617	stem system	Sproßsystem *n*
8618	stem tuber	Sproßknolle *f*
8619	stenohaline	stenohalin
8620	stenohygric	stenohyger
8621	stenothermic, stenothermous	stenotherm
8622	steppe plant	Steppenpflanze *f*
8623	stercobilin	Sterkobilin *n*
8624	stereome	Stereom *n*
8625	stereomicroscope	Stereomikroskop *n*
8626	stereotyped movements	Bewegungsstereotypien *f pl*
8627	sterigma	Sterigma *n*
8628	sterile	steril
8629	sterility	Sterilität *f*
8630	sterility gene	Sterilitätsgen *n*
8631	sterilization	Sterilisation *f*
8632	to sterilize	sterilisieren
8633	sternal rib → true rib	—
8634	sternum, breast bone	Brustbein *n*
8635	steroid	Steroid *n*
8636	steroid hormone	Steroidhormon *n*
8637	sterol	Sterin *n*, Sterol *n*

No.	Français	Español
8574	stade *m* de maturité	estado *m* de madurez
8575	eau *f* stagnante	agua *f* estancada
8576	colorer	colorear, teñir
8577	colorant *m*, matière *f* colorante, substance *f* ~	colorante *m*, sustancia *f* ~, materia *f* ~
8578	colorabilité *f*	colorabilidad *f*
8579	colorable	coloreable, tingible
8580	coloration *f*	coloración *f*, tinción *f*
8581	méthode *f* colorante, ~ de colorisation	método *m* de coloración, ~ de tinción
8582	pédoncule *m*, tige *f*	pedúnculo *m*, tallo *m*
8583	œil *m* pédonculé	ojo *m* pedunculado
8584	—	—
8585	étamine *f*	estambre *m*
8586	poil *m* staminal	pelo *m* estaminal
8587	fleur *f* étaminée, ~ staminifère, ~ mâle	flor *f* estaminada, ~ masculina
8588	staminode *m*	estaminodio *m*
8589	déviation *f* standard, écart-type *m*, écart *m* quadratique moyen	desviación *f* standard, ~ tipo, ~ típica, ~ cuadrática media
8590	erreur *f* standard, ~ moyenne	error *m* medio
8591	préparation *f* standardisée	preparación *f* standard
8592	solution *f* standard	solución *f* standard
8593	type *m* standard	tipo *m* standard
8594	«standing crop»	cultivo *m* estable, población *f* vegetal
8595	étrier *m* *anat.*	estribo *m* *anat.*
8596	gel *m* d'amidon	gel *m* de almidón
8597	grain *m* d'amidon	grano *m* de almidón
8598	réserve *f* amylacée	reserva *f* de almidón
8599	état *m* d'agrégation	estado *m* de agregación
8600	état *m* d'excitation	estado *m* de excitación, ~ excitatorio
8601	—	—
8602	noyau *m* stationnaire	núcleo *m* estacionario
8603	phase *f* stationnaire	fase *f* estacionaria
8604	statoblaste *m*	estatoblasto *m*
8605	statocyste *m*	estatocisto *m*
8606	statolithe *m*	estatolito *m*
8607	—	—
8608	équilibre *m* dynamique	equilibrio *m* dinámico
8609	acide *m* stéarique	ácido *m* esteárico
8610	stèle *f*	estela *f*
8611	tige *f*	tallo *m*
8612	apex *m* de la tige	ápice del tallo, ~ ~ vástago
8613	—	—
8614	bourgeon *m* caulinaire	yema *f* caulinar
8615	faisceau *m* caulinaire	haz *m* caulinar
8616	—	—
8617	appareil *m* caulinaire	sistema *m* caulinar
8618	tubercule *m* caulinaire	tubérculo *m* caulinar
8619	sténohalin	estenohalino
8620	sténohygrique	estenohídrico
8621	sténotherme	estenotermo, estenotérmico, estenotermal
8622	plante *f* des steppes	planta *f* esteparia
8623	stercobiline *f*	estercobilina *f*
8624	stéréome *m*	estereoma *m*
8625	microscope *m* stéréoscopique	estereomicroscopio *m*
8626	mouvements *m pl* stéréotypés	movimientos *m pl* estereotipados
8627	stérigma *m*	esterigma *m*
8628	stérile	estéril
8629	stérilité *f*	esterilidad *f*
8630	gène *m* de stérilité	gen *m* de esterilidad
8631	stérilisation *f*	esterilización *f*
8632	stériliser	esterilizar
8633	—	—
8634	sternum *m*	esternón *m*
8635	stéroïde *m*	esteroide *m*
8636	hormone *f* stéroïde	hormona *f* esteroide
8637	stérol *m*	esterol *m*, esterina *f*

8638	STH → somatotrophic hormone	—
8639	stigma[1] *bot.*	Narbe *f bot.*
8640	stigma[2], spiracle *zool.*	Stigma *n*, Atemloch *n*
8641	stigma[3] → eye-spot	—
8642	stilt-root, buttress-root	Stelzwurzel *f*
8643	stimulant	Stimulans *n*
8644	stimulating effect	Reizwirkung *f*
8645	stimulating electrode, exciting ~	Reizelektrode *f*
8646	stimulating substance	Reizstoff *m*
8647	stimulation	Reizung *f*
8648	stimulus (*pl* stimuli)	Reiz *m*
8649	stimulus intensity, ~ strength	Reizintensität *f*, Reizstärke *f*
8650	stimulus parameter	Reizparameter *m*
8651	stimulus perception	Reizperzeption *f*, Reizwahrnehmung *f*
8652	stimulus quality	Reizqualität *f*
8653	stimulus response	Reizantwort *f*, Reizreaktion *f*
8654	stimulus-response relationship	Reiz-Reaktions-Beziehung *f*
8655	stimulus situation	Reizsituation *f*
8656	stimulus source, source of stimulation	Reizquelle *f*, Reizursache *f*
8657	stimulus strength → ~ intensity	—
8658	stimulus summation	Reizsummation *f*
8659	stimulus threshold	Reizschwelle *f*
8660	sting	Stachel *m zool.*
8661	stinging *ins.*	stechend *ins.*
8662	stinging capsule → nematocyst	—
8663	stinging cell → cnidoblast	—
8664	stinging hair	Brennhaar *n*
8665	stink gland, repugnatorial ~	Stinkdrüse *f*
8666	stipule	Nebenblatt *n*, Stipel *f*
8667	stirrup (bone) → stapes	—
8668	stock	Pfropfunterlage *f*
8669	stock culture	Mutterkultur *f*
8670	stock plant → parent plant	—
8671	stock solution	Stammlösung *f*
8672	stolon, runner	Ausläufer *m*, Stolon *m*
8673	stoma (*pl* stomata)	Spaltöffnung *f*, Stoma *n* (*pl* Stomata)
8674	stomach	Magen *m*
8675	stomachic → gastric	—
8676	stomatic(al) transpiration	stomatäre Transpiration *f*
8677	stomochord	Stomochord *n*
8678	stomodaeum	Stomodaeum *n*, Vorderdarm *m embr.*
8679	stone (of fruit)	Stein(kern) *m*
8680	stone canal, madreporic ~	Steinkanal *m*
8681	stone cell → sclereid	—
8682	stone fruit, drupe	Steinfrucht *f*
8683	storage capacity	Speicherungsvermögen *n*
8684	storage organ	Speicherorgan *n*
8685	storage root	Speicherwurzel *f*, Nährwurzel *f*
8686	storage starch → reserve starch	—
8687	storage tissue	Speichergewebe *n*
8688	story of vegetation	Vegetationsstufe *f*
8689	strain[1] *(bacteria etc.)*	Stamm *m*
8690	strain[2] → variety	—
8691	strangulation	Durchschnürung *f*
8692	stratification	Schichtung *f*
8693	stratified, stratose, layered	geschichtet
8694	stratified epithelium	Schichtepithel *n*
8695	stratum (*pl* strata), layer	Schicht *f*
8696	streak culture	Strichkultur *f*
8697	stream pollution → river pollution	—
8698	streptomycin	Streptomyzin *n*, Streptomycin *n*
8699	stretch reflex, myotatic ~	Dehnungsreflex *m*
8700	striate	gestreift

8638	—	—
8639	stigmate *m*	estigma *m*
8640	stigmate *m* trachéen, spiracle *m*	estigma *m*, orificio *m* respiratorio, espiráculo *m*
8641	—	—
8642	racine *f* échasse	raíz *f* zanco, ~ fúlcrea
8643	stimulant *m*	estimulante *m*
8644	effet *m* stimulateur	efecto *m* estimulante, acción *f* excitante
8645	électrode *f* active, ~ stimulatrice	electrodo *m* estimulante
8646	substance *f* stimulante	sustancia *f* estimulante, ~ estimuladora
8647	stimulation *f*	estimulación *f*
8648	stimulus *m* (*pl* stimuli)	estímulo *m*
8649	intensité *f* du stimulus	intensidad *f* del estímulo
8650	paramètre *m* du stimulus	parámetro *m* del estímulo
8651	perception *f* du stimulus	percepción *f* del estímulo
8652	qualité *f* du stimulus	calidad *f* del estímulo, ~ estimulante
8653	réaction *f* d'excitation	respuesta *f* a un estímulo
8654	relation *f* stimulus-réponse	relación *f* (de) estímulo-respuesta
8655	situation *f* stimulante	situación *f* estimulante
8656	cause *f* stimulante, source *f* de stimulus	fuente *f* de estímulo
8657	—	—
8658	sommation *f* des stimuli, addition *f* ~ ~	sumación *f* de estímulos
8659	seuil *m* de stimulation	umbral *m* de estimulación
8660	aiguillon *m*, piquant *m*; dard *m ins.*	aguijón *m*
8661	piqueur *ins.*	picador *ins.*
8662	—	—
8663	—	—
8664	poil *m* urticant	pelo *m* urticante
8665	glande *f* répugnatoire	glándula *f* repugnatoria
8666	stipule *f*	estípula *f*
8667	—	—
8668	porte-greffe *m*	patrón *m*, portainjerto *m*
8669	culture-stock *f*	cultivo *m* stock
8670	—	—
8671	solution-mère *f*, solution-étalon *f*	solución *f* patrón
8672	stolon *m*, coulant *m*	retoño *m*, estolón *m*
8673	stomate *m*	estoma *m*
8674	estomac *m*	estómago *m*
8675	—	—
8676	transpiration *f* stomatique	transpiración *f* estomática
8677	stomocorde *f*	estomocordio *m*, estomocorda *f*
8678	stomodœum *m*, stomodéum *m*	estomodeo *m*
8679	noyau *m bot.*	hueso *m*; carozo *m* (LA)
8680	canal *m* madréporique, ~ hydrophore	conducto *m* pétreo, ~ hidróforo
8681	—	—
8682	drupe *f*, fruit *m* à noyau	drupa *f*, fruto *m* de hueso; ~ de carozo (LA)
8683	pouvoir *m* d'accumulation	poder *m* de acumulación
8684	organe *m* de réserve, ~ de stockage	órgano *m* de reserva, ~ reservante, ~ de almacenamiento
8685	racine *f* de réserve	raíz *f* nutricia, ~ nutrífera
8686	—	—
8687	tissu *m* de réserve	tejido *m* de reserva, ~ reservante
8688	étage *m* de végétation	piso *m* de vegetación
8689	souche *f*	cepa *f*, estirpe *f*
8690	—	—
8691	étranglement *m*	estrangulación *f*
8692	stratification *f*	estratificación *f*
8693	stratifié	estratificado
8694	épithélium *m* stratifié	epitelio *m* estratificado
8695	strate *f*, couche *f*, assise *f*	estrato *m*, capa *f*
8696	culture *f* en stries	cultivo *m* en estría
8697	—	—
8698	streptomycine *f*	estreptomicina *f*
8699	réflexe *m* d'allongement	reflejo *m* de alargamiento, ~ de elongación, ~ miotático
8700	strié	estriado

8701	striated muscle, striped ~, voluntary ~	quergestreifter Muskel *m*
8702	striation	Streifung *f*
8703	stridulating organ	Stridulationsorgan *n*, Zirporgan *n*
8704	stridulation	Stridulation *f*, Zirpen *n*
8705	strobile → cone *bot.*	—
8706	strobiliform, strobilate	zapfenartig, zapfenförmig
8707	strobilization, strobilation	Strobilation *f*
8708	stroma	Stroma *n*
8709	structural change, ~ modification	Strukturveränderung *f*, ~modifikation *f*
8710	structural gene, structure ~	Strukturgen *n*
8711	structural feature	Baumerkmal *n*, Organisationsmerkmal *n*
8712	structural heterozygosis	Strukturheterozygotie *f*
8713	structural hybrid	Strukturhybride *f*
8714	structural plan, body ~	Bauplan *m*
8715	structural protein	Strukturprotein *n*
8716	struggle for existence	Kampf *m* ums Dasein
8717	strychnin(e)	Strychnin *n*
8718	stylar canal	Griffelkanal *m*
8719	style *bot.*	Griffel *m* *bot.*
8720	stylet *zool.*	Stechborste *f*
8721	subcellular	subzellulär
8722	subclass	Unterklasse *f*
8723	subclimax	Subklimax *f*
8724	subculture	Abimpfung *f*, Subkultur *f*, Nachkultur *f*, Unterkultur *f*
8725	subcutaneous tissue, subcutis, hypoderm	Unterhaut(binde)gewebe *n*, Subkutis *f*, Hypoderm *n*
8726	subdivision *tax.*	Unterabteilung *f* *tax.*
8727	suber, suberous → cork, corky	—
8728	suberin	Suberin *n*, Korkstoff *m*
8729	suberification, suberization	Verkorkung *f*
8730	to suberificate, to suberize	verkorken
8731	subfamily	Unterfamilie *f*
8732	subgene	Subgen *n*
8733	subgenus	Untergattung *f*
8734	subimago, pseudoimago	Subimago *f*
8735	subitaneous egg → summer egg	—
8736	subkingdom	Unterreich *n*
8737	sublethal, semilethal	subletal
8738	sublimate	Sublimat *n*
8739	subliminal, subthreshold	unterschwellig
8740	sublit(t)oral zone	Sublitoral *n*
8741	submersed plant	submerse Pflanze *f*, Unterwasserpflanze *f*
8742	submersion culture, submerged ~	Submerskultur *f*
8743	submicron, hypomicron	Submikron *n*
8744	submicroscopic	submikroskopisch
8745	submissive gesture *eth.*	Demutsgebärde *f* *eth.*
8746	suborder	Unterordnung *f*
8747	subphylum (*pl* subphyla)	Unterstamm *m*
8748	subpopulation	Unterpopulation *f*
8749	subsere, secondary sere, ~ succession	Subserie *f*
8750	subsidiary cell → auxiliary cell	—
8751	subspecies	Unterart *f*, Subspezies *f*
8752	substitute association	Ersatzgesellschaft *f*
8753	substitution	Substitution *f*
8754	substitution crossing → grading up	—
8755	substrate, substratum	Substrat *n*; Unterlage *f*
8756	substrate specificity	Substratspezifität *f*
8757	subterranean animal, ground ~	Bodentier *n*, Subterrantier *n*
8758	subterranean shoot, underground ~	Erdsproß *m*
8759	subthreshold → subliminal	—
8760	subulate, awl-shaped	pfriemförmig
8761	subunit	Untereinheit *f*
8762	succession *ecol.*	Sukzession *f* *ecol.*

8701	muscle *m* strié, ~ volontaire	músculo *m* estriado, ~ voluntario
8702	striation *f*	estriado *m*, estriación *f*
8703	organe *m* stridulant	órgano *m* estridulador, ~ estridulante
8704	stridulation *f*	estridulación *f*
8705	—	—
8706	strobiliforme	estrobiliforme
8707	strobilisation *f*	estrobilación *f*, estrobilización *f*
8708	stroma *m*	estroma *m*
8709	modification *f* de structure	modificación *f* estructural, alteración *f* ~
8710	gène *m* structural, ~ de structure	gen *m* estructural
8711	caractère *m* d'organisation	carácter *m* de organización
8712	hétérozygose *f* de structure	heterocigosis *f* de estructura
8713	hybride *m* de structure, ~ structurel	híbrido *m* de estructura
8714	plan *m* de structure	plan *m* estructural
8715	protéine *f* de structure	proteína *f* estructural
8716	lutte *f* pour la (sur)vie	lucha *f* por la existencia
8717	strychnine *f*	estricnina *f*
8718	canal *m* stylaire	canal *m* estilar
8719	style *m bot.*	estilo *m bot.*
8720	stylet *m zool.*	estilete *m zool.*
8721	subcellulaire	subcelular
8722	sous-classe *f*	subclase *f*
8723	subclimax *m*	subclímax *m*
8724	subculture *f*	subcultivo *m*
8725	tissu *m* sous-cutané, ~ hypodermique, hypoderme *m*	tejido *m* subcutáneo, hipodermis *f*
8726	sous-embranchement *m tax.*	subdivisión *f tax.*
8727	—	—
8728	subérine *f*	suberina *f*
8729	subérification *f*	suberificación *f*, suberización *f*
8730	subérifier	suberificar, suberizar(se)
8731	sous-famille *f*	subfamilia *f*
8732	sous-gène *m*	subgen *m*
8733	sous-genre *m*	subgénero *m*
8734	subimago *f*	subimago *m*, subadulto *m*
8735	—	—
8736	sous-règne *m*	subreino *m*
8737	sublét(h)al, semi-lét(h)al	subletal, semiletal
8738	sublimé *m*	sublimado *m*
8739	sousliminaire, infraliminaire, inférieur au seuil	subliminar, infraumbral, inferior al umbral
8740	zone *f* sublittorale	zona *f* sublitoral, ~ infralitoral
8741	plante *f* submergée, végétal *m* submergé	planta *f* (acuática) sumergida
8742	culture *f* de submersion	cultivo *m* de sumersión
8743	submicron *m*, ultramicron *m*	submicrón *m*, hipomicrón *m*
8744	submicroscopique	submicroscópico
8745	geste *m* de soumission, attitude ~ ~ *eth.*	actitud *f* de sumisión *eth.*
8746	sous-ordre *m*, infra-ordre *m*	suborden *m*
8747	sous-embranchement *m*	subfilum *m*, subfilo *m*
8748	sous-population *f*	subpoblación *f*
8749	sous-série *f*	subsere *f*, sucesión *f* secundaria
8750	—	—
8751	sous-espèce *f*	subespecie *f*
8752	association *f* de remplacement	asociación *f* de reemplazo
8753	substitution *f*	sustitución *f*
8754	—	—
8755	substrat *m*; substratum *m*	substrato *m*
8756	spécificité *f* envers du substrat, ~ pour le substrat	especificidad *f* frente al substrato
8757	animal *m* terricole, ~ du sol	animal *m* terrícola
8758	pousse *f* souterraine	vástago *m* subterráneo
8759	—	—
8760	subulé, alené	alesnado
8761	sous-unité *f*, subunité *f*	subunidad *f*
8762	succession *f ecol.*	sucesión *f ecol.*

8763	successional series → sere	—
8764	successional stage → seral stage	—
8765	succinic acid	Bernsteinsäure *f*
8766	succulence	Sukkulenz *f*
8767	succulent (plant)	Sukkulente *f*
8768	sucker[1]	Saugorgan *n*
8769	sucker[2]	Saugnapf *m*
8770	sucking *ins.*	saugend *ins.*
8771	sucking disc	Saugscheibe *f*
8772	sucking root	Saugwurzel *f*
8773	sucking tube, feeding ~	Saugrüssel *m*
8774	sucrase → invertase	—
8775	sucrose → saccharose	—
8776	suction *bot.*	Saugwirkung *f*, Saugeffekt *m* *bot.*
8777	suction force of the soil, absorptive capacity ~ ~ ~	Saugkraft *f* des Bodens
8778	sudation	Schweißabsonderung *f*
8779	sudoriferous gland → sweat gland	—
8780	suffrutex, under-shrub, half-shrub	Halbstrauch *m*
8781	sulphur bacteria → thiobacteria	—
8782	sulphur cycle (US: sulfur)	Schwefelkreislauf *m*
8783	summation *neur.*	Summation *f* *neur.*
8784	summer annual	Sommerannuelle *f*
8785	summer dormancy → aestivation	—
8786	summer egg, subitaneous ~	Sommerei *n*, Subitanei *n*
8787	summer green	sommergrün
8788	summer-spore → uredospore	—
8789	sun leaf, heliophyll	Sonnenblatt *n*, Lichtblatt *n*
8790	sun plant → heliophyte	—
8791	superclass	Überklasse *f*
8792	superdominance → overdominance	—
8793	superfamily	Überfamilie *f*
8794	superfemale, metafemale	Überweibchen *n*
8795	superficial tissue → surface tissue	—
8796	superf(o)etation	Überbefruchtung *f*
8797	supergene	Supergen *n*
8798	superior *bot.*	oberständig
8799	superior maxillary bone, upper jawbone	Oberkieferknochen *m*
8800	supermale	Übermännchen *n*
8801	supernormal stimulus	Überreiz *m*
8802	supernumerary, supernumery	überzählig
8803	superorder	Überordnung *f*
8804	superparasite → hyperparasite	—
8805	superphylum	Hauptstamm *m*
8806	superposition eye	Superpositionsauge *n*
8807	supersonics → ultrasonics	—
8808	superspecies	Superspezies *f*
8809	supplemental air → reserve air	—
8810	supplementary gene	Supplementärgen *n*
8811	supporting cell	Stützzelle *f*
8812	supporting tissue, mechanical ~	Festigungsgewebe *n*, Stützgewebe *n*, mechanisches Gewebe *n*
8813	suppressor	Suppressor *m*
8814	suppressor mutation	Suppressormutation *f*, kompensierende Mutation *f*
8815	supraliminal, above threshold	überschwellig
8816	supralit(t)oral zone	Supralitoral *n*
8817	suprarenal gland → adrenal gland	—
8818	supraspecies	Supraspezies *f*, Überart *f*
8819	supravital staining	Supravitalfärbung *f*
8820	surface culture	Oberflächenkultur *f*
8821	surface humus	Auflagehumus *m*
8822	surface tension	Oberflächenspannung *f*
8823	surface tissue, superficial ~	Abschlußgewebe *n*
8824	surface water	Oberflächenwasser *n*

8763	—	—
8764	—	—
8765	acide *m* succinique	ácido *m* succínico
8766	succulence *f*	suculencia *f*
8767	plante *f* succulente	planta *f* suculenta
8768	suçoir *m*, organe *m* de succion	órgano *m* de succión, ~ suctor
8769	ventouse *f*	ventosa *f*
8770	suceur *ins.*	chupador *ins.*
8771	disque *m* ventousaire	disco *m* de succión
8772	racine *f* absorbante	raíz *f* absorbente
8773	tube *m* suceur	tubo *m* chupador, ~ alimentador
8774	—	—
8775	—	—
8776	succion *f*	succión *f*, acción *f* aspirante
8777	succion *f* du sol, force *f* d'attraction (du sol pour l'eau)	poder de absorción (del suelo), fuerza *f* absorbente (del suelo)
8778	sudation *f*, sécrétion *f* sudoripare, ~ sudorale	sudación *f*, sudoración *f*
8779	—	—
8780	sous-arbrisseau *m*, suffrutescent *m*	sufrútice *m*
8781	—	—
8782	cycle *m* du soufre	ciclo *m* del azufre
8783	sommation *f* *neur.*	sumación *f* *neur.*
8784	annuelle *f* estivale, ~ d'été	anual *f* estival
8785	—	—
8786	œuf *m* immédiat	huevo *m* subitáneo, ~ de verano
8787	vert estival	verde en verano
8788	—	—
8789	feuille *f* de soleil, ~ de lumière	hoja *f* heliófila, ~ de luz, ~ asoleada
8790	—	—
8791	super-classe *f*	superclase *f*
8792	—	—
8793	superfamille *f*	superfamilia *f*
8794	superfemelle *f*	superhembra *f*
8795	—	—
8796	superfétation *f*	superfetación *f*, superfecundación *f*
8797	supergène *m*	supergen *m*
8798	supère	súpero
8799	(os *m*) maxillaire *m* supérieur	maxilar *m* superior
8800	supermâle *m*	supermacho *m*
8801	stimulus *m* superstimulant	estímulo *m* supranormal, super-estímulo *m*
8802	surnuméraire	supernumerario
8803	super-ordre *m*	superorden *m*
8804	—	—
8805	superembranchement *m*	superfílum *m*
8806	œil *m* à vision par superposition	ojo *m* de visión por superposición
8807	—	—
8808	super-espèce *f*	superespecie *f*
8809	—	—
8810	gène *m* supplémentaire	gen *m* suplementario
8811	cellule *f* de soutien	célula *f* de sostén
8812	tissu *m* de soutien, ~ mécanique	tejido *m* de sostén, ~ mecánico
8813	suppresseur *m*	supresor *m*
8814	mutation *f* suppressive	mutación *f* supresora
8815	susliminaire, supraliminaire, supérieur au seuil	supraliminar, supraumbral, superior al umbral
8816	supralittoral *m*, zone *f* supralittorale	zona *f* supralitoral
8817	—	—
8818	supra-espèce *f*	supraespecie *f*
8819	coloration *f* supravitale	coloración *f* supravital, tinción *f* ~
8820	culture *f* en surface	cultivo *m* superficial
8821	humus *m* de couverture	humus *m* de superficie
8822	tension *f* superficielle	tensión *f* superficial
8823	tissu *m* de revêtement, ~ superficiel	tejido *m* superficial
8824	eaux *f pl* de surface	aguas *f pl* superficiales

8825	survival	Überleben *n*
8826	survival curve	Überlebenskurve *f*
8827	survival index	Erhaltungsquotient *m*
8828	survival of fittest	Überleben *n* des Tüchtigsten
8829	survival rate	Überlebensrate *f*
8830	survival reduction	Verringerung *f* der Überlebensrate
8831	survival value	Überlebenswert *m*
8832	to survive	überleben
8833	susception	Suszeption *f*, Reizaufnahme *f*
8834	suspended matter	Schwebstoffe *m pl*
8835	suspensor	Suspensor *m*, Embryoträger *m*
8836	swallowing → deglutition	—
8837	swallowing reflex, palatal ~	Schluckreflex *m*, Gaumenreflex *m*
8838	swarm	Schwarm *m*
8839	to swarm	schwärmen
8840	swarm spore → zoospore	—
8841	swarming	Schwärmen *n*
8842	swarming impulse	Schwarmtrieb *m*
8843	sweat	Schweiß *m*
8844	sweat gland, sudoriferous ~	Schweißdrüse *f*
8845	to swell	schwellen, anschwellen; quellen
8846	swelling	Schwellung *f*; Quellung *f*
8847	swim bladder, air-bladder	Schwimmblase *f*
8848	swimming ability	Schwimmfähigkeit *f*
8849	swimming movement	Schwimmbewegung *f*
8850	swimming plate	Schwimmplatte *f*, Wimperplatte *f*
8851	symbiont, symbiote	Symbiont *m*
8852	symbiosis	Symbiose *f*
8853	symbiotic	symbio(n)tisch
8854	symmetrical plane, plane of symmetry	Symmetrieebene *f*
8855	sympathetic chain	sympathischer Grenzstrang *m*, sympathische Ganglienkette *f*
8856	sympathetic (nervous) system, sympathicus, orthosympathetic system	Sympathikus *m*, Orthosympathikus *m*, sympathisches Nervensystem *n*
8857	sympatric	sympatrisch
8858	sympetalous → gamopetalous	—
8859	symphysis	Symphyse *f*, Knochenfuge *f*
8860	sympodial	sympodial
8861	sympodium, false axis, pseud(o)axis	Sympodium *n*, Scheinachse *f*
8862	synangium	Synangium *n*
8863	synapse *neur.*	Synapse *f* *neur.*
8864	synapsis, pairing of chromosomes	Synapsis *f*, Chromosomenpaarung *f*
8865	synaptene → zygotene	—
8866	synaptic	synaptisch
8867	synaptic cleft	synaptischer Zwischenraum *m*
8868	synaptic delay	synaptische Verzögerung *f*
8869	synaptic vesicle	synaptisches Bläschen *n*
8870	synarthrosis	Synarthrose *f*
8871	syncarpous	synkarp
8872	synchorology	Synchorologie *f*, Gesellschaftsverbreitung *f*
8873	synchronous	synchron
8874	syncytium	Synzytium *n*
8875	syndesis	Syndese *f*
8876	syndesmosis	Syndesmose *f*
8877	synecology	Synökologie *f*
8878	synergia, synergy	Synergie *f*
8879	synergic, synergetic	synergetisch
8880	synergid, help-cell	Synergide *f*, Gehilfin *f*
8881	synergism	Synergismus *m*
8882	synergist	Synergist *m*
8883	syngameon	Syngameon *n*, Sammelart *f*
8884	syngamic, syngamous	syngam
8885	syngamy	Syngamie *f*

8825	survie *f*, survivance *f*	supervivencia *f*
8826	courbe *f* de survie	curva *f* de supervivencia
8827	indice *m* de survie	índice *m* de supervivencia
8828	survivance *f* des mieux adaptés, persistance *f* des plus aptes	supervivencia *f* del más apto
8829	taux *m* de survie, degré *m* ~ ~	tasa *f* de supervivencia *f*, porcentaje *m* ~ ~, grado *m* ~ ~
8830	diminution *f* de survie	reducción *f* de supervivencia
8831	valeur *f* de survie	valor *m* de supervivencia
8832	survivre	sobrevivir
8833	susception *f*	suscepción *f*
8834	corps *mpl* en suspension, matières *fpl* ~ ~	cuerpos *mpl* en suspensión
8835	suspenseur *m*	suspensor *m*
8836	—	—
8837	réflexe *m* de déglutition, ~ palatin	reflejo *m* deglutorio, ~ palatino
8838	essaim *m*; troupe *f*	enjambre *m*; banda *f*, bando *m*
8839	essaimer	enjambrar
8840	—	—
8841	essaimage *m*	enjambrazón *m*, enjambramiento *m*
8842	fièvre *f* d'essaimage	impulso *m* de enjambrazón
8843	sueur *f*	sudor *m*
8844	glande *f* sudoripare	glándula *f* sudorípara
8845	se renfler, se gonfler	hincharse
8846	renflement *m*, gonflement *m*	hinchazón *f*, hinchamiento *m*
8847	vessie *f* natatoire, ~ gazeuse	vejiga *f* natatoria, ~ aérea
8848	flottabilité *f*, capacité *f* de flottaison	flotabilidad *f*
8849	mouvement *m* natatoire	movimiento *m* natatorio, ~ de natación
8850	palette *f* natatoire, ~ ciliée	paleta *f* natatoria
8851	symbio(n)te *m*	simbionte *m*
8852	symbiose *f*	simbiosis *f*
8853	symbiotique	simbiótico
8854	plan *m* de symétrie	plano *m* de simetría
8855	chaîne *f* sympathique	cadena *f* (ganglionar) simpática
8856	sympathique *m*, système *m* nerveux sympathique, ~ orthosympathique	sistema *m* simpático, ~ ortosimpático
8857	sympatrique	simpátrico
8858	—	—
8859	symphyse *f*	sínfisis *f*
8860	sympodial	simpódico, simpodial
8861	sympode *m*, faux-axe *m*	simpodio *m*
8862	synange *m*	sinangio *m*
8863	synapse *f* *neur.*	sinapsis *f* *neur.*
8864	synapsis *f*, appariement *m* de chromosomes	sinapsis *f* *gen.*, apareamiento *m* de cromosomas
8865	—	—
8866	synaptique	sináptico
8867	espace *m* synaptique, interstice *m* ~	intersticio *m* sináptico
8868	intervalle *m* synaptique	retraso *m* sináptico
8869	vésicule *f* synaptique	vesícula *f* sináptica
8870	synarthrose *f*	sinartrosis *f*
8871	syncarpe	sincarpo, sincárpico
8872	distribution *f* des communautés	sincorología *f*
8873	synchrone	sincrónico
8874	syncytium *m*	sincitio *m*, sincicio *m*
8875	syndèse *f*	síndesis *f*
8876	syndesmose *f*	sindesmosis *f*
8877	synécologie *f*, écologie *f* associationnelle	sinecología *f*
8878	synergie *f*	sinergia *f*
8879	synénergétique	sinérgico, sinergético
8880	synergide *f*	sinérgida *f*
8881	synergisme *m*	sinergismo *m*
8882	synergiste *m*	sinergista *m*
8883	syngaméon *m*, espèce *f* collective	singámeon *m*, especie *f* colectiva
8884	syngame	singámico
8885	syngamie *f*	singamia *f*

8886	syngenesis	Syngenese *f*
8887	syngenetic, syngenic	syngenetisch
8888	synkaryon	Synkarion *n*
8889	synoecious, synoicous	synözisch, gemischtgeschlechtig
8890	synoecy, synoeky	Synökie *f*
8891	synovia, synovial fluid	Gelenkschmiere *f*, ~flüssigkeit *f*, Synovia *f*
8892	synovial → articular	—
8893	synovial bursa, ~ sac	Schleimbeutel *m*
8894	synovial membrane	Gelenkhaut *f*
8895	to synthesize	aufbauen
8896	synthetase → ligase	—
8897	syntrophic	syntroph
8898	syntrophism	Syntrophismus *m*
8899	synusia, synusium	Synusie *f*
8900	synzoospore	Synzoospore *f*
8901	syrinx	Syrinx *m/f*
8902	systematics	Systematik *f*
8903	systematist	Systematiker *m*
8904	systemic circulation, greater ~	Körperkreislauf *m*, großer Kreislauf *m*
8905	systemic mutation	Systemmutation *f*
8906	systole	Systole *f*
8907	tabular root	Brettwurzel *f*
8908	tachygenesis	Tachygenese *f*
8909	tactile	Tast...
8910	tactile corpuscle	Tastkörperchen *n*
8911	tactile hair	Tasthaar *n*
8912	tactile organ, touch ~	Tastorgan *n*, Fühlorgan *n*
8913	tactile receptor → tangoreceptor	—
8914	tactile sensation	Tastempfindung *f*
8915	tactile stimulus, touch ~	Tastreiz *m*
8916	tactual sense → sense of touch	—
8917	tadpole	Kaulquappe *f*
8918	taiga	Taiga *f*
8919	tail feather, rectrix (*pl* rectrices)	Schwanzfeder *f*, Steuerfeder *f*
8920	tail fin → caudal fin	—
8921	tailless → anurous	—
8922	tallow	Talg *m*
8923	tandem selection	Tandemselektion *f*
8924	tangoreceptor, touch receptor, tactile ~	Tastrezeptor *m*, Tangorezeptor *m*, Tastsinnesorgan *n*
8925	tannic acid	Gerbsäure *f*
8926	tannin	Tannin *n*, Gallusgerbsäure *f*
8927	tapetal cell	Tapetenzelle *f*
8928	tapetum, tapetal layer	Tapetum *n*, Tapetenschicht *f*
8929	tap-root	Pfahlwurzel *f*
8930	Tardigrada, tardigrades	Bärtierchen *npl*
8931	target theory, hit ~	Treffertheorie *f*
8932	tarsal bone	Fußwurzelknochen *m*
8933	tarsus	Fußwurzel *f*
8934	tartaric acid	Wein(stein)säure *f*
8935	taste	Geschmack *m*
8936	taste-bud	Geschmacksknospe *f*
8937	taste cell, ~ organ → gustatory cell, ~ organ	—
8938	taste receptor	Geschmacksrezeptor *m*
8939	taste stimulus → gustatory stimulus	—
8940	taste threshold	Geschmacksschwelle *f*
8941	taxis	Taxie *f*, Taxis *f* (*pl* Taxien)
8942	taxon	Taxon *n* (*pl* Taxa)
8943	taxonomic	taxonomisch
8944	taxonomist	Taxonom *m*
8945	taxonomy	Taxonomie *f*
8946	tear... see also lacrimal...	—
8947	tear pit, larmier	Tränengrube *f*

8886	syngenèse *f*	singénesis *f*, singenesia *f*
8887	syngénique	singenético
8888	syncaryon *m*, syncarion *m*	sincario *m*, sincarión *m*
8889	synoecique, synoïque	sinoico
8890	synécie *f*, synoecie *f*	sinecia *f*
8891	synovie *f*, liquide *m* synovial	sinovia *f*, líquido *m* sinovial
8892	—	—
8893	bourse *f* synoviale, ~ muqueuse	bolsa *f* sinovial
8894	membrane *f* synoviale	membrana *f* sinovial
8895	synthétiser	sintetizar
8896	—	—
8897	syntrophe	síntrofo
8898	syntrophisme *m*, syntrophie *f*	sintrofismo *m*
8899	synusie *f*	sinusia *f*
8900	synzoospore *f*	sinzoóspora *f*
8901	syrinx *f*, syringe *f*	siringe *f*
8902	systématique *f*	sistemática *f*
8903	systématicien *m*	sistemático *m*
8904	circulation *f* systémique, ~ générale, grande circulation *f*	circulación *f* mayor, ~ general
8905	mutation *f* systémique	mutación *f* sistémica
8906	systole *f*	sístole *f*
8907	racine-palette *f*	raíz *f* tabular
8908	tachygenèse *f*	taquigénesis *f*
8909	tactile	táctil
8910	corpuscule *m* tactile, ~ de tact	corpúsculo *m* táctil
8911	poil *m* tactile	pelo *m* táctil
8912	organe *m* tactile, ~ du toucher, ~ du tact	órgano *m* del tacto
8913	—	—
8914	sensation *f* tactile	sensación *f* táctil
8915	stimulus *m* tactile	estímulo *m* táctil
8916	—	—
8917	têtard *m*	renacuajo *m*
8918	taïga *f*	taiga *f*
8919	(plume *f*) rectrice *f*, ~ caudale	(pluma *f*) timonera *f*, ~ caudal
8920	—	—
8921	—	—
8922	suif *m*	sebo *m*
8923	sélection *f* en tandem	selección *f* en tandem
8924	tangorécepteur *m*, récepteur *m* tactile	tangiorreceptor *m*, receptor *m* táctil
8925	acide *m* tannique	ácido *m* tánico
8926	tan(n)in *m*	tanino *m*
8927	cellule *f* du tapétum	célula *f* del tapete
8928	tapétum *m*, tapis *m*, assise *f* nourricière	tapete *m*
8929	racine *f* pivotante, pivot *m*	raíz *f* axonomorfa
8930	tardigrades *mpl*	tardígrados *mpl*
8931	théorie *f* de la cible, ~ du choc, ~ de l'impact	teoría *f* del blanco
8932	os *m* tarsien	hueso *m* tarsiano, ~ tarsal
8933	tarse *m*	tarso *m*
8934	acide *m* tartrique	ácido *m* tartárico
8935	goût *m*	gusto *m*
8936	bourgeon *m* gustatif, corpuscule *m* ~	mamelón *m* gustativo, botón *m* ~
8937	—	—
8938	récepteur *m* du goût	receptor *m* gustativo, gustatorreceptor *m*
8939	—	—
8940	seuil *m* gustatif	umbral *m* gustativo
8941	taxie *f*, tactisme *m*	taxis *f*, taxia *f*, tactismo *m*
8942	taxon *m*	taxón *m*
8943	taxonomique, taxinomique	taxonómico, taxinómico
8944	taxonomiste *m*, taxinomiste *m*	taxónomo *m*, taxonomista *m*, taxinomista *m*
8945	taxonomie *f*, taxinomie *f*	taxonomía *f*, taxinomía *f*
8946	—	—
8947	larmier *m*	lagrimal *m*

8948	teat, dug	Zitze *f*
8949	tectology	Tectologie *f*
8950	tectorial membrane → Corti's membrane	—
8951	teething → dentition	—
8952	tegmentum → bud-scale	—
8953	tegument → integument	—
8954	telegony	Telegonie *f*
8955	telencephalon, endbrain	Telenzephalon *n*, Endhirn *n*
8956	Teleostei, teleosts	Teleostier *pl*, Knochenfische *m pl* *(im engeren Sinne)*
8957	tele(re)ceptor → distance receptor	—
8958	teleutospore, teliospore, winter-spore	Teleutospore *f*, Winterspore *f*, Dauerspore *f*
8959	telocentric	telozentrisch
8960	telolecithal	telolezithal
8961	telome	Telom *n*
8962	telome theory	Telomtheorie *f*
8963	telomere	Telomer *n*
8964	telophase	Telophase *f*
8965	telosynapsis, telosyndesis	Telosynapsis *f*, Telosyndese *f*
8966	telotaxis	Telotaxis *f*
8967	telson	Telson *n*
8968	temperate phage	temperierter Phage *m*, gemäßigter ~
8969	temperature coefficient, Q_{10}	Temperaturkoeffizient *m*, Q_{10}
8970	temperature gradient	Temperaturgefälle *n*
8971	temperature optimum	Temperaturoptimum *n*
8972	temperature receptor	Temperaturrezeptor *m*
8973	temperature-sensitive	temperaturempfindlich
8974	temple	Schläfe *f*
8975	temporal bone	Schläfenbein *n*
8976	temporal facilitation	zeitliche Bahnung *f*
8977	temporal lobe	Schläfenlappen *m*
8978	temporal summation	zeitliche Summation *f*
8979	temporary parasite	zeitweiliger Parasit *m*
8980	temporary wilting	zeitweiliges Welken *n*
8981	tendinous	sehnig
8982	tendinous sheath	Sehnenscheide *f*
8983	tendon	Sehne *f*
8984	tendon reflex, ~ jerk	Sehnenreflex *m*
8985	tendril	Ranke *f*
8986	tensile strength	Zugfestigkeit *f*
8987	tensor muscle	Spannmuskel *m*, Tensor *m*
8988	tentacle	Fangarm *m*, Tentakel *m/n*
8989	tepal	Tepal *n*, Perigonblatt *n*
8990	teratology	Teratologie *f*
8991	terminal *bot.*	endständig, gipfelständig
8992	terminal arborization	Nervenendverzweigung *f*
8993	terminal bud → apical bud	—
8994	terminal cell → end-cell	—
8995	terminal flower	Endblüte *f*
8996	terminal organ → end-organe	—
8997	terminal shoot	Gipfeltrieb *m*
8998	terminal stage	Endstadium *n*, Terminalphase *f*
8999	terminalization	Terminalisation *f*
9000	termones	Termone *pl*
9001	terpene	Terpen *n*
9002	terrestrial, land-dwelling, land-living	landbewohnend, landlebend, terrestrisch
9003	terrestrial animal, land-living ~	Landtier *n*, landlebendes Tier *n*
9004	terrestrial ecosystem	Landökosystem *n*
9005	terrestrial plant, land ~	Landpflanze *f*
9006	terricolous	erdbewohnend, bodenbewohnend
9007	territorial behavio(u)r, territoriality, territorialism	Revierverhalten *n*, Territorialverhalten *n*, Territorialität *f*
9008	territorial defense, territory ~	Revierverteidigung *f*
9009	territorial marking, territory ~	Reviermarkierung *f*
9010	territorial song	Reviergesang *m*

8948	tétine *f*, trayon *m*	teta *f*, pezón *m*
8949	tectologie *f*	tectología *f*
8950	—	—
8951	—	—
8952	—	—
8953	—	—
8954	télégonie *f*	telegonía *f*
8955	télencéphale *m*, cerveau *m* terminal	telencéfalo *m*, cerebro *m* terminal
8956	téléostéens *mpl*	teleósteos *mpl*
8957	—	—
8958	téleutospore *f* téliospore *f*, spore *f* d'hiver, ~ durable	teleutóspora *f*, telióspora *f*, espora *f* de invierno
8959	télocentrique	telocéntrico *m*
8960	télolécithe, télolécithique	telolecito, telolecítico
8961	télome *m*	teloma *m*
8962	théorie *f* du télome	teoría *f* telomática
8963	télomère *m*	telómero *m*
8964	télophase *f*	telofase *f*
8965	télosynapse *f*, télosyndèse *f*	telosinapsis *f*, telosíndesis *f*
8966	télotaxie *f*, télotactisme *m*	telotaxia *f*, telotactismo *m*
8967	telson *m*	telson *m*
8968	phage *m* tempéré	fago *m* temperado
8969	coefficient *m* de température, Q_{10}	coeficiente *m* de temperatura, Q_{10}
8970	gradient *m* de température	gradiente *m* de temperatura
8971	optimum *m* de température	óptimo *m* térmico
8972	récepteur *m* de température	receptor *m* de temperatura
8973	sensible à la température	sensible a la temperatura
8974	tempe *f*	sien *f*
8975	os *m* temporal	hueso *m* temporal
8976	facilitation *f* temporelle	facilitación *f* temporal
8977	lobe *m* temporal	lóbulo *m* temporal
8978	sommation *f* temporelle	sumación *f* temporal
8979	parasite *m* temporaire	parásito *m* temporal
8980	flétrissement *m* passager	marchitamiento *m* temporal
8981	tendineux	tendinoso
8982	gaine *f* tendineuse	vaina *f* tendinosa
8983	tendon *m*	tendón *m*
8984	réflexe *m* tendineux	reflejo *m* tendinoso
8985	vrille *f*	zarcillo *m*
8986	résistance *f* à la traction	resistencia *f* a la tracción, ~ al estiramiento
8987	muscle *m* tenseur	músculo *m* tensor
8988	tentacule *m*	tentáculo *m*
8989	tépale *m*	tépalo *m*
8990	tératologie *f*	teratología *f*
8991	terminal *bot.*	terminal *bot.*
8992	arborisation *f* terminale	arborización *f* terminal
8993	—	—
8994	—	—
8995	fleur *f* terminale	flor *f* terminal
8996	—	—
8997	pousse *f* terminale	brote *m* terminal
8998	stade *m* terminal	estad(i)o *m* final, fase *f* terminal
8999	terminalisation *f*	terminalización *f*
9000	termones *fpl*	termonas *fpl*, termones *mpl*
9001	térébène *m*, terpène *m*	terpeno *m*
9002	terrestre	terrestre
9003	animal *m* terrestre	animal *m* terrestre
9004	écosystème *m* terrestre	ecosistema *m* terrestre
9005	plante *f* terrestre	planta *f* terrestre
9006	terricole	terrícola
9007	comportement *m* territorial, territorialisme *m*, territorialité *f*	comportamiento *m* territorial, territorialidad *f*
9008	défense *f* territoriale, ~ du territoire	defensa *f* del territorio
9009	marquage *m* du territoire	marcación *f* del territorio
9010	chant *m* territorial	canto *m* territorial

9011	territory	Revier *n*, Territorium *n*
9012	tertiary consumer	Tertiärkonsument *m*
9013	Tertiary (period)	Tertiär *n*
9014	tertiary structure	Tertiärstruktur *f*
9015	tesselated epithelium → pavement ~	—
9016	test cross, ~ mating	Testkreuzung *f*, Testpaarung *f*
9017	test for complementation	Komplementationstest *m*
9018	test paper	Reagenzpapier *n*
9019	test plant	Testpflanze *f*, Kontrollpflanze *f*
9020	test tube	Reagenzglas *n*
9021	test-tube culture	Reagenzglaskultur *f*
9022	testa → seed coat	—
9023	testis, testicle	Hoden *m*, Testikel *m*, Testis *f*
9024	testosterone	Testosteron *n*
9025	tetrad	Tetrade *f*
9026	tetrad analysis	Tetradenanalyse *f*
9027	tetraploid	tetraploid
9028	tetraploidy	Tetraploidie *f*
9029	tetrapod, quadruped	Vierfüsser *m*, Tetrapode *m*
9030	tetrasomic	tetrasom
9031	tetraspore	Tetraspore *f*
9032	tetratype	Tetratyp *m*
9033	thalamus[1]	Sehhügel *m*, Thalamus *m*
9034	thalamus[2] → receptacle	—
9035	thallophyte	Thallophyt *m*, Thalluspflanze *f*, Lagerpflanze *f*
9036	thallose, thalloid	thallös
9037	thallus	Thallus *m*, Lager *n*
9038	theca	Theka *f* (*pl* Theken)
9039	thely(o)toky	Thelytokie *f*
9040	theory of catastrophes	Katastrophentheorie *f*, Kataklysmentheorie *f*
9041	theory of creation	Schöpfungstheorie *f*, Konstanztheorie *f*
9042	theory of evolution, evolutionary theory, evolutionism	Evolutionstheorie *f*, Transformismus *m*
9043	theory of germinal continuity, ~ of continuity of the germ plasm	Keimplasmatheorie *f*, Idioplasmatheorie *f*, Hypothese *f* von der Kontinuität des Keimplasmas
9044	theory of selection	Selektionstheorie *f*
9045	thermal... see also heat...	—
9046	thermal rigidity, heat rigo(u)r	Hitzestarre *f*, Wärmestarre *f*
9047	thermal sense	Wärmesinn *m*
9048	thermal sensitivity, sensitivity to heat	Hitzeempfindlichkeit *f*
9049	thermal stimulus	Wärmereiz *m*
9050	thermal unit	Wärmeeinheit *f*
9051	thermocline, metalimnion, discontinuity layer	thermokline Zone *f*, Metalimnion *n*, Sprungschicht *f*
9052	thermodynamic	thermodynamisch
9053	thermodynamics	Thermodynamik *f*
9054	thermogenesis → calorification	—
9055	thermogenic centre → heat centre	—
9056	thermolabile	thermolabil, hitzeunbeständig, wärmeunbeständig
9057	thermolysis	Thermolyse *f*
9058	thermonasty	Thermonastie *f*
9059	thermoperiodicity	Thermoperiodismus *m*
9060	thermophil(ic)	wärmeliebend, thermophil
9061	thermoreceptor	Wärmerezeptor *m*, Thermorezeptor *m*
9062	thermoregulation	Thermoregulation *f*, Wärmeregulation *f*
9063	thermoregulatory	wärmeregulierend
9064	thermoregulatory centre (US: center)	Wärmeregulationszentrum *n*
9065	thermostability	Wärmestabilität *f*, Wärmebeständigkeit *f*
9066	thermostable, thermostabile	thermostabil, wärmebeständig, hitzebeständig
9067	thermotaxis	Thermotaxis *f*
9068	thermotropism	Thermotropismus *m*

9011	territoire *m*	territorio *m*
9012	consommateur *m* tertiaire	consumidor *m* terciario
9013	tertiaire *m*	terciario *m*
9014	structure *f* tertiaire	estructura *f* terciaria
9015	—	—
9016	croisement *m* d'épreuve, ~ d'essai, accouplement *m* de testage, test-cross *m*	cruzamiento *m* de prueba, cruce *m* ~ ~
9017	test *m* de complémentarité	prueba *f* de complementación
9018	papier *m* indicateur	papel *m* reactivo
9019	plante-témoin *f*, plante-test *f*	planta *f* testigo
9020	tube *m* à essai, éprouvette *f*	tubo *m* de ensayo, probeta *f*
9021	culture *f* en tube	cultivo *m* en tubo (de ensayo)
9022	—	—
9023	testicule *m*	testículo *m*
9024	testostérone *f*	testosterona *f*
9025	tétrade *f*	tétrade *f*
9026	analyse *f* des tétrades	análisis *m* de (las) tétrades
9027	tétraploïde	tetraploide
9028	tétraploïdie *f*	tetraploidia *f*
9029	tétrapode *m*, quadrupède *m*	tetrápodo *m*, cuadrúpedo *m*
9030	tétrasomique	tetrasómico
9031	tétraspore *f*	tetráspora *f*
9032	tétratype *m*	tetratipo *m*
9033	thalamus *m*, couche *f* optique	tálamo *m* óptico
9034	—	—
9035	thallophytc *m*	talófito *m*, talofita *f*
9036	thallodique	taloso
9037	thallus *m*, thalle *m*	talo *m*
9038	thèque *f*	teca *f*
9039	thélytocie *f*, thélytokie *f*	telitoquía *f*, telitocia *f*
9040	théorie *f* des catastrophes	teoría *f* de las catástrofes, catastrofismo *m*
9041	créationnisme *m*, fixisme *m*	teoría *f* creacionista, ~ de la constancia, creacionismo *m*
9042	théorie *f* de l'évolution, ~ évolutionniste, évolutionnisme *m*, transformisme *m*	teoría *f* de la evolución, ~ evolutiva, evolucionismo *m*, transformismo *m*, teoría *f* transformista
9043	hypothèse *f* de la continuité germinale	teoría *f* de la continuidad del plasma germinal
9044	théorie *f* de la sélection (naturelle)	teoría *f* de la selección (natural)
9045	—	—
9046	rigidité *f* thermique, catalepsie *f* calorifique	rigidez *f* debida al calor
9047	sens *m* thermique	sentido *m* térmico
9048	thermosensibilité *f*, sensibilité *f* thermique	sensibilidad *f* térmica
9049	stimulus *m* thermique	estímulo *m* del calor, ~ térmico
9050	unité *f* calorique	unidad *f* calorífica
9051	thermocline *f*, métalimnion *m*, couche *f* de discontinuité	termoclina *f*, capa *f* de discontinuidad
9052	thermodynamique	termodinámico
9053	thermodynamique *f*	termodinámica *f*
9054	—	—
9055	—	—
9056	thermolabile	termolábil
9057	thermolyse *f*	termólisis *f*
9058	thermonastie *f*, thermonastisme *m*	termonastia *f*
9059	thermopériodisme *m*	termoperiodicidad *f*
9060	thermophile	termófilo, termofílico
9061	thermorécepteur *m*, récepteur *m* thermique	termorreceptor *m*, receptor *m* térmico
9062	thermorégulation *f*, régulation *f* thermique	termorregulación *f*, regulación *f* térmica
9063	thermorégulateur	termorregulador
9064	centre *m* thermorégulateur	centro *m* termorregulador
9065	thermostabilité *f*, stabilité *f* à la chaleur	termoestabilidad *f*, estabilidad *f* térmica
9066	thermostable, thermostabile	termo(e)stable, termo(e)stábil
9067	thermotaxie *f*, thermotactisme *m*	termotaxis *f*, termotaxia *f*
9068	thermotropisme *m*	termotropismo *m*

9069	therophyte	Therophyt *m*
9070	thiamin(e), aneurin, vitamin B_1	Thiamin *n*, Aneurin *n*, antineuritisches Vitamin *n*, Vitamin *n* B_1
9071	to thicken	eindicken, verdicken
9072	thickening	Verdickung *f*
9073	thigh-bone → femur	—
9074	thigmomorphosis	Thigmomorphose *f*
9075	thigmotaxis	Thigmotaxis *f*
9076	thigmotropism	Thigmotropismus *m*
9077	thin-layer chromatography	Dünnschichtchromatographie *f*
9078	thin section	Dünnschliff *m*
9079	thiobacteria, sulphur bacteria (US: sulfur ~)	Schwefelbakterien *f pl*
9080	thioctic acid, liponic ~	Thioktansäure *f*, Liponsäure *f*
9081	thionin(e)	Thionin *n*
9082	thoracic cavity, pleural ~	Brusthöhle *f*
9083	thoracic duct	Milchbrustgang *m*
9084	thoracic vertebra	Brustwirbel *m*
9085	thorax	Brustkorb *m*, Thorax *m*
9086	thorn, spine	Dorn *m*
9087	thread-cell → cnidoblast	—
9088	threadworms → nematods	—
9089	threat posture, ~ position	Drohstellung *f*, Drohhaltung *f*
9090	three-point test cross	Dreipunktversuch *m*
9091	threonine	Threonin *n*
9092	threshold of pain	Schmerzschwelle *f*
9093	threshold reduction, ~ lowering	Schwellenerniedrigung *f*
9094	threshold stimulus → liminal ~	—
9095	threshold substance	Schwellensubstanz *f*, Schwellenstoff *m*
9096	threshold value	Schwellenwert *m*
9097	thrombin	Thrombin *n*
9098	thrombocyte, blood platelet	Blutplättchen *n*, Thrombozyt *m*
9099	thrombogen, prothrombin	Thrombogen *n*, Prothrombin *n*
9100	thrombokinase	Thrombokinase *f*
9101	throw-back	Rückschlag *m*
9102	throwing-back → atavism	—
9103	thylose → tylose	—
9104	thymidine	Thymidin *n*
9105	thymidylic acid	Thymidylsäure *f*
9106	thymine	Thymin *n*
9107	thymonucleic acid	Thymonukleinsäure *f*
9108	thymus (gland)	Thymus *m*, Thymusdrüse *f*
9109	thyroid cartilage	Schildknorpel *m*
9110	thyroid (gland)	Schilddrüse *f*
9111	thyroid hormone	Schilddrüsenhormon *n*
9112	thyrotrop(h)ic hormone, thyrotrophin, thyroid-stimulating hormone, TSH	thyreotropes Hormon *n*, Thyreotropin *n*, TSH
9113	thyroxin(e)	Thyroxin *n*
9114	thyrsus, thyrse	Thyrsus *m*
9115	tibia, shin-bone	Schienbein *n*
9116	tidal volume	Atemvolumen *n*
9117	tigellum	Keimstengel *m*
9118	to tiller	bestocken
9119	tiller	Bestockungstrieb *m*
9120	tillering	Bestockung *f*
9121	tillering node	Bestockungsknoten *m*
9122	timberline → tree limit	—
9123	time factor	Zeitfaktor *m*
9124	time-sense	Zeitsinn *m*
9125	timer	Zeitgeber *m*
9126	tip *bot.* → apex	—
9127	tissue	Gewebe *n*

9069	thérophyte *m*	terófito *m*
9070	thiamine *f*, aneurine *f*, vitamine *f* antinévritique, vitamine *f* B_1	tiamina *f*, aneurina *f*, vitamina *f* antineurítica, vitamina *f* B_1
9071	épaissir	espesar
9072	épaississement *m*	engrosamiento *m*
9073	—	—
9074	thigmomorphose *f*	tigmomorfosis *f*
9075	thigmotaxie *f*, thigmotactisme *m*	tigmotaxis *f*, tigmotaxia *f*
9076	thigmotropisme *m*	tigmotropismo *m*
9077	chromatographie *f* en couches minces	cromatografía *f* en capa fina
9078	section *f* fine, coupe *f* mince	corte *m* delgado, preparación *f* delgada
9079	thiobactéries *fpl*, bactéries *fpl* du soufre, sulfobactéries *fpl*	tiobacterias *fpl*, bacterias *fpl* de azufre
9080	acide *m* thioctique, ~ liponique	ácido *m* tióctico, ~ tioctánico, ~ lipónico
9081	thionine *f*	tionina *f*
9082	cavité *f* thoracique, ~ pleurale	cavidad *f* torácica, ~ pleural
9083	canal *m* thoracique	conducto *m* torácico
9084	vertèbre *f* thoracique, ~ dorsale	vértebra *f* torácica
9085	thorax *m*, cage *f* thoracique	tórax *m*, caja *f* torácica
9086	épine *f*, piquant *m*	espina *f*
9087	—	—
9088	—	—
9089	posture *f* menaçante, ~ de menace, position *f* ~ ~, attitude *f* menaçante	postura *f* amenazadora, ~ de amenaza
9090	test-cross *m* à trois points	cruza *f* probadora de tres puntos
9091	thréonine *f*	treonina *f*
9092	seuil *m* de douleur	umbral *m* de(l) dolor, dintel *m* doloroso
9093	abaissement *m* du seuil	reducción *f* del umbral, descenso *m* ~ ~, ~ liminal
9094	—	—
9095	substance *f* à seuil	sustancia *f* liminal
9096	valeur *f* de seuil, ~ liminaire	valor *m* umbral, ~ liminal
9097	thrombine *f*	trombina *f*
9098	thrombocyte *m*, plaquette *f* sanguine	trombocito *m*, plaqueta *f* sanguínea
9099	thrombogène *m*, prothrombine *f*	trombógeno *m*, protrombina *f*
9100	thrombokinase *f*	tromboquinasa *f*
9101	individu *m* atavique; caractère *m* ~	carácter *m* atávico
9102	—	—
9103	—	—
9104	thymidine *f*	timidina *f*
9105	acide *m* thymidylique	ácido *m* timidílico
9106	thymine *f*	timina *f*
9107	acide *m* thymonucléique	ácido *m* timonucleico
9108	thymus *m*	timo *m*, glándula *f* tímica
9109	cartilage *m* thyroïde	cartílago *m* tiroides
9110	(glande *f*) thyroïde *f*	(glándula *f*) tiroides *f*
9111	hormone *f* thyroïdienne	hormona *f* tiroidea
9112	hormone *f* thyréotrope, thyréostimuline *f*, TSH	hormona *f* tirotrópica, tirotrofina *f*, TSH
9113	thyroxine *f*	tiroxina *f*
9114	thyrse *m*	tirso *m*
9115	tibia *m*	tibia *f*, espinilla *f*
9116	volume *m* courant	volumen *m* respiratorio, ~ de ventilación pulmonar
9117	tigelle *f*	tallo *m* embrional
9118	taller	(a)macollar, ahijar
9119	talle *f*	hijuelo *m*
9120	tallage *m*	macollamiento *m*, amacollado *m*, ahijamiento *m*
9121	nœud *m* de tallage	nudo *m* de macollamiento
9122	—	—
9123	facteur *m* temps	factor *m* tiempo
9124	sens *m* du temps	sentido *m* del tiempo
9125	indicateur *m* de temps	indicador *m* del tiempo
9126	—	—
9127	tissu *m*	tejido *m*

No.	English	German
9128	tissue culture	Gewebekultur *f*, Gewebezüchtung *f*
9129	tissue differentiation, histodifferentiation	Gewebedifferenzierung *f*
9130	tissue disintegration	Gewebezerfall *m*
9131	tissue extract	Gewebeextrakt *m*
9132	tissue fluid, intercellular ~, interstitial ~	Gewebeflüssigkeit *f*
9133	tissue hormone	Gewebehormon *n*
9134	tissue preparation	Gewebepräparat *n*
9135	tissue protein	Gewebeprotein *n*
9136	tissue section, histological ~	Gewebeschnitt *m*
9137	tissue specifity	Gewebespezifität *f*
9138	tissular, histic	Gewebe...
9139	to titrate	titrieren
9140	titration	Titration *f*, Titrierung *f*
9141	titration curve	Titrationskurve *f*
9142	titre; titer (US)	Titer *m*
9143	titrimetry	Titrimetrie *f*
9144	TNP → nicotinamide adenine dinucleotide phosphate	—
9145	tocopherol, antisterility vitamin, vitamin E	Tokopherol *n*, Antisterilitätsvitamin *n*, Fruchtbarkeitsvitamin *n*, Vitamin *n* E
9146	tolerance limit	Toleranzgrenze *f*
9147	tolerance value	Rückstandshöchstmenge *f*
9148	tonoplast, internal plasma membrane	Tonoplast *m*, innere Plasmahaut *f*
9149	tonus, tonicity, tone	Tonus *m*
9150	tool using *eth.*	Werkzeuggebrauch *m eth.*
9151	tooth root	Zahnwurzel *f*
9152	tooth socket → dental alveolus	—
9153	toothed → dentate	—
9154	top-cross	Top-Cross *n*
9155	top-crossing	Top-Crossing *n*
9156	topographical isolation	topographische Isolation *f*
9157	topotaxis	Topotaxie *f*
9158	torus *bot.*	Torus *m bot.*
9159	total cleavage → complete ~	—
9160	total metabolism	Gesamtumsatz *m*, Gesamtstoffwechsel *m*
9161	total nitrogen	Gesamtstickstoff *m*
9162	total respiration	Gesamtveratmung *f*
9163	totipotency, omnipotency	Totipotenz *f*, Omnipotenz *f*
9164	totipotent(ial)	totipotent, omnipotent
9165	touch organ, ~ stimulus → tactile organ, ~ stimulus	—
9166	touch receptor → tangoreceptor	—
9167	toxic, poisonous, venomous	giftig, toxisch
9168	toxic effect, poisonous ~	Giftwirkung *f*
9169	toxic gas, poison ~	Giftgas *n*
9170	toxicity, venomousness	Giftigkeit *f*, Toxizität *f*
9171	toxicologist	Toxikologe *m*
9172	toxicology	Toxikologie *f*
9173	toxigenic, toxicogenic	toxigen, toxogen
9174	toxin, toxic substance, ~ agent, poisonous substance	Giftstoff *m*, Toxin *n*
9175	trabant, satellite	Trabant *m*, Satellit *m*
9176	trabecula	Trabekel *f*
9177	trace element, minor ~, micro-element	Spurenelement *n*
9178	tracer	Tracer *m*
9179	tracer method	Tracer-Methode *f*
9180	trachea[1], windpipe	Luftröhre *f*, Trachea *f*
9181	trachea[2] *zool.*	Trachee *f zool.*
9182	trachea[3] *bot.*, wood vessel	Trachee *f bot.*
9183	tracheal gill	Tracheenkieme *f*

9128	culture *f* de tissu, ~ tissulaire	cultivo *m* de tejido
9129	différenciation *f* des tissus, ~ tissulaire	diferenciación *f* de tejidos, ~ hística
9130	désintégration *f* des tissus, ~ tissulaire	desintegración *f* de los tejidos
9131	extrait *m* tissulaire	extracto *m* de tejido
9132	liquide *m* interstitiel	líquido *m* tisular, ~ hístico, ~ intersticial, ~ intercelular
9133	hormone *f* du tissu	hormona *f* tisular
9134	préparation *f* histologique	preparación *f* de tejido, ~ histológica
9135	protéine *f* tissulaire	proteína *f* hística
9136	coupe *f* de tissu, ~ histologique	corte *m* de tejido, ~ histológico, sección *f* hística
9137	spécifité *f* tissulaire	especifidad *f* de los tejidos
9138	tissulaire	tisular, hístico
9139	titrer	titular
9140	titrage *m*, titration *f*	titulación *f*
9141	courbe *f* de titrage	curva *f* de titulación
9142	titre *m*	título *m*
9143	titrimétrie *f*	análisis *m* por titulación
9144	—	—
9145	tocophérol *m*, vitamine *f* d'antistérilité, ~ de fertilité, ~ E	tocoferol *m*, vitamina *f* antiestéril, ~ de la fecundidad, ~ E
9146	limite *f* de tolérance	límite *m* de tolerancia
9147	teneur *f* en résidus tolérée, ~ résiduelle maximale	tolerancia *f* residual
9148	tonoplaste *m*	tonoplasto *m*, membrana *f* plasmática interna
9149	tonus *m*, tonicité *f*	tono *m*, tonicidad *f*
9150	usage *m* de l'outil *eth.*	utilización *f* de utensilios *eth.*
9151	racine *f* dentaire	raíz *f* dentaria
9152	—	—
9153	—	—
9154	top-cross *m*	cruzas *f pl* radiales
9155	top-crossing *m*	cruzamiento *m* radial
9156	isolement *m* topographique	aislamiento *m* topográfico
9157	topotaxie *f*, topotactisme *m*	topotaxia *f*, topotactismo *m*
9158	torus *m* *bot.*	toro *m* *bot.*
9159	—	—
9160	métabolisme *m* total	metabolismo *m* total
9161	azote *m* total	nitrógeno *m* total
9162	respiration *f* totale	respiración *f* total
9163	totipotence *f*, omnipuissance *f*	totipotencia *f*
9164	totipotent, omnipotent	totipotente
9165	—	—
9166	—	—
9167	toxique; vénéneux (plantes), venimeux (animaux)	tóxico, venenoso
9168	effet *m* toxique	efecto *m* tóxico
9169	gaz *m* toxique	gas *m* tóxico, ~ venenoso
9170	toxicité *f*, venimosité *f*	toxicidad *f*, venenosidad *f*
9171	toxicologue *m*	toxicólogo *m*
9172	toxicologie *f*	toxicología *f*
9173	toxigène	toxígeno, toxicógeno
9174	toxine *f*, substance *f* toxique	toxina *f*, sustancia *f* tóxica, ~ venenosa
9175	trabant *m*, satellite *m*	trabante *m*, satélite *m*
9176	trabécule *f*	trabécula *f*
9177	élément-trace *m*, oligo-élément *m*, micro-élément *m*, élément *m* mineur	elemento *m* traza, oligoelemento *m*, microelemento *m*
9178	traceur *m*, indicateur *m*	trazador *m*, marcador *m*, sustancia *f* trazadora
9179	méthode *f* des traceurs	método *m* de los trazadores
9180	trachée *f*, trachée-artère *f*	tráquea *f*
9181	trachée *f* *zool.*	tráquea *f* *zool.*
9182	trachée *f*, vaisseau *m* de bois *bot.*	tráquea *f* *bot.*
9183	branchie *f* trachéenne, trachéo-branchie *f*	branquia *f* traqueal

9184	tracheal respiration	Tracheenatmung *f*
9185	tracheal system	Tracheensystem *n*
9186	tracheal trunk	Tracheenstamm *m*
9187	Tracheata	Tracheaten *pl*, Antennaten *pl*, Röhrenatmer *mpl*
9188	tracheid	Tracheide *f*
9189	tracheole	Tracheole *f*
9190	tracheophyte	Tracheophyt *m*
9191	training	Dressur *f*
9192	training experiment	Dressurversuch *m*
9193	transaminase	Transaminase *f*
9194	transamination	Transaminierung *f*
9195	trans-configuration, trans-arrangement	Trans-Konfiguration *f*
9196	transcription *gen.*	Transkription *f gen.*
9197	transduction	Transduktion *f*
9198	transduction clone	Transduktionsklon *m*
9199	transfer of genetic material	Übertragung *f* des genetischen Materials
9200	transfer of information	Informationsübertragung *f*
9201	transfer RNA, t-RNA, soluble RNA	Transfer-RNS *f*, t-RNS, lösliche RNS *f*
9202	transferability → transmissibility	—
9203	transferase	Transferase *f*
9204	transformation *gen.*	Transformation *f gen.*
9205	transfusion tissue	Transfusionsgewebe *n*
9206	transgenation → gene mutation	—
9207	transgression, transgressive segregation	Transgression *f*
9208	transitional epithelium	Übergangsepithel *n*
9209	transitional moor, transition peat	Übergangsmoor *n*
9210	transition(al) stage, transitory ~	Übergangsstadium *n*
9211	translation *gen.*	Translation *f gen.*
9212	translocation	Translokation *f*
9213	translocation heterozygosis	Translokationsheterozygotie *f*
9214	translucent	lichtdurchlässig
9215	transmethylation	Transmethylierung *f*
9216	transmissibility, transferability	Übertragbarkeit *f*
9217	transmissible, transferable	übertragbar
9218	to transmit to	vererben
9219	transmitter, ~ substance, neurotransmitter, chemical mediator	Überträgersubstanz *f*, Mittlersubstanz *f*, Transmitter *m*, Neurotransmitter *m*, chemischer Mittler *m*
9220	transmutation	Transmutation *f*
9221	transphosphorylation	Transphosphorylierung *f*
9222	transpiration	Transpiration *f*
9223	transpiration rate	Transpirationsrate *f*, Transpirationsintensität *f*
9224	transplant, graft	Transplantat *n*
9225	to transplant, to graft	transplantieren, überpflanzen, verpflanzen
9226	transplantation, grafting	Transplantation *f*, Überpflanzung *f*, Verpflanzung *f*
9227	transport of substances	Stofftransport *m*
9228	transspecific	transspezifisch
9229	transudate	Transsudat *n*
9230	transudation	Transsudation *f*
9231	transverse process → diapophysis	—
9232	transverse section → cross section	—
9233	transverse septum → cross wall	—
9234	transverse striation, cross ~	Querstreifung *f*
9235	traumatin, traumatinic acid	Traumatin *n*, Traumatinsäure *f*
9236	traumatropism	Traumatotropismus *m*
9237	tree crown	Baumkrone *f*
9238	tree-dwelling → arboreal	—
9239	tree-fern	Baumfarn *m*
9240	tree limit; timberline (US)	Baumgrenze *f*

9184	respiration *f* trachéenne	respiración *f* traqueal
9185	système *m* trachéen	sistema *m* traqueal
9186	tronc *m* trachéen	tronco *m* traqueal
9187	trachéates *mpl*, antennates *mpl*	traqueados *mpl*, antenados *mpl*
9188	trachéide *f*	traqueida *f*
9189	trachéole *f*	traquéolo *m*, traqueola *f*
9190	trachéophyte *m*	traqueofito *m*
9191	dressage *m*	adiestramiento *m*, amaestramiento *m*
9192	expérience *f* de dressage	experiencia *f* de adiestramiento, ensayo *m* ~ ~
9193	transaminase *f*	transaminasa *f*
9194	transamination *f*	transaminación *f*
9195	configuration *f* trans	disposición *f* trans, transconfiguración *f*
9196	transcription *f gen.*	transcripción *f gen.*
9197	transduction *f*	tra(n)sducción *f*
9198	clone *m* de transduction	clon *m* de tra(n)sducción
9199	transmission *f* du matériel génique	transmisión *f* genética, transferencia *f* del material genético
9200	transmission *f* de l'information, transfert *m* ~ ~	transmisión *f* de información, transferencia *f* ~ ~
9201	ARN *m* de transfert, transfert-ARN *m*, t-ARN *m*, ARN *m* soluble	ARN *m* de transferencia, ARN *m* soluble
9202	—	—
9203	transférase *f*	transferasa *f*
9204	transformation *f gen.*	transformación *f gen.*
9205	tissu *m* de transfusion	tejido *m* de transfusión
9206	—	—
9207	transgression *f*, ségrégation *f* transgressive	transgresión *f*, segregación *f* transgresiva
9208	épithélium *m* de passage	epitelio *m* de transición
9209	tourbière *f* de transition	turbera *f* de transición
9210	stade *m* transitoire	estado *m* de transición
9211	translation *f gen.*	tra(n)slación *f gen.*
9212	translocation *f*	tra(n)slocación *f*
9213	hétérozygotie *f* de translocation	heterocigosis *f* de tra(n)slocación
9214	translucide	translúcido
9215	transméthylation *f*	transmetilación *f*
9216	transmissibilité *f*, transférabilité *f*	transferibilidad *f*
9217	transmissible, transférable	transmisible, transferible
9218	transmettre à	transmitir (hereditariamente) a
9219	substance *f* médiatrice, neuromédiateur *m*, neurotransmetteur *m*, médiateur *m* chimique	sustancia *f* transmisora, neurotransmisor *m*, transmisor *m* químico, intermediario *m* ~
9220	transmutation *f*	transmutación *f*
9221	transphosphorylation *f*	transfosforilación *f*
9222	transpiration *f*	transpiración *f*
9223	taux *m* de transpiration	intensidad *f* de transpiración, ~ transpiratoria
9224	transplant *m*, greffe *f*, greffon *m*	injerto *m*
9225	transplanter	trasplantar, injertar
9226	transplantation *f*, greffage *m*	trasplante *m*, injerto *m*
9227	transport *m* des substances	transporte *m* de sustancias
9228	transspécifique	transespecífico
9229	transsudat *m*	tra(n)sudado *m*
9230	transsudation *f*	tra(n)sudación *f*
9231	—	—
9232	—	—
9233	—	—
9234	striation *f* transversale	estriación *f* transversal, estriado *m* ~
9235	traumatine *f*, acide *m* traumati(ni)que	traumatina *f*
9236	trauma(to)tropisme *m*	trauma(to)tropismo *m*
9237	cime *f* d'arbre	copa *f* del árbol
9238	—	—
9239	fougère *f* arborescente	helecho *m* arborescente
9240	limite *f* d'arbres, ~ des arbres	límite *m* del arbolado, ~ de la vegetación arbórea

9241	tree ring → annual ring	—
9242	tree stratum, ~ layer	Baumschicht *f*
9243	trematodes	Trematoden *pl*, Saugwürmer *mpl*
9244	trephone	Trephon *n*
9245	triad	Triade *f*
9246	trial-and-error learning	Versuch- und Irrtum-Methode *f*
9247	Triassic (period)	Trias *f*
9248	tribe *tax.*	Tribus *f* *tax.*
9249	tricarboxylic cycle → Krebs cycle	—
9250	trichocyst	Trichozyste *f*
9251	trichogyne	Trichogyne *f*
9252	trichome, plant hair	Trichom *n*, Pflanzenhaar *n*
9253	trigeminal nerve	Trigeminus *m*, Drillingsnerv *m*
9254	trihybrid	trihybrid
9255	trihybridism	Trihybridie *f*
9256	trimerical	trimer
9257	trimery	Trimerie *f*
9258	trimorphism	Trimorphismus *m*, Dreigestaltigkeit *f*
9259	trioecious, trioikous	triözisch, dreihäusig
9260	triose phosphate	Triosephosphat *n*
9261	tripeptide	Tripeptid *n*
9262	triphosphopyridine nucleotide → nicotinamide adenine dinucleotide phosphate	—
9263	triple cross	Dreifachkreuzung *f*, Dreilinienkreuzung *f*
9264	triplet	Triplett *n*
9265	triplet code	Triplett-Kode *m*
9266	triplex	triplex
9267	triploidy	Triploidie *f*
9268	tripton	Tripton *n*
9269	trisomic	trisom
9270	trisomy	Trisomie *f*
9271	trivalent	Trivalent *n*
9272	t-RNA → transfer RNA	—
9273	trochlear nerve	Trochlearis *m*, Augenrollnerv *m*
9274	troglobionts → cave fauna	—
9275	trophallaxis	Trophophyllaxe *f*, Nahrungsaustausch *m*
9276	trophic	trophisch
9277	trophic level	trophische Ebene *f*
9278	trophobiosis	Trophobiose *f*, Ernährungssymbiose *f*
9279	trophoblast	Trophoblast *m*
9280	trophocyte → nutritive cell	—
9281	trophology	Ernährungswissenschaft *f*
9282	trophotaxis	Trophotaxis *f*
9283	trophotropism	Trophotropismus *m*
9284	tropic curvature	tropistische Krümmung *f*
9285	tropical rain forest	tropischer Regenwald *m*
9286	tropism	Tropismus *m*
9287	tropophyte	Tropophyt *m*
9288	tropotaxis	Tropotaxis *f*
9289	true fungi, Eumycophyta	echte Pilze *mpl*, höhere ~, Eumyzeten *mpl*
9290	true rib, sternal ~	echte Rippe *f*
9291	trunk[1]	Stamm *m*, Baumstamm *m*
9292	trunk[2] → proboscis	—
9293	trypsin	Trypsin *n*
9294	trypsinogen	Trypsinogen *n*
9295	tryptone	Trypton *n*
9296	tryptophan(e)	Tryptophan *n*
9297	TSH → thyrotrophic hormone	—
9298	tube-foot → ambulacrum	—
9299	tuber	Knolle *f*
9300	tuber formation	Knollenbildung *f*
9301	tuberous → bulbous	—
9302	tuberous plant	Knollenpflanze *f*, Knollengewächs *n*

9241	—	—
9242	strate *f* arborescente	estrato *m* arbóreo
9243	trématodes *mpl*	trematodos *mpl*
9244	tréphone *f*	trefona *f*
9245	triade *f*	triada *f*
9246	méthode *f* des essais et erreurs, apprentissage *m* par essais et erreurs	método *m* de ensayo y error
9247	trias *m*	triásico *m*
9248	tribu *f* *tax.*	tribu *f* *tax.*
9249	—	—
9250	trichocyste *m*	tricocisto *m*
9251	trichogyne *m*	tricógina *f*, tricógino *m*
9252	trichome *m*	tricoma *m*
9253	nerf *m* trijumeau	nervio *m* trigémino
9254	trihybride	trihíbrido
9255	trihybridisme *m*	trihibridismo *m*
9256	trimérique	trimérico
9257	trimérie *f*	trimería *f*
9258	trimorphisme *m*	trimorfismo *m*
9259	trioïque	trióico
9260	triose-phosphate *m*	triosa *f* fosfato
9261	tripeptide *m*	tripéptido *m*
9262	—	—
9263	croisement *m* triple	cruzamiento *m* triple, ~ de tres líneas
9264	triplet *m*	triplete *m*
9265	code *m* par triplets	código *m* de tripletes
9266	triplex	triplexo
9267	triploïdie *f*	triploidia *f*
9268	tripton *m*	tripton *m*
9269	trisomique	trisómico
9270	trisomie *f*	trisomia *f*
9271	trivalent *m*	trivalente *m*
9272	—	—
9273	nerf *m* trochléen, ~ pathétique	nervio *m* troclear, ~ patético
9274	—	—
9275	trophallaxie *f*, échange *m* de nourriture	trofalaxis *f*, trofofilaxia *f*, intercambio *m* alimenticio
9276	trophique	trófico
9277	niveau *m* trophique	nivel *m* trófico
9278	trophobiose *f*, symbiose *f* nutritionnelle	simbiosis *f* nutritiva
9279	trophoblaste *m*	trofoblasto *m*
9280	—	—
9281	trophologie *f*	trofología *f*
9282	trophotaxie *f*, trophotactisme *m*	trofotaxia *f*, trofotaxis *f*
9283	trophotropisme *m*	trofotropismo *m*
9284	courbure *f* tropique, incurvation *f* ~	curvatura *f* trópica, encorvadura *f* ~
9285	forêt *f* ombrophile tropicale	bosque *m* de lluvia tropical, ~ lluvioso ~
9286	tropisme *m*	tropismo *m*
9287	tropophyte *m*	tropófito *m*
9288	tropotaxie *f*, tropotactisme *m*	tropotaxia *f*, tropotaxis *f*
9289	eumycètes *mpl*	eumicetos *mpl*, hongos *mpl* verdaderos, ~ superiores
9290	côte *f* vraie, ~ sternale	costilla *f* verdadera, ~ esternal
9291	tronc *m*	tronco *m*
9292	—	—
9293	trypsine *f*, tryptase *f*	tripsina *f*
9294	trypsinogène *m*	tripsinógeno *m*
9295	peptone *f* trypsique	triptona *f*
9296	tryptophane *m*	triptófano *m*
9297	—	—
9298	—	—
9299	tubercule *m*	tubérculo *m*
9300	tubérisation *f*, tubération *f*	tuberización *f*
9301	—	—
9302	plante *f* tubéreuse, ~ à tubercule	planta *f* tuberosa

9303	tubiform floret, tubular ~	Röhrenblüte *f*
9304	tubular, tubulate, tubiform	röhrenförmig
9305	tubular trachea	Röhrentrachee *f*
9306	tundra	Tundra *f*
9307	Tunicata, tunicates, Urochordata, urochordates	Manteltiere *npl*, Tunikaten *pl*
9308	Turbellaria, turbellarians, eddyworms	Strudelwürmer *mpl*
9309	turgescent, turgid	turgeszent
9310	turgor, turgescence, turgidity	Turgor *m*, Turgeszenz *f*
9311	turgor movement	Turgorbewegung *f*
9312	turgor pressure	Turgordruck *m*
9313	turio(n)	Turione *f*
9314	turnover number	Wechselzahl *f*
9315	turnover rate	Umsatzrate *f*, „turnover"-Rate *f*
9316	tusk	Stoßzahn *m*, Hauer *m*
9317	twig	Zweig *m*
9318	twin genes → pseudoalleles	—
9319	twin species, geminate ~	Zwillingsarten *fpl*
9320	twining plant → volubilate plant	—
9321	twinning, twin pregnancy	Zwillingsträchtigkeit *f*
9322	two-point test cross	Zweipunktversuch *m*
9323	tylose, tylosis, thylose	Thylle *f*
9324	tympanal organ	Tympanalorgan *n*
9325	tympanic cavity	Paukenhöhle *f*
9326	tympanum, ear drum	Trommelfell *n*
9327	type	Typ(us) *m*, Standard *m*
9328	type specimen → holotype	—
9329	typogenesis	Typogenese *f*
9330	typolysis	Typolyse *f*
9331	typostasis	Typostase *f*
9332	tyrosine	Tyrosin *n*
9333	tyrothricin	Tyrothrizin *n*
9334	ubiquinone	Ubichinon *n*
9335	ubiquist	Ubiquist *m*
9336	ulna → cubitus	—
9337	ultracentrifugation	Ultrazentrifugation *f*
9338	ultracentrifuge	Ultrazentrifuge *f*
9339	ultrafilter	Ultrafilter *m*
9340	ultrafiltration	Ultrafiltrierung *f*
9341	ultramicroscope	Ultramikroskop *n*
9342	ultramicroscopic(al)	ultramikroskopisch
9343	ultramicrosome	Ultramikrosom *n*
9344	ultramicrotome	Ultramikrotom *n*
9345	ultrasonics, ultrasonic sound, supersonics	Ultraschall *m*, Überschall *m*
9346	ultrastructure	Ultrastruktur *f*
9347	ultrathin section	Ultradünnschnitt *m*
9348	ultraviolet absorption	UV-Absorption *f*
9349	ultraviolet irradiation, UV irradiation	Ultraviolettbestrahlung *f*, UV-Bestrahlung *f*
9350	ultraviolet light, UV light	ultraviolettes Licht *n*, UV-Licht *n*
9351	ultraviolet microscopy	Ultraviolettmikroskopie *f*, UV-Mikroskopie *f*
9352	ultraviolet radiation	Ultraviolettstrahlung *f*, UV-Strahlung *f*
9353	ultraviolet rays	ultraviolette Strahlen *mpl*, UV-Strahlen *mpl*
9354	umbel	Dolde *f*
9355	umbellate	doldig
9356	Umbelliferae, umbelliferous plants	Doldengewächse *npl*, Umbelliferen *fpl*
9357	umbelliferous	doldentragend
9358	umbelliform	doldenförmig
9359	umbilical, omphalic	Nabel...
9360	umbilical circulation	Nabelschnurkreislauf *m*, Umbilikalkreislauf *m*
9361	umbilical cord, navel ~	Nabelschnur *f*, Nabelstrang *m*
9362	umbilicus → navel	—

9303	fleur *f* tubuleuse, ~ tubulée	flor *f* tubulosa, ~ tubiforme
9304	tubulaire, tubuliforme	tubular, tubiforme
9305	trachée *f* tubulaire	tráquea *f* tubular
9306	toundra *f*	tundra *f*
9307	tuniciers *mpl*, urocordés *mpl*	tunicados *mpl*, urocordados *mpl*
9308	turbellariés *mpl*	turbelarios *mpl*
9309	turgescent	turgescente, turgente, túrgido
9310	turgescence *f*	turgencia *f*
9311	mouvement *m* de turgescence	movimiento *m* turgente, ~ de turgencia
9312	pression *f* de turgescence	presión *f* de turgencia, ~ túrgida
9313	turion *m*	turión *m*
9314	nombre *m* de renouvellement, ~ ~ rotation	número *m* de cambio
9315	taux *m* de renouvellement, vitesse *f* de recyclage	velocidad *f* de retorno, ~ ~ recambio
9316	défense *f*	defensa *f*, colmillo *m*
9317	rameau *m*	ramo *m*
9318	—	—
9319	espèces *fpl* jumelles	especies *fpl* gemelas
9320	—	—
9321	gémellité *f*, gestation *f* gémellaire	gemelaridad *f*, gestación *f* gemelar
9322	test-cross *m* à deux points	cruza *f* probadora de dos puntos
9323	thylle *m*, t(h)yllose *f*	tílide *f*, tilo *m*, tilis *f*
9324	organe *m* tympanal, ~ tympanique	órgano *m* timpánico
9325	caisse *f* du tympan	caja *f* del tímpano
9326	tympan *m*, membrane *f* tympanique	tímpano *m*, membrana *f* timpánica
9327	type *m*	tipo *m*
9328	—	—
9329	typogenèse *f*	tipogénesis *f*
9330	typolyse *f*	tipólisis *f*
9331	typostase *f*	tipóstasis *f*
9332	tyrosine *f*	tirosina *f*
9333	tyrothricine *f*	tirotricina *f*
9334	ubiquinone *f*	ubiquinona *f*
9335	ubiquiste *m*	ubicuo *m*
9336	—	—
9337	ultracentrifugation *f*	ultracentrifugación *f*
9338	ultracentrifugeuse *f*	ultracentrífuga *f*
9339	ultrafiltre *m*	ultrafiltro *m*
9340	ultrafiltration *f*	ultrafiltración *f*
9341	ultramicroscope *m*	ultramicroscopio *m*
9342	ultramicroscopique	ultramicroscópico
9343	ultramicrosome *m*	ultramicrosoma *m*
9344	ultramicrotome *m*	ultramicrótomo *m*
9345	ultrason(s) *m* (*pl*)	ultrasonido *m*, sonido *m* silencioso
9346	ultrastructure *f*	ultraestructura *f*
9347	section *f* ultrafine, coupe *f* ~	sección *f* ultrafina
9348	absorption *f* de l'ultraviolet, ~ en UV	absorción *f* ultravioleta
9349	irradiation *f* aux UV	irradiación *f* ultravioleta, ~ UV
9350	lumière *f* ultraviolette, ultraviolet *m*	luz *f* ultravioleta, UV *m*
9351	microscopie *f* ultraviolette, ~ à UV	microscopia *f* de luz ultravioleta
9352	rayonnement *m* ultraviolet	radiación *f* ultravioleta, ~ UV
9353	rayons *mpl* ultraviolets	rayos *mpl* ultraviolados, ~ ultravioleta
9354	ombelle *f*	umbela *f*
9355	ombellé	umbelado
9356	ombellifères *fpl*	umbelíferas *fpl*
9357	ombellifère	umbelífero
9358	ombelliforme	umbeliforme
9359	ombilical	umbilical
9360	circulation *f* ombilicale	circulación *f* umbilical
9361	cordon *m* ombilical, funicule *m* ~	cordón *m* umbilical
9362	—	—

9363	uncinate, hooked	hakenförmig
9364	unconditioned reflex	unbedingter Reflex *m*
9365	undergrowth, underwood, understory	Unterwuchs *m*
9366	under-shrub → suffrutex	—
9367	undulating membrane	undulierende Membran *f*
9368	undulating movement	Schlängelbewegung *f*
9369	unfertilized	unbefruchtet
9370	Ungulata, ungulates, hoofed animals	Huftiere *n pl*
9371	unguligrade	Zehenspitzengänger *m*
9372	uniaxial → monaxial	—
9373	unicellular, monocellular, one-celled, single-celled	einzellig
9374	unicellular culture	Einzellkultur *f*
9375	unicellular organism, unicell	Einzeller *m*
9376	uniflagellate, unimastigote, monomastigote	eingeißelig
9377	uniflorous	einblütig
9378	unifoliate → monophyllous	—
9379	unigerminal, monogerminal	einkeimig
9380	unilateral inheritance	einseitige Vererbung *f*, unilaterale ~
9381	unilocular, monolocular	einfächerig, unilokulär
9382	uninucleate → mononuclear	—
9383	uniovular → monozygotic	—
9384	unipolar, monopolar	einpolig, unipolar, monopolar
9385	uniqueness of the individual	Einmaligkeit *f* des Individuums
9386	unisexual	eingeschlechtig, unisexuell
9387	unisexuality	Eingeschlechtigkeit *f*
9388	unistrate, unistratose	einschichtig
9389	unit membrane	Einheitsmembran *f*, Elementarmembran *f*
9390	univalent	Univalent *n*
9391	univalve	einschalig
9392	universality of the genetic code	Universalität *f* des genetischen Kodes
9393	unmyelinated → amyelinated	—
9394	unsaturated	ungesättigt
9395	uperization	Uperisation *f*, Ultrapasteurisation *f*
9396	upper eyelid	Oberlid *n*
9397	upper jaw → maxilla	—
9398	upper jaw bone → superior maxillary bone	—
9399	upper lip	Oberlippe *f*
9400	upper palea, superior ~	Vorspelze *f*
9401	uracil	Urazil *n*
9402	urbanization	Urbanisation *f*, Verstädterung *f*
9403	urea	Harnstoff *m*
9404	urea cycle → ornithine cycle	—
9405	urease	Urease *f*
9406	uredospore, summer-spore	Uredospore *f*, Sommerspore *f*
9407	ureter	Harnleiter *m*, Ureter *m*
9408	urethra	Harnröhre *f*, Urethra *f*
9409	uric acid	Harnsäure *f*
9410	uridylic acid	Uridylsäure *f*
9411	urinary bladder	Harnblase *f*
9412	urinary tract	Harnwege *m pl*
9413	urine	Harn *m*, Urin *m*
9414	uriniferous tubule	Harnkanälchen *n*
9415	urobilin	Urobilin *n*
9416	Urochordata → Tunicata	—
9417	urogenital system, genitourinary ~	Urogenitalsystem *n*
9418	uronic acid	Uronsäure *f*
9419	uropyge, uropygium	Bürzel *m*, Stert *m*, Sterz *m*
9420	uropygial gland, preen ~, oil ~	Bürzeldrüse *f*
9421	useful animal	Nützling *m*, Nutztier *n*
9422	useful insect, beneficial ~	Nutzinsekt *n*
9423	useful plant	Nutzpflanze *f*
9424	uterine cavity	Uterushöhle *f*
9425	uterine tube → Fallopian tube	—

9363	unciforme, uncinulé	unciforme
9364	réflexe *m* inconditionné, ~ non conditionné	reflejo *m* incondicionado
9365	sous-bois *m*, sous-étage *m*	sotobosque *m*
9366	—	—
9367	membrane *f* ondulante	membrana *f* ondulante
9368	mouvement *m* ondulatoire	movimiento *m* ondulante
9369	non fertilisé	no fertilizado
9370	ongulés *m pl*	ungulados *m pl*
9371	onguligrade *m*	ungulígrado *m*
9372	—	—
9373	unicellulaire, monocellulaire	unicelular, monocelular
9374	culture *f* unicellulaire	cultivo *m* unicelular
9375	organisme *m* unicellulaire	organismo *m* unicelular
9376	uniflagellé, monoflagellé, monomastigote	uniflagelado, monomastigoto
9377	uniflore, à fleur unique	unifloro
9378	—	—
9379	unigerminal, monogerminal, monogerme	unigerminal, monogerminal
9380	hérédité *f* unilatérale	herencia *f* unilateral
9381	uniloculaire, monoloculaire	unilocular, monolocular
9382	—	—
9393	—	—
9384	unipolaire	unipolar
9385	unicité *f* de l'individu	unicidad *f* del individuo
9386	unisexuel, unisexué, monosexuel	unisexual
9387	unisexualité *f*	unisexualidad *f*, unisexualismo *m*
9388	unistratifié, monostratifié	uni(e)stratificado, monoestratificado
9389	membrane *f* unité, ~ unitaire, ~ de base, ~ type	membrana *f* unidad, ~ unitaria, ~ elemental
9390	univalent *m*	univalente *m*
9391	univalve	univalvo
9392	universalité *f* du code génétique	universalidad *f* del código genético
9393	—	—
9394	non saturé	no saturado
9395	upérisation *f*	uperisación *f*
9396	paupière *f* supérieure	párpado *m* superior, pálpebra *f* ~
9397	—	—
9398	—	—
9399	lèvre *f* supérieure	labio *m* superior
9400	glumelle *f* supérieure	glumela *f* superior
9401	uracile *m*	uracilo *m*
9402	urbanisation *f*	urbanización *f*
9403	urée *f*	urea *f*
9404	—	—
9405	uréase *f*	ureasa *f*
9406	urédospore *f*, spore *f* d'été	uredóspora *f*, espora *f* de verano
9407	urètre *m*	uréter *m*
9408	uretère *m*	uretra *f*
9409	acide *m* urique	ácido *m* úrico
9410	acide *m* uridylique	ácido *m* uridílico
9411	vessie *f* urinaire	vejiga *f* urinaria
9412	voies *f pl* urinaires	vías *f pl* urinarias
9413	urine *f*	orina *f*
9414	tube *m* urinifère	túbulo *m* urinífero
9415	urobiline *f*	urobilina *f*
9416	—	—
9417	appareil *m* urogénital, ~ génito-urinaire	sistema *m* urogenital, aparato *m* genitourinario
9418	acide *m* uronique	ácido *m* urónico
9419	croupion *m*	uropigio *m*
9420	glande *f* uropygienne	glándula *f* uropigia
9421	animal *m* utile	animal *m* útil
9422	insecte *m* utile, auxiliaire *m*	insecto *m* útil, ~ benéfico, ~ beneficioso
9423	plante *f* utile	planta *f* útil
9424	cavité *f* utérine	cavidad *f* uterina
9425	—	—

9426	uterus, matrix	Uterus *m*, Gebärmutter *f*, Fruchthalter *m*
9427	utilization rate *ecol.*	Nutzungsrate *f ecol.*
9428	utilization time *neur.*	Nutzzeit *f neur.*
9429	UV... → ultraviolet...	—
9430	uvea	Traubenhaut *f*
9431	vacuolar membrane	Vakuolenmembran *f*
9432	vacuolar sap	Vakuolensaft *m*
9433	vacuole	Vakuole *f*
9434	vacuome, vacuolar system	Vakuom *n*
9435	vacuum	Vakuum *n*
9436	vacuum activity	Leerlaufhandlung *f*
9437	vagina	Scheide *f*, Vagina *f*
9438	vagus (nerve), pneumogastric ~	Vagus *m*
9439	valine	Valin *n*
9440	valvar, valval	valvat, klappig
9441	valve	Klappe *f*, Schalenklappe *f*
9442	variability	Variabilität *f*
9443	variable	Variable *f*
9444	variance	Varianz *f*, Streuung *f*
9445	variant	Variante *f*
9446	variation	Variation *f*
9447	variegate	panaschiert
9448	variegation	Variegation *f*, Panaschierung *f*, Scheckung *f*
9449	variety, strain; cultivar	Sorte *f*, Abart *f*, Spielart *f*, Varietät *f*; Cultivar *n*, Cultigen *n*
9450	Varolius' bridge → pons Varolii	—
9451	to vary	abarten
9452	vascular	vaskular, Gefäß...
9453	vascular bundle	Leitbündel *n*, Gefäßbündel *n*
9454	vascular cambium	Leitbündelkambium *n*
9455	vascular cryptogams	Gefäßkryptogamen *f pl*
9456	vascular cylinder	Gefäßzylinder *m*
9457	vascular membrane	Gefäßhaut *f*
9458	vascular plant	Gefäßpflanze *f*
9459	vascular strand	Gefäßstrang *m*
9460	vascular system	Gefäßsystem *n*, Leitbündelsystem *n*
9461	vascular tissue	Gefäßgewebe *n*, Leitbündelgewebe *n*
9462	vascularization	Vaskularisierung *f*, Gefäßversorgung *f*, Gefäßbildung *f*
9463	vascularized	gefäßreich
9464	vasoconstriction	Gefäßverengung *f*
9465	vasodilatation	Gefäßerweiterung *f*
9466	vasomotor nerves	Vasomotoren *pl*
9467	vasopressin, pitressin, adiuretin, antidiuretic hormone	Vasopressin *n*, Adiuretin *n*
9468	vector	Überträger *m*
9469	vegetable protein	pflanzliches Eiweiß *n*
9470	vegetal, vegetable	pflanzlich
9471	vegetal pole, vegetative ~	vegetativer Pol *m*
9472	vegetation	Vegetation *f*
9473	vegetation belt	Vegetationsgürtel *m*
9474	vegetation map	Vegetationskarte *f*
9475	vegetation type, vegetational ~	Vegetationstyp *m*
9476	vegetation unit	Vegetationseinheit *f*
9477	vegetation zone, vegetational ~	Vegetationsgebiet *n*
9478	vegetational cover → plant cover	—
9479	vegetational survey	Vegetationsaufnahme *f*
9480	vegetative	vegetativ
9481	vegetative cone	Vegetationskegel *m*, Vegetationsspitze *f*
9482	vegetative nervous system → autonomic ~ ~	—
9483	vegetative nucleus	vegetativer Kern *m*
9484	vegetative organ	Vegetationsorgan *n*
9485	vegetative period	Vegetationsperiode *f*, Vegetationszeit *f*
9486	vegetative point, growing ~	Vegetationspunkt *m*, Wachstumspunkt *m*

9426	utérus *m*, matrice *f*	útero *m*, matriz *f*
9427	degré *m* d'utilisation *ecol.*	proporción *f* de utilización *ecol.*
9428	temps *m* utile *neur.*	tiempo *m* útil, ~ de utilización *neur.*
9429	—	—
9430	uvée *f*	úvea *f*
9431	membrane *f* vacuolaire	membrana *f* vacuolar
9432	suc *m* vacuolaire	líquido *m* vacuolar
9433	vacuole *f*	vacuola *f*, vacúolo *m*
9434	vacuome *m*, système *m* vacuolaire, appareil *m* ~	vacuoma *m*, sistema *m* vacuolar
9435	vide *m*	vacío *m*
9436	activité *f* (à) vide	acto *m* en el vacío
9437	vagin *m*	vagina *f*
9438	nerf *m* vague, ~ pneumogastrique	(nervio *m*) vago *m*, ~ neumogástrico
9439	valine *f*	valina *f*
9440	valvaire	valvar
9441	valve *f*	valva *f*
9442	variabilité *f*	variabilidad *f*
9443	variable *f*	variable *f*
9444	variance *f*	varianza *f*
9445	variante *f*	variante *f*
9446	variation *f*	variación *f*
9447	panaché	variegado
9448	panachure *f*	variegación *f*
9449	variété *f*; cultivar *m*	variedad *f*
9450	—	—
9451	varier	variar
9452	vasculaire	vascular
9453	faisceau *m* vasculaire, ~ conducteur	haz *m* vascular, ~ conductor
9454	cambium *m* vasculaire	cámbium *m* vascular
9455	cryptogames *mpl* vasculaires	criptógamas *fpl* vasculares
9456	cylindre *m* vasculaire	cilindro *m* vascular
9457	membrane *f* vasculaire	membrana *f* vascular
9458	plante *f* vasculaire, végétal *m* ~	planta *f* vascular
9459	cordon *m* vasculaire	cordón *m* vascular
9460	système *m* vasculaire, appareil *m* ~	sistema *m* vascular
9461	tissu *m* vasculaire	tejido *m* vascular
9462	vascularisation *f*	vascularización *f*
9463	vascularisé	vascularizado
9464	vasoconstriction *f*	vasoconstricción *f*
9465	vasodilatation *f*	vasodilatación *f*
9466	nerfs *mpl* vasomoteurs	nervios *mpl* vasomotores
9467	vasopressine *f*, adiurétine *f*, hormone *f* antidiurétique	vasopresina *f*, adiuretina *f*, hormona *f* antidiurética
9468	vecteur *m*	vector *m*
9469	protéine *f* végétale	proteína *f* vegetal
9470	végétal	vegetal
9471	pôle *m* végétal, ~ végétatif	polo *m* vegetativo
9472	végétation *f*	vegetación *f*
9473	ceinture *f* de végétation	cintura *f* de vegetación
9474	carte *f* de végétation	mapa *m* de vegetación
9475	type *m* de végétation	tipo *m* de vegetación
9476	unité *f* de végétation	unidad *f* de vegetación
9477	zone *f* de végétation	zona *f* de vegetación, región *f* ~ ~
9478	—	—
9479	relevé *m* de végétation	inventario *m* de vegetación
9480	végétatif	vegetativo
9481	cône *m* végétatif	cono *m* vegetativo
9482	—	—
9483	noyau *m* végétatif	núcleo *m* vegetativo
9484	organe *m* végétatif	órgano *m* vegetativo
9485	période *f* de végétation	período *m* vegetativo
9486	point *m* végétatif, ~ de végétation, ~ de croissance	punto *m* vegetativo

9487	vegetative propagation	vegetative Vermehrung *f*
9488	vegetative reproduction → asexual ~	—
9489	vegetative rest	Vegetationsruhe *f*
9490	vein	Vene *f*
9491	velamen	Wurzelhülle *f*, Velamen *n*
9492	venation → nervation	—
9493	venom → poison	—
9494	venomous, venomousness → toxic, toxicity	—
9495	venomous spine	Giftstachel *m*
9496	venous	venös
9497	venous sinus → sinus venosus	—
9498	venous valvula	Venenklappe *f*
9499	ventral → abdominal	—
9500	ventral cord	Bauchmark *n*; Strickleiternervensystem *n*
9501	ventral fin, pelvic ~	Bauchflosse *f*
9502	ventral root, motor ~, anterior ~	ventrale Rückenmarkswurzel *f*, motorische ~
9503	ventral shield → plastron	—
9504	ventricidal	ventrizid, bauchnahtspaltig
9505	ventricle (of the heart)	Herzkammer *f*, Ventrikel *m*
9506	vermiform	wurmförmig
9507	vermiform appendix	Wurmfortsatz *m*
9508	vernalization, yarovization, jarovization	Vernalisation *f*, Jarowisation *f*
9509	vernation, prefoliation	Knospenlage *f*, Vernation *f*
9510	vertebra	Wirbel *m*
9511	vertebral	Wirbel...
9512	vertebral arch	Wirbelbogen *m*
9513	vertebral body	Wirbelkörper *m*
9514	vertebral canal → spinal canal	—
9515	vertebral column, spinal ~, spine, rachis, backbone	Wirbelsäule *f*, Rückgrat *n*
9516	vertebral foramen	Wirbelloch *n*
9517	Vertebrata, vertebrates, Craniata, craniates	Wirbeltiere *npl*, Vertebraten *mpl*, Kranioten *pl*
9518	verticil, whorl	Wirtel *m*, Quirl *m*
9519	verticillaster	Scheinquirl *m*
9520	verticillate, whorled	wirtelig, quirlständig
9521	vesicular, vesicle-like	blasenförmig
9522	vessel *bot., anat.*	Gefäß *n bot., anat.*
9523	vessel element	Gefäßelement *n*
9524	vestibule *anat.*	Vorhof *m anat.*
9525	vestigial organ → rudimentary organ	—
9526	viability	Lebensfähigkeit *f*
9527	viable	lebensfähig
9528	vibrio	Vibrio *m* (*pl* Vibrionen), Kommabazillus *m*
9529	vicariad (species)	Vikariant *m*, vikariierende Art *f*
9530	villus (*pl* villi)	Zotte *f*
9531	vinegar bacteria → Acetobacter	—
9532	viral genetics	Virengenetik *f*
9533	viral protein	Virusprotein *n*
9534	virescence	Vireszenz *f*, Vergrünung *f*
9535	virgen forest, primeval ~	Urwald *m*
9536	virological	virologisch
9537	virologist	Virologe *m*
9538	virology	Virologie *f*, Virusforschung *f*
9539	virosis → virus disease	—
9540	virulency, virulence	Virulenz *f*
9541	virulent	virulent
9542	virulent phage → intemperate phage	—
9543	virus (*pl* viruses)	Virus *n* (*pl* Viren)
9544	virus culture	Viruskultur *f*
9545	virus disease, virosis	Viruskrankheit *f*, Virose *f*
9546	virus strain	Virusstamm *m*
9547	viscera	Eingeweide *pl*, Viszera *pl*

9487	multiplication *f* végétative	propagación *f* vegetativa
9488	—	—
9489	repos *m* de la végétation	reposo *m* vegetativo, vegetación *f* latente
9490	veine *f*	vena *f*
9491	velamen *m*, voile *m* de racine	velamen *m*
9492	—	—
9493	—	—
9494	—	—
9495	épine *f* venimeuse, aiguillon *m* venimeux, piquant *m* ~	espina *f* venenosa, aguijón *m* venenoso
9496	veineux	venoso
9497	—	—
9498	valvule *f* veineuse	válvula *f* de las venas
9499	—	—
9500	cordon *m* ventral	cordón *m* ventral
9501	nageoire *f* ventrale, ~ abdominale	aleta *f* abdominal, ~ pelviana
9502	racine *f* ventrale, ~ motrice	raíz *f* ventral, ~ motora, ~ anterior
9503	—	—
9504	ventricide	ventricida
9505	ventricule *m* (du cœur)	ventrículo *m* (cardíaco)
9506	vermiforme	vermiforme
9507	appendice *m* vermiforme	apéndice *m* vermiforme
9508	vernalisation *f*, jarovisation *f*, printanisation *f*	vernalización *f*, yarovización *f*
9509	vernation *f*, préfoliation *f*, préfoliaison *f*	vernación *f*, prefoliación *f*
9510	vertèbre *f*	vértebra *f*
9511	vertébral	vertebral
9512	arc *m* vertébral	arco *m* vertebral
9513	corps *m* vertébral	cuerpo *m* vertebral
9514	—	—
9515	colonne *f* vertébrale, épine *f* dorsale, rachis *m*	columna *f* vertebral, ~ espinal, espina *f* dorsal, espinazo *m*, raquis *m*
9516	trou *m* vertébral	agujero *m* vertebral
9517	vertébrés *mpl*	vertebrados *mpl*, craniotas *mpl*
9518	verticille *m*	verticilo *m*
9519	pseudo-verticille *m*	verticilastro *m*
9520	verticillé	verticilado
9521	vésiculaire	vesicular, vesiculiforme
9522	vaisseau *m* *bot.*, *anat.*	vaso *m* *bot.*, *anat.*
9523	élément *m* vasculaire	elemento *m* vaso, ~ vascular
9524	vestibule *m* *anat.*	vestíbulo *m* *anat.*
9225	—	—
9526	viabilité *f*	viabilidad *f*
9527	viable	viable
9528	vibrion *m*, bacille-virgule *m*	vibrio *m*, bacilo *m* coma
9529	vicariant *m*, espèce *f* vicariante	vicario *m*, vicariante *m*, especie *f* vicaria
9530	villosité *f*	vellosidad *f*
9531	—	—
9532	génétique *f* virale, virogénétique *f*	genética *f* vírica
9533	protéine *f* virale	proteína *f* vírica, ~ viral, viruproteína *f*
9534	virescence *f*	virescencia *f*
9535	forêt *f* vierge	selva *f* virgen, bosque *m* ~
9536	virologique	virológico
9537	virologiste *m*, virologue *m*	virólogo *m*, virologista *m*
9538	virologie *f*	virología *f*
9539	—	—
9540	virulence *f*	virulencia *f*
9541	virulent	virulento
9542	—	—
9543	virus *m*	virus *m*
9544	culture *f* de virus	cultivo *m* de virus, ~ vírico
9545	maladie *f* à virus, virose *f*	enfermedad *f* de virus, ~ vírica, ~virótica, virosis *f*
9546	souche *f* virale	cepa *f* de virus
9547	viscères *mpl*	vísceras *fpl*

9548	visceral, splanchnic	Eingeweide..., viszeral
9549	visceral arch	Viszeralbogen *m*
9550	visceral ganglion	Eingeweideganglion *n*
9551	visceral nervous system → autonomic ~ ~	—
9552	visceral organ	Eingeweideorgan *n*
9553	viscerocranium, splanchnocranium	Gesichtsschädel *m*, Eingeweideschädel *m*, Viszerokranium *n*
9554	visceroskeleton, visceral skeleton, splanchnoskeleton	Viszeralskelett *n*
9555	viscosity	Zähflüssigkeit *f*, Viskosität *f*
9556	viscous	zähflüssig, viskos, viskös
9557	visual ability, ~ faculty	Sehvermögen *n*
9558	visual acuity	Sehschärfe *f*
9559	visual cell	Sehzelle *f*
9560	visual centre (US: center)	Sehzentrum *n*
9561	visual cortex	Sehrinde *f*
9562	visual field	Gesichtsfeld *n*
9563	visual pigment	Sehpigment *n*, Sehfarbstoff *m*
9564	visual purple → rhodopsin	—
9565	visual sense	Gesichtssinn *m*
9566	visual stimulus, optical ~	Gesichtsreiz *m*
9567	vital activity	Lebenstätigkeit *f*
9568	vital capacity	Vitalkapazität *f*, Atmungsgröße *f*
9569	vital dye	Vitalfarbstoff *m*
9570	vital force, ~ energy	Lebenskraft *f*
9571	vital microscopy	Lebendgewebemikroskopie *f*
9572	vital optimum	Lebensoptimum *n*
9573	vital phenomenum	Lebenserscheinung *f*
9574	vital staining	Vitalfärbung *f*
9575	vitalism, vistalistic theory	Vitalismus *m*
9576	vitalist	Vitalist *m*
9577	vitality	Vitalität *f*
9578	vitamin	Vitamin
9579	vitamin A → axerophthol	—
9580	vitamin B_1 → thiamine	—
9581	vitamin B_2 → riboflavin	—
9582	vitamin B_6 → pyridoxine	—
9583	vitamin B_{12} → cobalamine	—
9584	vitamin B complex	Vitamin B-Komplex *m*
9585	vitamin C → ascorbic acid	—
9586	vitamin D_2 → calciferol	—
9587	vitamin E → tocopherol	—
9588	vitamin H → biotin	—
9589	vitamin K → phylloquinone	—
9590	vitamin P → antipermeability vitamin	—
9591	vitamin content	Vitamingehalt *m*
9592	vitamin deficiency	Vitaminmangel *m*
9593	vitamin requirements, ~ needs	Vitaminbedarf *m*
9594	vitellarium → yolk gland	—
9595	vitellin	Vitellin *n*
9596	vitelline duct → yolk duct	—
9597	vitelline membrane, oolemma	Dotterhaut *f*, Oolemma *n*
9598	vitelline vein	Dottervene *f*
9599	vitellogenesis	Dotterbildung *f*
9600	vitellus → yolk	—
9601	vitreous body	Glaskörper *m*
9602	vitreous humo(u)r	Glaskörperflüssigkeit *f*
9603	viviparity	Viviparie *f*, Lebendgebären *n*
9604	viviparous	vivipar, lebendgebärend
9605	vivisection	Vivisektion *f*
9606	vocal cord	Stimmband *n*
9607	vocalization	Lautäußerung *f*, Lautgebung *f*
9608	volatile oil → essential oil	—
9609	volatilization	Verflüchtigung *f*

9548	viscéral, splanchnique	visceral, esplácnico
9549	arc *m* viscéral	arco *m* visceral
9550	ganglion *m* viscéral	ganglio *m* visceral
9551	—	—
9552	organe *m* viscéral	órgano *m* visceral
9553	splanchnocrâne *m*	esplacnocráneo *m*
9554	viscérosquelette *m*, squelette *m* viscéral	esqueleto *m* visceral, esplacno(e)squeleto *m*
9555	viscosité *f*	viscosidad *f*
9556	visqueux	viscoso
9557	capacité *f* visuelle, faculté *f* ~	capacidad *f* visual
9558	acuité *f* visuelle	agudeza *f* visual
9559	cellule *f* visuelle	célula *f* visual
9560	centre *m* visuel, ~ optique	centro *m* visual, ~ óptico
9561	cortex *m* visuel	cortex *f* visual
9562	champ *m* visuel	campo *m* visual
9563	pigment *m* oculaire	pigmento *m* visual
9564	—	—
9565	sens *m* de la vue	sentido *m* de la vista
9566	stimulus *m* visuel	estímulo *m* visual, ~ óptico
9567	activité *f* vitale	actividad *f* vital
9568	capacité *f* (pulmonaire) vitale	capacidad *f* vital
9569	colorant *m* vital	colorante *m* vital
9570	force *f* vitale	fuerza *f* vital
9571	microscopie *f* vitale	microscopia *f* vital
9572	optimum *m* vital	óptimo *m* vital
9573	phénomène *m* vital	fenómeno *m* vital
9574	coloration *f* vitale	coloración *f* vital, tinción *f* ~
9575	vitalisme *m*	vitalismo *m*
9576	vitaliste *m*	vitalista *m*
9577	vitalité *f*	vitalidad *f*
9578	vitamine *f*	vitamina *f*
9579	—	—
9580	—	—
9581	—	—
9582	—	—
9583	—	—
9584	complexe *m* vitaminique B	complejo *m* vitamínico B
9585	—	—
9586	—	—
9587	—	—
9588	—	—
9589	—	—
9590	—	—
9591	teneur *f* en vitamines	contenido *m* vitamínico, ~ en vitaminas
9592	déficience *f* vitaminique, carence *f* de vitamine, manque *m* ~ ~	deficiencia *f* vitamínica, carencia *f* de vitaminas
9593	besoins *m pl* vitaminiques, ~ en vitamines, exigences *f pl* ~ ~	requerimientos *m pl* vitamínicos, necesidades *f pl* vitamínicas
9594	—	—
9595	vitelline *f*	vitelina *f*
9596	—	—
9597	membrane *f* vitelline	membrana *f* vitelina
9598	veine *f* vitelline	vena *f* vitelina
9599	vitellogenèse *f*	vitelogénesis *f*
9600	—	—
9601	corps *m* vitré	cuerpo *m* vítreo
9602	humeur *f* vitrée	humor *m* vítreo
9603	viviparité *f*	viviparidad *f*
9604	vivipare	vivíparo
9605	vivisection *f*	vivisección *f*
9606	corde *f* vocale	cuerda *f* vocal
9607	émission *f* vocale	vocalización *f*, emisión *f* de sonidos
9608	—	—
9609	volatilisation *f*	volatilización *f*

9610 to volatilize — sich verflüchtigen
9611 volubilate plant, twining ~ — Schlingpflanze *f*, Windepflanze *f*
9612 voluntary muscle → striated muscle — —
9613 volutin — Volutin *n*
9614 vomer, ploughshare bone — Pflugscharbein *n*
9615 vulva — Vulva *f*

9616 walking leg — Laufbein *n*, Gangbein *n*, Schreitbein *n*
9617 wandering cell, migratory ~, planocyte — Wanderzelle *f*
9618 warm-blooded, h(a)ematothermal — warmblütig
9619 warm-blooded animal — Warmblüter *m*
9620 warning call, ~ signal → alarm call, ~ signal — —
9621 warning colo(u)r, aposematic ~ — Warnfarbe *f*
9622 warning colo(u)ration, aposematic ~ — Warnfärbung *f*, Warntracht *f*

9623 waste, refuse — Abfall *m*, Abfälle *mpl*
9624 waste disposal — Abfallbeseitigung *f*

9625 waste product — Abfallprodukt *n*
9626 waste substance — Abfallstoff *m*
9627 waste treatment, utilization of waste — Abfallverwertung *f*
9628 waste water → sewage — —
9629 water absorption — Wasseraufnahme *f*, Wasserabsorption *f*
9630 water balance — Wasserbilanz *f*

9631 water conduction *bot.* — Wasserleitung *f* *bot.*
9632 water content — Wassergehalt *m*
9633 water culture — Wasserkultur *f*
9634 water deficiency, shortage of water — Wassermangel *m*

9635 water dispersal → hydrochory — —
9636 water economy — Wasserhaushalt *m*
9637 water elimination — Wasserabgabe *f*
9638 water-gland — Wasserdrüse *f*
9639 water holding capacity — Wasserkapazität *f* (des Bodens)
9640 water impermeability — Wasserundurchlässigkeit *f*
9641 water impermeable, ~ impervious — wasserundurchlässig
9642 water insolubility — Wasserunlöslichkeit *f*
9643 water insoluble, insoluble in water — wasserunlöslich
9644 water loss — Wasserverlust *m*
9645 water needs, ~ requirements — Wasserbedarf *m*
9646 water of imbibition — Quellungswasser *n*
9647 water permeability — Wasserdurchlässigkeit *f*
9648 water permeable, pervious to water — wasserdurchlässig
9649 water pollution — Wasserverschmutzung *f*, Wasserverseuchung *f* Gewässerverschmutzung *f*
9650 water purification — Wasserreinigung *f*
9651 water resources, aquatic ~ — Wasserreserven *fpl*, Wasservorräte *mpl*
9652 water rise *bot.* — Wasseranstieg *m* *bot.*
9653 water sample — Wasserprobe *f*
9654 water softener — Wasserenthärter *m*
9655 water solubility — Wasserlöslichkeit *f*
9656 water soluble, hydrosoluble — wasserlöslich
9657 water stoma — Wasserspalte *f*
9658 water storing — Wasserspeicherung *f*
9659 water-storing tissue — Wasserspeichergewebe *n*
9660 water supply — Wasserversorgung *f*
9661 water treatment — Wasseraufbereitung *f*
9662 water vapo(u)r — Wasserdampf *m*
9663 water vascular system — Wassergefäßsystem *n*, Ambulakralsystem *n*

9664 watery solution → aqueous solution — —
9665 wave length — Wellenlänge *f*
9666 wax coating — Wachsüberzug *m*, Wachsbelag *m*
9667 wax gland — Wachsdrüse *f*
9668 weathering — Verwitterung *f*

9610	se volatiliser	volatilizar(se)
9611	plante *f* volubile	planta *f* voluble, ~ enredadera
9612	—	—
9613	volutine *f*	volutina *f*
9614	vomer *m*	vómer *m*
9615	vulve *f*	vulva *f*
9616	patte *f* ambulatoire, ~ locomotrice	pata *f* ambulatoria, ~ locomotora
9617	cellule *f* migratrice, planocyte *m* (inusité)	célula *f* errante, planocito *m*
9618	à sang chaud	de sangre caliente
9619	animal *m* à sang chaud	animal *m* de sangre caliente
9620	—	—
9621	couleur *f* avertisseuse, ~ aposématique	color *m* advertidor
9622	pigmentation *f* avertisseuse, ~ aposématique	coloración *f* de alerta, ~ de advertencia, ~ aposemática
9623	déchets *mpl*, ordures *fpl*	desechos *mpl*, desperdicios *mpl*
9624	évacuation *f* des déchets	eliminación *f* de (los) desperdicios, evacuación *f* de desechos
9625	produit *m* de déchet	producto *m* de desecho
9626	substance *f* de déchet	sustancia *f* de desecho
9627	traitement *m* des déchets, utilisation *f* ~ ~	aprovechamiento *m* de (los) desperdicios
9628	—	—
9629	absorption *f* d'eau	absorción *f* de agua
9630	équilibre *m* hydrique, balance *f* ~, bilan *m* ~	equilibrio *m* hídrico, ~ acuoso, balance *m* del agua
9631	conduction *f* d'eau *bot.*	conducción *f* de agua *bot.*
9632	teneur *f* en eau, taux *m* aqueux	contenido *m* hídrico, ~ de agua, ~ en agua
9633	aquiculture *f*	cultivo *m* acuático
9634	manque *m* d'eau, carence *f* en eau, déficit *m* hydrique	falta *f* de agua, escasez *f* ~ ~, penuria *f* ~ ~, déficit *m* hídrico, deficiencia *f* hídrica
9635	—	—
9636	économie *f* d'eau	economía *f* hídrica
9637	élimination *f* d'eau	eliminación *f* de agua
9638	glande *f* à eau	glándula *f* acuosa
9639	capacité *f* de rétention d'eau	capacidad *f* de retención de agua
9640	imperméabilité *f* à l'eau	impermeabilidad *f* al agua
9641	imperméable à l'eau	impermeable al agua
9642	insolubilité *f* dans l'eau	insolubilidad *f* en agua
9643	insoluble dans l'eau	insoluble en agua
9644	perte *f* d'eau, ~ en eau, déperdition *f* ~ ~	pérdida *f* de agua
9645	besoins *mpl* en eau, exigences *fpl* hydriques	necesidades *fpl* de agua
9646	eau *f* d'imbibition	agua *f* de imbibición
9647	perméabilité *f* à l'eau	permeabilidad *f* al agua
9648	perméable à l'eau	permeable al agua
9649	pollution *f* des eaux, contamination *f* ~ ~	polución *f* del agua, contaminación *f* ~ ~ (o: de las aguas), ~ hídrica
9650	épuration *f* de l'eau	depuración *f* de las aguas
9651	ressources *fpl* aquatiques, réserve *f* d'eau	recursos *mpl* acuáticos, ~ hídricos
9652	ascension *f* de l'eau *bot.*	ascensión *f* del agua *bot.*
9653	échantillon *m* d'eau	muestra *f* de agua
9654	adoucisseur *m* d'eau	reblandecedor *m* de agua
9655	hydrosolubilité *f*, solubilité *f* dans l'eau	solubilidad *f* en agua
9656	hydrosoluble, soluble dans l'eau	hidrosoluble, soluble en agua
9657	stomate *m* aquifère	estoma *m* acuífero
9658	stockage *m* d'eau, accumulation *f* ~	almacenamiento *m* de agua
9659	tissu *m* aquifère	tejido *m* acuífero
9660	approvisionnement *m* en eau	abastecimiento *m* de agua, suministro *m* ~ ~
9661	traitement *m* des eaux	tratamiento *m* de las aguas
9662	vapeur *f* d'eau	vapor *m* de agua
9663	appareil *m* aquifère, système *m* ~, appareil *m* ambulacraire	sistema *m* vascular acuoso, ~ ambulacral, ~ hidrovascular, aparato *m* acuífero
9664	—	—
9665	longueur *f* d'ondes	longitud *f* de onda
9666	enduit *m* de cire	cubierta *f* cérea
9667	glande *f* cirière, ~ cérigène	glándula *f* de cera
9668	métamorphisme *m*	meteorización *f*, disgregación *f*

9669	web	Schwimmhaut *f*
9670	wet weight, humid ~	Feuchtgewicht *n*
9671	white blood corpuscle → leucocyte	—
9672	white matter	weiße Substanz *f*
9673	whole mount (preparation)	Ganzpräparat *n*, Totalpräparat *n*
9674	whorl → verticil	—
9675	wild type	Wildtyp *m*
9676	to wilt	welken, verwelken
9677	wilted	welk, verwelkt
9678	wilting	Welken *n*
9679	wilting coefficient	Welkungskoeffizient *m*, Welkekoeffizient *m*
9680	wilting point	Welkepunkt *m*
9681	wind dispersal → anemochory	—
9682	wind erosion	Winderosion *f*
9683	windpipe → trachea[1]	—
9684	wind pollination → anemophily	—
9685	wing	Flügel *m*
9686	wing venation	Flügelgeäder *n*
9687	winged → alate	—
9688	winged fruit, samara	Flügelfrucht *f*
9689	wingless → apterous	—
9690	winter annual	Winterannuelle *f*
9691	winter bud → hibernaculum	—
9692	winter dormancy	Winterruhe *f*
9693	winter egg, resting ~, dormant ~	Winterei *n*, Dauerei *n*, Latenzei *n*
9694	winter green	wintergrün
9695	winter hardiness → cold resistance	—
9696	winter quarter	Winterquartier *n*
9697	winter rigidity, cold rigo(u)r	Kältestarre *f*
9689	winter-spore → teleutospore	—
9699	wintering → overwintering	—
9700	wintering ground	Überwinterungsplatz *m*
9701	wisdom-tooth	Weisheitszahn *m*
9702	Wolffian body → mesonephros	—
9703	Wolffian duct → mesonephric duct	—
9704	wood	Holz *n*
9705	wood parenchyma	Holzparenchym *n*
9706	wood vessel → trachea[3]	—
9707	woody, ~ plant → ligneous, ~ plant	—
9708	woody tissue, ligneous ~	Holzgewebe *n*
9709	wound hormone, necrohormone	Wundhormon *n*, Nekrohormon *n*
9710	wrist	Handgelenk *n*
9711	xanthine	Xanthin *n*
9712	xanthophyll	Xanthophyll *n*
9713	X-chromosome	X-Chromosom *n*
9714	xenia	Xenie *f*
9715	xenogamous	xenogam
9716	xenogamy, cross fertilization	Xenogamie *f*, Fremdbestäubung *f*, Kreuzbestäubung *f*
9717	xeromorphic, xeromorphous	xeromorph
9718	xeromorphy	Xeromorphie *f*
9719	xerophil(ous)	xerophil
9720	xerophyte, xerophilous plant	Xerophyt *m*, Trockenpflanze *f*
9721	xerosere	Xeroserie *f*
9722	X-irradiation	Röntgenbestrahlung *f*
9723	X-ray analysis	Röntgenstrahlanalyse *f*
9724	X-ray crystallography	Röntgenkristallographie *f*
9725	X-ray diffraction	Röntgen(strahl)beugung *f*
9726	X-ray dose, roentgen dose	Röntgenstrahlendosis *f*
9727	X-rays, roentgen rays	Röntgenstrahlen *mpl*
9728	xylem	Xylem *n*, Holzteil *m*
9729	xylol	Xylol *n*

9669	palmure *f*, membrane *f* natatoire	membrana *f* natatoria
9670	poids *m* humide	peso *m* húmedo
9671	—	—
9672	substance *f* blanche	sustancia *f* blanca
9673	préparation *f* entière	preparación *f* entera, preparado *m* entero
9674	—	—
9675	type *m* sauvage	tipo *m* silvestre, ~ salvaje
9676	se faner, se flétrir	marchitarse
9677	fané, flétri	marchito
9678	fanaison *f*, flétrissement *m*	marchitez *f*, marchitamiento *m*
9679	coefficient *m* de fanaison	coeficiente *m* de marchitamiento
9680	point *m* de fanaison, ~ ~ flétrissement	punto *m* de marchitamiento
9681	—	—
9682	érosion *f* éolienne	erosión *f* eólica
9683	—	—
9684	—	—
9685	aile *f*	ala *f*
9686	nervation *f* alaire, veination *f* ~	nerviación *f* alar, venación *f* ~
9687	—	—
9688	fruit *m* ailé, samare *f*	fruto *m* alado, sámara *f*
9689	—	—
9690	annuelle *f* invernale, ~ d'hiver	anual *f* invernal, ~ hibernante
9691	—	—
9692	repos *m* hivernal, léthargie *f* hivernale, dormance *f* ~, latence *f* ~	reposo *m* invernal, letargo *m* hibernal
9693	œuf *m* d'hiver, ~ de repos, ~ de résistance	huevo *m* en reposo, ~ de invierno, ~ durable
9694	vert hivernal	verde en invierno
9695	—	—
9696	quartier *m* d'hiver	cuartel *m* de invierno, ~ ~ invernada
9697	catalepsie *f* frigorifique, ~ de froid	rigidez *f* debida al frío
9698	—	—
9699	—	—
9700	lieu *m* d'hivernage, aire *f* ~, zone ~	área *f* de invernación
9701	dent *f* de sagesse	muela *f* de juicio
9702	—	—
9703	—	—
9704	bois *m*	madera *f*, leño *m*
9705	parenchyme *m* ligneux	parénquima *m* leñoso
9706	—	—
9707	—	—
9708	tissu *m* ligneux	tejido *m* lignificado
9709	hormone *f* de blessure, ~ de cicatrisation, nécrohormone *f*	hormona *f* traumática, ~ de (las) heridas, necrohormona *f*
9710	poignet *m*	muñeca *f*
9711	xanthine *f*	xantina *f*
9712	xanthophylle *m*	xantofila *f*
9713	chromosome *m* X	cromosoma *m* X
9714	xénie *f*	xenia *f*
9715	xénogame	xenógamo
9716	xénogamie *f*	xenogamia *f*
9717	xéromorphe	xeromorfo
9718	xéromorphie *f*	xeromorfismo *m*
9719	xérophile	xerófilo
9720	xérophyte *m*, plante *f* xérophile	xerófito *m*, xerófita *f*, planta *f* xerófila
9721	série *f* xérophile	xerosere *f*, xeroserie *f*
9722	irradiation *f* aux rayons X	irradiación *f* con rayos X
9723	analyse *f* aux rayons X	análisis *m* roentgenográfico
9724	cristallographie *f* aux rayons X	cristalografía *f* por medio de rayos X
9725	diffraction *f* de rayons X	difracción *f* de los rayos X
9726	dose *f* de rayons X	dosis *f* de rayos X
9727	rayons *mpl* X, ~ de Roentgen	rayos *mpl* X, ~ (de) Roentgen
9728	xylème *m*	xilema *m*
9729	xylol *m*	xilol *m*

9730	xylophagous, wood-eating	holzfressend, xylophag
9731	xylose	Xylose *f*, Holzzucker *m*
9732	yarovization → vernalization	—
9733	Y-chromosome	Y-Chromosom *n*
9734	yeast	Hefe *f*
9735	yeast autolysate	Hefeautolysat *n*
9736	yeast extract	Hefeextrakt *m*
9737	yellow enzyme → flavoprotein	—
9738	yellow body → corpus luteum	—
9739	yellow spot, ~ macula, macula lutea	gelber Fleck *m*
9740	yolk, vitellus	Dotter *m*, Vitellus *m*
9741	yolk cell	Dotterzelle *f*
9742	yolk duct, vitelline ~, vitelloduct	Dottergang *m*
9743	yolk gland, vitellarium	Dotterstock *m*, Vitellarium *n*
9744	yolk plug	Dotterpfropf *m*
9745	yolk sac	Dottersack *m*
9746	young (animal)	Jungtier *n*, Junge *n*
9747	young stage → juvenile stage	—
9748	zeatin	Zeatin *n*
9749	zero point mutation	Nullpunktmutation *f*
9750	zoidiogamic, zoidiogamous	tierblütig
9751	zoidiogamic plant	Tierblütler *m*
9752	zoidiogamy	Tierblütigkeit *f*, Zoidiogamie *f*, Zoidiophilie *f*
9753	zona pellucida	Zona pellucida
9754	zonal electrophoresis, zone ~	Zonenelektrophorese *f*
9755	zonation	Zonation *f*, Zonierung *f*, Zonenabgrenzung *f*
9756	zone of hybridization	Bastardierungszone *f*
9757	zoochlorella	Zoochlorelle *f*
9758	zoochory, animal dispersal	Zoochorie *f*, Tierverbreitung *f* (des Samens)
9759	zooflagellates, zoomastigophores	Zooflagellaten *pl*
9760	zoogamete, planogamete	Zoogamet *m*, Planogamet *m*
9761	zoogenetics, animal genetics	Tiergenetik *f*
9762	zoogeography, animal geography	Tiergeographie *f*, Zoogeographie *f*
9763	zoogl(o)ea	Zoogloea *f*
9764	zooid	Zooid *n*
9765	zoological	zoologisch
9766	zoologist	Zoologe *m*
9767	zoology	Zoologie *f*, Tierkunde *f*
9768	zooparasite, parasitic animal	Zooparasit *m*, Schmarotzertier *n*, tierischer Schmarotzer *m*
9769	zoophagous	zoophag
9770	zoophyte	Zoophyt *m*, Pflanzentier *n*
9771	zooplankton, animal plankton	Zooplankton *n*
9772	zoosporangium	Zoosporangium *n*
9773	zoospore, planospore, swarm spore	Zoospore *f*, Schwärmspore *f*, Schwärmzelle *f*
9774	zoosterol	Zoosterin *n*
9775	zootechnics, zootechny	Tierzuchtlehre *f*
9776	zootomy, animal anatomy	Zootomie *f*, Tieranatomie *f*
9777	zootoxin, animal toxin	Zootoxin *n*, Tiergift *n*
9778	zooxanthella	Zooxanthelle *f*
9779	zwitterion	Zwitterion *n*
9780	zygapophysis	Gelenkfortsatz *m*, Zygapophyse *f*
9781	zygoma, zygomatic arch	Jochbogen *m*
9782	zygomatic bone, jugal (~), malar (~)	Jochbein *n*, Backenknochen *m*
9783	zygomorphic, monosymmetrical	zygomorph, monosymmetrisch
9784	zygomorphy	Zygomorphie *f*
9785	zygosis	Zygose *f*, Zygosis *f*
9786	zygosome	Zygosom *n*
9787	zygospore	Zygospore *f*

9730	xylophage	xilófago
9731	xylose *m*	xilosa *f*
9732	—	—
9733	chromosome *m* Y	cromosoma *m* Y
9734	levure *f*	levadura *f*
9735	autolysat *m* de levure	autolisado *m* de levadura
9736	extrait *m* de levure	extracto *m* de levadura
9737	—	—
9738	—	—
9739	tache *f* jaune, macule *f* ~	mácula *f* lútea
9740	vitellus *m*, lécithe *m*	vitelo *m*, lecito *m*, yema *f*
9741	cellule *f* vitelline	célula *f* vitelina
9742	canal *m* vitellin, vitelloducte *m*	viteloducto *m*
9743	glande *f* vitellogène, vitellarium *m*	glándula *f* vitelina
9744	bouchon *m* vitellin	tapón *m* de vitelo
9745	sac *m* vitellin	saco *m* vitelino
9746	jeune *m*	animal *m* joven
9747	—	—
9748	zéatine *f*	zeatina *f*
9749	mutation *f* de point zéro	mutación *f* de punto cero
9750	zoïdiogame	zoidiófilo, zoidiógamo
9751	plante *f* zoïdiogame	zoidiófila *f*
9752	zoïdiogamie *f*	zoidiogamia *f*, zoidiofilia *f*
9753	zone *f* pellucide, membrane *f* ~	zona *f* pelúcida
9754	électrophorèse *f* zonale	electroforesis *f* de zona
9755	zonation *f*	zonación *f*
9756	zone *f* d'hybridation	zona *f* de hibridación
9757	zoochlorelle *f*	zooclorela *f*
9758	zoochorie *f*, dispersion *f* par les animaux, dissémination *f* zoochore	zoocoria *f*, dispersión *f* zoocora
9759	zooflagellés *mpl*	zooflagelados *mpl*, zoomastigóforos *mpl*
9760	zoogamète *m*, planogamète *m*	zoogameto *m*, planogameto *m*, zoogámeta *m*, planogámeta *m*
9761	zoogénétique *f*, génétique *f* animale	zoogenética *f*, genética *f* animal
9762	zoogéographie *f*	zoogeografía *f*
9763	zooglée *f*	zooglea *f*
9764	zooïde *m*	zooide *m*
9765	zoologique	zoológico
9766	zoologiste *m*, zoologue *m*	zoólogo *m*
9767	zoologie *f*	zoología *f*
9768	zooparasite *m*, animal *m* parasite	zooparásito *m*, animal *m* parásito
9769	zoophage	zoófago
9770	zoophyte *m*	zoófito *m*
9771	zooplancton *m*	zooplancton *m*, plancton *m* animal
9772	zoosporange *m*	zoosporangio *m*
9773	zoospore *f*, planospore *f*, spore *f* flagellée	zoóspora *f*, planóspora *f*, espora *f* móvil
9774	zoostérol *m*	zooesterol *m*, esterol *m* animal
9775	zootechnie *f*	zootecnia *f*
9776	zootomie *f*	zootomía *f*
9777	zootoxine *f*, toxine *f* animale	zootoxina *f*
9778	zooxanthelle *f*	zooxantela *f*
9779	zwittérion *m*, ion *m* hybride, ~ bipolaire	zwitterion *m*, ion *m* híbrido, ~ hermafrodita, ~ anfótero
9780	zygapophyse *f*, apophyse *f* articulaire	cigapófisis *f*, apófisis *f* articular
9781	zygoma *m*, arc *m* zygomatique	cigoma *m*, arco *m* cigomático
9782	os *m* zygomatique, (~) malaire *m*, pommette *f*	malar *m*, pómulo *m*
9783	zygomorphe	cigomorfo, zigomorfo, monosimétrico
9784	zygomorphie *f*	cigomorfismo *m*
9785	zygose *f*	cigosis *f*, zigosis *f*
9786	zygosome *m*	zigosoma *m*
9787	zygospore *f*	cigospora *f*, zigóspora *f*

9788	zygote	Zygote *f*
9789	zygote nucleus	Zygotenkern *m*
9790	zygotene, synaptene	Zygotän *n*, Synaptän *n*
9791	zygotic	zygotisch
9792	zymase	Zymase *f*
9793	zymogen, proenzyme, proferment, enzyme precursor	Zymogen *n*, Proenzym *n*, Proferment *n*
9794	zymogen(ic) granules	Zymogengranula *pl*, Zymogenkörnchen *npl*
9795	zymology	Zymologie *f*, Gärungskunde *f*

9788	zygote *m*	cigoto *m*, zigoto *m*
9789	noyau *m* zygotique	núcleo *m* cigótico
9790	zygotène *m*, synaptène *m*	cigóteno *m*, zigóteno *m*
9791	zygotique	cigótico, zigótico
9792	zymase *f*	cimasa *f*, zimasa *f*
9793	zymogène *m*, proenzyme *m*, proferment *m*	cimógeno *m*, zimógeno *m*, proenzima *f*, profermento *m*, enzima *f* precursor
9794	granules *mpl* zymogènes, grains *mpl* de zymogène	gránulos *mpl* de cimógeno
9795	zymologie *f*	cimología *f*

DEUTSCH
Alphabetisches Register

A

AAM 4690
Abart 9449
abarten 9451
Abbau 1262
abbauen 1258
Abbauprodukt 1263
Abbaustoffwechsel 1459
Abdomen 1
abdominal 2
Abduktor 8
Abduzens 7
Aberration 9
Aberrationsrate 10
Abfall 9623
Abfallbeseitigung 9624
Abfälle 9623
Abfallprodukt 9625
Abfallstoff 9626
Abfallverwertung 9627
Abgriff 2409
Abimpfung 8724
Abiogenese 8537
abiogenetisch 12
abiotisch 13
Abklatschpräparat 4581
Abkömmlinge 7326
Ablaichen 8425
ablaichen 8423
Ableitung 2409
Ablenkungsverhalten 2297
Ablesen des genetischen Kodes 1843
abnorm 15
Abnormität 16
abscheiden 3195
Abschirmung 8038
Abschlußgewebe 8823
Abschuppung 2436
absondern 8057
Absonderung 8059
Absonderungsgewebe 8066
Absorbens 27
absorbieren 26
absorbierend 28
Absorption 29
Absorptionsfähigkeit 35
Absorptionskurve 32
Absorptionsmittel 27
Absorptionsspektrum 33
Absorptionsvermögen 35
abstammen von 2423
Abstammung 2426
Abstammungslehre 2427
Abstammungslinie 5182
Abstammungsnachweis 6673
Abstammungstheorie 2427
absterben 2515
Absterben 2516
Abstrich 8330
Abstrichpräparat 8333
Abszisin 21
Abszisin II 2680
Abszission 22
Abteilung *(tax.)* 2652
abtöten 2469
Abundanz 37
Abwasser 8210
Abwässer 8210
Abwasseraufbereitung 8213
Abwasserbeseitigung 8211
Abwasserreinigung, biologische 1083
Abwasserreinigungsanlage 7483
Abwasserverwertung 8213
Abwehrstellung 2325
Abwehrstoff 2327
Abweichung 2468
mittlere quadratische 8589
abwerfen (Pflanzenteile) 20
Abwerfen (von Pflanzenteilen) 22
Abyssal 40
abyssal 38
abyssisch 38
Abziehmuskel 8
Achäne 68
achlamydeisch 69
Achromasie 71
Achromatin 74
achromatisch 72
Achselknospe 876
achselständig 875
Achsenfaden 870
Achsenplasma 882
Achsenskelett 872
Achsenstab 883
Achsenstruktur 873
Achsenzylinder 879
ACTH 181
Adap(ta)tion 134
adaptiv 139
Adaptormolekül 138
Adduktor 149
Adelphogamie 150
Adenin 151
Adenohypophyse 153
adenoid 154
Adenosindiphosphat 155
Adenosindiphosphorsäure 155
Adenosinmonophosphat 156
Adenosinmonophosphorsäure 156
Adenosintriphosphat 157
Adenosintriphosphorsäure 157
Adenylsäure 156
Aderhaut 1687
Adermin 7500
Aderung 6038
Adhäsion 161
Adhäsionsfähigkeit 164
Adhäsionskultur 4084
Adiurctin 9467
Adjuvans 170
Adoleszenz 171
ADP 155
Adrenalin 179
adrenergisch 180
Adrenokortikotropin 181
Adrenosteron 182
Absorbens 184
adsorbieren 183
Adsorption 185
Adsorptions-Chromatographie 187
Adsorptionsfähigkeit 186
Adsorptionsmittel 184
Adsorptionsvermögen 186
adult 188
Adulttier 189
Adventivembryonie 193
Adventivknospe 192
Adventivpflanze 194
Adventivwurzel 195
Aecidie 197
Aecidienspore 196
Aerenchym 198
aerob 203
Aerobier 202
Aerobiologie 204
Aerobiont 202
Aerobiose 205
Aerophyt 206
Aeroplankton 207
Aerotaxis 208
Aerotropismus 209
afferent 214
Afferenz 213
Affinität 215
After 629
Afterdrüse 451
agam 226
Agamet 225
Agamobium 227
Agamogenese 228
Agamogonie 229
Agamont 230
Agamospermie 232
Agamospezies 231
Agar(-Agar) 233
Agarblock 234
Agargel 236
Agarkultur 235
Agarnährboden 237
Agarplatte 238
Agenesie 245
Agens 246
Agglutinat 248
Agglutination 250
Agglutinationsreaktion 251
agglutinierbar 247
agglutinieren 249
Agglutinin 252
Agglutinogen 253
Aggregat 254
Aggregationsgrad 2344
Aggregatzustand 8599
Aggressionstrieb 257
Aggressionsverhalten 256
agmatoploid 258
Agmatoploidie 259
Agnathen 260
Agonist 261
Agrobiologie 262
Ahn 472
Ahnentafel 3721
A-Horizont 263
Ährchen 8492
Ähre 8490
ährenförmig 8489
Akinese 273
akinetisch 274
Akklimatisation 53
akklimatisieren 54
Akklimatisierung 53
Akkommodation 55
Akranier 98
akrogyn 101
akrokarp 99
akropetal 102
Akrosom 103
akrozentrisch 100
Aktin 105
aktinomorph 107
Aktinomorphie 108
Aktinomyzin 110
Aktionskatalog 3126
Aktionspotential 114
Aktionsspektrum 115
Aktionsstrom 113
Aktionssystem 116
Aktivator 123
Aktivatorenzym 120
aktivieren 117
Aktivierung 121

anemophil 488
Anemophilie 490
Anemotaxis 491
aneuploid 492
Aneuploidie 493
Aneurin 9070
Anfangsstadium 4685
anfärbbar 8579
anfärben 8576
angeboren 1979
angepaßt 135
Angepaßtheit 137
Angiologie 495
Angiospermen 497
Angriffstrieb 257
Anhang 662
Anhydrid 500
Anion 531
Anisogamet 532
Anisogamie 533
Anisogenie 534
Anisophyllie 535
Anisoploidie 536
anisotrop 537
Anisotropie 538
Anlage 7306
Anneliden 540
Annuelle 543
Anode 549
Anodenstrahlen 550
anomal 552
Anomalie 553
anorganisch 4700
Anorthogenese 554
Anöstrus 551
Anoxybiose 448
anpaaren 5408
Anpassung 134
anpassungsfähig 133
Anpassungsfähigkeit 132
Anpassungsfärbung 141
Anpassungshormon 143
Anpassungsmerkmal 140
Anpassungsreaktion 145
Anpassungsverlust 2285
Anpassungswert 146
anreichern 2976
Anreicherung 2978
Anreicherungshorizont 1014
Anreicherungskultur 2979
Anreicherungsversuch 2980
Ansatzstelle 4714
ansäuern 90
Ansäuerung 89
anschwellen 8845
Antagonismus 556
Antagonist 557
Antennaten 9187
Antenne 558
Antennendrüse 559
Antennula 560
Anthere 564
Antheridium 565
Antherozoid 566
Anthese 567
Anthoxanthin 571
Anthozoen 572
Anthozyan(in) 568
Anthozyanidin 569
Anthropobiologie 573
Anthropochorie 574
anthropogen 576
Anthropogenese 575
Anthropogenie 577
Anthropogeographie 578
Anthropographie 579
Anthropoiden 580
Anthropologie 581
Anthropometrie 582
Anthropomorphismus 583
Anthropomorphologie 584
Anthropo-Ökologie 4399
Antiauxin 585
Antibiont 586
Antibiose 587
Antibiotikum 588
Antidot 593
antidrom 594
Antienzym 595
Antiferment 595
Antifertilizin 596
Antigen 597
Antigen-Antikörper-Reaktion 598
Antigenität 599
Antigibberellin 600
Antihormon 602
antiklin 590
Antikodon 591
Antikörper 589
Antimetabolit 603
antimorph 604
Antimutagen 605
Antipellagravitamin 6117
Antiperistaltik 609
Antipode 611
Antiseptikum 616
antiseptisch 616
Antiserum 617
Antisterilitätsvitamin 9145
Antithrombin 619
Antitoxin 621
antitoxisch 620
Antitrypsin 622
Antivitamin 623
Antwortreaktion 7813
Anuren 627
Anus 629
Anziehmuskel 149
Aorta 630
Aortenbogen 631
apetal 632
Apex 633
Apfelsäure 5346
Aphyllie 637
Apikaldominanz 640
Apikalknospe 638
Apikalmeristem 642
Aplanospore 643
Aplasie 644
Apoenzym 649
Apogamie 650
apokarp 645
Apokarpie 646
Apomeiose 651
apomiktisch 652
Apomixis 653
Aponeurose 654
Apophyse 655
Aporogamie 656
Aposporie 658
Apothezium 659
Appetenzverhalten 664
Apposition 665
Appositionsauge 666
Appositionswachstum 667
Apterygoten 669
Äquationsteilung 3096
Äquatorialebene 3098
Äquatorialkörper 3097
Äquatorialplatte 3099
Arachniden 678
Arachnoidea 679
Arachnologie 680
Arbeitskern 5533
Arbeitsteilung 2653
Archaikum 684
Archallaxis 686
Archäozoikum 685
Archegoniaten 687
Archegonium 688
Archenteron 690
Archenzephalon 689
Archespor 691
Archetyp 692
Archibenthal 693
Archiblast 694
Archipallium 695
Archiplasma 696
Areal 7600
  disjunktes 2599
  diskontinuierliches 2599
  geschlossenes 2017
  kontinuierliches 2017
Arealkarte 2643
Arealkunde 1688
Arealtyp 699
Arginase 701
Arginin 702
Argininphosphat 703
Arillus 705
Arm... 1236
Armfüsser 1237
Armleuchteralgen 1610
aromatisch 708
aromatisieren 710
Aromatisierung 709
Aromorphose 711
Arrhenotokie 715
Art 8430
  dominante 2670
  vikariierende 9529
Art... 8447
Artareal 8431
Artbastard 4780
Artbildung 8429
Artdichte 2384
Artefakt 716
arteigen 8446
artenarm 7182
Artengruppe 8436
Artenkreis 717
artenreich 8442
Artenvielfalt 8434
Artenzusammensetzung 8432
Arterenol 6179
arterhaltend 8440
Arterhaltung 8439
Arterie 721
arteriell 719
Arteriole 720
Arterkennung 8441
Artfremder 6173
Artgenosse 1997
artgleich 1996
Arthrologie 722
Arthropoden 723
Arthrospore 724
Artioploidie 736
Artisolierung 8437
Artkreuzung 4781
Artname 8448
Artreinheit 7487
artspezifisch 8443
Artspezifität 8444
Artstruktur 8445

Deutsch

arttypisch 8446
Aschelminthen 6010
Aschenanalyse, mikroskopische 5609
Aschenbild 8531
asexuell 749
Askogon 743
Askomyzeten 744
Askorbinsäure 745
Askospore 746
Askus 747
Asparagin 751
Asparaginsäure 752
Assimilat 759
Assimilation 757
genetische 3736
Assimilationsfähigkeit 762
Assimilationsfarbstoff 758
Assimilationsgewebe 756
Assimilationsprodukt 759
Assimilationsquotient 763
Assimilationsrate 760
Assimilationsstärke 761
assimilierbar 754
assimilieren 755
Assoziation 765
Assoziationsfeld 766
Assoziationslernen 768
Assoziationszentrum 767
Ast 1251
Aster 769
Ästheten 210
Ästivation 212
Astknoten 4981
Astrosphäre 773
Astrozyt 772
Asynapsis 774
Asyndese 774
Aszendent 472
Aszendenz 738
Atavismus 775
atavistisch 776
Atem ... 7789
Atembewegung 7802
Atemfrequenz 3583
Atemgas 7797
Atemhöhle *(bot.)* 4516
Atemloch 8640
Atemminutenvolumen 7455
Atemorgan 7803
Atemröhre 7808
Atemvolumen 9116
Atemwege 7805
Atemwurzel 7067
Atemzentrum 7791
Äthanol 3130
Äthylalkohol 3130
Äthylen 3131
atmen 1265
Atmung 7788
thorakale 2091
Atmungs ... 7789
Atmungsferment 7796
Atmungsgröße 9568
Atmungsintensität 7798
Atmungskette 7792
Atmungskettenphosphorylation 6491
Atmungsorgan 7803
Atmungspigment 7804
Atmungsquotient 7806
Atmungsregulation 7794
Atmungsstoffwechsel 7800
Atmungssystem 7809
Atmungsverlust 7799
Atomenergie 781
Atomgewicht 785
Atomkern 783
Atommüll 7560
Atomzeitalter 780
ATP 157
atrich 788
Atrium 789
atrop 791
Atrophie 790
Attrappenversuch 5729
aufbauen 8895
aufforsten 216
Aufforstung 217
aufgießen 4665
Aufgußtierchen 4666
aufhellen 1773
Aufhellung 1775
Aufhellungsmittel 1776
Auflagehumus 8821
Auflösungsvermögen *(mikr.)* 7785
Aufnahme, elektronenmikroskopische 2834
Aufnahmefläche 7968
aufspalten 8529
Aufspaltung 3424
Aufspaltungsverhältnis 5489
Aufwuchs 6744
Augapfel 3283
Auge *(bot.)* 1311
schlafendes 2676
treibendes 3969
Augenbecher 6361
Augenbläschen 6364
Augenfleck 3288
Augenhöhle 6371
Augenkeil 6331
Augenlid 3285
Augenmuskelnerv 6281
seitlicher 7
Augenrollnerv 9273
Augenstiel 3289
Ausatmung 3243
Ausatmungsluft 3245
ausbrüten 4612
ausdauernd 6713
Ausdrucksbewegung 3251
Ausdünstung 6775
Ausfall, radioaktiver 7554
ausfällbar 7231
ausfällen 7233
Ausfällung 7235
ausflocken 3463
Ausflockung 3464
Ausführungsgang 3198
Ausgangspopulation 6392
ausgestorben 3265
Ausläufer 8672
Auslese 8092
natürliche 5989
auslesen 8090
Auslesezüchtung 8098
Auslöschen *(Reiz)* 3267
Auslösemechanismus, angeborener 4690
auslösen 7711
Auslöser 7712
Auslösereiz 7714
ausmerzen 2852
Ausmerzen 2853
Ausprägung, phänotypische 6817
Ausprägungsgrad 3252
Ausrottung 3256
ausscheiden 3195
Ausscheidung 3196
Ausscheidungs ... 3197
Ausscheidungsgang 3198
Ausscheidungsorgan 3199
Ausscheidungsprodukt 3200
ausschlüpfen 4109
Ausschlüpfen 4110
Ausschwitzung 3279
Außenfaktor 3258
Außenkelch 1365
Außenmedium 3260
Außenreiz 3262
Außenschmarotzer 2783
Außenskelett 3216
aussterben 3266
Aussterben 3267
aussterbend 991
Ausstoß(ungs)reaktion 3253
Ausstrahlung 4849
ausstreichen 8331
Ausstrich 8330
Ausstrichkultur 8332
Ausstrichpräparat 8333
Ausstülpung 3151
Austausch der Atemgase 7795
Austrocknung 2437
Auswuchs 3193
Auszählung 2096
Autoadaptation 812
Autoalloploidie 813
Autoantigen 814
Autobivalent 815
Autochorie 817
autochthon 818
autogam 821
Autogamie 8112, 8115
autogen 825
Autogenese 823
autogenisch 824
Autoheteroploidie 827
Autoimmunisierung 828
Autoinkompatibilität 8114
Autokatalyse 816
Autoklav 819
Autökologie 811
Autolysat 829
Autolyse 830
Automatie 831
Automixis 832
Automutagen 833
Autonomie 835
Autoorientierung 836
Autoparthenogenese 837
Autoplantat 848
Autoplastik 849
autoplastisch 838
Autopolyploidie 840
Autoradiographie 841
Auto(re)duplikation 8109
Autoreproduktion 8119
Autosom 843
Autosyndese 844
Autosynthese 845
Autotomie 846
Autotoxin 847
autotroph 851
Autotrophie 850
Autotropismus 852
autözisch 820
Auxanographie 854
Auxiliarzelle 855
Auxin 856
Auxospore 858
auxotroph 859

B

Beiknospe 47
beimpfen 4697
Beinnerv 50
Beintaster 6674
Beischilddrüse 6611
Beizmittel 5805
Bekämpfung der Umweltverschmutzung 613
Bekräftigung *(eth.)* 7707
belaubt 3513
Belaubung 3510
Belebtschlamm 118
Belebtschlammverfahren 119
Belegknochen 5473
Benthal 1007
benthisch 1006
Benthos 1008
beringen 7880
Beringung 1134
Bernsteinsäure 8765
Berührungsreiz 2009
besamen 4711
Besamung 4712
 künstliche 734
Beschädigungsversuch 2324
Beschwichtigungsgeste 661
Beseitigung von Umweltverschmutzungen 2406
Bestand 8594
Bestandteil 2000
bestäuben 7107
Bestäuber 7108
Bestäuberinsekt 7109
Bestäubung 7110
Bestäubungstropfen 7111
Bestimmungsschlüssel 4530
bestocken 9118
Bestockung 9120
Bestockungsknoten 9121
Bestockungstrieb 9119
bestrahlen 4848
Bestrahlung 4849
Bestrahlungsmenge 4850
Betastrahlen 1012
Betriebsstoffwechsel 2965
Beugemuskel 3453
Beuger 3453
Beugereflex 3454
Beute 7285
Beuteerwerb 7286
Beutefang 7286
Beuteltiere 5387
Beutetier 7285
Bevölkerungsbiologie 7185
Bevölkerungsdichte 7186
Bevölkerungsdruck 7194
Bevölkerungsexplosion 7190
Bevölkerungspyramide 7195
Bevölkerungswachstum 7193
beweglich 5828
Beweglichkeit 5829
Bewegung, amöboide 412
 lokomotorische 5230
 morphogenetische 5812
 nastische 5984
 nyktinastische 6266
Bewegungs ... 4966
Bewegungsapparat 5228
bewegungsempfindlich 4963
Bewegungsenergie 4967
Bewegungsgefühl 4962
Bewegungslosigkeit 273
Bewegungsnorm 3431
Bewegungsorgan 5227
Bewegungsstereotypien 8626
Bewegungswahrnehmung 5846
Beweis, experimenteller 3233
bewimpert 1740
Bewimperung 1743
bewurzelt 7548
Bewurzelung 7549
Bezahnung 2393
B-Horizont 1014
Biegungsfestigkeit 1004
Bienengift 992
Bienenkönigin 7520
Bienne 1016
Bierhefe 993
Biestmilch 1901
Bilateralfurchung 1022
Bilateralsymmetrie 1023
bilateralsymmetrisch 1024
Bildungsdotter 6348
Bildungsgewebe 5495
 primäres 7292
Bilineurin 1656
Biliprotein 1033
Bilirubin 1034
Biliverdin 1035
Bindegewebe 1991
Bindegewebsknorpel 3388
Bindegewebszelle 3389
Bindehaut 1989
Bindung, energiereiche 2969
Binokularmikroskop 1040
Binominalverteilung 1042
Bioblast 1048
Biochemie 1054
Biochemiker 1053
biochemisch 1051
Biochore 1055
Biodynamik 1065
Bioenergetik 1066
Biogen 1067
Biogenese 1068
biogenetisch 1069
Biogenie 1072
Biogeochemie 1074
Biogeographie 1075
Biokatalysator 1050
Biokatalyse 1049
Bioklimatologie 1057
Biokybernetik 1098
Biologe 1086
Biologie 1087
biologisch 1076
biologisch abbaufähig 1063
Biolumineszenz 1088
Biolyse 1089
Biom 1091
Biomasse 1090
Biomechanik 1092
Biometeorologie 1093
Biometrie 1095
Biometrik 1095
biometrisch 1094
Biomolekül 1096
Bion 1097
Bionik 1098
Bionomie 1099
Biont 1100
Biophore 1101
Biophysik 1103
Biophysiker 1102
Biopsie 1104
Biorhythmus 1105
Bios 1106
Biosom 1108
Biosoziologie 1107
Biosphäre 1109
Biostatistik 1095
Biosstoffe 1106
Biosynthese 1111
Biosystematik 1112
Biotest 1047
Biotin 1117
biotisch 1113
Biotop 1118
Biotyp 1119
Biozid 1056
Biozönologie 1058
Biozönose 1059
Biozönotik 1058
biozönotisch 1060
Biozyklus 1062
Bipedie 1123
bipolar 1125
Bipolarität 1126
Bisexualität 1141
bisexuell 1140
Biuretreaktion 1142
Bivalent 1143
Bivalven 1145
Bläschen, pinozytäres 6954
 synaptisches 8869
Bläschenauge 6364
Bläschenfollikel 3919
Blasenauge 6364
blasenförmig 9521
Blasenkeim 1160
Blastem 1146
Blastoderm 1151
Blastodiskus 1152
blastogen 1154
Blastogenese 1153
Blastokolin 1147
Blastomer(e) 1155
Blastomyzet 1156
Blastoporus 1157
Blastospore 1159
Blastozöl 1148
Blastozyste 1149
Blastozyt 1150
Blastula 1160
Blastulabildung 1161
Blatt ... 3509
Blatt 5066
 zusammengesetztes 1950
Blattabsorption 5068
Blattabwurf 5067
Blattachsel 5069
Blattader 5090
Blattaderung 5091
Blattanlage 5080
Blattansatz 5078
blattartig 3505
Blattbasis 5070
Blattbildung 3510
Blattdorn 5084
Blättermagen 6327
Blattfall 5076
Blattfläche 5085
Blattfüßer 1257
Blattfußkrebse 1257
Blattgrün 1646
Blattgrund 5070
Blatthäutchen 5167
Blattkiemer 3139
Blattknospe 5072
blattlos 636
Blattlosigkeit 637
Blattlücke 5077
Blattnarbe 5082
Blattnerv 5090
Blattpflanze 3508

Blattpolster 5074
Blattranke 5087
Blattrippe 5081
Blattrosette 7914
Blattscheide 5083
Blattspindel 7530
Blattspitze 5088
Blattspreite 5071
Blattspur 5089
Blattsteckling 5075
Blattstellung 6880
Blattstiel 6781
Blattstielblatt 6878
Blattwedel 3591
Blaualgen 2210
Blepharoplast 1165
Blinddarm 1338
Blinzhaut 6122
Blockierung, genetische 3737
blühen 3481
Blühen 3489
Blühhormon 3491
Blühinduktion 4642
Blumenkrone 2062
Blumentiere 572
Blut ... 4050
Blutausstrich 1192
Blutbild 1183
blutbildend 4063
Blutbildung 4062
Blutdruck 1187
Blüte 3482, 3489
  männliche 8587
  weibliche 6959
Bluteiweiß 1188
Blütenachse 3468
Blütenanlage 3475
Blütenbecher 4478
Blütenbildung 3486
Blütenblatt 3473
Blütenboden 7632
Blütendiagramm 3469
Blütenformel 3471
Blütenhülle 6717
Blütenkelch 1368
Blütenknäuel 3868
Blütenknospe 3483
Blütenköpfchen 1395
Blütenkörbchen 1395
Blütenöffnung 567
Blütenorgan 3474
Blütenpflanzen 8472
Blütensproß 3493
Blütenstand 4656
  begrenzter 2332
  unbegrenzter 4623
Blütenstandachse 4657
Blütenstaub 7096
Blütenstiel 6670, 6678
blütentragend 3477
Blutersatzlösung 6890
Blütezeit 3492
Blutfaktor
Blutfarbstoff 1184
Blutfaserstoff 3384
Blutgefäß 1200
Blutgefäß-System 1199
Blut-Gehirn-Schranke 1170
Blutgerinnsel 1174
Blutgerinnung 1173
Blutgruppe 1179
Blutgruppen-Antigen 1180
Blutgruppenbestimmung 1181
Blutinsel 1182
Blutkieme 1178
Blutkörperchen 1175
  rotes 3109
  weißes 5123
Blutkreislauf 1172
Blutlakune 1193
Blutpigment 1184
Blutplasma 1185
Blutplättchen 9098
Blutprobe 1189, 1197
blutsaugend 4046
Blutsenkung 1190
Blutserum 1191
Blutstammzelle 4052
Blutstrom 1176
Blutung *(bot.)* 1162
Blutvolumen 1201
Blutzelle 4055
Blutzirkulation 1172
Blutzucker 1195
Blutzuckergehalt 3888
Bodenbakterien 8353
Bodenbedingungen 8355
bodenbewohnend 9006
Bodenbildung 8360
Bodenbiologie 8354
Bodenbrüter 3961
Bodenerhaltung 8356
Bodenfauna 8359
Bodenfließen 8382
bodenfremd 333
Bodenhorizont 8361
Bodenklimax 2789
Bodenkunde 8372
Bodenlebewelt 2793
Bodenmikrobiologie 8362
Bodenmikrofauna 8363
Bodenmikroflora 8364
Bodenmikroorganismen 8365
Bodenmüdigkeit 8358
Bodennister 3961
Bodenorganismen 2793
Bodenprobe 8371
Bodenprofil 8368
Bodenreaktion 8370
Bodenschutz 8369
bodenständig 818
Bodentier 8757
Bodentyp 8373
Bodenverschmutzung 8367
Bodenverseuchung 8357
Bodenzeiger 4630
Bogengänge 8127
Borke 947
Borste 1287
Borstenhaar 1287
Borstenwürmer 7125
borstig 1288
Botanik 1232
Botaniker 1231
botanisch 1230
Boten-RNS 5522
Bouillonkultur 1233
Brachiopoden 1237
Brachyblast 1238
Brachymeiosis 1239
Brackwasser 1241
bradytelisch 1245
Braktee 1242
Brakteole 1244
Branchialbogen 3828
Branchialskelett 1256
Branchie 3827
Braunalgen 6789
Brechungsindex 7684
breitblättrig 5054
Breitenwachstum 3980
Brennhaar 8664
Brenztraubensäure 7506
Brettwurzel 8907
Bronchie 1291
Bronchiole 1290
Bruchfrucht 5235
Bruch-Fusions-Brücken-Zyklus 1260
Bruch-Hypothese 1259
Bruch-Reunions-Bivalent 1261
Brücke *(anat.)* 7181
Brücken-Fragment-Konfiguration 1284
Brunft 6297
brunften 6298
Brunst 6297
brunsten 6298
brünstig (sein) 6298
Brunstlosigkeit 551
Brunstzeit 6296
Brunstzyklus 6295
Brust ... 6667
Brustatmung 2091
Brustbein 8634
Brustdrüse 5357
Brustfell 7056
Brustflosse 6668
Brusthöhle 9082
Brustkorb 9085
Brustwarze 5354
Brustwirbel 9084
Brut 1292, 4613
Brutablösung 8254
Brutbecher 1956
Brutbeutel 5388
Brutdauer 4618
brüten 4612
Brüten 4613
  in Kolonien 1275
Brutfleck 4616
Brutfürsorge 6622
Brutkleid 6239
Brutknospe 1295, 3695
Brutkolonie 1272
Brutkörbchen 1956
Brutkörper 7344
Brutparasit 1278
Brutparasitismus 1279
Brutpflege 1427
Brutpflegeverhalten 6623
Brutplatz 1270
Brutrevier 1282
Brutschrank 4619
Bruttasche 1300
Bruttoproduktion *(ecol.)* 3958
Bruttrieb 4615
Brutverhalten 4614
Brutzeit 1281
Brutzwiebel 1329
Bryologie 1306
Bryozoen 1308
Buchlunge 1227
Bukettstadium 1235
Bulbille 1329
Bündelkambium 3311
Burdo 1332
Bürstensaum 1304
Bürzel 9419
Bürzeldrüse 9420
Busch 8273
Büschelwurzel 3312
Buttersäure 1334
Buttersäuregärung 1335
Byssus 1336

Deutsch

C

^{14}C 7564
^{14}C-Test 7566
Caecum 1338
Caenogenese 1339
Carapax 1398
Carcinologie 1419
Casparyscher Streifen 1453
Chalaza 1598
Chalazogamie 1599
Chamäphyten 1600
Charakterart 1609
Charophyceen 1610
Chasmogamie 1611
Chasmophyten 1612
Chelat 1615
Chelatbildner 1616
Chelatbildung 1617
Chelatisierung 1617
Chelator 1616
Chelatverbindung 1615
Chelizeraten 1618
Chelizere 1619
chemoautotroph 1621
Chemonastie 1622
Chemorezeption 1623
Chemorezeptor 1624
Chemosynthese 1625
chemotaktisch 1626
Chemotaxis 1627
chemotroph 1628
Chemotropismus 1629
Chiasma 1631
  kompensierendes 1928
  komplementäres 1939
  optisches 6360
Chiasmafrequenz 1632
Chiasmainterferenz 1633
Chiasmatypie 1634
Chilopoden 1635
Chimäre 1636
Chinin 7524
Chinon 7525
Chitin 1638
chitinhaltig 1639
Chitinhülle 1640
Chitinmembran 1640
chitinös 1639
Chitinplatte 1641
Chlamydospore 1643
Chlorenchym 1644
Chlorokruorin 1645
Chlorophyceen 1647
Chlorophyll 1646
chlorophyllfrei 70
Chlorophyllkörnchen 1648
Chloroplast 1648
Choane 1649
Choanenfische 1650
Choanoflagellaten 1652
Choanozyte 1651
Cholesterin 1655
Cholin 1656
cholinergisch 1657
Cholinesterase 1658
Chondrin 1662
Chondriokont 1663
Chondriom 1664
Chondriomer 1665
Chondriosom 5709
Chondriosphäre 1667
Chondroblast 1668
Chondroklast 1669
Chorda dorsalis 6182
Chordamesoderm 1677
Chordaten 1678
Chordatiere 1678
Chordotonalorgan 1679
Chorioallantois 314
Chorion 1681
Choriongonadotropin 1683
Chorionhöhle 1682
Chorionzotte 1684
choripetal 2480
C-Horizont 1686
Chorologie 1688
chromaffin 1689
Chromatide 1691
Chromatidenaberration 1692
Chromatidenbrücke 1693
Chromatideninterferenz 1694
Chromatin 1695
Chromatinkörper 1696
chromatisch 1690
Chromatogramm 1697
Chromatographie 1699
chromatographieren 1698
Chromatometer 1700
Chromatophor 1701
Chromatoplasma 1702
Chromomer 1705
Chromomeren ... 1706
Chromonema 1708
Chromoplast 1709
Chromoproteid 1710
Chromosom 1713
  homologes 4356
Chromosomen ... 1711
Chromosomenaberration 1714
Chromosomenarm 1715
Chromosomenbestand 1717
Chromosomenbruch 1716
Chromosomenduplikation 1718
Chromosomenkarte 1719
Chromosomenkartierung 5370
Chromosomenmutation 1712
Chromosomenpaar 1721
Chromosomenpaarung 8864
Chromosomensatz 1723
Chromosomenschenkel 1715
Chromosomensegment 1722
Chromosomentheorie der Vererbung 1724
Chromosomentrennung 2609
Chromosomenummusterung 7731
Chromosomenverdoppelung 1718
Chromosomenzahl 1720
Chromosomenzahlenbastard 6238
Chromosomenzahlenkonstanz 6237
Chromosomin 1725
Chromozentrum 1703
Chromozyt 6933
Chronaxie 1726
Chrysalide 7468
Chylus 1729
Chylusbildung 1731
Chylusgefäß 1730
Chymosin 7727
Chymotrypsin 1734
Chymus 1732
Cis-Konfiguration 1757
Cis-Trans-Effekt 1758
Cis-Trans-Test 1759
Cistron 1760
Coecum 1338
Colchizin 1866
Colinearität 1880
Corium 2418
Corpora allata 2064
Corpora cardiaca 2065
Corpora pedunculata 5911
Corpus luteum 2068
Corpus-luteum-Hormon 2069
Corpus striatum 2070
Cortex, motorischer 5832
Cortisches Organ 2077
Crossing-over 2141
Crossing-over-Modifikator 2142
Crossing-over-Prozentsatz 2146
Cross(ing)-over-Suppressor 2144
Crossing-over-Wert 2146
Cross-over-Einheit 2145
Cultigen 9449
Cultivar 9449
Cyathium 2211
Cyto ... siehe Zyto ...

D

dachziegelig 4547
Damm 6736
Dammarharz 2269
Darm 4799
Darm ... 4791
Darmbein 4540
Darmflora 4793
Darmhöhle 4012
Darmkanal 4792
Darmlymphe 1729
Darmparasit 2989
Darmsaft 4794
Darmschmarotzer 2989
Darmtrakt 4796
Darmwand 4798
Darmzotte 4797
Darwinismus 2275
Dauerei 9693
Dauerform 7819
Dauergesellschaft 6757
Dauergewebe 6760
Dauerhumus 8573
Dauermodifikation 2276
Dauermyzel 8034
Dauerpräparat 6759
Dauerreiz 2015
Dauerspore 8958
dauerwarm 4352
Dauerwelke 6762
Dauerzahn 6761
Daune 2701
Deadaptation 2285
Deckblatt 1242
Deckelkapsel 7507
Deckfeder 2020
Deckflügel 2860
Deckglas 2106
Deckknochen 5473
Deckmembran 2076
Deckschuppe 2107
Deckspelze 5106
Deckungsgrad 2345
Dedifferenzierung 2320
Defäkation 2323
Degeneration 2340
degenerieren 2338
Dehiszenz 2352
dehnbar 2554
Dehnbarkeit 2553
Dehnungsreflex 8699
Dehydrase 2358

doppelbrechend 1137
Doppelbrechung 1136
Doppel-Crossing-over 2693
Doppeldolde 1952
Doppeldominant 2694
Doppelfüßer 2580
Doppelhelix 2696
Doppelhybrid 2410
Doppelkreuzung 2692
Doppelspirale 2696
doppelt gefiedert 1124
Dormin 2680
Dorn 9086
Dornfortsatz 6081
dornig 8507
dorsal 2681
Dorsiventralität 2686
Dosiseffekt 2688
Dosis-Effekt-Kurve 2689
Dosiskompensation 2687
Dotter 9740
dotterarm 6323
Dotterbildung 9599
Dottergang 9742
Dotterhaut 9597
dotterlos 293
Dotterpfropf 9744
Dottersack 9745
Dotterstock 9743
Dottervene 9598
Dotterzelle 9741
DPN 6118
Drang 2706
Dreifachkreuzung 9263
Dreigestaltigkeit 9258
dreihäusig 9259
Dreilinienkreuzung 9263
Dreipunktversuch 9090
Dressur 9191
Dressurversuch 9192
Drifttheorie 2013
Drillingsnerv 9253
Drohhaltung 9089
Drohne 2708
Drohstellung 9089
Druck, osmotischer 6420
Druckdifferenz 7279
Druckfestigkeit 1953
Druckgefälle 7279
Druckrezeptor 7280
Drüse 3850
  akzessorische 49
  apokrine 647
  ekkrine 2748
  endokrine 2910
  exokrine 3208
  holokrine 4316
  inkretorische 2910
  innersekretorische 2910
Drüsen ... 3851
drüsenähnlich 154
Drüsenepithel 3853
Drüsengewebe 3857
Drüsenhaar 3854
Drüsenhypophyse 153
Drüsenmagen 7418
Drüsenschuppe 3855
Drüsentätigkeit 8061
Drüsenzelle 3852
drüsig 3858
Duftdrüse 6283
Duftmarke 6285
Duftspur 6286
Duftstoff 6284
Dunkelfeld 2271
Dunkelfeldmikroskopie 2272
Dunkelkammer 2274
Dunkelreaktion 2273
Dünndarm 8329
Dünnschichtchromatographie 9077
Dünnschliff 9078
Dunstglocke 4113
Duodenum 2728
duplex 2729
Duplikat 2730
Duplikation 2732
duplizieren 2731
Durchblutung 1196
durchlässig 6765
Durchlässigkeit 6764
Durchlaßzelle 6644
Durchlüftungsgewebe 198
Durchschnürung 8691
Dürre 2710
dürrefest 2713
Dürrefestigkeit 2712
Dürrehärte 2712
dürreresistent 2713
Dürreresistenz 2712
Dyade 2739
Dyas 6768
Dysploidie 2742
dystroph 2743

E

Ebene, trophische 9277
Echinodermen 2753
Echolotung 2754
Echoortung 2754
Eckzahn 1385
Ectoprocta 2785
edaphisch 2788
Edaphon 2793
Edelreis 3923
EEG 2826
Effektor 2795
efferent 2797
Efferenz 2796
Egel 4287
Ei 2799
Eiablage 6465
Eiablageplatz 6466
Eiapparat 2801
Eibefruchtung 6343
Eibildung 6461
Eichel *(anat.)* 939
Eieralbumin 6446
Eier legen 6464
eierlegend 6463
Eierschale 2807
Eierstock 6452
eiförmig 6470
Eigenschaft 1605
  erworbene 96
Eignung 3427
Eihaut 2804
Eihülle 2804
Eikern 2805
Eiklar 283
Eilegeapparat 6467
Eileiter 3301, 6460
einachsig 5751
Einatmung 4719
Einatmungsluft 4721
einbetten 2862
Einbettung 2863
Einbettungsmasse 2864
Einbettungsmittel 2864
einblättrig 5788
einblütig 9377
eindicken 9071
Einehe 5769
einehig 5768
eineiig 5800
einfächerig 9381
Einfachkreuzung 8297
einfrüchtig 5754
eingehen 2515
Eingehen 2516
eingeißelig 9376
Ein-Gen-ein-Enzym-Hypothese 6337
eingeschlechtig 9386
Eingeschlechtigkeit 9387
eingestaltig 5783
Eingeweide 9547
Eingeweide ... 9548
Eingeweideganglion 9550
Eingeweidelehre 8523
Eingeweidenerv 8521
Eingeweideorgan 9552
Eingeweideschädel 9553
einhäusig 5764
Einhäusigkeit 5765
Einheitsmembran 9389
einjährig 541
einkapseln, sich 2897
Einkapseln 2898
einkeimblättrig 5761
einkeimig 9379
einkernig 5784
Einkrümmung 4620
Einmaligkeit des Individuums 9385
Einmieter 4701
Einnistung 6124
einpolig 9384
einsamig 5796
einschalig 9391
einschichtig 9388
Einschlußkörper 4597
Einschnürung 2003
Einsichtverhalten 4716
einspeicheln 4702
Einstülpung 4811
Einzelfrucht 8384
Einzelkornstruktur 8298
Einzeller 7396, 9375
einzellig 9373
Einzellkultur 9374
Eiröhre 6450
Eirollbewegung 2806
Eischale 2807
Eischlauch 6450
Eisenbakterien 4847
Eiszeit 3846
Eiszeitrelikt 3847
Eiweiß 7370
  einfaches 8293
  konjugiertes 1951
  körperfremdes 3550
  pflanzliches 9469
  tierisches 522
  zusammengesetztes 1951
Eiweißabbau 7371
eiweißartig 7385
Eiweißbedarf 7380
Eiweißgehalt 7374
Eiweißhülle 7372
Eiweißminimum 7377
Eiweißmolekül 7378

Eiweißreserve 7381
Eiweißstickstoff 7379
Eiweißstoff 7382
Eiweißstoffwechsel 7376
Eiweißverbindung 7373
Eizelle 6478
Eizellenbildung 6344
Ejakulat 2810
Ejakulation 2811
Ekdyse 5839
Ekdyson 2751
EKG 2822
Ektoblast 2779
Ektoderm 2779
Ektoenzym 2780
Ektogenese 2781
ektolezithal 2782
Ektoparasit 2783
Ektoplasma 2784
Ektoskelett 3216
Ektospore 3218
Ektotoxin 3224
ektotroph 2786
Elaioplast 2813
Elaiosom 2813
Elasmobranchier 2814
Elastin 2815
Elatere 2816
Elektrode 2824
Elektrodialyse 2825
Elektroenzephalogramm 2826
Elektroenzephalographie 2827
Elektrokardiogramm 2822
Elektrokardiographie 2823
Elektrolyt 2828
Elektromyogramm 2829
Elektromyographie 2830
Elektron 2831
Elektronenbild 2833
Elektronenfluß 3480
Elektronenmikroskop 2835
Elektronenmikroskopie 2836
Elektronenstrahl 2838
Elektronenstrahlung 2837
Elektronenträger 2832
Elektronentransport 2829
Elektronentransportkette 2840
Elektronen-Transport-Partikel 2841
Elektronentransportsystem 2842
Elektronik 2843
Elektroosmose 2844
Elektrophorese 2845
Elektrophysiologie 2846
Elektroplatte 2847
Elektroplaxe 2847
Elektroreiz 2818
Elektrotitration 2819
Elektrotonus 2848
Elektrotropismus 3640
Elementarkörperchen 2850
Elementarmembran 9389
Elementarteilchen 2851
Elimination 2853
Elitepflanzen 8091
Ellbogenhöcker 6305
Elle 2181
Ellipsoidgelenk 2854
Elter 6617
  männlicher 5342
  mütterlicher 3351
  väterlicher 5342
  weiblicher 3351
elterlich 6621
Elterngeneration 6625
Elternzelle 6618
Eluat 2857
Elution 2858
Emaskulation 1457
Embolie 2865
Embryo 2866
Embryogenese 2870
Embryogenie 2870
Embryoid 2871
Embryologe 2872
Embryologie 2873
embryonal 2875
Embryonalentwicklung 2877
Embryonalgewebe 2882
Embryonalhülle 2879
Embryonalknoten 2874
Embryonalschild 2880
Embryonalstadium 2881
Embryonalzelle 2867
Embryophyten 2883
Embryosack 2868
Embryosackmutterzelle 2869
Embryoträger 8835
Embryotrophe 2884
Emergenz 2885
EMG 2829
Emission *(env.)* 2886
Empfängerbakterie 7640
Empfängerzelle 7634
Empfängnishügel 3368
Empfindlichkeit 8155
Empfindung 8145
Empfindungsschwelle 8161
Emulgator 2889
emulgieren 2888
Emulgierung 2887
Emulsin 2890
Emulsion 2891
Emulsoid 2892
Endblüte 8995
Enddarm 4279, 7316
endemisch 2901
Endemismus 2903
Endemit 2902
endergonisch 2904
Endhandlung 2007
Endhirn 8955
Endknospe 638
Endobiose 2905
Endodermis 2915
Endoenzym 2916
Endoferment 2916
endogam 2917
Endogamie 2918
endogen 2919
Endogenese 2921
Endokard(ium) 2907
Endokarp 2908
Endokranium 2909
Endokrinologe 2912
Endokrinologie 2913
Endolymphe 2922
Endometrium 2923
Endomitose 2924
Endomixis 2925
Endomysium 2926
Endoneurium 2927
Endoparasit 2928
Endopeptidase 2929
Endophyt 2930
Endoplasma 2931
endoplasmatisch 2932
Endopolyploidie 2934
Endoprocta 2935
Endorgan 2937
Endosmose 2940
Endosom 2941
Endosperm 2942
Endospor(ium) 2944
Endospore 2943
Endost 2945
Endostom 2946
Endostyl 2947
Endosymbiose 2948
Endothel(ium) 2951
Endothelzelle 2950
endotherm 2952
Endothezium 2949
Endotoxin 2953
endotroph 2954
Endplatte 2955
  motorische 5833
Endplattenpotential 2956
Endprodukt 2957
Endstadium 8998
endständig 8991
Endwirt 2333
Endzelle 2900
Energide 2959
Energie, kinetische 4967
  potentielle 7224
Energiebedarf 2961
Energieerzeugung 2967
Energiefluß 2962
energieliefernd 3202
Energieniveau 2963
Energiepotential 2966
Energiequelle 2970
energiereiche Bindung 2969
Energiespeicherung 2971
Energiestoffwechsel 2965
Energieübertragung 2972
Energieumformung 2973
Energieverbrauch 2960
energieverbrauchend 2904
Energieverlust 2964
Engholz 5042
Engramm 2975
entarten 2338
Entartung 2340
entblättert 2336
Entblätterung 2337
Entdifferenzierung 2321
Entelechie 2981
Enteramin 8195
Enterogastron 2987
Enterokinase 2988
Enterokrinin 2986
Enteropeptidase 2988
Enterozöl 2985
Enterozoon 2989
entfärben 2307
Entfärbung 2306
entgiften 2447
Entgiftung 2448
Enthalpie 2990
enthärten *(Wasser)* 8351
enthemmen 2605
Enthemmung 2606
entkalken 2292
Entkalkung 2291
Entkeimungsfilter 915
entkernen 3002
Entkernung 3003
Entladung 2593
Entladungsfrequenz 2594
entlaubt 2336
Entlaubung 2337
Entlaubungsmittel 2335

entmannen 1456
Entmannung 1457
entnervieren 2374
Entnervierung 2375
Entoblast 2914
Entoderm 2914
entomogam 2997
Entomogamie 2999
Entomologe 2994
Entomologie 2995
Entomophilie 2999
Entoparasit 2928
Entoplasma 2931
Entoskelett 2939
Entozoon 3000
Entparaffinierung 2298
Entropie 3001
entsalzen 2420
Entsalzung von Meerwasser 2421
Entsalzungsanlage 2422
entsäuern 2284
Entsäuerung 2283
entschlüsseln 2304
Entschlüsselung (des gen. Kodes) 2305
entschmutzen 2405
Entschmutzung 2406
entschwefeln 2441
Entschwefelung 2440
entseuchen 2313
Entseuchung 2315
Entseuchungsanlage 2316
Entseuchungsmittel 2314
Entwässerung 2355
Entwicklung 2457
  postnatale 7217
  pränatale 7270
  regressive 7700
Entwicklungsbeschleunigung 45
Entwicklungsbiologie 2459
entwicklungsfähig 2456
Entwicklungsfähigkeit 2460
Entwicklungsgeschichte 6338
entwicklungsgeschichtlich 6339
Entwicklungsgeschwindigkeit 2465
Entwicklungskräfte 3164
Entwicklungsmechanik 2463
Entwicklungsphase 6738
Entwicklungsphysiologie 2463
Entwicklungsprozeß 2464
Entwicklungsrate 2465
Entwicklungsreihe 3168
Entwicklungsstadium 2466
Entwicklungsstillstand 713
Entwicklungsstörung 2461
Entwicklungsverzögerung 2361
Entwicklungszyklus 3162
Enzephalisierung 2896
Enzym 3029
  adaptives 142
  induzierbares 142
  konstitutives 2002
  repressibles 7738
Enzymaktivität 3030
enzymatisch 3028
Enzymhemmer 3035
Enzymhemmung 3034
Enzyminduktion 3033
Enzyminhibitor 3035
Enzymkatalyse 3031
Enzymologie 3043
Enzymprotein 3037
Enzymreaktion 3038
Enzymrepression 3039
Enzymspezifität 3040
Enzym-Substrat-Komplex 3041
Enzymsynthese 3042
Enzystierung 2898
Eosin 3045
eosinophil 3047
Eozän 3044
Eozoikum 685
Ependym 3048
Ephapse 3049
Epharmonie 3050
Ephedrin 3051
Ephemere 3052
Epiblast 2779
Epibolie 3055
Epidemiologie 3059
Epidermis 3061
Epidermiszelle 3060
epigäisch 3064
Epigastrium 3063
Epigenese 3065
Epigenetik 3066
epigyn 3068
Epikard(ium) 3056
Epikarp 3205
Epikotyl 3058
Epilimnion 3069
Epilith 7934
Epimorphose 3071
Epimysium 3072
Epinastie 3073
Epineurium 3075
Epipelagial 3076
epiphyll 3077
Epiphyse 3079, 3080
Epiphysenknorpel 3078
Epiphyten 3082
epiphytisch 3083
Episit 7245
Episitie 7244
Episitismus 7244
Episom 3085
Epistasie 3087
Epithel 3092
  kubisches 2180
  prismatisches 1913
Epithelgewebe 3091
Epithelkörperchen 6611
Epithelzelle 3090
Epithem 3093
Epithezium 3088
epizoisch 3094
Epizoon 3095
EPSP 3187
Erbanalyse 3735
Erbänderung 4185
Erbanlage 4186
erbbedingt 4183
Erbbild 3765
Erbeigenschaft 4192
Erbeinheit 4193
erben von 4670
Erbfaktor 3696
Erbfehler 4188
Erbgang 2099
Erbgefüge 4187
Erbgut 4191
Erbhygiene 3137
Erbinformation 4190
Erbkoordination 3431
Erblehre 3748
erblich 4183
Erblichkeit 4194
Erblichkeitsgrad 4194
Erbmaterial 4191
Erbmerkmal 4186
Erbplasma 3787
Erbschaden 4188
Erbträger 990
Erbtyp(us) 3765
Erdaltertum 6535
erdbewohnend 9006
Erdfrüchtigkeit 3773
Erdmittelalter 5521
Erdneuzeit 1546
Erdpflanzen 3777
Erdschürfepflanze 4160
Erdsproß 8758
Erdwendigkeit 3779
Erdzeitalter 3776
Erepsin 3101
Erfolgsorgan 2795
Ergänzungsmännchen 1938
Ergastoplasma 3102
Ergin 3103
Ergon 3103
Ergosom 7166
Ergosterin 3105
Ergotinin 3106
Erhaltungsquotient 8827
Erholungsraum *(env.)* 5105
Erkundungsverhalten 3248
Ermüdbarkeit 3325
Ermüdung 3324
Ernährung 6253
Ernährungsfaktor 6256
Ernährungsgewohnheit 3529
Ernährungsphysiologie 6893
Ernährungssymbiose 9278
Ernährungsweise 5726
Ernährungswissenschaft 9281
Erneuerung 7694
Erneuerungsknospe 7725
Erneuerungssproß 4696
Erosion 3107
erregbar 3182
Erregbarkeit 3181
Erregung 3183
erregungsfähig 3182
Erregungsfähigkeit 3181
Erregungshöhe 3189
Erregungsleitung 1967
Erregungsmuster 3184
Erregungsniveau 3189
Erregungsschwelle 3185
Erregungsstrom 3186
Erregungssubstanz 3188
Erregungszustand 8600
Ersatzgesellschaft 8752
Ersatzknochen 1446
Erscheinungsbild 6814
Erscheinungstyp 6814
erstarren 8381
Erstarrung 8380
Erstmünder 7412
erwachsen 188
Erwachsenenstadium 191
Erythroblast 3108
Erythrozyt 3109
Essigsäure 60
Essig(säure)bakterien 63
Essigsäuregärung 61
Ester 3116
Esterase 3117
Etheogenese 3123
Ethogramm 3126
Ethologie 3129
ethologisch 3127
Etiolement 3132
Euchromatin 3134
euchromatisch 3133

Fetus 3504
Feuchtgewicht 9670
Feuchtigkeitsgehalt 5737
Feuchtigkeitsgrad 5738
feuchtigkeitsliebend 4470
Feuchtigkeitsmesser 4468
Feuchtpflanzen 4471
Feuchtwald 5736
Feulgenreaktion 3375
F-Faktor 3365
F_1-Generation 3377
F_2-Generation 3378
fibrillär 3383
Fibrille 3382
Fibrin 3384
Fibrinogen 3385
Fibrinolysin 3386
Fibroblast 3387
Fibroin 3390
Fibrozyt 3389
Fibula 3395
Fieder 6949
Fiederblatt 6951
Fiederblattstiel 6782
fiederförmig 6952
fiedernervig 6953
Filament 3403
Filialgeneration 3406
  erste 3377
  zweite 3378
Filopodien 3409
filtern 3410
Filterpapier 3412
Filterung 3414
Filtrat 3413
Filtration 3414
filtrieren 3410
Filtrierer *(zool.)* 3411
Filtrierpapier 3412
Fischbank 8015
Fischbrut 3421
Fischkunde 4525
Fischschwarm 8015
fixieren 3429
Fixierung 3430
Fixier(ungs)flüssigkeit 3433
Fixier(ungs)mittel 3432
Flachauge 3288
flächenständig 5021
Flachmoor 5251
Flachsproß 1763
Flachwurzler 8253
Flagellaten 3437
Flagellum 3442
Flammenzelle 3444
Flaum 2700
Flaumfeder 2701
flaumig 5030
Flavin 3446
Flavinadenindinukleotid 3447
Flavinenzym 3451
Flavinmononukleotid 3448
Flavon 3449
Flavonoid 3450
Flavoprotein 3451
Flechte 5131
Flechtenkunde 5133
Flechtenstärke 5132
Flechtgewebe 7046
Fleck, blinder 1167
  gelber 9739
fleischfressend 1431
Fleischfresser 1430
fleischig 3452
Flexor 3453
Fließend-Wasser-Ökosystem 3495
Fließgewässer 3494
Fließgleichgewicht 8608
Flimmer 1744
Flimmerbewegung 1737
Flimmerepithel 1742
Flimmertrichter 6035
Flimmerzelle 1741
Flockenbildung 3464
flockig 3465
Flora 3467
Florenelement 3470
Florengebiet 3476
Florenkunde 3479
Florenreich 3472
Florigen 3491
Floristik 3479
Flosse 3415
Flossenstrahl 3416
Fluchtreaktion 3112
Fluchttrieb 3111
Fluchtverhalten 3110
Flügel 9685
Flügelfrucht 9688
Flügelgeäder 9686
Flügelkiemer 7442
flügellos 668
Flügellose 669
Flügelmuskel 279
flugfähig 1387
Flugfähigkeit 3456
Flugfeder 3457
Flughaut 6648
Fluglärm 267
Flugmuskel 3458
flugunfähig 3459
Fluktuation 7191
Fluoreszenz 3496
Fluoreszenzmikroskopie 3497
Flüssigkeitskultur 5210
Flußökosystem 3498
Flußverschmutzung 7888
FMN 3448
Folgeblatt 5991
Folgemeristem 8052
Follikel 3515
Follikelhormon 6294
Follikelreifungshormon 3518
Follikelsprung 6475
Follikelzelle 3517
Folsäure 3511
Fontanelle 3522
Foramen 3546
Formaldehyd 3563
Formation 3564
  retikulare 7830
Formensinn 3561
Formentrennung 8178
Formkonstanz 3560
Formsehen 3562
forschen (über) 7768
Forscher 7773
Forschung 7767
Forschungsarbeit 7772
Forschungsgebiet 7769
Forschungslaboratorium 7770
Forschungsprogramm 7771
Forstbotanik 3555
Fortbewegung 5226
Fortbewegungsfähigkeit 5229
Fortbewegungsorgan 5227
Fortbewegungsweise 5571
fortpflanzen (sich) 7742
Fortpflanzung 7745
  geschlechtliche 8247
Fortpflanzung, sexuelle 8247
  ungeschlechtliche 750
  vegative 750
Fortpflanzungsbiologie 7751
fortpflanzungsfähig 7744
Fortpflanzungsfähigkeit 7753
Fortpflanzungsgemeinschaft 7748
Fortpflanzungskörper 7752
Fortpflanzungsorgan 7759
Fortpflanzungsphysiologie 7761
Fortpflanzungspotential 7762
Fortpflanzungsrate 7746
Fortpflanzungsschranke 7749
Fortpflanzungsstadium 7763
Fortpflanzungssystem 7764
Fortpflanzungstrakt 7764
Fortpflanzungstrieb 7757
Fortpflanzungsverhalten 7750
Fortpflanzungsvermögen 7753
Fortpflanzungsweise 5573
Fortpflanzungszeit 7760
Fortpflanzungszelle 7754
Fortpflanzungszyklus 7756
fossil 3566
Fossil 3566
fossilführend 3567
Fossilisation 3568
fossilisieren 3569
fötal 3500
Fötalisation 3503
Fötus 3504
Fovea 3572
Fraktionskollektor 3573
freier Raum 660
Freiheitsgrad 2346
frei(kron)blättrig 2480
Freilandversuch 3398
freisetzen *(Energie etc.)* 5130
Freizeitgebiet *(env.)* 5105
Fremdbefruchtung 338
Fremdbestäubung 338, 9716
Fremdkörper 3549
Fremdstoff 3551
Fremdzucht 6442
Frequenzmodulation 3581
Frequenzverteilung 3580
Freßtrieb 3342
Freßverhalten 3341
Freßzelle 6791
Frischgewicht 3589
Frischpräparat 3584
Froschlurche 627
Frosthärte 3596
Frostresistenz 3596
Frostschaden 3595
Frucht 3603
  zusammengesetzte 1947
Frucht bilden 3600
Frucht tragen 3600
Fruchtabwurf 3607
Fruchtansatz 3608
Fruchtauge 3606
fruchtbar 3362
Fruchtbarkeit 3363
Fruchtbarkeitsvitamin 9145
Fruchtbarkeitsziffer 3364
Fruchtbecher 2196
Fruchtbildung 3599
Fruchtblatt 1438
Fruchtboden 4519
früchtefressend 3602
Fruchtfall 3607
Fruchtfleisch 7460
Fruchthalter 1440, 9426
Fruchthof, dunkler 697

Fruchthof, heller 698
Fruchtholz 3611
Fruchthülle 2879
Fruchtknoten 6452
Fruchtkörper 3605
Fruchtkuchen 219
Fruchtschuppe 8072
Fruchtstand 4662
Fruchtstiel 3609
fruchttragend 3598
Fruchtwand 6722
Fruchtwasser 407
Fruchtwassersack 408
Fruchtzucker 3601
Frühholz 7296
Frühlingsholz 7296
frühreif 7239
Frühreife 7240
Fruktifikation 3599
fruktifizieren 3600
Fruktose 3601
Frustel 3612
FSH 3518
Fühler 558
Fühlerlose 1618
Fühlorgan 8912
Führer *(Vogelzug)* 5064
Fukoxanthin 3617
Fumarase 3618
Fumarsäure 3619
Fungi 3624
  imperfecti 3625
Fungizid 3626
Funikulus 3627
Funktionseinheit 3620
Funktionsgen 1760
Funktionswechsel 1604
furchen (sich) 1778
Furchung 1779
Furchungsebene 1784
Furchungshöhle 1148
Furchungskern 1783
Furchungsspindel 1785
Furchungsteilung 1782
Furchungszelle 1781
Fusion, zentrische 1553
Fusionskern 3632
fußlos 648
Fußwurzel 8933
Fußwurzelknochen 8932
Futterplatz 3344

G

gabelig 2507
gabelteilig 2507
Gabelteilung 2509
Gabelung 2509
Galaktolipid 3634
Galaktose 3636
Galle 1025
Gallen ... 1030
Gallenabsonderung 1031
Gallenblase 3638
Gallenfarbstoff 1028
Gallengang 1027
  gemeinsamer 1654
Gallenpigment 1028
Gallensalz 1029
Gallensäure 1026
Gallensekretion 1031
Gallert(e) 3684
gallertig 3688
Gallusgerbsäure 8926
Galvanotaxis 3639
Galvanotropismus 3640
Gamet 3643
Gametangiogamie 3641
Gametangium 3642
Gameten ... 3644
Gametenkern 3645
gametisch 3644
Gametogamie 3647
gametogen 3649
Gametogenese 3648
Gametogenie 3648
Gametogonie 3656
Gametophor 3651
Gametophyt 3652
Gametozyt 3646
Gammaglobulin 3653
Gammastrahlen 3654
Gamobium 3655
Gamogonie 3656
Gamon 3657
Gamont 3658
gamopetal 3659
Gamophase 3660
gamosepal 3661
Gamotropismus 3662
Gangbein 9616
Ganglienkette, sympathische 8855
Ganglienzelle 3664
Ganglion 3663
Ganoidschuppe 3665
Ganzpräparat 9673
ganzrandig 2991
Ganzschmarotzer 4324
gären 3355
gärfähig 3356
Gärröhrchen 3358
Gärung 3357
  alkoholische 289
Gärungsbakterien 3359
Gärungskunde 9795
gärungsverhindernd 625
Gasaustausch 3669
  respiratorischer 7795
Gaschromatographie 3668
Gastral ... 3671
Gastralhöhle 3670
Gastralraum 1851
gastrisch 3671
Gastropoden 3676
Gastrovaskularraum 3677
Gastrozöl 3674
Gastrula 3678
Gastrulation 3679
Gattung 3769
Gattungs ... 3730
Gattungsbastard 4748
Gattungskreuzung 4749
Gattungsname 3770
Gattungstyp 3732
Gaumen 6541
  harter 4099
  weicher 8348
Gaumen ... 6537
Gaumenbein 6538
Gaumenmandel 6540
Gaumenreflex 8837
Gaumensegel 6542
Geäder 6038
Gebärmutter 9426
Gebiß 2394
Geburt 1138
Geburtenrate 1139
Geburtenziffer 5985
Gedächtnisinhalt 2975
gedreht *(Knospendeckung)* 2019
gefaltet *(Knospenlage)* 7059
Gefäß 9522
Gefäß ... 9452
Gefäßbildung 9462
Gefäßbündel 9453
Gefäßelement 9523
Gefäßerweiterung 9465
Gefäßgewebe 9461
Gefäßhaut 1687, 9457
Gefäßkryptogamen 9455
Gefäßlehre 495
gefäßlos 6177
Gefäßpflanze 9458
gefäßreich 9463
Gefäßstrang 9459
Gefäßsystem 9460
Gefäßverengung 9464
Gefäßversorgung 9462
Gefäßzylinder 9456
Gefieder 7062
gefiedert 6950
gefingert 2547
geflügelt 280
gefranst 4996
Gefriermikrotom 3578
Gefrierpunkt 3579
Gefrierschnitt 3597
Gefriertrocknung 5285
gegabelt 1021
Gegenfüßlerzelle 611
Gegengift 593
Gegenhormon 602
gegenständig 6358
Gehilfin 8880
Gehirn 1246
Gehirn ... 1579
Gehirnhaut, harte 2733
  weiche 6929
Gehirnnerv 2116
Gehirnrückenmarksflüssigkeit 1587
Gehirnwindung 4031
Gehölzkunde 2373
Gehör 4116
Gehör ... 796
Gehörbläschen 809
Gehörgang, äußerer 801
  innerer 801
Gehörkapsel 799
Gehörknöchelchen 2746
Gehörsinn 805
Gehörstein 6441
Geigerzähler 3680
Geißel 3442
Geißelalgen 6902
Geißelbewegung 3436
Geißelepithel 1742
geißelförmig 3441
Geißelinfusorien 3437
geißellos 788
Geißeltierchen 3437
Geißelzelle 1742, 3439
Geitonogamie 3682
Geitonokarpie 3681
gekammert 1374
gekerbt 2126
Gekröse 5504
Gel 3683
gelappt 5222
gelartig 3688
Gelatine 3684
gelatinieren 3686
Gelatinierung 3685

H

Haftorgan 163
Haftscheibe 162
Haftvermögen 164
Haftwurzel 1767
Hagelschnur 1598
Haiartige 2814
hakenförmig 9363
Halbaffe 7355
Halbantigen 4093
Halbchiasma 4072
Halbchromatide 4073
Halbchromosom 4074
halbdurchlässig 8139
Halbdurchlässigkeit 8138
Halbflügler 4165
halbflüssig 8131
Halbmond, grauer 3954
Halbmutante 4076
Halbparasit 4164
Halbrasse 4077
Halbschmarotzer 4164
Halbseitenzwitter 4079
Halbstrauch 8780
Halbwertzeit 4075
biologische 1079
Halbwüste 8129
Haliplankton 4082
Halm 2183
halophil 4080
Halophyten 4081
Haloplankton 4082
Hals ... 1591
Halsschlagader 1436
Halswirbel 1592
Haltere 4083
Häm 4037
Hämagglutination 4038
Hämagglutinin 4039
Hämalbogen 4040
Hämatin 4041
Hämatogenese 4044
Hämatologie 4045
hämatophag 4046
Hämatopoese 4062
Hämatoxylin 4048
Hämerythrin 4049
Hämin 4051
Hammer *(anat.)* 5347
Hämoblast 4052
Hämodynamik 4056
Hämoglobin 4057
Hämogramm 1183
Hämolymphe 4059
Hämolyse 4061
Hämolysin 4060
Hämozöl 4053
Hämozyanin 4054
Hämozyt 4055
Handflügler 1637
handförmig 6552
handförmig gelappt 6553
handförmig geteilt 6554
Handgelenk 9710
Handlungskette 8189
Handwurzel 1443
Handwurzelknochen 1437
Hängetropfenkultur 4084
Haplobiont 4085
haplochlamyd(eisch) 4086
haploid 4087
Haploidie 4088
Haplont 4089
Haplophase 4090
Haplosis 4091
haplostemon 4092
Hapten 4093
Hapteren 4094
Haptonastie 4096
Haptotropismus 4097
Harn 9413
Harnblase 9411
Harnkanälchen 9414
Harnleiter 9407
Harnröhre 9408
Harnsäure 9409
Harnstoff 9403
Harnstoffzyklus 6394
Harnwege 9412
hartblättrig 8030
Härtegrad 4103
härten 4101, 8381
Hartholz 2734
Hartlaubgewächse 8031
Härtung 4102, 8380
Harz 7781
Harzgang 7782
Harzkanal 7782
Hauer 9316
Hauptgen 6322
Hauptnährstoff 5316
Hauptnerv 5662
Hauptstamm 8805
Haupttrieb 5065
Hauptwirt 2333
Hauptwurzel 5332
Haustier 2662
Haustierwerdung 2665
Haustorie 4111
Haut 8314
Haut ... 2200
Hautatmung 2203
Hautdrüse 8316
häuten (sich) 5838
Hautfalte 8315
Hautflügler 4475
häutig 5480
Hautknochen 5473
Hautpapille 2415
Hautpflegehandlungen 3957
Hautpigment 8317
Hautsinnesorgan 2202
Hautskelett 3216
Hauttalg 8048
Häutung 3282, 5839
Häutungsdrüse 5841
Häutungsflüssigkeit 5840
Häutungshormon 2751
Hautzahn 2390
Hefe 9734
Hefeautolysat 9735
Hefeextrakt 9736
Hefepilz 1156
Heimfinden 4335
Heimkehrvermögen 4335
Heimrevier 4327
Heiratskreis 4875
Hektokotyl 4139
heliophil 4146
heliophob 4147
Heliophyt 4149
Heliotropismus 6867
Helix 4151
Helix-Struktur 4140
Hellfeld 1285
Hellfeldmikroskopie 1286
Helminthologie 4152
Helophyt 4153
Helotismus 4154
Hemichordaten 4159
Hemikaryon 4161
Hemikryptophyt 4160
Hemimetabolen 4162
Hemimetabolie 4163
Hemiparasit 4164
Hemipteren 4165
hemitrop 4166
Hemizellulose 4158
hemizygot 4167
hemmend 4676
Hemmer 4675
Hemmfaktor 4678
Hemmstoff 4681
Hemmsubstanz 4681
Hemmsynapse 4682
Hemmung, kompetitive 1934
nichtkompetitive 6155
reziproke 7644
Hemm(ungs)faktor 4678
Hemmungsreaktion 4680
Hemmwirkung 4677
Hensenscher Knoten 7302
Heparin 4170
Herbar(ium) 4176
Herbivore 4178
Herbizid 4177
Herde 3466, 4181
in Herden lebend 3952
Herdentrieb 4182
Heritabilität 4194
Herkogamie 4180
Hermaphrodit 4196
hermaphroditisch 4197
Hermaphroditismus 4199
Herpetologie 4200
Herrentiere 7297
Herz ... 1421
Herzbeutel 6721
Herzbeutelhöhle 6720
Herzfrequenz 3582
Herzinnenhaut 2907
Herzkammer 9505
Herzklappe 4120
Herzkranzgefäß 2063
Herzminutenvolumen 4119
Herzmuskel 5950
Herzohr 810
Herzrhythmus 1425
Herzschlag 4118
Herztätigkeit 1422
Herzvorhof 789
Herzzyklus 1423
Heteroallel 4201
Heteroauxin 4638
heterochlamyd(eisch) 4205
Heterochromatie 4208
Heterochromatin 4207
heterochromatisch 4206
Heterochromosom 4209
heterodont 4211
Heterogamet 532
heterogametisch 4215
Heterogamie 533
heterogen(isch) 4220
heterogenetisch 4218
Heterogenie 4221
Heterogenität 4217
Heterogenote 4219
Heterogonie 4222
heterogyn 4224
heterokarp 4203
Heterokaryon 4225
Heterokaryose 4226
Heterokinese 4227
heterokont 4228
heterolog 4229

Heterologie 4230
Heterolyse 4231
heteromer 4232
Heterometabolen 4162
Heterometabolie 4163
heteromorph 4234
Heteromorphismus 4235
Heteromorphose 4236
heteronom 4237
Heterophyllie 4238
heterophytisch 4239
Heteroplantat 4223
Heteroplasmonie 4240
Heteroplastie 4241
heteroploid 4242
Heteroploidie 4243
Heteropyknose 4244
Heterosis 4245
Heterosiszüchtung 4262
heterosomal 4246
heterospor 4247
Heterosporie 4248
heterostyl 4249
Heterostylie 4250
Heterosynapsis 4251
Heterosyndese 4252
Heterothallie 4254
heterothallisch 4253
heterotrich 4255
heterotroph 4257
Heterotrophie 4256
Heterotyp 4258
heterozentrisch 4204
Heterözie 4213
heterözisch 4212
heterozygot 4261
Heterozygote 4260
Heterozygotie 4259
Heterozyste 4210
Heuaufguß 4112
hexaploid 4263
Hexaploidie 4264
Hexokinase 4266
Hexose 4267
Hibernakel 4268
Hilfszelle 855
Hilum 4277
Hinterhaupt 6278
Hinterhauptsbein 6275
Hinterhauptshöcker 6276
Hinterhauptslappen 6277
Hinterhirn 5569
Hinterhorn 2683
Hinterleib 1
Hirn 1246
Hirn ... 1579
Hirnanhangdrüse 4511
Hirnbläschen 1250
Hirnhäute 5491
Hirnkammer 1584
Hirnmantel 6551
Hirnnerv 2116
Hirnreizung, elektrische 2820
Hirnrinde 1581
Hirnschädel 6087
Hirnschale 2114
Hirnstamm 1249
Hirnventrikel 1584
Hirnwindung 4031
Hirudin 4286
Histamin 4288
Histidin 4289
Histiozyt 4290
Histochemie 4292
histochemisch 4291
Histogen 4294
Histogenese 4295
Histogramm 4296
Histologe 4299
Histologie 4300
histologisch 4297
Histolyse 4301
Histon 4303
Historadiographie 4305
hitzebeständig 9066
Hitzeempfindlichkeit 9048
hitzeresistent 4135
Hitzeresistenz 4134
Hitzestabilität 4137
Hitzestarre 9046
hitzeunbeständig 9056
Hochblatt 4522
Hochmoor 4274
Hochzeitsflug 6240
Hochzeitskleid 6239
Hoden 9023
Hodenkanälchen 8137
Hodensack 8039
Hoftüpfel 1228
Höhenstufe 368
Höhenwachstum 3981
Höhenzuwachs 3981
Höherentwicklung 449
höhlenbewohnend 1493
Höhlenbrüter 1492
Höhlenfauna 1491
Höhlentiere 1491
Hohlorgan 4311
Hohlspindel 4312
Hohltiere 1850
Hohlvene 1490
holandrisch 4307
Holarktis 4308
Holobasidie 4313
holoblastisch 4314
Holoenzym 4317
hologam 4319
Hologamet 4818
Hologamie 4320
hologyn 4321
Holometabolen 4322
Holometabolie 4323
Holoparasit 4324
Holostandard 4326
holotrich 4325
Holozän 4315
Holz 9704
Holzfaser 5163
holzfressend 9730
Holzgewebe 9708
holzig 5159
Holzparenchym 9705
Holzpflanzen 5160
Holzstoff 5165
Holzteil 9728
Holzzucker 9731
Hominiden 4336
Homoallel 4337
Homochromie 4339
homodont 4340
homogam 4343
homogametisch 4342
Homogamie 4344
homogen 4347
Homogenat 4345
homogenetisch 4348
Homogenität 4346
Homogentisinsäure 4349
homoiolog 4329
Homokaryon 4353
Homokaryose 4354
homolezithal 4355
Homologie 4357
Homomerie 4358
homomorph 4359
Homöokinese 4328
homöolog 4329
Homöosis 4330
Homöostasis 4331
homöotherm 4352
homöotypisch 4334
homophytisch 4360
Homoplantat 4350
homoplastisch 4361
Homoploidie 4362
homopolar 4363
Homopteren 4364
homosomal 4365
homostyl 4368
Homostylie 4369
Homosynapsis 4332
Homothallie 4371
homothallisch 4370
Homotyp 4372
Homotypie 4373
homozentrisch 4338
homozygot 4376
Homozygote 4374
Homozygotie 4375
Horde 4381
Hör ... 796
Hören 795
Hormogonium 4382
Hormon 4386
  adrenokortikotropes 181
  follikelstimulierendes 3518
  gonadotropes 3911
  laktotropes 5011
  luteinisierendes 5269
  luteotrop(h)es 5011
  melanophorenstimulierendes 4759
  thyreotropes 9112
  zwischenzellenstimulierendes 5269
hormonal 4383
hormonbildend 4387
hormonell 4383
Hormontätigkeit 4384
Horn 4388
Hornbildung 2060
Hörnerv 802
Hornhaut 2056
Hornhautreflex 2057
hornig 2058
Hornplatte 4389
Hornschicht 2059
Hornsubstanz 4950
Hörorgan 803
Hörreiz 806
Hörschärfe 797
Hörschwelle 807
Hörvermögen 798
Hörzelle 800
Huf 4378
Hufeisenwürmer 6826
Hüft ... 8019
Hüftbein 4283
Hüfte 4282
Hüftgelenk 4285
Hüftgelenkpfanne 58
Huftiere 9370
Hüftnerv 8020
Hüllchen 4822
Hüllfrucht 2224

Hüllkelch 4823
Hüllspelze 3881
Hülse 5103
Hülsenfrüchtler 5103
Humanbiologie 4398
Humangenetik 4400
Humanökologie 4399
Humerus 4402
Humifizierung 4406
Huminsäure 4403
Humus 4407
  milder 5679
  saurer 7610
Humusbildung 4406
Hundertfüßer 1635
Hyalin 4408
Hyaloplasma 4410
Hyaluronsäure 4411
hybrid 4412
Hybrid(e) 4413
  numerische 6238
Hybrid(en)schwarm 4417
Hybridensterilität 4416
Hybridenzone 4419
Hybridisierung 4421
  introgressive 4807
Hybridismus 4420
Hybridität 4420
Hydathode 4423
Hydratation 4424
hydratisieren 4425
Hydratisierung 4424
hydrieren 4443
Hydrierung 4444
Hydrobiologie 4427
hydrobiologisch 4426
Hydrochorie 4431
Hydrogamie 4452
Hydrogel 4434
Hydrogenase 4442
Hydrokortison 4433
Hydrokultur 4456
Hydrolase 4445
Hydrolysat 4446
Hydrolyse 4447
hydrolysieren 4449
hydrolytisch 4448
Hydronastie 4450
hydrophil 4451
Hydrophilie 4452
hydrophob 4453
Hydrophyten 4455
Hydroponik 4456
Hydroserie 4457
Hydrosol 4458
Hydrosphäre 4460
Hydrotaxis 4461
Hydrotropismus 4462
Hydroxyd 4463
Hydroxylamin 4465
Hydroxylgruppe 4464
Hydroxylierung 4466
Hydrozoen 4467
Hydrozöl 4432
Hygrometer 4468
Hygronastie 4469
hygrophil 4470
Hygrophyten 4471
hygroskopisch 4472
Hygroskopizität 4473
Hymenium 4474
Hymenopteren 4475
Hyoidbogen 4476
Hypanthium 4478
Hyperchimäre 4479
Hyperchromatizität 4480
Hyperchromatose 4480
Hypermetabolie 4481
Hypermetamorphose 4481
hypermorph 4482
Hypermorphose 4483
Hyperparasit 4484
Hyperplasie 4485
hyperploid 4486
Hyperploidie 4487
Hyperpolarisation 4488
Hypertelie 4489
Hypertonie 4491
hypertonisch 4490
Hypertrophie, kompensatorische 1931
Hyphe 4492
Hyphomyzeten 4493
Hypobasidie 7310
Hypoblast 2914
Hypoderm(is) 4498, 8725
hypogäisch 4500
Hypogastrium 4499
Hypogenes(i)e 4501
Hypoglossus 4502
hypogyn 4503
Hypokotyl 4496
Hypolimnion 4504
hypomorph 4506
Hyponastie 4507
hypophyll 4508
Hypophyse 4511
Hypophysenhinterlappen 7211
Hypophysenhinterlappenhormon 7212
Hypophysenhormon 4509
Hypophysenstiel 4510
Hyopophysenvorderlappen 561
Hypophysenvorderlappenhormon 562
Hypoplasie 4512
hypoploid 4513
Hypoploidie 4514
Hypostase 4515
Hypothalamus 4517
Hypothallus 4518
Hypothese von der Kontinuität des Keimplasmas 9043
Hypothezium 4519
hypotonisch 4520
Hypoxanthin 4521
Hysterese 4523

I

Ichthyologie 4525
ICSH 5269
Id 4527
Idant 4528
Idioblast 4531
Idiochromatin 4532
Idiogramm 4533
Idioplasma 3787
Idioplasmatheorie 9043
Idiotyp 4535
Idiovariation 4536
Idiozom 4537
IES 4638
Ileum 4538
Imaginalhäutung 190
Imaginalscheibe 4542
Imago 4544
imbrikat 4547
immergrün 3157
Immersionsobjektiv 4550
Immission *(env.)* 4553
immun 4554
Immunbiologie 4562
immunisieren 4561
Immunisierung 4559
Immunität 4557
  aktive 124
  erworbene 97
  passive 6645
Immunitätszüchtung 4558
Immunkörper 4555
Immunkörperunterdrückung 4571
Immunmechanismus 4569
Immun(o)chemie 4563
Immunoelektrophorese 4564
Immunoglobulin 4565
Immunologe 4567
Immunologie 4568
immunologisch 4566
Immun(o)reaktion 4556
Immun(o)toleranz 4572
impermeabel 4578
Impermeabilität 4577
impfen 4697
Impfkultur 4699
Impfung 4698
Implantat 4579
Imponiergehaben 2622
Impulsfrequenz 4584
Impulsleitung 1968
inaktivieren 4586
Inaktivierung 4587
Indikator, ökologischer 2762
Indikatorpflanze 4630
Individualauslese 4636
Individualdistanz 4635
Individualpotenz 7274
Individuenzahl 6236
Indol(yl)essigsäure 4637
Indolyl-3-Essigsäure 4638
Induktion, embryonale 2878
Induktionsfähigkeit 4644
Induktionsstoff 4645
Induktor 4645
Indusium 4646
Industrieabfälle 4648
Industriemelanismus 4647
Infloreszens 4656
Information, genetische 3742
Informationstheorie 4658
Informationsübertragung 9200
Infrarotlicht 4659
Infrarotstrahlen 4660
Infraschall 4661
Infundibulum 4664
Infusionstierchen 4666
Infusorien 4666
Infusorienerde 4667
Inhibitor 4675
Inhibitorgen 4674
inhibitorisch 4676
Initiale 4683
Initialzelle 4683
inkompatibel 4600
Inkompatibilität 4598
Inkompatibilitätsfaktor 4599
Inkret 4607
Inkretion 4608
Inkretionsstoff 4607
inkretorisch 4609
Inkrustation 4611
Inkubationszeit 4617

J

K

Deutsch

Kollaterale 1886
Kollembolen 1888
Kollenchym 1889
Kollodium 1890
Kolloid 1891
kolloid(al) 1892
Kolon 1895
absteigendes 2425
aufsteigendes 740
Kolonie 1896
Koloniebildung 1897
Koloniebrüter 1135
Kolonienbrüten 1275
Kolorimeter 1899
Kolorimetrie 1900
Kolossalfaser 3822
Kolostrum 1901
Kolumella 1911
Kombination 1914
Kombinationseignung 1915
Kombinationsfähigkeit 1915
Kommabazillus 9528
Kommensale 1917
Kommensalismus 1918
Kommissur 1919
Kommunikation 1921
Kompartimentisierung 1926
Kompatibilität 1927
Kompensationspunkt 1929
Kompensator 1930
Kompetenz 1932
Komplementärgen 1940
Komplementärkern 1941
Komplementärluft 1937
Komplementation 1942
Komplementationstest 9017
Komplementbindung 1935
Komplementbindungsreaktion 1936
Komplexauge 1949
Konchin 1959
Konchiolin 1959
Konchylie 1958
Konchyolin 1959
Konditionalfaktor 1962
Konditionierung, instrumentelle 4727
klassische 1769
konduplikativ *(Knospenlage)* 1972
Konfliktverhalten 1978
Konidie 1982
Konidienträger 1981
Konidiophor 1981
Konifere 1983
Koniferen 1985
konisch 1977
Konjugant 1986
Konjugation 1988
Konkurrenz, innerartliche 4803
interspezifische 4779
intraspezifische 4803
zwischenartliche 4779
Konnektiv 1990
Konsoziation 1995
Konstanztheorie 9041
Konstitutionstyp 2001
Konsubspezies 2005
Konsument *(ecol.)* 2006
Kontaktpunkt 2008
Kontaktreiz 2009
Kontinentalschelf 2014
Kontinentalsockel 2014
Kontinentalverschiebungstheorie 2013
Kontinuität des Keimplasmas 2016
kontort *(Knospendeckung)* 2019
kontraktil 2021
Kontraktilität 2024
Kontraktionsfähigkeit 2024
Kontrollkreuzung 1613
Kontrollpflanze 9019
Kontrolltier 2025
Kontrollversuch 2026
Konturfeder 2020
Konvergenz 2029
Konversion 2031
Konzentrationsgefälle 1955
Konzentrationsgradient 1955
Konzeptakel 1956
Koordination 2032
Koordinationszentrum 2033
Koorientierung 2034
Kopepoden 2035
Kopf... 1570
Kopfbrust(stück) 1576
Kopfdarm 4114
Kopffüßer 1575
Kopfindex 1571
Kopfnerv 2116
Kopfschild 4115
Kopfschlagader 1436
Kopfskelett 3207
Kopiewahlhypothese 2042
Kopp(e)lung 5190
Kopp(e)lungsgruppe 5191
Kopp(e)lungsphase 2098
Koprophage 2036
Kopulation 2038
Kopulationsorgan 2041
Kopulationsschlauch 3370
kopulieren 2037
Korallenriff 2044
Korallentiere 572
Korbblütler 1946
Korbzelle 977
Ko-Repressor 2045
Kork 2048
Korkbildungsgewebe 6806
Korkgewebe 2052
korkig 2053
Korkkambium 6806
Korkpore 5109
Korkschicht 2051
Korkstoff 8728
Korkzelle 2050
Kormophyten 2055
Kormus 2054
körnerfressend 3934
Kornfrucht 1451
Korngröße 3927
körnig 3935
Korolle 2062
Koronararterie 2063
Körperbau 1211
Körperbautyp 2001
Körperchen 2071
Körperflüssigkeit 1208
Körpergewicht 1213
Körperhöhle 1206
Körperkreislauf 8904
Körpertemperatur 1212
Körperwärme 1209
Körperzelle 8393
Korrelation 2072
Korrelationsindex 2073
Korrelationskoeffizient 2073
Kortex 1581, 7879
Kortikoid 2079
Kortikosteroid 2080
Kortikosteron 2081
Kortikotropin 181
Kortin 2083
Kortisol 4433
Kortison 2085
Kosmopolit 2088
Kotfresser 2036
Kotyledone 2094
Kovarianz 2105
Kovariation 2105
Koxaldrüse 2110
Kragengeißelzelle 1651
Kralle 1772
Kraniologie 2118
Kraniometrie 2119
Kranioten 9517
krankheitserregend 6653
Krankheitserreger 6654
Krankheitskeim 2601
Kratzer 43
Kraut 4173
krautfressend 4179
Krautfresser 4178
krautig 4174
Krautschicht 4175
Kreatin 2122
Kreatinphosphat 2123
Krebse 2156
krebserregend 1418
Krebsforschung 1419
Krebszyklus 4982
Kreide 2127
Kreislauf, fötaler 3501
großer 8904
kleiner 7456
plazentärer 3501
Kreislaufsystem 1754
Kreuzbein 7943
Kreuz(bein)wirbel 7942
Kreuzbestäubung 338, 9716
Kreuzblütler 2152
kreuzen 2134
kreuzgegenständig 2318
Kreuzung 2133
inkongruente 4603
interspezifische 4781
kongruente 1980
reziproke 7643
Kreuzungsreaktivierung 2140
Kreuzungsversuch 2137
Kreuzungszüchtung 2136
Kriechbewegung 2121
Kriechpflanze 2125
Kriechtiere 7765
kristallin 2170
Kristallisation 2173
kristallisierbar 2172
kristallisieren 2174
Kristallographie 2175
Kron(en)blatt 6780
Kropf 2131
Kropfmilch 2132
Krümelstruktur 2145
Krummdarm 4538
Krümmung, tropistische 9284
Krümmungsbewegung 2197
Krustazeen 2156
Krustenflechte 2157
Kryophyt 2159
Kryoplankton 2160
Kryoskopie 2161
Kryostat 2162
kryptogam 2164

Deutsch

L

Letalfaktor, bedingter 1963
Letalität 5121
Letalmutation 5120
Leuchtbakterien 5263
Leuchtorgan 5264
Leukoplast 5124
Leukozyt 5123
  eosinophiler 3046
  polymorphkerniger 3938
Leuzin 5122
Lezithin 5100
LH 5269
Liane 5127
Lias 5128
Lichenin 5132
Lichenologie 5133
Licht, polarisiertes 7094
  ultraviolettes 9350
Lichtblatt 8789
lichtbrechend 7688
Lichtbrechung 7682
Lichtbrechungsvermögen 7687
lichtdurchlässig 9214
lichtempfindlich 5155
Lichtempfindlichkeit 5156
Lichtenergie 5148
Lichtintensität 5149
Lichtmikroskop 5150
Lichtquant 6850
Lichtquelle 5157
Lichtreaktion 5152
Lichtreiz 5158
Lichtrezeptor 6859
Lichtsinnesorgan 6375
Lichtstreuung 5154
lichtundurchlässig 6353
Lichtwahrnehmung 6852
Lidschlußreflex 3286
Ligament 5146
Ligase 5147
Lignin 5165
Ligula 5167
limnetisch 5176
Limnion 5177
Limnologie 5178
Limnoplankton 5179
Linie, männliche 5341
  mütterliche 3350
  reine 7480
  väterliche 5341
  weibliche 3350
Linienbegründer 3570
Linienzucht 5180
Linin 74
linksdrehend 5125
linkswindend 8299
Linolensäure 5193
Linolsäure 5192
Linse 2171
Linsenanlage 5108
linsenförmig 5110
Lipase 5194
Lipid 5195
Lipochrom 5198
Lipoid 5201
Lipolyse 5202
Liponsäure 9080
Lipopolysac(c)harid 5205
Lipoprotein 5206
Lippe *(bot.)* 4985
Lippen ... 4986
Lippentaster 4987
Lithophyt 5212
Lithoserie 5213
Lithosphäre 5214
lithotroph 5215
Litoral(zone) 5216
Lockruf 1351
Lockstoff 792
Locus 5232
Log-Phase 3250
Lokalrasse 5224
lokulizid 5231
Lophophor 5250
löslich 8387
Löslichkeit 8385
löslich machen 8386
Lösung, kolloide 1894
  wässerige 677
Lösungsenzym 5296
Lösungsmittel 8390
Lösungsprodukt 5286
LT 5011
Luftatmung 265
Luftfeuchtigkeit 777
Luftkammer 266
Luftpflanze 206
Luftröhre 9180
Luftsack 271
Luftspalte 270
Luftsproß 201
Luftstickstoff 778
Luftverschmutzung 269
Luftverseuchung 269
Luftwurzel 200
lumbal 5259
Lumen 5261
Lumineszenz 5262
Lunge 5265
Lungen ... 7453
Lungenalveole 7454
Lungenatmung 7459
Lungenbläschen 7454
Lungenfell 7458
Lungenfische 2585
Lungenkreislauf 7456
Lungenlappen 7457
Lupe 5331
Lurch 418
Lutein 5268
Luteotrophin 5011
Luziferin 5258
Lyase 5271
Lykopin 5272
Lymphdrüse 5277
Lymphe 5274
Lymphgefäß 5280
Lymph(gefäß)system 5279
Lymphgewebe 5282
Lymphherz 5276
Lymphknoten 5277
Lymphkreislauf 5275
Lymphozyt 5281
Lymphraum 5278
Lyochrom 3446
Lyoferment 5284
Lyophilisierung 5285
Lysat 5286
Lyse 5289
Lysergsäure 5287
lysieren 5298
lysigen 5290
Lysin 5288
Lysis 5289
lysogen 5290
Lysogenie 5291
Lysosom 5292
Lysotyp 5293
Lysozym 5294
lytisch 5295

M

Madreporenplatte 5330
Magen 8674
Magen ... 3671
Magendarm ... 3675
Magendrüse 3672
Magenmund 1420
Magensaft 3673
Makroevolution 5305
Makrofauna 5307
Makroflora 5308
Makrogamet 5309
Makrogamie 4320
Makroglia 5310
Makroklima 5303
Makrokonsument 5304
Makromer 5311
Makromethode 5312
Makromolekül 5313
Makromutation 5314
Makronährstoff 5316
Makronukleus 5315
Makrophage 5317
Makrophyll 5318
Makrophylogenese 5319
makroskopisch 5320
Makrosmat 5321
Makrosporangium 5322
Makrospore 5323
Makrosporogenese 5324
Makrosporophyll 5325
Makrotom 5327
Malako(zoo)logie 5336
Maleinsäure 5343
Malonsäure 5348
Malpighische Gefäße 5351
Malpighisches Körperchen 5349
Maltase 5352
Maltobiose 5353
Maltose 5353
Malzzucker 5353
Mamille 5354
Mamma 5357
Mammalogie 5356
Mandibel 5358
Mandibulata 5360
Mangelkrankheit 2330
Mangelmutante 6254
Mangelsymptom 2331
Männchen 5339
männlich 5338
Männlichkeit 5390
Mannose 5361
Mantel *(zool.)* 5362
Mantelchimäre 6726
Mantelhöhle 5363
Manteltiere 9307
Manubrium 5364
Mark 5446
  verlängertes 5447
markhaltig 5948
Markhirn 5945
Markhöhle 5449
markieren (mit) 4984
Markierung 5383
Markierungsgen 5382
Markierungsmethode 5384
Markierungsstoff 5381
Markierungs- und Wiederfang-
  methode 5380
Markkanal 5448
marklos 434
Markraum 5449

Markscheide 5947
Markstrahl 5452
Markstrahlzelle 5453
Markzelle 5450
Marsupialier 5387
Marsupium 5388
Massenauslese 5397
Masseneffekt 5393
Massenkreuzung 7126
Massenkultur 5392
Massenmutation 5395
Massenselektion 5397
Massenspektrograph 5398
Massenvermehrung 5396
Massenwanderung 5394
Massenwechsel 7191
Massenwirkung 5391
Massenwirkungsgesetz 5057
Mastax 5401
Mastdarm 7657
Mastigophoren 3437
Mastzelle 5400
Matrix 5418
  zytoplasmatische 4410
Matrixbrücke 5420
Matrizen-RNS 5522
matroklin 5421
Matroklinie 5422
Maulbeerkeim 5819
Maulbrüter 5843
Mauser 5839
mausern 5838
Maxillardrüse 5433
Maxillartaster 5434
Maxille 5430
Mazeration 5299
Mechanorezeptor 5441
Mediane 5443
Medianebene 5443
Mediastinum 5445
Medium, festes 8379
  flüssiges 5211
Medulla 5446
Medullarplatte 5451
Meduse 5456
Meeresbiologie 5374
Meeresboden 8041
Meeresfauna 5376
Meeresflora 5377
Meereskunde 6279
Meeresökologie 5375
Meeresplankton 4082
Meerestiefwasser 8042
Meerestier 5373
Meerwasser 8043
Meerwasserentsalzung 2421
Meerwasserverschmutzung 7118
Megaevolution 5458
Megasporangium 5322
Megaspore 5323
Megasporogenese 5324
Megasporophyll 5325
Mehrfachaustausch 1948
mehrfächerig 5869
Mehrfachreaktivierung 5887
mehrjährig 7065
Mehrjährigkeit 6712
mehrkernig 5880
mehrsamig 7170
mehrschichtig 5877
mehrteilig 5881
mehrzellig 5870
Meidereaktion 864
Meiose 5460
Meioseprodukt 5463
Meiospore 5461
meiotisch 5462
Meiozyte 5459
Melanin 5464
Melanismus 5465
Melanophor 5467
Melanotonin 5469
Melanotropin 4759
Melatonin 5469
Melitose 7589
Melitriose 7589
Membran 5472
  biologische 1080
  halbdurchlässige 8140
  semipermeable 8140
  undulierende 9367
membranartig 5480
Membranelle 5479
Membranfilter 5481
Membrankapazität 5474
Membranpermeabilität 5476
Membranpotential 5477
Membranresistenz 5478
Membranstrom 5475
Mendelgesetze 5483
Mendelismus 5490
Mendeln 5490
Mendelpopulation 5488
Mendelsche Gesetze 5483
  erstes 5484
  zweites 5485
  drittes 5486
Mendelscher Faktor 5487
Mendelvererbung 6640
Meningen 5491
Menotaxis 5492
Menschenaffen 580
Menschenkunde 581
Menschwerdung 575
Merikarp 5493
Meriklinalchimäre 5494
Meristem 5495
  primäres 7292
  sekundäres 8052
Merkmal 1605
  erworbenes 96
  genotypisches 3767
  phänotypisches 6816
  qualitatives 7511
  quantitatives 1163
  unabhängiges 4626
Merkmalsänderung 1802
Merkmalsgefälle 1803
Merkmalsgradient 1803
Merkmalskonstanz 1606
Merkmalsverschiebung 1607
meroblastisch 5496
Merogamet 5497
Merogamie 5498
Merogonie 5499
Meromixis 5500
Merozygote 5501
Mesenchym 5503
Mesenterium 5504
Mesenteron 5505
Mesenzephalon 5502
Mesoblast 5510
Mesoderm 5510
Mesogloea 5511
Mesokarp 5507
Mesokotyl 5509
Mesomitose 5512
Mesonephros 5514
mesophil 5515
Mesophyll 5516
Mesophyt 5517
Mesothel 5518
mesotroph 5519
Mesozoikum 5521
Mesozöl 5508
Mesozoon 5520
Messenger-RNS 5522
Messung 5439
Metabiose 5523
metabolisieren 5544
Metabolismus 5541
Metabolit 5542
metachromatisch 5548
Metagenese 5551
Metakinese 5552
Metalimnion 9051
Metamer 5554
Metamerie 5555
Metamitose 5556
Metamorphose 5558
  unvollkommene 4163
  unvollständige 4163
  vollkommene 4323
  vollständige 4323
metamorphosieren 5557
Metanephros 5559
Metaphase 5560
Metaphasepaarungsindex 5561
Metaphyten 5562
Metaplasie 5563
Metaxenie 5566
Metaxylem 5567
metazentrisch 5547
Metazöl 5549
Metazoon 5568
Metenzephalon 5569
Methionin 5570
Methylgruppe 5574
Methylierung 5575
Metöstrus 5576
Mevalonsäure 5577
Migration 5668
Migrationsverhalten 5670
Mikroanalyse 5579
Mikroautoradiographie 5580
Mikrobe 5582
mikrobiell 5583
Mikrobiologe 5587
Mikrobiologie 5588
mikrobiologisch 5586
Mikrobion 5582
mikrobisch 5583
mikrobizid 5585
Mikrochemie 5590
Mikrochirurgie 5653
mikrochirurgisch 5654
Mikrochromosom 5591
Mikrodissektion 5596
Mikroelektrode 5597
Mikroevolution 5600
Mikrofauna 5601
Mikrofibrille 5602
Mikroflora 5603
Mikrofossil 5604
Mikrogamet 5605
Mikrogen 5606
Mikroglia 5607
Mikroinjektion 5610
Mikrokinematographie 5592
Mikroklima 5593
Mikrokonsument 5594
Mikromanipulation 5611
Mikromanipulator 5612
Mikromer 5613
Mikrometer 5614

Mikrometerschraube 5616
Mikromethode 5615
Mikromilieu 5599
Mikromutation 5618
Mikron 5619
Mikronährstoff 5622
Mikronukleus 5621
mikrooperativ 5654
Mikroorganismus 5623
Mikropaläontologie 5624
Mikrophag 5625
Mikrophotographie 5626
Mikrophyll 5627
Mikrophylogenese 5628
Mikropipette 5629
Mikroprotein 5630
Mikropyle 5631
Mikroradiographie 5632
Mikroskop 5633
Mikroskopie 5640
Mikroskopiker 5639
mikroskopisch 5636
mikroskopisch untersuchen 3177
Mikrosmat 5641
Mikrosom 5642
Mikrospektrophotometer 5644
Mikrospektrophotometrie 5645
Mikrospezies 5643
Mikrosphäre 5646
Mikrosporangium 5647
Mikrospore 5648
Mikrosporophyll 5650
Mikrosporozyt 7101
Mikrosubspezies 5652
Miktrotechnik 5655
Miktrotom 5656
Mikrotomie 5657
Mikrotubuli 5658
Mikroveraschung 5609
Mikrovilli 5659
Mikrowaage 5581
Mikrozentrum 5589
Mikrozotten 5659
Mikrozyste 5595
Milch *(zool.)* 5684
Milchabsonderung 5681
Milchalbumin 5002
Milchbrustgang 9083
Milchdrüse 5357
Milcher 5687
Milchgang 5009
Milchner 5687
Milchröhre 5052
Milchsaft 5051
Milchsaftgefäß 5052
Milchsaftzelle 5053
Milchsäure 5006
Milchsäurebakterien 5007
Milchsäuregärung 5008
Milchsekretion 5681
Milchzahn 5683
Milchzucker 5013
Milieu, inneres 4764
Milz 8526
Milz ... 8527
Mimese 5688
mimetisch 5689
Mimik 3290
Mimikry 5693
  Batessche 986
  Müllersche 5868
Mineralernährung 5697
Mineralisation 5699
Mineralnährstoff 5696
Mineral(o)kortikoid 5700
Mineralsalz 5698
Mineralstoff 5694
Mineralstoffwechsel 5695
Minimalareal 5702
Minimalmedium 5701
Minimumgesetz 5059
Miozän 5705
mischbar 5707
Mischbarkeit 5706
mischerbig 4261
Mischerbigkeit 4259
Mischkultur 5721
Mischpopulation 5722
Mißbildung 5345
Mißteilung 5708
Mitochondrium 5709
Mitose 5711
Mitoseapparat 5714
Mitosegift 5718
Mitosehemmung 5717
Mitoseindex 5716
Mitosespindel 5719
Mitosezyklus 5715
Mitosom 5712
mitotisch 5713
Mitralklappe 5720
Mittel, pilztötendes 3626
Mitteldarm 5505
Mittelfell 5444
Mittelfellraum 5445
Mittelfuß 5565
Mittelfußknochen 5564
Mittelhand 5546
Mittelhandknochen 5545
Mittelhirn 5502
Mittellamelle 5664
Mittelnerv *(bot.)* 5662
Mittelohr 5663
Mittelrippe 5662
mittelständig 5442
Mittler, chemischer 9219
Mittlersubstanz 9219
Mixoploidie 5723
Mizelle 5578
Modellversuch 5729
Moder 6642
Modifikation 5730
Modifikationsfaktor 5732
Modifikationsgen 5732
Modifikator 5732
modifizierend 5731
Modulation 5733
Modulator 5734
Molar 5739
Molekül 5746
molekular 5740
Molekularbiologie 5741
Molekulargefüge 5744
Molekulargenetik 5742
Molekulargewicht 5745
Molekularstruktur 5744
Molekülmodell 5743
Mollusken 5747
Monade 5749
Monaster 5750
Moneren 5752
Monide 5753
Monochasium 5757
monochlamyd 5758
monofaktoriell 5767
monogam 5768
Monogamie 5769
monogen 5771
Monogenese 5770
Monogenie 5772
Monogonie 5774
Monohaploidie 5775
monohybrid 5776
Monohybridie 5777
monokarp 5754
Monokaryon 5778
monoklin 5759
monokotyl 5761
Monokotyle(done)n 5760
monomer 5781
monomorph 5783
monoöstrisch 5766
monophag 5785
monphyletisch 5786
Monophylie 5787
Monoploidie 5789
monopodial 5790
Monopodium 5791
monopolar 9384
Monosac(c)harid 5793
Monose 5793
Monosom 5794
Monosomie 5795
monosymmetrisch 9783
Monotremen 5798
monotrich 5799
monozentrisch 5756
Monözie 5765
monözisch 5764
monozygot 5800
monozyklisch 5762
Monozyt 5763
Monsunwald 5802
Mooskunde 1306
Moospflanzen 1307
Moosstärke 5132
Moostierchen 1308
Morgan-Einheit 5806
Morphallaxis 5807
Morphin 5808
Morphium 5808
Morphogenese 5809
morphogenetisch 5810
Morpholaxis 5807
Morphologie 5814
morphologisch 5813
Morphoplasma 5815
Morphose 5816
Mortalität 5817
Morula 5819
Morulabildung 5820
Morulatier 5520
Mosaikei 5822
Mosaikentwicklung 5821
Motilität 5829
Motoneuron 5835
Motorcortex 5832
Motorik 5837
m-RNS 5522
MSH 4759
Mukopolysac(c)harid 5855
Mukoprotein 3876
Mull 5679
Müll 2663
Müllbeseitigung 2664
Mülldeponie 2727
Müllersche Mimikry 5868
Müllerscher Gang 5867
Müllverbrennung 7691
Müllverbrennungsanlage 7692
Multienzymkomplex 5873
multifaktoriell 7130
Multigen 5876
multipolar 5891
Multivalent 5892

N

Deutsch

Nebenhoden 3062
Nebenkern 6590
Nebenniere 175
Nebennierenhormon 176
Nebennierenmark 177
Nebennierenmarkhormon 178
Nebennierenrinde 173
Nebennierenrindenhormon 174
Nebenschilddrüse 6611
Nebenschilddrüsenhormon 6610
Nebenwurzel 5050
Nebenzelle 855
Negativfärbung 6008
Nekrobiose 5998
Nekrohormon 9709
Nekrophage 6001
Nekrose 6002
Nektar 6003
Nektardrüse 6004
Nektarium 6004
Nekton 6009
Nemathelminthen 6010
Nematizid 6011
Nematoden 6013
Nematozid 6011
Nematozyste 6012
Neoblast 6015
Neocortex 6016
Neodarwinismus 6017
Neogäa 6019
Neogen 6020
Neolamarckismus 6021
Neolithikum 6022
neomorph 6023
Neopallium 6024
Neostandard 6028
Neotenie 6026
Neotenin 4920
Neotropis 6027
Neozoikum 1546
Nephridium 6031
Nephron 6033
Nephroporus 6034
Nephrostom 6035
Nephrotom 6036
Nephrozöl 6032
neritisch 6037
Nerv 6039
  motorischer 5834
  sensibler 8172
Nervatur 6038
Nerven ... 6054
Nervenbahn 6053
Nervenendigung 6044
Nervenendverzweigung 8992
Nervenerregung 6056
Nervenfaser 6045
Nervengeflecht 6049
Nervengewebe 6059
Nervenimpuls 6046
Nervenknoten 6048
Nervenlehre 6099
Nervenleitung 6042
Nervenmark 5946
Nervennetz 6074
Nervenphysiologie 6103
Nervenplexus 6049
Nervenring 6050
Nervenstamm 6052
Nervenstrang 6043
Nervensystem 6058
  autonomes 834
  parasympathisches 6608
  peripheres 6742
  sympathisches 8856
Nervensystem, vegetatives 834
  viszerales 834
Nerventätigkeit 6055
Nervenverzweigung 6057
Nervenwurzel 6051
Nervenzelle 6040
  motorische 5835
Nervenzentrum 6041
nervlich 6083
Nesselkapsel 6012
Nesseltiere 1819
Nesselzelle 1820
Nestablösung 7719
Nestbau 6062
Nestbautrieb 6064
Nestbauverhalten 6063
Nestflüchter 6126
Nesthocker 6125
Nestling 6070
Nestmaterial 6069
Nestverteidigung 6065
Nettoproduktion *(ecol.)* 6071
Netz *(anat.)* 6330
netzartig 7829
Netzauge 1949
Netzflügler 6105
netzförmig 7829
Netzgefäß 7832
Netzhaut 7838
Netzhautzapfen 7839
Netzmagen 7837
netznervig 6072
Netzverdickung 7831
Neubildung 6018
Neukombination 7648
Neumünder 2453
neural 6073
Neuralachse 6082
Neuralbogen 6074
Neuralfalte 6077
Neuralkanal 6075
Neuralleiste 6076
Neuralplatte 6079
Neuralrinne 6078
Neuralrohr 6080
Neurilemma 8018
Neurit 879
Neuroblast 6086
neuroendokrin 6090
Neuroepithel 6091
Neurofibrille 6092
neurogen 6094
Neuroglia 3860
Neurohormon 6096
Neurohypophyse 6097
Neurokranium 6087
Neurokrinie 6088
Neurolemma 8018
Neurologe 6098
Neurologie 6099
Neuron 6040
  sensorisches 8173
Neuronentheorie 6101
Neurophysiologe 6102
Neurophysiologie 6103
Neuropil 6104
Neuropteren 6105
Neurosekret 6106
Neurosekretion 6106
neurosekretorisch 6107
neurosensorisch 6108
Neurotransmitter 9219
neurovegetativ 6110
Neurozyt 6040
Neurula 6111
Neurulation 6112
Neuston 6113
Neutron 6114
neutrophil 6115
Neuzüchtung 1277
Niazin 6121
Niazinamid 6117
nicht allel 6153
nichtlaubabwerfend 6157
Nichtreduktion 6169
nichtsäurefest 85
Nichtschwellensubstanz 6185
Nichtschwester-Chromatide 6172
nichtseptiert 6170
Nichtumkehrbarkeit der Evolution 4851
nichtvirulent 862
nichtzellig 56
Nickhaut 6122
Nidamentaldrüse 6123
Nidation 6124
Niederblatt 1469
Niedermoor 5251
Niederschlag, radioaktiver 7554
Niederungsmoor 5251
Niere 4959
Nieren ... 7721
Nierenbecken 7723
nierenförmig 7726
Nierenkanälchen 7724
Nierenkelch 7722
Nikotin 6120
Nikotinamid-Adenin-Dinukleotid 6118
Nikotinamid-Adenin-Dinukleotidphosphat 6119
Nikotinsäure 6121
Nikotin(säure)amid 6117
Nische, ökologische 2764
nisten 6061
Nisten 6068
Nistmaterial 6069
Nistplatz 6067
Nitrifikanten 6132
Nitrifikation 6130
nitrifizieren 6131
Nomenklatur, binäre 1043
Nomenklaturregeln, internationale 4769
Non-Disjunktion 6158
Nonparental-Dityp-Tetrade 6166
Noradrenalin 6179
Notogäa 6183
notomorph 6184
Notruf *(eth.)* 2641
Nozizeptor 6145
Nucellus 6190
Nuklearreaktion, Feulgensche 3375
Nuklease 6212
Nuklein 6215
Nukleinisierung 6216
Nukleinsäure 6214
Nukleohiston 6218
Nukleoid 6219
Nukleolareinschnürung 6220
Nukleolarzone 6221
Nukleolonema 6222
Nukleolus 6223
Nukleolus-Organisator-Region 6224
Nukleoplasma 6225
Nukleoprotein 6227
Nukleosid 6228
Nukleosom 4940

Nukleotid 6230
Nukleus 6231
nulliplex 6233
nullisom 6234
Nullisomie 6235
nullosom 6234
Nullosomie 6235
Nullpunktmutation 9749
Nuß 6242
Nußfrucht 6242
Nüstern 6181
Nutation 6243
Nutritismus 6244
Nutzinsekt 9422
Nützling 9421
Nutzpflanze 9423
Nutztier 9421
Nutzungsrate 9427
Nutzzeit 9428
Nuzellus 6190
Nyktinastie 6266
nyktinastisch 6265
Nymphe 6267
Nymphenhäutung 6268
Nymphenstadium 6269

O

obdiplostemon 6271
Oberarmknochen 4402
Oberbauch 3063
Oberboden 263
Oberflächenkultur 8820
Oberflächenspannung 8822
Oberflächenwasser 8824
Oberhaut 3061
Oberhautzelle 3060
oberirdisch 3064
oberirdische Teile 199
Oberkiefer 5430
Oberkieferknochen 8799
Oberlid 9396
Oberlippe 4994 *(ins.)*, 9399
Oberschenkelknochen 3353
oberständig 8798
Objektiv 6272
Objekttisch 5635
Objektträger 5634
Odontoblast 6282
Ohr, äußeres 3257
inneres 4691
Ohr ... 6438
Ohrbläschen 809
Ohrenschmalz 1590
Ohrmuschel 1957
Ohrspeicheldrüse 6634
Ohrtrompete 3147
ökobiotisch 2755
ökogeographisch 2758
Ökokatastrophe 3013
ökoklimatisch 2757
Ökologe 2769
Ökologie 2770
ökologisch 2759
Ökomorphose 2771
Ökophysiologie 2772
Ökospezies 2773
Ökosphäre 2774
Ökosystem 2775
Ökoton 2776
Ökotyp(us) 2777
Okular 3287
Okulomotorius 6281
Öl, ätherisches 3115
Olein 6307
Oleinsäure 6306
Olfaktorius 6312
Ölgang 6300
Oligodendroglia 6321
Oligogen 6322
oligolezithal 6323
Oligosac(c)harid 6324
oligotroph 6326
Oligozän 6319
Ölimmersion 6302
Ölkörperchen 2813
Ölpest 6303
Ölsäure 6306
Ölteppich 6304
Ölverschmutzung 6303
ombrophil 6328
Ommatidium 6331
Ommochrom 6332
omnipotent 9164
Omnipotenz 9163
omnivor 6335
Omnivore 6334
Önozyte 6287
Ontogenese 6338
ontogenetisch 6339
Ontogenie 6338
oogam 6342
Oogamie 6343
Oogenese 6344
Oogonium 6345
Ookinesis 6346
Oolemma 9597
Ooplasma 6348
Oosom 6349
Oosphäre 6350
Oospore 6351
Ootide 6352
Oozentrum 6469
Oozyte 6341
Operator(gen) 6355
Operkulum 6356
Operon 6357
Opsonin 6359
Optikus 6362
Optimaltemperatur 6369
Ordnung 6372
Ordovizium 6373
Organ, elektrisches 2817
lauterzeugendes 8407
organähnlich 6386
organbildend 6384
Organbildung 6383
Organell(e) 6377
Organisationsebene 6381
Organisationshöhe 6381
Organisationsmerkmal 8711
Organisationszentrum 6380
Organisator 6382
organisch 6378
Organismus 6379
Organkultur 6374
Organlehre 6387
organogen 6384
Organogenese 6383
Organographie 6385
organoid 6386
Organologie 6387
organotroph 6388
Organsystem 6376
Orientierung 6389
Orientierungsbewegung 6390
Orientierungsreaktion 6391
Orientierungssinn 8149
Ornithin 6393
Ornithinzyklus 6394
Ornithogamie 6400
Ornithologe 6396
Ornithologie 6397
ornithophil 6398
Ornithophilie 6400
Orotsäure 6401
orthodrom 6402
Orthoevolution 6404
Orthogenese 6403
Orthoploidie 6405
Orthopteren 6406
Orthosympathikus 8856
orthotrop 6407
Ortstreue 6820
osmiophil 6410
Osmiumsäure 6409
Osmiumtetroxid 6411
Osmolarität 6412
Osmometer 6413
Osmoregulation 6415
Osmorezeptor 6414
Osmose 6416
Osmotaxis 6417
osmotisch 6418
Ösophagus 6288
Osphradium 6423
Ossein 6424
Ossifikation 6426
Osteoblast 6429
Osteogenese 6433
Osteoklast 6430
Osteokranium 6431
Osteologie 6434
Osteometrie 6435
Osteozyt 6432
Ostiolum 6436
Ostium 6437
Östradiol 6289
Östralzyklus 6295
Östran 6290
Östriol 6292
Östrogen 6293
Östron 6294
Östrus 6297
Oszillograph 6408
Otolith 6441
Outbreeding 6442
Ovalbumin 6446
Ovar 6452
Ovarialhormon 6449
Ovariole 6450
Ovariotestis 6451
Ovarium 6452
Ovidukt 6460
ovipar 6463
Oviparie 6462
Ovogenese 6344
Ovovitellin 6472
ovovivipar 6474
Ovoviviparie 6473
Ovozentrum 6469
Ovulation 6475
Ovum 2799
Oxalazetat 6479
Oxalessigsäure 6479
Oxalsäure 6480
Oxybiose 205
Oxydant 6482
Oxydase 6483
Oxydation 6485
Oxydationsmittel 6482
Oxydations-Reduktionspotential 6487

Oxydationsteich 6486
Oxydator 6482
oxydierbar 6481
oxydieren 6484
Oxydierung 6485
Oxydoreduktase 6492
Oxydoreduktion 6493
Oxygenase 6503
Oxygenierung 6504
Oxygenisation 6504
Oxyhämoglobin 6505
Oxytozin 6506
Ozeanographie 6279
Ozelle 6280

P

Paarbildung 6515
paaren 5408, 6514
Paarhufer 735
paarig 6517
paarig gefiedert 6632
Paarkernphase 2552
Paarung 5411
paarungsbereit 7624
Paarungsorgan 2051
Paarungspartner 5407
Paarungsruf 5413
Paarungstyp 5417
Paarungsverhalten 5412
Paarungszeit 5416
Paarzeher 735
Pachytän 6508
Pädogamie 6510
Pädogenese 6511
paläarktisch 6520
Palade-Granula 7875
Palä(o)anthropologie 6521
Paläobiologie 6522
Paläobotanik 6523
Paläogen 6528
Paläogeographie 6529
Paläoklimatologie 6525
Paläolithikum 6530
Paläoökologie 6526
Paläontologe 6531
Paläontologie 6532
Paläophytologie 6523
Paläotropis 6534
Paläozän 6524
Paläozoikum 6535
Paläozoologie 6536
Palenzephalon 6544
Palingenese 6546
Palisadengewebe 6548
Palisadenparenchym 6548
Palisadenzelle 6547
Pallium 6551
Palmitinsäure 6557
Palpus 6558
Palynologie 6559
panaschiert 9447
Panaschierung 9448
Pangen 6564
Pangenesis 6565
Pankreas 6560
Pankreassaft 6561
Pankreatin 6562
Pankreozymin 6563
Panmixie 6567
Pansen 7927
Panspermie 6568
Panspermie-Theorie 2661
Pantothensäure 6569
Panzer 1398
Papain 6570
Papierchromatographie 6571
Papierelektrophorese 6572
Papille 6573
Pappus 6574
Parabasalapparat 6575
Parabiose 6576
Paraffinblock 6580
Paraffineinbettung 6581
Paraffinschnitt 6582
parakarp 6577
Parallelmutation 6583
parallelnervig 6584
Parameter 6586
Paramitose 6587
Paramylon 6588
Paranukleus 6590
Paraphyse 6591
Parapinealorgan 6631
Paraplasma 6593
Parapodium 6594
Parapophyse 6595
Parasexualität 6598
parasexuell 6596
Parasit 6599
  fakultativer 3297
  obligater 6273
  ständiger 6758
  zeitweiliger 8979
parasitär 6600
Parasitenbefall 6603
Parasitie 6602
parasitieren 6604
parasitisch 6600
Parasitismus 6602
Parasitologe 6606
Parasitologie 6607
Parasympathikus 6608
Parasynapsis 6609
Parasyndese 6609
Parathormon 6610
Paratyp(us) 6612
Paravariation 6613
parazentrisch 6579
Parazoon 6614
Parenchym 6615
Parenchymzelle 6616
parental 6621
Parental-Dityp 6624
Parentalgeneration 6625
Parentaltyp 6626
parietal 6627
Parietalauge 6629
Parietalorgan 6631
parökisch 6633
Parthenogenese 6636
Parthenokarpie 6635
Partialdruck 6638
PAS-Reaktion 6739
Pasteurisation 6646
pasteurisieren 6647
Pasteurisierung 6646
pathogen 6653
Pathogenität 6655
patroklin 6656
Patroklinie 6657
Paukenhöhle 9325
Paurometabolie 6659
PD 6624
Pedipalpus 6674
Pedogenese 8360
Pedologie 8372
Pektase 6664
Pektin 6666
Pektinesterase 6664
Pektinsäure 6665
Pelagial 6683
pelagisch 6682
Pelorie 6686
Pendelbewegung 6694
Penetranz 6695
  unvollständige 4602
  vollständige 1945
Penis 6698
Penizillin 6697
Penizill(in)säure 6696
Pentosan 6699
Pentose 6700
Pepsin 6701
Pepsinogen 6702
Peptid 6704
Peptidase 6703
Peptidbindung 6705
Peptidhormon 6707
Peptidkette 6706
Pepton 6708
Peptonisierung 6709
Perenne 6714
perennierend 6713
Perforatorium 103
Perianth(ium) 6717
Periblast 6718
Periblem 6719
Perichondrium 6724
Periderm 6728
Peridie 6729
Peridineen 2564
Perigon(ium) 6730
Perigonblatt 8989
perigyn 6731
Perikard(ium) 6721
Perikardialhöhle 6720
Perikarp 6722
Perikaryon 6732
periklin 6725
Periklinalchimäre 6726
Perilymphe 6733
Perimetrium 6734
Perimysium 6735
Perineum 6736
Perineurium 6737
Periode, adulte 191
  sensible 8158
Periodizität 6740
Periost 6741
Peripheriewanderung 6743
Periplasma 6745
Perisperm 6747
Perispor 6748
Peristaltik 6750
Peristase 6751
Peristom 6752
Perithezium 6753
Peritonealhöhle 3
Peritoneum 6755
peritrich 6756
perizentrisch 6723
Perizykel 6727
Perjodsäure-Schiff-Reaktion 6739
Perm 6768
permeabel 6765
Permeabilität 6764
  selektive 8101
Permeabilitätsfaktor 610
Permease 6766
Permeation 6767
Permutation 6770

Phytozönose 6990
Phytozoon 6928
Pigment 6931
  akzessorisches 51
Pigmentbecher 6934
Pigmentfarbe 6938
Pigmentfleck 6937
Pigmenthormon 4759
pigmentieren 6932
Pigmentierung 6939
Pigmentschicht 6935
Pigmentwanderung 6936
Pigmentzelle 6933
Pilze 3624
  echte 9289
  höhere 9289
  unvollständige 3625
Pilzfaden 4492
Pilzhut 6941
Pilzkörper 5911
Pilzkunde 5941
Pilzwurzel 5943
Pinealorgan 6948
Pinozytose 6955
Pinozytosebläschen 6954
Pinzette 6946
Pionierpflanze 6956
Pipette 6957
Pistillum 6958
Pituizyt 6965
Plagiogeotropismus 6973
plagiotrop 6974
Plagiotropismus 6975
Plakoidschuppe 2390
Planation 6977
Plankter 6980
Plankton 6978
Planktonfresser 6981
planktonisch 6979
Planktonorganismus 6980
Planktont 6980
Planogamet 9760
Planula 7013
Plaque 7014
Plasma 7015
Plasmaeinschluß 2253
Plasmaeiweiß 7020
Plasmagen 7017
Plasmahaut 7019
  äußere 7018
  innere 9148
Plasmalemma 7018
Plasmamembran 7019
Plasmaprotein 7020
Plasmaströmung 7409
plasmatisch 7021
Plasmavererbung 2254
Plasmaverschmelzung 7027
Plasmazelle 7016
Plasmazirkulation 2216
Plasmid 7023
Plasmin 7024
Plasmodesma 7025
Plasmodium 7026
Plasmogamie 7027
Plasmolyse 7028
Plasmon 7029
Plasmosom 7030
Plasmotomie 7031
Plasmotyp 7029
Plast 7032
Plastid(e) 7033
Plastidenmutation 7035
Plastidenvererbung 7034
Plastidenvorstadium 7350
Plastidom 7036
Plastidotyp 7036
Plastochinon 7040
Plastogamie 7037
Plastogen 7038
Plastom 7036
Plastomer 1665
Plastosom 5709
Plattenepithel 6660
Plattenkultur 7044
plattieren 7043
Plattwürmer 7045
Plazenta 6968
Plazentalier 6970
Plazentatiere 6970
Plazentation 6971
Pleiochasium 7047
pleiotrop 7049
Pleiotropie 7050
Pleistozän 7051
Plektenchym 7046
pleomorph 7052
Pleomorphismus 7053
Pleopode 7054
Plerom 7055
Pleura 7056
Pleuston 7058
plikativ *(Knospenlage)* 7059
Pliozän 7060
Plumula 7063
plurifaktoriell 7130
poikilotherm 7072
Pol, animaler 520
  vegetativer 9471
Polarisation 7090
Polarisationsfilter 7091
Polarisationsmikroskop 7092
polarisieren 7093
Polarität 7089
Polfeld 7095
Polkappe 7084
Polkapsel 7085
Polkern 7086
Polkörperchen 7083
Pollen 7096
Pollenanalyse 7097
Pollendiagramm 7099
pollenführend 7112
Pollenkammer 7098
Pollenkern 7100
Pollenmutterzelle 7101
Pollensack 7102
Pollenschlauch 7105
Pollensterilität 7103
pollentragend 7112
Pollenübertragung 7104
Pollinarium 7106
Pollinium 7113
Polozyt 7083
Polplatte 7087
Polsterpflanzen 2198
Polstrahlen 7088
Polyandrie 7122
polyandrisch 7121
Polycrossmethode 7126
Polyembryonie 7128
Polyenergide 7129
polygam 7131
Polygamie 7132
polygen(isch) 7135
Polygen 7133
Polygenese 7134
Polygenie 7136
polygyn 7137
Polygynie 7138
Polyhaploidie 7139
polyhybrid 7140
Polyhybridie 7141
polykarp 7123
Polykaryozyt 7143
polymer 7145
Polymerase 7144
Polymergen 7146
Polymerie 7147
Polymerisation 7148
polymorph 7150
Polymorphismus 7151
Polynukleotid 7152
Polynukleotidkette 7153
polyöstrisch 7154
Polyp 7155
Polypeptid 7156
Polypeptidkette 7157
polyphag 7159
Polyphagie 7158
polyphän 7049
Polyphänie 7050
polyphyletisch 7161
Polyphylie 7162
polyploid 7163
Polyploidie 7165
Polyploidisierung 7164
Polyribosom 7166
Polysac(c)harid 7167
polysom 7168
Polysom 7166
Polysomie 7169
Polyspermie 7171
Polystelie 7172
polytän 7173
Polytänie 7174
polytop 7175
Polytopie 7176
polytroph 7177
polytypisch 7178
polyzentrisch 7124
polyzyklisch 7127
Population 7183
Populationsanalyse 7184
Populationsbiologie 7185
Populationsdichte 7186
Populationsdruck 7194
Populationsdynamik 7187
Populationsgenetik 7192
Populationsgleichgewicht 7189
Populationsgröße 7196
Populationsökologie 7188
Populationswachstum 7193
Populationswellen 7197
P/O-Quotient 7198
Pore 7199
Porenkapsel 7200
Porenvolumen 7201
Porenwasser 4788
Poriferen 7202
Porogamie 7203
Porphyrin 7204
Portalgefäß 7206
Positionseffekt 7207
Positionspseudoallel 7208
Postadaptation 7209
postembryonal 7210
postganglionär 7214
Postglazial 7215
Postheterokinese 7216
Postnatalentwicklung 7217
Postreduktion 7218
postsynaptisch 7219
Potential, biotisches 1116

Potential, elektrisches 2821
exzitatorisches postsynaptisches 3187
hemmendes postsynaptisches 4679
Potentialdifferenz 7227
Potentiometer 7225
Potenz *(gen.)* 7222
Po(te)tometer 7226
Präadaptation 7228
Präaquation 7218
Prachtkleid 2103
Prädetermination 7251
Präformation 7256
Präformationstheorie 7257
präganglionär 7258
Prägung 4582
Präheterokinese 7263
Präimmunität 7268
Präkambrium 7229
Präkozitätstheorie 7241
Prämolar 7267
Prämunität 7268
Prämutation 7269
Pränatalentwicklung 7270
Präparat 7271
mikroskopisches 5638
Präparationstechnik 5572
Präparator 7272
präparieren 7273
Präpariermikroskop 2625
Präpariernadel 5620
Präpotenz 7274
Präputium 7275
Präreduktion 7276
Präsentationszeit 7278
Präsenz-Absenz-Hypothese 7277
präsumtiv 7282
präsynaptisch 7283
Präzipitat 7234
Präzipitation 7235
Präzipitin 7236
Präzipitinreaktion 7237
Pregnandiol 7260
Primärblatt 7291
Primärkonsument 7290
Primärproduktion 7293
Primärstruktur 7295
Primaten 7297
Primitivknoten 7302
Primitivrasse 5986
Primitivrinne 7300
Primitivstreifen 7303
Primordialknochen 1446
Primordium 7306
Priserie 7307
Proband 7309
Probasidie 7310
Probeentnahme 7969
Probefläche 7968
Proben entnehmen 7967
ziehen 7967
Prochromosom 7314
Proctodaeum 7316
Produktionsbiologie 7318
Produktionsrate 7319
Produzent *(ecol.)* 7317
Proembryo 7320
Proenzym 9793
Proferment 9793
Proflavin 7322
Profundal 7323
progam 7324
Progenese 7325
Progesteron 7328
Progestin 7328
Prokambium 7313
Prolaktin 5011
Prolamin 7330
Prolan 1683
A 3518
B 5269
Proliferation 7334
Prolin 7335
Promeristem 7305
Prometaphase 7337
Prometaphasestreckung 7338
Promitose 7339
Promyzel 7340
Pronephros 7341
Pronukleus 7342
Proöstrus 7343
Propansäure 7349
Prophage 7345
Prophase 7346
Prophaseindex 7347
Propionsäure 7349
Proplastide 7350
Propositus 7309
Proprio(re)zeptor 7351
Prosenchym 7354
Prosenzephalon 7353
Prospezies 4595
Prostaglandin 7356
Prostata 7357
Protamin 7359
Protandrie 7361
protandrisch 7360
Protease 7363
Proteid 7369
Protein 7370
kreuzreagierendes 2147
Protein ... 7385
proteinartig 7385
Proteinase 7384
Proteinfaktor, tierischer 523
Proteinfaser 7375
Proteinhormon 7388
Proteinhülle 7372
Proteinmolekül 7378
Proteinoid 7386
Proteinoplast 7387
proteinspaltend 7390
Proteinsynthese 7383
Proteohormon 7388
Proteolyse 7389
proteolytisch 7390
Proterandrie 7361
proterandrisch 7360
proterogyn 7401
Proterogynie 7402
Prothallium 7394
Prothrombin 9099
Protisten 7396
niedere 5752
Protistenkunde 7397
Protistologie 7397
Protochlorophyll 7399
protogyn 7401
Protogynie 7402
Proton 7403
Protonema 7404
Protonephridium 7405
Protophyt(e) 7406
Protoplasma 7407
Protoplasmaströmung 7409
protoplasmatisch 7408
Protoplast 7410
Protostele 7411
Protostomier 7412
prototroph 7413
Protoxylem 7415
Protozephalon 7398
Protozoenkunde 7416
Protozoologie 7416
Protozoon 7417
Provitamin 7419
Psammon 7421
Psammophyten 7422
Pseudoallele 7423
Pseudoallelie 7424
Pseudobranchie 7426
Pseudochrysalis 7436
Pseudodominanz 7428
Pseudogamie 7429
Pseudohaploidie 7430
Pseudohermaphroditismus 7431
Pseudomixis 7429
Pseudoparenchym 7433
Pseudopodium 7434
Pseudoreduktion 7437
Pteridophyten 7439
Pteridospermen 7440
Pterin 7441
Pterygoten 7444
Ptyalin 7445
puberal 7446
pubertär 7446
Pubertät 7447
Puff 7450
Puffbildung 7451
Puffer 1318
Puffergen 1322
Pufferkapazität 1321
Pufferlösung 1323
puffern 1319
Puffersubstanz 1324
Puffersystem 1325
Pufferung 1326
Pufferungsvermögen 1321
Pufferwirkung 1320
pulmonar 7453
Pulpahöhle 7461
Puls 7464
Pulsen 7463
Pulsieren 7463
pulverisiert 7466
Punkt, isoelektrischer 4865
Punktauge 6280
Punktmutation 3714
Puparium 7471
Pupille 7474
Pupillenreflex 7475
pupipar 7476
Puppe 7468
bedeckte 6274
freie 3178
gemeißelte 3178
Puppenhülle 7469
Puppenstadium 7470
Purin 7485
Purinbase 7486
Putzbewegungen 3957
Pyknidium 7493
Pyknose 7494
Pylorus 7495
Pyramidenbahn 7497
Pyramiden(bahn)kreuzung 2319
Pyramidenzelle 7496
Pyrenoid 7498
Pyridin 7499
Pyridoxin 7500
Pyrimidin 7501

Pyrimidinbase 7502
Pyrrol 7503
Pyrrolring 7504
Pyruvat 7505

Q

Q_{10} 8969
quadriplex 7508
Quadrivalent 7509
Qualle 5456
Quant 7515
Quantenbiologie 7516
Quantenevolution 7517
Quartär 7518
Quaternärstruktur 7519
quellen 8845
Quellung 4545, 8846
Quellungsdruck 4546
Quellungswasser 9646
Quellwasser 8564
Querfortsatz 2493
Querscheibe *(gen.)* 946
Querschnitt 2148
Querstreifung 9234
Querwand 2150
Querzitrin 7521
Quetschpräparat 2155
Quirl 9518
quirlständig 9520
Quotient, respiratorischer 7806

R

Rabenschnabelbein 2043
Rachen 6800
Rachen ... 6796
Rachenhöhle 6798
Rachenmandel 6799
Rädertierchen 7919
Radialsymmetrie 7537
radialsymmetrisch 7538
radiär 7536
radiärsymmetrisch 7538
Radiation, adaptive 144
Radikante 7547
Radikula 7550
radioaktiv 7551
Radioaktivität 7561
Radioautographie 841
Radiobiologie 7563
Radiochemie 7567
Radiochromatogramm 7568
Radiochromatographie 7569
Radioelement 7571
Radioindikator 7555
Radioisotop 7573
  Einbau von ~en 4604
Radiokarbonmethode 7566
Radiokohlenstoff 7564
Radiolarien 7574
Radiomimetika 7577
radiomimetisch 7576
Radionuklid 7578
Radioökologie 7570
Radula 7588
Raffinose 7589
Randomisation 7599
randständig 5372
Rangordnung 7602
Ranke 8985
Ranvierscher Knoten 7603
Ranvierscher Schnürring 7603
Raphe 7604
Rasse 7526
  biologische 1084
  geographische 3775
  ökologische 2765
  physiologische 6887
  reine 7478
Rassenbastard 7534
Rassenbildung 7533
Rassenkreis 7847
Rassenkreuzung 7532
Raubbau 2438, 7249
Räuber 7245
Räuber-Beute-Beziehung 7246
räuberisch 7247
Räubertum 7244
Raubfisch 7250
Raubtiere 1429
Raubvogel 1130
Raubwirtschaft 7249
rauhblättrig 753
Raumordnung 8417
Raumorientierung 8413
Raumplanung 8417
Raumsinn 8414
Rautenhirn 7863
razemös 7528
Reagens 7628
Reagenzglas 9020
Reagenzglaskultur 9021
Reagenzpapier 9018
reagieren auf 7614
  (auf einen Reiz) 7812
Reaktion 7615
  aufgeschobene 2363
  enzymkatalisierte 3032
  immunologische 4556
Reaktion (auf einen Reiz) 7813
reaktionsbereit 7625
Reaktionsdauer 7618
Reaktionsfähigkeit 1390, 7623
Reaktionsgeschwindigkeit 7619
Reaktionskette 7617
Reaktionsnorm 6180
Reaktionspotential 3160
Reaktionsrate 7619
Reaktionsratengen 7608
Reaktionszeit 7620
Reaktionszentrum 7616
reaktivieren 7621
Reaktivierung 7622
rechtsdrehend 2474
rechtswindend 2475
Recon 7653
Redox 6493
Redoxase 6492
Redoxpotential 6487
Redoxreaktion 6488
Redoxsystem 6489
Reduktant 7663
Reduktase 7665
Reduktionsmittel 7663
Reduktionspotential 7664
Reduktionsteilung 7666
Redukton 7667
Reduplikation 7668
  identische 8109
Reduzent 2309
Reflex 7669
  bedingter 1964
  unbedingter 9364
Reflexantwort 7677
Reflexbahn 7676
Reflexbewegung 7675
Reflexbogen 7671
Reflexhemmung 7674
Reflexkette 1597
Reflexologe 7678
Reflexologie 7679
Reflexwirkung 7670
Reflexzentrum 7672
Refraktärphase 7685
Refraktärstadium 7686
Refraktärzeit 7685
Refugialgebiet 7689
Refugium 7689
Regelkreis 7701
Regelsystem 3339
Regenbogenhaut 4845
Regeneration 7694
regenerationsfähig 1389
Regenerationsfähigkeit 7695
Regenerationsvermögen 7695
regenerieren 7693
regenliebend 6328
Regenpflanze 6329
Regenwald 7590
  tropischer 9285
Regenzeit 7591
Regression 7696
  lineare 5185
  nichtlineare 6160
  partielle 6639
Regressionsgleichung 7698
Regressionskoeffizient 7697
regressiv 7699
Regulation 7702
Regulationsei 7703
Regulationsfaktor 7704
Regulationsmechanismus 7706
Regulatorgen 7705
Reibplatte 7588
Reibzunge 7588
reif 5427
Reife 5429
Reifegrad 2348
Reifei 6352
reifen 5428
Reifestadium 8574
Reifeteilung 5424
Reifezeit 5425
Reifung 5423
Reifungsphase 5426
Reifungsteilung 5424
Reihenexperiment 8187
reinerbig 4376
Reinerbigkeit 4375
reinigen 7484
Reinkultur 7479
Reinpräparat 7482
reinrassig 7477
Reinzucht 7481
Reißzahn 1428
Reiz 8648
  adäquater 159
  auslösender 7714
  elektrischer 2818
  haptischer 2009
Reizantwort 8653
Reizaufnahme 8833
reizbar 4853
Reizbarkeit 4852
Reizelektrode 8645
Reizempfindlichkeit 8160
Reizintensität 8649
Reizleitung 1969
Reizmengengesetz 7646

Reizparameter 8650
Reizperzeption 8651
Reizqualität 8652
Reizquelle 8656
Reizreaktion 8653
Reiz-Reaktions-Beziehung 8654
Reizschwelle 8659
Reizsituation 8655
Reizstärke 8649
Reizstoff 8646
Reizsummation 8658
Reizsummengesetz 5055
Reizung 8647
Reizursache 8656
Reizwahrnehmung 8651
Reizwirkung 8644
Rekapitulation 7629
Rekapitulationstheorie 7630
rekombinant 7647
Rekombination 7648
Rekombinationsanalyse 7649
Rekombinationseinheit 7652
Rekombinationshäufigkeit 7650
Rekombinationsprozentsatz 7651
Rekombinationswert 7651
Rekon 7653
Rektum 7657
Releaserfaktor 7713
Relikt 7718
Reliktendemismus 7715
Reliktenfauna 7716
Reliktenflora 7717
Rennin 7727
Reoxydation 7728
reoxydieren 7729
Reparation 7730
Repellent 7732
Replika-Technik 7734
Replikation 7735
semikonservative 8128
Replikationseinheit 7736
Repolarisation 7737
Repression 7739
Repressor 7740
Repressor-Gen 7741
Reproduktion 7745
Reproduktionsrate 7746
Reptilien 7765
RES 7836
Reservat 5993
Reserveeiweiß 7775
Reserveluft 7774
Reservematerial 7777
Reservestärke 7776
Reservestoff 7777
Reservevolumen, expiratorisches 7774
inspiratorisches 1937
Residualgebiet 7689
Residualluft 7779
resistent 7784
Resistenz 7783
Reistenzzüchtung 1274
resorbieren 7786
Resorption 7787
respiratorisch 7789
Respirometer 7811
Restitution 7826
Restitutionskapazität 7825
Restluft 7779
Restmeristem 4737
Reststickstoff 7815
Resupination 7827
Resynthese 7828
Retikularformation 7830
Retikulin 7833
Retikulinfaser 7834
Retikuloendothel 7836
Retikulozyt 7835
Retikulum, endoplasmatisches 2933
sarkoplasmatisches 7985
Retina 7838
Retraktor 7841
Retranslokation 7842
Revier 9011
Reviergesang 9010
Reviermarkierung 9009
Revierverhalten 9007
Revierverteidigung 9008
Rezeptakulum 7632
Rezeptor 7633
Rezeptorbakterie 7640
Rezeptororgan 7635
Rezeptorpotential 7636
Rezeptorzelle 7634
rezessiv 7637
Rezessivität 7639
reziprok 7642
Reziprozitätsgesetz 5484
Rhabdom 7844
Rhabdomer 7845
Rhachis 7530
Rheobase 7846
Rheogameon 7847
Rheotaxis 7848
Rheotropismus 7849
Rhesusfaktor 7850
Rhizodermis 7855
Rhizogenese 7903
Rhizoid 7856
Rhizom 7857
Rhizomorph 7858
Rhizopoden 7859
Rhizosphäre 7860
Rhodophyzeen 7861
Rhodopsin 7862
Rhombenzephalon 7863
Rhythmus, endogener 2920
zirkadianer 1748
Riboflavin 7868
Ribonuklease 7869
Ribonukleinsäure 7870
lösliche 9201
ribosomale 7874
Ribonukleoprotein 7871
Ribonukleosid 7872
Ribose 7873
Ribosom 7875
Ribulose 7876
Ribulosediphosphat 7877
Richtungskörper 7083
Riech ... 6309
Riechen 6308
Riechhirn 7852
Riechlappen 6311
Riechnerv 6312
Riechorgan 6313
Riechstoff 6284
Riesenaxon 3819
Riesenchromosom 3821
Riesen(nerven)faser 3822
Riesenwuchs 3825
Riesenzelle 3820
Rinde 7879
Rindenparenchym 2075
ringartig 545
Ringchromosom 7883
Ringelborke 7881
Ringelung *(bot.)* 3844
Ringelwürmer 540
Ringerlösung 7885
ringförmig 545
Ringgefäß 548
Ringkanal 7882
Ringknorpel 2128
Ringmuskel 1751
Ringtracheide 547
Ringverdickung 546
Rippe 7866
echte 9290
falsche 3304
freie 3462
Rippenfell 2090
Rippenknorpel 2089
Rippenquallen 2178
Rispe 6566
Ritualisierung 7887
RNS 7870
lösliche 9201
ribosomale 7874
RNS-Protein-Granula 7875
Rodentizid 7894
Rogen 7895
Rog(e)ner 8424
Rohhumus 7610
Rohproteinwert 2153
Röhrenatmer 9187
Röhrenblüte 9303
röhrenförmig 9304
Röhrenknochen 5236
Röhrentrachee 9305
Rohrzucker 7940
Röntgenbestrahlung 9722
Röntgeneinheit 7896
Röntgenkristallographie 9724
Röntgenstrahlanalyse 9723
Röntgen(strahl)beugung 9724
Röntgenstrahlen 9727
Röntgenstrahlendosis 9726
Röntgenstrahlung 7898
Rosettenpflanze 7915
rosettenständig 7916
Rotalgen 7861
Rotationskreuzung 7918
Rottange 7861
Rückbestäubung 902
Rückbildung 4826
Rücken ... 2681
Rückenflosse 2682
Rückenmark 8496
Rückenmarkskanal 8494
Rückenmarksnerv 8498
Rückenmarkstiere 2453
Rückenmarkswurzel, dorsale 2684
motorische 9502
sensible 2684
ventrale 9502
Rückensaite 6182
Rückenschild 2685
rückenspaltig 5231
Rückgrat 9515
Rückkoppelung 3337
Rückkoppelungshemmung 3340
rückkreuzen 2135
Rückkreuzung 895
alternierende 2129
Rückkreuzungselter 897
Rückkreuzungsverhältnis 898
Rückkreuzungszüchtung 896
Rückmutation 7843
Rückregulierung 3338
Rückschlag 9101
Rückschlagsbildung 775
Rückstände 7780

Rückstände von Schädlingsbekämpfungsmitteln 6779
Rückstandshöchstmenge 9147
Rückzieher(muskel) 7841
Rudel 6509
Ruderalpflanze 7923
Rudiment 7924
rudimentär 7925
Rudimentärorgan 7926
Ruhekern 7820
Ruheknospe 7816
Ruheperiode 7823
Ruhepotential 7821
Ruhespore 7822
Ruhestadium 7823
Ruhezelle 7817
Ruhezustand 7824
Rundmäuler 2217
Rüssel 7312
Rüsseltiere 7311

S

Sac(c)harase 4815
Sac(c)harid 7938
Sac(c)harimeter 7939
Sac(c)harobiose 7940
Sac(c)harose 7940
sackförmig 7941
Saftanstieg 741
Saftstrom 7972
Sagittalebene 7945
Sagittalschnitt 7946
Saisondimorphismus 8044
Saisonperiodik 8046
Saisonrhythmik 8046
Saitenorgan 1679
Salizylsäure 7948
Saltation 7964, 7965
Salzdrüse 7960
Salzgehalt 7950
salzhaltig 7949
Salzhaltigkeit 7950
Salzhaushalt 7957
salzig 7949
Salzkonzentration 7958
salzliebend 4080
Salzlösung 7963
Salzpflanzen 4081
Salzregulierung 7962
Salzsäure 4430
Salzwasserplankton 4082
Samen 8070, 8124
Samenanlage 6476
Samenbehälter 8134
Samenbläschen 8135
Samenblatt 2094
Sameneiweiß 284
Samenerguß 2811
Samenfaden 8474
Samenfarne 7440
Samenflüssigkeit 8133
Samengang 2812
Samenkanälchen 8137
Samenkapsel 8073
Samenkorn 3926
Samenleiste 6968
Samenleiter 2338
Samenmantel 705
Samennaht 7604
Samenpflanzen 8472
Samenröhrchen 8137
Samenruhe 8076
Samenschale 8074
Samenschuppe 8072
Samenstiel 3627
Samenstrang 8462
Samentasche 8134
samentragend 8136
Samenträger 8471
Samenverbreitung 8075
Samenzelle 8474
Samenzellenbildung 8468
Sammelart 8883
Sammelfrucht 5886
Sandpflanzen 7422
Saprobie 7973
Saprobiont 7973
saprogen 7974
Sapropel 7975
Saprophage 7976
Saprophyt 7977
Saprophytismus 7978
Saprozoon 7979
Sarkolemm 5955
Sarkomer 7983
Sarkoplasma 7984
Sarkosom 7986
Sat-Chromosom 7989
Satellit 9175
Satellitenchromosom 7989
Sattelgelenk 7944
Sättigung 7991
Sättigungseffekt 7992
Sättigungsgrad 2351
SAT-Zone 6221
Sauerstoffaufnahme 6495
Sauerstoffbedarf 6500
  biochemischer 1052
Sauerstoffentzug 2397
Sauerstoffgehalt 6498
Sauerstoffmangel 6501
Sauerstoffschuld 6499
Sauerstoff(über)träger 6496
Sauerstoffverbrauch 6497
Sauerstoffversorgung 6502
Sauerstoffzufuhr 6502
Saugeffekt 8776
saugend *(Insekt)* 8770
Säugetiere 5355
Säugetierkunde 5356
Saugfortsatz 4111
Saugkraft des Bodens 8777
Saugnapf 8769
Saugorgan 8768
Saugrüssel 8773
Saugscheibe 8771
Saugwert 6422
Saugwirkung 8776
Saugwürmer 9243
Saugwurzel 8772
Säulenchromatographie 1912
Säure, salpetrige 6143
Säureagglutination 88
Säureamid 80
Säure-Basen-Gleichgewicht 81
säurebeständig 83
säurefest 83
Säurefestigkeit 86
Säuregehalt 91
Säuregrad 2343
säureliebend 93
säurelöslich 87
Säurezeiger 92
Sauropsiden 7994
Savanne 7995
Schachtelhalmartige 8480
Schädel 2120
Schädel ... 2111
Schädelbasis 960
Schädelhöhle 2115
Schädelkapazität 2113
Schädelkapsel 2114
Schädelknochen 2112
Schädellehre 2118
Schädellose 98
Schädelmeßlehre 2119
Schadinsekt 4703
schädlich 6187
Schädlichkeit 6147
Schädlichkeitsschwelle 2270
Schädling 6777
Schädlingsbekämpfung, biologische 1082
Schädlingsbekämpfungsmittel 6778
Schadstoff 6189
Schafhaut 403
Schalendrüse 5433, 8257
Schalenfrüchte 8256
Schalenhaut 8258
Schalenklappe 9441
Schalfrucht 1451
Schallmauer 8405
Schallquelle 8410
Schallreiz 8411
Schallsignal 95
Schallwahrnehmung 8406
Schallwelle 8412
Schaltneuron 4770
Schaltzelle 4770
Schambein 7449
Schambeinfuge 7448
Schar 3466
Scharniergelenk 4281
Schattenblatt 8320
schattenliebend 8318
Schattenpflanze 8319
Schauapparat 8272
Scheckung 9448
Scheibenblüte 2592
Scheide 9437
Scheidewand 8182
scheidewandbrüchig 8181
scheidewandspaltig 8180
Scheinachse 8861
Scheinfrucht 7427
Scheinfüßchen 7434
Scheingewebe 7433
Scheinpuppe 7436
Scheinquirl 9519
Scheintod 441
Scheinzwittertum 7431
Scheitel 633
Scheitelauge 6629
Scheitelbein 6628
Scheitellappen 6630
Scheitelzelle 639
Schere 1614
Schicht 8695
  monomolekulare 5782
Schichtepithel 8694
Schichtung 8692
Schienbein 9115
Schildblatt 6688
Schilddrüse 9110
Schilddrüsenhormon 9111
schildförmig 8040
Schildknorpel 9109
Schizogamie 8008
schizogen 8010
Schizogenese 8009
Schizogonie 8011

Schizomyzeten 8012
Schizont 8013
Schlafbewegung 6266
Schläfe 8974
Schläfenbein 8975
Schläfenlappen 8977
Schlagader 721
Schlamm, belebter 118
Schlammbelebungsverfahren 119
schlammfressend 5175
Schlängelbewegung 9368
Schlauchbefruchtung 8304
Schlauchpilze 744
Schlauchwürmer 6010
Schleim 5863
schleimabsondernd 5854
Schleimabsonderung 5865
schleimartig 5851, 5858
Schleimbeutel 8893
Schleimdrüse 5860
Schleimhaut 5861
schleimig 5851, 5858
Schleimpilze 5969
Schleimscheide 5862
Schleimstoff 5852
Schleimzelle 5859
Schleuderfrucht 944
Schließfrucht 4624
Schließhaut 1811
Schließmuskel 8486
Schließzelle 4000
Schlingpflanze 9611
Schlittenmikrotom 8324
Schlucken 2341
Schluckreflex 8837
Schlund 6800
Schlund . . . 6796
Schlundkanal 6797
Schlundnerv 3870
schlüpfen 4109
Schlüpfen 4110
Schlüsselbein 1771
Schlüsselreiz 8280
schmalblättrig 499
schmarotzen 6604
schmarotzend 6600
Schmarotzer 6599
  pflanzlicher 6601
  tierischer 9768
Schmarotzerpflanze 6601
Schmarotzertier 9768
Schmarotzertum 6602
Schmecken 4003
Schmelzpunkt 5471
Schmelzschuppe 3665
Schmerzempfindung 6513
Schmerzreiz 6144
Schmerzrezeptor 6512
Schmerzschwelle 9092
Schmerzsinn 8150
Schmetterlinge 5111
Schmutzstoff 7114
Schnabelkerfe 4165
Schnecke *(anat.)* 1840
Schnecken 3676
Schneegrenze 8337
Schneidezahn 4596
schnellwachsend 3313
schnellwüchsig 3313
Schnittfläche 2199
Schnürring, Ranvierscher 7603
Schnurwürmer 6014
Schöpfungstheorie 9041
Schoß 8263
Schoßfuge 7448
Schößling 8263
Schötchen 8287
Schote 8289
Schraubel 1229
schraubenförmig 4144
Schraubengefäß 4143
Schraubentracheide 8514
Schraubenverdickung 8513
schraubig 4144
Schreckreaktion 3590
Schreckstoff 278
Schreitbein 9616
Schrittmacherpotential 6507
Schulter . . . 8003
Schulterblatt 8002
Schultergelenk 8271
Schultergürtel 8270
Schuppe 7998
schuppenartig 8569
Schuppenblatt 8000
Schuppenborke 7999
Schuppenepithel 8570
schuppenförmig 8569
schuppig 8568
Schüttelbecher 4030
Schutzfärbung 7366
Schutzgewebe 7368
Schutzhülle 7367
Schwachlichtpflanze 8319
Schwämme 7202
Schwammgewebe 8536
schwammig 8534
Schwammnadel 8532
Schwammparenchym 8538
Schwammtiere 7202
schwanger 7261
Schwangerschaft 7259
Schwangerschaftshormon 2069
Schwankungen, jahreszeitliche 8045
Schwannsche Scheide 8018
Schwannsche Zelle 8017
Schwanzfeder 8919
Schwanzflosse 1481
schwanzlos 628
Schwanzwirbel 1482
Schwarm 8838
Schwarmbildung *(Fische)* 8016
schwärmen 8839
Schwärmen 8841
Schwärmspore 9773
Schwarmtrieb 8842
Schwärmzelle 9773
Schwebepflanze 3461
Schwebstoffe 8834
Schwefelbakterien 9079
Schwefelkreislauf 8782
Schweiß 8843
Schweißabsonderung 8778
Schweißdrüse 8844
Schwelle, absolute 25
schwellen 8845
Schwellenanhebung 7592
Schwellenerhöhung 7592
Schwellenerniedrigung 9093
Schwellenreiz 5171
Schwellenstoff 9095
Schwellensubstanz 9095
Schwellenwert 9096
Schwellgewebe 1495
Schwellkörper 1494
Schwellung 8846
Schwerefeld 3943
Schwerereiz 3946
Schweresinnesorgan 3945
Schweresteinchen 8606
Schwerewahrnehmung 3942
Schwerkraft 3944
Schwerkraftreiz 3946
Schwesterchromatide 8308
Schwiele 1353
Schwimmbewegung 8849
Schwimmblase 8847
Schwimmblatt 3460
Schwimmfähigkeit 8848
Schwimmfüßler 6556
Schwimmhaut 9669
Schwimmplatte 8850
Schwimmvogel 6556
Schwingkölbchen 4083
Schwingungsschreiber 6408
Schwingungszähler 8021
Schwungfeder 7720
Sedimentationsgeschwindigkeit 8069
Sedimentationskonstante 1999
Seenkunde 5178
Segmentallopolyploidie 8081
Segmentaustausch 8082
Segregation 8087
Sehen, binokulares 1041
Sehfarbstoff 8563
Sehgrube 3572
Sehhügel 9033
Sehkeil 6331
Sehne 8983
Sehnenhaut 654
Sehnenreflex 8984
Sehnenscheide 8982
Sehnerv 6362
sehnig 8981
Sehorgan 6375
Sehpigment 9563
Sehpurpur 7862
Sehrinde 9561
Sehschärfe 9558
Sehstab 7044
Sehvermögen 9557
Sehzelle 9559
Sehzentrum 9560
Seidendrüse 8290
Seismonastie 8089
Seitenkette 8275
Seitenknospe 876
Seitenlinienorgan 5048
Seitenverzweigung 5049
Seitenwurzel 5050
Sekret 8059, 8065
Sekretbehälter 8063
sekretieren 8057
Sekretin 8058
Sekretion 8059
  äußere 3209
  innere 4608
  exokrine 3209
Sekretionsgewebe 8066
Sekretkörperchen 8062
sekretorisch 8060
Sekretzelle 8064
Sektorialchimäre 8067
Sekundäreinschnürung 6220
Sekundärkonsument 8051
Sekundärstruktur 8056
selbstbefruchtend 8110
Selbstbefruchter 8113
Selbstbefruchtung 8112
Selbstbestäubung 8115
Selbstdifferenzierung 8106
Selbsterhaltungstrieb 4723
Selbsterneuerung 842

Deutsch

selbstfertil 8110
Selbstfertilität 8111
Selbstfruchtbarkeit 8111
Selbstimmunisierung 828
Selbstkompatibilität 8111
Selbstregulierung 8117
Selbstreinigung 8116
Selbstreproduktion 8119
selbstreproduktiv 8120
selbststeril 8121
Selbststerilität 8122
Selbststeuerung 8117
Selbstung 8112
Selbstunverträglichkeit 8114
Selbstverdauung 8107
Selbstverdoppelung 8109
Selbstverstümmelung 846
Selektion 8092
  disruptive 2623
  gerichtete 2589
  kanalisierende 1380
  natürliche 5989
  rekurrente 7658
  stabilisierende 8572
Selektionsdifferential 8094
Selektionsdruck 8102
Selektionsgipfel 8100
Selektionskoeffizient 8093
Selektionstheorie 9044
Selektionsvorteil 8097
Selektionswert 8103
selektiv 8096
Selektivität 8104
Selektivnährboden 8099
semiarid 8126
Semidominanz 8130
semipermeabel 8139
Semipermeabilität 8138
semisteril 8142
Semisterilität 8143
Seneszenz 8144
Senkungsgeschwindigkeit 8069
Senkwasser 3947
sensibilisieren 8163
Sensibilisierung 8162
Sensibilisierungswirkung 8164
Sensibilität 8155
Sensillum 8157
sensoriell 8166
sensorisch 8166
Sensorium 8165
septiert 8179
septifrag 8181
septizid 8180
Septum 8182
Seralstufe 8184
Serie *(ecol.)* 8186
  allelomorphe 321
Serienschnitt 8188
Serin 8191
Serodiagnostik 8200
Serologe 8192
Serologie 8193
serös 8196
Serosa 8194
Serotonin 8195
Serum 8198
Serumalbumin 8199
Serumprotein 8201
Sesambein 8202
sessil 8203
Seston 8204
Seta 8208
Seuchenlehre 3059
Sexduktion 8222
sexual 8238
Sexualdimorphismus 8221
Sexualhormon 8227
Sexualindex 8236
Sexualität 8249
Sexuallockstoff 8239
Sexualorgan 8235
Sexualverhalten 8240
Sexualzelle 3643
sexuell 8238
sezernieren 8057
sezieren 2624
Sezieren 2627
Shift 8259
Shikimisäure 8260
Sichel *(bot.)* 2703
sichelförmig 3300
Siebbein 3125
Siebplatte 8277
Siebröhre 8278
Siebröhrenelement 8279
Siebröhrenglied 8279
Siebröhrenzelle 8279
Siebzelle 8276
Siedepunkt 1215
Signalreiz 8280
Silifikation 8285
silifizieren 8286
Silur 8291
simplex 8294
Simultanadaptation 8295
Simultanmutation 8296
Sinnes ... 8166
Sinnesborste 8167
Sinneseindruck 8171
Sinnesepithel 8169
Sinneshaar 8170
Sinnesnerv 8172
Sinnesnervenzelle 8173
Sinnesorgan 8154
Sinnesphysiologie 8174
Sinnesreiz 8176
Sinneszelle 8168
Sinus 8300
Sinusdrüse 8301
Sipho 8303
Siphonogamie 8304
Siphonostele 8306
Sippe 1766
Sippenzüchtung 3306
Sitzbein 4854
sitzend 8068
Skelett 8312
Skelettmuskel 8311
Sklereide 8027
Sklerenchym 8028
Sklerit 8029
Sklerophyllen 8031
Skleroprotein 8032
Sklerotium 8034
Sklerotom 8035
Skolopalorgan 8036
Skoloparium 8036
Skrotum 8039
Sohlengänger 7011
Sol 8374
Solbildung 8377
Solenozyt 8378
Solifluktion 8382
solitär 8383
Solut 8389
Soma 8391
somatisch 8392
Somatogamie 8394
somatogen 8395
Somatologie 8396
Somatopleura 8397
Somatotropin 8398
Somazelle 8393
Somit 8400
Sommerannuelle 8784
Sommerei 8786
sommergrün 8787
Sommerholz 5042
Sommerschlaf 211
Sommerspore 9406
Sonnenblatt 8789
Sonnenenergie 8375
sonnenliebend 4146
Sonnenpflanze 4149
Sonnenstrahlung 8376
Soredium 8402
Sorption 8403
Sorte 9449
Sortenabbau 2445
Sorus 8404
Soziabilität 8344
Sozialattraktion 8339
Sozialparasitismus 8343
Sozialverhalten 8340
Soziation 8345
Sozietät 8346
Spadix 8415
Spaltalgen 2210
spaltbar 3423
Spaltbarkeit 3422
Spaltbein 1786
spalten 8529
spalten, sich 8086
Spaltfrucht 8007
Spaltfuß 1786
Spaltkapsel 2353
Spaltöffnung 8673
Spaltpflanze 8014
Spaltpilze 8012
Spaltprodukt 3425
Spaltung 3424, 8087
Spaltungsgeneration 3406
Spaltungsgesetz 5485
Spannmuskel 8987
spatelförmig 8421
spatelig 8421
Spatha 8416
Spätholz 5042
Speiche 7587
Speichel 7951
Speichelabsonderung 7955
Speicheldiastase 7445
Speicheldrüse 7954
Speicheldrüsenchromosom 7953
Speichelfluß 7956
Speichergewebe 8687
Speicherorgan 8684
Speicherstoff 7777
Speicherungsvermögen 8683
Speicherwurzel 8685
Speiseröhre 6288
Spektralanalyse 8450
Spektrophotometer 8451
Spektrophotometrie 8452
Spektrum, biologisches 1085
Spelze 3881
Spender 2672
Spenderbakterie 2673
Spenderzelle 2674
Sperma 8455
Spermakern 8458
Spermatide 8465
Spermatium 8466
spermatogen 8469

Spermatogenese 8468
Spermatogonium 8470
Spermatophore 8471
Spermatophyten 8472
Spermatozoid 8473
Spermatozoon 8474
Spermatozyt 8467
Spermaträger 8471
Spermazelle 8456
spermienbildend 8469
Spermienflagellum 8463
Spermienkern 8458
Spermin 8475
Spermiogenese 8476
Spermium 8474
Sperreaktion 3667
Sperren 3666
Sperrmustereffekt 1169
Speziation 8429
Spezies 8430
spezifisch 8447
Spezifität 8449
Sphagnum 6662
Sphärom 8483
Sphäroprotein 8484
Sphärosom 8485
Sphingomyelin 8487
Sphingosin 8488
Sphinkter 8486
Spielart 9449
Spike 8491
Spinalganglion 8497
Spinalkanal 8494
Spinalnerv 8498
Spindel, achrome 73
Spindelachse 8500
Spindelfaser 8501
spindelförmig 3630
Spindelgift 8502
Spindelpol 8503
Spinndrüse 8506
Spinnenkunde 680
Spinnentiere 678
Spinnwarze 8505
Spinnwebenhaut 679
Spiraculum 8509
Spiralfurchung 8511
Spiralisation 8515
Spiraltracheide 8514
Spirem 8516
Spirille 8517
Spirochäte 8518
Spirometer 8519
Spitze 633
Spitzenpotential 8491
Spitzenwachstum 641
Splanchnologie 8523
Splanchnopleura 8524
Splint(holz) 7980
Spodogramm 8531
Spongin 8533
Sporangienträger 8539
Sporangium 8540
Spore 8541
Sporenbehälter 8540
Sporen bilden 8561
sporenbildend 8554
Sporenbildung 8562
Sporenblatt 8557
Sporenkapsel 8543
Sporenlager 4474
Sporenmutterzelle 8546
Sporenpflanzen 2165
Sporenschlauch 747
Sporentierchen 8559
sporentragend 8549
Sporenträger 8556
Sporenverbreitung 8544, 8562
Sporidie 8548
sporogen 8554
Sporogenese 8553
Sporogon 8555
Sporogonie 8553
Sporokarp 8550
Sporophyll 8557
Sporophyt 8558
Sporozoon 8559
Sporozyste 8551
Sporozyt 8546
Sport 8560
Sporulation 8562
Sprachzentrum *(anat.)* 8453
Spreublatt 6543
sprießen 8262
Springfrucht 2354
Springschwänze 1888
Spritzloch 8509
Sproß 8263, 8611
Sproßachse 1489
sprossen 8262
Sproßknolle 8618
Sproßknospe 7063
Sproßkonidie 1159
Sproßmutation 8560
Sproßpflanzen 2055
Sproßpilz 1156
Sproßscheitel 8612
Sproßspitze 8612
Sproßsystem 8617
Sprossung 1317
Sprungschicht 9051
Spurenelement 9177
Squalen 8567
Stäbchen *(Netzhaut)* 7891
stäbchenförmig 933
Stachel 7287, 8660
stachelfrüchtig 42
Stachelhäuter 2753
stachelig 7289
Stachelzelle 7288
Stamen 8585
Staminodium 8588
Stamm 6885, 8611, 8689, 9291
Stamm ... 1486
Stammbaum 6671, 6883
Stammbaumselektion 6672
Stammbaumverzweigung 1764
Stammbaumzüchtung 6672
stammblütig 1484
Stammblütigkeit 1485
Stammesentwicklung 6881
Stammesgeschichte 6881
stammesgeschichtlich 6882
Stammesreihe 6876
Stammform 7299
Stammhirn 1249
Stammknospe 8614
Stammlösung 8671
Stammtier 3570
Standard 9327
Standardabweichung 8589
Standardfehler 8590
Standardlösung 8592
Standardnährboden 965
Standardpräparat 8591
Standardtyp 8593
Ständer 972
Ständerpilze 970
Ständerspore 971
standing crop 8594
Standort 4032
Stärke, tierische 3892
stärkeführend 437
Stärkegel 8596
stärkehaltig 437
Stärkekorn 8597
Stärkenzuwachs 3983
stärkereich 437
Stärkereserve 8598
Starklichtpflanze 4149
Starre 7878
Starrheit 7878
Stationärkern 8602
Statoakustikus 802
Statoblast 8604
Statolith 8606
Statozyste 8605
Staubbeutel 564
Staubblatt 8585
Staubblatthaar 8586
Staubblüte 8587
Staubfaden 3403
Staubgefäß 8585
Staude 6714
Stearinsäure 8609
Stechborste 8720
stechend *(Insekt)* 8661
Steigbügel *(anat.)* 8595
Stein *(bot.)* 8679
Steinfrucht 8682
Steinkanal 8680
Steinkern 8679
Steinpflanze 5212
Steinzelle 8027
Steißbein 1839
Steißwirbel 1838
Stele 8610
Stelzwurzel 8642
Stempel 6958
Stempelblüte 6959
Stempeltechnik 7734
Stengel 8611
Stengel ... 1486
Stengelbündel 8615
stengelfrüchtig 1488
Stengelglied 4772
Stengelknospe 8614
Stengelknoten 6148
stengellos 44
stengeltreibend 1483
stengelumfassend 433
stenohalin 8619
stenohyger 8620
stenotherm 8621
Steppenpflanze 8622
Sterbetafel 5143
Sterbeziffer 2290
Sterblichkeit 5817
Stereom 8624
Stereomikroskop 8625
Sterigma 8627
steril 8628
Sterilisation 8631
sterilisieren 8632
Sterilität 8629
Sterilitätsgen 8630
Sterin 8637
Sterkobilin 8623
Sternstrahlen 7088
Steroid 8635
Steroidhormon 8636
Sterol 8637
Stert 9419
Stetigkeit 1998
Steuerfeder 8919

Deutsch

Steuerung, hormonale 4385
STH 8398
Stichkultur 8571
Stichprobe 7598
Stickstoffabfälle 6142
Stickstoffbase 6134
Stickstoffbilanz 6133
Stickstoffbinder 6139
Stickstoffbindung 6138
Stickstoffentzug 2379
Stickstoff-Fixierung 6138
Stickstoffgleichgewicht 6137
stickstoffhaltig 6141
Stickstoffkreislauf 6136
Stickstoff-Stoffwechsel 6140
Stickstoffverbindung 6135
Stickstoffzyklus 6136
Stiel 8582, 8611
Stielauge 8583
Stigma 3288, 8640
Stimmband 9606
Stimmritze 3871
Stimmung *(eth.)* 5803
Stimulans 8643
Stinkdrüse 8665
Stipel 8666
Stirnauge 6280
Stirnbein 3592
Stirnhöhle 3594
Stirnlappen 3593
Stoff, gelöster 8389
Stoffaufnahme 4668
Stoffaustausch 3179
Stoffkreislauf 1752
Stofftransport 9227
Stoffwanderung 5847
Stoffwechsel 5541
  intermediärer 4754
  oxydativer 6490
Stoffwechsel ... 5525
Stoffwechselabbauprodukt 1460
stoffwechselaktiv 5540
Stoffwechselantagonist 5527
Stoffwechselendprodukt 5529
Stoffwechselenergie 5530
Stoffwechselgift 5531
Stoffwechselgleichgewicht 6889
Stoffwechselinhibitor 5531
Stoffwechselprodukt 5542
Stoffwechselprozeß 5535
Stoffwechselrate 5537
Stoffwechselreaktion 5538
Stoffwechselschlacken 5539
Stoffwechselstörung 5528
Stoffwechseltätigkeit 5526
Stoffwechselumsatz 5537
Stoffwechselvorgang 5535
Stoffwechselweg 5534
Stoffwechselzwischenprodukt 5532
Stolon 8672
Stoma 8673
Stomochord 8677
Stomochordaten 4159
Stomodaeum 8678
Stoßzahn 9316
Strahlen, ionisierende 4844
  kosmische 2087
  radioaktive 7558
  ultraviolette 9353
Strahlenbiologie 7563
Strahlenblüte 7612
strahlendurchlässig 7586
Strahlendurchlässigkeit 7535
strahlenempfindlich 7583
Strahlenempfindlichkeit 7584
Strahlenflosser 111
Strahlengenetik 7572
strahlengeschützt 7613
Strahlenkörper 1736
Strahlenpilz 109
strahlenresistent 7582
Strahlenresistenz 7581
Strahlenschutz 7545
strahlensicher 7613
Strahlentierchen 7574
strahlenundurchlässig 7580
Strahlenundurchlässigkeit 7579
Strahlenwirkung 7611
strahlig 7536
Strahlung 7540
  radioaktive 7557
Strahlungen, mitogenetische 5710
Strahlungsdosis 7542
Strahlungsenergie 7539
Strahlungsfeld 7544
Strahlungsökologie 7570
Strahlungsquelle 7546
Strahlungsschaden 7575
Strauch 8273
strauchartig 3614
Strauchflechte 3615
strauchförmig 3614
Strauchschicht 1305
Strecker 3254
Streckreflex 3255
Streckungswachstum 2855
Streckungszone 2856
streifennervig 6584
Streifung 8702
Streptomyzin 8698
Streßhormon 143
Streufrucht 2354
Streuung 2618, 9444
Strichkultur 8696
Strickleiternervensystem 9500
Stridulation 8704
Stridulationsorgan 8703
Strobilation 8707
Strobilus 1974
Stroma 8708
Strömchentheorie 5223
Strudelwürmer 9308
Strukturgen 8710
Strukturheterozygotie 8712
Strukturhybride 8713
Strukturmodifikation 8709
Strukturprotein 8715
Strukturveränderung 8709
Strychnin 8787
Stuhlentleerung 2323
Stummelfüßer 6340
Stützblatt 1242
Stützgewebe 8812
Stützwurzel 7352
Stützzelle 8811
Suberin 8728
Subgen 8732
Subimago 8734
Subitanei 8786
Subklimax 8723
Subkultur 8724
Subkutis 8725
subletal 8737
Sublimat 8738
Sublitoral 8740
Submerskultur 8742
Submikron 8743
submikroskopisch 8744
Subserie 8749
Subspezies 8751
Substanz, graue 3955
  weiße 9672
Substanzanlagerung 665
Substitution 8753
Substrat 8755
Substratspezifität 8756
Subterrantier 8757
subzellulär 8721
Sukkulente 8767
Sukkulenz 8766
Sukzession 8762
  ökologische 2766
Sukzessionsserie 8186
Summation 8783
  räumliche 8420
  zeitliche 8978
Sumpfpflanze 4153
Superdominanz 6454
Supergen 8797
Superpositionsauge 8806
Superspezies 8808
Supplementärgen 8810
Suppressor 8813
Suppressormutation 8814
Supralitoral 8816
Supraspezies 8818
Supravitalfärbung 8819
Suspensor 8835
Süßwasser 3585
Süßwasserkunde 5178
Süßwasserpflanze 3588
Süßwasserplankton 5179
Süßwassertier 3586
Suszeption 8833
Symbiont 8851
symbio(n)tisch 8853
Symbiose 8852
Symmetrie, radiäre 7537
Symmetrieebene 8854
Sympathikus 8856
sympatrisch 8857
sympetal 3659
Symphyse 8859
sympodial 8860
Sympodium 8861
Synangium 8862
Synapse *(neur.)* 8863
Synapsis *(gen.)* 8864
Synaptän 9790
synaptisch 8866
Synarthrose 8870
Synchorologie 8872
synchron 8873
Syndese 8875
Syndesmose 8876
synergetisch 8879
Synergide 8880
Synergie 8878
Synergismus 8881
Synergist 8882
syngam 8884
Syngameon 8883
Syngamie 8885
Syngenese 8886
syngenetisch 8887
Synkarion 8888
synkarp 8871
Synökie 8890
Synökologie 8877
Synovia 8891
synözisch 8889
Synthetase 5147
syntroph 8897
Syntrophismus 8898

Synusie 8899
Synzoospore 8900
Synzytium 8874
Syrinx 8901
System, limbisches 5169
ökologisches 2775
retikulo-endotheliales 7836
zellfreies 1513
Systematik 8902
botanische 7008
Systematiker 8903
Systemmutation 8905
Systole 8906
Szintillationszähler 8021

T

Tachygenese 8908
tagblühend 4157
Tagesperiodik 2649
Tagesrhythmik 2649
Tagesrhythmus 1748
Tagtier 2647
Taiga 8918
Talg 8922
Talgdrüse 8047
Talgsäure 8609
Tandemselektion 8923
Tangorezeptor 8924
Tannin 8926
Tapetenschicht 8928
Tapetenzelle 8927
Tapetum 8928
tarnen 1376
Tarnfärbung 2163
Tarntracht 2163
Tarnung 1377
Tast ... 8909
Tastempfindung 8914
Taster 6558
Tasthaar 8911
Tastkörperchen 8910
Tastorgan 8912
Tastreiz 8915
Tastrezeptor 8924
Tastsinn 8153
Tastsinnesorgan 8924
Taupunkt 2471
Taxie 8941
Taxis 8941
Taxon 8942
Taxonom 8944
Taxonomie 8945
taxonomisch 8943
Tectologie 8949
Tegment 1316
Teilareal 2610
Teilblättchen 3512
Teile, oberirdische 199
Teilfrucht 5493
Teilung, meiotische 5460
mitotische 5711
Teilungsebene 2654
Teilungsvermehrung 8025
Teilungszone 2655
Telegonie 8954
Telenzephalon 8955
Teleostier 8956
Teleutospore 8958
telolezithal 8960
Telom 8961
Telomer 8963
Telomtheorie 8962
Telophase 8964
Telosynapsis 8965
Telosyndese 8965
Telotaxis 8966
telozentrisch 8959
Telson 8967
temperaturempfindlich 8973
Temperaturgefälle 8970
Temperaturkoeffizient 8969
Temperaturoptimum 8971
Temperaturrezeptor 8972
Tensor 8987
Tentakel 8988
Tepal 8989
Teratologie 8990
Terminalisation 8999
Terminalknospe 638
Terminalphase 8998
Termone 9000
Terpen 9001
terrestrisch 9002
Territorialität 9007
Territorialverhalten 9007
Territorium 9011
Tertiär 9013
Tertiärkonsument 9012
Tertiärstruktur 9014
Testa 8074
Testikel 9023
Testis 9023
Testkreuzung 9016
Testosteron 9024
Testpaarung 9016
Testpflanze 9019
Tetrade 9025
Tetradenanalyse 9026
tetraploid 9027
Tetraploidie 9028
Tetrapode 9029
tetrasom 9030
Tetraspore 9031
Tetratyp 9032
Thalamus 9033
Thallophyt 9035
thallös 9036
Thallus 9037
Thalluspflanze 9035
Theka 9038
Thelytokie 9039
Thermodynamik 9053
thermodynamisch 9052
thermolabil 9056
Thermolyse 9057
Thermonastie 9058
Thermoperiodismus 9059
thermophil 9060
Thermoregulation 9062
Thermorezeptor 9061
thermostabil 9066
Thermotaxis 9067
Thermotropismus 9068
Therophyt 9069
Thiamin 9070
Thigmomorphose 9074
Thigmonastie 4096
Thigmotaxis 9075
Thigmotropismus 9076
Thioktansäure 9080
Thionin 9081
Thorax 9085
Threonin 9091
Thrombin 9097
Thrombogen 9099
Thrombokinase 9100
Thrombozyt 9098
Thylle 9323
Thymidin 9104
Thymidylsäure 9105
Thymin 9106
Thymonukleinsäure 9107
Thymus(drüse) 9108
Thyreotropin 9112
Thyroxin 9113
Thyrsus 9114
Tiefenwasserzone 7323
Tiefsee ... 38
Tiefseefauna 39
Tiefwurzler 2321
Tier, dauerwarmes 4351
höheres 4285
landlebendes 9003
wechselwarmes 7071
Tieranatomie 9776
Tierart 526
tierblütig 9750
Tierblütigkeit 9752
Tierblütler 9751
Tiergenetik 9761
Tiergeographie 9762
Tiergesellschaft 524
Tiergift 9777
Tierkunde 9767
Tierleben 514
Tierökologie 507
Tierphysiologie 518
Tierreich 513
Tierschmarotzer 516
Tierschutz 521
Tiersoziologie 525
Tierstamm 517
Tierstock 1896
Tiersystematik 528
Tierverbreitung (des Samens) 9758
Tierversuch 3241
Tierwanderung 515
Tierwelt 530
Tierzelle 505
Tierzucht 504
Tierzüchter 503
Tierzuchtlehre 9775
Tintendrüse 4687
Tintensack 4688
Titer 9142
Titration 9140
Titrationskurve 9141
titrieren 9139
Titrierung 9140
Titrimetrie 9143
Tochterchromatide 2278
Tochterchromosom 2279
Tochtergeneration 3406
erste 3377
zweite 3378
Tochterkern 2280
Tochterzelle 2277
Todesrate 2290
Tokopherol 9145
Toleranzdosis, genetische 3746
Toleranzgrenze 9146
Tönnchen 7471
Tönnchenpuppe 1832
Tonoplast 9148
Tonus 9149
Top-Cross 9154
Top-Crossing 9155
Topotaxis 9157
Torf 6661
Torfmoos 6662
Torus 9158

Totalfurchung 1943
Totalpräparat 9673
totipotent 9164
Totipotenz 9163
toxigen 9173
Toxikologe 9171
Toxikologie 9172
Toxin 9174
toxisch 9167
Toxizität 9170
toxogen 9173
TPN 6119
Trabant 9175
Trabantenchromosom 7989
Trabekel 9176
Tracer 9178
radioaktiver 7559
Tracer-Methode 9179
Trachea 9180
Tracheaten 9187
Trachee 9181, 9182
Tracheenatmung 9184
Tracheenkieme 9183
Tracheenlunge 1227
Tracheenstamm 9186
Tracheensystem 9185
Tracheide 9188
Tracheole 9189
Tracheophyt 9190
trächtig 7261
Trächtigkeit 3817
Trächtigkeitsdauer 3818
Tragblatt 1242
Träger 1444
kolloidaler 649
Trägersubstanz 1444
Tragezeit 3818
Tragknospe 3606
Tränenbein 4997
Tränendrüse 4999
Tränengrube 8947
Tränenkanal 4998
Tränennasengang 5001
Tränensack 5000
Transaminase 9193
Transaminierung 9194
Transduktion 9197
abortive 18
Transduktionsklon 9198
Transferase 9203
Transfer-RNS 9201
Transformation *(gen.)* 9204
Transformismus 9042
Transfusionsgewebe 9205
Transgenation 3714
Transgression 9207
Trans-Konfiguration 9195
Transkription 9196
Translation 9211
Translokation 9212
insertionale 4715
reziproke 7645
Translokationsheterozygotie 9213
Transmethylierung 9215
Transmitter 9219
Transmutation 9220
Transphosphorylierung 9221
Transpiration 9222
kutikuläre 2205
stomatäre 8676
Transpirationsintensität 9223
Transpirationsrate 9223
Transplantat 9224
Transplantation 9226
autoplastische 849
transplantieren 9225
Transport, aktiver 127
Transposition 8259
transspezifisch 9228
Transsudat 9229
Transsudation 9230
Transversalgeotropismus 6973
Traube 7527
Traubenhaut 9430
Traubenzucker 3878
traubig 7528
Traumatin 9235
Traumatinsäure 9235
Traumatotropismus 9236
Treffertheorie 8931
Trematoden 9243
Trennungsschicht 23
Trennungszone 24
Trephon 9244
Treppengefäß 7997
Treppentracheide 7996
Triade 9245
Trias 9247
Tribus *(tax.)* 9248
Trichogyne 9251
Trichom 9252
Trichozyste 9250
trichterförmig 4663
Trieb 2706, 8565
Triebhandlung 4724
Trigeminus 9253
trihybrid 9254
Trihybridie 9255
Trikarbonsäurezyklus 4982
trimer 9256
Trimerie 9257
Trimorphismus 9258
Trinkwasser 2704
Trinkwasserversorgung 2705
Triosephosphat 9260
triözisch 9259
Tripeptid 9261
Triphosphopyridindinukleotid 6119
Triplett 9264
Triplett-Kode 9265
triplex 9266
Triploidie 9267
Tripton 9268
trisom 9269
Trisomie 9270
Trivalent 9271
t-RNS 9201
Trochlearis 9273
Trockenextrakt 2716
Trockenfrucht 2718
Trockengewicht 2722
Trockenheit 2710
Trockenmasse 2721
Trockennährboden 2715
Trockenperiode 2711
Trockenpflanze 9720
Trockenpräparat 2719
Trockenresistenz 2712
Trockenstarre 441
Trockensubstanz 2721
Trockenwald 2717
Trockenzeit 2720
Trocknen 2437
Troglobionten 1491
Trommelfell 9326
Tröpfchenkultur 2709
trophisch 9276
Trophobiose 9278
Trophoblast 9279
Trophophyllaxe 9275
Trophotaxis 9282
Trophotropismus 9283
Trophozyt 6255
Tropismus 9286
Tropophyt 9287
Tropotaxis 9288
Trugdolde 2218
trugdoldig 2219
Trypsin 9293
Trypsinogen 9294
Trypton 9295
Tryptophan 9296
TSH 9112
Tundra 9306
Tunikaten 9307
Tüpfel 6960
Tüpfelgefäß 6964
turgeszent 9309
Turgeszenz 9310
Turgor 9310
Turgorbewegung 9311
Turgordruck 9312
Turione 9313
„turnover“-Rate 9315
Tympanalorgan 9324
Typ(us) 9327
Typogenese 9329
Typolyse 9330
Typostase 9331
Tyrosin 9332
Tyrothrizin 9333

U

Überart 8818
Überbefruchtung 8796
Überdauerungsform 7819
Überdominanz 6454
Überfamilie 8793
Übergangsepithel 9208
Übergangsmoor 9209
Übergangsstadium 9210
Überhitzung 6455
überimpfen 4697
Überklasse 8791
Überkreuzvererbung 2130
Überkreuzzüchtung 2129
überleben 8832
Überleben 8825
Überleben des Tüchtigsten 8828
Überlebenschance 1602
Überlebenskurve 8826
Überlebensrate 8829
Verringerung der ~ 8830
Überlebenswert 8831
Übermännchen 8800
Überordnung 8803
Überparasit 4484
Überpflanzen 3082
überpflanzen 9225
Überpflanzung 9226
Überreiz 8801
Überschall 9345
Überschallknall 8401
überschwellig 8815
Übersprungbewegung 2621
Übersprunghandlung 2621
übertragbar 9217
Übertragbarkeit 9216
Überträger 9468
Überträgersubstanz 9219

Übertragung des genetischen
Materials 9199
Übertragungsweise 5728
Übervölkerung 6457
Überweibchen 8794
überwintern 6458
Überwinterung 6459
Überwinterungsknospe 4268
Überwinterungsorgan 4270
Überwinterungsplatz 9700
überzählig 8802
Ubichinon 9334
Ubiquist 9335
Uhr, biologische 1078
innere 1078
physiologische 1078
Ulna 2181
Ultradünnschnitt 9347
Ultrafilter 9339
Ultrafiltrierung 9340
Ultramikroskop 9341
ultramikroskopisch 9342
Ultramikrosom 9343
Ultramikrotom 9344
Ultrapasteurisation 9395
Ultraschall 9345
Ultrastruktur 9346
Ultraviolettbestrahlung 9349
Ultraviolettmikroskopie 9351
Ultraviolettstrahlung 9352
Ultrazentrifugation 9337
Ultrazentrifuge 9338
Umbelliferen 9356
Umbilikalkreislauf 9360
Umgebungstemperatur 376
Umorientierung 7661
Umsatzrate 9315
umsetzbar 5543
umsetzen 5544
Umwelt 3004
Umwelt ... 3005
Umweltänderung 3009
umweltangepaßt 136
Umweltanpassung 3006
Umweltbedingungen 3010
Umweltbiologie 3008
Umwelteinfluß 3017
Umweltfaktor 3015
Umweltforscher 3027
Umweltgesetzgebung 3018
Umwelthygiene 3016
Umweltkatastrophe 3013
Umweltlärm 375
Umweltplanung 3020
Umweltpolitik 3021
Umweltqualität 3024
Umweltreiz 3026
Umweltschutz 3023
Umweltspezialist 3027
Umweltstörung 3014
Umweltveränderung 3007
Umweltverbesserung 4583
Umweltverhältnisse, Verschlechterung der ~ 3012
Umweltverschmutzer 7114, 7116
Umweltverschmutzung 3011
Umweltverseuchung 3011
Umweltwiderstand 3025
Unabhängigkeitsgesetz 5486
unbefruchtet 9369
unbelebt 5145
unbeweglich 6163
Unbrunst 551
undurchlässig 4578
Undurchlässigkeit 4577
unerregbar 4650
Unerregbarkeit 4649
unfruchtbar 4654
Unfruchtbarkeit 4655
ungesättigt 9394
ungeschlechtlich 749
ungiftig 6175
Ungiftigkeit 6176
ungleicherbig 4261
Ungleicherbigkeit 4259
unifaktoriell 5767
Uniformitätsgesetz 5484
unilokulär 9381
unipolar 9384
unisexuell 9386
Univalent 9390
Universalität des genetischen
Kodes 9392
Unkrautvernichtungsmittel 4177
unlöslich 4718
Unlöslichkeit 4717
unmischbar 4552
Unpaarhufer 6749
unpaarig 4573
unpaarig gefiedert 4575
Unpaarzeher 6749
unpaarzehig 4574
unreif 4548
Unreife 4549
unschädlich 4694
Unschädlichkeit 4695
unseptiert 6170
unspezifisch 6174
Unterabteilung 8726
Unterart 8751
Unterbauch 4499
Unterboden 1014
Untereinheit 8761
Unterentwicklung 4512
Unterfamilie 8731
Untergattung 8733
Untergrund 1686
Unterhaut(binde)gewebe 8725
Unterhautfettgewebe 167
Unterkiefer 5358
Unterklasse 8722
Unterkultur 8724
Unterlage 8755
Unterleib 1
Unterlid 5252
Unterlippe 4988 *(ins.)*, 5254
Unterordnung 8746
Unterpopulation 8748
Unterreich 8736
Unterscheidungsmerkmal 2640
Unterschiedsschwelle 2522
unterschwellig 8739
Unterstamm 8747
unterständig 4652
Untersuchung, experimentelle 3239
mikroskopische 5637
Unterwasserpflanze 8741
Unterwuchs 9365
unverdaulich 4634
Unverdaulichkeit 4633
unverträglich 4600
Unverträglichkeit 4598
Uperisation 9395
Urazil 9401
Urbanisation 9402
Urdarm 690
Urdarmdach 1677
Urease 9405
Uredospore 9406
Ureizelle 6345
Ureter 9407
Urethra 9408
Urform 7299
Urgeschichte 7264
Urgestein 6619
Urharnsack 316
Urhirn 689
Uridylsäure 9410
Urin 9413
Urinsekten 669
Urmeristem 7305
Urmollusken 428
Urmund 1157
Urmünder 7412
Urmundlippe 1158
Urniere 5514
Urnierengang 5513
Urobilin 9415
Urogenitalsystem 9417
Uronsäure 9418
Ursamenzelle 8470
Ursegment 5963
Ursegmentstiel 6036
Ursprungspopulation 6392
Urstele 7411
Urtierchen 7417
Urtyp 692
Urwald 9535
Urwirbel 8400
Urzeugung 8537
Uterus 9426
Uterushöhle 9424
UV-Absorption 9348
UV-Bestrahlung 9349
UV-Licht 9350
UV-Mikroskopie 9351
UV-Strahlen 9353
UV-Strahlung 9352

V

Vagina 9437
Vagus 9438
Vagusstoff 66
Vakuole 9433
kontraktile 2023
pulsierende 2023
Vakuolenmembran 9431
Vakuolensaft 9432
Vakuom 9434
Vakuum 9435
Valenz, ökologische 2768
Valin 9439
valvat 9440
Variabilität 9442
alternative 2600
diskontinuierliche 2600
fluktuierende 2018
kontinuierliche 2018
qualitative 2600
quantitative 2018
Variable 9443
Variante 9445
Varianz 9444
Varianzanalyse 459
Variation 9446
Variationsbereich 7601
Variationsbreite 7601
Variationskoeffizient 1849
Variegation 9448
Varietät 9449
vaskular 9452
Vaskularisierung 9462

Vasomotoren 9466
Vasopression 9467
väterlich 6651
Vatertier 8307
Vegetation 9472
Vegetationsaufnahme 9479
Vegetationseinheit 9476
Vegetationsgebiet 9477
Vegetationsgürtel 9473
Vegetationskarte 9474
Vegetationskartierung 5371
Vegetationskegel 9481
Vegetationskunde 6921
Vegetationsorgan 9484
Vegetationsperiode 9485
Vegetationspunkt 9486
Vegetationsruhe 9489
Vegetationsspitze 9481
Vegetationsstufe 8688
Vegetationstyp 9475
Vegetationszeit 9485
vegetativ 9480
Velamen 9491
Vene 9490
Venenklappe 9498
Venensinus 8302
venös 9496
Ventilationsgröße 7455
Ventrikel 9505
ventrizid 9504
Veraschung 4593
Verästelung 683
Verband 327
Verbänderung 3310
Verbreitungseinheit 2494
Verbreitungsgebiet 7600
Verbreitungsgrenze 5172
Verbrennungswärme (physiologische) 1916
verdauen 2538
verdaulich 2540
Verdauung 2541
Verdauungsapparat 2546
Verdauungsdrüse 2543
Verdauungsenzym 2542
Verdauungsferment 2542
Verdauungskanal 304
Verdauungsorgan 2545
Verdauungssaft 2544
Verdauungssystem 2546
verdicken 9071
Verdickung 9072
verdoppeln (sich) 2731
Verdoppelung 2732
Verdoppelungszeit 2699
Verdrängungskreuzung 3921
verdünnen 2556
Verdünnung 2557
Verdünnungsmittel 2555
verdunsten 3152
Verdunstung 3153
  und Transpiration 3155
Verdunstungsverlust 3154
Vereisung 3849
vererbbar 4671
vererben 9218
vererblich 4671
Vererbung 4672
  alternative 367
  einseitige 9380
  erworbener Eigenschaften 4673
  extrachromosomale 2254
  geschlechtsgebundene 8234
  mendelnde 6640
  nichtmendelnde 6162
Vererbung, plasmatische 2254
  quantitativer Merkmale 1164
  unilaterale 9380
  zytoplasmatische 2254
Vererbungsgesetze 5062
Vererbungslehre 3748
Vererbungsmerkmal 4186
verestern 3119
Veresterung 3118
Verflächung 6977
verflüchtigen (sich) 9610
Verflüchtigung 9609
verflüssigen 5209
Verflüssigung 5208
vergärbar 3355
vergären 3356
Vergärung 3357
Vergeilung 3132
vergesellschaftet 764
Vergesellschaftung 765
Vergletscherung 3849
Vergrünung 9534
Verhalten 994
  tierisches 502
  umorientiertes 7661
Verhaltensanalyse 998
Verhaltensflexibilität 995
Verhaltensforscher 3128
Verhaltensforschung 3129
  vergleichende 1925
Verhaltensgenetik 996
Verhaltensinventar 3126
Verhaltensmerkmal 999
Verhaltensmuster 997
Verhaltensphysiologie 1000
Verhaltensweise 997
Verhärtung 4102
verholzen 5164
Verholzung 5162
verhornen 2061
Verhornung 2060
verjüngen 7693
Verjüngung 7694, 7708
verkalken 1349
Verkalkung 1347
verkieseln 8286
Verkieselung 8285
Verklonung 1808
verknöchern 6427
Verknöcherung 6426
Verknorpelung 1651
verkorken 8730
Verkorkung 8729
Verletzungsversuch 2324
vermehren (sich) 5890
Vermehrung durch Knospenbildung 3694
  vegetative 9487
vermehrungsfähig 7744
Vermehrungsfähigkeit 5888
Vermehrungsgeschwindigkeit 5889
Vermehrungsrate 5889
Vernalisation 9508
Vernation 9509
verpflanzen 9225
Verpflanzung 9226
verpuppen 7472
Verpuppung 7473
versauern *(Boden)* 90
Versauerung *(Boden)* 89
Verschiedenartigkeit 4217
verschiedengestaltig 4234
verschiedengriff(e)lig 4249
Verschiedengriff(e)ligkeit 4250
verschlüsseln *(gen. Kode)* 1842
Verschlüsselung 1844
verschmelzen 3629
Verschmelzung 3631
verschmutzen 2011, 7115
Verschmutzung 2012, 7117
Verschmutzungsgrad 2349
Verschmutzungsquelle 7119
verseuchen 2011
Verseuchung 2012
  radioaktive 7553
Verseuchungsmittel 2010
Verstädterung 9402
Verständigung *(eth.)* 1921
versteinern 6785
Versteinerung 6784
Versteinerungskunde 6532
Versuch 3225
Versuchsanordnung 3237
Versuchsbedingungen 3231
Versuchsergebnis 3236
Versuchslaboratorium 2458
Versuchsmethode 3234
Versuchsperson 3240
Versuchspflanze 3235
Versuchsreihe 8190
Versuchsserie 8190
Versuchsstadium 3238
Versuchstier 3228
Versuchszüchtung 3230
Versuch-und-Irrtum-Methode 9246
Vertebraten 9517
Verteidigungsreaktion 2326
Verteilungschromatographie 6641
Verträglichkeit 1927
Vertrocknung 2437
verunreinigen 7115
Verunreinigung 4585, 7117
Verunreinigungsgrad 2349
verwachsen 1961
verwachsenblättrig 3659
Verwachsung 1960
Verwandtschaft 7709
Verwandtschaftsgrad 2350
Verwandtschaftskoeffizient 1848
Verwandtschaftszucht 4589
verwelken 9676
verwelkt 9677
verwesen 2308
Verwesung 2310
verwildern 7931
Verwilderung 7933
Verwitterung 9668
Verzögerung, synaptische 8868
Verzögerungsphase 5016
verzweigen (sich) 7594
Verzweigung 7593
  dichotome 2508
  gabelige 2508
Vibrio 9528
Vielborster 7125
Vielfüßer 5964
vielgeiß(e)lig 5875
vielkammerig 5869
vielkernig 5880
viellappig 5878
vielsamig 7170
vielschichtig 5877
vielteilig 5881
Vielteilung 5885
Vielzeller 5871
vielzellig 5870
Vielzelligkeit 5872
Vierfüßer 9029
Vikariant 9529

W

X

Y

Z

Zwischenraum 4783
synaptischer 8867
Zwischenrippen ... 4742
Zwischenscheibe 4734
Zwischenstadium 4757
Zwischenstoffwechsel 4754
Zwischenwirbelscheibe 4790
Zwischenwirt 4755
Zwischenzelle 4785
Zwischenzellraum 4740
Zwitter 4196
Zwitterblüte 4198
Zwitterdrüse 6451
zwitterig 4197
Zwitterion 9779
Zwittertum 4199
Zwittrigkeit 4199
Zwölffingerdarm 2728
Zyanin 2208
Zygapophyse 9780
zygomorph 9783
Zygomorphie 9784
Zygose 9785
Zygosom 9786
Zygospore 9787
Zygotän 9790
Zygote 9788
Zygotenkern 9789
zyotisch 9791
zyklisch 2213
Zyklomorphose 2215
Zyklosis 2216
Zyklostomen 2217
Zyklus, biogeochemischer 1073
parasexueller 6597
Zylinderepithel 1913
Zymase 9792
Zymogen 9793
Zymogengranula 9794
Zymogenkörnchen 9794
Zymologie 9795
zymös 2219
Zyste 2220
Zystein 2221
Zystenbildung 2898
Zystide 2222
Zystin 2223
Zystokarp 2224
Zystolith 2225
Zytidylsäure 2226
Zytochrom 2231
Zytode 2232
Zytodiärese 2233
Zytogen 2234
Zytogenese 2235
Zytogenetik 2238
Zytogenetiker 2237
zytogenetisch 2336
Zytogonie 2239
Zytokinese 2240
Zytokinin 2241
Zytologe 2243
Zytologie 2244
zytologisch 2242
Zytolymphe 2245
Zytolyse 2247
Zytolysin 2246
Zytom 2248
Zytomorphose 2249
Zytoplasma 2251
zytoplasmatisch 2252
Zytopyge 2256
Zytosin 2257
Zytosom 2258
Zytostom 2259
Zytotaxonomie 2260
Zytotoxin 2261
Zytotrophoblast 2262
Zytotropismus 2263
Zytotubuli 5658

FRANÇAIS

Index Alphabétique

A

abaissement du seuil 9093
abdomen 1
abdominal 2
abeille mâle 2708
  mère 7520
aberration 9
  chromatidique 1692
  chromosomique 1714
abiogenèse 8537
abiogénétique 12
abiotique 13
abomasum 17
abondance 37
  de la ponte 1817
abscissine 21
  II 2680
abscission 22
  des feuilles 5067
  des fruits 3607
absorbant 27, 28
absorber 26
absorption 29
  d'eau 9629
  foliaire 5068
  d'oxygène 6495
  de l'ultraviolet 9348
  en UV 9348
abyssal 38
abysse 40
acanthocarpe 42
acanthocéphales 43
acaryote 6164
acaudé 628
acaule 44
accélération 45
accepteur 46
  (d') hydrogène 4436
acclimatation 53
acclimater 54
accommodation 55
accouplement 5411
  de bases 961
  génétique 3713
  au hasard 7597
  de testage 9016
accoupler 5408
s'accoupler *(chromosomes)* 6514
accoutumance 4034
accroissement diamétral 3983
accumulation d'aliments 3541
  d'eau 9658
  d'énergie 2971
  de nourriture 3541
acellulaire 56
acentrique 57
acéphales 1145
acétabule 58
acétaldéhyde 59
acétobactéries 63
acétone 64
acétylation 65
acétylcholine 66
achaine 68
achène 68
achlamydé 69
achromasie 71
achromatine 74
achromatique 72
achrome 1910
achromique 1910
aciculaire 77
acide abscissique 2680
acide acétique 60
  acétyle-acétique 62
  adénosine diphosphorique 155
  adénosine monophosphorique 156
  adénosine triphosphorique 157
  adénylique 156
  d'aldéhyde 291
  aldéhydique 291
  alginique 300
  aminé 395
  amino-acétique 3891
  aminobenzoïque 399
  ascorbique 745
  aspartique 752
  azoteux 6143
  biliaire 1026
  butyrique 1334
  caféique 1341
  carbonique 1411
  carboxylique 1417
  cétonique 4955
  chlorhydrique 4430
  cinnamique 1746
  citrique 1761
  cytidylique 2226
  désoxyribonucléique 2399
  diaminé 2485
  diaminopimélique 2486
  folique 3511
  formique 3565
  fumarique 3619
  gibbérellique 3823
  gluconique 3875
  glutamique 3882
  glutarique 3884
  glycérique 3889
  glycolique 3895
  gras 3326
  guanylique 3999
  homogentisique 4349
  humique 4403
  hyaluronique 4411
  indol(yl)acétique 4637
  3-indolyl-acétique 4638
  isocitrique 4863
  lactique 5006
  linoléique 5192
  linolénique 5193
  liponique 9080
  lysergique 5287
  maléique 5343
  malique 5346
  malonique 5348
  mévalonique 5577
  muramique 5893
  nicotinique 6121
  nucléique 6214
  oléique 6306
  orotique 6401
  osmique 6409
  oxalacétique 6479
  oxalique 6480
  palmitique 6557
  pantothénique 6569
  pectique 6665
  pénicillique 6696
  phosphatidique 6829
  phosphoglycérique 6832
  phosphorique 6836
  propionique 7349
  pyruvique 7506
  ribonucléique 7870
  salicylique 7948
  shikimique 8260
acide silicique 8284
  stéarique 8609
  succinique 8765
  tannique 8925
  tartrique 8934
  thioctique 9080
  thymidylique 9105
  thymonucléique 9107
  traumati(ni)que 9235
  uridylique 8410
  urique 8409
  uronique 9418
acidification 89
acidifier 90
acidiphile 93
acidité 91
acido-nonrésistant 85
acidophile 93
acidorésistance 86
acido-résistant 83
acido-soluble 87
aciforme 77
acinèse 273
acinétique 274
acraniens 98
acrocarpe 99
acrocentrique 100
acrogyne 101
acropète 102
acrosome 103
acte consommatoire 2007
  instinctif 4724
ACTH 181
actine 105
actinomorphe 107
actinomorphie 108
actinomycète 109
actinomycine 110
actinoptérygiens 111
action des gènes 3697
  génique 3697
  glandulaire 8061
  inhibitrice 4677
  de masse 5391
  réflexe 7670
  rétrograde 3337
  sécrétoire 8061
  tampon 1320
activateur 123
activation 121
  des acides aminés 396
  du gène 3698
activer 117
activité cambiale 1370
  cardiaque 1422
  de cour 2104
  enzymatique 3030
  génique 3699
  hormonale 4384
  métabolique 5526
  nerveuse 6055
  (à) vide 9436
  vitale 9567
actomyosine 129
acuité auditive 797
  visuelle 9558
acyclique 130
acylation 131
adaptabilité 132
adaptable 133
adaptatif 139
adaptation 134
  chromatique 141
  à l'environnement 3006
  au milieu 3006

Français

adaptation simultanée 8295
adaptativité 132
adapté 135
au milieu 136
adaptif 139
adaptivité 132
addition des stimuli 8658
adducteur 149
adelphogamie 150
adénine 151
adénocyte 3852
adénohypophyse 153
adénoïde 154
adénosine diphosphate 155
monophosphate 156
triphosphate 157
adermine 7500
adhérence 161
adhésion 161
adhésivité 164
adipeux 165
adipolytique 5199
adiurétine 9467
adjuvant 170
ADN 2399
adolescence 171
adoucir (eau) 8351
adoucisseur d'eau 9654
ADP 155
adrénaline 179
adrénergique 180
adrénine 179
adrénocorticotrophine 181
adrénostérone 182
adsorbant 184
adsorber 183
adsorption 185
adulte 188, 189
aérenchyme 198
aérobie 202, 203
aérobiologie 204
aérobiose 205
aéroplancton 207
aérotactisme 208
aérotropisme 209
afférence 213
afférent 214
affinité 215
after-discharge 221
agame 226
agamète 225
agamobium 227
agamoespèce 231
agamogenèse 228
agamogonie 229
agamonte 230
agamospermie 232
agar(-agar) 233
âge atomique 780
agénésie 245
agent 246
de chélation 1616
de contamination 2010
décontaminant 2314
de dispersion 2619
émulsionnant 2889
mutagène 5914
oxydant 6482
pathogène 6654
pollinisateur 7108
polluant 7114
agents radiomimétiques 7577
agglutinable 247
agglutinat 248
agglutination 250
agglutination acide 88
acidique 88
agglutiner 249
agglutinine 252
agglutinogène 253
agitateur 4030
agitation migratoire 5677
agmatoploïde 258
agmatoploïdie 259
agnathes 260
agnathostomes 260
agoniste 261
agrégat 254
cellulaire 1500
agrobiologie 262
AIA 4638
aiguille *(bot.)* 76
aguillon 7287, 8660
venimeux 9495
aguillonné 7289
aile 9685
ailé 280
aine 3956
air alvéolaire 370
complémentaire 1937
expiratoire 3245
inspiratoire 4721
de réserve 7774
résiduel 7779
supplémentaire 7774
aire associative 766
continue 2017
discontinue 2599
disjointe 2599
de distribution 7600
-échantillon 7968
des espèces 8431
germinative 3794
d'hivernage 9700
minima 5702
de nidification 6067
opaque 697
pellucide 698
relicte 7689
de répartition 7600
aisselle de la feuille 5069
akène 68
double akène 2477
akinésie 273
alanine 275
albinisme 281
albinos 282
albumen 283, 284
albumine 285
albuminoïde 286
alcalimètre 308
alcalimétrie 309
alcalin 310
alcalinité 311
alcaloïde 312
alcool éthylique 3130
aldéhyde 290
acétique 59
formique 3563
aldostérone 292
alécithe 293
alécithique 293
aléné 8760
aleurone 294
alevin 3421
algocepteur 6512
algologie 301
algonkien 302
algue 298
algues bleues 2210
algues brunes 6789
filamenteuses 3405
rouges 7861
vertes 1647
alimentation 6253
allaesthétique 313
allanto-chorion 314
allantoïde 316
allantoïne 315
allèle 317
allèles multiples 5882
allélique 318
allélisme 319
multiple 5883
allélomorphe 317, 318
allélomorphisme 319
allélotaxie 322
allélotype 323
allergène 324
allergie 326
allergique 325
alliance 327
allobiose 328
allocarpie 329
allochroïque 330
allochroïsme 331
allochrone 332
allochtone 333
allocyclie 334
allodiploïde 335
allodiploïdie 336
allogame 337
allogamie 338
allogène 339, 341
allogénique 340
allohaploïde 342
allohaploïdie 343
allohétéroploïde 344
allohétéroploïdie 345
alloiogenèse 346
allolysogénique 347
allométrie 348
allongement cellulaire 1510
allopatrique 349
alloplasme 350
allopolyploïde 351
allopolyploïdie 352
segmentaire 8081
allosome 4209
allosomique 354
allosyndèse 356
allosyndétique 357
allotétraploïde 358
allotétraploïdie 359
allotriploïde 360
allozygote 361
alluvium 4315
altération du milieu 3007
alternance de(s) générations 365
d'hôte 4391
alterne *(bot.)* 362
alvéole 374
dentaire 2386
pulmonaire 7454
ambiance sonore 375
ambient 3005
ambisexué 377
ambisexuel 377
ambivalent 378
ambocepteur 379
ambulacre 381
améiose 383
améiotique 384
amélioration de l'environnement 4583

aménagement de l'environnement 3020
paysagiste 5028
du territoire 8417
amentacées 386
amétaboles 669
amétabolique 388
amibe 409
amibocyte 410
amiboïde 411
amicron 389
amicroscopique 390
amide 391
acide 80
nicotinique 6117
amidon animal 3892
d'assimilation 761
de réserve 7776
amination 392
amine 393
amino-acide 395
aliphatique 307
essentiel 3114
nonessentiel 6159
aminoacide-décarboxylase 397
aminoacide-oxydase 398
amitose 401
ammoniac 402
amnios 403
amniotes 404
amniotique 405
amorphe 413
AMP 156
amphiaster 416
amphibie 419
amphibiens 418
amphibies 418
amphicarpe 420
amphicaryon 424
amphidiploïde 358
amphidiploïdie 359
amphigastres 422
amphigonie 423
amphimictique 425
amphimixie 426
amphi-mull 6642
amphimutation 427
amphineures 428
amphiploïde 429
amphitène 430
amphitoquie 431
amphogène 432
amplexicaule 433
amyélinique 434
amygdale palatine 6540
pharyngienne 6799
amylacé 437
amylase 436
salivaire 7445
amylifère 437
amylopectine 438
amylose 440
anabiose 441
anabolie 444
anabolique 442
anabolisme 443
anadrome 445
anaérobie 446, 447
anaérobionte 446
anaérobiose 448
anagenèse 449
analogie 455
analogique 453
analogue 453, 454
analyse 458
analyse du comportement 998
des facteurs 3296
factorielle 3296
génétique 3735
pollinique 7097
de population 7184
aux rayons X 9723
de la recombinaison 7649
du sang 1197
spectrale 8450
des tétrades 9026
de variance 459
analyser 456
analyseur 457
anamniés 460
anamniotes 460
anaphase 461
anaphorèse 463
anaphylaxie 464
anastomose 466
anastomoser 465
anatomie 469
végétale 6986
anatomiste 468
anatonose 470
anatrope 471
ancêtre 472
androcée 474
androdioécie 473
androgamète 5605
androgène 475
androgenèse 476
androgénie 476
androgyne 477
androgynie 478
androhermaphrodite 479
andromérogonie 480
androsome 481
androsporange 5647
androspore 5648
androstérone 484
anélectrotonus 485
anémochorie 486
anémogame 488
anémogamie 490
anémophile 488
anémophilie 490
anémotactisme 491
anémotaxie 491
aneuploïde 492
aneuploïdie 493
aneurine 9070
angiologie 495
angiosperme 496
angiospermes 497
angle de divergence 498
angustifolié 499
anhydrase carbonique 1412
anhydride 500
carbonique 1407
animal aquatique 671
diurne 2647
domestique 2662
dulçaquicole 3586
exotique 3223
d'expérience 3228
expérimental 3228
fondateur 3570
hibernant 4272
de laboratoire 4989
macrosmatique 5321
marin 5373
microsmatique 5641
nocturne 6146
nuisible 6777
animal parasite 9768
poïkilotherme 7071
à sang chaud 9619
à sang froid 1867
du sol 8757
supérieur 4275
témoin 2025
terestre 9003
terricole 8757
tête de ligne 3570
utile 9421
venimeux 7080
anion 531
anisogamète 532
anisogamie 533
anisogénie 534
anisophyllie 535
anisoploïdie 536
anisotrope 537
anisotropie 538
anneau annuel 544
de Balbiani 940
de croissance 3992
nerveux 6050
pyrrolique 7504
annélation 3844
annélides 540
annuel 541
annuelle 543
estivale 8784
d'été 8784
d'hiver 9690
invernale 9690
annulaire 545
anode 549
anoestrus 551
anomalie 16, 553
anormal 15, 552
anorthogenèse 554
anoure 628
anoures 627
anoxybionte 446
anoxybiose 448
anoxybiotique 447
antagonisme 556
ionique 4831
antagoniste 557
métabolique 5527
antécambrien 7229
antéhypophyse 561
antennates 9187
antenne 558
antennule 560
anthère 564
anthéridie 565
anthérozoïde 566
anthèse 567
anthocyanidine 569
anthocyan(in)e 568
anthophylle 3473
anthoxanthine 571
anthozoaires 572
anthropobiologie 573
anthropogène 576
anthropogenèse 575
anthropogénie 577
anthropogéographie 578
anthropographie 579
anthropoïdes 580
anthropologie 581
biologique 573
morphologique 584
anthropométrie 582
anthropomorphisme 583
anthropomorphologie 584

antiauxine 585
antibiose 587
antibiote 586
antibiotique 588
anticlinal 590
anti-codon 591
anticorps 589
antidote 593
antidromique 594
antienzyme 595
antifertilisine 596
antigène 597
  de groupe sanguin 1180
antigénéité 599
antigibbérelline 600
antihormone 602
antimétabolite 603
antimorphe 604
antimutagène 605
antipéristaltisme 609
antipode 611
antiseptique 616
anti-sérum 617
antithrombine 619
antitoxine 621
antitoxique 620
antitrypsine 622
antivitamine 623
antizymotique 625
anucléaire 6164
anucléé 6164
anus 629
aorte 630
apétale 632
apex 633
  de racine 7911
  de la tige 8612
aphylle 636
aphyllie 637
aplanospore 643
aplasie 644
apocarpe 645
apocarpie 646
apode 648
apoenzyme 649
apogamie 650
apoméiose 651
apomictique 652
apomixie 653
aponévrose 654
apophyse 655
  articulaire 9780
  épineuse 6081
  mastoïde 5406
  transverse 2492
aporogamie 656
aposporie 658
apothécie 659
appareil ambulacraire 9663
  aquifère 9663
  caulinaire 8617
  circulatoire 1754
  compte-globules 2097
  conducteur 1965
  digestif 2546
  excréteur 3201
  génital 3756
  génito-urinaire 9417
  de Golgi 3905
  locomoteur 5228
  mitotique 5714
  parabasal 6575
  de ponte 2801
  radical 7910
  reproducteur 7764
appareil respiratoire 7809
  sécréteur 8063
  urogénital 9417
  vacuolaire 9434
  vasculaire 9460
appariement 5411
  de chromosomes 8864
apparier 5408
appel sexuel 5413
appendice 662
  vermiforme 9507
apport calorique 1359
  nutritif 3543
  d'oxygène 6502
apposition 665
apprentissage 5097
  associatif 768
  par essais et erreurs 9246
  latent 5045
approvisionnement en eau 9660
  en eau potable 2705
apte à évoluer 2456
  à la germination 3809
  à la procréation 7744
  à se reproduire 7744
  à voler 1387
aptère 668
aptérygotes 669
aptitude à la combinaison 1915
  locomotrice 5229
aquatique 670
aqueux 674
aquiculture 9633
arachnides 678
arachnoïde 679
arachnologie 680
arborescent 681
arboricole 682
arborisation 683
  terminale 8992
arbre à feuilles aciculaires 1983
  à feuilles caduques 2303
  feuillu 2303
  généalogique 6883
  phylogénétique 6883
arbrisseau 8273
arbuste 8273
arbustif 3614
arc aortique 631
  branchial 3828
  hémal 4040
  hyoïdien 4476
  mandibulaire 5359
  neural 6074
  (de) réflexe 7671
  vertébral 9512
  viscéral 9549
  zygomatique 9781
archallaxie 686
archéen 684
archégone 688
archégoniates 687
archencéphale 689
archentère 690
archentéron 690
archéozoïque 685
archéspore 691
archétype 692
archiblaste 694
archipallium 695
archiplasme 696
archoplasme 696
aréolé 700
arête 866
arginase 701
arginine 702
  -phosphate 703
arille 705
ARN 7870
  m 5522
  -matrice 5522
  -messager 5522
  ribosomal 7874
  ribosomique 7874
  soluble 9201
  de transfert 9201
aromatique 708
aromatisation 709
aromatiser 710
aromorphose 711
arrangement cis 1757
  linéaire des gènes 5184
arrêt de croissance 714
  de développement 713
arrhénotocie 715
arrière-cerveau 5945
arrière-faix 219
arrière-gorge 6800
artefact 716
artère 721
  coronaire 2063
artériel 721
artériole 720
arthrologie 722
arthropodes 723
arthrospore 724
articulaire 725
articulation 732
  coxofémorale 4285
  ellipsoïde 2854
  de la hanche 4285
  scapulo-humérale 8271
  en selle 7944
articulées 8480
articulés 729
artiodactyles 735
artioploïdie 736
ascendance 738
ascendant 472
ascension de l'eau 9652
  de la sève 741
ascogone 743
ascomycètes 744
ascospore 746
asexué 749
asexuel 749
asparagine 751
aspérifolié 753
asque 747
assèchement 2437
assimilable 754
assimilat 759
assimilation 757
  du carbone 1403
  carbonée 1403
  du gaz carbonique 1408
  génétique 3736
assimiler 755
assise 8695
  de Caspary 1453
  cellulaire 1516
  génératrice 1369
  nourricière 8928
  pilifère 6942
  subéro-phellodermique 6806
association climacique 1798
  migratrice 5673
  permanente 6757
  de remplacement 8752
  (végétale) 765

associé 764
assortiment indépendant (des gènes) 4625
aster 769
astrocyte 772
astrosphère 773
asynapsis 774
atavique 776
atavisme 775
atoxique 6175
ATP 157
atriche 788
atrium 789
atrophie 790
atrophique 791
attitude de défense 2325
menaçante 9089
de soumission 8745
attractif 792
attraction sociale 8339
aubier 7980
auditif 796
audition 795
auricule 810
autécologie 811
autoadaptation 812
auto-alloploïdie 813
auto-antigène 814
autobivalent 815
autocatalyse 816
autochtone 818
autoclave 819
autodifférenciation 8106
autodigestion 8107
autoduplication 8109
auto-épuration 8116
autofécondant 8110
autofécondation 8112
autofécondité 8111
autofertile 8110
autofertilité 8111
autogame 821
autogamie 8112, 8115
autogène 825
autogenèse 823
autogénique 824
autogreffe 848
autohétéroploïdie 827
auto-immunisation 828
auto-incompatibilité 8114
autoïque 820
autolysat 829
de levure 9735
autolyse 830
automatisme 831
automixie 832
automutagène 833
autonomie 835
auto-orientation 836
autoparthénogenèse 837
autoplastique 838
autopollinisation 8115
autopolyploïdie 840
autoradiographie 841
autorégénération 842
autorégulation 8117
auto-réplication 8109
autoreproductible 8120
autoreproduction 8119
autosome 843
autosyndèse 844
autosynthèse 845
autostérile 8121
autostérilité 8122
autotomie 846
autotransplant 848
autotransplantation 849
autotrophe 851
autotrophie 850
autotropisme 852
auxanographie 854
auxiblaste 5301
auxiliaire 9422
auxine 856
auxocyte 857
auxospore 858
auxotrophe 859
auxotrophique 859
avantage sélectif 8097
avifaune 861
avirulent 862
avitaminose 863
axe caulinaire 1489
floral 3468
fusorial 8500
de l'inflorescence 4657
longitudinal 5243
neural 6082
axérophtol 869
axilemme 874
axillaire 875
axocoele 878
axone 879
géant 3819
axoplasme 882
axostyle 883
azoïque 884
azotation 885
azote atmosphérique 778
protéique 7379
total 9161
azoté 6141
azygote 886

B

baccifère 887
bacciforme 888
bacillaire 890
bacillariophycées 2499
bacille 893
-virgule 9528
bacilliforme 892
backcross 895
bactéricide 916, 917
bactérie 931
donatrice 2673
donneuse 2673
hôte 4392
réceptrice 7640
bactérien 903
bactéries acétiques 63
dénitrifiantes 2378
du fer 4847
de fermentation 3359
fixatrices d'azote 6139
lactiques 5007
lumineuses 5263
nitrifiantes 6132
des nodosités 6150
des nodules 6150
photogènes 5263
de putréfaction 7490
du sol 8353
du soufre 9079
bactériforme 918
bactériochlorophylle 919
bactériocine 920
bactériogénétique 909
bactériologie 923
bactériologique 921
bactériologiste 922
bactériologue 922
bactériolyse 925
bactériolysine 924
bactériolytique 926
bactériophage 927
bactériostatique 928
bactériotoxine 929
bactériotropine 6359
bactéroïde 932
baguage 1134
baguer 7880
baie 1009
balance azotée 6133
génique 3700
hydrique 9630
des sels 7957
balancier 4083
ballast 943
banc de poissons 8015
bande 946 *(gen.)*, 3466, 6509
médiane 4758
migratrice 5675
d'oiseaux 1128
bandelette germinative 3782
bang supersonique 8401
barbe 866
barrière de diffusion 2528
hémato-encéphalique 1170
base analogue 957
azotée 6134
du crâne 960
foliaire 5070
purique 7486
pyrimidique 7502
basichromatine 968
basicité 969
baside 972
basidiomycètes 970
basidiospore 971
basifuge 973
basilaire 974
basipète 976
basiphile 978
basique 964
basophile 978
basophilie 979
basophilisme 979
bassin 6693
bassinet 7723
bâtonnet 7891
batraciens 418
battement cardiaque 4118
du cœur 4118
baume 945
du Canada 1379
béhaviorisme 1001
béhavioriste 1002
benthique 1006
benthos 1008
besoin nutritif 6251
d'oxygène 6500
de protéine 7380
en substances nutritives 6251
besoins alimentaires 3534
en calories 1360
en eau 9645
énergétiques 2961
en vitamines 9593
vitaminiques 9593
bicaryon 2551
biennal 1015

biennale 1016
bifide 1017
biflagellé 1018
biflore 1019
bifolié 1020
bifurqué 1021
bilan azoté 6133
  de chaleur 4123
  hydrique 9630
bile 1025
biliaire 1030
bilineurine 1656
biliprotéine 1033
bilirubine 1034
biliverdine 1035
bilobé 1036
bilobulaire 1036
biloculaire 1037
binucléaire 1046
binucléé 1046
bioblaste 1048
biocatalyse 1049
biocatalyseur 1050
biocénose 1059
biocénotique 1058, 1060
biochimie 1054
biochimique 1051
biochimiste 1053
biochore 1055
biocide 1056
bioclimatologie 10517
biocoenose 1059
biocoenotique 1058
biocybernétique 1098
biodégradable 1063
biodynamique 1065
bioélectronique 1098
bioénergétique 1066
bioessai 1047
biogène 1067
biogenèse 1068
biogénétique 1069
biogénie 1072
biogéochimie 1074
biogéographie 1075
  végétale 6907
biologie 1087
  cellulaire 1501
  du développement 2459
  de l'environnement 3008
  humaine 4398
  marine 5374
  moléculaire 5741
  des populations 7185
  de la production 7318
  des quanta 7516
  de radiation 7563
  de la reproduction 7751
  sociale 1107
  du sol 8354
  végétale 6897
biologique 1076
biologiste 1086
bioluminescence 1088
biolyse 1089
biomasse 1090
biome 1091
biomécanique 1092
biométéorologie 1093
biométrie 1095
biométrique 1094, 1095
biomolécule 1096
bion 1097
bionique 1098
bionomie 1099
biophore 1101
biophysicien 1102
biophysique 1103
biopsie 1104
bios 1106
biosome 1108
biosphère 1109
biostatistique 1095
biosynthèse 1111
biosystématique 1112
biote 1100
biotine 1117
biotique 1113
biotope 1118
biotype 1119
biovulaire 2656
bipartition 1120
bipède 1121, 1122
bipédie 1123
bipenné 1124
bipolaire 1125
bipolarité 1126
biréfringence 1136
biréfringent 1137
bisexualité 1141
bisexué 1140
bisexuel 1140
bivalent 1143
  «rupture-réunion» 1261
bivalve 1144
bivalves 1145
blanc d'œuf 283
blastème 1146
blastocèle 1148
blastocoele 1148
blastocyste 1149
blastocyte 1150
blastoderme 1151
blastodisque 1152
blastogène 1154
blastogenèse 1153
blastokoline 1147
blastomère 1155
blastomycète 1156
blastopore 1157
blastospore 1159
blastula 1160
blastulation 1161
blépharoplaste 1165
blindage 8038
bloc d'agar 234
  de gélose 234
  de gènes 1168
  de paraffine 6580
blocage génétique 3737
bois 9704
  d'automne 5042
  de cœur 2734
  parfait 2734
  primaire 7296
  de printemps 7296
  tardif 5042
boisement 217
boiser 216
boîte crânienne 2114
  de Petri 6783
bol alimentaire 303
bonnet 7837
à bord entier 2991
bordure en brosse 1304
botanique 1230, 1232
  forestière 3555
  systématique 7008
botaniste 1231
bouchon vitellin 9744
bouclier céphalique 4115
  (dorsal) 2685
  ventral 7042
boues activées 118
  à globigérines 3864
bouillon de culture 6247
bourdon 1332
bourgeon 1311
  accessoire 47
  adventif 192
  apical 638
  axillaire 876
  caulinaire 8614
  dormant 7816
  floral 3483
  foliaire 5072
  à fruits 3606
  gustatif 8936
  latent 7816
  de membres 5168
  de remplacement 7725
  terminal 638
  en vie ralentie 7816
bourgeonnement 1317
bourgeonner 1312
bourse copulatrice 3754
  muqueuse 8893
  synoviale 8893
boursouflure chromosomique 7450
bouteille à culture 2189
bouton 1311
  embryonnaire 2874
  à fleurs 3483
  floral 3483
boutonner 1312
bouture de feuille 5075
brachial 1236
brachiopodes 1237
brachyblaste 1238
brachyméiose 1239
bractée 1242
bractée-écaille 2107
bractéole 1244
bradytélique 1245
branche 1251
  fruitière 3611
branchial 1254
branchie 3827
  sanguine 1178
  trachéenne 9183
branchiopodes 1257
bras chromosomique 1715
brévistylé 5651
bronche 1291
bronchiole 1290
bruit de fond 900
bruits ambiants 375
  émis par la navigation aérienne 267
bryologie 1306
bryophytes 1307
bryozoaires 1308
  vrais 2785
buccal 1309
buissonnant 3614
bulbe 1327
  olfactif 7852
  pileux 4065
  rachidien 5447
bulbeux 1330
bulbille 1329
byssus 1336

C

C_{14} 7564
cacogenèse 1337
cadre de Caspary 1453
caducifolié 2300
caduque 2299
caenogenèse 1339
cage thoracique 9085
caillette 17
caillot 1831
de sang 1174
caisse du tympan 9325
cal 1355
calamus 1342
calciférol 1346
calcification 1347
calcifier (se) 1349
calice 1368
rénal 7722
calicule 1365
callose 1352
callosité 1353
callus 1355
calorie 1358
grande ~ 4961
petite ~ 3928
calorification 1362
calorimètre 1363
calorimétrie 1364
calotte nucléaire 6191
polaire 7084
calyptre 7901
calyptrogène 1367
cambium 1369
fasciculaire 3311
interfasciculaire 4743
vasculaire 9454
cambrien 1372
camouflage 1377
camoufler 1376
campylotrope 1378
canal annulaire 7882
biliaire 1027
cholédoque 1654
déférent 2328
digestif 304
éjaculateur 2812
excréteur 3198
galactophore 5009
génital 3751
gommifère 4002
hydrophore 8680
intestinal 4792
lacrymal 4998
lacrymo-nasal 5001
lactifère 5009
madréporique 8680
médullaire 5448
de Müller 5867
naso-lacrymal 5001
neural 6075
oléifère 6300
pharyngien 6797
rachidien 8494
résinifère 7782
stylaire 8718
thoracique 9083
vertébral 8494
vitellin 9742
de Wolff 5513
canalicule rénal 7724
canaux semi-circulaires 8127
cancérigène 1418
cancérologie 1419
canine 1385
cannibalisme 1386
capable d'évolution 2456
de germer 3809
de régénérer 1389
de voler 1387
capacité absorbante 35
d'absorption 35
d'adhésion 164
d'adsorption 186
d'apprentissage 5098
d'assimilation 762
auditive 798
crânienne 2113
d'échange de bases 959
évolutive 2460
fécondante 3373
de flottaison 8848
germinative 3806
inductive 4644
de membrane 5474
pulmonaire 7455
(pulmonaire) vitale 9568
de réaction 1390
à réagir 7623
régénérative 7695
reproductrice 7753
de résolution 7785
(de rétention) au champ 3397
de rétention d'eau 9639
visuelle 9557
vitale 9568
de vol 3456
capillaire 1391, 1392
capitule 1395
capsule 1397
articulaire 726
bactérienne 904
déhiscente 2353
otique 799
polaire 7085
poricide 7200
séminale 8073
sporifère 8543
surrénale 175
capture de nourriture 3528
des proies 7286
caractère 1605
acquis 96
adaptatif 140
atavique 9101
du comportement 999
distinctif 2640
éthologique 999
génotypique 3767
héréditaire 4186
indépendant 4626
mêlé 1163
d'organisation 8711
phénotypique 6816
qualitatif 7511
quantitatif 1163
sexuel 8242
sexuel secondaire 8055
caractéristique 1605
acquise 96
carapace 1398
dorsale 2685
carbone 1402
radioactif 7564
carbonifère 1413
carboxylase 1415
carboxylation 1416
carbure d'hydrogène 4428
carcinogène 1418
carcinologie 1419
cardia 1420
cardiaque 1421
carence alimentaire 3539
en eau 9634
en substances nutritives 6249
de vitamine 9592
carnassier 1431
carnassière 1428
carnassiers 1429
carnivore 1430, 1431
carnivores 1429
caroténase 1433
carotène 1434
carotide 1436
carotinoïde 1435
carpe 1443
carpelle 1438
carpogone 1439
carpophore 1440
carpospore 1441
carposporophyte 1442
carte chromosomique 1719
factorielle 3744
génétique 3744
génique 3744
de répartition 2643
de végétation 9474
cartilage 1445
articulaire 727
aryténoïde 737
de conjugaison 3078
costal 2089
cricoïde 2128
épiphysaire 3078
fibreux 3388
hyalin 4409
thyroïde 9109
cartilagineux 1447
cartographie de la végétation 5371
caryocinèse 5711
caryogamie 4927
caryogène 4928
caryogenèse 4929
caryogramme 1533
caryologie 4932
caryolymphe 6207
caryolyse 4934
caryomère 4935
caryoplasme 6225
caryoplasmique 6226
caryopse 1451
caryorrhexie 4939
caryosome 4940
caryostasie 4941
caryosystématique 2260
caryothèque 6202
caryotine 4943
caryotype 4944
caséine 1452
cassure de chromosomes 1716
d'insertion 4713
caste 1454
castrat 1455
castration 1457
castrer 1456
(femelle) 8428
catabolisme 1459
protéique 7371
catabolite 1460
catadrome 1461
catalase 1462
catalepsie 1463
calorifique 9046

Français

catalepsie frigorifique 9697
de froid 9697
catalyse 1464
enzymatique 3031
catalyser 1466
catalyseur 1467
catalytique 1465
cataphase 4946
cataphorèse 1468
cataphylle 1469
catastrophe écologique 3013
catéchine 1470
catécholamine 1471
catélectrotonus 1472
caténation 1473
cathepsine 1474
cathode 1475
cation 1478
caulescent 1483
cauliflоré 1484
cauliflorie 1485
caulinaire 1486
caulocarpe 1488
cause stimulante 8656
cavernicole 1493
cavernicoles 1491
cavité abdominale 3
à air 266
amniotique 406
articulaire 728
branchiale 3831
chorionique 1682
coelomique 1852
corporelle 1206
cotyloïde 2093
crânienne 2115
gastrale 3670
gastrique 3670
intestinale 4012
médullaire 5449
nasale 5978
orbitaire 6371
palléale 5363
pelvienne 6690
péricardique 6720
péritonéale 3
pharyngée 6798
pleurale 9082
pulpaire 7461
de segmentation 1148
thoracique 9082
utérine 9424
cécidie 1497
ceinture pectorale 8270
pelvienne 6692
scapulaire 8270
thoracique 8270
de végétation 9473
cellobiose 1535
cellulaire 1537
cellule 1499
adipeuse 5200
animale 505
annexe 1924
antipode 611
apicale 639
auditive 800
auxiliaire 855
bactérienne 905
basale 951
caliciforme 3903
cambiale 1371
cartilagineuse 1671
ciliée 1741
cellule á cils vibratiles 1741
à collerette 1651
-compagne 1924
compte-globules 2097
conjonctive 3389
à corbeille 977
criblée 8276
donatrice 2674
embryonnaire 2867
endothéliale 2950
-engrais 5400
épidermique 3060
à épine 7288
épineuse 7288
épithéliale 3090
équatoriale 3097
fille 2277
flagellée 3439
à flagelles 3439
-flamme 3444
folliculaire 3517
ganglionnaire 3664
géante 3820
germinale 3783
germinative 3783
glandulaire 3852
gliale 3861
du goût 4005
gustative 4005
internodale 4771
hôte 4393
interstitielle 4785
laticifère 5053
libérienne 981
médullaire 5450
mère 5825
mère des grains de pollen 7101
mère du sac embryonnaire 2869
mère de spores 8546
migratrice 9617
motrice 5835
á mucus 5859
muqueuse 5859
musculaire 5897
nerveuse 6040
névroglique 3861
nourricière 6255
nutritive 6255
-œuf 6478
osseuse 6432
palissadique 6547
en panier 977
parenchymateuse 6616
parentale 6618
pierreuse 8027
pigmentaire 6933
plasmatique 7016
pyramidale 7496
quiescente 7817
des rayons médullaires 5453
réceptrice 7634
au repos 7817
reproductrice 7754
sanguine 4055
de Schwann 8017
scléreuse 8027
sécrétrice 8064
de segmentation 1781
sensorielle 8168
sexuelle 3643
somatique 8393
de soutien 8811
spermatique 8456
stomatique 4000
subéreuse 2050
cellule du tapétum 8927
terminale 2900
traversée 6644
du tube criblé 8279
urticante 1820
végétale 6989
visuelle 9559
vitelline 9741
cellulose 1541
cendres d'os 1217
cénobe 1856
cénobie 1856
cénogenèse 1339
cénozoïque 1546
centre actif 125
associatif 767
d'association 767
calorifiant 4124
de coordination 2033
génétique 3701
germinal 3784
du langage 8453
nerveux 6041
optique 9560
organisateur 6380
d'organisation 6380
de réaction 7616
réflexe 7672
respiratoire 7791
thermique 4124
thermorégulateur 9064
visuel 9560
centrifugation 1555
différentielle 2518
fractionnée 3574
en gradients de densité 2382
centrifuge 1554
centrifuger 1556
centrifugeur 1557
centrifugeuse 1557
centriole 1558
centripète 1559
centrique 1552
centrochromatine 1560
centrodesmose 1561
centrogène 1562
centrolécithe 1563
centrolécithique 1563
centromère 1564
centroplasme 1567
centrosome 1568
centrosphère 1569
céphaline 1572
céphalique 1570
céphalisation 1573
céphalocordés 98
céphalopodes 1575
céphalothorax 1576
cercle d'espèces 717
de régulation 7701
tentaculaire 5250
cérébelleux 1577
cérébral 1579
cérébralisation 1585
cérébroside 1586
cerne 544
certation 1589
certificat d'ascendance 6673
cerumen 1590
cerveau 1588
antérieur 7353
intermédiaire 2517
moyen 5502
postérieur 5569
terminal 8955

circulation fœtale 3501
générale 8904
grande ~ 8904
lymphatique 5275
ombilicale 9360
petite ~ 7456
placentaire 3501
portale 7205
porte 7205
pulmonaire 7456
du sang 1172
sanguine 1172
de la sève 7972
des substances 1752
systémique 8904
circumnutation 1755
cirre 1756
cistron 1760
clade 6885
cladode 1763
cladogenèse 1764
clan 1766
clarification 1775
classe 1768
clavicule 1771
clé de détermination 4530
cléistocarpe 1787
cléistogame 1788
cléistogamie 1789
climax 1797
édaphique 2789
cline 1803
clinostat 1804
clinotaxie 4977
clivage 1779
cliver 1778
cloaque 1805
cloche de brume 4113
cloison nasale 5980
séparatrice 8182
transversale 2150
cloisonné 1374, 8179
cloisonnement 1375
clonage 1808
clonal 1807
clone 1809
de transduction 9198
clonique 1807
clostridium 1812
cnidaires 1819
cnidoblaste 1820
cnidocil 1821
cnidocyste 6012
coacervat 1824
coacervation 1825
coadaptation 1826
coagulabilité 1827
coagulable 1828
coagulation 1830
sanguine 1173
coaguler 1829
coagulum 1831
coalescence 1960
coalescent 1961
coaptation 1826
cobalamine 1835
cocaïne 1836
coccus 1837
coccyx 1839
cochlée (auditive) 1840
cocon 1841
codage 1844
code chevauchant 6456
dégénéré 2339
génétique 3738
code non chevauchant 6165
par triplets 9265
coder 1842
codification 1844
codominance 1845
codominant 1846
codon 1847
coecum 1338
coefficient d'agrégation 2344
de consanguinité 4590
de corrélation 2073
de diffusion 2529
de fanaison 9679
de fécondité 3364
de parenté 1848
de régression 7697
de sélection 8093
de température 8969
de variation 1849
coelentérates 1850
coelentérés 1850
coelentéron 1851
coelome 1852
externe 3206
extraembryonnaire 3206
coelomoducte 1855
coenobe 1856
coenocyte 1857
coeno-espèce 1859
coenogamète 1858
coenzyme 1860
cœur branchial 1255
lymphatique 5276
cofacteur 1861
cohésion du groupe 3966
coiffe 7901
coïncidence 1864
coït 1865
colchicine 1866
coléoptères 1876
coléoptile 1877
coléorhize 1879
colinéarité 1880
collagène 1881, 1882
collatérale 1886
collecte de la nourriture 3528
collecteur de fractions 3573
collemboles 1888
collenchyme 1889
collet de la racine 7902
collodion 1890
colloïdal 1892
colloïde 1891
côlon 1895
ascendant 740
descendant 2425
colonie 1896
bactérienne 906
de ponte 1272
reproductrice 1272
colonne vertébrale 9515
colorabilité 8578
colorable 8579
colorant 8577
acide 82
basique 966
nucléaire 6209
vital 9569
coloration 1909, 8580
dissimulatrice 2163
de Gram 3932
intravitale 4804
négative 6008
nucléaire 6210
coloration protectrice 7366
secondaire 224
supravitale 8819
vitale 9574
colorer 8576
colorimètre 1899
colorimétrie 1900
colostrum 1901
columelle 1911
combinaison 1914
de gènes 3702
combiné azoté 6135
commensal 1917
commensalisme 1918
commissure 1919
communication 1921
communauté 1922
biotique 1059
climacique 1798
forestière 3556
de reproduction 7748
végétale 6990
comparaison mère-fille 2268
compartimentation 1926
compatibilité 1927
compensateur 1930
compensation de dosage 2687
compétence 1932
compétition alimentaire 1933
interspécifique 4779
intraspécifique 4803
complémentation 1942
complexe enzyme-substrat 3041
de gènes 3703
vitaminique B 9584
comportement 994
agressif 256
d'agression 256
alimentaire 3341
d'alimentation 3341
animal 502
d'appariement 5412
d'appétence 664
appétitif 664
de combat 3402
de copulation 2040
copulatoire 2040
de cour 2102
de distraction 2297
explorateur 3248
de fuite 3110
génésique 7750
d'incubation 4614
instinctif 4725
migrateur 5670
nidificateur 6063
nuptial 2102
de parade 2102
parental 6623
de reproduction 7750
sexuel 8240
en situations conflictuelles 1978
social 8340
territorial 9007
de toilette 3957
composacées 1946
composé carboné 1404
cellulaire 1505
protéique 7373
composées 1946
composition en espèces 8432
compréhension brusque 4716
comptage 2096
compteur (de) Geiger(-Müller) 3680

compteur de scintillations 8021
conassociation 1995
concentration en ions 4838
des ions (d')hydrogène 4441
saline 7958
conceptacle 1956
conchyoline 1959
concrescence 1960
concurrence alimentaire 1933
interspécifique 4779
intraspécifique 4803
conditionnement classique 1769
instrumentel 4727
conditions ambiantes 3010
climatiques 1792
édaphiques 8355
environnantes 3010
d'existence 5136
expérimentales 3231
du milieu 3010
pédologiques 8355
vitales 5136
conducteur 5064
conductibilité 1971
calorifique 4125
thermique 4125
conduction d'eau 9631
de l'excitation 1967
de l'influx 1968
nerveuse 6042
saltatoire 7965
du stimulus 1969
conductivité 1971
conduit alvéolaire 371
auditif (externe, interne) 801
excréteur 3198
génital 3751
lacrymal 4998
mésonéphrique 5513
nasal 5979
paramésonéphrique 5867
conduite instinctive 4725
condupliquée *(vernation)* 1972
condyle 1973
occipital 6276
cône 1974
d'attraction 3368
rétinien 7839
végétatif 9481
configuration cis 1757
«fragment-pont» 1284
trans 9195
conformation du corps 1211
congénère 1996, 1997
congénital 1979
conidie 1982
conidiophore 1981
conidiospore 1982
conifère 1983
conifères 1985
coniforme 1977
conique 1977
conjoint 5407
conjonctive 1989
conjugaison 1988
conjugant 1986
connectif 1990
consanguinité 4589
conservation de l'espèce 8439
de la nature 1993
du sol 8356
consommateur *(ecol.)* 2006
primaire 7290
secondaire 8051
tertiaire 9012
consommation énergétique 2960
d'oxygène 6497
constance des caractères 1606
cellulaire 1504
des espèces 1998
de forme 3560
numérique des chromosomes 6237
constante de diffusion 2530
de dissociation 2633
de sédimentation 1999
constituant 2000
cellulaire 1505
constitution héréditaire 4187
constriction 2003
nucléolaire 6220
secondaire 6220
construction du nid 6062
consubespèce 2005
contaminant 7114
contamination 2012
des eaux 9649
radioactive 7553
du sol 8357
contaminer 2011
continuité germinale 2016
contractibilité 2024
contractile 2021
contractilité 2024
contraction musculaire 5905
contre-poison 593
convergence 2029
conversion 2031
coordination 2032
héréditaire 3431
coorientation 2034
copépodes 2035
coprophage 2036
copulation 2038
copuler 2037
coque 1837
coquille 1958
calcaire 1343
(de l'œuf) 2807
coracoïde 2043
coralliaires 572
corbeille branchiale 3829
corde dorsale 6182
vocale 9606
cordés 1678
cordon ombilical 9361
spermatique 8462
vasculaire 9459
ventral 9500
corépresseur 2045
coriacé 2046
cormophytes 2055
cormus 2054
corne 4388
postérieure 2683
corné 2058
cornée 2056
cornification 2060
corolle 2062
corpora allata 2064
corpora cardiaca 2065
corps adipeux 3315
calleux 2067
caverneux 1494
cellulaire 1502
de chromatine sexuelle 8216
chromatinien 1696
ciliaire 1736
élémentaire 2850
étranger 3549
corps de Golgi 3906
graisseux 3315
gras 3315
jaune 2068
pédonculés 5911
pinéal 3081
reproducteur 7752
de sécrétion 8062
strié 2070
en suspension 8834
vertébral 9513
vitré 9601
de Wolff 5514
corpuscule 2071
de Barr 8216
basal 950
central 1568
de chromatine 1696
épithélial 6611
gustatif 8936
huileux 2813
intracellulaire 4597
de Malphigi 5349
polaire 7083
reproducteur 3605
de tact 8910
tactile 8910
corrélation 2072
cortex 7879
cérébral 1581
moteur 5832
surrénal 173
visuel 9561
corticoïde 2079
corticostéroïde 2080
corticostérone 2081
cortico-stimuline 181
cortico-surrénale 173
corticotrophine 181
cortine 2083
cortisol 4433
cortisone 2085
corymbe 2086
cosmopolite 2088
cosse 5103
côte 7866
asternale 3304
fausse ~ 3304
de feuille 5081
flottante 3462
sternale 9290
vraie 9290
côtelé 2092
cotyle 2093
cotylédon 2094
couche 8695
d'abscission 23
adipeuse 3318
d'aleurone 296
cellulaire 1516
cornée 2059
de discontinuité 9051
germinative 5350
lipidique 3318
monomoléculaire 5782
optique 9033
pigmentaire 6935
séparatrice 23
subéreuse 2051
coulant 8672
couleur aposématique 9621
avertisseuse 9621
coupe à blanc 1774
blanche 1774
à la celloïdine 1536

Français

D

Français

E

Français

environnement 3004
environnementaliste 3027
enzymatique 3028
enzyme 3029
d'activation 120
activatrice 120
adap(ta)tif 142
constitutif 2002
inductible 142
lytique 5296
de perméation 6766
protéolytique 7363
répressible 7738
respiratoire 7796
enzymologie 3043
éocène 3044
éosine 3045
éosinocyte 3046
éosinophile 3047
éozoïque 685
épaississement 9072
annelé 546
réticulé 7831
spiralé 8513
épaissir 9071
épendyme 3048
éphapse 3049
épharmonie 3050
éphédrine 3051
éphémère 3052
épi 8490
épibolie 3055
épicardium 3056
épicarpe 3205
épicotyle 3058
épidémiologie 3059
épiderme 3061
de l'anthère 3221
de la racine 7855
épididyme 3062
épigastre 3063
épigé 3064
épigenèse 3065
épigénétique 3066
épiglotte 3067
épigyne 3068
épillet 8492
épilimnion 3069
épimorphose 3071
épimysium 3072
épinastie 3073
épine 9086
dorsale 9515
foliaire 5084
venimeuse 9495
épinèvre 3075
épinéphrine 179
épineux 8507
épiphylle 3077
épiphyse 3080, 3081
épiphyte 3083
épiphytes 3082
épiphytique 3083
épiploon 6330
épisome 3085
épisperme 8074
épistasie 3087
épithécium 3088
épithélium 3092
cilié 1742
germinal 3797
cubique 2180
cylindrique 1913
glandulaire 3853
de passage 9208
épithélium pavimenteux 6660
sensoriel 8169
squameux 8570
stratifié 8694
vibratile 1742
épithème 3093
épizoaire 3095
épizoïque 3094
époque du frai 8426
glaciaire 3846
épreuve de la descendance 7327
éprouvette 9020
épuration biologique des eaux
résiduaires 1083
de l'eau 9650
équation de régression 7698
équilibre acido-basique 81
azoté 6137
démographique 7189
de diffusion 2531
dynamique 8608
écologique 2760
génétique 3741
génique 3700
hydrique 9630
ionique 4837
métabolique 6889
mutationnel 5932
osmotique 6119
physiologique 6889
équipement chromosomique 1717
génétique 8206
équivalent roentgenien 7896
ère géologique 3776
érepsine 3101
éreptase 3101
ergastoplasme 3102
ergine 3103
ergostérine 3105
ergostérol 3105
ergotinine 3106
érosion 3107
éolienne 9682
erpétologie 4200
erreur moyenne 8590
standard 8590
érythroblaste 3108
érythrocyte 3109
espace intercellulaire 4740
intermicellaire 4760
libre apparent 660
lymphatique 5278
sanguin 1193
synaptique 8867
vert 3951
vital 5220
espèce 8430
agame 231
animale 526
caractéristique 1609
collective 8883
dominante 2670
hôte 4396
indicatrice 1609
naissante 4595
qui perpétue l'espèce 8440
-relique 7718
type 1609
végétale 7007
vicariante 9529
espèces jumelles 9319
espérance de (sur)vie 5138
essai biologique 1047
de croisement 2137
d'élevage 1273
essai d'enrichissement 2980
de laboratoire 4991
de modèle 5729
essaim 8838
migrateur 5675
essaimage 8841
essaimer 8839
ester 3116
estérase 3117
d'acétylcholine 67
estérification 3118
estérifier 3119
esthètes 210
estivation 211, 212
estomac 8674
glandulaire 7418
masticateur 3845
musculaire 3845
étage altitudinal 368
de végétation 8688
étaler en frottis 8331
étamine 8585
étape successive 8184
état d'adaptation 137
adulte 191
d'agrégation 8599
d'excitation 8600
quiescent 7824
de repos 7824
de vie latent 5043
établissement de cartes 5369
de cartes chromosomiques 5370
étang d'oxydation 6486
s'éteindre 3266
éteint 3265
étendue de variation 7601
éthanol 3130
éthéogenèse 3123
éther-sel 3116
ethmoïde 3125
éthogramme 3126
éthologie 3129
éthologique 3127
éthologiste 3128
étiolement 3132
étranger à l'espèce 6173
étranglement 8691
annulaire 7603
de Ranvier 7603
être en chaleur 6298
en fleurs 3481
en rut 6298
vivant 5219
étrier 8595
étude comparée du comportement
1925
du comportement 3129
expérimentale 3239
eucaryotes 3138
euchromatine 3134
euchromatique 3133
euchromosome 3135
eugénique 3136, 3137
eugénisme 3137
eulamellibranches 3139
eumycètes 9289
euploïde 3142
euploïdie 3143
euryhalin 3144
euryphage 3145
eurytherme 3146
eurythermique 3146
eutélie 1504
euthériens 6970
eutrophe 3149

Français

eutrophique 3149
eutrophisation 3150
évacuation des déchets 9624
fécale 2323
des ordures ménagères 2664
évagination 3151
évaporation 3153
évaporer 3152
évapotranspiration 3155
évocateur 3159, 7712
évocation 3158
évolutif 3174
évolution 3161
des communautés 1923
convergente 2030
divergente 2651
en ligne droite 6404
orthogénétique 6404
progressive 449
du quantum 7517
régressive 7700
évolutionnisme 9042
évolutionniste 3173
évolutivité 2460
examen microscopique 5637
examiner au microscope 3177
excipulum 3180
excitabilité 3181
excitable 3182
excitant 3188
excitation 3183
nerveuse 6056
ex-conjugant 3191
excréments 3192
excreta 3194
excrétér 3195
excréteur 3197
excrétion 3196
excrétoire 3197
excroissance 3193
exergonique 3202
exigences alimentaires 3534
énergétiques 2961
hydriques 9645
nutritionnelles 3534
en vitamines 9593
exine 3203
exocarpe 3205
exocoelome 3206
exoderme 3210
exogamie 3211
exogastrulation 3212
exogène 3213
exopeptidase 3214
exoptérygotes 4162
exosmose 3217
exosporium 3219
exosquelette 3216
exostome 3220
exothermique 3222
exotoxine 3224
expectance de (sur)vie 5134
expérience 3225
de dressage 9192
d'élevage 1273
d'isolement 4879
de laboratoire 4991
de labyrinthe 5437
de plein champ 3398
en série 8187
témoin 2026
expérimental 3227
expérimentateur 3242
expérimentation sur des animaux 3241
expérimentation sur le terrain 3398
expérimenter 3226
expert en matière d'environnement 3027
expiration 3243
explant 3246
explantation 3247
exploitation abusive 2438, 7249
explosion démographique 7190
nucléaire 6197
exposant d'hydrogène 4439
expression génétique 3708
phénotypique 6817
du sexe 8224
expressions faciales 3290
expressivité (du gène) 3252
exsudat 3278
exsudation 3279
de la plante (après la taille) 1162
extenseur 3254
extermination 3256
extéroceptif 3263
extéro(ré)cepteur 3264
extinction 3267
extracellulaire 3269
extrachromosomique 3270
extraire 3272
extrait 3271
aqueux 675
cellulaire 1511
de levure 9736
sec 2716
tissulaire 9131
extrapolation 3273
extra-pyramidal 3274
extrémité radiculaire 7911
extrémités 3275
extrorse 3277
exuviation 3282
exuvie 3280
larvaire 5039

F

faciès 3292
facilitation 3293
spatiale 8418
temporelle 8976
façon de vivre 5725
facteur alimentaire 6256
ambiant 3015
biotique 1115
climatique 1793
complémentaire 1940
conditionnel 1962
de croissance 3977
déchaînant 7713
édaphique 2791
d'évolution 3163
externe 3258
extrinsèque 1835
F 3365
héréditaire 3696
d'incompatibilité 4599
inhibiteur 4678
d'intensification 4731
intrinsèque 4805
kappa 4925
lét(h)al 5119
de libération 7713
limitant 5173
facteur limitatif 5173
mendélien 5487
du milieu 3015
de mutation 5927
de motivation 5830
motivationnel 5830
nutritif 6256
de protéine animale 523
régulateur 7704
Rh 7850
Rhésus 7850
temps 9123
faculté d'adaptation 132
d'apprentissage 5098
germinative 3806
visuelle 9557
FAD 3447
faire la cour à la femelle 2100
faisceau caulinaire 8615
conducteur 9453
foliaire 5073
vasculaire 9453
falciforme 3300
famille 3305
fanaison 9678
permanente 6762
fané 9677
faner, se 9676
fascia 3309
fasciation (rubanée) 3310
fatigabilité 3325
fatigue 3324
du sol 8358
faune 3330
abyssale 39
des cavernes 1491
marine 5376
des oiseaux 861
relique 7716
du sol 8359
fausse côte 3304
faux-axe 8861
faux-bourdon 2708
faux-fruit 7427
faux jumeaux 2657
F-duction 8222
fèces 3299
fécond 3362
fécondabilité 3373
fécondable 3366
fécondation 3367
croisée 338
double ~ 2695
externe 3259
interne 4763
féconder 3371
fécondité 3363
feed-back 3337
femelle 3348, 3349
féminité 3352
fémur 3353
fenêtre foliaire 5077
ovale 6445
ronde 7921
fente branchiale 3834
ferment 3029
digestif 2542
jaune 3451
fermentation 3357
acétique 61
alcoolique 289
butyrique 1335
lactique 5008
fermenter 3355
fermentescible 3356

Français

formation de bancs 8016
de colonies 1897
du couple 6515
d'embryons adventifs 193
des espèces 8429
de fleurs 3486
de groupes 3968
de puffs 7451
de races 7533
réticulaire 7830
(végétale) 3564
forme, en ~ de bâtonnet 933
biologique 5139
de croissance 3978
cultivée 2185
de culture 2185
de durée 7819
de jeunesse 4919
larvaire 5037
originale 7299
primitive 7299
de souche 7299
formule dentaire 2387
florale 3471
hématologique 1183
sanguine 1183
fosse nasale 5978
fossile 3566
caractéristique 4627
fossilifère 3567
fossilisation 3568
fossiliser (se) 3569
fougère arborescente 9239
fougères 7443
à graines 7440
fourchu 1021
fovéa 3572
fractionnement cellulaire 1512
fragmentation nucléaire 6199
frai 3421, 8422, 8425
frayer 8423
frayère 8427
free-martin 3576
fréquence des chiasmata 1632
du cœur 3582
de décharge 2594
des gènes 3710
génique 3710
des influx 4584
de manifestation 6695
de recombinaison 7650
respiratoire 3583
fronde 3591
frontal 3592
frottis 8330
de sang 1192
fructifère 3598
fructifiant 3598
fructification 3599
fructifier 3600
fructose 3601
frugivore 3602
fruit 3603
agrégé 1947
ailé 9688
capsulaire 1396
composé 1947
déhiscent 2354
explosif 944
indéhiscent 4624
lomentacé 5235
multiple 5886
à noyau 8682
à pépin 7180
sec 2718
fruit solitaire 8384
fruits à coques 8256
frustule 3612
frutescent 3614
FSH 3518
fucoxanthine 3617
fumarase 3618
funicule 3627
ombilical 9361
spermatique 8462
fuseau achromatique 73
central 1551
creux 4312
mitotique 5719
musculaire 5901
nucléaire 6208
de segmentation 1785
fusiforme 3630
fusion 3631
cellulaire 1514
centrique 1553
nucléaire 6200
fusionner 3629

G

gaine foliaire 5083
mucilagineuse 5862
à myéline 5947
myélinique 5947
de Schwann 8018
tendineuse 8982
galactine 5011
galactolipide 3634
galactose 3636
galbule 3637
galle 1497
galvanotactisme 3639
galvanotaxie 3639
galvanotropisme 3640
gamétange 3642
gamétangie 3641
gamète 3643
gamétique 3644
gamétocyte 3646
gamétogamie 3647
gamétogenèse 3648
gamétogénie 3648
gamétogonie 3656
gamétophore 3651
gamétophyte 3652
gammaglobuline 3653
gamobium 3655
gamogonie 3656
gamone 3657
gamonte 3658
gamopétale 3659
gamophase 3660
gamosépale 3661
gamotropisme 3662
gangliocyte 3664
ganglion 3663
basal 952
cérébral 1582
lymphatique 5277
spinal 8497
viscéral 9550
garniture chromosomique 1723
gastéropodes 3676
gastrique 3671
gastrocèle 3674
gastro-intestinal 3675
gastrula 3678
gastrulation 3679
gaz carbonique 1407
respiratoire 7797
toxique 9169
géitonocarpie 3681
géitonogamie 3682
gel 3683
d'agar 236
d'amidon 8596
de gélose 236
gélatine 3684
gélatineux 3688
gélatinisant 3687
gélati(ni)sation 3685
gélatiniser 3686
gélation 3685
gélifiant 3687
gélification 3685
gélifier (se) 3686
gélose 233
nutritive 6246
gémellité 9321
gemmation 3692
gemme 1295
gemmifère 3693
gemmiparité 3694
gemmule 3695
gencive 4001
gène 3696
-clé 6322
complémentaire 1940
contrôleur 7608
de différenciation sexuelle 2521
inhibiteur 4674
d'isolation 4880
lét(h)al 5119
majeur 6322
marqueur 5382
mineur 7133
modificateur 5732
modifieur 5732
-mutateur 5934
nucléaire 6201
opérateur 6355
polymère 7146
régulateur 7705
répresseur 7741
de stérilité 8630
structural 8710
de structure 8710
supplémentaire 8810
tampon 1322
générateur 3727
génératif 3727
génération 3723
F_1 3377
F_2 3378
filiale 3406
filiale, deuxième 3378
filiale, première 3377
P_1 6625
des parents 6625
sexuée 8244
spontanée 8537
générique 3730
généritype 3732
gènes jumeaux 7423
genèse 3733
genetic drift 3740
généticien 3747
génétique 3734, 3748
animale 9761
bactérienne 909
du comportement 996
évolutive 2462

génétique humaine 4400
moléculaire 5742
des populations 7192
végétale 6904
virale 9532
généticiste 3747
génique 3749
génitalias 3757
géniteur 6617, 6621
génocentre 3701
génocopie 3758
génome 3759
génomère 3761
génomique 3762
génonème 3763
génosome 3764
génotype 3765
génotypique 3766
génovariation 3714
genre 3769
géobiologie 3771
géocarpie 3773
géographie botanique 6907
géophytes 3777
géotactisme 3778
géotaxie 3778
géotropisme 3779
germarium 3791
germe 3780
pathogène 2601
germen 3787
germer 3781
germicide 3792
germinal 3793
germinatif 3809
germination 3805
ralentie 2362
retardée 2362
gérontologie 3812
gésier 3845
gestagène 3813
gestaltisme 3816
gestaltiste 3814
gestaltpsychologie 3815
gestalt-théorie 3816
gestante 7261
gestation 3817
gémellaire 9321
geste d'apaisement 661
de soumission 8745
gibbérelline 3824
gigantisme 3825
gigantocyte 3820
gingival 3842
ginglyme 4281
glaciation 3849
gland 939
glande 3850
accessoire 49
anale 451
antennaire 559
apocrine 647
cérigène 9667
cirière 9667
coquillière 8257
coxale 2110
cutanée 8316
dermique 8316
digestive 2543
à eau 9638
eccrine 2748
endocrine 2910
exocrine 3208
fileuse 8506
gastrique 3672
glande génitale 3908
holocrine 4316
incrétoire 2910
lacrymale 4999
lactéale 5357
linguale 5187
mammaire 5357
maxillaire 5433
à mucus 5860
de mue 5841
muqueuse 5860
nectaire 6004
nectarifère 6004
nidamentaire 6123
du noir 4687
odorante 6283
odoriférante 6283
pinéale 3081
pituitaire 4511
répugnatoire 8665
salivaire 7954
sébacée 8047
à sécrétion externe 3208
à sécrétion interne 2910
à sel 7960
séricigène 8290
sexuelle 3908
du sinus 8301
sinusaire 8301
sudoripare 8844
surrénale 175
thyroïde 9110
uropygienne 9420
venimeuse 7077
à venin 7077
vitellogène 9743
glandulaire 3851
glanduleux 3858
gliadine 3862
globe oculaire 3283
globine 3865
globoïde 3866
globule blanc (du sang) 5123
rouge (du sang) 3109
du sang 1175
globuline 3867
glomérule floral 3868
glossien 5186
glotte 3871
glucagon 3872
glucide 1399
glucocorticoïde 3874
glucoprotéine 3876
glucosamine 3877
glucose 3878
glucose-phosphate 3879
glucoside 3899
glume 3881
glumelle 6543
inférieure 5106
supérieure 9400
glutamine 3883
glutathion 3885
glutéline 3886
gluten 3887
glycémie 3888
glycérine 3890
glycérol 3890
glycine 3891
glycocolle 3891
glycogène 3892
glycogenèse 3893
glycogénolyse 3894
glycolyse 3896
glycométabolisme 3897
glycoprotéine 3876
glycoside 3899
gnathique 5431
gnathostomes 3902
golgiocinèse 3907
gonade 3908
gonadostimuline 3911
gonadotrope 3910
gonadotrophine 3911
A 3518
B 5269
chorionique 1683
placentaire 1683
gonadotrophique 3910
gone 3912
gonflement 8846
gonfler, se 8845
gonidie 3913
gonie 3912
gonimoblaste 3914
gonochorisme 3915
gonocyte 3916
gonoducte 3917
gonomérie 3918
gousse 5103
articulée 730
lomentacée 730
goût 8935
goutte de pollinisation 7111
gouttelette lipidique 3317
gouttière neurale 6078
primitive 7300
gradation 5396
gradient de concentration 1955
de densité 2381
de pression 7279
de température 8970
grain d'aleurone 295
d'amidon 8597
basal 953
de pollen 7100
graine 3926
grains de Palade 7875
ribonucléoprotéiques 7875
de zymogène 9794
graisse 3314
de dépôt 2407
graisseux 165
gramme-calorie 3928
gramnégatif 3929
grampositif 3930
grana 3933
grand noyau 5315
grande calorie 4961
circulation 8904
grandeur de la population 7196
granivore 3934
granulation 3937
granulé 3935
granuler 3936
granules de zymogène 9794
granuleux 3935
granulocyte 3938
grappe 7527
gravicepteur 3945
gravide 7261
gravitation 3944
greffage 9226
nucléaire 6211
greffe 9224
autoplastique 848
hétérologue 4223
homologue 4350
homoplastique 4350
greffon 3923, 9224

Français

grégaire 3952
grégarisme 4182
griffe 1772
à gros grains 1833
gros intestin 5034
grosse 7261
  molécule 5313
grossesse 7259
groupe aminé 400
  amino 400
  carboxyle 1414
  carboxylique 1414
  d'espèces 8436
  hydroxyle 4464
  de liaison 5191
  méthyle 5574
  prosthétique 7358
  sanguin 1179
groupement 3968
  aminé 400
  amino 400
  en bancs 8016
  carboxyle 1414
  carboxylique 1414
  hydroxyle 4464
  méthyle 5574
  prosthétique 7358
guanidine 3997
guanine 3998
guide 5064
gustatif 4004
gustation 4003
guttation 4013
gymnocarpe 4014
gymnospermes 4016
gymnospermique 4015
gynandrie 4018
gynandromorphe 4019
  biparti 4079
gynandromorphisme 4020
gynécée 4017
gynéphore 4026
gynodioïque 4022
gynogenèse 4023
gynomérogonie 4024
gynomonoïque 4025
gynosporange 5322
gynospore 5323
gynostège 4028
gynostème 4029

H

habit nuptial 6239
habitat 4032
habituation 4034
habitude alimentaire 3529
habitus 4035
halophile 4080
halophytes 4081
halotolérant 3144
haltère 4083
hanche 4282
haplobionte 4085
haplobiontique 4085
haplochlamydé 4086
haploïde 4087
haploïdie 4088
haplonte 4089
haplophase 4090
haplose 4091
haplostémone 4092
haptène 4093
haptères 4094
haptonastie 4096
haptotropisme 4097
haustorie 4111
hectocotyle 4139
hélice, double 2696
hélicoïde 4144
héliophile 4146
héliophobe 4147
héliotropisme 6867
hélix 4151
helminthologie 4152
hélophyte 4153
hélotisme 4154
hémal 4050
hématie 3109
hème 4037
hémagglutination 4038
hémagglutinine 4039
hématine 4041
hématogenèse 4044
hématologie 4045
hématophage 4046
hématopoïèse 4062
hématopoïétique 4063
hématosine 4041
hématoxyline 4048
hémérythrine 4049
hémicaryon 4161
hémi-cellulose 4158
hémicryptophyte 4160
hémiligule rayonnante 7612
hémimétaboles 4162
hémine 4051
hémiparasite 4164
hémiperméabilité 8138
hémiperméable 8139
hémiptères 4165
hémisphère cérébral 1583
hémitrope 4166
hémizygote 4167
hémoblaste 4052
hémocoele 4053
hémocyanine 4054
hémocyte 4055
hémodynamique 4056
hémoglobine 4057
hémogramme 1183
hémolymphe 4059
hémolyse 4061
hémolysine 4060
hémopoïèse 4062
hémopoïétique 4063
héparine 4170
hépatique 4171
hépatiques 4172
herbacé 4174
herbe 4173
herbicide 4177
herbier 4176, 7927
herbivore 4178, 4179
hercogamie 4180
héréditaire 4183
hérédité 4672
  alternative 2130
  alternée 367
  des caractères acquis 4673
  (cyto)plasmique 2254
  extrachromosomique 2254
  à facteurs multiples 1164
  liée au sexe 8234
  mendélienne 6640
  non-chromosomique 2254
  non-mendélienne 6162
  plastidique 7034
hérédité du sexe 8229
  sexuelle 8229
  unilatérale 9380
héritabilité 4194
hériter de 4670
hermaphrodite 4196, 4197
hermaphrodi(ti)sme 4199
herpétologie 4200
hétéroallèle 4201
hétéroauxine 4638
hétérocarpe 4203
hétérocaryon 4225
hétérocaryose 4226
hétérocentrique 4204
hétérochlamydé 4205
hétérochromatine 4207
hétérochromatique 4206
hétérochromatisme 4208
hétérochromosome 4209
hétérocinèse 4227
hétérocinésie 4227
hétéroconté 4228
hétérocyste 4210
hétérodonte 4211
hétéroécie 4213
hétéroécique 4212
hétérogamète 532
hétérogamie 533
hétérogamétique 4215
hétérogène 4220
hétérogénéité 4217
hétérogénétique 4218
hétérogénie 4221
hétérogénique 4220
hétérogénote 4219
hétérogonie 4222
hétérogreffe 4223
hétérogyne 4224
hétéroïque 4212
hétérologie 4230
hétérologue 4229
hétérolyse 4231
hétéromère 4232
hétérométaboles 4162
hétéromorphe 4234
hétéromorphie 4235
hétéromorphisme 4235
hétéromorphose 4236
hétéronome 4237
hétérophyllie 4238
hétérophytique 4239
hétéroplasmonie 4240
hétéroplastie 4241
hétéroploïde 4242
hétéroploïdie 4243
hétéroprotéine 1951
hétéropycnose 4244
hétérosis 4245
hétérosomal 4246
hétérosporé 4247
hétérosporie 4248
hétérostylé 4249
hétérostylie 4250
hétérosynapse 4251
hétérosyndèse 4252
hétérothallie 4254
hétérothallique 4253
hétérothallisme 4254
hétérotherme 7072
hétérotriche 4255
hétérotrophe 4257
hétérotrophie 4256
hétérotrophique 4257
hétérotype 4258
hétérozygose 4259

Français

hydrotactisme 4461
hydrotaxie 4461
hydrotropisme 4462
hydroxyde 4463
hydroxylamine 4465
hydroxylation 4466
hydrozoaires 4467
hygiène de l'environnement 3016
hygromètre 4468
hygronastie 4469
hygrophile 4470
hygrophytes 4471
hygroscopicité 4473
hygroscopique 4472
hyménium 4474
hyménoptères 4475
hyoïde 4477
hypanthium 4478
hyperchimère 4479
hyperchromatose 4480
hypermétamorphose 4481
hypermorphe 4482
hypermorphose 4483
hyperparasite 4484
hyperplasie 4485
hyperplastie 4485
hyperploïde 4486
hyperploïdie 4487
hyperpolarisation 4488
hypertélie 4489
hypertonicité 4491
hypertonie 4491
hypertonique 4490
hypertrophie compensatoire 1931
hyphe 4492
hypnospore 7822
hypoblaste 2914
hypocotyle 4496
hypoderme 4498, 8725
hypogastre 4499
hypogé 4500
hypogenèse 4501
hypogénésie 4501
hypogyne 4503
hypolimnion 4504
hypomorphe 4506
hyponastie 4507
hypophylle 4508
hypophyse 4511
hypoplasie 4512
hypoploïde 4513
hypoploïdie 4514
hypostase 4515
hypothalamus 4517
hypothalle 4518
hypothécie 4519
hypothèse des alvéoles 372
  de la cassure d'abord 1259
  de la choix de copie 2042
  de la continuité germinale 9043
  «un gène, un enzyme» 6337
  de la panspermie 2661
  sur la réplication du gène 3719
hypotonique 4520
hypoxanthine 4521
hystérésis 4523

I

ichthyologie 4525
idante 4528
ide 4527
identification des congénères 8441
idioblaste 4531
idiochromatine 4532
idiogramme 4533
idioplasme 3787
idiotype 4535
idiovariation 4536
idiozome 4537
iléon 4538
ilion 4540
îlot sanguin 1182
îlots de Langerhans 4856
ilotisme 4154
imago 4544
imbibition 4545
imbriqué 4547
imitateur 5690
immature 4548
immaturité 4549
immersion d'huile 6302
immiscible 4552
immission 4553
immunisateur 4555
immunisation 4559
immuniser 4561
immunisé 4554
immunité 4557
  acquise 97
  active 124
  passive 6645
immunobiologie 4562
immunochimie 4563
immunoélectrophorèse 4564
immunoglobuline 4565
immunologie 4568
immunologique 4566
immunologiste 4567
immunomécanisme 4569
immunoréaction 4556
immunosuppression 4571
immunotolérance 4572
impair 4573
imparidigité 4574
imparipenné 4575
imperméabilité 4577
  à l'eau 9640
imperméable 4578
  à l'eau 9641
implant 4579
implantation 6124
impression sensorielle 8171
imprinting 4582
impulsion 2706
  agressive 257
  combative 2707
  de couver 4615
  migratoire 5669
impureté 4585
inactivation 4587
inactiver 4586
inanimé 5145
inbreeding 4589
incapable de voler 3459
incinération 4593
  de déchets 7691
incisive 4596
inclure 2862
inclusion 2863
  cellulaire 1515
  dans la paraffine 6581
incolore 1910
incompatibilité 4598
incompatible 4600
incorporation de radioisotopes
  4604
incrément 4607
incrétion 4608
incrétoire 4609
incrustation 4611
incubateur 4619
  buccal 5843
  oral 5843
incubation 4613
incuber 4612
incurvation 4620
  tropique 9284
indépendant de la densité 2383
indicateur 9178
  de l'acidité (du sol) 92
  de chaux 6996
  écologique 2762
  radioactif 7555
  de temps 9125
indice d'appariement méta-
  phasique 5561
  céphalique 1571
  de corrélation 2073
  de générations 3724
  d'isolation 4881
  mitotique 5716
  photosynthétique 6863
  des prophases 7347
  de réfraction 7684
  de survie 8827
indigérable 4634
indigestibilité 4633
indigestible 4634
individu atavique 9101
  autofécondant 8113
  à autoreproduction 8113
  souche 3570
inducteur 4645
induction embryonnaire 2878
  enzymatique 3033
  de floraison 4642
  de mutations 4643
indusie 4646
inexcitabilité 4649
inexcitable 4650
infécond 4654
infécondabilité 949
infécondité 4655
infère 4652
inférieur au seuil 8739
infertile 4654
infertilité 4655
inflorescence 4656
  définie 2332
  indéfinie 4623
influence du milieu 3017
influencé par le sexe 8228
influx nerveux 6046
information génétique 3742
  héréditaire 4190
infraliminaire 8739
infra-ordre 8746
infrason(s) 4661
infrutescence 4662
infundibuliforme 4663
infundibulum 4664
infuser 4665
infusion de foin 4112
infusoires 4666
ingestion 4668
  des aliments 3530
inguinal 4669
inhibiteur 4675
  de croissance 3985
  d'enzymes 3035
  de la germination 3808
  métabolique 5531

inhibiteur mitotique 5718
inhibition compétitive 1934
  de croissance 3984
  enzymatique 3034
  de la germination 3807
  mitotique 5717
  non compétitive 6155
  réciproque 7644
  (de) réflexe 7674
  par rétrocontrôle 3340
initiale 4683
inné 1979
innervation 4693
innerver 4692
innocuité 4695
inoculation 4698
inoculer 4697
inoculum 4699
inoffensif 4694
inorganique 4700
inquiétude migratrice 5677
inquilin 4701
insaliver 4702
insecte nuisible 4703
  parfait 4544
  pollinisateur 7109
  utile 9422
insectes 4706
  amétaboliques 669
  sociaux 8342
insecticide 4707
insectivore 4708, 4709
insémination 4712
  artificielle 734
inséminer 4711
insertion foliaire 5078
  musculaire 5899
insight 4716
insolubilité 4717
  dans l'eau 9642
insoluble 4718
  dans l'eau 9643
inspiration 4719
installation de décontamination 2316
instinct 4722
  alimentaire 3342
  combatif 2707
  de la conservation 4723
  de couver 4615
  de fuite 3111
  génésique 7757
  grégaire 4182
  maternel 5410
  migrateur 5669
  nidificateur 6064
  de reproduction 7757
  sexuel 8243
insuline 4728
intégument 4729
intensité d'assimilation 760
  liminaire 7846
  lumineuse 5149
  de production 7319
  respiratoire 7798
  du stimulus 8649
interaction 4733
  des gènes 3712
interbande 4734
interchangement de segments 8082
intercinèse 4752
intercostal 4742
interférence 4744
  du centromère 1566
  chiasmatique 1633
interférence chromatidique 1694
interférone 4747
interlocking 4753
intermédine 4759
intermittent 4761
interneurone 4770
intéro(ré)cepteur 4773
interphase 4774
intersexe 4776
intersexualité 4777
intersexué 4776
interspécifique 4778
interstérilité 4782
interstice 4783
  intercellulaire 4740
  synaptique 8867
interstitiel 4784
intervalle entre les générations 3725
  synaptique 8868
intestin 4799
  antérieur 3548
  céphalique 4114
  grêle 8329
  gros ~ 5034
  moyen 5505
  postérieur 4279, 7316
  primitif 690
intestinal 4791
intine 4800
intracellulaire 4801
intraspécifique 4802
introgression 4806
introrse 4808
intussusception 4809
inuline 4810
invagination 4811
inversion 4813
  du sexe 8237
invertase 4815
invertébrés 4816
invertine 4815
investigateur 7773
investigation 7767
inviabilité 4818
inviable 6178
in vitro 4820
in vivo 4821
involucelle 4822
involucre 4823
involuté 4825
involution 4826
iode 4828
iodé 4827
iodifère 4827
ion 4830
  bipolaire 9779
  hybride 9779
  (d')hydrogène 4440
ionisation 4842
ioniser 4843
iris 4845
irradiation 4849
  aux rayons X 9722
  aux UV 9349
irradier 4848
irréversibilité de l'évolution 4851
irrigation sanguine 1196
irritabilité 4852
irritable 4853
ischion 4854
isidie 4855
isoagglutinine 4857
isoallèle 4858
isoanticorps 4859
isoantigène 4860
isochromosome 4862
isodiamétrique 4864
isoénergétique 4861
isoenzyme 4866
isogame 4868
isogamète 4867
isogamie 4869
isogène 4870
isogénie 4872
isogénique 4870
isogénomique 4871
iso-ionie 4873
isokonté 4874
isolat 4875
isolation 4878
isolécithe 4882
isolécithique 4882
isolement 4878
  biogénétique 1070
  écologique 2763
  des espèces 8437
  géographique 3774
  des lignées 8178
  phénologique 6812
  reproductif 7758
  sexuel 7758
  spatial 8419
  topographique 9156
isoler 4876
isoleucine 4883
isomar 4894
isomérase 4886
isomère 4884, 4887
isomérie 4888
isomérisme 4888
isométrie 4890
isométrique 4889
isomorphe 4891
isomorphisme 4892
isophane 4894
isophène 4895
isophénique 4895
isophénogamie 4896
isopolyploïde 4897
isoprène 4898
isoprénoïde 4899
isosmotique 4902
isosporé 4366
isosporie 4367
isotonicité 4901
isotonie 4901
isotonique 4902
isotope 4903
  du carbone 1409
  radioactif 7573
isotrope 4906
isotype 4907
isotypie 4908
isozyme 4866

J

jabot 2131
jarovisation 9508
jéjunum 4914
jeu chromosomique 1723
jeune 9746
  oiseau au nid 6070
jointure 732
jumeaux biovulaires 2657
  dizygotiques 2657
  faux ~ 2657

Français

jumeaux monozygotes 5801
uniovulaires 5801
univitellins 5801
vrais 5801
jurassique 4918

K

kamptozoaires 2935
kératine 4950
kératinisation 4951
kératiniser 4952
kilocalorie 4961
kinase 4964
kinétine 4969
kinesthésie 4962
kinesthésique 4963
kinétochore 1564
kinétogenèse 4972
kinétoplaste 4974
kinine 2241
kyste 2220

L

labelle 4985
labferment 7727
labial 4986
labium 4988
laboratoire d'expériences 2458
de recherches 7770
labre 4994
labrocyte 5400
labyrinthe membraneux 5482
osseux 1223
lacinié 4996
lactalbumine 5002
lactase 5003
lactation 5004
lacthormone 5011
lactoflavine 7868
lactoglobuline 5012
lactose 5013
lacune aérifère 270
foliaire 5077
sanguine 1193
lag phase 5016
lait du jabot 2132
laitance 5685
laite 5685
lamarckisme 5017
lame branchiale 3830
porte-objet 5634
lamelle 5018
intergranaire 4751
mitoyenne 5664
moyenne 5664
lamellibranches 1145
laminaire 5021
lancéolé 5024
lanugineux 5030
lanugo 5031
larmier 8947
larve 5036
larynx 5041
latence 5043
hivernale 9692
latent 5044
latex 5051
laticifère 5052
latifolié 5054
leader 5064
lécheur *(insecte)* 5032
lécithe 9740
lécithine 5100
lectotype 5101
lecture du code génétique 1843
législation relative à l'environnement 3018
légume 5103
légumineuses 5104
lemma 5106
lenticelle 5109
lenticulaire 5110
lenticulé 5110
lépidoptères 5111
leptonème 5112
leptosporangié 5113
leptotène 5114
lésion par chaleur 4130
expérimentale 2324
lest 943
lét(h)al 5117
balancé 936
conditionné 1963
lét(h)alité 5121
de l'hybride 4414
léthargie estivale 211
hivernale 9692
leucine 5122
leucocyte 5123
leucoplaste 5124
lévogyre 5125
lèvre *(bot.)* 4985
blastoporique 1158
inférieure 5254
supérieure 9399
lévulose 3601
levure 9734
de bière 993
de boulangerie 934
LH 5269
liaison 5190
double ~ 2691
génétique 3713
peptidique 6705
riche en énergie 2969
liane 5127
lias 5128
liber 5129
libérer *(énergie etc.)* 5130
lichen 5131
crustacé 2157
foliacé 3414
fruticuleux 3615
lichénine 5132
lichénologie 5133
lié au sexe 8233
liège 2048
liégeux 2053
liénal 8527
liénique 8527
lieu de frai 8427
d'hivernage 9700
de nidification 6067
de ponte 6466
ligament 5146
ligase 5147
ligne primitive 7303
lignée cellulaire 1517
consanguine 5183
évolutive 3165
femelle 3350
généalogique 5181
germinale 3789
mâle 5341
lignée maternelle 3350
paternelle 5341
phylétique 6876
pure 7480
lignées d'élite 8091
ligneux 5159
lignification 5162
lignifier 5164
lignine 5165
ligule 5167
limbe (foliaire) 5071
limitante cellulaire 1503
limitation au sexe 8230
limite d'arbres 9240
de distribution 5172
forestière 3557
des neiges 8337
de tolérance 9146
limité à un sexe 8231
limivore 5175
limnétique 5176
limnivore 5175
limnologie 5178
limnoplancton 5179
line-breeding 5180
lingual 5186
linine 74
linkage 5190
au sexe 8232
lipase 5194
lipide 5195
lipochrome 5198
lipocyte 5200
lipoïde 5201
lipolyse 5203
lipolytique 5199
lipopolyoside 5205
lipopolysaccharide 5205
lipoprotéine 5206
liposolubilité 3321
liposoluble 3322
liquéfaction 5208
liquéfier 5209
liquide amniotique 407
céphalorachidien 1587
cérébrospinal 1587
coelomique 1853
corporel 1208
à culture 2190
de culture 6252
exuvial 5840
interstitiel 9132
organique 1208
de Ringer 7885
séminal 8133
spermatique 8133
synovial 8891
lithophyte 5212
lithosphère 5214
lithotrophe 5215
littoral 5216
livrée nuptiale 6239
lobe antérieur (de l'hypophyse) 561
frontal 3593
occipital 6277
olfactif 6311
pariétal 6630
postérieur (de l'hypophyse) 7211
temporal 8977
lobé 5222
lobulaire 5222
lobule pulmonaire 7457
localisation des gènes 5225

localisation génique 5225
locataire 4701
locomotion 5226
loculicide 5231
locus 5232
loge pollinique 7098
loi d'action de masse 5057
  de l'assortiment indépendant 5486
  biogénétique (fondamentale) 1071
  contre la contamination de l'environnement 612
  de la disjonction indépendante 5486
  de la dominance ou de l'uniformité (des hybrides) 5484
  de Haeckel 1071
  de Liebig 5059
  de Mendel, première 5484
  de Mendel, seconde 5485
  de Mendel, troisième 5486
  du minimum 5059
  de pureté des caractères 5486
  de la quantité de stimulus 7646
  du rapport inverse entre la masse corporelle de l'organisme et son métabolisme 4812
  de la réciprocité 7646
  de (la) ségrégation 5485
  de la sommation hétérogène des stimuli 5055
  du tout ou rien 353
lois de l'hérédité 5062
  de Mendel 5483
  mendéliennes 5483
lombaire 5259
lombes 5234
longévité 5242
longicaude 5240
longistylé 5236
longueur d'ondes 9665
lophophore 5250
lot chromosomique 1723
loupe 5331
LTH 5011
luciférine 5258
lumen 5261
lumière 5261
  infrarouge 4659
  polarisée 7094
  ultraviolette 9350
luminescence 5262
lutéine 5268
luté(in)ostimuline 5269
lutte biologique (contre les parasites) 1082
  contre le bruit 6151
  contre la pollution 613
  pour la (sur)vie 8716
lyase 5271
lycopène 5272
lycopodinées 5273
lymphe 5274
lymphocyte 5281
lyochrome 3446
lyodiastase 5284
lyophilisation 5285
lysat 5286
lyse 5289
lyser 5298
lysigène 5290
lysine 5288
lysogène 5290
lysogénicité 5291
lysogénie 5291
lysogénique 5290
lysosome 5292
lysotype 5293
lysozyme 5294
lytique 5295

M

macération 5299
mâchoire 4911
  inférieure 5358
  supérieure 5430
macrobite 5241
macroblaste 5301
macrocarpe 5302
macroclimat 5303
macroconsommateur 5304
macro-élément 5316
macroévolution 5306
macrofaune 5307
macroflore 5308
macrogamète 5309
macroglie 5310
macromère 5311
macrométhode 5312
macromolécule 5313
macromutation 5314
macronoyau 5315
macronucléus 5315
macrophage 5317
macrophylogenèse 5319
macroscopique 5320
macrosmatique 5321
macrosporange 5322
macrospore 5323
macrotome 5327
macule jaune 9739
main préhensile 3940
maintien de l'espèce 8439
malacologie 5336
maladie de carence 2330
  carentielle 2330
  à virus 9545
malaire 9782
mâle 5338, 5339
  complémentaire 1938
malformation 5345
maltase 5352
maltose 5353
mamelon 5354
mammalogie 5356
mammifères 5355
  placentaires 6970
mandibulates 5360
mandibule 5358
mangeur de plancton 6981
mannose 5361
manque d'eau 9634
  de nourriture 3539
  d'oxygène 6501
  de vitamine 9592
manteau 5362
  des hémisphères 6551
manubrium 5364
marginal 5372
marquage 5383
  isotopique 4904
  du territoire 9009
marque odorante 6285
marquer (à, par) 4984
marqueur 5381
marsupiaux 5387
marsupie 5388
marteau 5347
masculin 5338
masculinité 5390
masse d'inclusion 2864
masséter 5399
mastax 5401
masticateur *(insecte)* 1630
mastication 5402
mastigophores 3437
mastocyte 5400
mastoïde 5406
mastzelle 5400
matériaux du nid 6069
matériel génique 3745
maternel 5409
matière active 126
  alimentaire 3531
  colorante 8577
  grasse 3328
  polluante 7114
  de réserve 7777
  sèche 2721
matières fécales 3299
  en suspension 8834
matrice 5418, 9426
  cellulaire 1518
matroclinal 5421
matroclinie 5422
maturation 5423
maturité 5429
  sexuelle 8245
maxillaire 4911, 5431
  inférieur 5358
  supérieur 5430, 8799
maxille 5430
maxillipède 5436
méat auditif (externe, interne) 801
mécanique embryonnaire 2463
mécanisme d'action 5721
  inné de déclenchement 4690
  d'isolement 4877
  régulateur 7706
  de régulation 7706
méchano-récepteur 5441
médian 5442
médiastin 5445
médiateur chimique 9219
médulle 5446
médullo-surrénale 177
méduse 5456
mégaévolution 5458
mégaphylle 5318
mégasporange 5322
mégaspore 5323
mégasporocyte 2869
mégasporogenèse 5324
mégasporophylle 5325
méiocyte 5459
méiose 5460
méiospore 5461
méiotique 5462
mélanine 5464
mélanisme 5465
  industriel 4647
mélanophore 5467
mélanostimuline 4759
mélanotonine 5469
mélanotropine 4759
mélitose 7589
mélitriose 7589
membrane 5472
  alaire 6648

Français

membrane basale 963
  de base 9389
  basilaire 975
  biologique 1080
  cellulaire 1519
  cellulosique 1542
  chitineuse 1640
  chorio-allantoïdienne 314
  coquillière 8258
  déciduale 2299
  de fécondation 3369
  limitante 5174
  muqueuse 5861
  natatoire 9669
  nictitante 6122
  nucléaire 6202
  obturante 1811
  ondulante 9367
  pellucide 9753
  plasmatique 7019
  protectrice 7367
  semiperméable 8140
  synoviale 8894
  tectrice 2076
  tympanique 9326
  type 9389
  unitaire 9389
  unité 9389
  vacuolaire 9431
  vasculaire 9457
  vitelline 9597
membranelle 5479
membraneux 5480
membre d'une espèce étrangère 6173
membres 3275
mendélisme 5490
méninges 5491
ménotaxie 5492
mensuration 5439
mère 2267
méricarpe 5493
méristème 5495
  apical 642
  intercalaire 4737
  primaire 7292
  primordial 7305
  secondaire 8052
méroblastique 5496
mérogamète 5497
mérogamie 5498
mérogonie 5499
méromixie 5500
mérozygote 5501
mésencéphale 5502
mésenchyme 5503
mésentère 5504
mésoblaste 5510
mésocarpe 5507
mésocoele 5508
mésocotyle 5509
mésoderme 5510
mésoglée 5511
mésomitose 5512
mésonéphros 5514
mésophile 5515
mésophylle 5516
mésophyte 5517
mésothélium 5518
mésotrophe 5519
mésozoaire 5520
mésozoïque 5521
mesurage 5439
métabiose 5523
métabolique 5525
métaboliquement actif 5540
métabolisable 5543
métaboliser 5544
métabolisme 5541
  azoté 6140
  basal 955
  cellulaire 1520
  énergétique 2965
  glucidique 1400
  intermédiaire 4754
  lipidique 3319
  minéral 5695
  oxydatif 6490
  protidique 7376
  respiratoire 7800
  total 9160
métabolite 5542
  intermédiaire 5532
métacarpe 5546
métacentrique 5547
métachromatique 5548
métacinésie 5552
métacoele 5549
métagenèse 5551
métalimnion 9051
métamère 5554
métamérie 5555
métamérisme 5555
métamitose 5556
métamorphisme 9668
métamorphose 5558
  complète 4323
  directe 4163
  incomplète 4163
  indirecte 4323
  paurométabole 6659
métamorphoser 5557
métanéphros 5559
métaphase 5560
métaphytes 5562
métaplasie 5563
métatarse 5565
métaxénie 5566
métaxylème 5567
métazoaire 5568
métencéphale 5569
méthionine 5570
méthode du ^{14}C 7566
  colorante 8581
  de colorisation 8581
  des essais et erreurs 9246
  expérimentale 3234
  de goutte 2709
  de Gram 3931
  de marquage 5384
  des réplications 7734
  des traceurs 9179
méthylation 5575
métoestrus 5576
mettre expérimentalement en évidence 2368
micelle 5578
micro-aiguille 5620
micro-analyse 5579
microautoradiographie 5580
microbalance 5581
microbe 5582
microbicide 5585
microbien 5583
microbiologie 5588
  du sol 8362
microbiologique 5586
microbiologiste 5587
microcentre 5589
microchimie 5590
microchirurgical 5654
microchirurgie 5653
microchromosome 5591
microcinématographie 5592
microclimat 5593
microconsommateur 5594
microcyste 5595
microdissection 5596
micro-électrode 5597
micro-élément 9177
microenvironnement 5599
micro-espèce 5643
micro-évolution 5600
microfaune 5601
  du sol 8363
microfibrille 5602
microflore 5603
  du sol 8364
microfossile 5604
microgamète 5605
microgène 5606
microglie 5607
micrographie électronique 2834
microhabitat 5608
micro-incinération 5609
micro-injection 5610
micromanipulateur 5612
micromanipulation 5611
micromère 5613
microméthode 5615
micromètre 5614
micromutation 5618
micron 5619
micronoyau 5621
micronucléus 5621
microorganisme 5623
microorganismes du sol 8365
micropaléontologie 5624
microphage 5625
microphotographie 5626
microphylle 5627
microphylogenèse 5628
micropipette 5629
microprotéine 5630
micropyle 5631
microradiographie 5632
microscope 5633
  binoculaire 1040
  à contraste de phase 6802
  à dissection 2625
  électronique 2835
  optique 5150
  photonique 5150
  polarisant 7092
  à polarisation 7092
  stéréoscopique 8625
microscopie 5640
  électronique 2836
  à fluorescence 3497
  fluorescente 3497
  à fond clair 1286
  à fond noir 2272
  interférante 4746
  ultraviolette 9351
  à UV 9351
  vitale 9571
microscopique 5636
microscopiste 5639
microsmatique 5641
microsome 5642
micro-sous-espèce 5652
microspectrophotomètre 5644
microspectrophotométrie 5645
microsphère 5646
microsporange 5647

microspore 5648
microsporocyte 7101
microsporophylle 5650
microtechnique 5655
microtome 5656
  à congélation 3578
  à glissière 8324
microtomie 5657
microtubules 5658
microvillosités 5659
MID 4690
migration 5668
  d'animaux 515
  cellulaire 1521
  massive 5394
  d'oiseaux 1129
  pigmentaire 6936
migrer 5665
milieu à agar 237
  ambiant 3004
  de culture 2191
  (de culture) basal 965
  de culture sec 2715
  de culture sélectif 8099
  environnant 3004
  extérieur 3260
  gélosé 237
  d'inclusion 2864
  intérieur 4764
  liquide 5211
  minimal 5701
  nutritif 6250
  solide 8379
mime 5690
mimer 5691
mimèse 5688
mimétique 5689
mimétisme 5693
  batésien 986
  müllérien 5868
mimique 3290
minéralisation 5699
minéralocorticoïde 5700
minimum de protéine 7377
miocène 5705
miscibilité 5706
miscible 5707
misdivision 5708
mise en évidence expérimentale 3233
  à fleurs 3486
  à fruit 3608
  en réserve d'énergie 2971
mitochondrie 5709
mito-inhibiteur 5718
mito-inhibition 5717
mitose 5711
  amphiastrale 417
  anastérale 467
mitosome 5712
mitotique 5713
mixoploïdie 5723
mobile 5828
mode d'action 5724
  de locomotion 5571
  de nutrition 5726
  de reproduction 5573
  de transmission 2099, 5728
  de vie 5725
modèle de comportement 997
  moléculaire 5743
moder 6642
modificateur 5731
  de crossing-over 2142
modification 5730
modification durable 2276
  héréditaire 4185
  du milieu 3007
  de structure 8709
modifieur 5732
modulateur 5734
modulation 5733
  de fréquence 3581
moelle 5446
  allongée 5447
  épinière 8496
  osseuse 1219
molaire 5739
moléculaire 5740
molécule 5746
  adaptrice 138
  en chaîne 1595
  grosse ~ 5313
  protéique 7378
mollusques 5747
monade 5749
monaster 5750
monaxial 5751
monde animal 530
  végétal 7010
monécie 5765
monères 5752
monide 5753
monocarpique 5754
monocaryon 5778
monocentrique 5756
monocellulaire 9373
monochlamydé 5758
monocline 5759
monocotylédone 5761
monocotylédones 5760
monocyclique 5762
monocyte 5763
monoécie 5765
monoecique 5764
monoestrien 5766
monofactoriel 5767
monoflagellé 9376
monogame 5768
monogamie 5769
monogamique 5768
monogenèse 5770
monogénie 5772
monogénique 5771
monogerme 9379
monogerminal 9379
monogonie 5774
monohaploïdie 5775
monohybride 5776
monohybridisme 5777
monoïque 5764
monoloculaire 9381
monomastigote 9376
monomérique 5781
monomorphe 5783
mononucléaire 5784
monophage 5785
monophosphate d'adénosine 156
monophylétique 5786
monophylie 5787
monophylle 5788
monoploïdie 5789
monopode 5791
monopodial 5790
monopodique 5790
monosaccharide 5793
monose 5793
monosexuel 9386
monosome 5794
monosomie 5795
monosperme 5796
monostratifié 9388
monotrèmes 5798
monotriche 5799
monozygote 5800
montage expérimental 3237
montée de la sève 741
mor 7610
mordant 5805
morgan 5806
morphallaxie 5807
morphine 5808
morphologie 5814
  histologique 4302
  végétale 6998
morphologique 5813
morphoplasme 5815
morphose 5816
mort cellulaire 1507
mortalité 5817
morula 5819
morulation 5820
motilité 5829
motoneurone 5835
motricité 5837
mousses 5824
mouvement amiboïde 412
  anaphasique 462
  antipéristaltique 609
  ciliaire 1737
  de courbure 2197
  expressif 3251
  flagellaire 3436
  de glissement 3863
  d'incurvation 2197
  instinctif 4726
  intentionnel 4732
  locomoteur 5230
  morphogène 5812
  natatoire 8849
  nyctinastique 6266
  ondulatoire 9368
  d'orientation 6390
  pendulaire 6694
  périphérique 6743
  pigmentaire 6936
  rampant 2121
  (de) réflexe 7675
  respiratoire 7802
  de turgescence 9311
mouvements de cour 2104
  de déplacement 2621
  péristaltiques 6750
  stéréotypés 8626
mucilage 5850
mucilagineux 5851
mucine 5853
mucipare 5854
mucopolyoside 5855
mucopolysaccharide 5855
mucosité 5863
mucus 5863
mue 5839
  imaginale 190
  larvaire 5038
  nymphale 6268
muer 5838
mull 5679
multicellulaire 5870
multicellularité 5872
multifactoriel 7130
multigène 5876
multilobaire 5878
multilobulaire 5878
multinucléaire 5880

Français

multinucléé 5880
multipartiel 5881
multiplication cellulaire 1529
  massive 5396
  végétative 9487
multiplier, se 5890
multipolaire 5891
multistratifié 5877
multivalent 5892
muni de plusieurs flagelles 5875
muqueuse 5861
muqueux 5858
mur du son 8405
mûr 5427
mûrir 5428
muscle 5895
  abducteur 8
  adducteur 149
  alaire 279
  antagoniste 557
  cardiaque 5950
  ciliaire 1738
  circulaire 1751
  extenseur 3254
  fléchisseur 3453
  involontaire 8336
  lisse 8336
  longitudinal 5247
  masséter 5399
  masticateur 5399
  squelettique 8311
  strié 8701
  tenseur 8987
  du vol 3458
  volontaire 8701
musculaire 5903
musculature 5910
mutabilité 5912
mutable 5913
mutagène 5914, 5918
mutagenèse 5917
mutagénicité 5919
mutagénique 5918
mutant 5920
  auxotrophe 6254
  nain 2736
  nutritionnel 6254
mutante 5920
mutation 5924
  bourgeonneuse 8560
  de bourgeons 8560
  chromosomique 1712
  double 427
  factorielle 3714
  génique 3714
  de génomes 3760
  lét(h)ale 5120
  majeure 5314
  parallèle 6583
  plastidiale 7035
  plastidique 7035
  au point 3714
  de point zéro 9749
  ponctuelle 3714
  provoquée 4639
  renversée 7843
  en retour 7843
  reverse 7843
  simultanée 8296
  spontanée 8538
  suppressive 8814
  systémique 8905
mutationnel 5931
mutationnisme 5930
muter 5923
muton 5935
mutualisme 5936
mycélium 5937
mycétologie 5941
mycétome 5939
mycétozoaires 5969
mycobactérie 5940
mycologie 5941
mycophytes 3624
mycor(r)hize 5943
mycotrophie 5944
myélencéphale 5945
myéline 5946
myélinique 5948
myélinisé 5948
myoblaste 5949
myocarde 5950
myofibrille 5952
myogène 5953
myoglobine 5954
myolemme 5955
myologie 5956
myomère 5957
myomètre 5958
myoplasme 5959
myosine 5961
myotome 5963
myriapodes 5964
myrmécophile 5965
myrmécophyte 5966
myxamibe 5967
myxoflagellés 5968
myxomycètes 5969

N

NAD 6118
NADP 6119
nageoire 3415
  abdominale 9501
  adipeuse 166
  anale 450
  caudale 1481
  dorsale 2682
  pectorale 6668
  ventrale 9501
nanisme 5972
nannoplancton 5973
nappe d'huile 6304
  phréatique 3965
  souterraine 3965
narine 5974
nasal 5976
naseaux 6181
nasopharynx 5982
nastie 5984
nastique 5983
nastisme 5984
natalité 5985
néarctique 5996
nebenkern 6590
nécrobiose 5998
nécrocytose 1507
nécrohormone 9709
nécrophage 6001
nécrose 6002
  cellulaire 1507
nectaire 6004
nectar 6003
necton 6009
némathelminthes 6010
nématicide 6011
nématocyste 6012
nématodes 6013
némertes 6014
némertiens 6014
néoblaste 6015
néocortex 6016
néo-darwinisme 6017
néoformation 6018
néogée 6019
néogène 6020
néo-lamarckisme 6021
néolithique 6022
néomorphe 6023
néopallium 6024
néoténie 6026
néotype 6028
néozoïque 1546
néphridie 6031
néphridiopore 6034
néphrique 7721
néphrocoele 6032
néphron 6033
néphrostome 6035
néphrotome 6036
nerf 6039
  abducens 7
  accessoire 50
  auditif 802
  crânien 2116
  facial 3291
  glosso-pharyngien 3870
  hypoglosse 4502
  moteur 5834
  moteur oculaire (commun) 6281
  oculaire externe 7
  olfactif 6312
  optique 6362
  pathétique 9273
  pneumogastrique 9438
  rachidien 8498
  sciatique 8020
  sensitif 8172
  sensoriel 8172
  spinal 8498
  splanchnique 8521
  trijumeau 9253
  trochléen 9273
  vague 9438
nerfs vasomoteurs 9466
néritique 6037
nervation 6038
  alaire 9686
  foliaire 5091
  à ~ réticulée 6072
nerveux 6054
nervure centrale 5662
  foliaire 5090
  médiane 5662
nervuré 2092
à nervures parallèles 6584
neural 6073
neurilemme 8018
neurite 879
neuroblaste 6086
neurocentre 6041
neurocrâne 6087
neurocrinie 6088
neurocyte 6040
neuroendocrine 6090
neuro-épithélium 6091
neuro-fibrille 6092
neurogéne 6094
neurohormone 6096
neurohypophyse 6097
neurologie 6099
neurologiste 6098

O

Français

P

pectine 6666
pectoral 6667
pédicelle 6670
pedigree 6671
pédipalpe 6674
pédoclimax 2789
pédogenèse 6511, 8360
pédologie 8372
pédoncule 6678, 8582
  fructifère 3609
  oculaire 3289
  optique 3289
pédonculé 6679
pelage 6681
pélagique 6682
pélécypodes 1145
pélorie 6686
pelté 8040
pelvien 6689
pelvis 6693
pénétrance 6695
  complète 1945
  incomplète 4602
pénicilline 6697
pénis 6698
penne 3457, 6949, 7720
penné 6950
penniforme 6952
penninervé 6953
pentosanne 6699
pentose 6700
pénurie d'oxygène 6501
pépin 4953
pepsine 6701
pepsinogène 6702
peptidase 6703
peptide 6704
peptone 6708
  trypsique 9295
peptonisation 6709
perception 6710
  acoustique 8406
  de la gravité 3942
  du mouvement 5846
  olfactive 6314
  sonore 8406
  des sons 8406
  du stimulus 8651
perceptivité 6711
perdre *(feuilles etc.)* 20
père 8307
pérennance 6712
pérennant 6713
pérenne 6713
pérennité 6712
perforateur 103
périanthe 6717
périblaste 6718
périblème 6719
péricarde 6721
péricarpe 6722
péricaryon 6732
péricentrique 6723
périchondre 6724
périclinal 6725
péricycle 6727
périderme 6728
péridiniens 2564
péridium 6729
périgone 6730
périgyne 6731
périlymphe 6733
périmétrium 6734
périmysium 6735
périnée 6736
périnèvre 6737
période biologique de demi-valeur 1079
  des chaleurs 6296
  de croissance 3972
  de couvaison 1281
  de demi-valeur 4075
  de duplication 2699
  d'essai 3238
  de floraison 3492
  du frai 8426
  géologique 3776
  de gestation 3818
  glaciaire 3846
  d'incubation 1281, 4617
  interglaciaire 4750
  juvénile 4922
  de latence 5046
  de maturité 5425
  post-glaciaire 7215
  quiescente 7823
  de réaction 7618
  réfractaire 7685
  de repos 7823
  de reproduction 7760
  du rut 6296
  de sécheresse 2711
  sensible 8158
  de végétation 9485
périodicité 6740
  annuelle 542
  diurne 2649
  journalière 2649
  saisonnière 8046
périoste 6741
péripates 6340
périphyton 6744
périplasme 6745
périsperme 6747
périspore 6748
périssodactyles 6749
péristaltisme 6750
péristase 6751
péristome 6752
périthèce 6753
périthécium 6753
péritoine 6755
péritriche 6756
perméabilité 6764
  cellulaire 1526
  à l'eau 9647
  membranaire 5476
  sélective 8101
perméable 6765
  à l'eau 9648
perméase 6766
perméation 6767
permien 6768
permutation 6770
péroné 3395
peroxydase 6771
perpétuation de l'espèce 8439
persistance 6772
persistant 6773
perspiration 6775
perte calorique 4131
  d'eau 9644
  énergétique 2964
  d'énergie 2964
  par évaporation 3154
  par respiration 7799
perturbation métabolique 5528
  du milieu 3014
peste huileuse 6303
pesticide 6778
pétale 6780
à pétales concrescents 3659
pétiole 6781
pétiolule 6782
petite calorie 3928
  circulation 7456
pétrification 6784
pétrifier, se 6785
peuplement mélangé 5722
pH 6788
phage 927
  intempéré 4730
  tempéré 8968
  virulent 4730
phagocyte 6791
phagocyter 6792
phagocytose 6793
phanérogames 8472
phanérophyte 6795
pharyngien 6796
pharynx 6800
  branchial 6801
phase d'accouplement 2098
  d'attraction 2098
  de croissance 3993
  de développement 6738
  exponentielle 3250
  initiale 4685
  de jeunesse 4922
  de latence 5016
  logarithmique 3250
  de maturation 5426
  réfractaire 7685
  stationnaire 8603
phelloderme 6805
phellogène 6806
phène 6807
phénocopie 6808
phénocytologie 6809
phénogénétique 6810
phénologie 6813
phénoloxydase 6811
phénomène vital 9573
phénotype 6814
  mutant 5921
phénotypique 6815
phénylalanine 6818
phéophycées 6789
phéophytes 6789
phéro(r)mone 6819
phloème 6821
phobotactisme 6822
phobotaxie 6822
phonation 6823
phonorécepteur 6824
phorésie 6825
phoronidiens 6826
phosphatase 6827
phosphatide 6828
phospholipide 6828
phosphomonoestérase 6827
phosphoprotéide 6834
phosphorescence 6835
phosphorolyse 6837
phosphorylase 6838
phosphorylation 6840
  de la chaîne respiratoire 6491
  oxydative 6491
phosphoryler 6839
photobactéries 5263
photobiologie 6842
photochimie 6844
photochimique 6843
photocinèse 6845
photologie 6842

photolyse 6846
photométrie 6847
photomorphose 6849
photon 6850
photonastie 6851
photoperception 6852
photopériode 6853
photopériodisme 6854
photophile 6855
photophore 5264
photophosphorylation 6857
  cyclique 2214
  non cyclique 6156
photoréactivation 6858
photorécepteur 6859
photosensibilité 5156
photosensible 5155
photosynthèse 6861
photosynthétique 6862
phototactisme 6864
phototaxie 6864
phototonus 6865
phototrophique 6866
phototropisme 6867
phragmoplastc 6868
phycobiline 6869
phycocyanine 6870
phycoérythrine 6871
phycologie 301
phycomycètes 6873
phycophéine 6874
phycoxanthine 3617
phylétique 6875
phylloclade 6877
phyllode 6878
phylloquinone 6879
phyllotaxie 6880
phyllotrachée 1227
phylogenèse 6881
phylogénétique 6882, 6884
phylogénie 6881
phylogénique 6882
phylum 6885
  animal 517
physiologie 6892
  animale 518
  cellulaire 1527
  du comportement 1000
  du développement 2463
  nerveuse 6103
  de la nutrition 6893
  de la reproduction 7761
  sensorielle 8174
  végétale 6919
physiologique 6886
physiologiste 6891
  végétal 7002
physique biologique 1103
phytine 6894
phyto-albumine 6895
phytobenthos 6896
phytobiologie 6897
phytocénose 6990
phytochimie 6899
phytochimique 6898
phytochrome 6900
phytoécologie 6993
phytoflagellés 6902
phytogène 6905
phytogenèse 6903
phytogénéticien 6987
phytogénétique 6904
phytogénie 6903
phytogénique 6905
phytogéographe 6906
phytogéographie 6907
  floristique 3479
phyto-globuline 6908
phytographie 6909
phytohormone 6910
phytol 6912
phytologie 1232
phytologue 1231
phytopaléontologie 6523
phytoparasite 6601
phytopathologie 6919
phytophage 6917, 6918
phytoplancton 6920
phytosociologie 6921
phytostérol 6922
phytotomie 6986
phytotoxicité 6925
phytotoxine 6926
phytotoxique 6924
phytotron 6927
phytozoaire 6928
pic de sélection 8100
pièce florale 3474
pièces buccales 5845
pied ambulacraire 381
  fendu 1786
pie-mère 6929
pigment 6931
  accessoire 51
  assimilateur 758
  biliaire 1028
  colorant 6938
  cutané 8317
  oculaire 9563
  respiratoire 7804
  sanguin 1184
  végétal 7003
pigmentation 6939
  aposématique 9622
  avertisseuse 9622
  cryptique 2163
pigmenter 6932
pileux 6943
pilorhize 7901
pilosité 6944
pince 1614
pincette 6946
pinné 6950
pinocytose 6955
pipette 6957
piquant 7287, 8660, 9086
  venimeux 9495
piqueur *(insecte)* 8661
piste odorante 6286
pistil 6958
pitocine 6506
pituicyte 6965
pivot 8929
placenta 6968
placentaires 6970
placentation 6971
placode cristallinienne 5108
plagiogéotropisme 6973
plagiotrope 6974
plagiotropisme 6975
plan de clivage 1784
  de division 2654
  équatorial 3098
  médian 5443
  d'organisation biologique 1081
  sagittal 7945
  de structure 8714
  de symétrie 8854
planation 6977
plancton 6978
plancton aérien 207
  dulçaquicole 5179
  marin 4082
  végétal 6920
planctonique 6979
planctonophage 6981
planification de l'environnement 3020
planocyte 9617
planogamète 9760
planospore 9773
plante 6984
  adventive 194
  aéricole 206
  aérophyte 206
  anémophile 489
  annuelle 543
  biennale 1016
  bisannuelle 1016
  à bulbe 1331
  bulbeuse 1331
  calcicole 1345
  calcifuge 1348
  carnivore 4710
  cultivée 2186
  de culture 2186
  désertique 2431
  d'eau douce 3588
  à enracinement profond 2321
  exotique 3223
  d'expérience 3235
  à feuilles 3508
  à fibre 3380
  à floraison nocturne 6128
  flottante 3461
  grimpante 1800
  héliophile 4149
  héméropériodique 5237
  -hôte 4395
  hybride 4415
  indicatrice 4630
  indifférente 2282
  indigène 4632
  inférieure 5256
  insectivore 4710
  de journée courte 8265
  de journée longue 5237
  ligneuse 5160
  des marais 4153
  -mère 6620
  naine 2737
  nourricière 3533
  nutritive 3533
  nyctipériodique 8265
  ombrophile 6329
  palustre 4153
  parasite 6601
  pérenne 6714
  photoapériodique 2282
  pionnière 6956
  racinée 7547
  à racines traçantes 8253
  radicante 7547
  rampante 2125
  de rocher 7934
  en rosette 7915
  rudérale 7923
  rupestre 7934
  rupicole 7934
  sarmenteuse 7987
  sciaphile 8319
  silicicole 8288
  de soleil 4149
  de sombre 8319
  des steppes 8622

Français

plante submergée 8741
succulente 8767
supérieure 8319
-témoin 9019
terrestre 9005
-test 9019
toxique 7081
à tubercule 9302
tubéreuse 9302
umbrophile 8319
utile 9423
vasculaire 9458
vénéneuse 7081
vivace 6714
volubile 9611
xérophile 9720
zoïdiogame 9751
plantes amentifères 386
aquatiques 4455
à fleurs 8472
à graines 8472
halophiles 4081
pulviniformes 2198
plantigrade 7011
plantule 7012
planule 7013
plaque 7014
d'agar 238
ambulacraire 380
basale 956
cellulaire 1528
chitineuse 1641
cornée 4389
à culture 2192
équatoriale 3099
de gélose 238
incubatrice 4616
madréporique 5330
médullaire 5451
neurale 6079
osseuse 1224
de Petri 6783
polaire 7087
terminale 2955
terminale motrice 5833
plaquer 7043
plaquette sanguine 9098
plasma 7015
sanguin 1185
plasmacyte 7016
plasmagène 7017
plasmalemme 7018
plasma-protéine 7020
plasmatique 7021
plasmazelle 7016
plasme germinal 3787
germinatif 3787
plasmide 7023
plasmine 7024
plasmique 7021
plasmocyte 7016
plasmode 7026
plasmodesme 7025
plasmogamie 7027
plasmolyse 7028
commençante 4594
limite 4594
plasmon(e) 7029
plasmosome 7030
plasmotomie 7031
plasmotype 7029
plaste 7032
amylifère 439
chlorophyllien 1648
plastide 7033
plastidome 7036
plastidotype 7036
plastogamie 7037
plastogène 7038
plastome 7036
plastoquinone 7040
plastosome 5709
plastron 7042
plateau continental 2014
plathelminthes 7045
platine (porte-objet) 5635
plectenchyme 7046
plein de barbes (ou arêtes) 707
pléiotropie 7050
pléiotropique 7049
pléistocène 7051
pléomorphe 7052
pléomorphisme 7053
pléopode 7054
pléotropie 7050
plérome 7055
pleuston 7058
plèvre 7056
costale 2090
médiastinale 5444
pulmonaire 7458
plexus nerveux 6049
pli dermique 8315
neural 6077
pliocène 7060
plissé *(vernation)* 7059
plumage 7062
juvénil 4921
plume 3333
caudale 8919
de contour 2020
de couverture 2020
rectrice 8919
tectrice 2020
voilière 3457
plumule 2701, 7063
pluriannuel 7065
pluricellulaire 5870
plurifactoriel 7130
pluriloculaire 5869
plurinucléé 5880
pluripartiel 5881
pluristratifié 5877
pneumatophore 7067
poche branchiale 3838
d'encre 4688
incubatrice 1300
marsupiale 5388
du noir 4688
poecilotherme 7072
pogonophores 7070
poids atomique 785
corporel 1213
frais 3589
humide 9670
moléculaire 5745
sec 2722
vif 5221
poignet 9710
poïkilotherme 7072
poil 4064
absorbant 7904
glanduleux 3854
radiculaire 7904
sensitif 8170
sensoriel 8170
staminal 8586
tactile 8911
urticant 8664
poilu 6943
point cardinal 1426
de compensation 1929
de congélation 3579
de contact 2008
de croissance 9486
d'ébullition 1215
de fanaison 9680
de fanaison permanente 6763
de flétrissement 9680
de flétrissement permanent 6763
de fusion 5471
d'insertion 4714
is(o)électrique 4865
de rosée 2471
végétatif 9486
de végétation 9486
pointe 633
poison 7075
d'enzymes 3035
fusorial 8502
mitotique 5718
végétal 6926
poisson carnassier 7250
laité 5687
migrateur 5666
œuvé 8424
prédateur 7250
rogué 8424
poissons cartilagineux 1660
osseux 6428
polarisation 7090
polariser 7093
polarité 7089
pôle animal 520
du fuseau 8503
germinal 3798
végétal 9471
végétatif 9471
politique de l'environnement 3021
pollen 7096
pollinarie 7106
pollinide 7113
pollinie 7113
pollinifère 7112
pollinisateur 7108
pollinisation 7110
aquatique 4452
croisée 338
par l'eau 4452
entomophile 2999
ornithophile 6400
en retour 902
polliniser 7107
polluant 7114
polluer 7115
pollueur 7116
pollution 7117
de l'air 269
atmosphérique 269
des cours d'eau 7888
des eaux 9649
de l'environnement 3011
des fleuves 7888
par les hydrocarbures 6303
de la mer 7118
du milieu 3011
radioactive 7553
du sol 8367
polocyte 7083
polyallélie 5883
polyandrie 7122
polyandrique 7121
polycarpien 7123
polycarpique 7123
polycaryocyte 7143

pression de sélection 8102
de turgescence 9312
présure 7727
présynaptique 7283
prêt à réagir 7625
preuve expérimentale 3233
primates 7297
primordium 7306
foliaire 5080
principe actif 126
de la thermodynamique 5060
printanisation 9508
prise des aliments 3530
d'essai 7969
privation de nourriture 3525
probabilité de fertilisation 7308
proband 7309
probasidie 7310
proboscidiens 7311
proboscis 7312
procambium 7313
procartilage 7230
procédé des boues activées 119
de sélection 1276
processus de développement 2464
évolutif 3167
métabolique 5535
vital 5140
prochromosome 7314
procréation 7745
proctodéum 7316
producteur *(ecol.)* 7317
production brute 3958
énergétique 2967
d'énergie 2967
nette 6071
primaire 7293
produit d'assimilation 759
catabolique 1460
contaminant 2010
de déchet 9625
de décomposition 2311
décontaminant 2314
de dégradation 1263
de démolition 1263
détergent 2443
d'excrétion 3200
final 2957
de fission 3425
du gène 3716
intermédiaire 4756
méiotique 5463
métabolique 5542
de scission 3425
sécrété 8065
de sécrétion 8065
terminal 2957
terminal du métabolisme 5529
proembryon 7320
proenzyme 9793
proferment 9793
profil pédologique 8368
du sol 8368
proflavine 7322
progame 7324
progenèse 7325
progéniture 7326
progestérone 7328
progestine 7328
programme de recherche(s) 7771
proie 7285
prolactine 5011
prolamine 7330
prolan 1683
prolifération 7334
prolifération cellulaire 1529
proline 7335
prométaphase 7337
promitose 7339
promycélium 7340
pronéphros 7341
pronucléus 7342
procœstrus 7343
propagule 7344
prophage 7345
prophase 7346
proplastide 7350
proportion mendélienne 5489
de rétrocroisement 898
des sexes 8236
propositus 7309
propre à l'espèce 8446
propriété héréditaire 4192
proprio(ré)cepteur 7351
prosencéphale 7353
prosenchyme 7354
prosimien 7355
prostaglandine 7356
prostate 7357
protamine 7359
protandre 7360
protandrie 7361
protandrique 7360
protéase 7363
protection des animaux 521
des eaux 7365
de l'environnement 3023
de la nature 1993
des oiseaux 1132
des paysages (naturels) 5029
des plantes 7005
contre les radiations 7545
radiologique 7545
du sol 8369
protégé contre les radiations 7613
protéide 7369
protéinacé 7385
protéinase 7384
protéine 7370
animale 522
bactérienne 5630
cellulaire 1538
composée 1951
conjugée 1951
enzymatique 3037
étrangère 3550
fibrillaire 8032
globulaire 8484
musculaire 5900
nucléaire 6227
de réserve 7775
sérique 8201
simple 8293
de structure 8715
tissulaire 9135
végétale 7007, 9496
virale 9533
protéinique 7385
protéinoïde 7386
protéique 7385
protéoclastique 7390
protéohormone 7388
protéolyse 7389
protéolytique 7390
protéométabolisme 7376
protéomolécule 7378
protéoplaste 7387
protérandre 7360
protérandrie 7361
protérandrique 7360
protérogyne 7401
protérogynie 7402
prothalle 7394
prothrombine 9099
protidique 7385
protistes 7396
protistologie 7397
protocéphalon 7398
protochlorophylle 7399
protogyne 7401
protogynie 7402
proton 7403
protonéma 7404
protonéphridie 7405
protophyte 7406
protoplasmatique 7408
protoplasme 7407
protoplasmique 7408
protoplaste 7410
protostèle 7411
protostomiens 7412
prototrophe 7413
protovertèbre 8400
protoxylème 7415
protozoaire 7417
protozoologie 7416
protubérance annulaire 7181
provitamine 7419
psammon 7421
psammophytes 7422
pseudoallèle de position 7208
pseudo-allèles 7423
pseudo-allélisme 7424
pseudobranchie 7426
pseudo-chrysalide 7436
pseudo-dominance 7428
pseudogamie 7429
pseudo-haploïdie 7430
pseudo-hermaphrodi(ti)sme 7431
pseudomixie 7429
pseudo-nymphe 7436
pseudo-parenchyme 7433
pseudopode 7434
pseudo-polyploïdie 7435
pseudo-réduction 7437
pseudo-verticille 9519
psilophytes 7438
psychologie de la forme 3815
ptéridophytes 7439
ptéridospermées 7440
ptérine 7441
ptérobranches 7442
ptérygotes 7444
ptyaline 7445
pubère 7446
puberté 7447
pubis 7449
puff 7450
pullulation 5396
pullelement 5396
pulmonaire 7453
pulpe 7460
dentaire 2388
pulsation 7463
pulsion 2706
migratoire 5669
sexuelle 8243
pulvérisé 7466
pupaison 7473
puparium 7471
pupation 7473
pupe 7468
coarctée 1832
pupille 7474
pupipare 7476

Q

R

Français

réarrangement chromosomique 7731
reboisement 7627
reboiser 7626
récapitulation 7629
réceptacle (floral) 7632
  séminal 8134
récepteur 7633
  cutané 2202
  de distance 2637
  du froid 1870
  du goût 8938
  nociceptif 6145
  olfactif 6315
  de pression 7280
  tactile 8924
  de température 8972
  thermique 9061
récessif 7637
récessivité 7639
recherche 7767
  de nourriture 3537
rechercher (sur) 7768
rechercheur 7773
récif coralliaire 2044
  corallien 2044
récipient de culture 2189
réciproque 7642
recombinaison 7648
  des gènes 3717
recombinant 7647
récon 7653
reconnaissance spécifique 8441
recroisement 895
recroiser 2135
rectrice 8919
rectum 7657
redirection 7661
redox 6493
réductase 7665
réducteur 2309, 7663
réduction chromatique 7666
réductone 7667
réduplication 7668
  des gènes 3718
réflexe 7669
  d'allongement 8699
  d'axone 880
  conditionné 1964
  cornéen 2057
  de déglutition 8837
  d'extension 3255
  de flexion 3457
  inconditionné 9364
  non conditionné 9364
  palatin 8837
  palpébral 3286
  photomoteur 7475
  de posture 7220
  pupillaire 7475
  rotulien 4980
  tendineux 8984
réflexes en série 1597
réflexologie 7679
réflexologiste 7678
réflexologue 7678
réfraction 7682
réfringence 7687
réfringent 7688
refuge glaciaire 3847
régénération 7694
régénérer, se 7693
régénérescence 7694
régime alimentaire 5726
région alimentaire 3344
région faunistique 3332
  florale 3476
  holarctique 4308
  d'insertion 4714
  néotropicale 6027
  de l'organisateur nucléolaire 6224
  paléotropicale 6534
  polaire 7095
règles de l'hérédité 5062
Règles Internationales de Nomenclature 4769
règne animal 513
  végétal 6997
régressif 7699
régression 7696
  linéaire 5185
  non-linéaire 6160
  partielle 6639
régulateur de croissance 3990
régulation 7702
  de croissance 3989
  hormonale 4385
  osmotique 6415
  respiratoire 7794
  par retour 3338
  saline 7962
  thermique 9062
rein 4959
reine (des abeilles) 7520
rejet(on) 8565
relais entre couveurs 8254
relation caryoplasmique 6217
  enzyme-gène 3706
  nucléo-plasm(at)ique 6217
  nutritive 6257
  prédateur-proie 7246
  stimulus-réponse 8654
relations alimentaires 3343
relaxation musculaire 5907
relève au nid 7719
relevé cartographique de la végétation 5371
  de végétation 9479
relique glaciaire 3848
remaniement chromosomique 7731
rémige 7720
rénal 7721
renflement 8846
renfler, se 8845
renforcement *(eth.)* 7707
réniforme 7726
rennine 7727
réoxydation 7728
réoxyder 7729
réparation 7730
répartition par âge 242
  au hasard 7595
  du travail 2653
repli cutané 8315
réplication 7735
  de l'ADN 2659
  semi-conservative 8128
répolarisation 7737
répondre (à un stimulus) 7812
réponse 7813
  réflexe 7677
  retardée 2363
repos estival 211
  germinatif 2675
  hivernal 9692
  de la végétation 9489
répresseur 7740
répression 7739
répression enzymatique 3039
  fermentative 3039
reproduction 7745
  asexuée 750
  panmictique 6567
  par rétrocroisement 896
  sexuée 8247
  végétative 750
reproductivité 7753
reproduire, se 7742
  par voie sexuée 7743
reptation 2121
reptiles 7765
répulsif 7732
réseau 7837
  capillaire 1393
  nerveux 6047
réserve adipeuse 5197
  alimentaire 3535
  amylacée 8598
  avienne 1133
  biologique 5993
  d'eau 9651
  glucidique 1401
  grasse 5197
  lipidique 5197
  naturelle 5993
  nourricière 3535
  nutritive 3535
  protéique 7381
  protidique 7381
réservoir de sécrétion 8063
résidence écologique 4032
résidus 7780
  alimentaires 3536
  des pesticides 6779
  radioactifs 7560
résine 7781
  dammar 2269
  échangeante d'ions 4834
résineux 1983
résistance 7783
  aux acides 86
  à la chaleur 4134
  de l'environnement 3025
  à la flexion 1004
  au froid 1871
  au gel 3596
  membranaire 5478
  du milieu 3025
  à la pression 1953
  à la sécheresse 2712
  à la traction 8986
résistant 7784
  aux acides 83
  à la chaleur 4135
  à la sécheresse 2713
résorber 7786
résorption 7787
respiration 7788
  abdominale 5
  aérienne 265
  branchiale 3840
  cellulaire 1539
  costale 2091
  cutanée 2203
  diaphragmatique 2490
  diffusive 2535
  pulmonaire 7459
  totale 9162
  trachéenne 9184
respiratoire 7789
respirer 1265
respiromètre 7811
ressources aquatiques 9651

restituabilité 7825
restitution 7826
résultat des expériences 3236
expérimental 3236
résupination 7827
resynthèse 7828
retard de croissance 3991
du développement 2361
réticulaire 7829
réticulé 7829
réticuline 7833
réticulocyte 7835
réticulo-endothélium 7836
réticulum endoplasm(at)ique 2933
sarcoplasmique 7985
rétine 7838
rétinerve 6072
retombées radioactives 7554
retour à l'état sauvage 7933
au gîte 4335
à l'habitat 4335
au nid 4335
retourner à l'état sauvage 7931
rétracteur 7841
retranslocation 7842
rétroaction 3337
rétrocontrôle 3338
rétrocroisement 895
rétroinhibition 3340
rétrorégulation 3338
réunion par bancs 8016
réversion sexuelle 8237
rhabdome 7844
rhabdomère 7845
rhéobase 7846
rhéogaméon 7847
rhéotactisme 7848
rhéotaxie 7848
rhéotropisme 7849
rhinencéphale 7852
rhinopharynx 5982
rhizogenèse 7903
rhizoïde 7856
rhizome 7857
rhizomorphe 7858
rhizopodes 7859
rhizosphère 7860
rhodophycées 7861
rhodophytes 7861
rhodopsine 7862
rhombencéphale 7863
rhytidome 947
annulaire 7881
écailleux 7999
riboflavine 7868
ribonucléase 7869
ribonucléoprotéine 7871
ribonucléoside 7872
ribose 7873
ribosome 7875
ribulose 7876
ribulosediphosphate 7877
riche en espèces 8442
rigidité 7878
thermique 9046
ritualisation 7887
robe de noce 6239
roche-mère 6619
rodenticide 7894
rogue 7895
rongeurs 7893
rosette de feuilles 7914
foliacée 7914
en rosette 7916
rotifères 7919
rotule 4979
route de migration 5678
rudiment 7924
rudimentaire 7925
rumen 7927
ruminant 7928
rumination 7930
ruminer 7929
rupture chromosomique 1716
rut 6297
rythme d'activité 128
biologique 1105
cardiaque 1425
circadien 1748
diurne 1748
endogène 2920
nycthéméral 1748
rythmicité 6740

S

sabot 4378
sac aérien 271
amniotique 408
coelomique 1854
embryonnaire 2868
lacrymal 5000
pollinique 7102
pulmonaire 1227
vitellin 9745
saccharase 4815
saccharide 7938
saccharimètre 7939
saccharose 7940
sacciforme 7941
sacculaire 7941
sacrum 7943
sagitté 7947
saison d'accouplement 5416
d'appariement 5416
de croissance 3972
de floraison 3492
des pluies 7591
pluvieuse 7591
reproductrice 7760
sèche 2720
salin 7949
salinité 7950
salivation 7956
salive 7951
saltation 7964
salure 7950
samare 9688
sang, à ∼ chaud 9618
à ∼ froid 1868
sanguin 4050
sans germes 3785
queue 628
saprobionte 7973
saprogène 7974
sapropèle 7975
saprophage 7976
saprophyte 7977
saprophytisme 7978
saprozoïte 7979
sarcolemme 5955
sarcomère 7983
sarcoplasme 7984
sarcosome 7986
SAT-chromosome 7989
satellite 9175
saturation 7991
saturé 7990
sauropsidés 7994
savane 7995
scapulaire 8003
schéma d'action 116
schème d'excitation 3184
schizocarpe 8007
schizogamie 8008
multiple 5885
schizogène 8010
schizogenèse 8009
schizogonie 8011
schizomycètes 8012
schizonte 8013
schizophyte 8014
sciaphile 8318
sciaphyte 8319
sciatique 8019
science du sol 8372
scinder 8529
scission 3424
scissiparité 8025
sclérenchyme 8028
sclérite 8029
sclérophylle 8030
sclérophytes 8031
scléroprotéine 8032
sclérote 8034
sclérotique 8026
sclérotome 8035
scrotum 8039
scutiforme 8040
sébum 8048
sécerner 8057
sécheresse 2710
seconde loi de Mendel 5485
sécréter 8057
sécréteur 8060
sécrétine 8058
sécrétion 8059
biliaire 1031
endocrine 4608
exocrine 3209
interne 4608
lactée 5681
muqueuse 5865
salivaire 7955
sudorale 8778
sudoripare 8778
sécrétoire 8060
section fine 9078
longitudinale 5248
sagittale 7946
en série 8188
transversale 2148
ultrafine 9347
sédentaire 8068
sédimentation globulaire 1190
segment chromosomique 1722
segmentation 1779
bilatérale 1022
spirale 8511
totale 1943
ségrégation 8087
transgressive 9207
ségréger 8086
séismonastie 8089
sel biliaire 1029
minéral 5698
nutritif 6259
sélectif 8096
sélection 1269, 8092
animale 504
canalisante 1380
clonale 1808
consanguine 4589

Français

synchrone 8873
syncytium 8874
syndèse 8875
syndesmose 8876
synécie 8890
synécologie 8877
synénergétique 8879
synergide 8880
synergie 8878
synergisme 8881
synergiste 8882
syngame 8884
syngaméon 8883
syngamie 8885
syngenèse 8886
syngénique 8887
synoecie 8890
synoecique 8889
synoïque 8889
synovial 725
synovie 8891
synthèse enzymatique 3042
  de(s) protéine(s) 7383
  protéique 7383
synthétase 5147
synthétiser 8895
syntrophe 8897
syntrophie 8898
syntrophisme 8898
synusie 8899
synzoospore 8900
syringe 8901
syrinx 8901
systématicien 8903
systématique 8902
  animale 528
  botanique 7008
système acellulaire 1513
  aquifère 9663
  circulatoire 1754
  de classification 1770
  conducteur 1965
  écologique 2775
  d'espaces intercellulaires 4738
  excréteur 3201
  gastro-vasculaire 3677
  lacunaire 5014
  limbique 5169
  lymphatique 5279
  multienzymatique 5873
  musculaire 5908
  nerveux 6058
  nerveux autonome 834
  nerveux central 1550
  nerveux périphérique 6742
  nerveux orthosympathique 8856
  nerveux sympathique 8856
  nerveux végétatif 834
  nerveux viscéral 834
  organe 6376
  oxydo-réducteur 6489
  parasympathique 6608
  porte 7205
  radiculaire 7910
  redox 6489
  à régulation 3339
  réticulo-endothélial 7836
  de rétrocontrôle 3339
  sanguin 1199
  tampon 1325
  trachéen 9185
  transporteur d'électrons 2842
  vacuolaire 9434
  vasculaire 9460
systole 8906

T

table de mortalité 5143
  de survie 5143
tableau généalogique 3721
tache aveugle 1167
  germinale 3801
  jaune 9739
  oculaire 3288
  oculiforme 3288
  pigmentaire 6937
tachygenèse 8908
tact 8153
tactile 8909
tactisme 8941
taïga 8918
tallage 9120
talle 9119
taller 9118
tampon 1318
tamponnage 1326
tamponner 1319
tangorécepteur 8924
tan(n)in 8926
tapétum 8928
tapis 8928
tardigrades 8930
tare héréditaire 4188
tarière 6467
t-ARN 9201
tarse 8933
taux d'aberration 10
  d'accroissement 7609
  aqueux 9632
  de croissance 3988
  de crossing-over 2146
  métabolique 5537
  de mortalité 2290
  de multiplication 5889
  de mutation 5929
  de(s) naissance(s) 1139
  de recombinaison 7651
  de renouvellement 9315
  de reproduction 7746
  de survie 8829
  de transpiration 9223
taxie 8941
taxinomie 8945
taxinomique 8943
taxinomiste 8944
taxon 8942
taxonomie 8945
taxonomique 8943
taxonomiste 8944
technique du marquage-et-recapture 5380
  de préparation 5572
tectologie 8949
tégument 4729
  pupal 7469
  séminal 8074
télégonie 8954
télencéphale 8955
téléostéens 8956
téleutospore 8958
téliospore 8958
télocentrique 8959
télolécithe 8960
télolécithique 8960
télome 8961
télomère 8963
télophase 8964
télosynapse 8965
télosyndèse 8965
télotactisme 8966
télotaxie 8966
telson 8967
tempe 8974
température ambiante 376
  corporelle 1212
  interne 1212
  optimale 6369
  optimum 6369
  préférentielle 7253
temps de génération 3726
  d'impression 7278
  de latence 5046
  de présentation 7278
  de réaction 7620
  utile 9428
tendineux 8981
tendon 8983
teneur en carbone 1405
  en eau 9632
  en humidité 5737
  nutritive 6248
  en oxygène 6498
  en protéines 7374
  résiduelle maximale 9147
  en résidus tolérée 9147
  saline 7950
  en substances nutritives 6248
  en vitamines 9591
tension artérielle 1187
  partielle 6638
  superficielle 8822
tentacule 8988
tépale 8989
tératologie 8990
térébène 9001
terminaison axonique 881
  nerveuse 6044
terminal *(bot.)* 8991
terminalisation 8999
termones 9000
terpène 9001
terrain d'alimentation 3344
terre d'infusoires 4667
terrestre 9002
terricole 9006
territoire 9011
  de ponte 1282
  de reproduction 1282
territorialisme 9007
territorialité 9007
tertiaire 9013
test 8074
  avoine 860
  biologique 1047
  cis-trans 1759
  de complémentarité 9017
  de courbure 860
  des répliques 7734
testage de la descendance 7327
test-cross 9016
  à deux points 9322
  à trois points 9090
testicule 9023
testostérone 9024
têtard 8917
tétin 5354
tétine 8948
tétrade 9025
  ditype non-parentale 6166
tétraploïde 9027
tétraploïdie 9028
tétrapode 9029
tétrasomique 9030
tétraspore 9031

Français

tétratype 9032
tétroxyde d'osmium 6411
thalamus 7632, 9033
thalle 9037
thallodique 9036
thallophyte 9035
thallus 9037
thélytocie 9039
théorie de l'âge et de l'espace 240
béhavioriste 1001
des catastrophes 9040
cellulaire 1533
du choc 8931
chromosomique de l'hérédité 1724
de la cible 8931
des courants locaux 5223
de la dérive des continents 2013
de la descendance 2427
de l'équilibre de la détermination des sexes 935
de l'évolution 9042
évolutionniste 9042
de la forme 3816
de l'impact 8931
de l'information 4658
ionique 4841
de la précocité 7241
de la préformation 7257
de la présence et de l'absence 7277
de la récapitulation 7630
de la sélection (naturelle) 9044
du télome 8962
de la tension de cohésion 1862
thèque 9038
thermocline 9051
thermodynamique 9052, 9053
thermogenèse 1362
thermolabile 9056
thermolyse 9057
thermonastie 9058
thermonastisme 9058
thermopériodisme 9059
thermophile 9060
thermorécepteur 9061
thermorégulateur 9063
thermorégulation 9062
thermorésistance 4134
thermorésistant 4135
thermosensibilité 9048
thermostabile 9066
thermostabilité 9065
thermostable 9066
thermotactisme 9067
thermotaxie 9067
thermotropisme 9068
thérophyte 9069
thiamine 9070
thigmomorphose 9074
thigmonastie 4096
thigmotactisme 9075
thigmotaxie 9075
thigmotropisme 9076
thiobactéries 9079
thionine 9081
thorax 9085
thréonine 9091
thrombine 9097
thrombocyte 9098
thrombogène 9099
thrombokinase 9100
thylle 9323
thyllose 9323
thymidine 9104
thymine 9106
thymus 9108
thyréostimuline 9112
thyroïde 9110
thyroxine 9113
thyrse 9114
tibia 9115
tige 8582, 8611
florale 6678
pileuse 4069
pituitaire 4510
tigelle 9117
tissu 9127
adipeux 168
d'aération 198
aquifère 9659
assimilateur 756
cartilagineux 1673
caverneux 1495
cellulaire 1540
cicatriciel 1735
conducteur 1966
conjonctif 1991
embryonnaire 2882
épithélial 3091
érectile 1495
fibreux 3394
fondamental 3963
glandulaire 3857
graisseux 168
hypodermique 8725
ligneux 9708
lymphoïde 5282
mécanique 8812
méristématique 5495
nerveux 6059
nourricier 6261
osseux 1226
en palissade 6548
palissadique 6548
permanent 6760
protecteur 7368
de réserve 8687
de revêtement 8823
sécréteur 8066
sous-cutané 8725
de soutien 8812
spongieux 8536
subéreux 2052
superficiel 8823
de transfusion 9205
vasculaire 9461
végétal 7009
tissulaire 9138
titrage 9140
titration 9140
titre 9142
titrer 9139
titrimétrie 9143
tocophérol 9145
tolérance immunologique 4572
tonicité 9149
tonnelet 7471
tonoplaste 9148
tonus 9149
musculaire 5909
top-cross 9154
top-crossing 9155
topotactisme 9157
topotaxie 9157
tordue *(estivation)* 2019
torus 9158
totipotence 9163
totipotent 9164
toucher 8153
toujours vert 3157
toundra 9306
tourbe 6661
tourbière basse 5251
haute 4277
à sphaignes 6662
de transition 9209
toxicité 9170
toxicologie 9172
toxicologue 9171
toxigène 9173
toxine 9174
animale 9777
végétale 6926
toxique 7075, 9167
TPN 6119
trabant 9175
trabécule 9176
trace foliaire 5089
traceur 9178
isotopique 7559
radioactif 7559
trachéates 9187
trachée 9180, 9181, 9182
-artère 9180
tubulaire 9305
trachéide 9188
annelée 547
scalariforme 7996
spiralée 8514
trachéo-branchie 9183
trachéole 9189
trachéophyte 9190
tractus génital 3756
intestinal 4796
trait 1605
traitement par chaleur 4138
des déchets 9627
des eaux 9661
des eaux résiduaires 8213
par le froid 1875
transaminase 9193
transamination 9194
transcription 9196
transduction 9197
abortive 18
transférabilité 9216
transférable 9217
transférase 9203
transfert-ARN 9201
transfert d'électrons 2839
d'énergie 2972
de gènes 3720
de l'information 9200
transformation 9204
énergétique 2973
d'énergie 2973
transformer en corne 2061
en pupe 7472
transformisme 9042
transgénation 3714
transgression 9207
transit des substances 5847
translation 9211
translocation 9212
insertionnelle 4715
réciproque 7645
translucide 9214
transméthylation 9215
transmettre à 9218
transmissibilité 9216
transmissible 9217
héréditairement 4671
transmission d'énergie 2972
héréditaire 4672

U

V

Français

vacuole pulsatrice 2023
vacuome 9434
vagin 9437
vaisseau 9522
  annelé 548
  de bois 9182
  capillaire 1392
  chylifère 1730
  laticifère 5052
  lymphatique 5280
  ponctué 6964
  réticulé 7832
  sanguin 1200
  scalariforme 7997
  spiralé 4143
valence écologique 2768
valeur d'adaptation 146
  adaptative 146
  calorique 1356
  excitatoire 3189
  génétique 1283
  liminaire 9096
  du métabolisme de base 954
  nutritive 6262
  osmotique 6422
  (du) pH 6788
  protéique brute 2153
  sélective 8103
  de seuil 9096
  de survie 8831
valine 9439
valvaire 9440
valve 9441
valvule cardiaque 4120
  mitrale 5720
  veineuse 9498
vapeur d'eau 9662
variabilité 9442
  continue 2018
  discontinue 2600
  qualitative 2600
  quantitative 2018
variable 9443
variance 9444
variante 9445
variation 9446
  des caractères 1802
  clinale 1802
varier 9451
variété 9449
vasculaire 9452
vascularisation 9462
vascularisé 9463
vase de filtrage 8212
vasoconstriction 9464
vasodilatation 9465
vasopressine 9467
vecteur 9468
végétal 6984, 9470
  inférieur 5256
  submergé 8741
  supérieur 4276
  unicellulaire 7406
  vasculaire 9458
végétatif 9480
végétation 9472
veination alaire 9686
veine 9490
  cave 1490
  porte 7206
  vitelline 9598
veineux 9496
velamen 9491
vénation 6038
vénation foliaire 5091
vénéneux 9167
venimeux 9167
venimosité 9170
venin 7075
  d'abeille 992
ventouse 8769
ventral 2
ventre 1
ventricide 9504
ventricule cérébral 1584
  (du cœur) 9505
  succenturié 7418
vérification expérimentale 3233
vermiforme 9506
vernalisation 9508
vernation 9509
vers annelés 540
  plats 7045
vert estival 8787
  hivernal 9694
vertébral 9511
vertèbre 9510
  caudale 1482
  cervicale 1592
  coccygienne 1838
  dorsale 9084
  lombaire 5260
  sacrée 7942
  thoracique 9084
vertébrés 9517
verticille 9518
verticillé 9520
vésiculaire 9521
vésicule auditive 809
  biliaire 3638
  cérébrale 1250
  germinative 3803
  ophtalmique 6364
  optique 6364
  otique 809
  de pinocytose 6954
  pulmonaire 7454
  séminale 8135
  synaptique 8869
vessie gazeuse 8847
  natatoire 8847
  urinaire 9411
vestibule 9524
viabilité 9526
viable 9527
vibrion 9528
vicariant 9529
vide 9435
vie animale 514
  à ~ courte 8266
  à ~ longue 5241
  moyenne 5438
vieillir 239
vieillissement 8144
vigueur hybride 4245
villosité 9530
  choriale 1684
  intestinale 4797
virescence 9534
virogénétique 9533
virologie 9538
virologique 9536
virologiste 9537
virologue 9537
virose 9545
virulence 9540
virulent 9541
virus 9543
vis micrométrique 5616
viscéral 9548
viscères 9547
viscérosquelette 9554
viscosité 9555
vision colorée 1908
  des couleurs 1908
  des formes 3562
visqueux 9556
vitalisme 9575
vitaliste 9576
vitalité 9577
vitamine 9578
  A 896
  antihémorragique 6879
  antinévritique 9070
  antipellagreuse 6117
  d'antiperméabilité 610
  antirachitique 1346
  antiscorbutique 745
  d'antistérilité 9145
  antixérophtalmique 869
  B_1 9070
  B_2 7868
  B_6 7500
  B_{12} 1835
  C 745
  D_2 1346
  E 9145
  de fertilité 9145
  H 1117
  K 6879
  P 610
  PP 6117
vitellarium 9743
vitelline 9595
vitelloducte 9742
vitellogenèse 9599
vitellus 9740
vitesse de croissance 3988
  de diffusion 2534
  d'évolution 2465
  de germination 8454
  de multiplication 5889
  de réaction 7619
  de recyclage 9315
  de sédimentation 8069
vivipare 9604
viviparité 9603
vivisection 9605
voie de conduction 6652
  en ~ d'extinction 991
  métabolique 5534
  nerveuse 6053
  pyramidale 7497
  réflexe 7676
voies aériennes 7805
  génitales 8223
  respiratoires 7805
  urinaires 9412
voile du palais 6542
  palatin 6542
  de racine 9491
voilier 1387
vol nuptial 6240
volatilisation 9609
volatiliser, se 9610
volume courant 9116
  -minute du cœur 4119
  des pores 7201
  de réserve expiratoire 7774
  de réserve inspiratoire 1937
  respiratoire par minute 7455
  sanguin 1201
volutine 9613
vomer 9614

vrille 8985
  foliaire 5087
vulve 9615

X

xanthine 9711
xanthophylle 9712
xénie 9714
xénogame 9715
xénogamie 9716
xéromorphe 9717
xéromorphie 9718
xérophile 9719
xérophyte 9720
xylème 9728
xylol 9729
xylophage 9730
xylose 9731

Z

zéatine 9748
zoïdiogame 9750
zoïdiogamie 9752
zonation 9755
zone d'abscission 24
  d'allongement 2856
  aphotique 635
  archibenthique 693
  bathyale 988
  bathypélagique 989
  benthique 1007
  biologique 5144
  climatique 1795
  crépusculaire 2741
  de croissance 3971
  de détente 5105
  de division 2655
  dysphotique 2741
  d'élongation 2856
  épipélagique 3076
  euphotique 3141
  floristique 3476
  germinative 3794
  d'habitat 4327
  d'hivernage 9700
  d'hybridation 9756
  hybride 4419
  inter(co)tidale 4789
  limnétique 5177
  de loisir(s) 5105
  des marées 4789
  de nidification 1270
  nucléolaire 6221
  partielle de répartition 2610
  pélagique 6683
  pellucide 9753
  photique 6841
  pilifère 6942
  profonde 7323
  de reproduction 1270
  résidentielle 7778
  SAT 6221
  de séparation 24
  sublittorale 8740
  supralittorale 8816
  de végétation 9477
  verte 3951
zoocénotique 525
zoochlorelle 9757
zoochorie 9758
zooflagellés 9759
zoogamète 9760
zoogénétique 9761
zoogéographie 9762
zooglée 9763
zooïde 9764
zoologie 9767
zoologique 9765
zoologiste 9766
zoologue 9766
zooparasite 9768
zoophage 9769
zoophysiologie 518
zoophyte 9770
zooplancton 9771
zoosporange 9772
zoospore 9773
zoostérol 9774
zootechnicien 503
zootechnie 9775
zootomie 9776
zootoxine 9777
zooxanthelle 9778
zwittérion 9779
zygapophyse 9780
zygoma 9781
zygomorphe 9783
zygomorphie 9784
zygose 9785
zygosome 9786
zygospore 9787
zygote 9788
zygotène 9790
zygotique 9791
zymase 9792
zymogène 9793
zymologie 9795

ESPAÑOL
Índice Alfabético

A

abastecimiento de agua 9660
  de agua potable 2705
abayado 888
abdomen 1
abdominal 2
abeja madre 7520
  reina 7520
aberración 9
  cromatídica 1692
  cromosómica 1714
abertura branquial 3837
  cloacal 1806
  genital 3750
  nasal posterior 1649
abiogénesis 8537
abiogenético 12
abiótico 13
abisal 38
abísico 38
abismal 38
ablandar *(agua)* 8351
abomaso 17
abotonar 1312
abscisina 21
abscisión 22
  de los frutos 3607
  de las hojas 5067
absorbente 27, 28
absorber 26
absorción 29
  de agua 9629
  foliar 5068
  de oxígeno 6495
  ultravioleta 9348
abundancia 37
acantocárpico 42
acantocarpo 42
acantocéfalos 43
acario 6164
acaule 44
acción amortiguadora 1320
  aspirante 8776
  excitante 8644
  génica 3697
  hormonal 4384
  inhibidora 4677
  de masas 5391
  muscular 5904
  refleja 7670
  secretoria 8061
  -tampón 1320
aceite esencial 3115
  etéreo 3115
  volátil 3115
aceleración 45
acelular 56
acéntrico 57
acept(ad)or de hidrógeno 4436
aceptor 46
acetábulo 58
acetaldehído 59
acetilación 65
acetilcolina 66
acetona 64
acíclico 130
acícula 76
acicular 77
acidez 91
acidificación 89
acidificar 90
ácido acético 60
  acetoacético 62
ácido adenílico 156
  aldehídico 291
  algínico 300
  aminado 395
  aminoacético 3891
  aminobenzoico 399
  ascórbico 745
  aspártico 752
  biliar 1026
  butírico 1334
  cafeico 1341
  carbónico 1411
  carboxílico 1417
  cetónico 4955
  cevitámico 745
  cinámico 1746
  citidílico 2226
  cítrico 1761
  clorhídrico 4430
  desoxirribonucleico 2399
  diamínico 2485
  diaminopimélico 2486
  esteárico 8609
  fólico 3511
  fórmico 3565
  fosfatídico 6829
  fosfoglicérico 6832
  fosfórico 6836
  fumárico 3619
  giberélico 3823
  glicérico 3889
  glicólico 3895
  glucónico 3875
  glutámico 3882
  glutárico 3884
  graso 3326
  guanílico 3999
  hialurónico 4411
  homogentísico 4349
  homogentisínico 4349
  húmico 4403
  indol(il)acético 4637
  indol-3-acético 4638
  isocítrico 4863
  láctico 5006
  linoleico 5192
  linolénico 5193
  lipónico 9080
  lisérgico 5287
  maleico 5343
  málico 5346
  malónico 5348
  mevalónico 5577
  murámico 5893
  nicotínico 6121
  nitroso 6143
  nucleico 6214
  oleico 6306
  orótico 6401
  ósmico 6409
  oxalacético 6479
  oxálico 6480
  palmítico 6557
  pantoténico 6569
  péctico 6665
  penicilínico 6696
  pirúvico 7506
  propiónico 7349
  ribonucleico 7870
  salicílico 7948
  silícico 8284
  siquímico 8260
  succínico 8765
  tánico 8925
  tartárico 8934
ácido timidílico 9105
  timonucleico 9107
  tioctánico 9080
  tióctico 9080
  úrico 9409
  uridílico 9410
  urónico 9418
acidófilo 93
ácidorresistencia 86
ácidorresistente 83
acigoto 886
acilación 131
acinesia 273
acinético 274
aclamídeo 69
aclarado 1775
aclarador 1776
aclaramiento 1775
aclarante 1776
aclarar 1773
aclimatación 53
aclimatar 54
acomodación 55
acopado 2195
acoplamiento 5411
acoplar 5408
acranios 98
acrocárpico 99
acrocéntrico 100
acrógino 101
acromasia 71
acromático 72
acromatina 74
acrómico 1910
acropetal 102
acrosoma 103
ACTH 181
actina 105
actinomiceto 109
actinomicina 110
actinomorfia 108
actinomorfo 107
actinopterigios 111
actinozoos 572
actitud de distracción 2297
  de galanteo 2104
  de sumisión 8745
activación 121
  de los aminoácidos 396
  del gen 3698
activador 123
activar 117
actividad del cámbium 1370
  cardíaca 1422
  de desplazamiento 2621
  enzimática 3030
  de los genes 3699
  hormonal 4384
  metabólica 5526
  nerviosa 6055
  nidificadora 6063
  reorientada 7661
  vital 9567
acto consumatorio 2007
  instintivo 4724
  en el vacío 9436
actomiosina 129
acuático 670
acuícola 670
acuidad auditiva 797
acumulación de materiales
  alimenticios 3541
acuoso 674
adaptabilidad 132
adaptable 133

adaptación 134
  al (medio) ambiente 3006
  simultánea 8295
adaptado 135
  al medio ambiente 136
adaptativo 139
adaptividad 132
adaptivo 139
adelfogamia 150
adenina 151
adenocito 3852
adenohipófisis 153
adenoide 154
adenosina difosfato 155
  monofosfato 156
  trifosfato 157
adenoso 3858
adermina 7500
adherencia 161
adhesión 161
adhesividad 164
adiestramiento 9191
adiposo 165
adiuretina 9467
adjutor 170
ADN 2399
adolescencia 171
adorno vistoso (durante el
  cortejo) 2103
ADP 155
adrenalina 179
adrenérgico 180
adrenina 179
adrenocorticotropina 181
adrenosterona 182
adsorbente 184
adsorber 183
adsorción 185
aductor 149
adultez 191
adulto 188, 189
aerénquima 198
aerobio 202, 203
aerobiología 204
aerobiosis 205
aerófito 206
aeroplancton 207
aerotaxis 208
aerotropismo 209
aferencia 213
aferente 214
afilia 637
áfilo 636
afinidad 215
aforestación 217
aforestar 216
agalla 1497, 3827
agámeto 225
agámico 226
ágamo 226
agamobio 227
agamoespecie 231
agamogénesis 228
agamogonia 229
agamonte 230
agamospermia 232
agar(-agar) 233
  de cultivo 6246
  nutritivo 6246
agenesia 245
agénesis 245
agente 246
  aclarante 1776
  activador 123
  catalizador 1467
agente
  contaminante 2010, 7114
  de(s)contaminador 2314
  gelatinizante 3687
  gelificante 3687
  inductivo 4645
  mutagénico 5914
  oxidante 6482
  patógeno 6654
  polinizador 7108
  de polución 7114
  de quelación 1616
  quelante 1616
  reductor 7663
agentes radiomiméticos 7577
agitador giratorio 4030
aglutinable 247
aglutinación 250
  ácida 88
aglutinado 248
aglutinar 249
aglutinina 252
aglutinógeno 253
agmatoploide 258
agmatoploidia 259
agnatos 260
agotamiento del suelo 8358
agregado 254
agrobiología 262
agrupación 3968
agua blanda 8350
  capilar 1394
  de capilaridad 1394
  corriente 3494
  dulce 3585
  dura 4100
  estancada 8575
  fluente 3494
  del fondo del mar 8042
  de gravitación 3947
  gravitacional 3947
  de imbibición 9646
  intersticial 4788
  de manantial 8564
  de mar 8043
  potable 2704
  salobre 1241
aguas freáticas 3964
  residuales 8210
  subterráneas 3964
  superficiales 8824
agudeza auditiva 797
  visual 9558
aguijón 7287, 8660
  venenoso 9495
aguijonado 7289
agujero vertebral 9516
ahijamiento 9120
ahijar 9118
ahilamiento 3132
AIA 4638
aire alveolar 370
  complementario 1937
  espiratorio 3245
  inspirado 4721
  inspiratorio 4721
  de reserva 7774
  residual 7779
  suplementario 7774
aislamiento 4878
  biogenético 1070
  ecológico 2763
  espacial 8419
  de (las) especies 8437
  fenológico 6812
aislamiento
  genocitoplásmico 3704
  geográfico 3774
  de líneas 8178
  reproductivo 7758
  topográfico 9156
aislar 4876
ala 9685
alado 280
alaestético 313
alanina 275
alantoides 316
alantoína 315
alargamiento celular 1510
albinismo 281
albino 282
albumen 283, 284
albúmina 285
  de huevo 6446
  sérica 8199
  vegetal 6895
albuminoide 286
albura 7980
alcalimetría 309
alcalímetro 308
alcalinidad 311
alcalino 310
alcaloide 312
alcohol etílico 3130
aldehído 290
  acético 59
  fórmico 3563
aldosterona 292
alecítico 293
alecito 293
alélico 318
alelismo 319
  múltiple 5883
alelo 317
alelomórfico 318
alelomorfismo 319
alelomorfo 317
alelos múltiples 5882
alelotaxia 322
alelotaxis 322
alelotipo 323
alergeno 324
alergia 326
alérgico 325
alesnado 8760
aleta 3415
  abdominal 9501
  adiposa 166
  anal 450
  caudal 1481
  dorsal 2682
  pectoral 6668
  pelviana 9501
aleurona 294
aleuroplasto 7387
alevín 3421
alga 298
algas azules 2210
  filamentosas 3405
  pardas 6789
  pardo-doradas 1728
  rojas 7861
  verdeazules 2210
  verdes 1647
algesiorreceptor 6512
algología 301
algonquino 302
alianza 327
alimentación 6253
  de las crías 3346

almacenamiento de agua 9658
de alimento 3541
energético 2971
de energía 2971
almidón animal 3892
de asimilación 761
de reserva 7776
alobiosis 328
alocarpia 329
alociclia 334
alocigoto 361
alocroico 330
alocroismo 331
alocromasia 331
alócrono 332
alóctono 333
alodiploide 335
alodiploidia 336
alogamia 338
alógamo 337
alogenético 340
alógeno 339, 341
alohaploide 342
alohaploidia 343
aloheteroploide 344
aloheteroploidia 345
aloiogénesis 346
alolisogénico 347
alometría 348
alopátrico 349
aloplasma 350
alopoliploide 351
alopoliploidia 352
segmental 8081
alosíndesis 356
alosindético 357
alosoma 4209
alosómico 354
alotetraploide 358
alotetraploidia 359
alotriploide 360
alteración estructural 8709
del medio ambiente 3007
alternación de generaciones 365
alternado *(bot.)* 362
alternancia de generaciones 365
alvéolo 374
dentario 2386
pulmonar 7454
amacollado 9120
amacollar 9118
amaestramiento 9191
ambiental 3005
ambisexual 377
ámbito del hogar 4327
ambivalente 378
amboceptor 379
ambulacro 381
ameba 409
amebocito 410
ameboide 411
ameiosis 383
ameiótico 384
amentifloras 386
amento 1480
ametábolo 388
ametábolos 669
amiba 409
amicrón 389
amicroscópico 390
amida 391
de ácido 80
amielínico 434
amígdala faríngea 6799
palatina 6540
amiláceo 437
amilasa 436
salivar 7445
amilífero 437
amilopectina 438
amiloplasto 439
amilosa 440
amina 393
aminación 392
aminoácido 395
alifático 307
esencial 3114
no esencial 6159
-oxidasa 398
amitosis 401
amnios 403
amniotas 404
amniótico 405
amoniaco 402
amorfo 413
amortiguación 1326
amortiguador 1318
amortiguar 1319
AMP 156
amplexicaule 433
anabiosis 441
anabolia 444
anabólico 442
anabolismo 443
anádromo 445
anaerobio 446, 447
anaerobiosis 448
anáfase 461
anafilaxis 464
anaforesis 463
anagénesis 449
análisis 458
del comportamiento 998
espectral 8450
de factores 3296
factorial 3296
genético 3735
de población 7184
polínico 7097
roentgenográfico 9723
de la recombinación 7649
de sangre 1197
de (las) tétradas 9026
por titulación 9143
de varianza 459
analizador 457
analizar 456
analogía 455
analógico 453
análogo 453, 454
de base 957
anamnios 460
anamniotas 460
anastomosar 465
anastomosis 466
anatomía 469
vegetal 6986
anatomista 468
anatonosis 470
anátropo 471
androceo 474
androdioecia 473
androgénesis 476
andrógeno 475
androginia 478
andrógino 477
androhermafrodita 479
andromerogonia 480
androsoma 481
andróspora 5648
androsporangio 5647
androsterona 484
anelectrotono 485
anélidos 540
anemocoria 486
anemofilia 490
anemófilo 488
anemogamia 490
anemógamo 488
anemotactismo 491
anemotaxis 491
anestro 551
aneuploide 492
aneuploidia 493
aneurina 9070
anfiaster 416
anfibio 419
anfibios 418
anficario 424
anficárpico 420
anfidiploide 358
anfidiploidia 359
anfigastros 422
anfigonia 423
anfimíctico 425
anfimixia 426
anfimixis 426
anfimutación 427
anfineuros 428
anfiploide 429
anfiteno 430
anfitoquía 431
anfogénico 432
angiología 495
angiospermas 497
angiospermo 496
angiospermos 497
ángulo de divergencia 498
angustifolio 499
anhidrasa carbónica 1412
anhídrido 500
carbónico 1407
anidación 6068
anidar 6061
anillado 1134
anillamiento 1134
anillar 7880
anillo anual 544
de Balbiani 940
de crecimiento 3992
nervioso 6050
pirrólico 7504
animal acuático 671
conductor 5064
control 2025
dañino 6777
diurno 2647
doméstico 2662
dulciacuícola 3586
exótico 3223
de experimentación 3228
experimental 3228
hibernante 4272
huésped 514
joven 9746
de laboratorio 4989
macrosmático 5321
marino 5373
microsmático 5641
nocivo 6777
nocturno 6146
parásito 9768
de sangre caliente 9619
de sangre fría 1867
superior 4275

animal terrestre 9003
terrícola 8757
testigo 2025
útil 9421
venenoso 7080
animales coralíneos 572
anion 531
anisofilia 535
anisogameto 532
anisogamia 533
anisogenia 534
anisoploidia 536
anisotropia 538
anisotrópico 537
ano 629
ánodo 549
anomalía 553
anormal 15, 552
anormalidad 16
anortogénesis 554
anoxibionte 446
anoxibiosis 448
antagonismo 556
de iones 4831
iónico 4831
antagonista 557
metabólico 5527
antecesor 472
antena 558
antenados 9187
anténula 560
antepasado 472
antera 564
anteridio 565
anterozoide 566
antesis 567
antiauxina 585
antibionte 586
antibiosis 587
antibiótico 588
anticímico 625
anticlinal 590
anticodon 591
anticuerpo 589
antídoto 593
antidrómico 594
antienzima 595
antifertilizina 596
antigenicidad 599
antígeno 597
de grupo sanguíneo 1180
antigiberelina 600
antihormona 602
antimetabolito 603
antimorfo 604
antimutagen 605
antípoda 611
antiséptico 616
antisuero 617
antitóxico 620
antitoxina 621
antitripsina 622
antitrombina 619
antivitamina 623
antocianidina 569
antocianina 568
antofilo 3473
antófitos 8472
antoxantina 571
antozoarios 572
antozoos 572
antropobiología 573
antropocoria 574
antropogénesis 575
antropogenia 577
antropogénico 576
antropogeografía 578
antropografía 579
antropoides 580
antropología 581
biológica 573
antropometría 582
antropomorfismo 583
antropomorfología 584
anual 541, 543
estival 8784
hibernante 9690
invernal 9690
anucleado 6164
anular 545
anuro 628
anuros 627
aorta 630
aovar 6464
aparato acuífero 9663
genital 3756
genitourinario 9417
de Golgi 3905
locomotor 5228
mitótico 5714
ovular 2801
parabasal 6575
de reclamo 8272
apareamiento 5411
al azar 7597
de bases 961
de cromosomas 8864
aparear 5408
aparearse *(cromosomas)* 6514
apego al lugar 6820
apéndice 662
vermiforme 9507
apertura del pico 3666
apétalo 632
ápice 633
foliar 5088
radical 7911
del tallo 8612
del vástago 8612
aplanóspora 643
aplasia 644
apocarpia 646
apocárpico 645
apocarpo 645
ápodo 648
apoenzima 649
apofermento 649
apófisis 655
articular 9780
espinosa 6081
mastoides 5406
transversa 2492
apogamia 650
apomeyosis 651
apomíctico 652
apomixia 653
apomixis 653
aponeurosis 654
aporogamia 655
aportación calórica 1359
aporte alimenticio 3543
de oxígeno 6502
aposición 665
aposporia 658
apotecio 659
aprendizaje 5097
asociativo 768
latente 5045
apresamiento 7286
aprovechamiento de las aguas
residuales 8213
de (los) desperdicios 9627
apterigoto 668
apterigotos 669
áptero 668
aptitud adaptiva 132
de combinación 1915
combinatoria 1915
locomotora 5229
apto para la fecundación 3366
para la reproducción 7744
aquenio 68
arácnidos 678
aracnoides 679
aracnología 680
árbol acicufolio 1983
deciduo 2303
filogenético 6883
genealógico 6883
de hojas caducas 2303
arbóreo 682
arborescente 681
arborícola 682
arboriforme 681
arborización 683
terminal 8992
arbustivo 3614
arbusto 8273
arcaico 684
arcalaxis 686
arco aórtico 631
branquial 3828
cigomático 9781
hemal 4040
hioideo 4476
mandibular 5359
neural 6074
reflejo 7671
vertebral 9512
visceral 9549
área de asociación 766
continua 2017
de cría 1270
discontinua 2599
de distribución 7600
disyunta 2599
embrionaria 3794
de especies 8431
germinativa 3794
de invernación 9700
mínima 5702
de muestra 7968
de nidificación 6067
opaca 697
pelúcida 698
polar 7095
de repartición 7600
de residencia 7778
transparente 698
areolado 700
arginasa 701
arginina 702
arilo 705
arista 866
aristado 707
ARN 7870
m 5522
matriz 5522
mensajero 5522
ribosomático 7874
soluble 9201
de transferencia 9201
aromático 708
aromatización 709

aromatizar 710
aromorfosis 711
arquegoniadas 687
arquegonio 688
arquencéfalo 689
arquenterio 690
arquenterón 690
arqueozoico 685
arquesporio 691
arquetipo 692
arquiblasto 694
arquipalio 695
arquiplasma 696
arrecife de coral 2044
  coralino 2044
arrenotoquía 715
arrosariado 7916
arrosetado 7916
artefacto 716
arterenol 6179
arteria 721
  carótida 1436
  coronaria 2063
arterial 719
arteriola 720
articulación 732
  de la cadera 4285
  condílea 2854
  coxofemoral 4285
  en charnela 4281
  elipsoidea 2854
  escapulohumeral 8271
  sellar 7944
articulados 729
articular 725
artiodáctilos 735
artioploidia 736
artrología 722
artrópodos 723
artróspora 724
asca 747
ascendencia 738
ascendiente 472
ascensión del agua 9652
  de la savia 741
asco 747
ascogonio 743
ascomicetos 744
ascospora 746
aserrado 8197
asexuado 749
asexual 749
asilvestrar 7931
asimilable 754
asimilación 757
  del carbono 1403
  del dióxido de carbono 1408
  genética 3736
asimilar 755
asinapsis 774
asindesis 774
asociación celular 1500
  migratoria 5673
  de reemplazo 8752
  (vegetal) 765
asociado 764
asparagina 751
asperifolio 753
asquelmintos 6010
asta 4388
áster 769
astrocito 772
astroesfera 773
atávico 776
atavismo 775
atoxicidad 6176
atóxico 6175
ATP 157
atracción social 8339
atractivo sexual 8239
atrayente 792
  sexual 8239
atrio 789
atriquio 788
atrofia 790
átropo 791
audición 795
auditivo 796
aurícula 810
autecología 811
autoadaptación 812
autoaloploidia 813
autoantígeno 814
autobivalente 815
autocatálisis 816
autoclave 819
autocompatibilidad 8111
autocoria 817
autóctono 818
autodepuración 8116
autodiferenciación 8106
autodigestión 8107
autoduplicación 8109
autoestéril 8121
autoesterilidad 8122
autofecundación 8112
autofecundante 8113
autofértil 8110
autofertilidad 8111
autofertilización 8112
autofertilizador 8113
autofertilizante 8113
autogamia 8112, 8115
autógamo 821
autogénesis 823
autogénico 824
autógeno 825
autoheteroploidia 827
autoico 820
autoincompatibilidad 8114
autoinjerto 848
autoinmunización 828
autolisado 829
  de levadura 9735
autolisis 830
automatismo 831
automixis 832
automutagen 833
autonomía 835
autoorientación 836
autopartenogénesis 837
autoplastia 849
autoplástico 838
autopolinización 8115
autopoliploidia 840
autopurificación 8116
autor de la polución 7116
autorradiografía 841
autorregeneración 842
autorregulación 8117
autorreplicación 8109
autorreproducción 8119
autorreproducible 8120
autorreproductor 8120
autosíndesis 844
autosíntesis 845
autosoma 843
autotomía 846
autotoxina 847
autotrasplante 849
autotrofia 850
autotrófico 851
autótrofo 851
autotropismo 852
auxanografía 854
auxina 856
auxocito 857
auxóspora 858
auxótrofo 859
ave altricial 6125
  nidícola 6125
  nidífuga 6126
  mientras está en el nido 6070
  migratoria 5671
  migrante 5671
  de presa 1130
  que anida en cavidades 1492
  que anida en el suelo 3961
  rapaz 1130
  de rapiña 1130
aves que anidan en colonias 1135
avifauna 861
avirulento 862
avitaminosis 863
axeroftol 869
axila de la hoja 5069
axilar 875
axilema 874
axocele 878
axolema 874
axón 879
  gigante 3819
axoplasma 882
axostilo 883
azoico 884
azotación 885
azúcar de caña 7940
  de fruto 3601
  de leche 5013
  de malta 5353
  sanguíneo 1195

B

baccífero 887
bacciforme 888
bacilar 890
bacilariofíceas 2499
bacilicultura 891
baciliforme 892
bacilo 893
  coma 9528
bacteria 931
  donadora 2673
  esférica 1837
  receptora 7640
bacterial 903
bacteriano 903
bacterias acéticas 63
  (ácido)lácticas 5007
  de azufre 9079
  desnitrificantes 2378
  edáficas 8353
  fermentativas 3359
  fijadoras de nitrógeno 6139
  fotógenas 5263
  lácticas 5007
  luminosas 5263
  nitrificantes 6132
  nodulares 6150
  de las nudosidades 6150
  de putrefacción 7490
  putrificantes 7490

C

Español

capaz de germinar 3809
de regeneración 1389
de reproducirse 7744
de volar 1387
capilar 1391, 1392
capítulo 1395
cápsula 1397
articular 726
auditiva 799
bacteriana 904
craneal 2114
dehiscente 2353
esporífera 8543
ótica 799
polar 7085
porífera 7200
seminal 8073
sinovial 726
suprarrenal 175
urticante 6012
captura del alimento 3528
de la presa 7286
capullo 1841
caracol *(anat.)* 1840
carácter 1605
de adaptación 140
adaptivo 140
adquirido 96
atávico 9101
cualitativo 7511
cuantitativo 1163
hereditario 4186
independiente 4626
mezclado 1163
de organización 8711
sexual 8242
sexual secundario 8055
característica 1605
adquirida 96
comportamental 999
etológica 999
fenotípica 6816
genotípica 3767
característico de la especie 8446
carbohidrato 1399
carbonífero 1413
carbono 1402
radiactivo 7564
carboxilación 1416
carboxilasa 1415
carcinógeno 1418
carcinología 1419
cardíaco 1421
cardias 1420
cardumen 8015
carencia en sustancias nutritivas 6249
de vitaminas 9592
carga genética 3743
cariocinesis 5711
cariogamia 4927
cariogen 4928
cariogénesis 4929
cariograma 4533
cariolinfa 6207
cariólisis 4934
cariología 4932
cariómero 4935
carioplasma 6225
cariópside 1451
cariopsis 1451
cariorrexis 4939
cariosoma 4940
cariostasis 4941
carioteca 6202
cariotina 4943
cariotipo 4944
carnicero 1431
carniceros 1429
carnívoro 1430, 1431
carnívoros 1429
carnoso 3452
carofíceas 1610
carofitos 1610
caroteno 1434
carotenoide 1435
carótida 1436
carotina 1434
carotinasa 1433
carotinoide 1435
carozo 8679
carpelo 1438
carpo 1443
carpóforo 1440
carpogonio 1439
carpopodio 1440
carpospora 1441
carposporófito 1442
cartilaginoso 1447
cartílago 1445
anular 2128
articular 727
aritenoides 737
artrodial 727
de conjunción 3078
costal 2089
cricoides 2128
epifisario 3078
fibroso 3388
hialino 4409
precursor 7230
tiroides 9109
cáscara del huevo 2807
casco 4378
caseína 1452
casmófitos 1612
casmogamia 1611
casquete nuclear 6191
polar 7084
radical 7901
casta 1454
castración 1457
castrado 1455
castrar 1456
(hembra) 8428
catabolismo 1459
proteico 7371
catabolito 1460
catádromo 1461
catafase 4946
catafilo 1469
cataforesis 1468
catalasa 1462
catalepsia 1463
catálisis 1464
biológica 1049
catalítico 1465
catalizador 1467
catalizar 1466
catástrofe ecológica 3013
catastrofismo 9040
catecolamina 1471
catelectrótono 1472
catenación 1473
catepsina 1474
catequina 1470
catión 1478
cátodo 1475
caudillo 5064
caulescente 1483
caulifloria 1485
caulífloro 1484
caulinar 1486
caulino 1486
caulocarpo 1488
cavernícola 1493
cavidad abdominal 3
amniótica 406
articular 728
branquial 3831
celomática 1852
coriónica 1682
corporal 1206
cotiloidea 2093
faríngea 6798
gástrica 3670
gastrovascular 3677
intestinal 4012
del manto 5363
medular 5449
nasal 5978
paleal 5363
pelviana 6690
pericardíaca 6720
peritoneal 3
pleural 9082
pulpar 7461
secretora 8063
de segmentación 1148
torácica 9082
uterina 9424
cecidio 1497
cefálico 1570
cefalina 1572
cefalización 1573
cefalocordados 98
cefalópodos 1575
cefalotórax 1576
celentéreos 1850
celenterio 1851
celenterón 1851
celo 6297
estar en ~ 6298
celobiosa 1535
celoma 1852
celomiducto 1855
célula 1499
acompañante 1924
adiposa 5200
anexa 1924
animal 505
antípoda 611
apical 639
auditiva 800
auxiliar 855
bacteriana 905
basal 951
caliciforme 3903
cambial 1371
cartilaginosa 1671
cebada 5400
en cesta 977
de cierre 4000
ciliada 1741
en collar 1651
corporal 8393
cribosa 8276
cuentaglóbulos 2097
dadora 2674
dentada 7288
donadora 2674
donante 2674
embrional 2867
en empalizada 6547
endotelial 2950

célula epidérmica 3060
epitelial 3090
errante 9617
espermática 8456
espinosa 7288
estomática 4000
flagelada 3439
flamígera 3444
folicular 3517
ganglionar 3664
genitora 6618
germinal 3783
germinativa 3783
gigante 3820
glandular 3852
glial 3861
grasa 5200
de guarda 4000
gustativa 4005
hija 2277
huésped 4393
huevo 6478
inicial 4683
intercalar 4770
internodal 4771
intersticial 4785
laticífera 5053
liberiana 981
límite 4210
llama 3444
madre 5825
madre de esporas 8546
madre del polen 7101
madre del saco embrional 2869
medular 5450
mucífera 5859
muscular 5897
nerviosa 6040
nerviosa sensorial 8173
neuroglial 3861
nutritiva 6255
oclusiva 4000
ósea 6432
parenquimática 6616
de paso 6644
pétrea 8027
pigmentaria 6933
piramidal 7496
plasmática 7016
progenitora 6618
de los radios medulares 5453
receptora 7634
de recuento 2097
en reposo 7817
reproductora 7754
sanguínea 4055
de Schwann 8017
secretora 8064
segmentada 1781
sensitiva 8168
sensorial 8168
sexual 3643
somática 8393
de sostén 8811
suberosa 2050
subsidiaria 855
del tapete 8927
terminal 2900
del tubo criboso 8279
urticante 1820
vegetal 6989
visual 9559
vitelina 9741
celular 1537
células eucarióticas 3138
celulosa 1541
ceniza ósea 1217
cenobio 1856
cenocito 1857
cenogameto 1858
cenogénesis 1339
cenospecie 1859
cenozoico 1546
céntrico 1552
centrífuga 1557
centrifugación 1555
diferencial 2518
fraccionada 3574
sobre gradientes de densidades 2382
centrifugadora 1557
centrifugar 1556
centrífugo 1554
centríolo 1558
centrípeto 1559
centro activo 125
de asociación 767
calorífico 4124
coordinador 2033
genético 3701
germinal 3784
del lenguaje 8453
nervioso 6041
óptico 9560
de organización 6380
de reacción 7616
reflejo 7672
respiratorio 7791
termogénico 4124
termorregulador 9064
visual 9560
centrocromatina 1560
centrodesma 1561
centrogen 1562
centrolecítico 1563
centrolecito 1563
centrómera 1564
centrómero 1564
centroplasma 1567
centrosfera 1569
centrosoma 1568
cepa 8689
de bacterias 912
mutante 5922
salvaje 3401
de virus 9546
cerda 1287
sensitiva 8167
cerebelar 1577
cerebelo 1578
cerebral 1579
cerebralización 1585
cerebro 1588
anterior 7353
intermedio 2517
medio 5502
posterior 5569
terminal 8955
cerebrósido 1586
ceremonia de instigación 2622
certación 1589
certificado de ascendencia 6673
cerumen 1590
cervical 1591
cesación del crecimiento 714
del desarrollo 713
cese del crecimiento 714
del desarrollo 713
cesta branquial 3829
cestodos 1593
cetoácido 4955
cetona 4954
cetosa 4956
cianina 2208
cianofíceas 2210
cianofitos 2210
ciático 8019
ciatio 2211
cibernética 2212
cicatriz foliar 5082
cíclico 2213
ciclo (del ácido) cítrico 4982
del ácido tricarboxílico 4982
alimentario 3524
alimenticio 3524
del azufre 8782
biogeoquímico 1073
del carbono 1406
cardíaco 1423
del celo 6295
estral 6295
del estro 6295
evolutivo 3162
del glioxalato 3900
glioxílico 3900
de Krebs 4982
de la materia 1752
mitótico 5715
del nitrógeno 6136
de la ornitina 6394
ovárico 6447
parasexual 6597
reproductor 7756
"ruptura-fusión-puente" 1260
sexual 8218
de la urea 6394
de vida 5137
vital 5137
ciclomorfosis 2215
ciclosis 2216
ciclóstomos 2217
ciego 1338
cieno activado 118
cifra de fertilidad 3364
cigapófisis 9780
cigoma 9781
cigomorfismo 9784
cigomorfo 9783
cigosis 9785
cigospora 9787
cigóteno 9790
cigótico 9791
cigoto 9788
ciliación 1743
ciliado 1740
ciliados 1739
cilindro central 1549
vascular 9456
cilindroeje 879
cilio 1744
cima 2218
bípara 2503
helicoidal 1229
unípara 5757
cimasa 9792
cimógeno 9793
cimología 9795
cimoso 2219
cincino 1745
cinesis 4965
cinestesia 4962
cinestético 4963
cinética 4968
cinético 4966
cinetogen 4971

Español

colorear 8576
colorimetría 1900
colorímetro 1899
colquicina 1866
columela 1911
columna espinal 9515
vertebral 9515
combinación 1914
de genes 3702
comedor por filtración 3411
comensal 1917
comensalismo 1918
comisura 1919
compañero sexual 5407
comparación madre-hija 2268
compartimentación 1926
compatibilidad 1927
compensación de dosificación 2687
compensador 1930
competencia 1932
alimentaria 1933
interespecífica 4779
intraespecífica 4803
complejo enzima-substrato 3041
de genes 3703
de Golgi 3905
multienzimático 5873
vitamínico B 9584
complementación 1942
complemento de genes 8206
genético 8206
componente celular 1505
comportamiento 994
agresivo 256
de alimentación 3341
animal 502
de apareamiento 5412
de apetencia 664
apetitivo 664
conflictivo 1978
de construcción del nido 6063
de cópula 2040
durante el cortejo 2102
desplazado 2621
epimelético 6623
de huida 3110
incubador 4614
instintivo 4725
de lucha 3402
migratorio 5670
nutricio 3341
reproductivo 7750
reproductor 7750
sexual 8240
en situaciones de conflicto 1978
social 8340
territorial 9007
composición en especies 8432
comprobación experimental 3233
compuestas 1946
compuesto carburado 1404
nitrogenado 6135
proteico 7373
comunicación 1921
comunidad 1922
béntica 1008
biótica 1059
climácica 1798
clímax 1798
final 1798
nemoral 3556
permanente 6757
reproductora 7748
vegetal 6990
concentración iónica 4838
de hidrogeniones 4441
de iones hidrógeno 4441
salina 7958
conceptáculo 1956
concrescencia 1960
concrescente 1961
concha 1958
calcárea 1343
condicionamiento clásico 1769
instrumental 4727
condiciones ambientales 3010
climáticas 1792
edáficas 8355
de existencia 5136
experimentales 3231
del medio 3010
de vida 5136
cóndilo 1973
occipital 6276
cóndrico 1447
condricties 1660
condrificación 1661
condrina 1662
condrioconto 1663
condrioma 1664
condriomero 1665
condriosfera 1667
condriosoma 5709
condroblasto 1668
condrocito 1671
condroclasto 1669
condrocráneo 1670
condroesqueleto 1675
condrogénesis 1672
condrología 1674
conducción de agua 9631
del estímulo 1969
de la excitación 1967
del impulso 1968
nerviosa 6042
saltatoria 7965
conducta 994
agresiva 256
combativa 3402
copulativa 2040
exploratoria 3248
de galanteo 2102
instintiva 4725
sexual 8240
conductibilidad 1971
calorífica 4125
conductismo 1001
conductista 1002
conductividad 1971
conducto alveolar 371
auditivo (externo, interno) 801
biliar 1027
biliar común 1654
circular 7882
deferente 2328
espinal 8494
excretor(io) 3198
eyaculador 2812
galactóforo 5009
genital 3751
gumífero 4002
hidróforo 8680
lactífero 5009
lagrimal 4998
mesonéfrico 5513
de Müller 5867
nasal 5979
nasolacrimal 5001
neural 6075
conducto oleífero 6300
pétreo 8680
raquídeo 8494
resinoso 7782
torácico 9083
vertebral 8494
de Wolff 5513
conductos semicirculares 8127
sexuales 8223
conduplicada *(vernación)* 1972
conectivo 1990
confección de mapas 5369
de mapas cromosómicos 5370
configuración de fragmento-puente 1284
congénere 1996, 1997
congénito 1979
cónico 1977
conidio 1982
conidióforo 1981
conífera 1983
coníferas 1985
coniforme 1977
conjugación 1988
conjugante 1986
conjuntiva 1989
conjunto cromosómico 1717
cono 1974
de fertilización 3368
retinal 7839
vegetativo 9481
conquiolina 1959
consanguinidad 4589
conservación de la(s) especie(s) 8439
de la naturaleza 1993
del suelo 8356
conservador de la especie 8440
consociación 1995
constancia de los caracteres 1606
celular 1504
de las especies 1998
de forma 3560
formal 3560
numérica de los cromosomas 6237
constante de difusión 2530
de disociación 2633
de sedimentación 1999
constitución corporal 1211
hereditaria 4187
constitutivo 2000
constituyente celular 1505
constricción 2003
nucleolar 6220
secundaria 6220
construcción del nido 6062
consubespecie 2005
consumidor *(ecol.)* 2006
primario 7290
secundario 8051
terciario 9012
consumo de energía 2960
de oxígeno 6497
contacto efáptico 3049
contador de escintilación 8021
(de) Geiger 3680
contaminación 2012
del agua 9649
ambiental 3011
atmosférica 269
hídrica 9649
marina 7118
petrolífera 6303
radiactiva 7553

Español

contaminación de los ríos 7888
sónica 6232
sonora 6232
del suelo 8357
contaminador 7116
contaminante 2010, 7114
contaminar 2011
contenido de agua 9632
en carbono 1405
hídrico 9632
en humedad 5737
nutritivo 6248
en oxígeno 6498
proteico 7374
en proteínas 7374
salino 7950
en sustancias nutritivas 6248
en vitaminas 9591
vitamínico 9591
conteo 2096
continuidad del plasma germinal 2016
contorta *(estivación)* 2019
contracción muscular 5905
contráctil 2021
contractilidad 2024
contraveneno 593
control biológico (de plagas) 1082
del crecimiento 3989
de la respiración 7794
por retroacción 3338
retroactivo 3338
de (los) ruidos 6151
convergencia 2029
conversión 2031
cónyuge 5407
coordinación 2032
coorientación 2034
copa del árbol 9237
copépodos 2035
coposo 3465
coprófago 2036
cópula 2038
copulación 2038
copular(se) 2037
coracoides 2043
corazón branquial 1255
linfático 5276
de la madera 2734
corcho 2048
corchoso 2053
cordados 1678
cordamesodermo 1677
cordón espermático 8462
espinal 8496
nervioso 6043
umbilical 9361
vascular 9459
ventral 9500
co-represor 2045
coriáceo 2046
corimbo 2086
corioalantoides 314
corion 1681, 2418
coripétalo 2480
cormo 2054
cormófitos 2055
córnea 2056
córneo 2058
cornificación 2060
cornificar(se) 2061
coroides 1687
corola 2062
corología 1688
corona dental 2151
corpora cardiaca 2065
corpúsculo 2071
de Barr 8216
central 1568
cromatínico 1696
elemental 2850
de inclusión 4597
de Malpighi(o) 5349
táctil 8910
correlación 2072
corriente de acción 113
estimulante 3186
excitadora 3186
de membrana 5475
plasmática 7409
sanguínea 1176
corte de celoidina 1536
por congelación 3597
delgado 9078
histológico 9136
a matarrasa 1774
en parafina 6582
en serie 8188
a tala rasa 1774
de tejido 9136
transversal 2148
cortejar 2100
cortejo 2101
cortex visual 9561
corteza 7879
adrenal 173
cerebral 1581
motora 5832
suprarrenal 173
corticoide 2079
corticosteroide 2080
corticosterona 2081
corticosuprarrenal 173
corticotropina 181
cortina 2083
cortisol 4433
cortisona 2085
cosmopolita 2088
costilla 7866
asternal 3304
esternal 9290
falsa 3304
flotante 3462
foliar 5081
verdadera 9290
cotila 2093
cotiledón 2094
cotilo 2093
covariación 2105
covarianza 2105
coxa 4282
coxal 8019
CR 7806
craneal 2111
craneano 2111
cráneo 2120
cartilaginoso 1670
craneología 2118
craneometría 2119
craniotas 9517
creación de nuevas variedades 1277
creacionismo 9041
creatina 2122
creatinina 2124
crecimiento 3973
en altura 3981
en anchura 3980
apical 641
por aposición 667
crecimiento demográfico 7193
diametral 3983
diferencial 2520
por elongación 2855
exponencial 3249
en grosor 3983
intercalar 4736
en latitud 3980
de ~ lento 8327
logarítmico 3249
en longitud 3982
longitudinal 3982
de (las) poblaciones 7193
de ~ rápido 3313
terminal 641
crenado 2126
cresta neural 6076
cretáceo 2127
cría 1269
en colonia 1275
experimental 3230
por líneas puras 7481
de selección 8098
selectiva 8098
criador 1268
de plantas 6987
crianza de líneas 5180
por líneas puras 7481
criba 8277
branquial 3839
criodesecación 5285
criófilo 2158
criófito 2159
crioplancton 2160
crioscopia 2161
criostato 2162
cripsis 1377
criptófito 2169
criptógamas 2165
vasculares 9455
criptógamo 2164
criptogenético 2166
criptógeno 2166
criptomería 2167
criptomerismo 2167
criptomitosis 2168
crisálida 7468
crisófitos 1728
criss-crossing 2129
cristalino 2170, 2171
cristalizable 2172
cristalización 2173
cristalizar 2174
cristalografía 2175
por medio de rayos X 9724
cromafín(ico) 1689
cromático 1690
cromátida 1691
hermana 8308
hija 2278
no hermana 6172
cromatina 1695
sexual 8215
cromatóforo 1701
cromatografía 1699
por adsorción 187
en capa fina 9077
en columna 1912
de gases 3668
de intercambio iónico 4833
en (sobre) papel 6571
de partición 6641
cromatografiar 1698
cromatograma 1697
cromatómetro 1700

Ch

D

deficiencia vitamínica 9592
déficit hídrico 9634
  de oxígeno 6501
  de la presión de difusión 2533
defoliación 2337
defoliado 2336
defoliante 2335
defosforilación 2401
degeneración 2340
  consanguínea 4591
degenerar 2338
deglución 2341
degradación 1262
  del medio ambiente 3012
  de variedades 2445
degradador 2309
degradar 1258
dehidrasa 2358
dehidrocorticosterona 2356
dehidroepiandrosterona 2357
dehidrogenación 2359
dehidrogenasa 2358
dehiscencia 2352
  longitudinal 5244
deleción 2365
deleterio 6187
demanda bioquímica de oxígeno 1052
  de oxígeno 6500
demo 2367
demoler 1258
demostrar experimentalmente 2368
dendrita 2371
dendrocronología 2372
dendrología 2373
denervación 2375
denervar 2374
densidad demográfica 7186
  de las especies 2384
  óptica 6365
  de (la) población 7186
dentado 2389
dentadura 2394
dental 2385
dentario 2385
dentición 2393
dentículo 2390
dentificación 2391
dentina 2392
denucleación 3003
denuclear 3002
dependiente de la densidad 2380
depigmentación 2402
deplasmólisis 2403
depolución 2406
depolucionar 2405
deposición de grasas 3316
depredador 7245
depresión consanguínea 4591
depuración de las aguas 9650
  biológica de aguas residuales 1083
deriva genética 3740
derivación 2409
dermatógeno 2416
dérmico 2200
dermis 2418
dermosqueleto 3216
derrepresión 2408
desacidificación 2283
desacidificar 2284
desadaptación 2285
desaferenciación 2286
desaferentización 2286
desalación del agua del mar 2421
desalar 2420
desalinar 2420
desalinización del agua del mar 2421
desalinizar 2420
desaminación 2289
desaminar 2288
desaminasa 2287
desarrollable 2456
desarrollo 2457
  embrionario 2877
  en mosaico 5821
  postnatal 7217
  prenatal 7270
desasimilación 1459
desastre ecológico 3013
descalcificación 2291
descalcificar 2292
descamación 2436
descarboxilación 2295
descarboxilasa 2294
  de aminoácidos 297
descarga 2593
  ulterior 221
descendencia 2426, 7326
descender de 2423
descendiente 2424
descendientes 7326
descenso liminal 9093
  del umbral 9093
desceración 2298
desciframiento (del código genético) 2305
descifrar (el código genético) 2304
descoloración 2306
descolorar 2307
descomponedor 2309
descomponerse 2308
descomposición 2310
descontaminación 2315
descontaminar 2313
desdoblamiento 1262, 2732
  proteico 7371
desdoblar(se) 1258, 2731
desecación 2437
  congelación 5285
desechos 9623
  atómicos 7560
  industriales 4648
  del metabolismo 5539
  nitrogenados 6142
  radiactivos 7560
desencadenador 7712
desencadenante 7712
desencadenar 7711
desensibilización 2429
desensibilizar 2430
deshidrasa 2358
deshidratación 2355
deshidrogenación 2359
deshidrogenasa 2358
desinfección 2604
desinfectante 2603
desinfectar 2602
desinhibición 2606
desinhibir 2605
desintegración 2608
  de los tejidos 9130
desintegrador 2309
desintegrarse 2607
desintoxicación 2448
desintoxicar 2447
deslaminación 2360
desmolasa 2433
desmona 2434
desmonte completo 1774
desmosoma 2435
desnaturalización 2369
desnaturalizar 2370
desnitrificación 2376
desnitrificar 2377
desnitrogenación 2379
desnucleinización 2395
desorden metabólico 5528
desovar 6464, 8423
desove 6465, 8425
desoxicorticosterona 2396
desoxidación 2397
desoxigenación 2397
desoxirribonucleasa 2398
desoxirribosa 2400
desparafinado 2298
desperdicios 9623
  industriales 4648
desplazamiento de caracteres 1607
  en masa 5394
despolarización 2404
despolución 2406
despolucionar 2405
desprenderse 20
desprendimiento 22
  de calor 4127
  de los frutos 3607
  de las hojas 5067
destilación 2639
destoxificación 2448
destoxificar 2447
desulfuración 2440
desulfurar 2441
desulfuricación 2440
desvernalización 2467
desviación 2468, 8259
  al azar 7596
  cuadrática media 8589
  standard 8589
  típica 8589
  tipo 8589
desvitalizar 2469
detención del crecimiento 714
detergente 2443
deterioración ambiental 3012
deterioro del medio ambiente 3012
determinación de la edad 241
  embrionaria 2876
  del grupo sanguíneo 1181
  del sexo 8219
  sexual 8219
determinante 2446
detrito 2449
detritóvoro 2450
deuteromicetos 3625
deuteroplasma 2455
deuterostomas 2453
deuterotocia 2454
deutolecito 2455
deutoplasma 2455
devónico 2470
dextrana 2472
dextrina 2473
dextrógiro 2474
dextrorrotatorio 2474
dextrorso 2475
dextrosa 3878
dextrosamina 3877
dextrovoluble 2475
diacinesis 2479
díada 2739
diáfisis 2491

diafragma 2489
diageotropismo 2478
diagrama floral 3469
polínico 7099
dialipétalo 2480
diálisis 2482
dializado 2481
dializar 2483
diamina 2484
diapausia 2487
diapédesis 2488
diapófisis 2492
diaquenio 2477
diaquinesis 2479
diartrosis 2493
diáspora 2494
diastasa 436
salivar 7445
diaster 2496
diástole 2497
diatomeas 2499
dibranquiados 2500
dicariófase 2552
dicario(n) 2551
dicarionte 2551
dicasio 2503
dicéntrico 2502
diclamídeo 2504
diclino 2510
dicogamia 2506
dicógamo 2505
dicotiledóneas 2511
dicotiledóneo 2512
dicotomía 2509
dicotómico 2507
dicótomo 2507
dictiocinesis 2513
dictiosoma 2514
diecia 2566
diencéfalo 2517
diente canino 1385
carnicero 1428
deciduo 5683
incisivo 4596
de leche 5683
molar 5739
permanente 6761
premolar 7267
temporal 5683
venenoso 7078
diestro 2567
diferencia de potencial 7223
de presión 7279
diferenciación 2524
celular 1508
hística 9129
sexual 8220
de tejidos 9129
diferencial de selección 8094
diferenciar 2523
difilético 2570
difosfopiridina-nucleótido 6118
difracción de los rayos X 9725
difundible 2526
difundir(se) 2525
difusión 2527
difusivo 2526
digamético 4215
digénico 2537
digerible 2540
digerir 2538
digestibilidad 2539
digestible 2540
digestión 2541
digitado 2547
digitígrado 2548
dihibridismo 2550
dihíbrido 2549
dilatabilidad 2553
dilatable 2554
dilución 2557
diluente 2555
diluir 2556
dimería 2560
dimérico 2559
dimerización 2558
dímero 2559
dimórfico 2561
dimorfismo 2562
estacional 8044
nuclear 6193
sexual 8221
dimorfo 2561
dinámica de las poblaciones 7187
dinitrofenol 2563
dinofíceas 2564
dinoflagelados 2564
dintel doloroso 9092
dioecia 2566
dioecismo 2566
dioico 2565
dióxido de carbono 1407
dipéptido 2568
diplobionte 2571
diplocario(n) 2577
diplocromosoma 2572
diplofase 2579
diplohaplonte 2573
diploide 2574
diploidia 2576
diploidización 2575
diplonte 2578
diplópodos 2580
diplosis 2581
diplosoma 2582
diplostémono 2583
diplóteno 2584
dipneustos 2585
dipnoos 2585
dípteros 2586
dirección evolutiva 3171
de la migración 5674
disacárido 2590
discernimiento 4716
disclímax 2595
disco adhesivo 162
germinativo 1152
imaginal 4542
intervertebral 4790
de succión 8771
disecar 2624
disección 2627
disector 7272
diseminación 8075
de esporas 8544
diseminulo 2494
disepimento 2630
disgregación 9668
disgregante 2309
disimétrico 2646
disimilación 2631
dislocación 2611
disociación 2632
disolvente 8390
disomia 2612
dispermia 2614
dispérmico 2613
dispermo 2613
dispersión 2618
de esporas 8544
dispersión luminosa 5154
de la luz 5154
de la semilla 8075
zoocora 9758
disploidia 2742
disponibilidad de alimento 3543
disposición *(eth.)* 5803
cis 1757
del ensayo 3237
de las hojas 6880
lineal de los genes 5184
trans 9195
dispuesto para el apareamiento 7624
a reaccionar 7625
distancia del centrómero 1565
diferencial 2519
entre los individuos 4635
de interferencia 4745
de mapa 5368
dístico 2638
distribución al azar 7595
binomial 1042
por edades 242
de frecuencias 3580
independiente (de los genes) 4625
distrófico 2743
disyunción 2609, 2610
alternativa 366
dítipo progenitor 6624
divergencia 2650
de caracteres 1607
diversidad de especies 8434
dividido en cámaras 1374
división 2652
amitótica 401
binaria 1120
celular 1509
ecuacional 3096
longitudinal 5245
de maduración 5424
meiótica 5460
mitótica 5711
nuclear 6194
reductora 7666
de segmentación 1782
del trabajo 2653
dizigótico 2656
doble dominante 2694
enlace 2691
hélice 2696
doctrina de la neurona 6101
domesticación 2665
dominancia 2666
alternativa 364
apical 640
incompleta 8130
dominante 2669
doble ~ 2694
ecológico 2761
dominigen 2671
donador 2672
de hidrógeno 4438
donante 2672
dormancia 2675
dormina 2680
dorsal 2681
dorsiventralidad 2686
dosis génica 3705
de irradiación 4850
letal 5118
permisible 3746

dosis de radiación 7542
de rayos X 9726
dotación cromosómica 1723
génica 3739
hereditaria 4191
DPD 2533
DPN 6118
drepanio 2703
drupa 8682
duodeno 2728
duplexo 2729
duplicación 2732
del ADN 2659
cromosómica 1718
duplicado 2730
duplicarse 2731
duración de la germinación 2735
de la gestación 3818
de la incubación 4618
de reacción 7618
de (la) vida 5142
duramadre 2733
duramen 2734

E

ecdisis 5839
ecdisona 2751
ECG 2822
ecidio 197
ecidióspora 196
eclosión 4110
hacer ~ 4109
eclosionar 4109
ecobiótico 2755
ecoclimático 2757
ecofisiología 2772
ecogeográfico 2758
ecolocalización 2754
ecología 2770
animal 507
fisiológica 2772
humana 4399
marina 5375
de población 7188
de la radiación 7570
vegetal 6993
ecológico 2759
ecólogo 2769
ecomorfosis 2771
economía hídrica 9636
ecosfera 2774
ecosistema 2775
de agua fluente 3495
fluvial 3498
terrestre 9004
ecospecie 2773
ecotipo 2777
ecotono 2776
ectoblasto 2779
ectodermo 2779
ectoenzima 2780
ectogenia 2781
ectolecítico 2782
ectolecito 2782
ectoparásito 2783
ectoplasma 2784
ectoproctos 2785
ectótrofo 2786
ectozoo 2783
ecuación de regresión 7698
edáfico 2788
edafología 8372
edafón 2793
EEG 2826
efarmonía 3050
efecto de Baldwin 941
cis-trans 1758
de combinación en bloque 1169
de la dosis 2688
estimulante 8644
de grupo 3967
inhibidor 4677
de masa 5393
de posición 7207
de la radiación 7611
de saturación 7992
sensibilizador 8164
tóxico 9168
efector 2795
efedrina 3051
eferente 2797
eficacia reproductora 3427
eficacidad 2794
eficiencia 2798
efímera 3052
eje caulinar 1489
floral 3468
del huso 8500
de la inflorescencia 4657
longitudinal 5243
ejemplar tipo 4326
elaioplasto 2813
elasmobranquios 2814
elastina 2815
elater 2816
elección del alimento 3538
electrocardiografía 2823
electrocardiograma 2822
electrodiálisis 2825
electrodo 2824
estimulante 8645
electroencefalografía 2827
electroencefalograma 2826
electroestímulo 2818
electrofisiología 2846
electroforesis 2845
sobre papel 6572
de zona 9754
electrolito 2828
electromiografía 2830
electromiograma 2829
electrón 2831
electrónica 2843
biológica 1098
electronografía 2833
electronomicroscopia 2836
electroosmosis 2844
electroplaca 2847
electrotono 2848
electrotropismo 3640
elemento faunístico 3331
floral 3470
florístico 3470
mineral 5694
radiactivo 7571
traza 9177
tubo-criboso 8279
vascular 9523
vaso 9523
eleoplasto 2813
elevación del umbral 7592
eliminación 2853
de agua 9637
de (las) aguas residuales 8211
de las basuras 2664
de (los) desperdicios 9624
eliminar 2852
élitro 2860
elongación celular 1510
prometafásica 7338
eluato 2857
elución 2858
emasculación 1457
emascular 1456
embarazada 7261
embarazo 7259
embolia 2865
embriofitas 2873
embriófitos 2873
embriogenia 2870
embriogénesis 2870
embrioide 2871
embriología 2873
experimental 2463
embriólogo 2872
embrión 2866
embrional 2875
embrionario 2875
embrionía adventicia 193
embriótrofo 2884
embudo ciliado 6035
emergencia 2885
emisión 2886
de sonidos 9607
emparejamiento 6515
de bases 961
emplazamiento del nido 6067
empollar 4612
emulsificación 2887
emulsificador 2889
emulsificar 2888
emulsina 2890
emulsión 2891
emulsionar 2888
emulsoide 2892
encefálico 1579
encefalización 2896
encéfalo 1246
encelarse 6298
encía(s) 4001
encorvadura 4620
de crecimiento 3975
trópica 9284
endémico 2901, 2902
endemismo 2903
conservativo 7715
endergónico 2904
endobiosis 2905
endoblasto 2914
endocardio 2907
endocarpo 2908
endocráneo 2909
endocrinología 2913
endocrinólogo 2912
endodermis 2915
endodermo 2914
endoenzima 2916
endoesqueleto 2939
endófito 2930
endogamia 2918
endogámico 2917
endogénesis 2921
endógeno 2919
endolinfa 2922
endometrio 2923
endomisio 2926
endomitosis 2924
endomixis 2925
endoneur(i)o 2927
endoparásito 2928
endopeptidasa 2929
endoplasma 2931

endoplasmático 2932
endoplásmico 2932
endopoliploidia 2934
endoproctos 2935
endopterigotos 4322
endosito 2928
endosmosis 2940
endosimbiosis 2948
endosoma 2941
endosperma 2942
endospermo 2942
endóspora 2943
endosporio 2944
endosqueleto 2939
endostilo 2947
endostio 2945
endóstoma 2946
endotecio 2949
endotelio 2951
endotérmico 2952
endotoxina 2953
endótrofo 2954
endurecer 4101
endurecimiento 4102
energía de activación 122
  alimenticia 3426
  atómica 781
  calórica 4128
  calorífica 4128
  cinética 4967
  lumínica 5148
  luminosa 5148
  metabólica 5530
  nuclear 781
  nutritiva 3426
  potencial 7224
  radiante 7539
  solar 8375
  trófica 3426
enérgida 2959
enfermedad por carencia 2330
  carencial 2330
  vírica 9545
  virótica 9545
  de virus 9545
engrama 2975
engrosamiento 9072
  anular 546
  espiral 8513
  helicoidal 8513
  reticulado 7831
enjambramiento 8841
enjambrar 8839
enjambrazón 8841
enjambre 8838
  migratorio 5675
enlace, doble 2691
  de hidrógeno 4437
  peptídico 6705
  péptido 6705
  rico en energía 2969
enocito 6287
enquistamiento 2898
enquistarse 2897
enriquecer 2976
enriquecimiento 2978
ensayo de adiestramiento 9192
  biológico 1047
entalpía 2990
entelequia 2981
enteramina 8195
entérico 4791
enterocele 2985
enterocrinina 2986
enterogastrona 2987
enteropeptidasa 2988
enteroquinasa 2988
enterosito 2989
enterozoo 2989
entero *(hoja)* 2991
entodermo 2914
entomófago 4709
entomofilia 2999
entomófilo 2997
entomógamo 2997
entomología 2995
entomólogo 2994
entozoario 3000
entozoo 3000
entrebanda 4734
entrecruzamiento 2141
  compuesto 1948
  doble 2693
entrenudo 4772
entropía 3001
enucleación 3003
enuclear 3002
envejecer 239
envejecimiento 8144
envoltura (extra)embrionaria 2879
  fetal 2879
  floral 6717
  larval 5039
  nuclear 6196
  pupal 7469
  quitinosa 1640
enzima 3029
  activador 120
  activante 120
  adap(ta)tiva 142
  constitutiva 2002
  digestiva 2542
  inducible 142
  isomerizante 4886
  lítica 5296
  precursor 9793
  proteolítica 7363
  represible 7738
  respiratoria 7796
enzimático 3028
enzimología 3043
enzimoproteína 3037
eoceno 3044
eosina 3045
eosinocito 3046
eosinófilo 3047
epéndimo 3048
epiblasto 2779
epibolia 3055
epicardio 3056
epicarpo 3205
epicotilo 3058
epidemiología 3059
epidermis 3061
epidídimo 3062
epifilo 3077
epífisis 3080, 3081
epífitas 3082
epifítico 3080
epífito 3080
epífitos 3082
epigastrio 3063
epigénesis 3065
epigenética 3066
epigeo 3064
epígino 3068
epiglotis 3067
epilimneo 3069
epimisio 3072
epimorfosis 3071
epinastia 3073
epinefrina 179
epineur(i)o 3075
epiplón 6330
episoma 3085
episperma 8074
epistasia 3087
epistasis 3087
epitecio 3088
epitelio 3092
  ciliado 1742
  cilíndrico 1913
  columnar 1913
  cúbico 2180
  escamoso 8570
  estratificado 8694
  germinal 3797
  glandular 3853
  neural 6091
  pavimentoso 6660
  plano 6660
  sensorial 8169
  de transición 9208
epitema 3093
epizoario 3095
epizoico 3094
epizoo 3095
época de apareamiento 5416
  de cría 1281
  de desove 8426
  de floración 3492
  geológica 3776
  glacial 3846
  de lluvia 7591
  de reproducción 7760
equilibrio ácido-base 81
  acuoso 9630
  demográfico 7189
  de difusión 2531
  dinámico 8608
  ecológico 2760
  genético 3741
  génico 3700
  hídrico 9630
  iónico 4837
  metabólico 6889
  mutacional 5932
  nitrogenado 6137
  osmótico 6419
  salino 7957
equinodermos 2753
equipo cromosómico 1717
  de genes 3739
era atómica 780
erepsina 3101
ereptasa 3101
ergastoplasma 3102
ergina 3103
ergosoma 7166
ergosterina 3105
ergosterol 3105
ergotinina 3106
eritroblasto 3108
eritrocito 3109
erosión 3107
  eólica 9682
error medio 8590
escama 7998
  del bulbo 1328
  del estróbilo 1976
  fructífera 8072
  ganoidea 3665
  gemaria 1316
  glandular 3855
  placoidea 2390

estanque de oxidación 6486
estar en celo 6298
estatoblasto 8604
estatocisto 8605
estatolito 8606
estela 8610
estenohalino 8619
estenohídrico 8620
estenotermal 8621
estenotérmico 8621
estenotermo 8621
éster 3116
esterasa 3117
  de acetilcolina 67
estercobilina 8623
estereoma 8624
estereomicroscopio 8625
esterificación 3118
esterificar 3119
esterigma 8627
estéril 8628
esterilidad 8629
  híbrida 4416
  del polen 7103
  polínica 7103
esterilización 8631
esterilizar 8632
esterina 8637
esterizar 3119
esternón 8634
esteroide 8635
esterol 8637
  animal 9774
estetos 210
estigma 8639, 8640
  ocular 3288
estilete 8720
estilo 8719
estimulación 8647
  continua 2015
  del crecimiento 3986
  eléctrica del cerebro 2820
  sostenida 2015
estimulante 8643
  del crecimiento 3987
estímulo 8648
  adecuado 159
  acústico 806, 8411
  ambiental 3026
  auditivo 806
  del calor 9049
  clave 8280
  de contacto 2009
  desencadenador 7714
  desencadenante 7714
  eléctrico 2818
  externo 3262
  del frío 1874
  de la gravedad 3946
  de gravitación 3946
  gustativo 4010
  háptico 2009
  liminar 5171
  lumínico 5158
  luminoso 5158
  llave 8280
  mínimo 5171
  nociceptivo 6144
  olfativo 6318
  óptico 9566
  sensitivo 8176
  sensorial 8176
  signo 8280
  supranormal 8801
  táctil 8915
estímulo térmico 9049
  umbral 5171
  visual 9566
estípula 8666
estirpe 1766, 8689
estivación 211, 212
estolón 8672
estoma 8673
  acuífero 9657
  aerífero 270
estomacal 3671
estómago 8674
estomocorda 8677
estomocordados 4159
estomocordio 8677
estomocordios 4159
estomodeo 8678
estradiol 6289
estrangulación 8691
  de Ranvier 7603
estrano 6290
estratificación 8692
estratificado 8693
estrato 8695
  de abscisión 23
  arbóreo 9242
  arbustivo 1305
  celular 1516
  córneo 2059
  germinativo 5350
  herbáceo 4175
  de Malpighi(o) 5350
estreptomicina 8698
estriación 8702
  transversal 9234
estriado 8700, 8702
  transversal 9234
estribo *(anat.)* 8595
estricnina 8717
estrina 6294
estriol 6292
estro 6297
estrobilación 8707
estrobiliforme 8706
estrobilización 8707
estróbilo 1974
  abayado 3637
estrógeno 6293
estroma 8708
estrona 6294
estructura alveolar 373
  axial 873
  celular 1531
  cuaternaria 7519
  del cuerpo 1211
  por edades 243
  de las especies 8445
  fina 3418
  fundamental 3623
  grumosa 2154
  en hélice 4140
  helicoidal 4140
  lamelar 5019
  molecular 5744
  monogranular 8298
  primaria 7295
  secundaria 8056
  terciaria 9014
estructuras vistosas 8272
estudio comparado del comportamiento 1925
  del comportamiento 3129
  experimental 3239
etanol 3130
etapa de crecimiento 3993
etapa de desarrollo 2466
  evolutiva 3169
  intermedia 4757
  reproductiva 7763
  seral 8184
  de sucesión 8184
eteogénesis 3123
etiolación 3132
etmoides 3125
etología 3129
  comparada 1925
etológico 3127
etólogo 3128
eucariotas 3138
eucromático 3133
eucromatina 3134
eucromosoma 3135
eugenesia 3137
eugenético 3136
eulamelibranquios 3139
eumicetos 9289
euploide 3142
euploidia 3143
eurifágico 3145
eurihalino 3144
euritérmico 3146
euritermo 3146
eutelia 1504
euterios 6970
eutroficación 3150
eutrófico 3149
evacuación de desechos 9624
evaporación 3153
evaporar 3152
evapotranspiración 3155
evocación 3158
evocador 3159
evolución 3161
  de las comunidades 1923
  convergente 2030
  del cuanto 7517
  divergente 2651
  en línea directa 6404
  ortogenética 6404
  progresiva 449
  regresiva 7700
evolucionismo 9042
evolucionista 3173
evolutivo 3174
examen microscópico 5637
examinar al microscopio 3177
excípulo 3180
excitabilidad 3181
excitable 3182
excitación 3183
  nerviosa 6056
exconjugante 3191
excrecencia 3193
excreción 3196
excrementos 3192
excreta 3194
exretar 3195
excretor(io) 3197
exento de gérmenes 3785
exergónico 3202
exigencias alimenticias 3534
  (de sustancias) nutritivas 6251
exina 3203
éxito evolutivo 3170
exocarpo 3205
exoceloma 3206
exocráneo 3207
exodermis 3210
exodermo 2779
exoenzima 2780

exoesqueleto 3216
exogamia 3211
exogastrulación 3212
exógeno 3213
exopeptidasa 3214
exoplasma 2784
exopterigotos 4162
exósmosis 3217
exóspora 3218
exosporio 3219
exosqueleto 3216
exóstoma 3220
exotecio 3221
exotérmico 3222
exotoxina 3224
expectativa de vida 5138
experiencia 3225
  de adiestramiento 9192
  de control 2026
  de cría 1273
  con modelos 5729
experimentación animal 3241
experimentador 3242
experimental 3227
experimentar 3226
experimento 3225
  de aislamiento 4879
  en animales 3241
  en el campo 3398
  de cría 1273
  de cruce 2137
  de cultivo 2188
  por defecto 2324
  de enriquecimiento 2980
  de laboratorio 4991
  seriado 8187
experto en (problemas del) medio ambiente 3027
explantación 3247
explante 3246
explosión demográfica 7190
  nuclear 6197
explotación abusiva 2438, 7249
exponente de hidrógeno 4439
expresión fenotípica 6817
  genética 3708
  génica 3708
  del sexo 8224
expresividad (génica) 3252
extender 8331
extensibilidad 2553
extensión 8330
  sanguínea 1192
extensor 3254
exterminación 3256
exterminio 3256
exteroceptivo 3263
exteroceptor 3264
extinción 3267
  en trance de ~ 991
  en vías de ~ 991
extinguido 3265
extinguirse 3266
extinto 3265
extracelular 3269
extracromosómico 3270
extracto 3271
  acuoso 675
  celular 1511
  de levadura 9736
  seco 2716
  de tejido 9131
extraer 3272
extraño a la especie 6173
extrapiramidal 3274
extrapolación 3273
extremidades 3275
extrorso 3277
exudación 3279
exudado 3278
exuvia 3280
exuviación 3282
eyaculación 2811
eyaculado 2810

F

facies 3292
facilitación 3293
  espacial 8418
  temporal 8976
factor alimentario 6256
  ambiental 3015
  antipelagroso 6117
  biótico 1115
  climático 1793
  complementario 1940
  condicional 1962
  de crecimiento 3977
  edáfico 2791
  de evolución 3163
  evolutivo 3163
  externo 3258
  extrínseco 1835
  F 3365
  de fertilidad 3365
  hereditario 3696
  de incompatibilidad 4599
  inhibitorio 4678
  de intensificación 4731
  intrínseco 4805
  kappa 4925
  letal 5119
  de liberación 7713
  limitador 5173
  limitante 5173
  mendeliano 5487
  modificador 5732
  múltiple 7133
  de mutación 5927
  de motivación 5830
  motivante 5830
  nutricional 6256
  PP 6117
  proteínico animal 523
  regulador 7704
  Rh 7850
  Rhesus 7850
  sexual 3365
  tiempo 9123
facultad de aprender 5098
  auditiva 798
  germinativa 3806
  reproductiva 7753
FAD 3447
fago 927
  intemperado 4730
  lítico 4730
  temperado 8968
  virulento 4730
fagocitar 6792
fagocito 6791
fagocitosis 6793
falciforme 3300
falso fruto 7427
falta de agua 9634
  de oxígeno 6501
familia 3305
fanerófita 6795
fanerófito 6795
fanerógamas 8472
fango activado 118
  de alcantarrilla 8212
  de globigerina 3864
faringe 6800
  branquial 6801
  nasal 5982
faríngeo 6796
fascia 3309
fasciación 3310
fase de acoplamiento 2098
  de atracción 2098
  de crecimiento 3993
  de desarrollo 6738
  dicariótica 2552
  diploide 2579
  de ensayo 3238
  estacionaria 8603
  exponencial 3250
  gerόntica 3811
  haploide 4090
  inicial 4685
  juvenil 4922
  larval 5040
  logarítmica 3250
  de maduración 5426
  de reposo 7823
  retardada 5016
  de retardo 5016
  retrasada 5016
  terminal 8998
fatiga 3324
fatigabilidad 3325
fauna 3330
  abisal 39
  aviar 861
  cavernícola 1491
  edáfica 8359
  marina 5376
  marítima 5376
  reliquial 7716
F-ducción 8222
fécula de liquen 5132
fecundabilidad 3373
fecundación 3367
  cruzada 338
  doble 2695
fecundar 3371
fecundidad 3363
fecundo 3362
feedback 3337
felema 2052
felodermis 6805
felodermo 6805
felógeno 6806
femenino 3348
femin(e)idad 3352
fémur 3353
fen 6807
fenilalanina 6818
fenocitología 6809
fenocopia 6808
fenogenética 6810
fenolase 6811
fenología 6813
fenómeno vital 9573
fenotípico 6815
fenotipo 6814
  mutante 5921
feofíceas 6789
feofitos 6789
fermentable 3356
fermentación 3357

fermentación acética 61
alcohólica 289
butírica 1335
láctica 5008
fermentar 3355
fermentescible 3356
fermento 3029
amarillo 3451
digestivo 2542
lab 7727
feromona 6819
ferredoxina 3361
ferrobacterias 4847
fértil 3362
fertilidad 3363
fertilizable 3366
fertilización 3367
cruzada 338
dispérmica 2614
doble 2695
externa 3259
interna 4763
fertilizar 3371
fertilizina 3372
fetal 3500
fetalización 3503
feto 3504
fibra 3379
argirófila 7834
colágena 1883
colateral 1886
gigante 3822
del huso 8501
leñosa 5163
liberiana 982
muscular 5906
nerviosa 6045
nerviosa gigante 3822
de proteína 7375
reticulínica 7834
fibrilar 3383
fibrilla 3382
muscular 5952
fibrina 3384
fibrinógeno 3385
fibrinolisina 3386
fibroblasto 3387
fibrocartílago 3388
fibrocito 3389
fibroína 3390
fibroso 3391
ficobilina 6869
ficocianina 6870
ficoeritrina 6871
ficofeína 6874
ficología 301
ficomicetos 6873
ficoxantina 3617
fidelidad asociativa 3396
al lugar 6820
fijación 3430
del complemento 1935
de nitrógeno 6138
fijador 3432
fijar 3429
filamento 3403
axial 870
branquial 3836
filamentoso 3404
filético 6875
filiación 2426
consanguínea 5183
genealógica 5182
filiforme 3404
filocladio 6877
filóclado 6877
filodio 6878
filogénesis 6881
filogenética 6884
filogenético 6882
filogenia 6881
filogénico 6882
filopluma 3408
filopodios 3409
filoquinona 6879
filotaxia 6880
filotaxis 6880
filotráquea 1227
filtración 3414
filtrado 3413
filtrar 3410
filtro bacteriano 915
impermeable para las bacterias 915
de membrana 5481
polarizador 7091
filum 6885
animal 517
fisibilidad 3422
fisible 3423
fisiología 6892
animal 518
celular 1527
del comportamiento 1000
del desarrollo 2463
de la nutrición 6893
de la reproducción 7761
sensorial 8174
vegetal 6919
fisiológico 6886
fisiólogo 6891
fisión 3424
nuclear 6198
fisiparidad 8025
fitina 6894
fitoalbúmina 6895
fitobentos 6896
fitobiología 6897
fitocenología 6921
fitocenosis 6990
fitocroma 6900
fitoecología 6993
fitófago 6917, 6918
fitofisiología 6919
fitofisiólogo 7002
fitoflagelados 6902
fitogénesis 6903
fitogenética 6904
fitogeneticista 6986
fitogenetista 6987
fitogenia 6903
fitógeno 6905
fitogeografía 6907
florística 3479
fitogeógrafo 6906
fitoglobulina 6908
fitografía 6909
fitohormona 6910
fitol 6912
fitología 1232
fitólogo 1231
fitopaleontología 6523
fitoparásito 6601
fitopatología 6916
fitoplancton 6920
fitoquímica 6899
fitoquímico 6898
fitoquinina 2241
fitorregulador 3990
fitosociología 6921
fitosterina 6922
fitosterol 6922
fitotaxonomía 7008
fitotomía 6986
fitotoxicidad 6925
fitotóxico 6924
fitotoxina 6926
fitotrón 6927
fitozoo 6928
flabelado 3435
flabeliforme 3435
flabelo 7854
flagelación 3440
flagelado 3438
flagelados 3437
flageliforme 3441
flagelo 3442
bacteriano 908
espermático 8463
flavina 3446
flavin-adenin-dinucleótido 3447
flavin-mononucleótido 3448
flavona 3449
flavonoide 3450
flavoproteína 3451
flecha 5065
flexibilidad del comportamiento 995
flexor 3453
floculación 3464
flocular 3463
floema 6821
flor 3482
entomófila 2998
estaminada 8587
femenina 6959
hermafrodita 4198
ligulada 5166
masculina 8587
ornitófila 6399
pistilada 6959
radial 7612
terminal 8995
tubiforme 9303
tubulosa 9303
flora 3467
intestinal 4793
marina 5377
marítima 5377
reliquial 7717
floración 3489
de ~ diurna 4157
de ~ nocturna 6264
florecer 3481
florescencia 3492
florífero 3477
florígeno 3491
florística 3479
flósculo discoide 2592
flotabilidad 8848
fluctuación de población 7191
fluctuaciones estacionales 8045
fluido celómico 1853
de la muda 5840
flujo de electrones 3480
energético 2962
de energía 2962
de genes 3709
de iones 4839
plasmático 7409
sanguíneo 1176
de savia 7972
fluorescencia 3496
FMN 3448
fobotactismo 6822

fobotaxia 6822
foco luminoso 5157
de polución 7119
foliáceo 3505
foliación 3510
foliar 3509
foliculina 6294
folículo 3515, 3516
de Graaf 3919
ovárico 3919
piloso 4067
folíolo 3512
folioso 3513
follaje 3506
fonación 6823
fondo genético 3715
marino 8041
oscuro 2271
sonoro 900
fonorreceptor 6824
fontanela 3522
foramen 3546
foresia 6825
foresis 6825
forma biológica 5139
de crecimiento 3978
de cultivo 2185
juvenil 4919
larval 5037
originaria 7299
primitiva 7299
de resistencia 7819
formación de colonias 1897
de las especies 8429
de flores 3486
del grupo 3968
de pareja 6515
de puffs 7451
de raíces 7903
de razas 7533
reticular 7830
(vegetal) 3564
formaldehído 3563
fómula dental 2387
floral 3471
foronídeos 6826
fosa craneal 2115
nasal 5978
fosfatasa 6827
fosfátido 6828
fosfato de arginina 703
de creatina 2123
fosfoarginina 703
fosfocreatina 2123
fosfolípido 6828
fosfolipina 6828
fosfolipoide 6828
fosfomonoesterasa 6827
fosfoproteína 6834
fosforescencia 6835
fosforilación 6840
de la cadena respiratoria 6491
fotosintética 6857
oxidativa 6491
fosforilasa 6838
fosforólisis 6837
fósil 3566
característico 4627
indicador 4627
fosilífero 3567
fosilización 3568
fosilizar(se) 3569
fotobiología 6842
fotocinesis 6845
fotófilo 4855
fotóforo 5264
fotofosforilación 6857
cíclica 2214
no cíclica 6156
fotólisis 6846
fotometría 6847
fotomicrografía 5626
fotomorfosis 6849
fotón 6850
fotonastia 6851
fotopercepción 6852
fotoperiodicidad 6854
fotoperiodismo 6854
fotoperíodo 6853
fotoquímica 6844
fotoquímico 6843
fotorreactivación 6858
fotorreceptor 6859
fotosensibilidad 5156
fotosensible 5155
fotosíntesis 6861
fotosintético 6862
fototactismo 6864
fototaxia 6864
fototaxis 6864
fototono 6865
fototrófico 6866
fotótrofo 6866
fototropismo 6867
fóvea 3572
fraccionamiento celular 1512
fragmentación nuclear 6199
fragmoplasto 6868
frasco de cultivo 2189
frecuencia cardíaca 3582
de descarga 2594
de genes 3710
génica 3710
de los impulsos 4584
de mutación 5929
de los quiasmas 1632
de recombinación 7650
respiratoria 3583
freemartin 3576
freza 8422, 8425, 8426
frezar 8423
frigidorreceptor 1870
fronda 3591
frontal 3592
frotis 8330
sanguíneo 1192
fructífero 3598
fructificación 3599
fructificar 3600
fructosa 3601
frugívoro 3602
frústulo 3612
frútice 8273
fruticoso 3614
fruto 3603
agregado 1947
alado 9688
capsular 1396
de carozo 8682
colectivo 5886
compuesto 1947
dehiscente 2354
espermobólico 944
falso ~ 7427
de hueso 8682
indehiscente 4624
múltiple 5886
de pepita 7180
seco 2718
simple 8384
frutos de cáscara 8256
FSH 3518
fucoxantina 3617
fuente de alimentación 3540
alimenticia 3540
contaminante 7119
energética 2970
de energía 2970
de estímulo 8656
de radiación 7546
del sonido 8410
sonora 8410
fuerza absorbente (del suelo) 8777
cohesiva 1863
de la gravedad 3944
vital 9570
fuerzas evolutivas 3164
fumarasa 3618
función génica 3711
fungicida 3626
funículo 3627
espermático 8462
fusiforme 3630
fusión 3631
celular 1514
céntrica 1553
nuclear 6200
fusionar(se) 3629

G

galactolípido 3634
galactolipina 3634
galactosa 3636
galantear 2100
gálbulo 3637
galvanotaxis 3639
galvanotropismo 3640
gámeta 3643
gametangio 3642
gametangiogamia 3641
gamético 3644
gameto 3643
gametocito 3646
gametofito 3652
gametóforo 3651
gametogamia 3647
gametogénesis 3648
gametogénico 3649
gammaglobulina 3653
gamobio 3655
gamófase 3660
gamogonia 3656
gamona 3657
gamonte 3658
gamopétalo 3659
gamosépalo 3661
gamotropismo 3662
ganglio 3663
basal 952
cerebral 1582
espinal 8497
linfático 5277
nervioso 6048
visceral 9550
gangliocito 3664
garfa 1772
garra 1772
gas respiratorio 7797
tóxico 9169
venenoso 9169
gasterópodos 3676
gástrico 3671

gastrocele 3674
gastrointestinal 3675
gastrópodos 3676
gástrula 3678
gastrulación 3679
geitonocarpia 3681
geitonogamia 3682
gel 3683
  de agar 236
  de almidón 8596
gelación 3685
gelatificación 3685
gelatina 3684
gelatinificar 3686
gelatinización 3685
gelatinizar 3686
gelatinoso 3688
gelificación 3685
gelificar 3686
gelosa 233
gema 1295
gemación 1317, 3692
gemelaridad 9321
gemelos biovulares 2657
  bivitelinos 2657
  dizigóticos 2657
  fraternos 2657
  idénticos 5801
  monocigóticos 5801
  uniovulares 5801
  univitelinos 5801
gemífero 3693
gémino 1143
gemiparidad 3694
gémula 3695
gen 3696
  de aislamiento 4880
  amortiguador 1322
  complementario 1940
  de control 7608
  de esterilidad 8630
  estructural 8710
  inhibidor 4674
  letal 5119
  marcador 5382
  mayor 6322
  menor 7133
  modificador 5732
  mutador 5934
  mutante 5934
  nuclear 6201
  operador 6355
  polímero 7146
  regulador 7705
  represor 7741
  sexual diferencial 2521
  suplementario 8810
gene 3696
genecología 3722
generación 3723
  espontánea 8537
  F_1 3377
  F_2 3378
  filial 3406
  filial, primera 3377
  filial, segunda 3378
  P_1 6625
  sexuada 8244
  sexual 8244
generador 3727
generativo 3727
genérico 3730
género 3769
generotipo 3732
genes gemelos 7423
génesis 3733
genética 3748
  animal 9761
  bacteriana 909
  del comportamiento 996
  conductista 996
  del desarrollo 2462
  humana 4400
  molecular 5742
  de (las) poblaciones 7192
  vegetal 6904
  vírica 9532
geneticista 3747
genético 3734
genetista 3747
génico 3749
genitales 3757
genitor 6617, 6621
  femenino 3351
  hembra 3351
  macho 5342
genocentro 3701
genocopia 3758
genoma 3759
genómero 3761
genómico 3762
genonema 3763
genosoma 3764
genotípico 3766
genotipo 3765
geobiología 3771
geobotánica 6907
geocarpia 3773
geófitos 3777
geografía botánica 6907
  vegetal 6907
geotactismos 3778
geotaxis 3778
geotropismo 3779
  transversal 6973
germario 3791
germen 3780
  patógeno 2601
germicida 3792
germígeno 3791
germinación 3805
  retardada 2362
germinal 3793
germinar 3781
germinativo 3809
gerontología 3812
gestación 3817
  gemelar 9321
gestágeno 3813
gestaltismo 3816
gestante 7261
gesto de apaciguamiento 661
  de destacar 2622
  pacificador 661
giberelina 3824
gigantismo 3825
gigantocito 3820
gimnocarpo 4014
gimnospermas 4016
gimnospermo 4015
gimnospermos 4016
ginandrismo 4018
ginandromorfismo 4020
ginandromorfo 4019
gineceo 4017
gingival 3842
gínglimo 4281
ginodioico 4022
ginóforo 4026
ginogénesis 4023
ginomerogonia 4024
ginomonoico 4025
ginóspora 5323
ginostegio 4028
ginostemo 4029
glaciación 3849
glande 939
glándula 3850
  accesoria 49
  acuosa 9638
  adrenal 175
  anal 451
  antenal 559
  apócrina 647
  de cera 9667
  cerrada 2910
  de la concha 8257
  coxal 2110
  cutánea 8316
  dérmica 8316
  digestiva 2543
  ecrina 2748
  endocrina 2910
  exocrina 3208
  gástrica 3672
  hiladora 8506
  holocrina 4316
  láctea 5357
  lagrimal 4999
  lingual 5187
  mamaria 5357
  maxilar 5433
  mucípara 5860
  mucosa 5860
  nectarífera 6004
  nidamental 6123
  odorífica 6283
  odorípara 6283
  paratiroides 6611
  parótida 6634
  pineal 3081
  pituitaria 4511
  prostática 7357
  protorácica 5841
  repugnatoria 8665
  salífera 7960
  salival 7954
  sebácea 8047
  de secreción externa 3208
  de secreción interna 2910
  secretora de la tinta 4687
  del seno 8301
  sericígena 8290
  sexual 3908
  sinusal 8301
  sudorípara 8844
  suprarrenal 175
  tímica 9108
  tiroides 9110
  tóxica 7077
  uropigia 9420
  venenosa 7077
  vitelina 9743
glandular 3851
glanduloso 3858
glía 3860
gliacito 3861
gliadina 3862
glicemia 3888
glicerina 3890
glicerol 3890
glicina 3891
glicocola 3891
glicogénesis 3893
glicogenia 3893

Español

glicógeno 3892
glicogenólisis 3894
glicólisis 3896
glicoproteína 3876
glicosamina 3877
glicósido 3899
globina 3865
globo ocular 3283
globoide 3866
globulina 3867
  gamma 3653
  vegetal 6908
glóbulo blanco (de la sangre) 5123
  rojo (de la sangre) 3109
  sanguíneo 1175
glomérulo floral 3868
glotis 3871
glucagón 3872
glúcido 1399
glucocorticoide 3874
glucógeno 3892
glucogenólisis 3894
glucometabolismo 3897
glucoproteína 3876
glucosa 3878
glucosafosfato 3879
glucosamina 3877
glucósido 3899
gluma 3881
glumela superior 9400
glutamina 3883
glutation(a) 3885
glutelina 3886
gluten 3887
gnático 5431
gnatóstomos 3902
golgiocinesis 3907
gónada 3908
gonadotrofina 3911
  A 3518
  B 5269
gonadotrópico 3910
gonadotropina 3911
  coriónica 1683
gonaducto 3917
gonidia 3913
gonidio 3913
gonimoblasto 3914
gonio 3912
gonócito 3916
gonocorismo 3915
gonomería 3918
gota polinizante 7111
gotita de grasa 3317
  lipídica 3317
gradiente de concentración 1955
  de densidad 2381
  de temperatura 8970
grado de acidez 2343
  de cobertura 2345
  de contaminación 2349
  de dureza 4103
  higrométrico 5738
  de humedad 5738
  de libertad 2346
  de madurez 2348
  de parentesco 2350
  salínico 7950
  de saturación 2351
  de supervivencia 8829
  de variación 7601
gramnegativo 3929
grampositivo 3930
grana 3933
granívoro 3934
grano 3926
  de aleurona 295
  de almidón 8597
  de clorofila 1648
  de ~ grueso 1833
  de polen 7100
  polínico 7100
granos de Palade 7875
granulación 3937
granulado 3935
granular 3935, 3936
gránulo basal 953
granulocito 3938
gránulos de cimógeno 9794
grasa 3314
  de depósito 2407
graso 165
grávida 7261
gravidez 7259
gravitación 3944
gregario 3952
gregarismo 4182
grito de alarma 276
grupo amino 400
  carboxílico 1414
  carboxilo 1414
  de especies 8436
  hidroxilo 4464
  de ligamiento 5191
  metilo 5574
  prostético 7358
  sanguíneo 1179
guanidina 3997
guanina 3998
guía 5064
  (del árbol) 5065
gusanos anillados 540
  de cabeza espinosa 43
  en cinta 6014
  planos 7045
  redondos 6010
gustación 4003
gustativo 4004
gustatorio 4004
gustatorreceptor 8938
gusto 8935
gutación 4013

H

hábitat 4032
hábito 4035
  alimenticio 3529
habituación 4034
hacer eclosión 4109
halófilo 4080
halofitas 4081
halófitos 4081
haloplancton 4082
halterio 4083
haplobionte 4085
haploclamídeo 4086
haplofase 4090
haploide 4087
haploidia 4088
haplonte 4089
haplosis 4091
haplostémono 4092
hapteno 4093
hapterios 4094
haptonastia 4096
haptotropismo 4097
haustorio 4111
haz caulinar 8615
  conductor 9453
  foliar 5073
  vascular 9453
heces 3299
hectocótilo 4139
helecho arborescente 9239
helechos 7443
  de semilla 7440
hélice 4151
  doble ~ 2696
helicoidal 4144
heliófilo 4146
heliófobo 4147
heliotropismo 6867
helmintología 4152
helófito 4153
helotismo 4154
hemaglutinación 4038
hemaglutinina 4039
hemático 4050
hematíe 3109
hematina 4041
hematocianina 4054
hematófago 4046
hematogénesis 4044
hematología 4045
hematopoyesis 4062
hematopoyético 4063
hematoxilina 4048
hembra 3349
hemeritrina 4049
hemicarión 4161
hemicelulosa 4158
hemicigoto 4167
hemicordados 4159
hemicordios 4159
hemicriptófito 4160
hemicromatida 4073
hemicromosoma 4074
hemimetabolia 4163
hemimetábolos 4162
hemina 4051
hemiparásito 4164
hemípteros 4165
hemisferio cerebral 1583
hemítropo 4166
hemizigótico 4167
hemo 4037
hemoaglutinación 4038
hemoaglutinina 4039
hemoblasto 4052
hemocele 4053
hemocianina 4054
hemocito 4055
hemodinámica 4056
hemoglobina 4057
hemograma 1183
hemolinfa 4059
hemolisina 4060
hemólisis 4061
hemopoyesis 4062
hendidura branquial 3834
heparina 4170
hepáticas 4172
hepático 4171
herbáceo 4174
herbario 4176
herbicida 4177
herbívoro 4178, 4179
hercogamia 4180
heredabilidad 4198
heredable 4183
heredar de 4670

hereditario 4183
herencia 4672
  alternativa 367, 2130
  de los caracteres adquiridos 4673
  citoplásmica 2254
  cruzada 2130
  ligada al sexo 8234
  mendeliana 6640
  mezclada 1164
  no mendeliana 6162
  plásmica 2254
  plastídica 7034
  del sexo 8229
  sexual 8229
  unilateral 9380
hermafrodita 4196, 4197
hermafroditismo 4199
herpetología 4200
heterecio 4212
heterecismo 4213
heteroalelo 4201
heteroauxina 4638
heterocarión 4225
heterocarionte 4225
heterocariosis 4226
heterocárpico 4203
heterocarpo 4203
heterocéntrico 4204
heterocigosis 4259
  de inversión 4814
  de tra(n)slocación 9213
heterocigótico 4261
heterocigoto 4260
heterocinesis 4227
heterocisto 4210
heteroclamídeo 4205
heteroconto 4228
heterocromático 4206
heterocromatina 4207
heterocromatismo 4208
heterocromosoma 4209
heterodonto 4211
heteroecismo 4213
heteroestilismo 4250
heterofilia 4238
heterofítico 4239
heterogamético 4215
heterogameto 532
heterogamia 533
heterogeneidad 4217
heterogéneo 4220
heterogenético 4218
heterogenia 4221
heterogénico 4220
heterógeno 4220
heterogenote 4219
heterógino 4224
heterogonia 4222
heteroico 4212
heteroinjerto 4223
heterólisis 4231
heterología 4230
heterólogo 4229
heterómero 4232
heterometabolia 4163
heterometábolos 4162
heteromorfia 4235
heteromórfico 4234
heteromorfismo 4235
heteromorfo 4234
heteromorfosis 4236
heterónomo 4237
heteropicnosis 4244
heteroplasmonia 4240
heteroplastia 4241
heteroploide 4242
heteroploidia 4243
heteroproteido 1951
heterosinapsis 4251
heterosíndesis 4252
heterosis 4245
heterosomal 4246
heterospóreo 4247
heterosporia 4248
heterósporo 4247
heterostilia 4250
heterostílico 4249
heterostilo 4249
heterotalia 4254
heterotálico 4253
heterotalismo 4254
heterotérmico 7072
heterotipo 4258
heterótrico 4255
heterotrofia 4256
heterotrófico 4257
heterótrofo 4257
heterozigosis 4259
  de estructura 8712
heterozigótico 4261
heterozigotismo 4259
heterozigoto 4260
hexaploide 4263
hexaploidia 4264
hexápodos 4706
hexoquinasa 4266
hexosa 4267
hialina 4408
hialoplasma 4410
hibernación 4271
hibernáculo 4268
hibernante 4272
hibernar 4269
hibridación 4421
  incongruente 4603
  interespecífica 4781
  intergenérica 4749
  interspecie 4781
  introgresiva 4807
hibridar 4422
hibridez 4420
hibridismo 4420
híbrido 4412, 4413
  derivado 2410
  doble 2410
  de estructura 8713
  de injerto 3924
  (inter)específico 4780
  (inter)genérico 4748
  interspecie 4780
  numérico 6238
  de razas 7534
hidátodo 4423
hidratación 4424
hidratar 4425
hidrato de carbono 1399
hidrobiología 4427
hidrobiológico 4426
hidrocarburo 4428
hidrocele 4432
hidrocoria 4431
hidrocortisona 4433
hidrocultivo 4456
hidrofilia 4452
hidrofílico 4451
hidrófilo 4451
hidrófitos 4455
hidrofóbico 4453
hidrogamia 4452
hidrogel 4434
hidrogenación 4444
hidrogenar 4443
hidrogenasa 4442
hidrogenión 4440
hidrógeno 4435
hidrolasa 4445
hidrólisis 4447
hidrolítico 4448
hidrolizado 4446
hidrolizar 4449
hidronastia 4450
hidroserie 4457
hidrosfera 4460
hidrosol 4458
hidrosoluble 9656
hidrotactismo 4461
hidrotaxis 4461
hidrotropismo 4462
hidróxido 4463
hidroxilación 4466
hidroxilamina 4465
hidrozoos 4467
hierba 4173
hifa 4492
hifomicetos 4493
hígado 5217
higiene ambiental 3016
  del medio ambiente 3016
higrófilas 4471
higrófilo 4470
higrófitos 4471
higrómetro 4468
higronastia 4469
higroscopicidad 4473
higroscópico 4472
hijuelo 9119
hilera 8505
hilo *(bot.)* 4277
himenio 4474
himenópteros 4475
hinchamiento 8846
hincharse 8845
hinchazón 8846
hioides 4477
hipanto 4478
hipercromatismo 4480
hipercromatosis 4480
hipermetamorfosis 4481
hipermórfico 4482
hipermorfosis 4483
hiperparásito 4484
hiperplasia 4485
hiperploide 4486
hiperploidia 4487
hiperpolarización 4488
hiperquimera 4479
hipertelia 4489
hipertonía 4491
hipertonicidad 4491
hipertónico 4490
hipertrofia compensatoria 1931
hipnóspora 7822
hipobasidio 7310
hipoblasto 2914
hipocótilo 4496
hipodermis 4498, 8725
hipodermo 4498
hipofilo 4508
hipófisis 4511
  glandular 153
hipogastrio 4499
hipogénesis 4501

hipogeo 4500
hipógino 4503
hipolimneo 4504
hipomicrón 8743
hipomórfico 4506
hipomorfo 4506
hiponastia 4507
hipoplasia 4512
hipoploide 4513
hipoploidia 4514
hipostasia 4515
hipostasis 4515
hipotálamo 4517
hipotalo 4518
hipotecio 4519
hipótesis de los alvéolos 372
  de la elección de la copia 2042
  de un gen por una enzima 6337
  de la genocopia 3719
  de "ruptura inicial" 1259
hipotónico 4520
hipoxantina 4521
hipsofilo 4522
hirudina 4286
hirudíneos 4287
histamina 4288
histeresis 4523
hístico 9138
histidina 4289
histiocito 4290
histocompatibilidad 4293
histofisiología 4304
histogénesis 4295
histogenia 4295
histógeno 4294
histograma 4296
histólisis 4301
histología 4300
histológico 4297
histólogo 4299
histomorfología 4302
histona 4303
histoquímica 4292
histoquímico 4291
historradiografía 4305
hoja 5066
  acicular 78
  adulta 5991
  asoleada 8789
  de ~ caduca 2300
  carpelar 1438
  compuesta 1950
  embrionaria 3786, 5080
  escamiforme 8000
  floral 3473
  flotante 3460
  foliácea 3507
  germinal 3786
  germinal externa 2779
  heliófila 8789
  juvenil 7291
  de ~ laciniada 4996
  de luz 8789
  peltada 6688
  de ~ persistente 6157
  pinnada 6951
  primordial 7291
  seminal 2094
  siguiente 5991
  de sombra 8320
  tectriz 1242
  vegetativa 3507
hojas, de ~ caedizas 2300
holándrico 4307
holobasidio 4313
holoblástico 4314
holoceno 4315
holoenzima 4317
hologameto 4318
hologamia 4320
hologámico 4319
holótipo 4319
hologínico 4321
holometábolos 4322
holoparásito 4324
holoproteido 8293
holótipo 4326
holotrico 4325
homeocinesis 4328
homeólogo 4329
homeosis 4330
homeosinapsis 4332
homeostasis 4331
homeotermo 4351, 4352
homeotípico 4334
homínidos 4336
homoalelo 4337
homocario 4353
homocarión 4353
homocariosis 4354
homocéntrico 4338
homocigosis 4375
homocigoto 4374, 4375
homocromía 4339
homodonto 4340
homofítico 4360
homogamético 4342
homogamia 4344
homógamo 4343
homogenato 4345
homogeneidad 4346
homogeneizado 4345
homogenético 4348
homógeno 4347
homoinjerto 4350
homolecítico 4355
homolecito 4355
homología 4357
homomeria 4358
homomórfico 4359
homoplástico 4361
homoploidia 4362
homopolar 4363
homópteros 4364
homosomal 4365
homosporia 4367
homósporo 4366
homostilia 4369
homostílico 4368
homostilo 4368
homotalia 4371
homotálico 4370
homotalismo 4371
homotérmico 4352
homotipia 4373
homotipo 4372
homozigosis 4375
homozigotismo 4375
homozigoto 4374
hongos 3624
  algoides 6873
  filamentosos 4493
  imperfectos 3625
  superiores 9289
  verdaderos 9289
horda 4381
horizonte A 263
  B 1014
  C 1686
  edáfico 8361
horizonte eluvial 263
  iluvial 1014
hormogonio 4382
hormona 4386
  adaptativa 143
  adrenocortical 174
  adrenocorticotropa 181
  adrenomedular 178
  antidiurética 9467
  cortical adrenal 174
  corticosuprarrenal 174
  de crecimiento 8398
  del cuerpo lúteo 2069
  esteroide 8636
  estimulante de las células intersticiales 5269
  estrógena 6293
  floral 3491
  florígena 3491
  folicular 6294
  foliculoestimulante 3518
  gonadal 8227
  gonádica 8227
  gonadotrófica 3911
  gonadotrópica 3911
  de (las) heridas 9709
  hipofisaria 4509
  inhibidora 602
  juvenil 4920
  lactógena 5011
  lactotropa 5011
  lútea 7328
  luteinizante 5269
  luteotrófica 5011
  luteótropa 5011
  masculina 475
  medular adrenal 178
  medulosuprarrenal 178
  melanocito-estimulante 4759
  de la muda 2751
  ovárica 6449
  paratiroides 6610
  peptídica 6707
  pituitaria 4509
  posthipofisaria 7212
  prehipofisaria 562
  proteínica 7388
  sexual 8227
  somatotropa 8398
  suprarrenal 176
  tiroidea 9111
  tirotrópica 9112
  tisular 9133
  traumática 9709
  vegetal 6910
hormonal 4383
hormonogénico 4387
hospedante 4390
  alternativo 363
  animal 512
  intermedio 4755
huesecillo del oído 2746
hueso 1216
  (bot.) 8679
  carpiano 1437
  cartilaginoso 1446
  de cartílago 1446
  corto 8264
  coxal 4283
  del cráneo 2112
  cutáneo 5473
  dérmico 5473
  esfenoides 8479
  etmoides 3125
  frontal 3592

hueso ilíaco 4283
lagrimal 4997
largo 5236
maxilar 5432
membranoso 5473
metacarpiano 5545
metatarsiano 5564
nasal 5977
occipital 6275
palatino 6538
parietal 6628
plano 3445
primordial 1446
sacro 7943
sesamoide 8202
de sustitución 1446
tarsal 8932
tarsiano 8932
temporal 8975
huésped 4390
accidental 52
alternativo 363
bacteriano 4392
definitivo 2333
intermediario 4755
huevas 7895
huevo 2799
durable 9693
de invierno 9693
en mosaico 5822
de regulación 7703
regulatorio 7703
en reposo 9693
subitáneo 8786
de verano 8786
humedad atmosférica 777
húmero 4402
humificación 4406
humor acuoso 676
vítreo 9602
humus 4407
ácido 7610
bruto 7610
estable 8573
suave 5679
de superficie 8821
huso acromático 73
central 1551
cóncavo 4312
mitótico 5719
muscular 5901
nuclear 6208
de segmentación 1785

I

ICSH 5269
ictiología 4525
id 4527
idante 4528
idioblasto 4531
idiocromatina 4532
idiograma 4533
idioplasma 3787
idiosoma 4537
idiótipo 4535
idiovariación 4536
ido 4527
íleon 4538
ilion 4540
imago 4544
imbibición 4545
imbricado 4547
impar 4573
imparidigitado 4574
imparipinnado 4575
impermeabilidad 4577
al agua 9640
impermeable 4578
al agua 9641
implantación 6124
implante 4579
impresión 4582
sensorial 8171
imprinting 4582
impulso 2706
de enjambrazón 8842
migratorio 5669
nervioso 6046
reproductor 7757
sexual 8243
impureza 4585
impuridad 4585
inactivación 4587
inactivar 4586
inanimado 5145
inbreeding 4589
incapacitación híbrida 4416
incineración 4593
de basuras 7691
incisivo 4596
incluir 2862
inclusión 2863
celular 1515
grasa 3327
lipídica 3327
en parafina 6581
del plasma 2253
incoloro 1910
incompatibilidad 4598
propia 8114
incompatible 4600
incorporación de radioisótopos 4604
increción 4608
incremento 4607
incretor(io) 4609
incrustación 4611
incubación 4613
incubador bucal 5843
incubadora 4619
incubar 4612
independiente de la densidad 2383
indicador de la acidez (del suelo) 92
de cal 6996
ecológico 2762
radiactivo 7555
del tiempo 9125
índice de aislamiento 4881
de apareamiento metafásico 5561
cefálico 1571
de crecimiento 3988
de metabolismo basal 954
mitótico 5716
de mortalidad 2290
de natalidad 1139
de polución 2349
de las profases 7347
de proteína bruta 2153
de refracción 7684
de reproducción 7746
de supervivencia 8827
indigerible 4634
indigestibilidad 4633
indigestible 4634
inducción embrionaria 2878
enzimática 3033
de la floración 4642
de mutaciones 4643
inductor 4645
indusio 4646
inervación 4693
inervar 4692
inexcitabilidad 4649
inexcitable 4650
infecundidad 949, 4655
infecundo 4654
inferior al umbral 8739
ínfero 4652
infértil 4654
infertilidad 4655
inflorescencia 4656
definida 2332
indefinida 4623
influencia ambiental 3017
del medio ambiente 3017
influido por el sexo 8228
información genética 3742
hereditaria 4190
infrasonido 4661
infraumbral 8739
infrutescencia 4662
infundibiliforme 4663
infundíbulo 4664
infundir 4665
infusión de heno 4112
infusorios 4666
ingestión 4668
de alimentos 3530
ingle 3956
inguinal 4669
inhibición competitiva 1934
del crecimiento 3984
enzimática 3034
de la germinación 3807
mitótica 5717
no competitiva 6155
recíproca 7644
del reflejo 7674
por retroacción 3340
inhibidor 4675, 4676
del crecimiento 3985
enzimático 3035
de la germinación 3808
metabólico 5531
inhibitorio 4676
inicial 4683
injertar 9225
injerto 3923, 9226
autoplástico 848
homoplástico 4350
inmadurez 4549
inmaduro 4548
inmaturo 4548
inmersión en aceite 6302
inmiscible 4552
inmisión 4553
inmóvil 6163
inmune 4554
inmunidad 4557
adquirida 97
activa 124
pasiva 6645
inmunización 4559
inmunizar 4561
inmunobiología 4562
inmunoelectroforesis 4564
inmunoglobulina 4565
inmunología 4568
inmunológico 4566

inmunólogo 4567
inmunoquímica 4563
inmunoreacción 4556
inmunosupresión 4571
innato 1979
innocuo 4694
inocuidad 4695
inoculación 4698
inocular 4697
inóculo 4699
inocuo 4694
inofensivo 4694
inorgánico 4700
inquietud motora en las aves migratorias 5677
inquilino 4701
insalivar 4702
insecticida 4707
insectívoro 4708, 4709
insecto adulto 4544
  beneficioso 9422
  benéfico 9422
  dañino 4703
  nocivo 4703
  perfecto 4544
  polinizador 7109
  útil 9422
insectos 4706
  sociales 8342
inseminación 4712
  artificial 734
inseminar 4711
inserción foliar 5078
  del músculo 5899
insight 4716
insolubilidad 4717
  en agua 9642
insoluble 4718
  en agua 9643
inspiración 4719
instinto 4722
  agresivo 257
  de alimentación 3342
  combativo 2707
  de conservación 4723
  genésico 7757
  gregario 4182
  de huida 3111
  de incubación 4615
  de lucha 2707
  maternal 5410
  materno 5410
  nidificador 6064
  de reproducción 7757
  reproductor 7757
  sexual 8243
insulina 4728
integumento 4729
intensidad del estímulo 8649
  lumínica 5149
  luminosa 5149
  respiratoria 7798
  de transpiración 9223
  transpiratoria 9223
interacción 4733
  factorial 3712
  de los genes 3712
  huésped-parásito 4394
intercambiador de iones 4835
  iónico 4835
intercambio alimenticio 9275
  de bases 958
  de calor 4129
  gaseoso 3669
  genético 3707
intercambio de iones 4832
  iónico 4832
  de material 3179
  múltiple 1948
  respiratorio 7795
  de segmentos 8082
  de sustancias 3179
intercierre 4753
intercinesis 4752
intercostal 4742
interespecífico 4778
interesterilidad 4782
interfase 4774
interferencia 4744
  del centrómero 1566
  cromatídica 1694
  quiasmática 1633
interferón 4747
interglaciación 4750
intermediario químico 9219
intermedina 4759
intermitente 4761
interneurona 4770
internodio 4772
interoceptor 4773
intersexo 4776
intersexuado 4776
intersexualidad 4777
intersticial 4784
intersticio 4783
  foliar 5077
  sináptico 8867
intervalo de generaciones 3725
intestinal 4791
intestino 4799
  anterior 3548
  branquial 6801
  cefálico 4114
  ciego 1338
  delgado 8329
  grueso 5034
  medio 5505
  posterior 4279
intina 4800
intracelular 4801
intraespecífico 4802
introgresión 4806
introrso 4808
intususcepción 4809
inulina 4810
invaginación 4811
invasión por parásitos 6603
inventario de vegetación 9479
invernación 6459
invernar 6458
inversión 4813
invertasa 4815
invertebrados 4816
invertina 4815
investigación 7767
investigador 7773
investigar (sobre) 7768
inviabilidad 4818
inviable 6178
in vitro 4820
in vivo 4821
involucela 4822
involucelo 4822
involución 4826
involucro 4823
involuto 4825
ion 4830
  anfótero 9779
  hermafrodita 9779
  híbrido 9779
ion hidrógeno 4440
ionización 4842
ionizar 4843
iris 4845
irradiación 4849
  con rayos X 9722
  ultravioleta 9349
  UV 9349
irradiar 4848
irreversibilidad de la evolución 4851
irrigación sanguínea 1196
irritabilidad 4852
irritable 4853
isidio 4855
isla hemática 1182
  sanguínea 1182
islotes de Langerhans 4856
isoaglutinina 4857
isoalelo 4858
isoantígeno 4860
isoconto 4874
isocromosoma 4862
isocuerpo 4859
isodiamétrico 4864
isoenergético 4861
isoenzima 4866
isófano 4894
isofénico 4895
isófeno 4894, 4895
isofenogamia 4896
isogameto 4867
isogamia 4869
isógamo 4868
isogenia 4872
isogénico 4870
isogenómico 4871
isoinjerto 4350
isoionía 4873
isolecítico 4882
isolecito 4882
isoleucina 4883
isomerasa 4886
isomerismo 4888
isómero 4885, 4887
isometría 4890
isométrico 4889
isomórfico 4891
isomorfismo 4892
isomorfo 4891
is(o)osmótico 4902
isopoliploide 4897
isopreno 4898
isoprenoide 4899
isosporia 4367
isósporo 4366
isotipia 4908
isotipo 4907
isotonia 4901
isotónico 4902
isótopo 4903
  del carbono 1409
  radiactivo 7573
isotrópico 4906
isótropo 4906
ísquion 4854

J

jerarquía de picoteo 6663
  social 8341
juego cromosómico 1723
jugo celular 1530

jugo digestivo 2544
entérico 4794
gástrico 3673
intestinal 4794
nuclear 6207
pancreático 6561
vegetal 7971
jurásico 4918

K

kainozoico 1546
kieselgur 4667
kilocaloría 4961
klinotaxia 4977

L

labelo 4985
laberinto membranoso 5482
óseo 1223
labial 4986
labio *(ins.)* 4988
del blastóporo 1158
inferior 5254
superior 9399
labor de investigación 7772
laboratorio de ensayos 2458
de experimentación 2458
de investigación 7770
labro 4994
labrocito 5400
laciniado 4996
lacrimación 1162
lactación 5004
lactalbúmina 5002
lactasa 5003
lactoflavina 7868
lactoglobulina 5012
lactosa 5013
lagrimal 8947
laguna sanguínea 1193
lamarckismo 5017
lamarquismo 5017
lamedor *(ins.)* 5032
lamelibranquios 1145
lámina branquial 3830
celular 1528
media 5664
laminal 5021
laminilla 5018
intergranular 4751
media 5664
lanceolado 5024
lanuginoso 5030
lanugo 5031
lapso vital 5142
laringe 5041
larva 5036
latencia 5043
latente 5044
látex 5051
laticífero 5052
latido cardíaco 4118
del corazón 4118
latifolio 5054
lecito 9740
lecitina 5100
lectotipo 5101
lectura del código genético 1843
lecha 5685
lechaza 5685
leche de buche 2132
de paloma 2132
legislación sobre el medio ambiente 3018
legumbre 5103
articulada 730
leguminosas 5104
lema 5106
lenticela 5109
lenticular 5110
leño 9704
amplio 7296
estrecho 5042
de otoño 5042
de primavera 7296
tardío 5042
temprano 7296
leñoso 5159
lepidópteros 5111
leptonema 5112
leptosporangiado 5113
leptóteno 5114
lesión por radiación 7575
letal 5117
condicionado 1963
equilibrado 936
letalidad 5121
del híbrido 4414
letargo 2675
estival 211
hibernal 9692
de las yemas 1314
leucina 5122
leucocito 5123
eosinófilo 3046
polimorfonuclear 3938
leucoplasto 5124
levadura 9734
de cerveza 993
de panificación 934
levantamiento fitocartográfico 5371
levógiro 5125
levovoluble 8299
levulosa 3601
ley de acción de masas 5057
anticontaminación 612
de la cantidad de estímulo 7646
de la distribución independiente 5486
de la dominancia o uniformidad (de los híbridos) 5484
fundamental biogenética 1071
de Mendel, primera 5484
de Mendel, segunda 5485
de Mendel, tercera 5486
del metabolismo inverso al tamaño 4812
del mínimo 5059
contra la polución del medio ambiente 612
de la segregación 5485
de la suma heterogénea de estímulos 5055
(de la) termodinámica 5060
del todo o nada 353
leyes hereditarias 5062
de la herencia 5062
de Mendel 5483
LH 5269
liana 5127
liasa 5271
liásico 5128
líber 5129
liberar *(energía etc.)* 5130
libre de gérmenes 3785
librea nupcial 6239
librillo 6327
libro 6327
licopeno 5272
licopina 5272
licópsidos 5273
licuación 5208
licuar 5209
licuefacción 5208
licuefacer 5209
lienal 8527
ligado al sexo 8233
ligamento 5146
ligamiento 5190
genético 3713
al sexo 8232
ligasa 5147
lignificación 5162
lignificarse 5164
lignina 5165
lígula 5167
limbo (foliar) 5071
limitación a un sexo 8230
limitado por el sexo 8231
límite del arbolado 9240
del bosque 3557
de distribución 5172
de sensibilidad 8161
de tolerancia 9146
de la vegetación arbórea 9240
limívoro 5175
limnético 5176
limnívoro 5175
limnología 5178
limnoplancton 5179
linaje de células 1517
evolutivo 3165
germinal 3789
lincaje genético 3713
línea de bosques 3557
consanguínea 5183
de descendencia 5182
evolutiva 3165
femenina 3350
genealógica 5181
intermedia 4758
masculina 5341
materna 3350
de nieve 8337
paterna 5341
primitiva 7303
pura 7480
linfa 5274
linfocito 5281
lingual 5186
linina 74
linkage 5190
liocromo 3446
lioenzima 5284
liofilización 5285
lipasa 5194
lípido 5195
lipocito 5200
lipoclástico 5199
lipocromo 5198
lipoide 5201
lipólisis 5202
lipolítico 5199
lipometabolismo 3319
lipopolisacárido 5205
lipoproteína 5206
liposolubilidad 3321
liposoluble 3322

liquen 5131
crustáceo 2157
foliáceo 3514
fruticuloso 3615
liquenina 5132
liquenología 5133
líquido amniótico 407
cefalorraquídeo 1587
de cultivo 2190
exuvial 5840
fijador 3433
hístico 9132
intercelular 9132
intersticial 9132
orgánico 1208
seminal 8133
sinovial 8891
tisular 9132
vacuolar 9432
lisado 5286
lisar 5298
lisígeno 5290
lisina 5288
lisis 5289
lisogenia 5291
lisogénico 5290
lisógeno 5290
lisosoma 5292
lisótipo 5293
lisozima 5294
lítico 5295
litófito 5212
litoral 5216
litoser(i)e 5213
litosfera 5214
litótrofo 5215
lobulado 5222
lobular 5222
lóbulo anterior (de la hipófisis) 561
frontal 3593
occipital 6277
olfativo 6311
parietal 6630
posterior (de la hipófisis) 7211
pulmonar 7457
temporal 8977
localización génica 5225
locomoción 5226
loculicida 5231
locus (génico) 5232
activo 125
lodo de aguas residuales 8212
lofóforo 5250
lomento 5235
lomo 5234
longevidad 5242
longevo 5241
longistilo 5326
longitud de onda 9665
LTH 5011
luciferina 5258
lucha contra el ruido 6151
antipolución 613
antipolutiva 613
por la existencia 8716
lugar de anidamiento 6067
de desove 8427
de puesta 6466
lumbar 5259
lumen 5261
luminescencia 5262
lúnula gris 3954
lupa 5331
luteína 5268
luteotrofina 5011
luz 5261
infrarroja 4659
polarizada 7094
ultravioleta 9350

LL

llamada de apareamiento 5413
de apuro 2641
de atracción 1351
sexual 5413

M

maceración 5299
macollamiento 9120
macollar 9118
macroblasto 5301
macrocarpo 5302
macroclima 5303
macroconsumidor 5304
macroevolución 5306
macrófago 5317
macrofauna 5307
macrofilo 5318
macrofilogénesis 5319
macroflora 5308
macrogámeta 5309
macrogameto 5309
macroglia 5310
macrómero 5311
macrométodo 5312
macromolécula 5313
macromutación 5314
macronúcleo 5315
macronutriente 5316
macronutrim(i)ento 5316
macroscópico 5320
macrospora 5323
macrosporangio 5322
macrostilo 5326
macrótomo 5327
mácula lútea 9739
macho 5339
complementario 1938
madera 9704
madre 2267
maduración 5423
madurar 5428
madurez 5429
sexual 8245
maduro 5427
malacología 5336
malar 9782
malformación 5345
maltasa 5352
maltosa 5353
mamalogía 5356
mamelón gustativo 8936
mamíferos 5355
placentarios 6970
mamila 5354
manada 6509
mancha ciega 1167
germinal 3801
incubadora 4616
ocular 3288
de pigmento 6937
mandíbula (inferior) 5358
superior 5430
mandibulados 5360
mano prensil 3940
manosa 5361
mantenimiento de la(s) especie(s) 8439
mantillo 4407
ácido 7610
estable 8573
suave 5679
manto 5362
manubrio 5364
mapa de área 2643
cromosómico 1719
factorial 3744
genético 3744
génico 3744
de vegetación 9474
marca olorosa 6285
marcación 5383
del territorio 9009
marcado 5383
isotópico 4904
con isótopos 4904
marcador 5381, 9178
radiactivo 7559
marcar (con) 4984
el celo 6298
marchitamiento 9678
temporal 8980
marchitarse 9676
marchitez 9678
permanente 6762
marchito 9677
marginal 5372
marsupiales 5387
marsupio 5388
martillo *(anat.)* 5347
masculinidad 5390
masculino 5338
masetero 5399
mástax 5401
masticación 5402
masticador 1630 *(ins.)*, 5399
mastigóforos 3437
mastocito 5400
mastoides 5406
mastzelle 5400
materia activa 126
colorante 8577
extraña 3551
fibrosa 943
grasa 3328
seca 2721
material genético 3745
hereditario 4191
del nido 6069
de reacción cruzada 2147
de reserva 7777
materiales nutritivos de reserva 3535
materias excrementicias 3192
fecales 3299
maternal 5409
materno 5409
matriz 5418, 9426
celular 1518
citoplasmática 4410
matroclinal 5421
matroclinia 5422
matroclino 5421
madurez 5429
maxila 5430
maxilar 4911, 5431
inferior 5358
superior 5430, 8799

maxilípedo 5436
máximo de selección 8100
meato auditivo (externo, interno) 801
intercelular 4740
mecanismo de acción 5724
de aislamiento 4877
aislante 4877
desencadenante innato 4690
inmunitario 4569
inmunológico 4569
de regulación 7706
regulador 7706
mecanorreceptor 5441
media luna gris 3954
raza 4077
mediano 5442
mediastino 5445
medición 5439
medio 5442
aclarante 1776
de agar 237
ambiente 3004
de cultivo 2191
(de cultivo) básico 965
de cultivo seco 2715
dispersante 2619
de dispersión 2619
dispersivo 2619
externo 3260
de inclusión 2864
interno 4764
líquido 5211
mínimo 5701
nutritivo 6250
selectivo 8099
sólido 8379
medula 5446
adrenal 177
espinal 8496
oblonga(da) 5447
ósea 1219
raquídea 8496
suprarrenal 177
medulado 5948
medusa 5456
megaevolución 5458
megafilo 5318
meganúcleo 5315
megaspora 5323
megasporangio 5322
megasporocito 2869
megasporofilo 5325
megasporogénesis 5324
meiocito 5459
meiosis 5460
meiospora 5461
meiótico 5462
mejora del medio ambiente 4583
melanina 5464
melanismo 5465
industrial 4647
melanóforo 5467
melanotrofina 4759
melatonina 5469
melitosa 7589
melitriosa 7589
membrana 5472
basal 963
basilar 975
biológica 1080
celular 1519
celulósica 1542
coclear 8258
corioalantoidea 314
membrana de Corti 2076
elemental 9389
(extra)embrionaria 2879
de fecundación 3369
fetal 2879
del huevo 2804
limitante 5174
mucosa 5861
natatoria 9669
nictitante 6122
nuclear 6202
obturante 1811
oclusiva 1811
ondulante 9367
ovular 2804
plasmática 7019
plasmática externa 7018
plasmática interna 9148
protectora 7367
quitinosa 1640
semipermeable 8140
serosa 8194
sinovial 8894
tectoria 2076
timpánica 9326
unidad 9389
unitaria 9389
vacuolar 9431
vascular 9457
vitelina 9597
membranela 5479
membranoso 5480
mendelismo 5490
meninges 5491
menotaxia 5492
mericarpo 5493
meristema 5495
apical 642
intercalar 4737
primario 7292
primordial 7305
secundario 8052
meristemo 5495
meroblástico 5496
merocigoto 5501
merogameto 5497
merogamia 5498
merogonia 5499
meromixia 5500
meromixis 5500
mesencéfalo 5502
mesénquima 5503
mesenterio 5504
mesenterón 5505
mesoblasto 5510
mesocarpo 5507
mesocéfalo 5502
mesocele 5508
mesocótilo 5509
mesodermo 5510
mesofílico 5515
mesofilo 5516
mesofita 5517
mesófito 5517
mesoglea 5511
mesohémera 2282
mesomitosis 5512
mesonefros 5514
mesotelio 5518
mesótrofo 5519
mesozoico 5521
mesozoo 5520
metabiosis 5523
metabólicamente activo 5540
metabólico 5525
metabolismo 5541
asimilador 443
de los azúcares 3897
basal 955
celular 1520
constructivo 443
destructivo 1459
energético 2965
glucídico 1400
graso 3319
hidrocarbonado 1400
intermediario 4754
lipídico 3319
mineral 5695
nitrogenado 6140
del nitrógeno 6140
oxidativo 6490
prote(ín)ico 7376
respiratorio 7800
total 9160
metabolito 5542
metabolizable 5543
metabolizar 5544
metacarpo 5546
metacele 5549
metacéntrico 5547
metacinesis 5552
metacromático 5548
metaestro 5576
metafase 5560
metafitas 5562
metáfitos 5562
metagénesis 5551
metamerismo 5555
metámero 5554
metamitosis 5556
metamorfosear 5557
metamorfosis 5558
completa 4323
directa 4163
incompleta 4163
indirecta 4323
metanefros 5559
metaplasia 5563
metatarso 5565
metaxenia 5566
metaxilema 5567
metazoario 5568
metazoo 5568
metencéfalo 5569
metionina 5570
método del carbono 14, 7566
de coloración 8581
de ensayo 3234
de ensayo y error 9246
experimental 3234
de las gotitas 2709
de Gram 3931
de marcado 5384
de marcado y recaptura 5380
de tinción 8581
de los trazadores 9179
meyosis 5460
meyótico 5462
micela 5578
micelio 5937
micetología 5941
micetoma 5939
micetozoos 5969
micobacteria 5940
micofitos 3624
micología 5941
micorriza 5943
micotrofia 5944
micra 5619

monogerminal 9379
monogonia 5774
monohaploidia 5775
monohibridismo 5777
monohíbrido 5776
monoico 5764
monolocular 9381
monomastigoto 9376
monomérico 5781
monomorfo 5783
mononucleado 5784
mononuclear 5784
monoploidia 5789
monopodial 5790
monopódico 5790
monopodio 5791
monos antropomorfos 580
monosa 5793
monosacárido 5793
monosimétrico 9783
monosoma 5794
monosomia 5795
monospermo 5796
monotremas 5798
monotrico 5799
mordiente 5805
morfalasis 5807
morfina 5808
morfogénesis 5809
morfogenético 5810
morfogenia 5809
morfogénico 5810
morfógeno 5810
morfología 5814
  vegetal 6998
morfológico 5813
morfoplasma 5815
morfosis 5816
morgan 5806
morir 2515
mortalidad 5817
mórula 5819
morulación 5820
motilidad 5829
motoneurona 5835
motricidad 5837
móvil 5828
movilización del pigmento 6936
movimiento ameboide 412
  anafásico 462
  antiperistáltico 609
  ciliar 1737
  de curvatura 2197
  deslizante 3863
  expresivo 3251
  flagelar 3436
  instintivo 4726
  de intención 4732
  intencional 4732
  de locomoción 5230
  locomotor 5230
  en masa 5394
  morfogenético 5812
  nástico 5984
  de natación 8849
  natatorio 8849
  nictinástico 6266
  ondulante 9368
  de orientación 6390
  pendular 6694
  periférico 6743
  peristáltico 6750
  reflejo 7675
  reorientado 7661
  reptante 2121
movimiento respiratorio 7802
  de rodar el huevo 2806
  de rotación del huevo 2806
  de sueño 6266
  de sustancias 5847
  de turgencia 9311
  turgente 9311
movimientos estereotipados 8626
  de limpieza 3957
mucilaginoso 5851
mucílago 5850
mucina 5853
mucíparo 5854
mucopolisacárido 5855
mucoproteína 3876
mucosa 5861
mucosidad 5863
mucoso 5858
mucus 5863
muda 5839
  imaginal 190
  larval 5038
  larvaria 5038
  ninfal 6268
mudar 5838, 5923
muela de juicio 9701
muerte celular 1507
  lenta 2516
muestra de agua 9653
  al azar 7598
  de sangre 1189
  de suelo 8371
muestreo 7969
multicelular 5870
multicelularidad 5872
multiflagelado 5875
multigen 5876
multilobular 5878
multinucleado 5880
multinuclear 5880
multiparcial 5881
multipartición 5885
multiplicación celular 1529
  en masa 5396
multiplicarse 5890
multipolar 5891
multivalente 5892
mundo animal 530
  vegetal 7010
muñeca 9710
muscular 5903
musculatura 5910
músculo 5895
  abductor 8
  aductor 149
  agonista 261
  alar 279
  antagonista 557
  cardíaco 5950
  ciliar 1738
  circular 1751
  esquelético 8311
  estriado 8701
  extensor 3254
  flexor 3453
  involuntario 8336
  liso 8336
  longitudinal 5247
  masticador 5399
  retractor 7841
  tensor 8987
  volador 3458
  voluntario 8701
  de vuelo 3458
musgos 5824
mutabilidad 5912
mutable 5913
mutación 5924
  de brotes 8560
  cromosómica 1712
  doble 427
  espontánea 8538
  génica 3714
  genómica 3760
  inducida 4639
  invertida 7843
  letal 5120
  mayor 5314
  paralela 6583
  plastidial 7035
  plastídica 7035
  provocada 4639
  de punto cero 9749
  puntual 3714
  reversiva 7843
  revertida 7843
  simultánea 8296
  sistémica 8905
  supresora 8814
  de yema 8560
mutacional 5931
mutacionismo 5930
mutagénesis 5917
mutagénico 5918
mutágeno 5914, 5918
mutante 5920
  auxotrófica 6254
  enano 2736
  nutricional 6254
mutar 5923
mutón 5935
mutualismo 5936

N

NAD 6118
NADP 6119
nanismo 5972
nanoplancton 5973
nar 5974
nasal 5976
nasofaringe 5982
nastia 5984
nástico 5983
natalidad 5985
neártico 5996
necesidades de agua 9645
  alimenticias 3534
  calóricas 1360
  energéticas 2961
  nutritivas 6251
  proteicas 7380
  de sustancias nutritivas 6251
  vitamínicas 9593
necrobiosis 5998
necrocitosis 1507
necrófago 6001
necrohormona 9709
necrosis 6002
néctar 6003
nectario 6004
necton 6009
néfrico 7721
nefridio 6031
nefridióporo 6034
nefrocele 6032
nefrón 6033
nefrona 6033

Español

nefrostoma 6035
nefrotomo 6036
nematelmintos 6010
nematicida 6011
nematocisto 6012
nematodos 6013
nemertinos 6014
neoblasto 6015
neocórtex 6016
neodarwinismo 6017
neoformación 6018
neogeno 6020
neogeo 6019
neolamarckismo 6021
neolítico 6022
neomorfo 6023
neopalio 6024
neotenia 6026
neotenina 4920
neotipo 6028
nerítico 6037
nervación 6038
  foliar 5091
nervado 2092
nervadura 6038
  foliar 5091
  de ~ paralela 6584
  de ~ reticulada 6072
nerviación 6038
  alar 9686
  foliar 5091
  de ~ reticulada 6072
nervio 6039
  abducens 7
  accesorio 50
  acústico 802
  auditivo 802
  central 5662
  ciático 8020
  craneal 2116
  espinal 8498
  esplácnico 8521
  facial 3291
  foliar 5090
  glosofaríngeo 3870
  hipogloso 4502
  medial 5662
  medio 5662
  motor 5834
  neumogástrico 9438
  ocular externo 7
  oculomotor 6281
  olfativo 6312
  óptico 6362
  patético 9273
  raquídeo 8498
  sensitivo 8172
  sensorial 8172
  trigémino 9253
  troclear 9273
  vago 9438
  visceral 8521
nervios vasomotores 9466
nervioso 6054
neumatóforo 7067
neural 6073
neurapófisis 6081
neuraxis 6082
néurico 6083
neurilema 8018
neurita 879
neuroblasto 6086
neurocentro 6041
neurocito 6040
neurocráneo 6087
neurocrinia 6088
neuroeje 6082
neuroendocrino 6090
neuroepitelio 6091
neurofibrilla 6092
neurofisiología 6103
neurofisiólogo 6102
neuroganglio 6048
neurogénico 6094
neuroglía 3860
neurohipófisis 6097
neurohormona 6096
neurología 6099
neurólogo 6098
neurona 6040
  de asociación 4770
  intermedia 4770
  internuncial 4770
  motora 5835
  sensitiva 8173
  sensorial 8173
neuropilo 6104
neurópteros 6105
neurosecreción 6106
neurosecretor 6107
neurosensorial 6108
neurotransmisor 9219
neurovegetativo 6110
neurula 6111
neurulación 6112
neuston 6113
neutrófilo 6115
neutrón 6114
niacina 6121
niacinamida 6117
nicotina 6120
nicotinamida 6117
  -adenina-dinucleótido 6118
  -adenina-dinucleótido-fosfato 6119
nictanto 6264
nictinastia 6266
nictinástico 6265
nicho ecológico 2764
nidación 6124
nidada 1292
nidarios 1819
nidificación 6068
  en colonia 1275
nidificar 6061
nidoblasto 1820
nidocilio 1821
nidocisto 6012
ninfa 6267
ninfalización 7473
ninfosis 7473
nitrificación 6130
nitrificar 6131
nitrogenado 6141
nitrógeno atmosférico 778
  no proteínico 7815
  proteico 7379
  residual 7815
  total 9161
nivel energético 2963
  de energía 2963
  de excitación 3189
  freático 3965
  de organización 6381
  de polución 2349
  trófico 9277
no ácido-resistente 85
  alélico 6153
  apareado 4573
  celular 56
no deciduo 6157
  disyunción 6158
  específico 6174
  fertilizado 9369
  medulado 434
  mielinizado 434
  miscible 4552
  móvil 6163
  nucleado 6164
  reducción 6169
  saturado 9394
  septado 6170
  tabicado 6170
  tóxico 6175
  vascular 6177
  viable 6178
  virulento 862
  viviente 5145
  volador 3459
nociceptor 6145
nocividad 6147
nocivo 6187
nódulo embrionario 2874
  linfático 5277
  de Ranvier 7603
nódulos radiculares 7906
nombre específico 8448
  genérico 3770
nomenclatura binaria 1043
  binomial 1043
noradrenalina 6179
norma de reacción 6180
notocorda 6182
notocord(i)o 6182
notogeo 6183
notomorfo 6184
nucela 6190
nucelo 6190
nucleado 6213
nucleasa 6112
nucleína 6215
nucleinación 6216
núcleo atómico 783
  (celular) 6231
  cigótico 9789
  complementario 1941
  emigrante 5676
  enérgico 5533
  espermático 8458
  estacionario 8602
  de fusión 3632
  gamético 3645
  generativo 3728
  hijo 2280
  (de) interfase 4775
  metabólico 5533
  ovular 2805
  polar 7086
  en reposo 7820
  segmentado 1783
  vegetativo 9483
nucleohistona 6218
nucleoide 6219
nucléolo 6223
nucleolonema 6222
nucleoplasma 6225
nucleoplasmático 6226
nucleoplásmico 6226
nucleoproteína 6227
nucleósido 6228
nucleótida 6230
nucleótido 6230
nudo *(bot.)* 4981
  (caulinar) 6148
  de Hensen 7302

O

osmolaridad 6412
osmómetro 6413
osmorreceptor 6414
osmorregulación 6415
ósmosis 6416
osmotactismo 6417
osmotaxis 6417
osmótico 6418
osteicties 6428
osteoblasto 6429
osteocito 6432
osteoclasto 6430
osteocráneo 6431
osteogénesis 6433
osteología 6434
osteometría 6435
ostíolo 6436
óstium 6437
ótico 6438
otocisto 809
otolito 6441
outbreeding 6442
ovado 6470
ovario 6452
ovariolo 6450
ovariotestis 6451
oviducto 6460
ovificación 6461
oviparidad 6462
ovíparo 6463
oviposición 6465
oviscapto 6467
ovoalbúmina 6446
ovocélula 6478
ovocentro 6469
ovocito 6341
ovogénesis 6344
ovogonio 6345
ovoide(o) 6470
ovonúcleo 2805
ovoplasma 6348
ovopositor 6467
ovotestis 6451
ovótida 6352
ovovitelina 6472
ovoviviparidad 6473
ovoviviparismo 6473
ovovivíparo 6464
ovulación 6475
óvulo 6476, 6478
oxibiosis 205
oxidable 6481
oxidación 6485
oxidante 6482
oxidar 6484
oxidasa 6483
  de aminoácidos 398
oxidorreducción 6493
oxidorreductasa 6492
oxigenación 6504
oxigenasa 6503
oxihemoglobina 6505
oxitocina 6506

P

pabellón (de la oreja) 1957
padre 8307
paidogamia 6510
paidogénesis 6511
paisaje natural 5988
  transformado por el hombre 2187
paladar 6541
  blando 8348
  duro 4099
palatal 6537
palatino 6537
pálea 6543
  inferior 5106
paleártico 6520
paleoantropología 6521
paleobiología 6522
paleobotánica 6523
paleoceno 6524
paleoclimatología 6525
paleoecología 6526
paleoencéfalo 6544
paleofitología 6523
paleógeno 6528
paleogeografía 6529
paleolítico 6530
paleoncéfalo 6544
paleontología 6532
  humana 6521
paleontólogo 6531
paleozoico 6535
paleozoología 6536
paleta natatoria 8850
paletilla 8002
palingenesia 6546
palingenia 6546
palinología 6559
palio 5362, 6551
palmado 6552
palmatilobulado 6553
palmatipartido 6554
palmeado 6552
palmípeda 6556
pálpebra 3285
  inferior 5252
  superior 9396
palpo 6558
  labial 4987
  maxilar 5434
páncreas 6560
pancreatina 6562
pancreozimina 6563
pángen 6564
pangénesis 6565
panícula 6566
panículo adiposo 167
panmixia 6567
panmixis 6567
panspermia 6568
panza 7927
papaína 6570
papel filtrante 3412
  -filtro 3412
  reactivo 9018
papila 6573
  anal 452
  dérmica 2415
  lingual 5188
papo 6574
paquíteno 6508
par 6517
  de cromosomas 1721
  cromosómico 1721
  de genes 6516
parabiosis 6576
paracárpico 6577
paracéntrico 6579
parada nupcial 2101
  sexual 2101
paráfisis 6591
paralelinervio 6584
parámetro 6586
parámetro del estímulo 8650
paramilo 6588
paramitosis 6587
paranúcleo 6590
paraplasma 6593
parapodio 6594
parápodo 6594
parapófisis 6595
parasexual 6596
parasexualidad 6598
parasíndesis 6609
parasitado 6605
parasitar 6604
parasitario 6600
parasitismo 6602
  de cría 1279
  social 8343
parásito 6599
  animal 516
  de cría 1278
  facultativo 3297
  forzado 6273
  intestinal 2989
  obligado 6273
  obligatorio 6273
  permanente 6758
  temporal 8979
  vegetal 7001
parasitología 6607
parasitólogo 6606
parathormona 6610
paratiroides 6611
paravariación 6613
parazoario 6614
parazoo 6614
pared abdominal 6
  celular 1534
  celulósica 1543
  divisoria 8182
  intestinal 4798
parénquima 6615
  aerífero 8535
  cortical 2075
  esponjoso 8535
  lagunoso 8535
  leñoso 9705
  liberiano 983
parental 6621
parentesco 7709
parición 1138
parietal 6627, 6628
paripinnado 6632
paroico 6633
parótida 6634
párpado 3285
  inferior 5252
  superior 9396
parque nacional 5992
parte constitutiva 2000
partenocarpia 6635
partenogénesis 6636
partes aéreas 199
  sexuales 3757
partícula alimenticia 3532
  elemental 2851
  fundamental 2851
  transportadora de electrones 2841
partículas ribonucleoproteicas 7875
parto 1138
paste(u)rización 6646
paste(u)rizar 6647

pata ambulatoria 9616
locomotora 9616
maxila 5436
patagio 6648
patela 4979
paternal 6651
paterno 6651
patogenicidad 6655
patógeno 6653
patrimonio genético 3739
hereditario 4191
patroclinia 6657
patroclino 6656
patrón 8668
paurometabolia 6659
pauta de conducta 997
peces cartilaginosos 1660
pulmonados 2585
óseos 6428
pecíolo (foliar) 6781
peciólulo 6782
pectasa 6664
pectina 6666
pectoral 6667
pedicelo (floral) 6670
pedigree 6671
pedipalpo 6674
pedogénesis 6511, 8360
pedología 8372
pedunculado 6679
pedúnculo 8582
(floral) 6678
fructífero 3609
hipofisario 4510
ocular 3289
pelágico 6682
pelaje 6681
pelecípodos 1145
pelo 4064
absorbente 7904
estaminal 8586
glandular 3854
radicular 7904
sensitivo 8170
sensorial 8170
táctil 8911
urticante 8664
peloria 6686
peltado 8040
peludo 6943
pelviano 6689
pélvico 6689
pelvis 6693
renal 7723
pena 3457
pene 6698
penetración completa 1945
(génica) 6695
incompleta 4602
penicilina 6697
penniforme 6952
penninervado 6953
penninervio 6953
pentosa 6700
pentosana 6699
penuria de agua 9634
pepa 4953
pepita 4953
pepsina 6701
pepsinógeno 6702
peptidasa 6703
péptido 6704
peptona 6708
peptonización 6709
percepción 6710
percepción acústica 8406
del estímulo 8651
de la gravedad 3942
del movimiento 5846
olfativa 6314
del sonido 8406
perceptibilidad 6711
pérdida de agua 9644
de calor 4131
de energía 2964
por evaporación 3154
respiratoria 7799
perecer 2515
perenne 6713
perennidad 6712
perennifolio 6157
perfil del suelo 8368
perforatorium 103
perianto 6717
periblasto 6718
periblema 6719
pericardio 6721
pericario 6732
pericarión 6732
pericarp(i)o 6722
pericéntrico 6723
periciclo 6727
periclinal 6725
pericondrio 6724
peridermis 6728
peridermo 6728
peridineas 2564
peridio 6729
perifiton 6744
perígino 6731
perigonio 6730
perilinfa 6733
perimetrio 6734
perimisio 6735
perineo 6736
perineur(i)o 6737
periodicidad 6740
anual 542
diaria 2649
diurna 2649
estacional 8046
período de calores 6296
del celo 6296
de crecimiento 3972
de desarrollo 6738
experimental 3238
de generación 3726
geológico 3776
de gestación 3818
glacial 3846
de incubación 1281, 4617
interglacial 4750
de latencia 5046
de letargo 7823
de maduración 5425
medio de vida 5438
postglacial 7215
de presentación 7278
refractario 7685
de reposo 7823
reproductor 7760
sensible 8158
sensitivo 8158
de sequía 2711
vegetativo 9485
de vida media 4075
periostio 6741
periplasma 6745
perisodáctilos 6749
perisperma 6747
perispermo 6747
perisporio 6748
peristalsis 6750
peristaltismo 6750
peristasis 6751
peristoma 6752
peritecio 6753
peritoneo 6755
peritrico 6756
perjudicial 6187
permeabilidad 6764
al agua 9647
celular 1526
de la membrana 5476
selectiva 8101
permeable 6765
al agua 9648
permeación 6767
permeasa 6766
pérmico 6768
permutación 6770
perniciosidad 6147
peroné 3395
peroxidasa 6771
perpetuación de la(s) especie(s) 8439
persistencia 6772
persistente 6773
perspiración 6775
perturbación del medio ambiente 3014
peso atómico 785
corporal 1213
fresco 3589
húmedo 9670
molecular 5745
seco 2722
vivo 5221
pesticida 6778
pétalo 6780
pétalos, de ~ concrescentes 3659
peto 7042
petrificación 6784
petrificarse 6785
pez (de)predador 7250
lechal 5687
migrante 5666
ovado 8424
rapaz 7250
pezón 5354, 8948
pezuña 1772
pH 6788
piamadre 6929
picador *(ins.)* 8661
picnidio 7493
picnosis 7494
pie ambulacral 381
bífido 1786
piel 8314
pieza floral 3474
piezas bucales 5845
pigmentación 6939
pigmentar 6932
pigmento 6931
accesorio 51
asimilador 758
asimilable 758
biliar 1028
cutáneo 8317
hemático 1184
respiratorio 7804
sanguíneo 1184
vegetal 7003
visual 9563
píleo 6941

Español

píloro 7495
pilorriza 7901
pilosidad 6944
piloso 6943
pinna 6949
pinnado 6950
pinnatinervio 6953
pinocitosis 6955
pinzas 6946
pipeta 6957
pirámide de las edades 7195
pirenoide 7498
piridina 7499
piridoxina 7500
pirimidina 7501
pirrol 7503
piruvato 7505
piso de altitud 368
  altitudinal 368
  de vegetación 8688
pista olorosa 6286
pistilo 6958
pitocina 6506
pituicito 6965
pituitaria 4511
  anterior 561
  posterior 7211
pixidio 7507
placa 7014
  de agar 238
  ambulacral 380
  basal 956
  celular 1528
  córnea 4389
  cribosa 8277
  de cultivo 2192
  ecuatorial 3099
  incubatriz 4616
  madrepórica 5330
  medular 5451
  neural 6079
  nuclear 3099
  ósea 1224
  de Petri 6783
  polar 7087
  quitinosa 1641
  terminal 2955
  terminal motora 5833
placenta 6968
placentación 6971
placentarios 6970
placoda del cristalino 5108
plaga animal 6777
plagiotropismo 6975
plagiótropo 6974
plaguicida 6778
plan estructural 8714
  de organización biológica 1081
planación 6977
planctófago 6981
plancton 6978
  aéreo 207
  animal 9771
  marino 4082
  vegetal 6920
planctónico 6979
planctonófago 6981
planctonte 6980
planificación del medio ambiente 3020
  territorial 8417
  del territorio 8417
plano de división 2654
  ecuatorial 3098
  mediano 5443
plano sagital 7945
  de segmentación 1784
  de simetría 8854
planocito 9617
planogámeta 9760
planogameto 9760
planóspora 9773
planta 6984
  (acuática) sumergida 8741
  adventicia 194
  de agua dulce 3588
  alimenticia 3533
  anemófila 489
  anual 543
  arrosetada 7915
  bienne 1016
  bulbífera 1331
  bulbosa 1331
  calcícola 1345
  calcífuga 1348
  carnívora 4710
  cultivada 2186
  depuradora 7483
  de desalación 2422
  de(s)contaminadora 2316
  desértica 2431
  desertícola 2431
  de día corto 8265
  de día largo 5237
  de día neutro 2282
  enana 2737
  de enraizamiento profundo 2321
  enredadera 9611
  esciófila 8319
  esteparia 8622
  exótica 3223
  experimental 3235
  fibrosa 3380
  de floración nocturna 6128
  flotante 3461
  fotoperiódicamente neutra 2282
  heliófila 4149
  híbrida 4415
  de hoja 3508
  hospedante 4395
  huésped 4395
  incineradora de basuras 7692
  indicadora 4630
  indígena 4632
  inferior 5256
  insectívora 4710
  leñosa 5160
  macrohémera 5237
  madre 6620
  microhémera 8265
  mirmecófila 5966
  nativa 4632
  noctiflora 6128
  nutriz 3533
  ombrófila 6329
  paludícola 4153
  palustre 4153
  parásita 6601
  parasitaria 6601
  perenne 6714
  pionera 6956
  radicante 7547
  de raíces superficiales 8253
  rastrera 2125
  (en) roseta 7915
  ruderal 7923
  rupestre 7934
  rupícola 7934
  saprófita 7977
planta silícola 8288
  de sombra 8319
  suculenta 8767
  sumergida 8741
  superior 4276
  terrestre 9005
  testigo 9019
  tóxica 7081
  trepadora 1800
  tuberosa 9302
  útil 9423
  vascular 9458
  venenosa 7081
  vivaz 6714
  voluble 9611
  xerófila 9720
  zarcillosa 7987
plantación de bosques 217
plantar bosques 216
plantas acuáticas 4455
  élites 8091
  esclerofilas 8031
  pulviniformes 2198
  selectas 8091
  de semilla 8472
  de semilla desnuda 4016
plantígrado 7011
plántula 7012
plánula 7013
plaqueta sanguínea 9098
plasma 7015
  germinal 3787
  sanguíneo 1185
plasmacito 7016
plasmagen(e) 7017
plasmalema 7018
plasmático 7021
plasmatogamia 7037
plasmazelle 7016
plásmico 7021
plasmidio 7023
plasmina 7024
plasmodesma 7025
plasmodio 7026
plasmogamia 7027
plasmólisis 7028
  incipiente 4594
  límite 4594
plasmón 7029
plasmosoma 7030
plasmótipo 7029
plasmotomía 7031
plastidio 7033
plastidiótipo 7036
plastidoma 7036
plasto 7032
plastogamia 7037
plastogen(e) 7038
plastoma 7036
plastoquinona 7040
plastosoma 5709
platelmintos 7045
platicultivo 7044
platina (microscópica) 5635
plecténquima 7046
plegado *(vernación)* 7059
pleiomorfismo 7053
pleiotropía 7050
pleiotrópico 7049
pleiotropismo 7050
pleistoceno 7051
pleocasio 7047
pleomórfico 7052
pleomorfismo 7053
pleópodo 7054

pleotropía 7050
pleotropismo 7050
pleotropo 7049
pleroma 7055
pleura 7056
costal 2090
mediastínica 5444
parietal 2090
pulmonar 7458
pleuston 7058
plexo nervioso 6049
pleyocasio 7047
pliegue germinativo 3755
gonadal 3755
neural 6077
plioceno 7060
pluma 3333
caudal 8919
cobertera 2020
timonera 8919
plumaje 7062
juvenil 4921
nupcial 6239
plumilla 7063
plumón 2700
plúmula 2701, 7063
pluricelular 5870
pluricelularidad 5872
plurienal 7065
plurilocular 5869
plurinucleado 5880
pluripartición 5885
pluristratificado 5877
pluvi(i)silva 7590
población 7183
consanguínea 4592
híbrida 4417
inicial 6392
de laboratorio 4992
mendeliana 5488
mezclada 5722
de origen 6392
original 6392
vegetal 8594
pobre en especies 7182
poder absorbente 35
de absorción 35
de absorción (del suelo) 8777
de acumulación 8683
de adsorción 186
amortiguador 1321
de cohesión 1863
fecundante 3373
germinativo 3804
patógeno 6655
raceador 7274
reproductor 7753
de resolución 7785
resolutivo 7785
resolvente 7785
tampón 1321
podredumbre 7488
pogonóforos 7070
poiquilotérmico 7072
poiquilotermo 7071, 7072
polaridad 7089
polarización 7090
polarizar 7093
polen 7096
polialelia 5883
poliandria 7122
poliándrico 7121
policario 5880
policariocito 7143
policárpico 7123
policasio 7047
policéntrico 7124
policíclico 7127
policruzamiento 7126
poliembrionía 7128
polienérgida 7129
poliéstrico 7154
polifactorial 7130
polifagia 7158
polífago 7159
polifenia 7050
polifilético 7161
polifiletismo 7162
poligamia 7132
polígamo 7131
poligen 7133
poligénesis 7134
poligenia 7136
poligénico 7135
poliginia 7138
poligínico 7137
polihaploidia 7139
polihibridismo 7141
polihíbrido 7140
poliholósido 7167
polimerasa 7144
polimería 7147
polimérico 7145
polimerismo 7147
polimerización 7148
polímero 7145
polimorfismo 7151
polimorfo 3938, 7150
polinario 7106
polinia 7113
polinífero 7112
polinio 7113
polinización 7110
cruzada 338
entomófila 2999
polinizante 7108
polinizar 7107
polinucleótido 7152
poliósido 7167
polipéptido 7156
poliploide 7163
poliploidia 7165
poliploidización 7164
pólipo 7155
poliquétidos 7125
poliquetos 7125
polirribosoma 7166
polisacárido 7167
polisoma 7166
polisomía 7169
polisómico 7168
polispermia 7171
polispermo 7170
polistelia 7172
politenia 7174
politénico 7173
política ambiental 3021
del medio ambiente 3021
politípico 7178
politópico 7175
politopismo 7176
politrófico 7177
polizoos 1308
polo animal 520
germinativo 3798
del huso 8503
vegetativo 9471
polocito 7083
poluante 7114
poluar 7115
polución 7117
del agua 9649
del agua del mar 7118
del aire 269
por hidrocarburos 6303
del mar 7118
del medio ambiente 3011
del suelo 8367
polucionar 7115
pollada 1292
pomo 7180
pómulo 9782
poner huevos 6464
en placa 7043
pool de genes 3715
porcentaje de aberración 10
de crossing-over 2146
de recombinación 7651
de supervivencia 8829
porfirina 7204
poríferos 7202
poro 7199
aerífero 270
areolado 1228
gustativo 4007
nuclear 6204
porogamia 7203
portador 1444
de electrones 2832
de la herencia 990
portainjerto 8668
portaobjetos 5634
posdescarga 221
posición defensiva 2325
posmaturación 223
postadaptación 7209
postcelo 5576
postembrionario 7210
postganglionar 7214
postheterocinesis 7216
posthipófisis 7211
pos(t)potencial 222
postreducción 7218
postsináptico 7219
postura 6465
de amenaza 9089
amenazadora 9089
potencia *(gen)* 7222
potencial de acción 114
de acción muscular 5896
en aguja 8491
biótico 1116
de demarcación 2366
de difusión 2532
eléctrico 2821
energético 2966
evocado 3160
evolutivo 3166
generador 3729
marcapaso 6507
de membrana 5477
de oxidorreducción 6487
de placa terminal 2956
postsináptico excitatorio 3187
postsináptico inhibitorio 4679
de punta 8491
de reacción 3160
de recepción 7636
receptor 7636
redox 6487
reductor 7664
de reposo 7821
reproductivo 7762
reproductor 7762
secundario 222

Español

potenciómetro 7225
po(te)tómetro 7226
preadaptación 7228
precámbrico 7229
precartílago 7230
precelo 7343
precipitable 7231
precipitación 7235
  radiactiva 7554
precipitado 7234
precipitante 7232
precipitar 7233
precipitina 7236
precocidad 7240
precoz 7239
precursor 7242
predación 7244
predador 7245
predatorio 7247
predatorismo 7244
predeterminación 7251
prefloración 212
prefoliación 9509
preformación 7256
preformismo 7257
preganglionar 7258
pregnandiol 7260
preheterocinesis 7263
prehipófisis 561
prehistoria 7264
prelínea 7303
premolar 7267
premunidad 7268
premutación 7269
preñada 7261
preñez 7259
preparación 7271
  aplastada 2155
  delgada 9078
  entera 9673
  por extensión 8333
  en fresco 3584
  histológica 9134
  húmeda 4404
  por impresión 4581
  microscópica 5638
  permanente 6759
  en seco 2719
  standard 8591
  de tejido 9134
preparado 7271
  entero 9673
  puro 7482
preparar 7273
prepotencia 7274
prepucio 7275
pre-reducción 7276
presa 1428, 7285
presere 7307
presináptico 7283
presión demográfica 7194
  de imbibición 4546
  de (in)migración 4551
  de mutación 5928
  osmótica 6420
  parcial 6638
  de población 7194
  radical 7907
  sanguínea 1187
  de selección 8102
  selectiva 8102
  de turgencia 9312
  túrgida 9312
presuntivo 7282
previtamina 7419
primates 7297
primera generación filial 3377
  ley de Mendel 5484
primordio 7306
  floral 3475
  foliar 5080
  plastídico 7350
principio activo 126
priser(i)e 7307
privación de alimentos 3525
probabilidad de fertilización 7308
  de sobrevivir 1602
  de supervivencia 1602
probando 7309
probasidio 7310
probeta 9020
probóscide 7312
proboscidios 7311
procámbium 7313
procedimiento de cría 1276
proceso de activación del fango 119
  de desarrollo 2464
  de evolución 3167
  evolutivo 3167
  del fango activo 119
  metabólico 5535
  vital 5140
procreación 7745
procromosoma 7314
proctodeo 7316
producción bruta 3958
  de energía 2967
  neta 6071
  primaria 7293
producto de asimilación 759
  asimilado 759
  de descomposición 2311
  de de(s)contaminación 2314
  de degradación 1263
  de desdoblamiento 1263
  de desecho 9625
  de excreción 3200
  final 2957
  final del metabolismo 5529
  de fisión 3425
  del gen 3716
  génico 3716
  intermedio 4756
  meiótico 5463
  metabólico 5542
  metabólico intermedio 5532
  secretorio 8065
productor *(ecol.)* 7317
proembrión 7320
proenzima 9793
proestro 7343
prófago 7345
profase 7346
profermento 9793
profilo 7348
proflavina 7322
progámico 7324
progénesis 7325
progenie 7326
progenitor 6617
  femenino 3351
  hembra 3351
  macho 5342
  recurrente 897
progesterona 7328
progestina 7328
programa de investigación 7771
prolactina 5011
prolamina 7330
prolán 1683
  A 3518
  B 5269
prole 7326
proliferación 7334
  celular 1529
promedio vital 5438
promeristema 7305
prometafase 7337
promicelio 7340
promitosis 7339
pronefros 7341
pronúcleo 7342
propagación vegetativa 9487
propágulo 7344
propiedad hereditaria 4192
propio de la especie 8446
propioceptor 7351
propiorreceptor 7351
proplastidio 7350
proporción de asimilación 760
  de crossing-over 2146
  fotosintética 6863
  de incremento 7609
  mendeliana 5489
  metabólica 5537
  de producción 7319
  de (los) sexos 8236
  de utilización 9427
prosencéfalo 7353
prosénquima 7354
prosimio 7355
prostaglandina 7356
próstata 7357
protalo 7394
protamina 7359
protandria 7361
protandro 7360
proteasa 7363
protección de (las) aguas 7365
  ambiental 3023
  de animales 521
  de aves 1132
  de espacios naturales 5029
  del medio ambiente 3023
  de la naturaleza 1993
  del paisaje (natural) 5029
  de las plantas 7005
  contra las radiaciones 7545
  del suelo 8369
protegido contra rayos 7613
proteico 7385
proteido 7369
proteína 7370
  animal 522
  celular 1538
  conjugada 1951
  enzimática 3037
  estructural 8715
  extraña 3550
  fibrosa 8032
  globular 8484
  hemática 1188
  hística 9135
  muscular 5900
  plasmática 7020
  de reserva 7775
  sanguínea 1188
  sérica 8201
  simple 8293
  vegetal 7006, 9469
  viral 9533
  vírica 9533
proteinasa 7384
proteínico 7385

proteinoide 7386
proteoclástico 7390
proteohormona 7388
proteólisis 7389
proteolítico 7390
proteometabolismo 7376
proteoplasto 7387
proterandria 7361
proterándrico 7360
proterandro 7360
proteroginia 7402
proterogínico 7401
proterógino 7401
proteróstomos 7412
protistas 7396
protistología 7397
protistos 7396
protocéfalo 7398
protoclorofila 7399
protocooperación 327
protófita 7406
protófito 7406
protoginia 7402
protógino 7401
protón 7403
protonefridio 7405
protonema 7404
protoplasma 7407
protoplasmático 7408
protoplásmico 7408
protoplasto 7410
protostela 7411
protótrofo 7413
protovértebra 8400
protoxilema 7415
protozoario 7417
protozoo 7417
protozoología 7416
protrombina 9099
protuberancia anular 7181
proventrículo 7418
provitamina 7419
prueba de la avena 860
cis-trans 1759
de complementación 9017
de la curvatura del coleóptilo 860
de descendencia 7327
experimental 3233
del laberinto 5437
de laboratorio 4991
de progenie 7327
sanguínea 1197
con simulacros 5729
psamófilas 7422
psamófitas 7422
psamon 7421
pseudoalelismo 7424
pseudoalelo de posición 7208
pseudoalelos 7423
pseudobranquia 7426
pseudocarpo 7427
pseudodominancia 7428
pseudogamia 7429
pseudohaploidia 7430
pseudohermafroditismo 7431
pseudomixis 7429
pseudoparénquima 7433
pseudopodio 7434
pseudópodo 7434
pseudopoliploidia 7435
pseudopupa 7436
pseudorreducción 7437
psicología de la gestalt 3815
psicólogo gestaltista 3814
psilópsidos 7438
PTE 2841
pteridofitas 7439
pteridófitos 7439
pteridospermas 7440
pterigoto 280
pterigotos 7444
pterina 7441
pterobranquios 7442
pterópsidos 7443
ptialina 7445
púber 7446
púbero 7446
pubertad 7447
pubis 7449
pudrir(se) 7491
puente cromatídico 1693
de hidrógeno 4437
de matriz 5420
de Varolio 7181
puesta 1816, 6465
ovular 6475
puff 7450
pulmón 5265
laminar 1227
pulmonar 7453
pulpa 7460
dentaria 2388
pulsación 7463
pulsión 2706
pulso 7464
pulverizado 7466
pulvínulo foliar 5074
punta 633
radicular 7911
punteadura *(bot.)* 6960
areolada 1228
punto cardinal 1426
ciego 1167
de compensación 1929
de congelación 3579
de contacto 2008
de ebullición 1215
de fusión 5471
de inserción 4714
isoeléctrico 4865
de marchitamiento 9680
de marchitez permanente 6763
oculiforme 3288
de rocío 2471
vegetativo 9486
pupa 7468
coartada 1832
exarada 3178
libre 3178
obtecta 6274
pupación 7473
pupario 7471
pupila 7474
pupíparo 7476
pureza de (las) especies 7487
purificación biológica de aguas residuales 1083
purificar 7484
purina 7485
púrpura visual 7862
putrefacción 7488
en estado de ~ 7492
putrefactivo 7489
putrefactor 7489
putrescencia 7488
putrescente 7492

Q

Q_{10} 8969
quela 1614
quelación 1617
quelado 1615
quelicerados 1618
quelícero 1619
queratina 4950
queratinización 4951
queratinizar 4952
quercitrina 7521
queta 1287
sensorial 8167
quetona 4954
quetosa 4956
quiasma 1631
de compensación 1928
complementario 1939
óptico 6360
quiasmatipia 1634
quilífero 1730
quilificación 1731
quilo 1729
quilópodos 1635
quimera 1636
mericlinal 5494
periclinal 6726
sectorial 8067
quimioautótrofo 1621
quimionastia 1622
quimiorrecepción 1623
quimiorreceptor 1624
quimiosíntesis 1625
quimiotáctico 1626
quimiotactismo 1627
quimiotaxis 1627
quimiotrófico 1628
quimiótrofo 1628
quimiotropismo 1629
quimo 1732
quimosina 7727
quimotripsina 1734
quinasa 4964
quinetina 4969
quinetócoro 1564
quinina 2241, 7524
quinona 7525
quirópteros 1637
quitina 1638
quitinoso 1639

R

rabdoma 7844
rabdómera 7845
rabdómero 7845
rabilargo 5240
racemoso 7528
racimo 7527
radiabilidad 7535
radiación 7540
adap(ta)tiva 144
calorífica 4133
electrónica 2837
nuclear 6205
radiactiva 7557
de rayos X 7898
roentgen 7898
solar 8376
ultravioleta 9352
UV 9352

radiaciones ionizantes 4844
radiactividad 7561
radiactivo 7551
radiado 7536
simétricamente 7538
radial 7536
radicación 7549
radicado 7548
radícula 7550
radio 7587
medular 5452
radioautografía 841
radiobiología 7563
radiocarbono 7564
radiocromatografía 7569
radiocromatograma 7568
radioelemento 7571
radiogenética 7572
radioindicador 7555
radioisótopo 7573
radiolarios 7574
radiolesión 7575
radiomimético 7576
radionúclido 7578
radiopacidad 7579
radio(o)paco 7580
radioquímica 7567
radiorresistencia 7581
radiorresistente 7582
radiosensibilidad 7584
radiosensible 7583
radiotransparente 7586
rádula 7588
rafe 7604
rafinosa 7589
raíz 7900
absorbente 8772
adherente 1767
adhesiva 1767
adventicia 195
aérea 200
aérea de sostén 7352
anterior 9502
axonomorfa 8929
capilar 4068
contráctil 2022
dentaria 9151
dorsal 2684
epigea 200
fasciculada 3312
fibrosa 3393
fúlcrea 8642
lateral 5050
motora 9502
nerviosa 6051
nutricia 8685
nutrífera 8685
posterior 2684
principal 5332
respiratoria 7067
secundaria 5050
sensitiva 2684
tabular 8907
trepadora 1801
ventral 9502
zanco 8642
rama 1251
fructífera 3611
ramificación 7593
dicotómica 2508
lateral 5049
nerviosa 6057
ramificar(se) 7594
ramo 9317
randomización 7599
rapaz 1130, 7247
rapidez de evolución 2465
de germinación 8454
raquis 7531, 9515
foliar 7530
rasgo de la conducta 999
distintivo 2640
rastro foliar 5089
rayo de la aleta 3416
de electrones 2838
medular 5452
rayos anódicos 550
del áster 7088
beta 1012
catódicos 1476
cósmicos 2087
gamma 3654
infrarrojos 4660
mitogenéticos 5710
mitógenos 5710
radiactivos 7558
(de) Roentgen 9727
ultraviolados 9353
ultravioleta 9353
X 9727
raza 7526
biológica 1084
climática 1794
ecológica 2765
fisiológica 6887
geográfica 3775
local 5224
media ~ 4077
original 5986
pura 7478
de ~ pura 7477
reabsorber 7786
reabsorción 7787
reacción 7615
en abrir el pico 3667
del ácido periódico de Schiff 6739
de adaptación 145
adaptativa 145
de aglutinación 251
de alarma 3590
antígeno-anticuerpo 598
del biureto 1142
en cadena 1596
coloreada 1905
defensiva 2326
enzimática 3038
enzimáticamente catalizada 3032
de evasión 864
de evitación 864
de expulsión 3253
de fijación del complemento 1936
de Gram 3931
de huida 3112
de inhibición 4680
inmunitaria 4556
inmunológica 4556
lumínica 5152
metabólica 5538
(nuclear) de Feulgen 3375
de orientación 6391
oscura 2273
de oxidorreducción 6488
PAS 6739
de precipitación 7237
de la precipitina 7237
redox 6488
de respuesta 7813
reacción retardada 2363
de seguimiento 3521
de seguir (a la madre) 3521
del suelo 8370
reaccionar a 7614
reactivación 7622
cruzada 2140
múltiple 5887
reactivar 7621
reactividad 7623
reactivo 7628
reactor nuclear 6206
reajuste cromosómico 7731
rebaño 3466, 4181
migrante 5667
reblandecedor de agua 9654
recapitulación 7629
receptáculo 7632
seminífero 8134
receptor 7633
cutáneo 2202
de distancia 2637
del dolor 6512
gravitatorio 3945
gustativo 8938
luminoso 6859
olfativo 6315
de presión 7280
del sonido 6824
táctil 8924
de temperatura 8972
térmico 9061
recesividad 7639
recesivo 7637
reciente 4315
recipiente secretorio 8063
recíproco 7642
recolección de los alimentos 3528
recombinación 7648
de genes 3717
génica 3717
recombinante 7647
recón 7653
reconocimiento entre las especies 8441
recto 7657
recuento 2096
recursos acuáticos 9651
hídricos 9651
red capilar 1393
nerviosa 6047
redecilla 7837
redox 6493
reducción cromática 7666
de supervivencia 8830
del umbral 9093
reductasa 7665
reductona 7667
reductor 7663
reduplicación 7668
de genes 3718
reflejo 7669
de alargamiento 8699
de axón 880
condicionado 1964
corneal 2057
deglutorio 8837
de elongación 8699
de extensión 3255
de flexión 3454
flexor 3454
incondicionado 9364

S

sensible a la luz 5155
  a la temperatura 8973
sensilio 8157
sensorial 8166
sensorio 8165
sentido auditivo 805
  del color 1906
  cromático 1906
  del dolor 8150
  del equilibrio 8147
  del espacio 8414
  estático 8147
  de formas 3561
  del gusto 4009
  del oído 805
  del olfato 6317
  de la orientación 8149
  de la posición 7221
  postural 7221
  del tacto 8153
  térmico 9047
  del tiempo 9124
  de la vista 9565
señal acústica 95
  advertidora 277
  preventiva 277
sépalo 8177
sépalos, de ~ concrescentes 3661
separación de cromosomas 2609
  longitudinal 5249
septado 8179
septicida 8180
septífrago 8181
septo 8182
sequedad 2710
sequía 2710
ser viviente 5219
  vivo 5219
sere 8186
serie alelomórfica 321
  de ensayos 8190
  evolutiva 3168
  experimental 8190
  de experimentos 8190
  filética 6876
  filogenética 3168
  primitiva 7307
  de soluciones alcohólicas 288
  sucesional 8186
serina 8191
seroalbúmina 8199
serodiagnóstico 8200
serología 8193
serólogo 8192
seroproteína 8201
serosa 8194
seroso 8196
serotonina 8195
sésil 8203
seston 8204
seta 8208
setáceo 1288
seudo... → pseudo...
sexo 8214
sexuado 8238
sexual 8238
sexualidad 8249
sexualmente maduro 8250
siempreverde 3157
sien 8974
sifón 8303
sifonogamia 8304
sifonostela 8306
silíceo 8281
silicificación 8285
silicificar 8286
silicua 8289
  articulada 731
silícula 8287
silúrico 8291
simbionte 8851
simbiosis 8852
  mutualista 5936
  nutritiva 9278
simbiótico 8853
simetría bilateral 1023
  radiada 7537
  radial 7537
simétricamente bilateral 1024
simpátrico 8857
simpétalo 3659
simplexo 8294
simpodial 8860
simpódico 8860
simpodio 8861
sin clorofila 70
  cola 628
  hojas 636
sinangio 8862
sinapsis 8863, 8864
  inhibidora 4682
sináptico 8866
sinartrosis 8870
sincario 8888
sincárpico 8871
sincarpo 8871
sincicio 8874
sincitio 8874
sincorología 8872
sincrónico 8873
síndesis 8875
sindesmosis 8876
sinecia 8890
sinecología 8877
sinergético 8879
sinergia 8878
sinérgico 8879
sinérgida 8880
sinergismo 8881
sinergista 8882
sínfisis 8859
  púbica 7448
singámeon 8883
singamia 8885
singámico 8884
singenesia 8886
singénesis 8886
singenético 8887
sinoico 8889
sinovia 8891
sinovial 725
sintasa 5271
síntesis enzimática 3042
  de proteínas 7383
  prote(ín)ica 7383
sintetasa 5147
sintetizar 8895
síntoma de deficiencia 2331
sintrofismo 8898
síntrofo 8897
sinusia 8899
sinzoóspora 8900
siringe 8901
sismonastia 8089
sistema de acción 116
  ambulacral 9663
  amortiguador 1325
  caulinar 8617
  circulatorio 1754
  de clasificación 1770
sistema conductor 1965
  digestivo 2546
  ecológico 2775
  eferente 2796
  de espacios intercelulares 4738
  de excreción 3201
  excretor 3201
  genital 3756
  hidrovascular 9663
  intercelular de espacios de aire 4738
  lagunar 5014
  libre de células 1513
  límbico 5169
  linfático 5279
  muscular 5908
  nervioso 6058
  nervioso autónomo 834
  nervioso central 1550
  nervioso parasimpático 6608
  nervioso periférico 6742
  nervioso vegetativo 834
  nervioso visceral 834
  orgánico 6376
  ortosimpático 8856
  de oxidorreducción 6489
  oxidorreductor 6489
  parasimpático 6608
  portal 7205
  radical 7910
  radicular 7910
  redox 6489
  de regulación 3339
  regulador 3339
  de retroacción 3339
  reproductivo 7764
  respiratorio 7809
  reticuloendotelial 7836
  simpático 8856
  tampón 1325
  transportador de electrones 2842
  traqueal 9185
  urogenital 9417
  vacuolar 9434
  vascular 9460
  vascular acuoso 9663
  vasculosanguíneo 1199
sistemática 8902
  animal 528
sistemático 8903
sístole 8906
sitio de puesta 6466
situación estimulante 8655
SNC 1550
sobrecalentamiento 6455
sobredominancia 6454
sobreevolución 4489
sobrevivir 8832
sociabilidad 8344
sociación 8345
social 8338
sociedad 8346
  animal 524
  de insectos 4705
sociología animal 525
  botánica 6921
  vegetal 6921
sol 8374
solación 8377
solenocito 8378
solidificación 8380
solidificar(se) 8381
soliflucción 8382

Español

solitario 8383
solubilidad 8385
  en agua 9655
solubilizar 8386
soluble 8387
  en ácido 87
  en agua 9656
solución acuosa 677
  amortiguadora 1323
  coloidal 1894
  colorante 1907
  de cultivo 6252
  nutritiva 6252
  patrón 8671
  de Ringer 7885
  salina 7963
  salina fisiológica 6890
  standard 8592
  tampón 1323
  yodada 4829
  de yodo 4829
soluto 8389
solvente 8390
soma 8391
somático 8392
somatogamia 8394
somatogénico 8395
somatógeno 8395
somatología 8396
somatopleura 8397
somatotrofina 8398
somatotropina 8398
somita 8400
somito 8400
sonido de alarma 276
  de llamada 1351
  silencioso 9345
soporte coloidal 649
sorción 8403
soredio 8402
soro 8404
sotobosque 9365
spike 8491
sport 8560
subadulto 8734
subcelular 8721
subclase 8722
subclímax 8723
subcultivo 8724
subdivisión 8726
súber 2048
suberificación 8729
suberificar 8730
suberina 8728
suberización 8729
suberizar(se) 8730
suberoso 2053
subespecie 8751
subfamilia 8731
subfilo 8747
subfllum 8747
subgen 8732
subgénero 8733
subida de la savia 741
subimago 8734
subletal 8737
sublimado 8738
subliminar 8739
submicrón 8743
submicroscópico 8744
suborden 8746
subpoblación 8748
subproductos metabólicos 5539
subreino 8736
subsere 8749
substrato 8755
  nutritivo 6260
subsuelo 1014
subunidad 8761
succión 8776
sucesión 8762
  ecológica 2766
  secundaria 8749
sucrasa 4815
sucrosa 7940
suculencia 8766
sudación 8778
sudor 8843
sudoración 8778
suelo marino 8041
suero 8198
  hemático 1191
  sanguíneo 1191
sufrútice 8780
sujeto (de experimentación) 3240
sumación 8783
  espacial 8420
  de estímulos 8658
  temporal 8978
suministro de agua 9660
  de oxígeno 6502
superclase 8791
superdominancia 6454
superespecie 8808
super-estímulo 8801
superfamilia 8793
superfecundación 8796
superfetación 8796
superficie cortada 2199
  de corte 2199
  foliar 5085
  del suelo 263
superfílum 8805
supergen 8797
superhembra 8794
superior al umbral 8815
supermacho 8800
supernumerario 8802
súpero 8798
superorden 8803
superparásito 4484
superpoblación 6457
supervivencia 8825
  del más apto 8828
supraespecie 8818
supraliminar 8815
suprarrenal 175
supraumbral 8815
supresor 8813
  de crossing-over 2144
surco neural 6078
  primitivo 7300
suscepción 8833
suspensión celular 1532
suspensor 8835
sustancia activa 126
  de alarma 278
  alimenticia 3531
  de atracción 792
  de atracción sexual 8239
  blanca 9672
  colorante 8577
  contaminadora 2010
  contaminante 7114
  de crecimiento 3995
  defensiva 2327
  de desecho 9626
  estimuladora 8646
  estimulante 8646
  excitante 3188
sustancia extraña 3551
  de fecundación 3657
  fijadora 3432
  fundamental 3962
  gris 3955
  inhibidora 4681
  inhibidora del crecimiento 3985
  intercelular 4741
  liminal 9095
  mucilaginosa 5852
  mutágena 5914
  natural 5990
  nociva 6189
  no liminar 6185
  nutritiva 6245
  nutritiva mineral 5696
  odorífera 6284
  olorosa 6284
  portadora 1444
  proteica 7382
  de reserva 7777
  tampón 1324
  tóxica 9174
  transmisora 9219
  trazadora 9178
  trazadora radiactiva 7559
  vagal 66
  vegetal en descomposición 6642
  venenosa 9174
sustancias residuales 7780
sustitución 8753

T

tabicación 1375
tabicado 8179
tabicamiento 1375
tabique divisorio 8182
  nasal 5980
  separatorio 8182
  transversal 2150
tabla genealógica 3721
  de vida 5143
táctil 8909
tactismo 8941
tacto 8153
taiga 8918
tala incontrolada 2438
tálamo 7632
  óptico 9033
talo 9037
talofita 9035
talófito 9035
taloso 9036
talud continental 2014
tallo 8582, 8611
  embrional 9117
  hipofisario 4510
  óptico 3289
  del pelo 4069
  principal 5065
tamaño de los granos 3927
  de las partículas 3927
  de la población 7196
tampón 1318
tamponar 1319
tangiorreceptor 8924
tanino 8926
tapete 8928
tapón de vitelo 9744
taquigénesis 8908
tara hereditaria 4188
tardígrados 8930

tarso 8933
tasa de crecimiento 3988
de crossing-over 2146
de fertilidad 3364
metabólica basal 954
de mortalidad 2290
de mutación 5929
de natalidad 1139
de reproducción 7746
de supervivencia 8829
taxia 8941
taxinomía 8945
taxinómico 8943
taxinomista 8944
taxis 8941
taxón 8942
taxonomía 8945
animal 528
botánica 7008
vegetal 7008
taxonómico 8943
taxonomista 8944
taxónomo 8944
teca 9038
técnica de preparación 5572
tectología 8949
tegmento 1316
tegumento 4729
larvario 5039
tejido 9127
acuífero 9659
adiposo 168
adulto 6760
aerífero 198
asimilador 756
cartilaginoso 1673
cavernoso 1495
celular 1540
cicatricial 1735
de conducción 1966
conductor 1966
conjuntivo 1991
embrionario 2882
en empalizada 6548
epitelial 3091
eréctil 1495
esponjoso 8536
fibroso 3394
fundamental 3963
glandular 3857
graso 168
lignificado 9708
linfoide 5282
mecánico 8812
nervioso 6059
nutricio 6261
nutritivo 6261
óseo 1226
protector 7368
de reserva 8687
reservante 8687
secretor 8066
de sostén 8812
subcutáneo 8725
suberoso 2052
superficial 8823
de transfusión 9205
vascular 9461
vegetal 7009
teleceptor 2637
telegonia 8954
telencéfalo 8955
teleósteos 8956
teleutóspora 8958
telióspora 8958
telitocia 9039
telitoquía 9039
telocéntrico 8959
telofase 8964
telolecítico 8960
telolecito 8960
teloma 8961
telómero 8963
telosinapsis 8965
telosíndesis 8965
telotactismo 8966
telotaxia 8966
telson 8967
temperatura ambiente 376
corporal 1212
interna 1212
óptima 6369
preferida 7253
tendinoso 8981
tendón 8983
tensión arterial 1187
superficial 8822
tentáculo 8988
teñir 8576
teoría del blanco 8931
de las catástrofes 9040
celular 1533
del circuito local 5223
de la constancia 9041
de la continuidad del plasma germinal 9043
creacionista 9041
cromosómica de la herencia 1724
de la deriva de los continentes 2013
de la descendencia 2427
de los desplazamientos continentales 2013
de la edad y área 240
del equilibrio de la determinación de los sexos 935
de la evolución 9042
evolutiva 9042
de la gestalt 3816
de la información 4658
iónica 4841
mutacionista 5930
de la neurona 6101
de la panspermia 2661
de la precocidad 7241
de la preformación 7257
de la presencia-ausencia 7277
de la recapitulación 7630
de la selección (natural) 9044
telomática 8962
de la tensión-cohesión 1862
transformista 9042
tépalo 8989
teratología 8990
tercera ley de Mendel 5486
terciario 9013
terminación nerviosa 6044
terminal 8991
axónico 881
terminalización 8999
termoclina 9051
termodinámica 9053
termodinámico 9052
termoestábil 9066
termoestabilidad 9065
termoestable 9066
termofílico 9060
termófilo 9060
termogénesis 1362
termolábil 9056
termólisis 9057
termonas 9000
termonastia 9058
termoperiodicidad 9059
termorreceptor 9061
termorregulación 9062
termorregulador 9063
termorresistencia 4134
termorresistente 4135
termostábil 9066
termostable 9066
termotaxia 9067
termotaxis 9067
termotropismo 9068
terófito 9069
terpeno 9001
terreno alimentario 3344
terrestre 9002
terrícola 9006
territorialidad 9007
territorio 9011
de cría 1282
testa 8074
testículo 9023
testosterona 9024
teta 8948
tétrade 9025
dítipo no progenitor 6166
tetraploide 9027
tetraploidia 9028
tetrápodo 9029
tetrasómico 9030
tetráspora 9031
tetratipo 9032
tetróxido de osmio 6411
tialina 7445
tiamina 9070
tibia 9115
tiempo de duplicación 2699
de latencia 5046
de presentación 7278
de reacción 7620
útil 9428
de utilización 9428
tierra de diatomeas 4667
silícea 4667
tigmomorfosis 9074
tigmonastia 4096
tigmotaxia 9075
tigmotaxis 9075
tigmotropismo 9076
tílide 9323
tilis 9323
tilo 9323
timidina 9104
timina 9106
timo 9108
timonera 8919
tímpano 9326
tinción 8580
complementaria 224
intravital 4804
negativa 6008
supravital 8819
vital 9574
tingible 8579
tiobacterias 9079
tionina 9081
tipo 9327
de apareamiento 5417
de área 699
climático 1796
constitucional 2001
copulante 5417

Español

tipo genitor 6626
salvaje 9675
silvestre 9675
de suelo 8373
standard 8593
de vegetación 9475
tipogénesis 9329
tipólisis 9330
tipóstasis 9331
tiroides 9110
tirosina 9332
tirotricina 9333
tirotrofina 9112
tiroxina 9113
tirso 9114
tisular 9138
titulación 9140
eléctrica 2819
titular 9139
título 9142
tocoferol 9145
tolerancia inmunológica 4572
residual 9147
toma de alimentos 3530
de muestras 7969
tomar muestras 7967
tonicidad 9149
tono 9149
muscular 5909
tonoplasto 9148
tonsila faríngea 6799
palatina 6540
topotactismo 9157
topotaxia 9157
tórax 9085
tornillo micrométrico 5616
toro *(bot.)* 9158
totipotencia 9163
totipotente 9164
toxicidad 9170
tóxico 7075, 9167
mitótico 5718
vegetal 6926
toxicógeno 9173
toxicología 9172
toxicólogo 9171
toxígeno 9173
toxina 9174
bacteriana 929
TPN 6119
trabante 9175
trabécula 9176
tracto genital 3756
intestinal 4796
nervioso 6053
reproductor 7764
traje nupcial 6239
trance, en ~ de extinción 991
transaminación 9194
transaminasa 9193
transconfiguración 9195
transcripción 9196
tra(n)sducción 9197
abortiva 18
transespecífico 9228
transferasa 9203
transferencia de energía 2972
de genes 3720
de información 9200
del material genético 9199
del polen 7104
transferibilidad 9216
transferible 9217
transformación 9204
energética 2973
transformación
de energía 2973
transformarse en crisálida 7472
en pupa 7472
transformismo 9042
transfosforilación 9221
transgenación 3714
tra(n)slación 9211
tra(n)slocación 9212
de inserción 4715
recíproca 7645
translúcido 9214
transmetilación 9215
transmisible 9217
hereditariamente 4671
por herencia 4671
transmisión de energía 2972
genética 9199
hereditaria 4672
hereditaria del sexo 8229
de información 9200
transmisor de la herencia 990
químico 9219
transmitir (hereditariamente) a
9218
transmutación 9220
transpiración 9222
cuticular 2205
estomática 8676
tra(n)splante nuclear 6211
transportador de electrones 2832
de oxígeno 6496
transporte activo 127
de electrones 2839
de iones 4836
de sustancias 9227
transudación 9230
transudado 9229
tráquea 9180, 9181, 9182
laminar 1227
tubular 9305
traqueados 9187
traqueida 9188
anular 547
fibriforme 3380
escalariforme 7996
espiralada 8514
helicoidal 8514
traqueofito 9190
traqueola 9189
traquéolo 9189
trasplantar 9225
trasplante 9226
nuclear 6211
trastorno de crecimiento 2645
de desarrollo 2461
ecológico 3014
metabólico 5528
trasudación 9230
trasudado 9229
tratamiento de las aguas 9661
de las aguas residuales 8213
por el calor 4138
(por el) frío 1875
traumatina 9235
trauma(to)tropismo 9236
traza foliar 5089
trazado de mapas 5369
trazador 9178
radiactivo 7559
trazar mapas *(cromosómicos etc.)*
5367
trefona 9244
trematodos 9243
treonina 9091
triada 9245
triásico 9247
tribu 9248
tricocisto 9250
tricógina 9251
tricógino 9251
tricoma 9252
trifosfopiridina-nucleótido 6119
trihibridismo 9255
trihíbrido 9254
trimería 9257
trimérico 9256
trimorfismo 9258
trioico 9259
triosa fosfato 9260
tripéptido 9261
triplete 9264
triplexo 9266
triploidia 9267
tripsina 9293
tripsinógeno 9294
triptófano 9296
tripton 9268
triptona 9295
trisomia 9270
trisómico 9269
trivalente 9271
trofalaxis 9275
trófico 9276
trofoblasto 9279
trofocito 6255
trofofilaxia 9275
trofología 9281
trofotaxia 9282
trofotaxis 9282
trofotropismo 9283
troglobios 1491
trombina 9097
trombocito 9098
trombógeno 9099
tromboquinasa 9100
trompa 7312
auditiva 3147
de Eustaquio 3147
de Falopio 3301
tronco 9291
cerebral 1249
encefálico 1249
nervioso 6052
traqueal 9186
tropel 4181
tropismo 9286
tropófito 9287
tropotaxia 9288
tropotaxis 9288
TSH 9112
tubérculo 9299
caulinar 8618
radicular 7912
tuberización 9300
tuberoso 1330
tubiforme 9304
tubo alimentador 8773
de copulación 3370
copulador 3370
criboso 8278
de cultivo 2194
chupador 8773
digestivo 304
de ensayo 9020
de fermentación 3358
germinal 3790
laticífero 5052
neural 6080
polínico 7105

tubo respiratorio 7808
seminífero 8137
uterino 3301
tubos de Malpighi(o) 5351
tubular 9304
túbulo renal 7724
urinífero 9414
tuétano 1219, 5446
tundra 9306
tunicados 9307
turba 6661
turbelarios 9308
turbera alta 4274
baja 5251
de esfagnales 6662
de transición 9209
turgencia 9310
turgente 9309
turgescente 9309
túrgido 9309
turión 9313

U

ubicuo 9335
ubiquinona 9334
ultracentrífuga 9338
ultracentrifugación 9337
ultraestructura 9346
ultrafiltración 9340
ultrafiltro 9339
ultramicroscópico 9342
ultramicroscopio 9341
ultramicrosoma 9343
ultramicrótomo 9344
ultrasonido 9345
umbela 9354
compuesta 1952
umbelado 9355
umbelíferas 9356
umbelífero 9357
umbeliforme 9358
umbilical 9359
umbral absoluto 25
auditivo 807
diferencial 2522
de(l) dolor 9092
de estimulación 8659
de excitación 3185
gustativo 8940
de nocividad 2270
de sensación 8161
umbrófilo 8318
unciforme 9363
ungulados 9370
ungulígrado 9371
uniaxial 5751
uniáxico 5751
unicelular 9373
unicidad del individuo 9385
unidad calorífica 9050
de crossing-over 2145
de dispersión 2494
funcional 3620
hereditaria 4193
de recombinación 7652
recombinacional 7652
de replicación 7736
roentgen 7896
de vegetación 9476
uniestratificado 9388
uniflagelado 9376
unifloro 9377
unifoliado 5788
unigerminal 9379
unilocular 9381
uninuclear 5784
uniovular 5800
unipolar 9384
unisexual 9386
unisexualidad 9387
unisexualismo 9387
unistratificado 9388
univalente 9390
univalvo 9391
universalidad del código genético 9392
uña 4378
uperisación 9395
uracilo 9401
urbanización 9402
urea 9403
ureasa 9405
uredóspora 9406
uréter 9407
uretra 9408
urobilina 9415
urocordados 9307
uropigio 9419
útero 9426
utilización de utensilios 9150
UV 9350
úvea 9430

V

vacío 9435
vacuola 9433
alimenticia 3544
contráctil 2023
digestiva 3544
pulsativa 2023
vacúolo 9433
vacuoma 9434
vagina 9437
vago 9438
vaina 5103
foliar 5083
medulada 5947
de mielina 5947
mielínica 5947
mucilaginosa 5862
de Schwann 8018
tendinosa 8982
valencia ecológica 2768
valina 9439
valor de adaptación 146
adaptativo 146
calórico 1356
genético 1283
liminal 9096
nutritivo 6262
osmótico 6422
del pH 6788
de recombinación 7651
selectivo 8103
de supervivencia 8831
umbral 9096
valva 9441
valvar 9440
válvula cardíaca 4120
mitral 5720
de las venas 9498
vapor de agua 9662
variabilidad 9442
continua 2018
variabilidad cualitativa 2600
cuantitativa 2018
discontinua 2600
variable 9443
variación 9446
clinal 1802
variante 9445
varianza 9444
variar 9451
variedad 9449
variegación 9448
variegado 9447
vascular 9452
vascularización 9462
vascularizado 9463
vaso 9522
anillado 548
anular 548
capilar 1392
escalariforme 7997
helicado 4143
helicoidal 4143
laticífero 5052
linfático 5280
punteado 6964
reticulado 7832
reticular 7832
sanguíneo 1200
vasoconstricción 9464
vasodilatación 9465
vasopresina 9467
vástago 8263
epigeo 201
subterráneo 8758
vector 9468
vegetación 9472
latente 9489
vegetal 6984, 9470
vegetativo 9480
vejiga aérea 8847
natatoria 8847
urinaria 9411
velamen 9491
velo del paladar 6542
palatino 6542
velocidad de conducción 1970
de difusión 2534
de evolución 2465
metabólica 5537
de multiplicación 5889
de reacción 7619
de recambio 9315
de retorno 9315
de sedimentación 8069
vellosidad 9530
corial 1684
coriónica 1684
intestinal 4797
vena 9490
cava 1490
porta 7206
vitelina 9598
venación 6038
alar 9686
foliar 5091
veneno 7075
de abeja 992
del huso 8502
mitótico 5718
vegetal 6926
venenosidad 9170
venenoso 9167
venoso 9496
ventaja selectiva 8097
ventana oval 6445

Español

X

Y

Z

zona de hibridación 9756
infralitoral 8740
de inserción 4714
limnética 5177
litoral 5216
nucleolar 6221
pelágica 6683
pelúcida 9753
pilífera 6942
profunda 7323
de recreo 5105
residencial 7778
SAT 6221
sublitoral 8740
supralitoral 8816
zona de vegetación 9477
verde 3951
zonación 9755
zooclorela 9757
zoocoria 9758
zooesterol 9774
zoófago 9769
zoófito 9770
zooflagelados 9759
zoogámeta 9760
zoogameto 9760
zoogenética 9761
zoogeografía 9762
zooglea 9763
zooide 9764
zoología 9767
zoológico 9765
zoólogo 9766
zoomastigóforos 9759
zooparásito 9768
zooplancton 9771
zoóspora 9773
zoosporangio 9772
zootecnia 9775
zootécnico 503
zootomía 9776
zootoxina 9777
zooxantela 9778
zwitterion 9779

Español